EIGHTH EDITION

The Earth Through Time

To see where we might be going,
we must understand where we have been.

Robert Tamarkin, 1993

EIGHTH EDITION

The Earth Through Time

HAROLD L. LEVIN

Professor of Geology
Washington University,
St. Louis, Missouri

JOHN WILEY & SONS, INC.

EXECUTIVE EDITOR	Ryan Flahive
MARKETING MANAGER	Jeffrey Rucker
SENIOR PRODUCTION EDITOR	Sandra Dumas
SENIOR DESIGNER	Madelyn Lesure
SENIOR MEDIA EDITORS	Thomas Kulesa/Lynn Pearlman
SENIOR PHOTO EDITOR	Jennifer MacMillan
SENIOR ILLUSTRATION EDITOR	Anna Melhorn
DEVELOPMENTAL EDITOR	Fred Schroyer
EDITORIAL ASSISTANT	Rachel Schneider

This book was typeset in 10/12 Janson at GGS Book Services and printed and bound by Von Hoffmann Corporation. The cover was printed by Lehigh Press Lithographers.

This book is printed on acid-free paper. ∞

To order books or for customer service please, call 1-800-CALL WILEY (225-5945).

Levin, Harold L.
The Earth Through Time, Eighth Edition

ISBN 13 978-0471-69743-5
ISBN 10 0-471-69743-5

Printed in the United States of America.

10 9 8 7 6 5 4 3 2 1

This book is dedicated to my wife Kay, who has cheerfully endured my preoccupation with preparing this new edition, and to Noah, Lillie, Eli, Mollie, Natalie, Emily, Caitlyn, Hannah, and Candis. May they have the wisdom to treat the Earth kindly.

Harold ("Hal") Levin began his career as a petroleum geologist in 1956 after receiving bachelor's and master's degrees from the University of Missouri and a doctorate from Washington University. His fondness for teaching brought him back to Washington University in 1962, where he is currently professor of geology and paleontology in the Department of Earth and Planetary Sciences. His writing efforts include authorship of seven editions of *The Earth Through Time*; four editions of *Contemporary Physical Geology*; *Essentials of Earth Science*; and co-authorship of *Earth: Past and Present*, as well as eight editions of *Laboratory Studies in Historical Geology*; *Life Through Time*; and, most recently, *Ancient Invertebrates and Their Living Relatives*.

For his courses in physical geology, historical geology, paleontology, sedimentology, and stratigraphy, Hal has received several awards for excellence in teaching. The accompanying photograph was taken during a lecture on life of the Cenozoic Era. The horse skull serves to illustrate changes in the teeth and jaws of grazing animals in response to the spread of prairies and savannahs during the Miocene and subsequent epochs.

The Earth Through Time is an excursion into the Earth's geologic past. The trip began about four and a half billion years ago, approaching the time when our planet had gathered most of its mass from a rotating cloud of dust, gases, and meteorites. From that time to the present, the Earth has undergone constant change. Ocean floors have expanded and moved, continents have broken apart, sea floors have been thrust skyward to form mountain ranges, and once peaceful landscapes have been disrupted by earthquakes and fiery floods of lava. Since about 3 billion years ago, life has existed on this dynamic planetary surface. Fossil remains of ancient life attest to their successes and failures in coping with perpetually changing conditions.

How the Earth and its inhabitants have changed provides a fascinating story. The story alone is sufficient reason for undertaking a study of the Earth's geologic past Historical geology, however, has value in many other ways as well. As a science, it informs us about the way science works, how discoveries are made, and how to tell the difference between valid and shoddy interpretations. Past events also relate to many problems we face today, including, global warming, environmental pollution, depletion of vital mineral resources, and the adverse effects of loss of animal and plant species. Understanding science, and realizing how each component of our past and present planet interacts with all other components will promote informed decisions about the many environmental issues now being debated in our governing bodies.

Determining the cause of past geologic events is a complex task. As an example, one might infer that the demise of a particular group of ancient animals resulted from cooling of the planet. This leads us to ask *why* cooling occurred. Was it the result of changes in the composition of the atmosphere, changes in the amount of radiation received from the sun, the upheaval of wind-diverting mountain ranges, changes in the Earth's axis of rotation, or shifts in the location of lands and seas? The answers come only after study of the reciprocal actions of all the Earth's systems: the solid Earth, the atmosphere, and the biosphere. This intergrated approach applies well both to the solution of contemporary problems and to an understanding of happenings in the remote geologic past. It is an approach embodied in this text.

As in earlier editions, this revision of *Earth Through Time* is designed to meet the requirements of the undergraduate student who has had little or no previous acquaintance with geology. However, for those exploring the possibility of an academic major in geology, the text will provide the initial background needed for advanced courses. Information about geologic materials and processes are included in *Earth Through Time* so that the book can be used either for a single, self-contained first course, or for the second course in a two semester sequence of Physical and Historical Geology.

THE EIGHTH EDITION

The preparation of a new edition provides many opportunities to refashion and improve explanations, to develop better illustrations, and to constructively expand, delete, and reorganize. The greatest effort in this revision has been to make the narrative more lively and engaging, to clearly convey the unique perspective and value of historical geology, and to improve the presentation so as to enhance the student's ability to retain essential concepts.

In this eighth edition, seventeen chapters have been organized under three major divisions. Part I. *Discovering Time and Deciphering Earth's Amazing History*, explains the methods used in reconstructing the Earth's geologic history, introduces contributions made by early geologists to the science, and describes how rocks are dated and then used to construct the geologic time scale. Part II, *Rocks and What They tell Us About Earth History* describes the nature and origin of Earth materials, and how rocks and fossils can reveal events of the geologic past. In Part III, *History of Planet Earth and Its Inhabitants* we explore the actual history of our planet from its fiery birth to the unfolding of the modern world.

WHAT KINDS OF PEDAGOGY DOES *EARTH THROUGH TIME* PROVIDE?

Earth Through Time has a variety of features to engage the student and promote learning.

- Technical terms are printed in **boldface** type the first time they are used, and clearly defined, both in the text and in the glossary.

- So that students will know what lies ahead, each chapter begins with a concise list of **Key Chapter Concepts**.

- Each chapter ends with a **Summary** of essential concepts. These provide students with a reality check. If the summary statement is not understood, it is a cue for the student to visit that topic in the chapter again.

- A list of **Key Terms** is provided. If the student does not remember the meaning of a term, it provides another opportunity to check their understanding of concepts developed in the chapter.

- **Questions for Review** allows students to test their understanding of material in a chapter and to further "process" what they have learned.

Caption questions occur beneath many photographs. These questions draw attention to geologic features depicted in the images, and clarify explanations in the text.

- Appendices in *Earth Through Time* include a **Classification of Living Things** that helps students place fossils described in the text in their correct taxonomic position, a **Glossary** containing a list of important terms and their definitions, Physiographic, political and bedrock geology **maps**, the **Periodic Table and Symbols for Chemical Elements**, **Convenient Conversion Factors**, an explanation of **Exponential or Scientific Notation**, a **Table of Rock Symbols**, and a table providing the composition and physical properties of **Common Rock-Forming Silicate Minerals**.

- Most chapters feature one or more boxes. The **Enrichment** boxes discuss some topics of general interest that relate to material in the chapter, and the **National Parks and Monuments** boxes provide an example of a location where the geology discussed can be directly seen.

- To help students correlate events they are reading about in the text to their actual time of occurrence, small time scales containing an age indicator accompany many photographs of rock formations and fossils.

► SUPPLEMENTS

The eighth edition of *The Earth Through Time* is accompanied by an extensive set of supporting materials. These include:

A **Student Study Guide** prepared by Harry Wagner of Victoria College. Designed to enhance students' understanding of the text, the guide includes Chapter Overviews, Learning Objectives for each chapter, Questions for Review, Key Terms, illustrations and maps.

An **Instructor's Manual and Test Bank** has been prepared by David T. King Jr., of Auburn University. To facilitate use of the text by instructors, the manual includes chapter outlines, questions that can be incorporated into examinations, and answers to questions in the textbook.

Computerized Test Banks are available in three disk versions: Windows, IBM, and Macintosh.

A set of 100 full-color **Overhead Transparencies** is provided for use in the laboratory or lecture hall, as well as a set of one-hundred 25-mm slides of illustrations and photographs.

Earth Sciences Instructor's Resource CD-ROM provides numerous images from this and other John Wiley & Sons geology texts.

John Wiley and Sons, Inc. may provide complementary instructional aids and supplementary packages to those adopters qualified under our adoption policy. Please contact your sales representative for more information.

► ACKNOWLEDGMENTS

Many changes in this edition are the result of incisive comments and suggestions of the diligent group of reviewers listed below:

Deniz Altin, *Georgia Perimeter College*

Brooks Ellwood, *Louisiana State University–Baton Rouge*

Pamela Gore, *Georgia Perimeter College*

Peter Isaacson, *University of Idaho*

Tim Kroeger, *Bemidji State University*

Mark Kulp, *University of New Orleans*

Paul Pause, *Midland College*

Kenneth Rasmussen, *Northern Virginia Community College–Annandale*

I would also like to extend my thanks to the following earth scientists who contributed as reviewers to the previous editions of the book:

Dennis Allen, *University of South Carolina–Aiken*

William Ausich, *Ohio State University*

David R. Berry, *California State Polytechnic University*

William Berry, *University of California–Berkeley*

Michael Bikerman, *University of Pittsburgh*

Mark Camp, *University of Toledo*

Roger J. Cuffey, *University of Pennsylvania*

James H. Darrell, *Georgia Southern University*

Larry E. David, *Washington State University*

William H. Easton, *University of Southern California*

George F. Englemann, *University of Nebraska at Omaha*

Stanley Fagerlin, *Southwest Missouri State College*

Cindy Ganes, *University of Saskatchewan*
John Gosse, *University of Kansas*
Jack Hall, *University of North Carolina at Wilmington*
Vicki Harder, *Santa Teresa, New Mexico*
John A. Howe, *Bowling Green State University*
Warren Huff, *University of Cincinnati*
John R. Huntsman, *University of North Carolina*
Allen Johnson, *West Chester University*
Gary D. Johnson, *University of South Dakota*
William M. Jordan, *Millersville University of Pennsylvania*
Roger Kaesler, *University of Kansas*
Ernst Kastning, *Radford University*
David T. King, Jr., *Auburn University*
Larry Knox, *Tennessee Technical University*
Peter Kresan, *University of Arizona*
Ralph L. Langenheim, Jr., *University of Illinois*
W. Brit Leatham, *California State University–San Bernardino*
Peter Leavens, *University of Delaware*
Joseph Lintz, *University of Nevada at Reno*
Daniel L. Lumsden, *Memphis State University*
Donald Marchand, *Old Dominion University*
William H. Mathews, III, *University of British Columbia*
Dewey McLean, *Virginia Polytechnic Institute*
Eldridge Moores, *University of California–Davis*
Peter Nielsen, *Keene State College*

Cathryn Newton, *Syracuse University*
Shannon O'Dunn, *Grossmont College*
Donald E. Owen, *Lamar University*
William Parker, *Florida State University*
John Pope, *Miami University*
Jennifer Smith Prouty, *Corpus Christi State University*
G.J. Retallack, *University of Oregon*
Thomas Roberts, *University of Kentucky*
Barun K. Sen Gupta, *Louisiana State University*
Thomas W. Small, *Frostburg University*
Leonard W. Soroka, *St. Cloud University*
Don Steinker, *Bowling Green State University*
Calvin H. Stevens, *San Jose State University*
James Stevens, *Lamar University*
David Thomas, *Washtenaw Community College*
Harry Wagner, *Victoria College*
Kenneth Van Dellen, *Macomb Community College*
Carl Vondra, *Iowa State University*
Peter Whaley, *Murray State University*
Lisa White, *San Francisco State University*
William Zinmeister, *Purdue University*
Thomas T. Zwick, *Eastern Montana College*

My thanks conclude with a special note of appreciation for the insightful guidance provided by our developmental editor, Fred Schroyer.

BRIEF CONTENTS

CHAPTER 1
The Science of Historical Geology 1

CHAPTER 2
Early Geologists Tackle History's
Mysteries 11

CHAPTER 3
Time and Geology 27

CHAPTER 4
Rocks and Minerals: Documents that
Record Earth's History 47

CHAPTER 5
The Sedimentary Archives 77

CHAPTER 6
Life on Earth: What Do Fossils Reveal? 117

CHAPTER 7
Plate Tectonics Underlies All Earth
History 162

CHAPTER 8
Earliest Earth: 2,100,000,000 Years of the
Archean Eon 208

CHAPTER 9
The Proterozoic: Dawn of a More Modern
World 243

CHAPTER 10
Early Paleozoic Events 267

CHAPTER 11
Late Paleozoic Events 295

CHAPTER 12
Life of the Paleozoic 327

CHAPTER 13
Mesozoic Events 373

CHAPTER 14
Life of the Mesozoic 405

CHAPTER 15
Cenozoic Events 451

CHAPTER 16
Life of the Cenozoic 487

CHAPTER 17
Human Origins 525

APPENDICIES A1

GLOSSARY G1

INDEX I1

CONTENTS

CHAPTER 1
The Science of Historical Geology 1

Why Study Earth History? **2**
Geology Lives in the Present and the Past **2**
A Way To Solve Problems: The Scientific Method **3**
Three Great Themes in Earth History **6**
What Lies Ahead? **8**

CHAPTER 2
Early Geologists Tackle History's Mysteries 1 1

The Intrigue of Fossils **12**
An Early Scientist Discovers Some Basic Rules **13**
European Researchers Unravel the Succession of Strata **15**
Neptunists and Plutonists Clash **15**
Uniformitarianism: James Hutton Recognizes that the Present is Key to the Past **16**
The Principle of Fossil Succession **18**
The Great Uniformitarianism–Catastrophism Controversy **18**
The Principle of Cross-cutting Relationships **19**
Evolution: How Organisms Change through Time **20**
Earth History in America **22**

CHAPTER 3
Time and Geology 2 7

Finding the Age of Rocks: Relative versus Actual Time **27**
A Scale of Geologic Time **28**
Actual Geologic Time: Clocks in the Rocks **32**
Radioactivity Provides a Way to Date Rocks **33**
What Occurs When Atoms Decay? **35**
The Principle Radioactive Timekeepers **38**

CHAPTER 4
Rocks and Minerals: Documents that Record Earth's History 4 7

Minerals as Documents of Earth History **48**
Minerals and Their Properties **48**
Common Minerals that Form Rocks **50**

Earth's Three Great Rock Families and How They Formed **55**
Igneous Rocks: "Fire-Formed" **56**
Sedimentary Rocks: Layered Pages of History **65**
Metamorphic Rocks: Changed without Melting **69**

CHAPTER 5
The Sedimentary Archives 7 7

Tectonic Setting is the Biggest Factor in Sediment Deposition **78**
Environments Where Deposition Occurs **80**
What Rock Color Tells Us **86**
What Rock Texture Tells Us **86**
ENRICHMENT   You Are the Geologist **89**
What Sedimentary Structures Tell Us **90**
What Four Sandstone Types Reveal About Tectonic Setting **95**
Limestones and How They Form **98**
Organizing Strata To Solve Problems **98**
Sea-level Change Means Dramatic Environmental Change **102**
Stratigraphy and the Correlating of Rock Bodies **103**
Unconformities: Something is Missing **105**
Depicting the Past **106**
GEOLOGY OF NATIONAL PARKS AND MONUMENTS
Grand Canyon National Park **113**

CHAPTER 6
Life on Earth: What Do Fossils Reveal? 1 1 7

Fossils: Surviving Records of Past Life **120**
ENRICHMENT   Amber, the Golden Preservative **123**
Figuring Out How Life is Organized **124**
Evolution: Continuous Changes in Life **128**
The Case for Evolution **136**
Fossils and Stratigraphy **139**
Fossils Indicate Past Environments 143
ENRICHMENT   Earbones Through the Ages **144**
How Fossils Indicate Paleogeography **149**
How Fossils Indicate Past Climates **152**
An Overview of the History of life **153**
Life on Other Planets: Are We Alone **156**

CHAPTER 7

Plate Tectonics Underlies All Earth History 1 6 2

Earthquake Waves Reveal Earth's Mysterious Interior 162
Earth's Internal Zones 163
Earth's Two Types of Crust 167
Broken, Squeezed, or Stretched Rocks Produce Geologic Structures 168
Plate Tectonics Theory Ties it All Together 171
Drifting Continents 174
Evidence for Continental Drift 175
Paleomagnetism: Ancient Magnetism Locked into Rocks 178
Today's Plate Tectonics Theory 181
What Happens at Plate Margins? 187
What Drives Plate Tectonics? 189
Verifying Plate Tectonics Theory 191
Thermal Plumes, Hotspots, and Hawaii 196
ENRICHMENT   Rates of Plates 197
Exotic Terranes 199
GEOLOGY OF NATIONAL PARKS AND MONUMENTS
Hawaii Volcanoes National Park 200

CHAPTER 8

Earliest Earth:2,100,000,000 Years of the Archean Eon 2 0 8

Earth in Context: A Little Astronomy 208
ENRICHMENT   The Origin of the Universe 213
A Solar System Tour, from Center to Fringe 214
Following Accretion, Earth Differentiates 219
The Primitive Atmosphere-Virtually No Oxygen 222
The Primitive Ocean and the Hydrologic Cycle 224
Origin of Precambrian "Basement" Rocks 224
The Origin of Life 230
GEOLOGY OF NATIONAL PARKS AND MONUMENTS
Voyageurs National Park 238
In Retrospect 239

CHAPTER 9

The Proterozoic: Dawn of a More Modern World 2 4 3

Highlights of the Paleoproterozoic(2.5 to 1.6 billion years ago) 245
ENRICHMENT   The 18.2-Hour Proterozoic Day 247
Highlights of the Mesoproterozoic(1.6 to 1.0 billion years ago) 249

ENRICHMENT   BIF: Civilization's Indispensable Treasure 250
Highlights of the Neoproterozoic(1.0 to 542 million years ago) 251
Proterozoic Rocks South of the Canadian Shield 253
ENRICHMENT   Heliotropic Stromatolites 254
Proterozoic Life 255

CHAPTER 10

Early Paleozoic Events 2 6 7

Dance of the Continent 269
Some Regions Tranquil, Others Active 269
Identifying the Base of the Cambrian 273
Early Paleozoic Events 273
Cratonic Sequences: The Seas Come in, the Seas Go Out 274
The Sauk and Tippecanoe Sequences 275
Way Out West: Events in the Cordillera 278
Deposition in the Far North 281
Dynamic Events in the Past 281
GEOLOGY OF NATIONAL PARKS AND MONUMENTS
Jasper National Park 282
ENRICHMENT   A Colossal Ordovician Ash Fall: Was it a Killer? 284
The Caledonian Orogenic Belt 288
ENRICHMENT   The Big Freeze in North Africa 290
Aspects of Early Paleozoic Climate 290

CHAPTER 11

Late Paleozoic Events 2 9 5

Epicontinental Sea on the Craton 298
Unrest in the West 303
To the East, a Clash of Contitnents 306
ENRICHMENT   The Wealth of Reefs 309
Sedimentation and Orogeny in the West 316
Europe During the Late Proterozoic 319
Gondwana During the Late Paleozoic 320
Climates of the Late Paleozoic 320
Mineral Productions of the Late Paleozoic 321
GEOLOGY OF NATIONAL PARKS AND MONUMENTS
Acadia National Park 322

CHAPTER 12

Life of the Paleozoic 3 2 7

Animals with Shells Proliferate—and so does Preservation 329
The Cambrian Explosion of Life: Amazing Fossil Sites in Canada and China 330

Continuing Diversification: Each Creature Found Its
  Ecological Niche 333
Protistans: Creatures of a Single Cell 334
Marine Invertebrates Populate the Seas 335
ENRICHMENT   The Eyes of Trilobites 349
Advent of the Vertebrates 353
The Rise of Fishes 354
Conodonts: Valuable but Enigmatic Fossils 359
Advent of Tetrapods 360
Plants of the Paleozoic 363
Mass Extinctions 366

CHAPTER 13
Mesozoic Events 373

The Breakup of Pangea 374
The Mesozoic in Eastern North America 375
The Mesozoic in Western North America 378
GEOLOGY OF NATIONAL PARKS AND MONUMENTS
Zion National Park 382
ENRICHMENT   Did Seafloor Spreading Cause Cretaceous
Epicontinental Seas? 388
GEOLOGY OF NATIONAL PARKS AND MONUMENTS
Grand Staircase-Escalante National Monument 394
The Tethys Sea in Europe 396
ENRICHMENT   Chunneling Through the
Cretaceous 399
Gondwana Events 399

CHAPTER 14
Life of the Mesozoic 405

Climate Controls It All 407
Mesozoic Invertebrates 409
Mesozoic Vertebrates 415
Dinosaurs: "Terrifying Lizards" 417
ENRICHMENT   Can We Bring Back the Dinosaurs? 424
GEOLOGY OF NATIONAL PARKS AND MONUMENTS
Dinosaur National Monument 426
Dinosuars: Cold-Blooded, Warm-Blooded,
  or Both? 429
Dinosaur Parenting 430
The Flying Archosaurs 431
A Return to the Sea 433
The Rise of Birds 437
The Mammalian Vanguard 437
Sea Plants and Phytoplankton 439
Land Plants 441
Late Cretaceous Catastrophe 443
ENRICHMENT   Is There a Bolide in Our Future? 446

CHAPTER 15
Cenozoic Events 451

The Tectonics-Climate Connection 452
Stability and Erosion along the North American
  Eastern Margin 454
Gulf Coast: Transgressing and Regressing Sea 455
Tectonics and Erosion in the Rockies 455
ENRICHMENT   Oil Shale 459
Creating the Basin and Range Province 460
GEOLOGY OF NATIONAL PARKS AND MONUMENTS
Badlands National Park 461
Colorado Plateau Uplift 464
Columbia Plateau and Cascades Volcanism 464
ENRICHMENT   Hellish Conditions in the Basin and Range
Province 464
Sierra Nevada and California 467
The New West Coast Tectonics 467
Meanwhile, Drama Overseas . . . 470
Big Freeze: The Pleistocene Ice Age 473
Why Did Earth's Surface Cool 480
Cenozoic Climates 483

CHAPTER 16
Life of the Cenozoic 487

Grasslands Expand, Mammals Respond 489
Plankton 490
Marine Invertebrates 492
Vertebrates 496
Mammals 499
Monotremes 501
Marsupials 501
Placental Animals 504
ENRICHMENT   How the Elephant Got Its Trunk 517
Demise of the Pleistocene Giants 521

CHAPTER 17
Human Origins 525

Primates 526
Modern Primates 527
The Prosimian Vanguard 528
The Early Anthropoids 529
The Australopithecine Stage and the Emergence of
  Hominids 532
The Homo Erectus Stage 537
ENRICHMENT   Being Upright: Good News, Bad News 538
Final Stages of Human Evolution 538
ENRICHMENT   Neandertal Ritual 540

Humans Arrive in the Americas 543
Population Growth 544

APPENDIX A
Classification of Living Things A1

APPENDIX B
Physiographic Provinces of the
United States A6

APPENDIX C
World Political Map A7

APPENDIX D
Periodic Table and Symbols for Chemical
Elements A9

APPENDIX E
Convenient Conversion Factors A12

APPENDIX F
Exponential or Scientific Notation A13

APPENDIX G
Rock Symbols A13

APPENDIX H
Bedrock Geology A14

APPENDIX I
Common Rock-Forming Silicate Minerals A15

GLOSSARY G1

INDEX I1

1

Q
T
K
J
Tr
P
M
D
S
O
Є
Pre-Є

***Fossil extracted from limestone of Jurassic age
(135–200 million years old).*** *This ammonite cephalopod
is an extinct mollusk, distantly related to today's
chambered nautilus. The "mother of paleontology," Mary
Anning (1799–1847), collected many of these fossils from
this locality at Lyme Regis on England's south coast.*

The Science of Historical Geology

To see where we might be going,
we must understand where we have been.
—Robert Tamarkin, 1993

O U T L I N E

▶ PART I—DISCOVERING TIME AND
  DECIPHERING EARTH'S AMAZING HISTORY
▶ WHY STUDY EARTH HISTORY?
▶ GEOLOGY LIVES IN THE PRESENT
  AND THE PAST
▶ A WAY TO SOLVE PROBLEMS:
  THE SCIENTIFIC METHOD
▶ THREE GREAT THEMES IN EARTH HISTORY
▶ WHAT LIES AHEAD?
▶ SUMMARY
▶ KEY TERMS
▶ QUESTIONS FOR REVIEW AND DISCUSSION
▶ WEB SITES

Key Chapter Concepts

- The study of events in the Earth's past can often be used to predict future events.
- The Earth and its inhabitants have undergone continuous change during the past 4.6 billion (4,600,000,000) years.
- Physical geology examines the structure, composition, and processes that affect the Earth today. Historical geology considers all past events on Earth.
- The scientific method is a way to find answers to questions and solve problems. It involves collection of information through observation and experimentation, formulation of tentative answers, and validation by testing.
- The three most pervasive themes in the history of Earth are the immensity of geologic time, plate tectonics, and organic evolution.

Welcome to the amazing history of our planet! Here you will discover many astonishing events of the past and learn how we came to understand them. You will learn the intriguing story of how life developed on Earth and how an extraordinary species evolved that is the only one capable of reading books like this: *us*.

Our planet formed about 4.6 billion years ago. Since that time, it has circled the sun like a small spacecraft observing a rather average star. Between about 300,000 and 150,000 years ago, a species of primate we call *Homo sapiens* (Latin: *wise human*) evolved on Earth. Unlike earlier animals, these creatures with oversized brains and nimble fingers asked questions about themselves and their surroundings. Their questioning has continued to the present day: *How did Earth form? Why do earthquakes occur? What lies beneath the land and below the ocean floor?*

Even ancient people sought answers to these questions. In frail wooden ships, they probed the limits of the known world, fearing that they might tumble from its edge or be consumed by dragons. Their descendants came to know the planet as an imperfect sphere, and they began to examine every obscure recess of its surface. In harsher regions, exploration proceeded slowly. It has been only within the last 100 years that humans have penetrated the deep interior of Antarctica. Today, except for a few areas of great cold or dense tropical forest, the continents are well charted. New frontiers for exploration now lie beneath the ocean and outward into space.

▶ WHY STUDY EARTH HISTORY?

Earth's spectacular history deserves to be closely examined, for it permits us to see the future. Events of the past will happen again. We owe it to ourselves and to our home planet to look carefully at those events and attempt to understand them.

From the time of its origin to the present day, Earth has undergone continuous modification. Continents have been flooded by vast inland seas. They also have ponderously drifted across the face of the globe and slowly collided with other landmasses to form lofty mountain ranges (Fig. 1-1). Massive glaciers have buried vast tracts of forest and prairie. Earth has witnessed recurrent earthquakes, rampant volcanism, catastrophic impacts of meteorites and asteroids, and major changes in the chemistry of the ocean and atmosphere. Along with these physical changes, life on Earth has also undergone change—sometimes slow, but occasionally swift and deadly.

All of these events of the geologic past have relevance to our lives today. By discovering why they occur, we can better predict the future. For example, we are carefully examining climatic trends of the past so we can better understand today's climatic changes. With knowledge of Earth's history, we can plan ahead. We can avoid further damage to this planetary haven in space that is our home. Aside from these concerns, an important reason to study Earth history is simply to better understand our favorite and unique planet and its amazing forms of life.

▶ GEOLOGY LIVES IN THE PRESENT AND THE PAST

For convenience, we divide the body of knowledge called geology into physical geology and historical geology. **Physical geology** studies Earth's materials (rocks and minerals) and the varied processes that occur both on the surface and deep in the interior. **Historical geology** addresses Earth's origin and evolution, changes in the distribution of lands and seas, the growth and reduction of mountains, and the succession of animals and plants through time. Historical geologists examine planetary materials and structures to discover how they came to exist. They observe the tangible *results* of past events and work backward in time to discover the *causes* of those events.

Geology primarily studies Earth, but its view has broadened to include other planets. This increase in scope is appropriate, because geologic knowledge is employed in interpreting the images of the surfaces of other planets and their moons, in estimating the power of volcanoes on Venus, and in identifying rocks and minerals from Earth's moon.

Geologists identify the minerals in meteorites (Fig. 1-2) to discover how Earth formed. With sophisticated instruments, they scrutinize images of planets or interpret data transmitted by space probes and planetary exploration rovers (Fig. 1-3). Still others busily unravel the structure of mountain ranges, attempt to predict hazards like earthquakes and volcanic eruptions (Fig. 1-4), or study the behavior of glaciers, streams, or underground water.

FIGURE 1-1 **The magnificent Canadian Rocky Mountains viewed from Malign Lake, British Columbia.** These mountains were initially raised over 80 million years ago. Their present appearance results from further uplift and erosional sculpting during subsequent geologic periods down to the present day.

FIGURE 1-2 **Stony meteorite resting on snow of the Antarctic Ice Cap.** The contrast of dark meteorites on white snow makes Antarctica a good place for collecting meteorites. The meteorite is composed of iron and magnesium silicates.

FIGURE 1-3 **NASA's robotic rover *Opportunity* on Mars in 2004.** Its instruments sampled soil and rock, photographed rock outcrops, and found solid evidence to support the hypothesis that Mars, now a global desert, once had flowing water.

Many "exploration geologists" search for fossil fuels and the metallic ores vital to our standard of living. This requires knowledge of both physical and historical geology. To understand where to find resources, exploration geologists draw on their knowledge of Earth history, astronomy, physics, chemistry, mathematics, and biology. For example, a petroleum geologist must understand the physics of moving fluids, the chemistry of oil and gas, and the biology of the fossils (Fig. 1-5) that are used to trace subsurface rock layers.

Because geology incorporates information from so many other scientific disciplines, it is an "eclectic" science—it draws on information from many sources. All sciences are eclectic to some degree, but geology is decidedly more so.

▶ A WAY TO SOLVE PROBLEMS: THE SCIENTIFIC METHOD

Geologists, both physical and historical, employ the same procedures used by scientists in other disciplines. Those procedures are called the **scientific method.**

FIGURE 1-4 **Geologist studies a disastrous mudflow resulting from the eruption of Mount St. Helens, Washington State, in 1980.** Mount St. Helens is in the background. The eruption devastated nearly 600 square kilometers and killed 57 people.

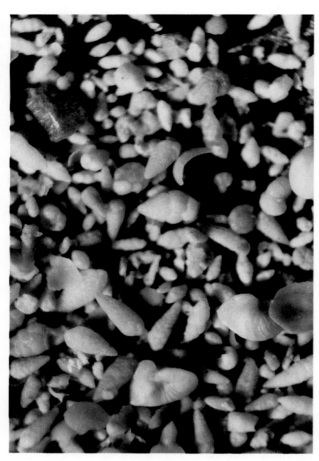

FIGURE 1-5 **Fossil shells of single-celled marine animals.** These distinctive shells of foraminifera are widely used to identify rock formations when drilling for oil and natural gas.

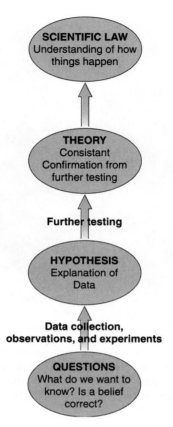

FIGURE 1-6 **Typical steps in the scientific method.** An initial question stimulates the collection of data. Scientists then study the data and build a hypothesis to answer the question. Through exhaustive testing, the hypothesis is accepted, revised, or rejected. A hypothesis that consistently answers questions rises to the level of a scientific theory. If the theory "works" in every known case over a long period, it can attain the status of a scientific law or principle.

The scientific method is a systematic way to find answers to questions, solutions to problems, and evidence to prove or disprove ideas and beliefs.

A scientific investigation often begins with a *question* (Figure 1-6). It proceeds to the collection of *data* (facts from observations and experiments), and is followed by the development of a ***hypothesis*** that fits all the data and is likely to account for observations in the future as well as the present. A hypothesis then is tested and critically examined by other scientists, so it subsequently may be confirmed, modified, or discarded. In some cases, several hypotheses may be proposed to explain the same set of data, and each is tested until the "best" one emerges. For example, the origin of the universe and the origin of life have each been the subject of several hypotheses.

A hypothesis that survives repeated challenges and is supported by accumulating favorable evidence may be elevated to a ***theory***. A theory has survived such intense scrutiny that it can be accepted with more confidence than a hypothesis. Examples are the theory of relativity, plate tectonics theory, evolutionary theory, and atomic theory.

It is important to understand that the term "theory" has very different meanings to scientists and to the public. To a scientist, a theory represents knowledge that has a very high probability of being correct. However, in common language, "theory" implies a lack of knowledge or a guess.

The search for scientific truth does not end with the formulation of a theory. Even after a theory has been firmly established, it must continue to survive rigorous testing derived from advances in science and technology that its author could not have foreseen. If a theory continues to triumph over every challenge, it can be raised to the level of a ***scientific law***, such as the law of gravitational attraction.

An Example of the Scientific Method

Applying the scientific method in historical geology, consider the following research into some curious features of the Mediterranean seafloor.

The questions. Several observations raised questions about the Mediterranean Sea's history:

1. Microscopic single-celled plants and animals living in the Mediterranean changed abruptly about 6 million years ago. Most of the older organisms were nearly wiped out. A few survived by migrating into the Atlantic. Somewhat later, the migrants returned, bringing new species with them. *What dramatic event happened? Why did the near extinction occur? Why did the migrants subsequently return?*

2. An enormous buried gorge extends seaward from the present course of the Rhone River (Fig. 1-7). Similar buried gorges had been found off the coast of North Africa. *What gave a stream sufficient power to erode such canyon-like features?*

3. A hard layer of sedimentary rock, detected by seismic instruments, lies 100 meters or so below the present sea floor. *What is the origin of this hard layer?*

4. Domelike rock structures exist deep beneath the Mediterranean seafloor. They were detected years earlier by echo-sounding instruments, but they had never been investigated by drilling. *Are they huge plumes of salt (called salt domes) like the ones that are common along the U.S. Gulf Coast? If so, what caused the precipitation of so much rock salt?*

In 1970, geologists Kenneth J. Hsu and William B.F. Ryan boarded the oceanographic research vessel *Glomar Challenger* to search for answers (Fig. 1-8). They drilled the Mediterranean seafloor to obtain samples. As drilling progressed, they recovered a sample from the surface of the hard layer. It consisted of pebbles of hardened sediment that had once been soft, deep-sea mud, plus granules of gypsum (a mineral commonly formed by the evaporation of seawater). However, not a single pebble was found to indicate that the sediment had been carried to the sea from surrounding land areas.

In the days following, samples of solid gypsum were repeatedly brought on deck as drilling penetrated the hard layer—clearly, it was a bed of gypsum. The composition and texture of the gypsum suggested it had formed by evaporation on desert flats. But sediment above and below the gypsum layer contained tiny marine fossils, indicating not a desert-like environment, but normal open-ocean conditions.

The Hypothesis. The time had come to *formulate a hypothesis that the Mediterranean Sea was once a desert.* Hsu and Ryan proposed that about 20 million years ago, the Mediterranean was a broad seaway linked to the Atlantic by narrow straits, like the present Strait of Gibraltar. Tectonic movements of Earth's crust closed the straits. Turned into a giant salt lake, the Mediterranean began to evaporate and shrink. Evaporation concentrated the various salts that were dissolved in the water, and this increasing salinity exterminated scores of marine species.

As evaporation continued, the remaining brine became so saturated that minerals dissolved in it were forced to precipitate—that is, they were forced to separate from the solution. This formed the hard layer of gypsum. Different salts precipitate at different rates, and the remaining brine in the central, deeper part of the basin was rich in sodium and chlorine ions, so the water evaporated to precipitate sodium chloride (table salt).

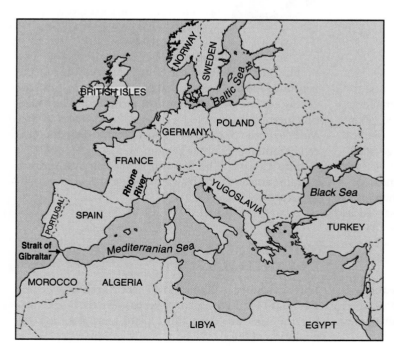

FIGURE 1-7 **Mediterranean region.** When the Mediterranean was a desert, the Rhone River no longer entered at sea level, but flowed down a steep slope, eroding a gorge over a kilometer deep that is now buried beneath sediment.

A

B

FIGURE 1-8 **Aboard the research ship** *Glomar Challenger*. (A) Roughnecks on the drill floor add lengths of drill pipe as drilling progresses deep into the Mediterranean seafloor. As they drill, sediment is forced into the pipe, like jamming a soda straw into mud. This sediment forms cylindrical "drill cores" that are removed from the pipe for study. (B) A geologist on board examines cores of sediment recovered during drilling operations. The drill cores reveal the layers on the Mediterranean's floor.

The dried-up Mediterranean had become a vast "Death Valley" 3000 meters deep. Streams entering the basin from Europe and Africa now had steep gradients, enabling them to erode spectacular gorges. Then, about 5.5 million years ago, new crustal movements caused the Strait of Gibraltar to open. The resulting deluge of Atlantic seawater cascaded down tremendous waterfalls into the vast desert, and began to refill the Mediterranean.

Initially, turbulent currents tore into the hardened salt flats, grinding them into the pebbles observed in the first sample taken by the *Glomar Challenger*. As the basin refilled, marine organisms returned, mostly migrants from the Atlantic. Soon layers of oceanic ooze were deposited above the old hard layer.

Long after the Mediterranean Basin was refilled, pressure from the weight of overlying sediments forced the salt to flow plastically upward to form salt domes like those on the U.S. Gulf Coast.

The questions about the faunal changes, the salt and gypsum deposits, the unusual pebbly sediment, and the deeply buried gorges were now answered. The scientific method had worked well, and *the hypothesis that the Mediterranean Sea was once a desert* could now be critically examined by other geologists.

The theory. The hypothesis has survived critical examination and is on its way to being accepted as a rigorous *theory*.

THREE GREAT THEMES IN EARTH HISTORY

Earth's history is like a novel, with grand, sweeping themes. The three major themes—intensely interacting themes—are deep time, plate tectonics, and the evolution of life.

Deep Time

Recognition of the immensity of geologic time is the single most important contribution to human knowledge made by geology. Geologists look back across a calculated 4.6 billion years (4,600,000,000) of Earth history, from our planet's chaotic birth to the present. Compared to the average duration of a human life, it is a span of time so huge as to be difficult to comprehend. It is not surprising, therefore, that our ancestors believed hills and valleys were changeless and eternal and that the planet originated only a few thousand years ago.

Eventually, we came to realize that the slow and relentless work of erosion reduces mountains to plains, that valleys are the result of long periods of erosion, and that sands and gravels produced by erosion have been turned to rock. These changes clearly required vast amounts of time.

But how much time? And which event preceded—or followed—another? To answer these questions, it was necessary to find the **absolute age** (actual age) of rocks in years. A way to do so was enabled by the discovery of *radioactivity* in 1896.

Certain atoms are unstable, causing them to be radioactive. This means they decay by expelling particles of their nuclei, converting themselves into stable "daughter" atoms, mostly of different elements. A well-known example is uranium, which is radioactive and continually expels particles. The rate of this decay can be accurately measured. The term **half-life** expresses the rate of decay. Half-life is the time required for one-half of the original quantity of radioactive atoms to decay.

As an example, the radioactive element uranium-235 has a half-life of approximately 704 million years. This means that after 704 million years pass, only half (50%) of the uranium-235 in a mineral will remain. After a second 704 million years pass, half of that half (25%) will have decayed. Thus, a rock that contains only 25% uranium-235 and 75% of the daughter atoms must be 1,408 million years old (704 + 704). Using this method, some Earth rocks have been calculated to be 4.04 billion years old (Fig. 1-9).

Before the discovery of radioactivity, geologists were able to determine only if particular layers or bodies of rock were *older* or *younger* than others. This determined a rock's **relative age**. Relative age determinations provided a framework in which to place events of the geologic past. Through relative dating, geologists developed a *geologic time scale*, and with dating through radioactive decay, that time scale was calibrated in actual years. We will examine how this was accomplished in Chapter 3.

Plate Tectonics

A significant number of events in both physical and historical geology are related to a grand unifying concept termed **plate tectonics**. "Tectonics," from the same Greek word as "architecture," refers to large-scale de-

FIGURE 1-9 **Geologist Samuel Bowring stands before Earth's oldest known rocks.** Named the Acasta Gneiss (pronounced "nice"), this most ancient of rocks outcrops in Canada's Northwest Territories. Radioactive dating methods indicate it is 4.04 billion years old, having survived 87% of Earth's 4.6-billion-year history.

formation of rocks in Earth's outer layers. The term "plate" is given to large slabs of Earth's lithosphere. The **lithosphere** is the rigid outer layer of Earth (roughly 100 km thick) that includes the **crust** as well as the uppermost part of the **mantle** (Fig. 1-10).

Earth's surface consists of seven large lithospheric plates and about twenty smaller ones. The plates rest on a plastic, easily deformed layer of the mantle called the **asthenosphere**. Probably because of heat-driven convectional flow in the asthenosphere, the plates move. They move almost imperceptibly, only millimeters per year and a few feet in a human lifetime.

Tectonic plates have well-defined edges or "margins." Where two or more plates move apart (diverge) from one another, the plate margins form **divergent boundaries**. Where plates converge, **convergent boundaries** occur. Where plates grind past one another, **transform boundaries** occur. You will meet these terms again in Chapter 7, where we examine plate tectonics in more detail.

Evolution of Life (Organic Evolution)

Plate tectonics is the "great unifying theory" that explains many physical phenomena in geology. In biology, evolution is the "great unifying theory" for understanding the history of life. Because of evolution, animals and plants living today are different from their ancestors. They have changed in appearance, in genetic characteristics, in the way they function, and in their body chemistry, apparently in response to changes in the environment and competition for food. Fortunately, fossils record these changes for us to study. Fossils are valuable indicators of the age of rocks.

Although Charles Darwin is credited for the concept of evolution, the idea began as early as 2600 years ago, in the seventh century BCE. We find it in the writ-

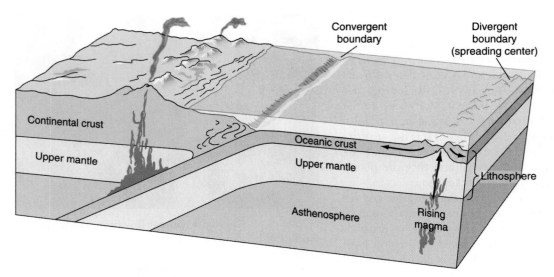

FIGURE 1-10 **The lithosphere is Earth's rigid outer shell.** It lies above the asthenosphere, and it includes the upper part of the mantle and both types of crust: continental and oceanic.

ings of the Greek philosopher Anaximander. But Darwin and his colleague Alfred R. Wallace were the first scientists to propose a hypothesis with convincing evidence. They also proposed a working mechanism for evolution, which Darwin called natural selection.

Natural selection is based on several important observations: that any given species produces more organisms than can survive to maturity; that variations exist among offspring; that offspring must compete for food and habitat; and that those lucky individuals with the most favorable variations are most likely to survive and pass their beneficial traits to the next generation. Scientists eventually came to understand genetics as the cause of these variations.

Darwin provided many lines of evidence for evolution. He cited the direct evidence of changes seen in fossils in successively younger strata. He noted that certain organs were fundamentally similar from species to species, but became modified to function differently, apparently to make the species more competitive. There were also useless organs in modern animals that clearly had a useful function in ancestral species but were "evolving out." (Human examples include the appendix and tailbone.) He further noted that animals that looked quite different as adults nevertheless had very similar-looking embryos.

There were many other lines of evidence as well, but none as compelling as subsequent work in genetics, biochemistry, and molecular biology. We now know that the biochemistry of closely related organisms is similar to, but distinctly unlike, that of their distant relatives. The sequence of amino acids in proteins and the characteristics of the famous DNA molecule are also most similar in closely related organisms. Discoveries such as these clearly indicate that animals and plants of each geologic era arose from earlier species by the

process we call organic evolution. (Organic means living, as opposed to physical evolution of Earth's rocks.) We will examine this important theory in Chapter 6.

▶ WHAT LIES AHEAD?

Your book is divided into three parts:

Part I—Discovering Time and Deciphering Earth's Amazing History gives you a broad perspective on our subject, introducing the pioneers of historical geology whose careful detective work and brilliant insights enabled them to unravel the geologic record.

Part II—Rocks and What They Tell Us about Earth History illustrates how rocks and fossils have recorded events in geologic history and how we have learned to read this wonderful archive. We examine sedimentary rocks, because often trapped in their sediment is the ancient animal and plant debris we call fossils. We study fossils because without them, nothing would be known about our planet's earlier inhabitants. You will learn about the evidence for plate tectonics and how this dynamic process has shaped Earth and influenced life throughout geologic time.

Part III—History of Earth and Its Inhabitants gives you our best chronology of physical and biological events on Earth since its origin 4.6 billion years ago. We describe the planet's birth as a primordial ball of cosmic debris and explain how our continents, oceans, and the atmosphere evolved and interacted over time. Here you can survey the vast panorama of animals and plants that populated bygone eras. Most recently, humans appeared on the scene, the first creatures we know to discover Earth's amazing history.

SUMMARY

- Earth is not a static ball of rock orbiting the sun. From the time of its origin 4.6 billion years ago to the present, it has been undergoing continuous change.
- Knowledge of events in the geologic past have relevance to conditions on Earth today and can be used to solve current and future problems.
- Geology is a science devoted to the study of Earth—its origin, history, composition, and properties. Geology has two interrelated branches. Physical geology focuses on processes within and on the surface of Earth, as well as Earth's chemical and physical features. Historical geology is the branch concerned with decoding the rock and fossil record of the planet's long history.
- The scientific method is a procedure by which scientists study problems and answer questions. The method usually begins with one or more questions, then collection of data. From this, scientists formulate a hypothesis that is supported by the data. It can then be tested for its validity.

With sufficient testing, a hypothesis can become elevated to a theory or even a law.
- Three great themes in historical geology are deep time, plate tectonics, and organic evolution.
- Geology's most important contribution to knowledge is recognition of the immensity of geologic time. Realization of deep time is fundamental to our understanding of Earth and our place within the universe.
- Plate tectonics describes large-scale movements and interactions of rigid plates of Earth's lithosphere. These movements and interactions control major geologic events, including mountain building, earthquakes, volcanism, and the configuration of continents and ocean basins.
- Organic evolution refers to changes that have occurred in organisms with the passage of time. It is fundamentally important for understanding humanity's relationship to Earth's biological realm, and it provides a basis for determining the relative age of rocks.

KEY TERMS

absolute age, p.7
asthenosphere, p.7
convergent boundary, p.7
crust, p.7
divergent boundary, p.7
half-life, p.7
historical geology, p.2
hypothesis, p.4
lithosphere, p.7

mantle, p.7
natural selection, p.8
physical geology, p.2
plate tectonics, p.7
relative age, p.7
scientific law, p.4
scientific method, p.3
theory, p.4
transform boundary, p.7

QUESTIONS FOR REVIEW AND DISCUSSION

1. Application of the scientific method may lead to the development of a hypothesis. The hypothesis may at a later stage be elevated to a theory (like the theory of plate tectonics). What must occur to validate a hypothesis in this way?

2. What is the lithosphere? How does it differ from the crust?

3. Define plate tectonics. Differentiate among convergent, divergent, and transform plate boundaries. What is the significance of plate tectonics in our lives today?

4. What is the importance of fossils to historical geology?

5. Why are fossils useful as evidence for evolution?

6. How old is our planet? What is the age of the oldest rocks found so far? Why is it unlikely that geologists will find rocks at Earth's surface that are as old as Earth itself?

7. Throughout geologic time, the climate and topography of particular regions have changed. Would these changes affect the course of evolution? Provide an example.

8. In determining the age of Earth, why is it important to determine the age of meteorites?

WEB SITES

The Earth Through Time Student Companion Web Site (www.wiley.com/college/levin) has online resources to help you expand your understanding of the topics in this chapter. Visit the Web Site to access the following:

1. Illustrated course notes covering key concepts in each chapter;

2. Online quizzes that provide immediate feedback;

3. Links to chapter-specific topics on the web;

4. Science news updates relating to recent developments in Historical Geology;

5. Web inquiry activities for further exploration;

6. A glossary of terms;

7. A Student Union with links to topics such as study skills, writing and grammar, and citing electronic information.

2

Q
T
K
J
Tr
Pr
P
M
D
S
O
€
Pre-€

Strata exposed at Siccar Point on the southeast coast of Scotland reveal a striking unconformity. *It was here that James Hutton perceived the significance of unconformities to Earth history.*

Early Geologists Tackle History's Mysteries

And some rin up hill and down dale,
knapping the chunky stanes to pieces wi' hammers,
like sae many road makers run daft. They say it is
to see how the world was made.
—Sir Walter Scott, *St. Ronan's Well*

Key Chapter Concepts

- Fossils are intriguing remnants of former life, without which a history of Earth would be incomplete.
- Nicholas Steno provides the basic principles of superposition, original horizontality, and original lateral continuity.
- John Strachey, Giovanni Arduino, Johann Lehmann, Georg Fuchsel, and Peter Simon Pallas recognize the general age relationships and nature of major groups of rock assemblages in Europe.
- Abraham G. Werner publishes the first great textbook of mineralogy, but errs in his belief that basalt was an oceanic deposit.
- James Hutton perceives the immensity of geologic time and understands the relation between processes of the present and of the geologic past.
- William Smith demonstrates the use of fossils for correlating strata.
- Georges Cuvier conceives the theory of catastrophism and stipulates that life on Earth underwent periodic extermination, followed by reappearance of entirely new life forms.
- Charles Lyell explains inclusions of one rock in another, and how rocks that cut into and across other rocks can be used to determine relative geologic age.
- Charles Darwin and Alfred R. Wallace conceive natural selection as a mechanism for evolution.
- Pioneers of American geology:
  - Louis Agassiz recognizes the evidence for the great Pleistocene Ice Age
  - James Hall understands the depositional and mountain-building history of the Appalachian Mountains
  - Ferdinand Hayden, John Powell, and Clarence King complete geologic and topographic surveys of the American West
  - Othniel C. Marsh and Edwin D. Cope collect and describe the multitude of extinct vertebrate animals that once populated North America

O U T L I N E

▶ THE INTRIGUE OF FOSSILS
▶ AN EARLY SCIENTIST DISCOVERS SOME BASIC RULES
▶ EUROPEAN RESEARCHERS UNRAVEL THE SUCCESSION OF STRATA
▶ NEPTUNISTS AND PLUTONISTS CLASH
▶ UNIFORMITARIANISM: JAMES HUTTON RECOGNIZES THAT "THE PRESENT IS KEY TO THE PAST"
▶ THE PRINCIPLE OF FOSSIL SUCCESSION
▶ THE GREAT UNIFORMITARIANISM–CATASTROPHISM CONTROVERSY
▶ THE PRINCIPLE OF CROSS-CUTTING RELATIONSHIPS
▶ EVOLUTION: HOW ORGANISMS CHANGE THROUGH TIME
▶ EARTH HISTORY IN AMERICA
▶ SUMMARY
▶ KEY TERMS
▶ QUESTIONS FOR REVIEW AND DISCUSSION
▶ WEB SITES

In 1795, William Smith, a surveyor in England was hired to determine the best route for a canal for coal barges. But before beginning work, Smith asked some preliminary questions: What kind of rocks would have to be excavated? Could he predict what rocks would be encountered along the route? Would the digging be easy or difficult?

As he walked along a creek, Smith studied every exposed edge of the layered rocks. At one point, he climbed down an embankment to closely examine a protruding ledge of rock. Oblivious to the water percolating into his shoes, he ran his hand across the rock surface. The shell of an ancient snail caught his eye. With hammer and chisel, Smith chipped his discovery from the rock and placed it in a canvas bag.

He recognized this rock with its signature fossil. He knew what rock stratum he would find below it and what lay above it. Only yesterday, he had seen this same layer two miles to the north. There was remarkable consistency here. Smith perceived that each stratum, known by its fossils and rock type, not only extended invisibly beneath the farms in the distance, but also maintained its place in the vertical succession of rock layers.

William Smith's discovery is just one of the great contributions to early geology that you will discover in this chapter.

THE INTRIGUE OF FOSSILS

The remains and traces of prehistoric life we call fossils have sparked the interest and imagination of people from before the advent of civilization to the present day. We think that fossils were prized possessions of Neandertals over 30,000 years ago, for fossils have been found among the artifacts of these heavy-browed cave dwellers.

We suspect that Neandertals believed that fossils held magical powers. But a more scientific interpretation was offered about 450 BCE by the Greek philosopher Herodotus. While traveling in Egypt and Libya, Herodotus saw fossil seashells in outcrops of sedimentary rock far from the sea and high above sea level. They were similar to those he had seen along the shores of the Mediterranean. From this, he concluded that the land he stood on was once beneath the sea.

In the centuries that followed, fossils of all kinds were observed and collected across Europe. Some were bones of ancient land animals, but the majority were fossil shells of marine creatures similar to those observed by Herodotus (Fig. 2-1). They indicated that seas had once covered the continent. But how could such a great flooding be reconciled with religious doctrine of the day? To many the answer seemed obvious. The fossil seashells and the rocks in which they were embedded were an affirmation of the biblical story of a great flood and Noah's Ark.

But why would ocean-dwelling creatures be exterminated by a flood of water, the very medium in which they lived? And how might we explain the evidence of many floods at different times and in different places? By the beginning of the 1800s, geologists had acquired a better understanding of the immensity of geologic time and were able to use methods of relative dating to derive a more valid interpretation of fossils. They now realized that fossils were the record of entire dynasties of living creatures that preceded the arrival of humans.

How Do Fossils Form?

Worldwide, sediment is continually deposited. Sediment includes mud and silt from streams, debris settling from ocean water onto the seafloor, dust or volcanic ash deposited by the wind, and chemical precipitates.

FIGURE 2-1 **Limestone containing 200-million-year-old fossil ammonites.** Ammonites are an extinct group of cephalopods related to the living chambered nautilus.

FIGURE 2-2 **Spider preserved in amber.**

Countless times in the geologic past, plants and animals—dead or living—have been covered by sediment and preserved as fossils.

For preservation, quick burial is needed. In the case of animals, if the creature has a hard skeleton or shell, all the better. After death, soft tissue will rot away as flesh, and other soft tissue are consumed by bacteria and scavengers. But shell and bone will be left to petrify in the gradually hardening matrix of sediment. This is why they are plentiful as fossils.

However, occasionally we are rewarded with evidence of soft parts. Small animals, especially insects, are preserved in the hardened resin ("amber") of conifers (Fig. 2-2). Carbonized imprints of jellyfish and worms can be seen in X-rays of rocks. Skin, hair, and stomach contents of Ice Age mastodons have been pickled in tar from oil seeps or frozen into glacial ice.

All of these preservations—as well as foot tracks, trails, and even holes dug by some animals—permit us to discern the history of life. The story is not complete, for new fossil discoveries are made daily. Often only a fragment of bone or shell remains, but therein lies the fascination. We become detectives, using our reason, imagination, and intuition to recreate the form and habits of a once-living thing.

► AN EARLY SCIENTIST DISCOVERS SOME BASIC RULES

Niels Stensen (1638–1687), a Danish physician, was widely recognized for his studies in anatomy. He moved to Italy, Latinized his name to Nicolaus Steno, and became physician to the Grand Duke of Tuscany. The duke was a generous employer, giving Steno ample time to tramp the countryside, visit quarries, and examine strata. His observations of sedimentary rocks led him to formulate three basic principles of historical geology: superposition, original horizontality, and original lateral continuity. All three are common-sense but valuable, and are described below.

Principle of Superposition

*The **principle of superposition** states that in any sequence of undisturbed strata, the oldest layer is at the bottom, and successively younger layers are successively higher.* This seems obvious, easily demonstrated by tossing sheets of paper on the floor: the one on the bottom was "deposited" first and therefore is the oldest. Yet Steno, based on his observations of strata in northern Italy, was the first to formally describe the concept applied to rock layers.

The fact that superposition is self-evident does not diminish the principle's importance in determining the relative age of strata, oldest to youngest. The superpositional relationship of strata is not always clear where layers are steeply tilted or even overturned (Fig. 2-3). In such instances, the geologist must examine the strata

FIGURE 2-3 **Principle of superposition.** Steeply dipping strata grandly exposed in the Himalayan Mountains. It is often difficult to recognize the original tops or bottoms of beds in strongly deformed sequences such as this. For example, at the snowy patch to the right of center, if you could see only this area and not the entire mountainside, would you be able to tell which layer was originally on the bottom?

FIGURE 2-4 **Principle of original horizontality.** This is a sequence of undisturbed, horizontal strata in Canyonlands National Park, Utah. The Colorado River in the foreground carved the deep gorge, exposing the strata in this natural geological laboratory. ◄ *Along this bend in the river, where would one find the youngest stratum? (Answers to questions appearing within figure legends can be found in the Student Study Guide.)*

for clues useful in recognizing their lowermost and uppermost layers. Useful clues in determining "which way was up" at the time of deposition are the way fossils lie in the rock and the evidence of mud cracks and ripple marks, which form on a surface.

Principle of Original Horizontality

The **principle of original horizontality** *states that sediment is deposited in layers that are originally horizontal.* Steno reasoned this to be so because most sedimentary particles settle out of water or air straight down, under the influence of gravity (Fig. 2-4). This is obvious when you pour a cup of sand into an aquarium filled with water; the result is a horizontal layer of sand. Today, if we see flat-lying strata, it is probably in its original horizontal position, untilted and unfolded. But if we see steeply inclined strata, this indicates crustal disturbance that occurred *after* deposition, altering the original horizontality (Fig. 2-5).

Principle of Original Lateral Continuity

The **principle of original lateral continuity** *states that a rock layer extends continuously in all directions until it thins out or encounters a barrier.* Steno's third principle recognizes that, when sediment is deposited on the floor of an ocean or a lake, it extends continuously in all directions until thinning at the margin of the basin. Again, pour a cup of sand into an aquarium, and this is how it behaves. The layer may end abruptly against some barrier to deposition, or grade laterally into a different kind of sediment (Fig. 2-6).

The significance of this in geology is that, if you observe an exposed cross-section of strata on a valley wall, you know that the strata, as originally deposited, will continue laterally on the other valley wall. Further, the

layer will continue for some distance in all directions. This can be determined by observing other valley walls, road cuts, quarries, mines, and drill cores from wells. If you do not observe lateral continuity, and the cause is not related to the reasons given earlier, then there may have been displacement of strata by faulting, or the layer may have been eroded away.

Today, we recognize that Steno's three principles are basic to the geologic specialty of **stratigraphy,** which is

FIGURE 2-5 **Steeply dipping shale and sandstone strata.** These are located in the Ouachita (*WASH-i-taw*) Mountains of Oklahoma.

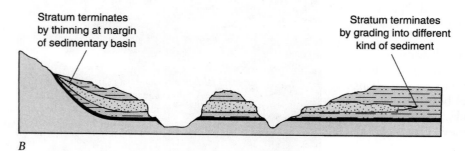

FIGURE 2-6   Principle of original lateral continuity. Cross-section **A** shows a sandstone layer deposited within a low-lying area or sedimentary basin. It received sediment that was eroded from surrounding uplands. **B** shows the same area after erosion has exposed the sandstone on hillsides. Note that the stratum continues horizontally until it stops at the margin of a basin or other obstruction, or until it grades into different sediment. Also, strata may terminate abruptly as a result of erosion or faulting, but originally they were laterally continuous. ✍ *What conditions in the environment of deposition may have caused the sandstone layer (stippled) to "pinch out" on the right side of section B?*

the study of layered rocks, including their texture, composition, arrangement, and correlation from place to place. Because stratigraphy enables geologists to sequence the events that are recorded in rocks, it is the key to understanding Earth's history.

EUROPEAN RESEARCHERS UNRAVEL THE SUCCESSION OF STRATA

The stratigraphic principles formulated by Steno were rediscovered several decades later by other European scientists. John Strachey (1671–1743) used the principles of superposition and original lateral continuity to decipher the strata of coal-bearing rock formations in two English counties, one in northern England and the other in the south. He clearly illustrated the sequence of formations encountered at the surface and in the coal mines. He described how horizontal strata rested upon the eroded edges of inclined older formations (what we now call an unconformity, described later).

Other naturalists developed a more global view of stratigraphy. In Italy, Giovanni Arduino (1713–1795) classified mountains according to their most abundant rock type. He defined "primary mountains" as those constructed of "crystalline rocks" (what we now call igneous and metamorphic). He recognized these "primary mountains" as likely to be the oldest in a mountain

range. His "secondary mountains" were constructed of layered consolidated fossiliferous rocks (what we now call sedimentary). Arduino's "tertiary mountains" were gravel, sand, and clay beds. He made a sensible start at classifying mountains and rocks.

The origin and history of mountain ranges also intrigued another geologist, Peter Simon Pallas. He traveled throughout Asia and carefully studied the Ural and Altai Mountains. Working in the Urals, he observed the change in rock assemblages as he traveled from the center to the flanks of the range. These observations helped him produce a geologic history of the Ural Mountains.

NEPTUNISTS AND PLUTONISTS CLASH

Historical geology might seem a quiet profession today, but its history has been filled with controversy and conflicting ideas. An early example was the collision of two hypotheses that attempted to explain the origin of rocks.

Professor Abraham Gottlob Werner (1749–1817) was an influential, eloquent, and enthusiastic lecturer at the Freiberg Mining Academy in Saxony (Germany). He published the first great mineralogy textbook, and many geologists used his methods to identify minerals and ores.

Werner insisted that all rocks were deposited or precipitated from a great ocean that once enveloped the entire planet. Because of this, Werner and his many followers were called *Neptunists* (after Neptune, the Roman god of the sea).

Werner envisioned his universal ocean as hot, steamy, and saturated with all the dissolved minerals needed to form the rocks of his oldest division, "primitive" rocks (the ones Arduino called "primary"). Werner thought that most of these rocks formed the cores of mountain ranges.

Werner's second stage of Earth history involved subsidence and cooling of the primitive ocean to resemble the ocean of today. Werner said this change was marked by deposition of fossil-bearing rock strata that lie above the primitive rocks. He called these "transition rocks," suggesting they were deposited when Earth became suitable for life.

Above his "transition rocks," Werner observed flat-lying sandstones, shales, coal beds, very fossiliferous limestones, and old lava flows. He added a final term, "alluvium," for the sand, gravel, and clay that usually rested on the "transition rocks."

Werner's ideas soon drew criticism. He failed to explain what had become of the immense volume of water that once covered Earth to a depth so great that all continents were submerged. An even greater problem was his insistence that the lava flows were deposited in the same manner as their enclosing limestones and shales. Other geologists showed field evidence of features that could have formed only in molten lava, demonstrating their volcanic origin.

Geologists with this opposing view were called *Plutonists* (after Pluto, the Roman god of the underworld). Plutonists said that "fire" (heat), rather than water, was the key to the origin of "primitive" igneous rocks. These rocks were not formed in Werner's steamy sea. James Hutton of Scotland, a prominent Plutonist, stated that rocks such as the lava flows and granite "formed in the bowels of the Earth of melted matter poured into rents and openings of the strata."

▶ UNIFORMITARIANISM: JAMES HUTTON RECOGNIZES THAT THE PRESENT IS KEY TO THE PAST

James Hutton (1726–1797), an Edinburgh physician and geologist (Fig. 2-7), saw Earth as a dynamic, ever-changing globe where new rocks, lands, and mountains arise slowly but continuously to balance their loss to erosion and weathering. His dynamic view opposed Werner's static concept of an Earth that had changed very little from its beginning. Hutton also believed that "the past history of our globe must be explained by what can be seen to be happening now." In other words, the present is key to the past.

FIGURE 2-7 **The present is key to the past.** James Hutton, Scottish physician, farmer, and geologist, recognized that studying present processes (weathering, erosion, deposition of sediment, volcanism, etc.) provides the means of interpreting ancient rocks. The idea became established as the principle of uniformitarianism.

Uniformitarianism

This simple yet powerful idea was later named *uniformitarianism* because it says that geologic processes are uniform through all time. Specifically, what is "uniform" are *the physical and chemical laws that govern nature*. Hence, uniformitarianism says that Earth's history may be deciphered by observing events today and assuming that the same phenomena have operated in the same way throughout geologic time.

These natural laws summarize all of our observational and experimental knowledge. They let us predict the conditions under which water becomes ice, the behavior of a volcanic gas when it is expelled at Earth's surface, or the effect of gravity on a grain of sand settling to the ocean floor. Uniform natural laws govern geologic processes such as weathering, erosion, glacial movement, earthquakes, volcanic eruptions, and the transport of sediment by streams.

Hutton's understanding of uniformitarianism was simple and logical. By observing geologic processes around him, he was able to infer the origin of features he discovered in rocks. For example, Hutton watched how waves wash onto the shore and produce ripple marks in the sand. From this he inferred that strata near Edinburgh that bore similar markings must have been shoreline deposits of an ancient sea. Those strata lay far inland from a coast, but Hutton was able to recognize that a sea once existed where Scottish sheep now graze.

Uniform, But . . .

The uniformitarianism concept works well for solving many geologic problems. However, the geologic past actually was sometimes quite *unlike* the present. For example, before Earth evolved its present oxygen-rich atmosphere, oxidation was not the major rock-weathering agent that it is today. Other chemical reactions weathered rocks instead. Also, life originated in a primordial, oxygenless atmosphere under conditions that have no present-day counterpart. Yet another example is meteorite bombardment of Earth's surface. Today, large meteorites that reach the surface are rare. But 3 billion years ago, large meteorites bombarded the surface mercilessly.

Many times in the geologic past, parts of continents stood higher above the oceans than they do today. This higher land elevation allowed for faster rates of erosion and harsher climates than were possible during the periods when lands were lower and partially covered with inland seas. Similarly, at one time or another in the geologic past, volcanism was much more extensive than it is today.

Nevertheless, ancient volcanoes disgorged gases and deposited lava and ash just as present-day volcanoes do. Modern glaciers are more limited in area than those of the recent geologic past, yet they still form erosional and depositional features that resemble those of their more ancient counterparts. All of this suggests that present events do indeed give us clues to the past.

At the same time, we must be constantly aware that in the past, the rates of change and intensity of processes often varied from those we are accustomed to seeing today, and that some events of long ago simply have no modern counterpart.

Actualism

To emphasize the importance of natural laws in the concept of uniformitarianism, many geologists prefer the term **actualism**. *Actualism* is the principle that natural laws governing both past and present processes on Earth have been the same. Hutton's friend John Playfair (1748–1819) provided an eloquent statement of actualism in 1802:

> Amid all the revolutions of the globe, the economy of Nature has been uniform, and her laws are the only thing that have resisted the general movement. The rivers and rocks, the seas and the continents have changed in all their parts; but the laws that describe those changes, and the rules to which they are subject, have remained invariably the same.

In addition to uniformitarianism, Hutton's *Theory of the Earth* (1785) brought together many concepts from naturalists who preceded him. He showed that rocks recorded events that had occurred over immense periods of time. He also demonstrated that Earth had experienced many episodes of mountain-raising upheaval, separated by quieter times of erosion and sedimentation. As he put it, there had been a "succession of former worlds."

Hutton saw a world of *cycles*. In these cycles, water sculpted Earth's surface and carried erosional detritus from the land, down the network of streams and rivers, and into the sea. The sediment in the sea became compacted into stratified rocks, and then, through the action of enormous forces, these sea-bottom rock layers were forced upward to form new lands. In this endless process, Hutton found "no vestige of a beginning, no prospect of an end." No longer could geologists compress all of Earth history into the short span suggested by the Old Testament of the Bible.

Unconformities

In his homeland of Scotland, at Siccar Point on the North Sea coast, Hutton observed a remarkable positioning of the rocks. As you can see in the chapter-opening photo, steeply inclined older strata are covered by flat-lying layers. It was clear to Hutton that the lower rocks had been deposited, then tilted, then partly removed by erosion, and finally the younger rocks were deposited on top.

The erosional surface separating the lower, sloping layers from the overlying horizontal beds indicated a time gap (hiatus) in the rock record. Because the flat-lying upper rocks did not "conform" to the tilted lower rocks, another geologist named the relationship an **unconformity**. Hutton was the first to understand and explain the geological significance of this feature.

An unconformity, or erosional gap in the rock record, can occur either between two flat-lying rock sequences or between two that lie at different angles, as at Siccar Point. Today, we call Hutton's site an *angular unconformity* because the lower beds are tilted at an angle to the upper beds. This and other unconformities provided Hutton with evidence for periods of erosion in his "succession of worlds."

During most of his career, Hutton's published reports attracted only modest attention, partly because his writing often was complex and difficult to follow. Much of the attention he did receive came from opponents who preferred Werner's views. But British scientists who valued Hutton's ideas convinced John Playfair, Hutton's friend and a professor of mathematics and natural philosophy, to publish a summary and commentary on Hutton's *Theory of the Earth*. By contrast, Playfair's text was lucid, engaging, and persuasive. Later geologists came to understand Hutton's ideas not through his own publications, but through Playfair's more "reader-friendly" versions.

Hutton's life was absorbed in investigating Earth. He scrutinized every rock exposure and soon became so familiar with certain strata that he could recognize them at different localities. But what he could not do was determine if dissimilar-looking strata at different locations were roughly equivalent in age. He had not discovered how to correlate beds that differed in composition and/or texture (lithology). This problem was soon to be resolved by William Smith.

▶ THE PRINCIPLE OF FOSSIL SUCCESSION

William Smith (1769–1839) was an English surveyor and engineer (Fig. 2-8). He was employed to locate routes of canals, to design drainage for marshes, and to restore springs. As he worked, he came to understand the principles of stratigraphy, for they were of immediate use to him. He knew that different types of stratified rocks occur in a definite order. He knew they can be identified by their lithology, the soils they form, and the fossils they contain. Thus, he predicted the types and thicknesses of rock that needed to be excavated for canals used to transport coal from mines to markets. It is small wonder that he acquired the nickname "Strata Smith."

Smith's use of fossils was particularly significant. Prior to his time, collectors rarely noted the precise rock layer from which a fossil was taken. But Smith carefully recorded the occurrence of fossils and quickly became aware that certain rock units could be identified by their fossil assemblages. He explained this relationship in *Strata Identified by Organized Fossils*. Smith used his knowledge to trace strata from place to place. Even if the layer he was tracing changed to a different composition (for example, from a very coarse sandstone to a finer siltstone), he knew the layers were "correlative" because of their similar fossils.

Ultimately, this knowledge led to formulation of the **principle of fossil succession**. This principle stipulates that the life forms of each age in Earth's long history are unique for particular periods, that fossils permit geologists to recognize contemporaneous deposits worldwide, and that fossils can be used to assemble scattered fragments in the rock record into a chronologic sequence.

Interestingly, Smith did not know *why* each rock unit had a particular fauna. It was over 40 years before Charles Darwin would publish *On the Origin of Species*. Today, we recognize that different animals and plants succeed one another over time because life evolves continuously. Because of this continuous change, or evolution, only rocks that formed during the same age can contain similar fossil assemblages.

Smith's success as a surveyor led him to all parts of England for consultation. On his many trips, he carefully recorded the types of rocks he saw and the fossils they contained. Armed with these notes, in 1815 he prepared a geologic map of England and Wales that is substantially accurate even today. In the 1830s, "Strata" Smith was declared the "father of English geology."

▶ THE GREAT UNIFORMITARIANISM–CATASTROPHISM CONTROVERSY

As Smith made his observations in England, scientists across the English Channel in France also were advancing the study of fossils. Baron Georges Léopold Cuvier (*KYOO-vee-yay*, 1769–1832), an expert in comparative anatomy, became the most respected vertebrate paleontologist of his day. Cuvier and his colleague Alexander Brongniart validated Smith's findings that fossils display a definite succession of types within a sequence of strata, and that this succession remains more or less constant even in widely separated locations.

Cuvier recognized that certain large groupings of strata were often separated by unconformities. Scanning from a lower group of strata across an unconformity to an overlying group, he often observed a dramatic change in animal fossils. From this, Cuvier concluded that the history of life was marked by frightful catastrophes involving flooding of the continents and crustal upheavals. This viewpoint came to be

FIGURE 2-8 William "Strata" Smith, "Father of English geology." Smith discovered that successive rock formations have distinctive fossils that can be used in correlation as indicators of relative age of rocks.

known as **catastrophism**. Cuvier considered the most recent of these catastrophic episodes to be the famous "Noah's flood" described in the Bible. He also thought that each catastrophe completely extinguished life forms and was followed by the appearance of new animals and plants.

But many geologists of the time, including the eminent Charles Lyell (more on him below), did not buy Cuvier's hypothesis. They thought that Earth's record of past life reflected continuous uniform change down through the ages (Hutton's uniformitarianism view). Thus began a controversy between catastrophism and uniformitarianism that rivaled the earlier Neptunist–Plutonist debates. Uniformitarians argued that seemingly abrupt changes in fossil fauna were actually caused by missing strata or other imperfections in the geologic record. They surmised that other apparent breaks in the fossil record were not sudden and that each animal group's fossil ancestors would be found in underlying beds.

Today, geologists accept that both uniformitarianism and catastrophism operate on our planet. Uniformity prevails in day-to-day geologic processes. But occasionally, catastrophic events punctuate the daily routine. Geologists recognize that rampant volcanism, asteroid impacts, or the onslaught of exceptionally harsh climatic conditions have indeed caused mass extinctions at various times in the geologic past.

THE PRINCIPLE OF CROSS-CUTTING RELATIONSHIPS

In the early 1800s, English geologist Sir Charles Lyell (1797–1875) authored a classic five-volume work, *Principles of Geology*. The first volume, published in 1830, expanded on Hutton's ideas, presented important geologic concepts of the day, and became an indispensable handbook for every English geologist.

FIGURE 2-9 **Sir Charles Lyell.** His five-volume work, *Principles of Geology*, expanded on Hutton's ideas, presented important geologic concepts of the day, and became an indispensable handbook for every English geologist.

For example, Lyell (Fig. 2-9) described how a geologic feature—like a mineral vein or a molten rock injected into other rocks—cuts across existing rock. This means that the penetrating rock must be younger than the rock that is being penetrated (Fig. 2-10). Put another way, the feature that is cut is older than the feature that does the cutting. This **principle of cross-cutting relationships** applies not only to rock bodies, but also to geologic structures like faults and unconformities.

FIGURE 2-10 **The principle of cross-cutting relationships.** The dark rock (basalt) cuts across the lighter-colored rock. The basalt was once molten, and in the molten state it penetrated the existing light-colored rock. Thus, the lighter-colored rock is older, and the basalt is younger.

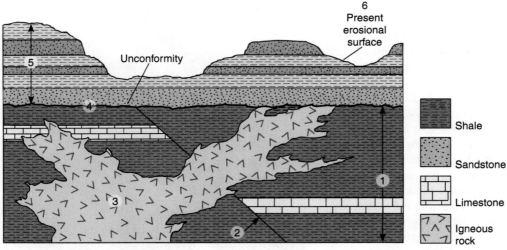

FIGURE 2-11 **Determining the sequence of geologic events from cross-cutting relationships and superposition.** From first to last, the sequence indicated in the cross-section is deposition of 1, then faulting 2, then intrusion of igneous rock mass 3, then erosion to produce the unconformity 4, then deposition of 5, and finally erosion 6. Strata labeled 1 are oldest, and strata labeled 5 are youngest.

Using the principle of cross-cutting relationships, let us reconstruct how the cross-section in Figure 2-11 came to be. We'll start with the oldest rock and build events toward the present.

1. First, the rocks in **1** were deposited. Note their "original horizontality."

2. Then, Earth movements broke the rocks at fault **2**. Note how the layer of limestone (brick pattern) is dramatically displaced by hundreds of feet diagonally. If it occurred suddenly, such a fault would generate a major earthquake. The fault obviously is younger than the original rocks it broke.

3. The orange mass of molten magma **3** intruded next, cutting across the earlier fault. So, **3** is younger than rocks **1** and fault **2**.

4. Next, a long period of erosion occurred, creating the unconformity **4**. The unconformity is younger than **1, 2**, and **3**.

5. New deposition began, forming rock layers **5**. The youngest of these layers, of course, is on top.

6. Many additional strata may have been deposited above **5**, but they have been eroded away, bringing us to the present erosional surface **6**.

We have just applied the basic principles of stratigraphy in the same manner that a geologist does.

Another of Lyell's *Principles* regards rock fragments. In Figure 2-12, along the granite-sandstone rock contact line, fragments of one rock appear in the other. These fragments are called **inclusions**. Lyell discerned that fragments within larger rock masses are older than the rock masses in which they are enclosed. Thus, whenever two rock masses are in contact, the one containing pieces of the other will be the younger of the two.

In Figure 2-12A, there are pebbles of granite (a coarse-grained igneous rock) within the sandstone (a fine-grained sedimentary rock). This tells us that the older, eroded granite fragments were incorporated into the younger sandstone as it formed. In contrast, in Figure 2-12B, the granite is younger, and it intruded as a melt into the older sandstone. Because there are sandstone inclusions in the granite, we know that the granite is the younger rock.

▶ EVOLUTION: HOW ORGANISMS CHANGE THROUGH TIME

William Smith and others recognized that strata often contain specific fossils and that a general progression toward more modern-looking shells occurs in higher (younger) strata. But why? Charles Darwin (1809–1882) provided a hypothesis that accounted for the changes seen in the fossil record.

Darwin (Fig. 2-13) had impressive knowledge of both biology and geology. This expertise enabled him to secure an unpaid position as a naturalist aboard the H.M.S. *Beagle* for a five-year mapping expedition around the world. On his return from the voyage in 1836, Darwin had volumes of notes to support his hypothesis of the evolution of organisms by natural selection.

Darwin's hypothesis was based on logical observations and conclusions. He observed that all living things tend to reproduce at prodigious rates. Yet, despite their reproductive potential to do so, no single

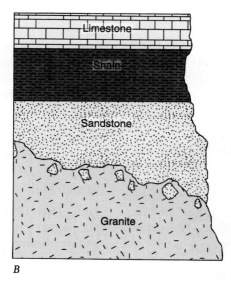

FIGURE 2-12 **Inclusions.**
(A) Granite inclusions in sandstone indicate that granite is the older unit. (B) Inclusions of sandstone in the granite indicate that sandstone is the older unit. ◀ *If the granite in (A) was found to be 150 million years old, and the shale above the sandstone 100 million years old, what can be stated about the age of the sandstone?*

group of organisms has overwhelmed Earth's surface. In fact, the size of any given animal or plant population remains fairly constant over long periods of time. Because of this, Darwin concluded that not all the individuals in a given generation can survive.

In addition, Darwin recognized that individuals of the same kind differ from one another in various features of morphology (form and structure) and physiology (organs and functions). He concluded that individuals with variations that are most favorable for survival and reproduction in a given environment would have the best chance of surviving and transmitting these same favorable traits to the next generation. Hence the term *natural selection.*

FIGURE 2-13 **Charles Darwin as a young man.** Darwin is shown shortly after he returned to England from his world voyage on the H.M.S. *Beagle.* Observations made during this voyage helped him formulate the concept of evolution by natural selection.

In those pioneering days, even the most brilliant scientists were handicapped by not knowing key facts or hearing about discoveries of others. In Darwin's case, he had no knowledge of genetics. Therefore, he did not know the *cause* of the variations in plants and animals that were so critical to his hypothesis. Gregor Mendel's 1865 report of experiments in heredity had escaped Darwin's attention. Nevertheless, in the decades following Darwin's death, geneticists established that the variability essential to Darwin's hypothesis of natural selection derives from new gene combinations that occur during reproduction and from genetic mutation.

Darwin may have been reluctant to face the controversy that his hypothesis would provoke, so he did not publish his findings immediately. He did, however, confide in Lyell and the great botanist Joseph Hooker, and these friends urged him to publish quickly, before someone else did. But Darwin procrastinated. Then, in 1858, a comparatively unknown young naturalist, Alfred Russel Wallace, sent Darwin a manuscript for review that contained the basic concepts of natural selection. Wallace had conceived of natural selection while on a biologic expedition to Indonesia. The idea came to him shortly after he read an essay on human overpopulation by Thomas Malthus.

Lyell and Hooker wanted Darwin to receive credit for his discovery and long years of research, so they arranged for immediate presentation of Darwin's work and Wallace's paper before the Linnaean Society, a prestigious group of natural scientists. Thus, the idea of natural selection was credited simultaneously to both scientists. Darwin now hastened to complete his book, *On the Origin of Species*, published in 1859.

By the time Darwin died in 1882, geologists everywhere were using their knowledge of evolution, biologic succession, superposition, and cross-cutting relationships to unravel Earth's extraordinary history.

EARTH HISTORY IN AMERICA

Geologists soon applied the knowledge gained in Europe on other continents. America proved to be an exciting natural laboratory in which to study Earth's history.

Louis Agassiz on Glaciers

Louis Agassiz (*AGG-uh-see*, 1807–1873), a Swiss paleontologist, arrived in North America in 1846. He became a Harvard University professor, founding the Harvard Museum of Comparative Zoology. Agassiz published important works on fossil fishes, mollusks, and echinoids. By 1859, he had become a well-known American scientist.

Before coming to America, Agassiz studied glaciers in the Alps. He proposed that immense ice sheets once covered much of North America and Eurasia. In his 1840 *Studies of Glaciers*, he wrote:

> The surface of Europe, adorned before by tropical vegetation and inhabited by troops of large elephants, enormous hippopotami, and gigantic carnivora, was suddenly buried under a vast mantle of ice, covering alike plains, lakes, seas, and plateaus.

Agassiz also found ample evidence in America for his "Ice Age hypothesis." Along the shores of Lake Superior, he showed his students bedrock with aligned scratches called *glacial striations*. These scratches were scoured by rocks frozen into the base of advancing ice, which "sandpapered" the rocks over which it flowed. He observed huge boulders transported by the ice from locations far to the north and deposited on the landscape, looking very out-of-place, as *glacial erratics*. On American prairies, Agassiz recorded *glacial moraines*, rock debris deposited by a melting glacier. And he correctly interpreted New York's elongated Finger Lakes as large depressions gouged by glacial erosion.

His "Ice Age" became recognized as the main event of the Pleistocene Epoch, beginning about 1.5 million years ago and ending only 8,000–10,000 years ago. During this frigid period, an estimated 40 million cubic miles of ice and snow blanketed one-third of the planet's land area.

James Hall's 7.5 miles of Strata

James Hall (1811–1898), a brilliant geologist and paleontologist, was appointed director of New York's first geological survey. Hall had plenty of strata to examine, for the total thickness of the stratigraphic sequence in New York, scattered from place to place, is a remarkable 40,000 feet (7.5 miles). Based on fossils, Hall knew these rocks had been deposited in shallow water. He correctly reasoned that the seafloor had subsided at the same time that the sediment that formed the rocks was deposited, and that subsequently the Appalachian Mountains were raised from what were once marine basins. Hall gained world fame for his knowledge of stratigraphy and his eight-volume *The Paleontology of New York*.

Western Geology

Geologic work in the United States shifted westward, often through surveys of the Western Territories mandated by Congress. A prominent leader in these expeditions was Ferdinand V. Hayden (1829–1887), known to natives as "he who picks up rocks running" (Fig. 2-14).

FIGURE 2-14 **The Hayden field party in the summer of 1870.** In this photo, taken near Red Buttes, Wyoming, Ferdinand Hayden is the bearded geologist seated at the center of the table.

In South Dakota Hayden surveyed the badlands, Black Hills, and territory along several major rivers. He was influential in convincing Congress to pass a bill authorizing establishment of the first U.S. national park, Yellowstone.

John Powell (1834–1902) led several geologic and geographic surveys of the West and directed the U.S. Geologic Survey. His greatest feat was a journey by boat through the Grand Canyon of the Colorado River in 1869 (Fig. 2-15). The Grand Canyon is among the most magnificent natural laboratories for stratigraphy on our planet, with 6,250 vertical feet (almost 1.2 miles) of rock layers exposed for study. These layers contain twelve major rock units representing two billion years (roughly one-third) of Earth's history.

Hayden, Powell, and others surveyed and drew maps, recording the geology, biology, archaeology, and paleontology of the country. Among their discoveries was a wealth of dinosaur and giant mammal bones in the rock formations of the western states.

The Dinosaur Rush

Two paleontologists who exploited these bony treasures were Othniel C. Marsh (1831–1899) and Edwin D. Cope (1850–1897). Marsh became the first professor of paleontology at Yale University and later founded the Peabody Museum of Natural History. Cope was a wealthy Quaker and taught at the University of Pennsylvania. To hasten their description of the abundant fossils in the West, both employed professional bone-hunters who excavated fossils with pick and shovel and shipped the bones back to Marsh in Connecticut and Cope in Philadelphia.

Unfortunately, Marsh and Cope were egotistical, jealous, and competitive. A bitter feud developed as each tried to surpass the other in naming and describing newly discovered vertebrates. Both made mistakes

FIGURE 2-15 **The Grand Canyon of the Colorado River.** You come face-to-face with the immensity of geologic time when viewing this most magnificent of natural features.

in their haste to be first in publishing descriptions of newly discovered fossil beasts.

Nevertheless, the great dinosaur rush led by Marsh and Cope provided thousands of specimens for study and museum exhibits, motivated research worldwide, enhanced our understanding of life during the Mesozoic and Cenozoic Eras, provided evidence for evolution, and established paleontology as a science with a spirit of discovery.

SUMMARY

- Fossils are the remains or traces of organisms from the geologic past preserved in rocks.
- During the 1600s and 1700s, pioneering geologists formulated principles that are the foundation of the science of historical geology.
- Nicholas Steno formulated three important principles used in the interpretation of sedimentary rocks. The *principle of superposition* states that, in layered sedimentary rocks, the age of individual strata decreases from bottom to top. The *principle of original horizontality* stipulates that because most sedimentary rocks are deposited more or less horizontally, where they are found sloping or folded, they have been deformed by strong crustal forces after deposition. The *principle of original lateral continuity* reminds us that strata extend laterally in all directions until they thin out because of non-deposition, encounter a barrier (such as a coastline) or merge into another stratum.
- In the 1700s, geologists in Europe classified mountains according to their most characteristic kind of rock. The term *primary* was assigned to mountains composed mostly of igneous and metamorphic rocks. *Secondary* mountains were mainly constructed of fossiliferous stratified sedimentary rocks. Unconsolidated sediments such as sands and gravels were given the name *Tertiary*.
- Abraham G. Werner wrote the first great mineralogy textbook. He supported the erroneous concept of Neptunism, proposing that all rocks were from an ancient sea.

- James Hutton recognized the immensity of geologic time. He explained the concept that was later to be named the principle of uniformitarianism. In opposition to Neptunism, Hutton was a Plutonist who recognized that layers of lava and granite solidified from molten rock. Although he did not invent the term *unconformity*, he recognized unconformities and understood the geologic events that produced them.
- Based on his studies of the order of strata near Bath, England, William Smith published *Strata Identified by Organized Fossils*. This work formed the basis for using fossils in correlation. Smith also produced the first geologic map of England and Wales.
- Georges Cuvier established vertebrate paleontology as a science. He favored the theory of catastrophism, which postulates periodic mass extinctions followed by the appearance of new faunas and floras.
- Charles Lyell was a great teacher and interpreter of geology. He is particularly known for *Principles of Geology*, in which he described the use of cross-cutting relationships and fragments (inclusions) for determining the relative age of rock bodies.

- Supported by his extensive research, Charles Darwin proposed that life on Earth has continuously evolved, and that new forms of life result from natural selection, which provides for the survival and reproduction of organisms best suited to their environment.
- After observing glaciers in the Alps, Louis Agassiz presented his hypothesis that continental ice sheets covered much of North America and Eurasia in the recent geologic past.
- James Hall proposed that the thick sequences of folded, shallow marine sedimentary rocks in the Appalachian Mountains indicate subsidence of the basin floor as sediment was added, followed by uplift and deformation that formed mountains.
- Ferdinand Hayden and John Powell were influential leaders in the geologic surveys of the Western United States. Powell is particularly remembered for his amazing 1869 exploration of the Grand Canyon of the Colorado River.
- Othniel Marsh and Edwin Cope organized and conducted expeditions into the American West to recover, classify, and describe hundreds of extinct vertebrates.

KEY TERMS

actualism, p.17

catastrophism, p.19

inclusions, p.20

principle of cross-cutting relationships, p.19

principle of fossil succession, p.18

principle of original lateral continuity, p.14

principle of original horizontality, p.14

principle of superposition, p.13

stratigraphy, p.14

unconformity, p.17

QUESTIONS FOR REVIEW AND DISCUSSION

1. Which of Steno's principles were used by William Smith in his studies of strata in England?

2. Smith found that there were different fossil groups in different rock layers. What are two or more reasons that might account for these differences?

3. What interpretation did Steno make when he observed strata that were sloping steeply into the ground or even vertical rather than horizontal? What interpretation would you make about a sequence of strata with *older* strata on top and *younger* strata on the bottom?

4. James Hutton visited an exposure of layered rocks in which the igneous rock basalt had been squeezed when molten between two beds of sedimentary rock. What features or evidence might Hutton have looked for that would indicate the once-molten origin of the basalt, as opposed to it having been deposited as sediment?

5. Match the description at right with the geologist at left.

____ John Powell	A. Pioneer mineralogist and
____ Charles Lyell	Neptunist
____ Abraham Werner	B. Proponent of catastrophism
____ Georges Cuvier	C. Uniformitarianism
____ Louis Agassiz	D. Original horizontality
____ James Hutton	E. Cross-cutting relationships
____ Charles Darwin	F. Natural selection
____ Nicholas Steno	G. Recognition of Ice Age
____ James Hall	H. Explorer of Grand Canyon
____ Othniel Marsh	I. From depositional basins to
	mountains
	J. Fossil vertebrates of
	Western U.S.
	K. Fossil-based correlation

6. From the photograph of the unconformity at Siccar Point at the beginning of this chapter, sketch five cross-sections indicating the geologic events that occurred: (1) deposition of the older Silurian strata, (2) deformation (folding) of those older strata, (3) erosion to produce the surface of unconformity, (4) deposition of the younger Devonian strata, and (5) further deformation to provide for slope (dip) in younger beds.

WEB SITES

The Earth Through Time Student Companion Web Site (www.wiley.com/college/levin) has online resources to help you expand your understanding of the topics in this chapter. Visit the Web Site to access the following:

1. Illustrated course notes covering key concepts in each chapter;

2. Online quizzes that provide immediate feedback;

3. Links to chapter-specific topics on the web;

4. Science news updates relating to recent developments in Historical Geology;

5. Web inquiry activities for further exploration;

6. A glossary of terms;

7. A Student Union with links to topics such as study skills, writing and grammar, and citing electronic information.

3

Q
T
K
J
Tr
Pr
P
M
D
S
O
Є
Pre-Є

Flat-lying sedimentary rocks of Permian, Triassic, and Jurassic age exposed along the Colorado River at Dead Horse Point, Utah. *Relative age of strata by superposition is displayed in these rocks, which record over 100 million years of Earth history.*

CHAPTER 3

Time and Geology

Each grain of sand, each minute crystal in the rocks about us is a tiny clock, ticking off the years since it was formed. It is not always easy to read them, and we need complex instruments to do it, but they are true clocks. The story they tell numbers the pages of Earth history.

—Patrick M. Hurley, noted geophysicist, 1959

Key Chapter Concepts

- Relative geologic time places events in the *order* in which they occurred *without reference to actual time in years*. Actual geologic time provides ages measured in years.
- Geochronologic units are units of pure time. Chronostratigraphic units are all the rocks deposited during a given time interval.
- The geologic time scale was constructed bit-by-bit from fossil-bearing stratigraphic sections in Europe. These became standard sections to which fossil-bearing strata in other locations could be compared worldwide.

We are curious about the great age of rocks and fossils. For this reason, newscasters unfailingly report that a recently discovered fossil or rock is "umpteen-zillion years old." And for this reason, geologists often are asked the age of rocks and fossils brought to them by students and collectors. When told that the samples have ages of ten-millions or even hundred-millions of years, people are pleased, but perplexed: "How can this geologist know the age of this specimen just by looking at it?" If you want the answer, read on.

The science of determining the age of rocks is called **geochronology**. It began over 400 years ago when Nicholas Steno described how the position of strata in a sequence could show the relative geologic age of the layers, usually with the youngest layer on top. As described in Chapter 2, this simple but important idea was expanded and refined by geologists in the 1700s and 1800s, who correlated strata from different locations that bore similar fossils. Outcrop by outcrop, they pieced together the rock sequences with their fossils, one above the other, until a standard geologic time scale based on relative ages was constructed.

FINDING THE AGE OF ROCKS: RELATIVE VERSUS ACTUAL TIME

There are two different frames of reference when dealing with geologic time. **Relative geologic dating** places geologic events, and the rocks representing

OUTLINE

► FINDING THE AGE OF ROCKS: RELATIVE VERSUS ACTUAL TIME
► A SCALE OF GEOLOGIC TIME
► ACTUAL GEOLOGIC TIME: CLOCKS IN THE ROCKS
► RADIOACTIVITY PROVIDES A WAY TO DATE ROCKS
► WHAT OCCURS WHEN ATOMS DECAY?
► THE PRINCIPAL RADIOACTIVE TIMEKEEPERS
► HOW OLD IS EARTH?
► SUMMARY
► KEY TERMS
► QUESTIONS FOR REVIEW AND DISCUSSION
► WEB SITES

those events, in *sequence* (what happened first, second, and so on). This is done without specifying the actual age in years. Relative geologic time simply tells us which event preceded or followed another, or which rock mass is older or younger than another.

With the discovery of radioactivity, **actual geologic dating** (or absolute dating) became possible. It expresses, in years, the approximate actual age of rocks or geologic events. These actual ages usually are determined from the rate of decay of radioactive elements in the rocks, and are typically expressed in hundred-thousands or millions of years. The actual dates quantify the relative dates, showing the duration of time periods and specifically when events occurred.

▶ A SCALE OF GEOLOGIC TIME

As early geologists were discovering the relationships among rocks, they had no way of knowing how many time units eventually would be in the completed geologic time scale. Nor could they know which fossils would be useful in correlation, or which new strata might be discovered at a future time in some distant part of the world. Consequently, the time scale grew piece-by-piece, in an uncoordinated manner. It was like a scavenger hunt to find jigsaw puzzle pieces one at a time, and then to assemble the pieces.

Consequently, time units were named inconsistently as they were discovered. Sometimes a unit's name was borrowed from local geography or from a mountain range in which rocks of a particular age were well exposed. Some units were named for ancient tribes of Welshmen, which seems odd until you learn that the rocks were studied in Wales. Sometimes the name was suggested by the kind of rock that predominated.

Overviewing the Time Scale

Refer to the Geologic Time Scale, Figure 3-1. The only name familiar to you might be "Jurassic," because of the classic dinosaur film, *Jurassic Park*. The other names may seem intimidating, but they will soon become just as familiar.

The column of rock layers on the left is hypothetical and generic; no such sequence literally exists. But its purpose is to convey the idea that the geologic time scale generally represents rock layers at various locations on Earth.

The scale itself is divided into Eons, Eras, Periods, and Epochs, which will be explained in the next section.

To the right of the scale are the time ranges for major plant and animal groups. Note that multicellular animals lacking backbones—the invertebrates—did not appear until around 1 billion years ago. Then came the fishes, land plants, and so on. Mammals and birds were the most recent to evolve, very roughly 200 million years ago.

Divisions in the Geologic Time Scale

This is a book about the history of Earth. Any history needs a time chart to which particular events or conditions can be referred. That is why we have a geologic time scale.

In history classes, boundaries on the time chart are placed at times of great changes in culture and government. On the geologic time scale, boundaries are placed where the fossil record reveals important changes in animals and plants. The divisions are somewhat like our calendar, which divides years into months, months into weeks, and weeks into days. The geologic time scale divides eons into eras, eras into periods, and periods into epochs.

Eons. Three **eons** span the entire sweep of Earth's history. The **Archean Eon** starts with Earth's origin 4.6 billion years ago. It includes a long period for which we have no rock record, because these early rocks were "recycled" long ago through the action of plate tectonics. The **Proterozoic Eon** spans the time interval from 2500 to 542 million years ago, and the **Phanerozoic Eon** covers the past 542 million years. For convenience, geologists often refer to the time before the Phanerozoic as the **Precambrian** (all of time prior to the Cambrian Period), which encompasses 87% of all geologic time.

Eras, Periods, Epochs, and Ages. Each eon is divided into **eras**. Their names give us a good opportunity to introduce Greek prefixes that are used in historical geology. Note on the time scale that some time interval names begin with *paleo-, meso-*, and so on. The prefixes mean: *eo* = dawn, *paleo* = ancient, *meso* = middle, *neo* = new, and *ceno* = recent.

The eras of the Phanerozoic Eon are the **Paleozoic** (ancient life), **Mesozoic** (middle life), and **Cenozoic** (recent life). The *paleo-* and *-zoon* of the Paleozoic refer to now-extinct ancient invertebrates, primitive fish, amphibians, and reptiles. The Mesozoic was dominated by dinosaurs. And the most recent era, the Cenozoic, is dominated by mammals and modern plants. We live in the Cenozoic, which began about 65 million years ago.

Eras are divided into shorter units called **periods**, and periods are divided into **epochs**. These divisions also are based on changes in life forms recorded by fossils. The terms represent increments of time called *geochronologic units*, which for convenience we will simply call **time units**.

The actual rocks formed or deposited during a specific time interval are called *chronostratigraphic units* or more simply **time-rock units**. Table 3-1 shows the time units that correspond to time-rock units. A **system**

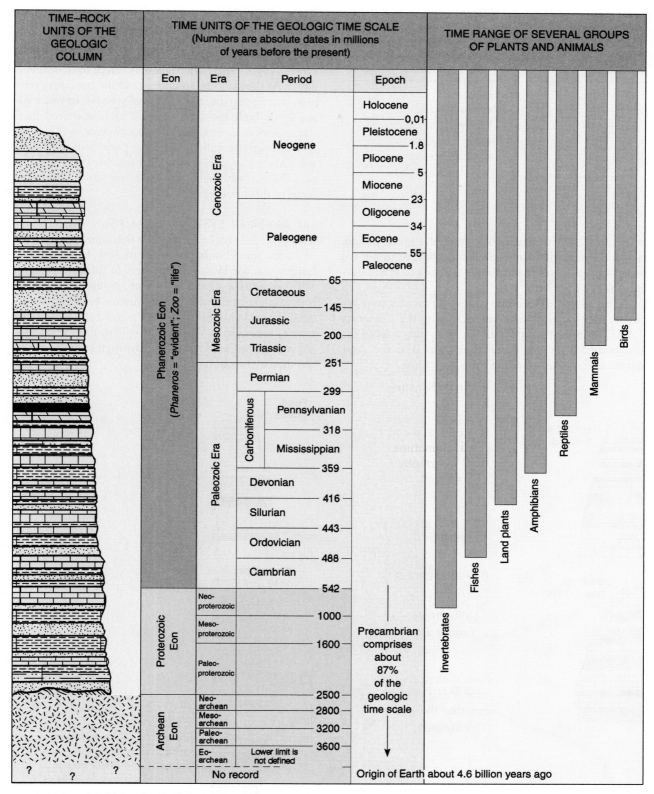

FIGURE 3-1 **Geologic Time Scale.** Following the principle of superposition, the scale is shown with the oldest time period at the bottom and youngest at the top. Note the time ranges shown for plant and animal groups. (See text for a walk-through of the time scale.)

TABLE 3-1 Hierarchy of Time (Geochronologic) and Time-Rock (Chronostratigraphic) Terms

Time Divisions	Equivalent Time-Rock Divisions
Eon	Eonothem
Era	Erathem
Period	System
Epoch	Series
Age	Stage
Chron	Zone (or chronozone)

refers to all of the actual rock units of a given period, whereas a **series** is the chronostratigraphic equivalent of an epoch, and a **stage** represents the rocks formed during an age.

As an example of the way these terms are used, we might say that climatic changes during the *Cambrian Period* are indicated by fossils found in the rocks of the *Cambrian System*. Note that time units bear the same names as the corresponding time-rock units.

Evolution of the Geologic Time Scale

As we discussed earlier, the geologic time scale was not originally conceived as a coherent whole, but rather evolved part-by-part as a result of research by many geologists. As they labored to understand the confusing relationships among the strata and fossils they observed, trial-and-error built the time scale. Widely scattered local sections became correlated to standard sections (Fig. 3-2).

Let us briefly walk through the periods of the Paleozoic, Mesozoic, and Cenozoic Eras, for these encompass the eon of "visible life" (Phanerozoic) that is the focus of this book.

THE CAMBRIAN SYSTEM In the 1830s, Adam Sedgwick (Fig. 3-3) named a sparsely fossiliferous sequence of rocks the **Cambrian System** (for Cambria, the Latin name for Wales). Exposures of these strata (Fig. 3-4) provided a standard section against which other rocks on Europe and other continents could be correlated. Worldwide, all rocks deposited during the same time as the rocks in Wales (542–488 million years ago) are recognized as "Cambrian" through comparison to the standard section in Wales.

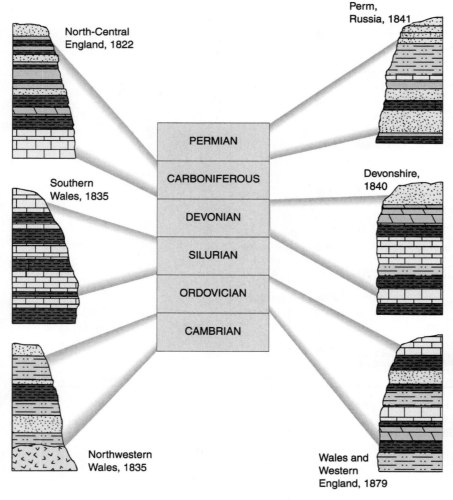

FIGURE 3-2 **The standard geologic time scale for the Paleozoic and other eras developed without benefit of a grand plan.** Instead, it developed by the compilation of "type sections" from various European locations for each of the periods. ◪ *What criteria at Devonshire demonstrated that these strata were younger than those in Wales?*

FIGURE 3-3 **Adam Sedgwick (1785–1873), one of the foremost geologists of the 1800s.** A geology professor at Cambridge University, Sedgwick defined the highly deformed system of rocks in northwestern Wales as the Cambrian System.

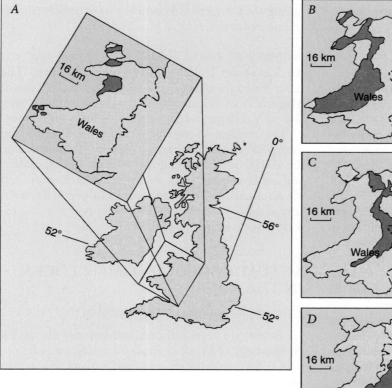

FIGURE 3-4 **Outcrop areas for some Paleozoic strata in Great Britain.** The systems shown are Cambrian (A), Ordovician (B), Silurian (C), and Devonian (D).

THE ORDOVICIAN AND SILURIAN SYSTEMS While Sedgwick labored over his Cambrian rocks in northern Wales, a former student, Sir Roderick Impey Murchison (1785–1871), studied fossil-bearing strata in southern Wales. Murchison named these rocks the **Silurian System**, for early inhabitants called the Silures. In 1835, the two men presented a paper, *On the Silurian and Cambrian Systems, Exhibiting the Order in Which the Older Sedimentary Strata Succeed Each Other in England and Wales*. This publication initiated the early Paleozoic time scale.

Sedgwick had not described distinctive fossils in the Cambrian, so it could not be recognized in other countries. So, Murchison said that Sedgwick's Cambrian was not a valid system. Murchison maintained that all fossiliferous strata between the Precambrian and the Devonian belonged within his Silurian System. This professional dispute was to sever their friendship.

Fossils were eventually found in the upper Cambrian, and they proved to be similar to those in Europe and North America. The Cambrian then became recognized outside England. Another geologist combined the upper part of Sedgwick's Cambrian and the lower part of Murchison's Silurian into a new system, the **Ordovician System** (after the Ordovices, an early Celtic tribe). Thus the first three systems of the Paleozoic were established.

THE DEVONIAN SYSTEM In 1839 Sedgwick and Murchison proposed the **Devonian System** for outcrops near Devonshire, England. They based their proposal on differences between fossils in rocks between the Carboniferous System and the underlying Silurian.

THE CARBONIFEROUS SYSTEM (MISSISSIPPIAN AND PENNSYLVANIAN) The term **Carboniferous System** was coined in 1822 for strata that included coal beds in north-central England. The name refers to the fact that coal consists largely of carbon. Europeans divided the system into the Lower Carboniferous and the Upper Carboniferous, the latter containing most of the workable coal seams. In North America, the broadly comparable systems are the **Mississippian** (Lower Carboniferous strata that are extensively exposed in the upper Mississippi River Valley) and **Pennsylvanian** (Upper Carboniferous strata that are heavily mined for coal in Pennsylvania).

THE PERMIAN SYSTEM The **Permian System** is named for Permia, an ancient Russian kingdom. In the 1840s, Murchison traveled across western Russia and was delighted to recognize fossils that matched England's Silurian, Devonian, and Carboniferous rocks. His establishment of the Permian System provides a fine example of the logic employed by early geologists in assembling the puzzle of the standard time scale. The Permian is the youngest period in the "ancient life" Paleozoic Era.

THE TRIASSIC, JURASSIC, AND CRETACEOUS SYSTEMS The "middle life" Mesozoic Era includes three periods—Triassic, Jurassic, and Cretaceous.

The **Triassic System** was named in 1834 by a German geologist for the trifold division of rocks of this age in Germany. The Triassic saw the rise of early mammals.

Another German scientist, Alexander von Humboldt (1769–1859), proposed the term **Jurassic** for strata of the Jura Mountains between France and Switzerland. Reptiles prevailed in the Jurassic, notably the dinosaurs. The flying reptiles that would become modern birds also appeared.

A Belgian geologist proposed the term **Cretaceous** (from the Latin *creta*, meaning "chalk") for rock outcrops in France, Belgium, and Holland. The famous White Cliffs of Dover, England, are composed of chalk. Although chalk beds are prevalent in some Cretaceous successions, the system is actually recognized on the basis of fossils. Dinosaurs still held center stage during the Cretaceous Period but were completely exterminated at the close of Mesozoic Era.

DIVISIONS OF THE CENOZOIC The "recent life" era, the Cenozoic, includes two geologic systems. The older **Paleogene** marks the beginning of the "Age of Mammals" when small horses, hornless rhinoceroses, and early rodents and primates made their appearance.

The younger **Neogene** includes the **Pleistocene Series**, composed primarily of deposits formed during the glacial ages. Mammoths, modern horses, camels, and saber-toothed cats roamed across Europe and North America. The most recent **Holocene Series** is composed of sediments and rocks formed after the retreat of the continental glaciers.

▶ ACTUAL GEOLOGIC TIME: CLOCKS IN THE ROCKS

Early Attempts to Determine Earth's Age

Geologists are never satisfied knowing only the *relative* geologic age of a rock or fossil. Scientific curiosity demands that we know its *actual* age in years. Numerous attempts were made to calculate Earth's age, some quite ingenious. Here is a brief summary.

Biblical Calculation. An early, unscientific attempt at quantitative dating was conducted in 1658 by James Ussher (1581–1656), an Irish archbishop and scholar. Archbishop Ussher calibrated the solar and lunar cycles against dates and events recorded in the Old Testament of the Judeo-Christian Bible, and calculated that Earth was created about 6000 years ago—on October 23, 4004 BCE. Another scholar placed the precise hour of creation at 9:00 A.M. The archbishop's date became inserted in many editions of the Bible and thus became widely accepted.

However, geologists in the 1800s showed that the archbishop's date could not be supported by objective scientific observation. They recognized that, to discover the true age of Earth or of individual rock bodies, they would have to find natural processes that continue at a constant rate *and* that leave a tangible record in the rocks. The evolution of life leaves a tangible record in the rocks, which became understood in the 1860s. Radioactive decay proceeds at a constant rate, and it was discovered in 1896.

Evolution of Fossils. Charles Lyell attempted to use fossils to determine Earth's actual age. He compared the amount of evolution exhibited by marine mollusks in successively younger Cenozoic strata with the amount that had occurred since the beginning of the Pleistocene Ice Age. Lyell estimated that 80 million years had elapsed since the beginning of the Cenozoic. He actually came astonishingly close to the correct figure, which modern evidence puts at 65.5 million years. For rocks older than Cenozoic, however, basing estimates on evolutionary rates is more difficult. Parts of the fossil record are missing, and the rates of evolution for some extinct organisms are poorly known.

Sediment Deposition Rate. In another attempt, geologists reasoned that if deposition rates could be determined for sedimentary rocks, they could estimate the time required for deposition of a given thickness of strata. Similar reasoning suggested that one could estimate total elapsed geologic time from the thickness of sediment transported annually to the oceans. The method was to divide the average annual thickness into the total thickness of sedimentary rock that had ever been deposited in the past.

But such estimates did not account for past differences in sedimentation rates or losses during episodes of erosion. Also, some very ancient sediments were no longer recognizable, having been converted to igneous and metamorphic rocks through tectonic action like mountain-building. Based on sedimentation rates, estimates of Earth's age ranged from a million years to over a billion.

Ocean Salinity. Sir Edmund Halley (1656–1742), best known for his comet discovery, devised another scheme for approximating Earth's age. Halley surmised that the ocean formed soon after the origin of the planet, and therefore it would be only slightly younger than the solid Earth. (By ocean, we mean the entire "world ocean," so-called because the Atlantic, Pacific, Indian, etc. are interconnected.)

He reasoned that the original ocean was pure water and not salty, and that subsequently salt derived from the weathering of rocks was brought to the sea by streams. Thus, if he knew the total amount of salt dissolved in the ocean and the amount added each year, he could calculate the ocean's age.

In 1899, the Irish geologist John Joly (1857–1933) attempted this calculation. From gauges placed at stream mouths, Joly estimated the amount of salt dumped in the oceans each year. Then, knowing both the salinity and volume of ocean water, he calculated the amount of salt already dissolved in the ocean. To estimate the ocean's age, he divided total ocean salt by the rate of salt addition each year. He came up with about 90 million years for the oceans to reach their present salinity.

This figure was off the mark by a factor of 50—the actual age of the world ocean is close to the age of Earth, which is 4.6 billion years. His error occurred because he could not account accurately for recycled salt and salt incorporated into clay minerals on the seafloors. Billions of tons of salt once in the sea had become evaporite deposits on land. Some of the salt being washed back to the sea is from these deposits. Nevertheless, Joly's calculations supported geologists who insisted that Earth was far older than a few million years.

The hypothesis of Earth's immense antiquity also was supported by Darwin and other biologists, who saw the need for hundreds of millions of years to accomplish the organic evolution apparent in the fossil record.

Cooling Rate. Curiously, the opinion of geologists and biologists that Earth was immensely old soon was challenged by Lord William Thomson Kelvin (1824–1907), an outstanding physicist. Kelvin assumed that Earth was once a sphere of molten rock, and that it had been cooling steadily since. Using the laws of heat convection and radiation, he estimated how long it would have taken Earth to cool from a hot mass to its present condition. He calculated that 24 to 40 million years was a reasonable age for the planet.

Kelvin's elegant mathematics were convincing, and the biologists and geologists had only inaccurate dating schemes and geologic intuition. However, new discoveries soon showed Kelvin's hypothesis to be in error.

Natural Radioactivity. With the 1896 detection of natural radioactivity by Henri Becquerel (1852–1908), scientists realized that Earth had its own built-in source of heat: radioactive decay. The planet was not cooling at a steady and predictable rate, as Kelvin thought. The discovery of radioactive decay also gave scientists a far more accurate way to date rocks and ultimately to determine the true age of Earth.

RADIOACTIVITY PROVIDES A WAY TO DATE ROCKS

To understand how atoms in rocks and minerals can reveal the numerical age of geologic events, we need a brief review of the nature of atoms.

Reviewing Atoms

An **atom** is the smallest particle of matter that can exist as a chemical **element** (oxygen, silicon, sodium, and so on). An individual atom consists of an extremely minute nucleus that is orbited by fast-moving **electrons** (Fig. 3-5). The electrons are farther from the nucleus, in proportion, than are the planets surrounding our Sun. Consequently, the atom consists primarily of empty space. However, electrons move so rapidly in this space around the nucleus that their repeated presence "fills" the space within their orbits, giving volume to the atom. The orbiting electrons, which have a negative electrical charge, also repel other atoms nearby. (Recall that like electrical charges repel one another.)

In the atom's nucleus are closely compacted particles called **protons**. A proton carries a positive electrical charge that is equal in strength to the negative electrical charge carried by an orbiting electron. Associated with protons in the nucleus are particles having the same mass as protons, but they are electrically neutral, and thus are called **neutrons**. Modern atomic physics has made us aware of still other particles in the nucleus, but they are not significant for our purpose here.

An atom's **atomic number** is the number of protons in its nucleus. Each of the 90 naturally occurring chemical elements has a unique atomic number. Atomic numbers start with 1 (for one proton) in hydrogen and increase to 92 (for 92 protons) in uranium (Table 3-2).

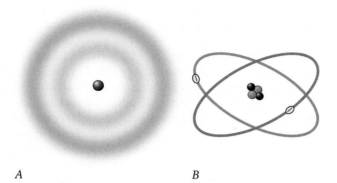

A   *B*

FIGURE 3-5   **Two models of atoms.** As depicted in (A), an atom is composed of negatively charged electrons that whirl around a positively charged nucleus. Within the cloud of rapidly orbiting electrons there is a high probability that a specific number of electrons will occupy certain zones or "shells." An atom of helium is depicted in (B). Helium has a nucleus containing two protons and two neutrons, and two orbiting electrons.

An atom's **atomic mass** approximately equals the sum of the masses of its protons and neutrons. (The mass of electrons is too slight to consider here.) Carbon-12 is used as the standard for comparison of mass. Setting the atomic mass of carbon at 12 makes the atomic mass of hydrogen, the lightest of elements, just a bit over 1 (1.008).

TABLE 3-2   **Number of Protons and Neutrons, and the Atomic Mass of Some Geologically Important Elements**

Element and Symbol	Atomic Number (Number of Protons in Nucleus)	Number of Neutrons in Nucleus	Mass Number
Hydrogen (H)	1	0	1
Helium (He)	2	2	4
Carbon-12 (C)*	6	6	12
Carbon-14 (C)	6	8	14
Oxygen (O)	8	8	16
Sodium (Na)	11	12	23
Magnesium (Mg)	12	13	25
Aluminum (Al)	13	14	27
Silicon (Si)	14	14	28
Chlorine-35 (Cl)*	17	18	35
Chlorine-37 (Cl)	17	20	37
Potassium (K)	19	20	39
Calcium (Ca)	20	20	40
Iron (Fe)	26	30	56
Barium (Ba)	56	82	138
Lead-208 (Pb)*	82	126	208
Lead-206 (Pb)	82	124	206
Radium (Ra)	88	138	226
Uranium-238 (U)	92	146	238

*When two isotopes of an element are given, the most abundant is starred.

An atom's mass number is the total of its protons and neutrons, rounded to the nearest whole number. Some atoms of the same substance have different mass numbers, like uranium-235 and uranium-238. Such variants are called **isotopes**. Isotopes are two or more varieties of the same element that have identical atomic numbers and chemical properties, but differ in mass numbers because they have a varying number of neutrons in the nucleus. Isotopes are critical to the age-dating of rocks.

We use a shorthand for designating each element and showing its mass number and atomic number. Using calcium as an example, the mass number is noted as a superscript preceding the symbol for the element, and the atomic number is placed beneath it as a subscript. Thus, $^{40}_{20}$Ca is the element calcium having a mass number of 40 and an atomic number of 20 (see Table 3-2 for data on geologically important elements).

WHAT OCCURS WHEN ATOMS DECAY?

Most atoms are stable, having an internal balance of charges between the positive protons and the negative neutrons. However, a number of elements have isotopes that are unbalanced, and therefore unstable. To achieve stability, they release particles or energy. This release of particles or energy is the natural **radioactivity** discovered by Henri Becquerel.

Unstable isotopes that are radioactive occur in elements such as uranium, radium, and thorium. Specific isotopes decay (release particles or energy) to form more stable isotopes of the same element, or to form other elements altogether. For example, given sufficient time, any individual uranium atom will eventually decay, through several stages, to the metal lead.

The Alpha, Beta, and Gamma of Decay

To better understand what we mean by radioactive decay, consider what happens to a radioactive element such as uranium-238 ($^{238}_{92}$U). Uranium-238 has a *mass* number of 238 (the total weight of the atom's protons and neutrons, with each having a mass of 1). Uranium has an *atomic* number (number of protons) of 92.

Sooner or later (and entirely spontaneously), an unstable uranium-238 atom will fire off a particle from its nucleus, called an **alpha particle**. Alpha particles are positively charged ions of the element helium. They have an atomic weight of 4 and an atomic number of 2. Thus, when the alpha particle is emitted, the resulting slightly lighter atom has a reduced atomic weight of 234 and a lowered atomic number of 90. This new atom, formed from its **parent element** by radioactive decay, is called a **daughter element**.

From the decay of the parent isotope, uranium-238, the daughter isotope, called thorium-234, is born (Fig. 3-6). A shorthand equation for this change is written:

$$^{238}_{92}\text{U} \rightarrow {}^{234}_{90}\text{Th} + 1\,{}^{4}_{2}\text{He}$$

However, this change is not the end of the process, for the nucleus of thorium-234 ($^{234}_{90}$Th) is also unstable. It eventually emits a **beta particle**. (A beta particle is an electron that is released when a neutron splits into a proton and an electron.)

Thus, from the unstable parent atom $^{234}_{90}$Th, the daughter element $^{234}_{91}$Pa (protactinium) is formed. In this case, the atomic number has been increased by 1.

A third kind of emission in the radioactive decay process is **gamma radiation**. Gamma rays are invisible electromagnetic waves (like radio waves) that have extremely high frequencies, higher than X-rays.

Why Radioactivity Lets Us Date Ancient Rocks with Confidence

In a nutshell, the rate of decay of radioactive isotopes is *uniform* and is *unaffected by changes in pressure, temperature, or the chemical environment*. Even in the most hellish environment a thousand miles underground, where temperatures and pressures are extreme, and over millions of years, *radioactive isotopes continue to decay at exactly the same rate*. This is the "multi-million-year clock" scientists needed to determine the age of rocks.

The mechanism works like this: Once a quantity of radioactive material (such as uranium, which is plentiful worldwide) has been incorporated into a growing mineral crystal, that fixed quantity will decay at a steady rate, with a known percentage of the atoms decaying in any given increment of time. Each radioactive isotope has its own mode of decay (alpha, beta, or gamma radiation, or a combination) and its own unique decay rate. As time passes, the quantity of the original (parent) isotope diminishes, and the number of daughter atoms increases, thereby indicating how much time has elapsed since the clock began its timekeeping.

The beginning, or "time zero," for any mineral containing radioactive material is the moment when the radioactive parent atoms became part of that mineral. Encased solidly within the crystal, the daughter elements cannot escape. The retention of daughter elements is critical, because, if any escape, it will throw off the count and give an inaccurate age.

"Counting" the quantities of parent and daughter isotopes in a rock sample usually is done with a *mass spectrometer*. This analytic instrument measures the atomic mass of isotopes. In the mass spectrometer, a sample is vaporized in a vacuum chamber, where it is

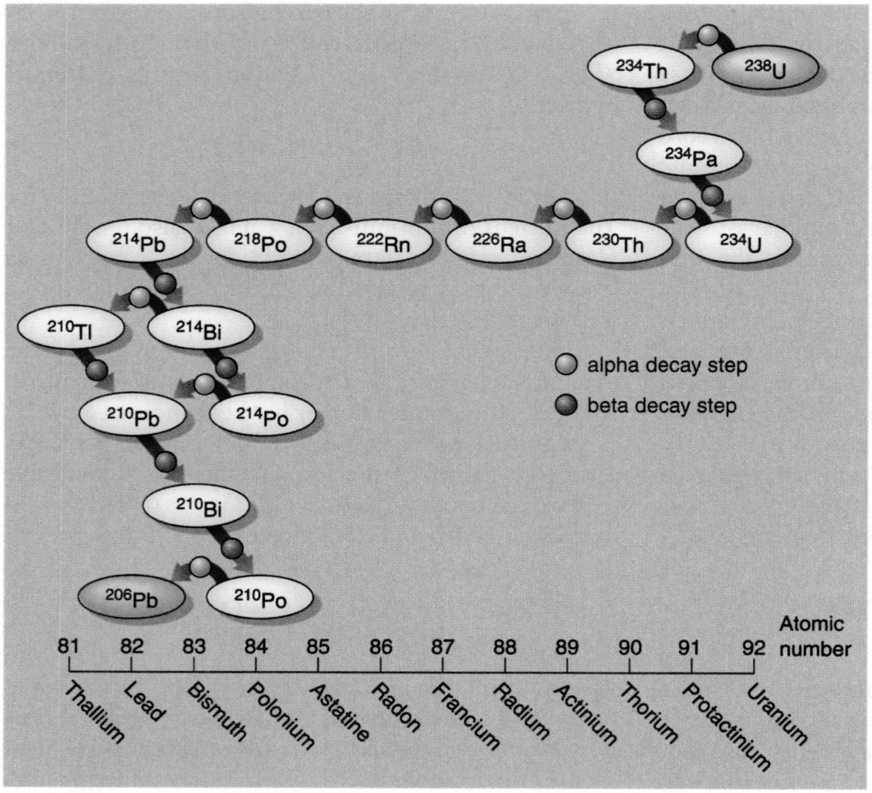

FIGURE 3-6 **Radioactive decay series.** This is for the isotope uranium-238 (^{238}U) to the isotope lead-206 (^{206}Pb).

bombarded by a stream of electrons. This bombardment knocks electrons off the atoms, turning them into positively charged ions. A stream of these positively charged ions passes through a magnetic field that deflects the stream. The degree of deflection varies with the charge-to-mass ratio. In general, the heavier the ion, the less it will be deflected.

Why Igneous Rocks Give the Most Trustworthy Dates

Of the three major families of rocks (igneous, metamorphic, and sedimentary), the igneous clan is by far the best for isotopic dating. First, igneous rocks crystallize from magma. So, when the crystals form, they "lock in" the atoms of radioactive elements. Second, if the crystals remain undisturbed, fresh samples of igneous rock are not likely to have lost any daughter atoms. This permits a trustworthy calculation of the number of years that have elapsed since the material cooled and solidified from a molten liquid mass to a solid rock.

In contrast, the minerals of sediments often are weathered and may have lost some of the daughter atoms. This makes age determination for sedimentary rocks problematic. Also, the age of a detrital grain in a sedimentary rock just gives the age of the original parent rock from which the grain was derived, not the age of the sedimentary rock itself.

As you might expect, metamorphic rocks are even more problematic, for they have undergone the trauma of heat and pressure, which may release daughter atoms. They require special care in interpretation. Thus, the age determined for a particular mineral could be when the rock first formed, but it also could be for any of several subsequent recrystallizations that occur when the rock becomes metamorphosed.

Once an age has been determined for a particular rock unit, that information can help determine the age of adjacent rocks. For example, in Figure 3-7, a geology student on a field trip points to a layer of weathered volcanic ash called **bentonite.** Crystals of the semiprecious mineral zircon in the bentonite contain uranium atoms, and they yield uranium-to-lead ratios indicating the layer is 453.7 million years old. Thus, strata below the bentonite are older than 453.7 million years, and those above are younger.

In Figure 3-8, a shale bed lies beneath a lava flow dated at 110 million years old and above another flow dated at 180 million years old. Therefore, the shale bed must have been deposited between 110 and 180 million years ago. Similarly, as shown in Figure 3-9, the age of a shale deposited on the erosional surface of a 480-million-year-old granite mass and covered by a

FIGURE 3-7 **Thin layer of altered volcanic ash (dark brown) between layers of Ordovician marine limestones.** ⬛ *Using a high-resolution uranium-lead method, the ash was found to be 453.7 million years old. What can now be stated about the age of the stratum beneath the ash?*

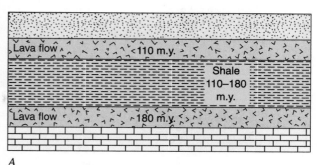

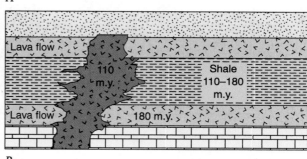

FIGURE 3-8 **Igneous rocks that have provided absolute radiogenic ages can often be used to date sedimentary layers.** (A) The shale is bracketed by two lava flows. (B) The shale lies above the older flow and is intruded by a younger igneous body (m.y. = million years).

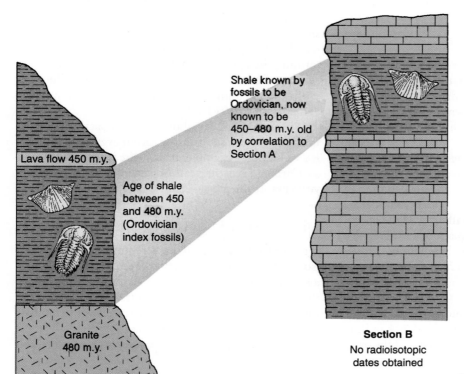

Section A
Some radioisotopic
dates obtained

Section B
No radioisotopic
dates obtained

FIGURE 3-9 **The actual age of rocks that cannot be dated isotopically can sometimes be ascertained by correlation.**

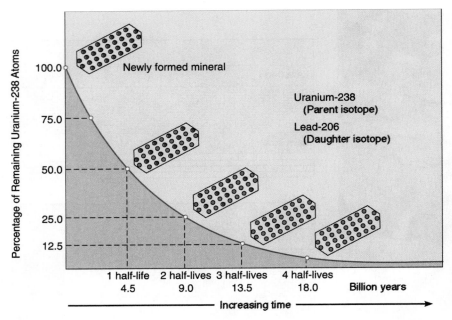

FIGURE 3-10 **Rate of radioactive decay of uranium-238 to lead-206.** During each half-life, half of the remaining amount of the radioactive element decays to its daughter element. In this simplified diagram, only the parent and daughter isotopes are shown. The assumption is that there was no contamination by daughter isotopes at the time the mineral formed. ◄ *If you were to draw a graph showing how many grains of sand passed through an hourglass each minute, how would the graph differ from the one depicted here?*

450-million-year-old lava flow must be between 450 and 490 million years old.

If there are fossils in the shale, they might then be used to assign the shale to a particular geologic system or series. Then, by correlation, the actual age determined at the initial locality (Fig. 3-9, Section A) could be assigned to formations at other locations (Fig. 3-9, Section B).

Half-Life

One cannot predict with certainty the moment of disintegration for any individual radioactive atom in a mineral. Experimenters have shown that more disintegrations per increment of time occur in the early stages of decay than in later stages (Fig. 3-10). Thus, we can statistically forecast what percentage of a large population of atoms will decay in a certain amount of time.

It is convenient to consider the time needed for half of the original quantity of atoms in a sample to decay. This span of time is termed the **half-life**. At the end of one half-life, half of the original quantity of radioactive element has decayed, but the other half has not. After a second half-life, half of what was left remains, or 1/4 of the original quantity. After a third half-life, only 1/8 remains, and so on.

Every isotope has its own unique half-life. Uranium-235, for example, has a half-life of 704 million years. Thus, if a sample contains 50 percent of the original amount of uranium-235 and 50 percent of its daughter product, lead-207, then that sample is 704 million years old. If the analyses indicate 25 percent of uranium-235 and 75 percent of lead-207, two half-lives

would have elapsed, and the sample would be 1408 million years old.

THE PRINCIPAL RADIOACTIVE TIMEKEEPERS

At one time, there were many more radioactive isotopes present on Earth than there are today. Many had short half-lives and have long since decayed to undetectably tiny quantities. Fortunately, for those who determine the ages of Earth's most ancient rocks, there remain a few long-lived radioactive isotopes. The most useful are uranium-238, uranium-235, rubidium-87, and potassium-40 (Table 3-3).

There are also a few short-lived radioactive elements that are used for dating more recent events. Carbon-14 is a well-known example. Its half-life is only 5730 years, so Carbon-14 dating can be accurate back to 50,000 years before the present.

Uranium-Lead Methods

Uranium-lead dating methods require that radioactive isotopes of uranium or thorium were incorporated into a rock at the time it formed. To determine the age of a mineral or rock sample, we must know the original number of parent isotope atoms as well as the number remaining at present. The original number of parent atoms should equal the sum of the current parent and daughter atoms.

We assume that no daughter atoms (lead) were present in the system when it formed. We also assume that the system has remained closed, so neither parent nor

TABLE 3-3 **Some of the More Useful Isotopes for Radioisotope Dating**

Parent Isotope	Half-Life†	Daughter Isotope	Source Materials
Carbon-14	5730 years	Nitrogen-14	Organic matter
Uranium-238	4.47 billion years	Lead-206	Zircon, uraninite, pitchblende
Uranium-235	704 million years	Lead-207	
Uranium-234	248 thousand years	Thorium-230	Lavas, coralline limestones
Thorium-232	14 billion years	Lead-208	
Rubidium-87	48.8 billion years	Strontium-87	Muscovite mica, potassium feldspar, biotite, glauconite, whole metamorphic or igneous rock
Potassium-40	1251 million years (1.251 billion years)	Argon-40 (and calcium-40)†	Muscovite, biotite, hornblende, whole volcanic rock, glauconite, potassium feldspar‡

†Half-life data from Steiger, R. H., and Jäger, E. 1977. Subcommission on geochronology: Convention on the use of decay constants in geo- and cosmochronology, *Earth and Planetary Science Letters* 36:359–362.
‡Although potassium-40 decays to argon-40 and calcium-40, only argon is used in the dating method because most minerals contain considerable calcium-40, even before decay has begun.

daughter atoms have been added or lost. The presence of any original lead in the mineral would cause the radiometric age to exceed the true age. Similarly, added or lost volumes of parent or daughter atoms can throw off the calculation significantly. Fortunately, geochemists are able to recognize original lead, if any is present, and make the needed corrections.

Because different isotopes decay at different rates, geochronologists simultaneously analyze two or three isotope pairs to cross-check their ages and thus detect errors. For example, if the ^{235}U/^{207}Pb (Pb is lead) radiometric ages and the ^{238}U/^{206}Pb ages from the same sample agree, then we can confidently assume that the age determination is valid. If they disagree, researchers must determine the cause and perhaps obtain new samples to test.

Isotopic ages that depend on uranium-lead ratios also may be checked against ages derived from the ratio of lead-207 to lead-206. The half-life of uranium-235 is much shorter than the half-life of uranium-238. So, the ratio of lead-207 (produced by the decay of uranium-235) to lead-206 will change regularly with age. This makes it an effective radioactive timekeeper (Fig. 3-11). This is called a lead-lead age, as opposed to a uranium-lead age.

The Potassium-Argon Method

Potassium and argon are another radioactive pair widely used for dating rocks. Potassium-40 occurs in many common minerals, like feldspar. About 11 percent of the potassium-40 in a mineral decays to argon-40 (by means of electron capture), which then may be retained within the parent mineral. The remaining

potassium-40 decays to calcium-40, but we cannot use this because the calcium-40 formed by decay cannot be distinguished from calcium-40 that may have originally been in the rock.

An advantage of argon-40 is that it is chemically inert—it does not combine with other elements. Argon-40 found in a mineral is very likely to have originated there following the decay of adjacent potassium-40 atoms in the mineral.

However, like all isotopic dating methods, potassium-argon has limitations. A sample can yield a valid age only if none of the argon, which is a gas, has leaked from the

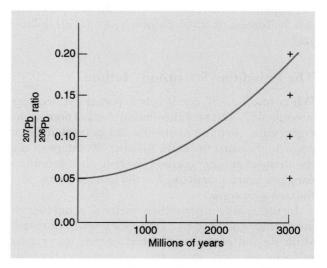

FIGURE 3-11 **Graph showing how the ratio of lead-207 to lead-206 can be used as a measure of age.**
☒ *What would be the age of a rock having a 207Pb/206Pb ratio of 0.15?*

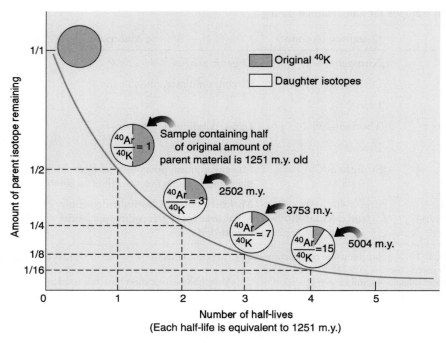

FIGURE 3-12 **Decay curve for potassium-40.** Potassium and argon are another radioactive pair widely used for dating rocks. Potassium-40 occurs in many common minerals, like feldspar. A sample can yield a valid age only if none of the argon gas has leaked from the mineral sample. Potassium-40's half-life is 1251 million years (1.251 billion years).

mineral sample. Leakage may indeed occur if the rock has experienced temperatures above about 125°C, which is not very hot and commonly occurs. In specific localities, the ages of rocks dated by this method reflect the last episode of heating, rather than the time of origin of the rock itself.

Potassium-40's half-life is 1251 million years (1.251 billion years). As illustrated in Figure 3-12, if the ratio of potassium-40 to daughter products is 1 to 1, then the age of the sample is 1251 million years (1.251 billion years). If the ratio is 1 to 3, then yet another half-life has elapsed, so the rock would have an isotopic age of two half-lives, or 2502 million years (2.502 billion years).

The Rubidium-Strontium Method

When rubidium-87 expels a beta particle, it becomes strontium-87. This provides a useful check of potassium-argon dates, because rubidium and potassium often occur in the same minerals. Another advantage of the rubidium-strontium method is that the strontium daughter isotope is not lost by mild heating events, as is the case with argon.

In the rubidium-strontium method, a number of samples are collected from the rock body to be dated. With the aid of the mass spectrometer, researchers calculate for each sample the amounts of radioactive rubidium-87, its daughter product strontium-87, and strontium-86. (Strontium-86 is an isotope not derived from radioactive decay.)

Then the ^{87}Rb/^{86}Sr ratio in each sample is plotted against the ^{87}Sr/^{86}Sr ratio (Fig. 3-13). From the points

on the graph, a straight line (isochron) is constructed. The slope of the isochron results from the fact that, with the passage of time, there is continuous decay of rubidium-87, which causes the rubidium-87/strontium-86 ratio to decrease. Conversely, the strontium-87/ strontium-86 ratio increases as strontium-87 is produced by the decay of rubidium-87. The older the

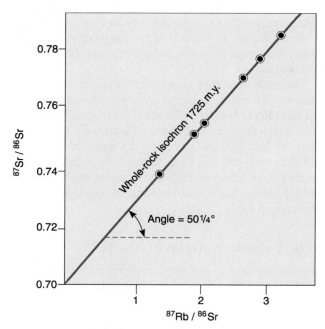

FIGURE 3-13 **Whole-rock rubidium-strontium isochron for a set of samples of a Precambrian granite body exposed near Sudbury, Ontario.**

rocks being investigated, the more the original isotope ratios will have been changed and the greater will be the inclination of the isochron. The degree of slope of the isochron thus permits a computation of the age of the rock.

How Carbon-14 Enters the Environment

All but a tiny amount of carbon on Earth is carbon-12, a stable atom. But another carbon isotope, carbon-14, provides an indispensable aid to archaeological research and is useful in deciphering very recent events in geologic history. Carbon-14's relatively short half-life—a mere 5730 years—means that most organic substances older than about 50,000 years cannot be dated.

Carbon-14 is created continuously in Earth's upper atmosphere. Its story is shown in Figure 3-14. It begins with cosmic rays, which are extremely high-energy particles (mostly protons) that come from space and bombard Earth continuously. Cosmic rays strike atoms in the upper atmosphere and split their nuclei into small particles, including neutrons. When one of these neutrons strikes the nucleus of a nitrogen-14 atom, the collision causes the nitrogen atom to emit a proton and capture a neutron. This is how carbon-14 is formed.

Carbon and oxygen atoms routinely combine to form carbon dioxide (CO_2). This happens regardless of whether a carbon atom is carbon-12 or carbon-14. The resulting carbon dioxide molecule is carried by wind and water currents and distributed anywhere around the globe. By this means, carbon-14 permeates the environment worldwide. Despite this wide distribution, the proportion of carbon-14 atoms in the atmosphere is very tiny—only one atom for every trillion carbon-12 atoms.

So, as photosynthetic plants draw in carbon dioxide to build tissue, a minute amount of that CO_2 contains carbon-14. Plants containing it are ingested by animals,

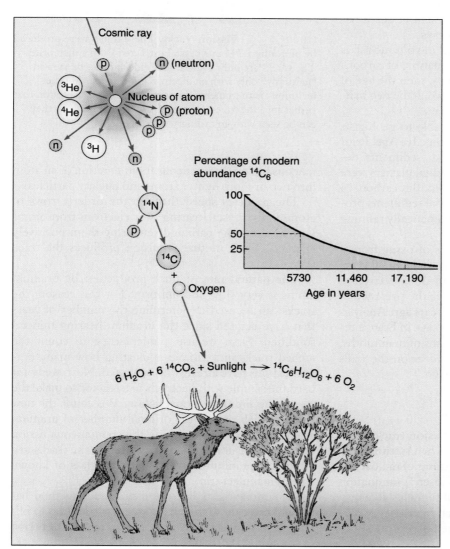

FIGURE 3-14 Carbon-14 is formed from nitrogen in the atmosphere. It combines with oxygen to form radioactive carbon dioxide and is then incorporated into all living things.

so the isotope becomes a part of animal tissue—including yours. Your body contains a tiny percentage of radioactive carbon-14 right now. Thus, since all life forms contain carbon, all life forms carry the carbon-14 needed for radioactive dating.

How Carbon-14 Dating Works

Eventually, each carbon-14 atom emits a beta particle and decays back to nitrogen-14. A plant that has removed CO_2 from the atmosphere, and an animal that eats that plant, both receive a share of carbon-14 proportional to that in the atmosphere. A state of equilibrium is reached in which the gain in newly produced carbon-14 is balanced by the loss from decay.

Unlike previously discussed dating schemes, carbon-14 dating is not determined from the ratio of parent to daughter isotopes. Rather, the age is estimated from the present ratio of the carbon-14 isotope to all other carbon isotopes in the sample. After an animal or plant dies, it no longer can add carbon-14 to itself, so the amount of carbon-14 present at death begins to diminish as the carbon-14 decays.

For example, if a pine tree buried in volcanic ash is measured to have 25 percent of the quantity of carbon-14 that exists in currently living pines, then the age of the wood (and the volcanic activity) would be two half-lives of 5730 years each, or 11,460 years.

The carbon-14 technique is valuable to geologists studying recent events of the Pleistocene Ice Age. Prior to the development of this method, sediments deposited by the last advance of continental glaciers were estimated to be about 25,000 years old. But carbon-14 dates for a peat layer beneath the glacial sediments provided an age of only 11,400 years, dramatically refining our knowledge of their true age.

In Figure 6–8, carbon-14 analysis of tissue from a baby mammoth indicates the animal died 44,000 years ago. The age of giant ground sloth dung recovered from a cave near Las Vegas indicates the presence of these great beasts in Nevada 10,500 years ago. Analysis of charcoal from the famous Lascaux cave in France reveals that the artists who drew pictures of mammoths, woolly rhinoceroses, bison, and reindeer on the walls of the cave lived about 15,000 years ago.

Fission Track Dating Method

In the early 1960s, nuclear particle **fission tracks** were discovered by electron microscope when scientists examined areas around presumed locations of radioactive particles embedded in mica. Closer examination showed the tracks to be really small tunnels—like bullet holes—that were produced when high-energy particles from uranium nuclei were fired off during spon-

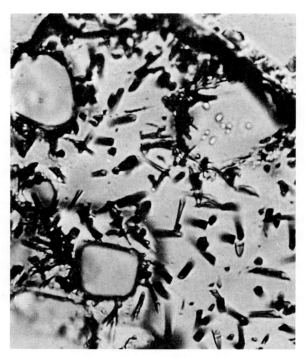

FIGURE 3-15 **Fission tracks.** These tracks were produced by plutonium-244 in a crystal extracted from a meteorite. The tracks are only a few atoms wide and can be viewed "naturally" only with an electron microscope. Another technique is to immerse the sample briefly in a solution that rushes into the tubes, enlarging the track tunnels so they can be seen with an ordinary microscope.

taneous fission (spontaneous fragmentation of an atom into two or more lighter atoms and nuclear particles).

The particles speed through the orderly rows of atoms in the crystal, tearing away electrons from atoms located along the path and rendering them positively charged. Their mutual repulsion produces the track (Fig. 3-15).

The natural rate of track production by uranium atoms is very slow and uniform. For this reason, the tracks can be used to determine the number of years that have elapsed since the uranium-bearing mineral solidified. First, we use a microscope to count the etched tracks in a sample, indicating how many uranium atoms have already disintegrated. Next, we bombard the sample with neutrons in a reactor to make the remaining uranium atoms fission. We count the new tracks, which reveals the original number of uranium atoms. Finally, we determine the spontaneous fission decay rate for uranium-238 by counting the tracks in a sample of uranium-bearing synthetic glass of known date of manufacture.

Geochronologists like the fission track method because it can date specimens from a few centuries to billions of years in age. The method helps to date the period between 100,000 and 1 million years ago—an interval for which neither carbon-14 nor potassium-argon methods are suitable. As with all radiometric techniques, however, there are caveats. If rocks have been subjected to high temperatures, tracks may heal and fade away.

▶ HOW OLD IS EARTH?

To find the age of Earth, one must first decide what event marked its "birth." Most geologists assume that "year 1" was when Earth had collected most of its present mass and had developed a solid crust. This, of course, was not literally in a single year but during a span of many thousands. Unfortunately, rocks that date from those earliest years have not been found. They have long since been altered and converted to other rocks by geologic processes, obliterating any trace of the original rocks.

The oldest materials known are grains of the mineral zircon from a sandstone in western Australia. The zircon grains are 4.2 billion years old. They probably eroded from nearby granitic rocks and were deposited by rivers, along with quartz and other detrital grains. Other very old rocks on Earth include 4.0-billion-year-old granites of southwestern Greenland, metamorphic rocks near the same age from Minnesota, and 4.04-billion-year-old rocks from the Northwest Territories of Canada.

Meteorites, possibly remnants of a shattered planet or asteroid that formed at about the same time as Earth, have provided uranium-lead and rubidium-strontium ages of about 4.6 billion years. From such data, and from estimates of how long it would take to produce the quantities of various lead isotopes now found on Earth, geochronologists accept with confidence that Earth is 4.6 billion years old.

Moon rocks substantiate this conclusion, with samples dating from 3.3 to about 4.6 billion years old. The older ages are for rocks collected on the lunar highlands, which may represent the original lunar crust. Certainly, the moons and planets of our solar system originated as a result of the same cosmic processes and at about the same time.

SUMMARY

- Early attempts to unravel Earth's history depended on relative dating of events. Using the stage of evolution indicated by fossils and superposition, geologists pieced together many separate stratigraphic sections to gradually construct a relative scale of geologic time.
- The largest increments of time in the geologic time scale are termed *eons*. From oldest to youngest, the eons are Archean, Proterozoic, and Phanerozoic. Eons are divided into *eras*, eras into *periods*, periods into *epochs*, and epochs into *ages*. These terms refer to intervals of time, and are therefore termed *geochronologic units* or *time units*.
- Geochronologic units are represented by bodies of rock formed during each chronologic time interval. These tangible bodies of rock are termed *time-rock units*. A time-rock *system* consists of rocks formed during a period. The term *series* is used for the rocks formed during an epoch, and *stage* is used for rocks formed during an age.
- A way to determine the actual age of rocks measured in years was achieved after the discovery of radioactivity at about the turn of the twentieth century. Scientists found that the rate of decay of radioactive elements is constant and can be measured. The proportion of the parent and daughter elements is used to indicate the age of the rock in years.
- The parent-daughter isotope pairs most widely used in the actual dating of rocks are uranium-238 to lead-206, uranium-235 to lead-207, thorium-232 to lead-208, potassium-40 to argon-40, rubidium-87 to strontium-87, and carbon-14 to nitrogen-14.
- Isotopic age determinations of meteorites, moon rocks, and terrestrial rocks indicate Earth is 4.6 billion years old.

KEY TERMS

actual geologic dating, p.28

alpha particle, p.35

Archean Eon, p.28

atom, p.34

atomic mass, p.34

atomic number, p.34

bentonite, p.36

beta particle, p.35

Cambrian System, p.31

Carboniferous System, p.32

Cenozoic, p.28

Cretaceous System, p.32

daughter element, p.35

Devonian System, p.32

electron, p.34

element, p.34

eon, p.28

epoch, p.28

era, p.28

fission tracks, p.42

gamma radiation, p.35

geochronology, p.27

half-life, p.38

Holocene Series, p.32

isotope, p.35

Jurassic System, p.32

Mesozoic, p.28

Mississippian System, p.32

Neogene Series, p.32

Ordovician System, p.32

Paleozoic, p.28

parent element, p.35

Pennsylvanian System, p.32

period, p.28

Permian System, p.32

Phanerozoic Eon, p.28

Pleistocene Series, p.32

Precambrian, p.28

Proterozoic Eon, p.28

proton, p.34

neutron, p.34

radioactivity, p.35

relative geologic dating, p.27

series, p.30

Silurian System, p.32

stage, p.30

system, p.28

time-rock unit (= chronostratigraphic unit), p.28

time unit (= geochronologic unit), p.28

Triassic System, p.32

QUESTIONS FOR REVIEW AND DISCUSSION

1. Explain the difference between a geochronologic division in the geologic time scale and a chronostratigraphic division.

2. Provide the actual geologic age of each of the following:
  a. Earth
  b. The beginning of the Proterozoic Eon
  c. The beginning of the Paleozoic Era
  d. The beginning of the Mesozoic Era
  e. The beginning of the Cenozoic Era
  f. The beginning of the Pleistocene Epoch

3. To make an age determination, why is it necessary to determine the amount of parent isotope still in a mineral as well as the amount of daughter isotope?

4. Why is the concept of half-life necessary? (Why not use "whole life"?)

5. Pebbles of a black igneous rock called basalt are present in a layer of sedimentary rock. The pebbles have an isotopic age of 300 million years. Is the rock younger or older than the pebbles contained within it?

6. How do isotopes of a given element differ from one another in regard to number of protons and neutrons in the nucleus?

7. How are dating methods involving decay of radioactive elements *unlike* methods for determining elapsed time that involve the funneling of sand through an hourglass?

8. If a small amount of argon-40 escaped from a zircon crystal, how would this affect the measurement of that crystal's age using the potassium-argon method?

9. What is the age of a sample of prehistoric mummified human skin that contains 12.5 percent of the original amount of carbon-14?

WEB SITES

The Earth Through Time Student Companion Web Site (www.wiley.com/college/levin) has online resources to help you expand your understanding of the topics in this chapter. Visit the Web Site to access the following:

1. Illustrated course notes covering key concepts in each chapter;

2. Online quizzes that provide immediate feedback;

3. Links to chapter-specific topics on the web;

4. Science news updates relating to recent developments in Historical Geology;

5. Web inquiry activities for further exploration;

6. A glossary of terms;

7. A Student Union with links to topics such as study skills, writing and grammar, and citing electronic information.

4

Radiating clusters of slender crystals of selenite gypsum (CaSO₄·2H₂O).

Rocks and Minerals: Documents that Record Earth's History

What stuff 'tis made of, whereof it is born,
I am to learn.
—William Shakespeare, *The Merchant of Venice*

If you have already taken a course in physical geology, you have learned about the "recording media" in which Earth's geologic history is preserved. These media are minerals, rocks, and the fossils embedded in rocks. A basic understanding of them is essential to your understanding of the material in this historical geology course.

If you have had a physical geology course, this chapter will be a brief review. If not, we present the basics of Earth materials to set the stage for understanding events that have shaped our planet.

Key Chapter Concepts

- A rock is a solid Earth material composed of one or more minerals. A mineral is a naturally occurring, solid, inorganic substance having a specific chemical composition and a crystal structure.

- Silicates form the most abundant crustal rocks. A silicate is a mineral that contains silicon and oxygen (like quartz), often combined with additional elements.

- Nonsilicate minerals include carbonates (like calcite and dolomite), chlorides (like halite), sulfates (like gypsum), and various sulfides (like pyrite) and oxides (like hematite).

- We group rocks into three families—igneous, sedimentary, and metamorphic—based on how they formed. However, a rock of any type may be transformed into any other type.

- Igneous rocks solidify from a cooling melt, and their texture reflects their cooling history.

- Intrusive igneous rocks solidify underground, whereas extrusive igneous rocks solidify above ground, under the atmosphere or the ocean. Three common intrusive rocks are granite, diorite, and gabbro. Their extrusive equivalents are rhyolite, andesite, and basalt.

- Sedimentary rocks are composed of sediment that has been consolidated by compaction, cementation, and crystallization.

O U T L I N E

▶ MINERALS AS DOCUMENTS OF EARTH HISTORY
▶ MINERALS AND THEIR PROPERTIES
▶ COMMON MINERALS THAT FORM ROCKS
▶ EARTH'S THREE GREAT ROCK FAMILIES AND HOW THEY FORMED
▶ IGNEOUS ROCKS: "FIRE-FORMED"
▶ SEDIMENTARY ROCKS: LAYERED PAGES OF HISTORY
▶ METAMORPHIC ROCKS: CHANGED WITHOUT MELTING
▶ SUMMARY
▶ KEY TERMS
▶ QUESTIONS FOR REVIEW AND DISCUSSION
▶ WEB SITES

- A metamorphic rock has been changed in texture and mineralogy from its parent rock (which can be igneous, sedimentary, or metamorphic) by heat, pressure, and chemically active gases or solutions.

FIGURE 4-1 **Two copper-bearing minerals, malachite and azurite.** The two green specimens are malachite, and the blue mineral on the right is azurite.

MINERALS AS DOCUMENTS OF EARTH HISTORY

Our way of life depends on minerals, from which we manufacture products. But minerals also provide valuable evidence about past events:

- Many minerals form within a narrow range of physical conditions. Thus, their presence (or absence) indicates what pressures and temperatures existed during the formation of mountain ranges and volcanoes.
- Some minerals develop primarily in ocean water. Thus, their presence is evidence of former seas.
- Some minerals form only under highly arid conditions. Their presence is evidence of former arid tropical belts.
- Magnetic minerals record the direction of Earth's magnetic field at the time the minerals formed, thus providing clues to drifting of continents, widening of oceans, and movement of our planet's magnetic poles.
- Minerals containing radioactive elements provide trustworthy natural calendars for determining the ages of rocks and structures.

Earth's crust is made of rocks. Every rock is composed of one or more minerals. Let us first review the definition of minerals and their distinctive properties.

MINERALS AND THEIR PROPERTIES

*A **mineral** is a naturally occurring, solid, inorganic substance having a specific chemical composition and a specific crystal structure.* Let us consider each part of that definition: A mineral cannot be manufactured; cannot be a liquid or gas; cannot be organic (living or once-living); cannot be a mixture of chemicals; and cannot have randomly arranged molecules.

Minerals are identified by several physical and chemical properties. Some of these properties can be recognized only with the use of microscopes or other special equipment. Fortunately, most common minerals can be identified from easily seen or measured properties such as color, streak, luster, cleavage, hardness, density, crystal form, and magnetism.

Color is a mineral's most conspicuous characteristic. Some minerals are always the same color, such as green in malachite and blue in azurite (Fig. 4-1), blue-green in turquoise (Fig. 4-2), or red in almandine gar-

net (Fig. 4-3). In other minerals, color can be altered by even small amounts of impurities. Quartz, for example, may be colorless, white, pink, purple, green, or blue. Thus, when identifying a mineral, don't just go by its color—consider its other properties, too.

Streak is the color of a mineral when it is ground to powder. The streak color often is quite different from the color of the overall sample. Streak is produced by rubbing the mineral across an unglazed porcelain *streak plate* and observing the color of the powdery "streak" that results. A streak's color can be surprising, as in the case of the mineral pyrite—a bright, gold-colored mineral that has a black streak (Fig. 4-4).

Luster is the way a mineral shines in reflected light. Minerals that reflect light like metals have a *metallic luster*, as in pyrite. Minerals that look glassy, like quartz, have a *vitreous luster*. The luster of diamonds is *adamantine*. Some minerals reflect light poorly, so their luster is *dull* or *earthy*.

FIGURE 4-2 **Turquoise, another copper-bearing mineral.** Turquoise is a hydrous copper aluminum phosphate mineral. Its blue-to-blue-green color has made it a prized gemstone for centuries.

FIGURE 4-3 **Crystalline mass of almandine garnet.** This metamorphic mineral is noted for its wine-red color.

Cleavage is the tendency of a mineral to break along certain directions in its crystal structure where the bonds between planes of ions are weakest. The number and direction of cleavage planes help us to identify the mineral. Minerals like mica and graphite, for example, have pronounced cleavage in a single direction. The common mineral calcite has pronounced cleavage in three directions, as you can see in Figure 4-5.

Hardness is a mineral's resistance to scratching by another substance of known hardness. The German mineralogist Frederick Mohs (1773–1839) developed the hardness scale that we still use today. Mohs

FIGURE 4-5 **Calcite.** Calcite is a common mineral in limestone. It has notable cleavage in three directions.

arranged ten common minerals in order of their increasing hardness—see Table 4-1. For example, if an unknown mineral can be scratched by quartz (hardness = 7), it must be softer than quartz. But if it scratches apatite (hardness = 5), it must be harder than apatite. Therefore, its hardness is between 5 and 7, so its hardness is comparable to orthoclase (hardness = 6). You can make a rough estimate of hardness by using a steel nail (hardness = 5), a penny (hardness = 3), or your fingernail (hardness = 2 to 2.5).

Density of a mineral is its mass per unit volume (density = mass/volume). Density is a reflection of the relative "heaviness" of a mineral. For example, the density of quartz is 2.65 grams per cubic centimeter, whereas gold has a high density of 19.3 grams per cubic centimeter.

Crystal form is the way a mineral grows, through the regular addition of ions to its surfaces from a surrounding rock melt (magma), solution, or gas. As a crystal grows, the mineral may acquire a characteristic

FIGURE 4-4 **Pyrite, often called "fool's gold," is known for its golden metallic luster.** But it is easily distinguished from the real thing. Pyrite has a black streak, whereas real gold has a true golden streak. And real gold can easily be shaped with a hammer, whereas pyrite will simply shatter when struck.

TABLE 4-1 **The Mohs Scale of Mineral Hardness***

Increasing hardness		
	1	Talc
	2	Gypsum
	3	Calcite
	4	Fluorite
	5	Apatite
	6	Orthoclase
	7	Quartz
	8	Topaz
	9	Corundum
	10	Diamond

*Mohs scale of mineral hardness widely used by geologists was formulated in 1822 by Frederich Mohs. The scale begins with talc, a very soft mineral, and continues to diamond, the hardest of all minerals.

FIGURE 4-6 **Magnetite.** This mineral is a natural magnet.

form, having crystal faces parallel to orderly intersecting layers of ions within the crystal. Perfect crystals are rare in nature, because crystals usually grow in crowded environments, in contact with one another. The result is an interlocking crystalline mass.

Magnetism, taste, flexibility, and feel. Some minerals have other easily identified characteristics. For example, magnetite attracts steel objects (Fig. 4-6). Halite tastes salty—it is what we call "rock salt." Mica is flexible. Talc has a soapy feel.

▶ COMMON MINERALS THAT FORM ROCKS

More than 3,000 minerals have been described. But only a handful compose common rocks, so we will focus on those. We divide rock-forming minerals into two broad groups, the silicates (which contain silicon)

and nonsilicates (which don't). Silicates are much more common, so we'll examine them first.

Historically, the silicates are the most primordial and fundamental group of minerals. Over 4 billion years ago, after Earth had formed as an aggregation of cosmic debris, it began to melt. The heat required for melting came largely from two sources, radioactive decay and gravitational energy. Denser elements like iron and nickel gravitated inward to form the core. Outside the core, less-dense silicates accumulated.

The planet's surface seethed with volcanic activity. The molten rock that poured from vents and fissures solidified to form the silicate minerals of igneous rocks. Once the planet had cooled enough to support a solid crust and to sustain a hydrosphere (ocean) and atmosphere, weathering of the primordial silicates and precipitation of these products of weathering produced nonsilicate minerals.

Silicate Minerals

As students with too much to learn and too little time, you will appreciate knowing that only 8 of Earth's 90 natural elements make up the bulk of the crust and minerals. There is no need to memorize the Periodic Table of the Elements. As shown in Table 4-2, the eight most-abundant elements are oxygen, silicon, aluminum, iron, calcium, sodium, potassium, and magnesium.

About 75% by weight of Earth's crust is composed of two elements, oxygen and silicon (Table 4-2). These elements usually occur in combination with the abundant metals aluminum, iron, calcium, sodium, potassium, and magnesium to form the silicate minerals. A single family of silicates, the **feldspars**, comprises over one-half of the material of the crust, and a single mineral species called **quartz** represents a sizable portion

TABLE 4-2 Abundances of Chemical Elements in Earth's Crust

Element and Symbol	Percentage by Weight	Percentage by Number of Atoms	Percentage by Volume
Oxygen (O)	46.6	62.6	93.8
Silicon (Si)	27.7	21.2	0.9
Aluminum (Al)	8.1	6.5	0.5
Iron (Fe)	5.0	1.9	0.4
Calcium (Ca)	3.6	1.9	1.0
Sodium (Na)	2.8	2.6	1.3
Potassium (K)	2.6	1.4	1.8
Magnesium (Mg)	2.1	1.9	0.3
All other elements	1.5	—	—
	100.0	100.0[†]	100.0[†]

Based on Mason, B. 1966. *Principles of Geochemistry.* New York: John Wiley & Sons. Note the high percentage of oxygen in Earth's crust.
[†]Includes only the first eight elements.

TABLE 4-3   **Common Rock-Forming Silicate Minerals**

Silicate Mineral	Composition	Physical Properties
Quartz	Silicon dioxide (silica, SiO_2)	Hardness of 7 (on scale of 1 to 10); will not cleave (fractures specific gravity: 2.65
Orthoclase feldspar group	Aluminosilicates of potassium	Hardness of 6.0–6.5; cleaves well in two directions; pink or white; specific gravity: 2.5–2.6
Plagioclase feldspar group	Aluminosilicates of sodium and calcium	Hardness of 6.0–6.5; cleaves well in two directions; white or gray; may show striations on cleavage planes; specific gravity: 2.6–2.7
Muscovite mica	Aluminosilicates of potassium and water	Hardness of 2–3; cleaves, perfectly in one direction, yielding flexible, thin plates; colorless; transparent in thin sheets; specific gravity: 2.8–3.0
Biotite mica	Aluminosilicates of magnesium, iron, potassium, and water	Hardness of 2.5–3.0; cleaves perfectly in one direction, yielding flexible, thin plates; black to dark brown; specific gravity: 2.7–3.2
Pyroxene group	Silicates of aluminum, calcium, magnesium, and iron	Hardness of 5–6; cleaves in two directions at 87° and 93°; black to dark green; specific gravity: 3.1–3.5
Amphibole group	Silicates of aluminum, calcium, magnesium, and iron	Hardness of 5–6; cleaves in two directions at 56° and 124°; black to dark green; specific gravity: 3.0–3.3
Olivine	Silicates of magnesium and iron	Hardness of 6.5–7.0; light green; transparent to translucent; specific gravity: 3.2–3.6
Garnet group	Aluminosilicates of iron, calcium, magnesium, and manganese	Hardness of 6.5–7.5; uneven fracture; red, brown, or yellow; specific gravity: 3.5–4.3

(Biotite mica, Pyroxene group, Amphibole group, and Olivine are bracketed as *Ferromagnesian minerals*)

of the remainder. The principal silicate minerals are shown in Table 4-3.

QUARTZ   Quartz is the most basic silicate mineral: just silicon and oxygen, SiO_2. It is familiar and common in many different rocks. Quartz is glassy, colorless, or white. It is relatively hard (see Mohs scale, Table 4-1) and will easily scratch glass.

When quartz has sufficient time and space to grow, it may form attractive hexagonal crystals (Fig. 4-7). More frequently, all possible crystal faces cannot develop because the orderly addition of atoms during crystal growth is restricted by adjacent growing crystals. The resulting texture looks granular and reflects many small points of light from the scattered crystal faces that did manage to grow. The texture is termed *crystalline*.

Crystals of quartz commonly grow in an igneous environment, crystallizing from magma. But they also may form in sedimentary environments, from silica-rich water solutions. Such "sedimentary quartz" forms common minerals like chert, flint, jasper, and agate. Chert (Fig. 4-8) is a dense, hard silicate mineral composed of sub-microcrystalline quartz. Flint is the popular name for a dark gray or black variety of chert. The

dark color results from carbon impurities. Jasper is chert that is colored red or red-brown by iron oxide. Agate is recognized by its bands of differing colors.

The origin of chert is a complex problem made even more difficult by the fact that different varieties are formed by somewhat different processes. Some cherts are replacements of earlier carbonate rocks. Others appear to have formed as a result of the solution and reprecipitation of silica derived from siliceous skeletal remains of organisms.

FELDSPAR   The feldspars (German: *field stone*) constitute about 60% of the total weight of Earth's crust. There are two major groups of feldspars, orthoclase and plagioclase.

All feldspars are aluminosilicates (made of aluminum, silicon, and oxygen). The orthoclase feldspars (Fig. 4-9) are *potassium* aluminosilicates (feldspars made of potassium, aluminum, silicon, and oxygen). The plagioclase feldspars are *sodium-calcium* aluminosilicates (Fig. 4-10). Members of the plagioclase group range from a calcium-rich anorthite ($CaAl_2Si_2O_8$) to sodium-rich albite ($NaAlSi_3O_8$). Between these two extremes, plagioclase minerals containing both sodium and calcium occur.

FIGURE 4-7 **A large crystal of quartz.** This specimen is 14 cm (about 5.5 inches) tall.

A

B

FIGURE 4-9 **Orthoclase.** (A) Crystals of orthoclase (potassium feldspar). (B) A cleavage fragment of orthoclase.

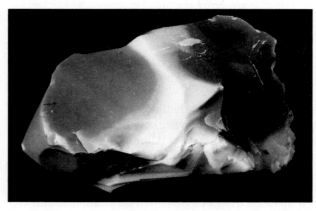

FIGURE 4-8 **Chert.** This variety called novaculite is used as a grinding stone because of its even texture.

The substitution of sodium for calcium is not random, and is governed by the temperature and composition of the parent magma or lava. Thus, by examining the feldspar content of a once-molten rock, it is possible to infer the physical and chemical conditions under which it originated.

Feldspars are nearly as hard as quartz and range in color from white or pink to bluish gray. They have pronounced cleavage in two directions, and the resulting flat, often rectangular surfaces are useful in identification. Potassium feldspars and plagioclase feldspars containing abundant sodium tend to occur in silica-rich rocks such as granite. Calcium-rich plagioclases are present in rocks that are dark-colored and rich in iron and magnesium, like the solidified lavas of Hawaii.

FIGURE 4-10 **The variety of plagioclase feldspar known as labradorite.** The mineral often displays beautiful blue and gold reflections as well as fine striations on cleavage planes.

FIGURE 4-11 **Muscovite mica, exhibiting its characteristic perfect cleavage in one plane.** The blade of the screwdriver is shown here separating pieces of the mineral along its cleavage plane.

MICA The **micas** are a family of silicate minerals easily recognized by their perfect and conspicuous cleavage along a single plane (Fig. 4-11). The two chief varieties are the colorless or pale-colored muscovite mica, which is a hydrous potassium aluminum silicate, and the dark-colored biotite mica, which also contains magnesium and iron. Both muscovite and biotite are common rock-forming silicates.

HORNBLENDE Hornblende is a glassy, black or very dark green mineral (Fig. 4-12). It is the most common member of a larger family of minerals called amphiboles, which have generally similar properties. Crystals of hornblende tend to be long and narrow.

Because of its content of iron and magnesium, hornblende (along with biotite, augite, and olivine) is designated a **ferromagnesian** or **mafic silicate mineral**.

AUGITE Just as hornblende is only one member of a family of minerals called amphiboles, **augite** is an important member of the pyroxene family, in which many other mineral species also occur. Like hornblende, it is a ferromagnesian mineral and thus dark-colored. An augite crystal is typically rather stumpy in shape, with pronounced cleavage developed along two planes that are nearly at right angles.

OLIVINE As you might guess, this glassy-looking iron and magnesium silicate often has an olive green color (Fig. 4-13). **Olivine** is present in dark rocks such as

basalt. Along with pyroxene and minor calcium plagioclase, it is also an important component of the ultramafic rock called peridotite. Peridotite is a prominent rock type in Earth's mantle, which is the rocky layer that lies beneath the crust.

CLAY MINERALS **Clay minerals** are silicates of hydrogen, aluminum, magnesium, iron, and potassium. Their basic structure is similar to that of mica, but individual flakes are extremely small (some are as small as

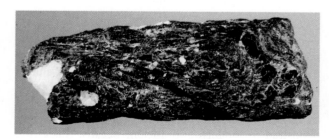

FIGURE 4-12 **Hornblende.** Specimen is 6 cm long. ☒ *Judging from its dark color, what metallic elements are likely to be present in this mineral?*

A

FIGURE 4-14 **Electron micrograph of the clay mineral kaolinite.** The flaky, stack-of-cards character of the clay crystals is a manifestation of their silicate sheet structure. (Magnified 32,000 times.)

B

FIGURE 4-13 **Olivine.** (A) is a specimen composed of intergrown crystals of olivine. (B) is a single crystal of olivine. When cut and polished, olivine crystals are marketed as the gemstone peridot.

viruses). This means that their mica-like form can be seen only with an electron microscope (Fig. 4-14). Clay minerals form as a result of weathering of other aluminosilicate minerals, such as feldspars.

Nearly 75% of the surface of continents is covered by clay minerals. They are the most abundant materials deposited in modern and ancient oceans. Shale, the most abundant of sedimentary rocks, is composed of about 50% clay. Clay minerals are an essential component of agricultural soils, serving to bind soil particles together and thus help to retain moisture and nutrients.

In geology, the term "clay" is used in two ways. A *clay mineral* is a hydrous aluminum silicate with mica-like form. But *clay* also designates any tiny particle of sediment that is less than 1/256 mm in diameter.

Nonsilicate Minerals

Approximately 8% of Earth's crust is composed of **nonsilicate minerals**. They are called nonsilicates for an obvious reason: They do not contain the silicon-oxygen structures that characterize silicate minerals. Nonsilicates include a host of carbonates, sulfides, sulfates, chlorides, and oxides. Carbonates, such as calcite and dolomite, are the most widespread.

CARBONATES: CALCITE AND DOLOMITE Calcite ($CaCO_3$) is the main constituent of limestone and marble. It forms in different ways—secreted as skeletal material by many invertebrate animals, precipitated directly from seawater, or deposited as dripstone from evaporating groundwater in caverns. **Calcite** is easily identified. It has rhombohedron-shaped cleaved fragments (see Fig. 4-15). Applying dilute hydrochloric acid to its surface will produce tiny bubbles of carbon dioxide (CO_2), which is liberated from the $CaCO_3$ by action of the acid.

Dolomite refers to either the magnesium-bearing carbonate mineral $CaMg(CO_3)_2$ or the rock that is composed largely of that mineral. In dolomite, calcium and magnesium occur in approximately equal proportions.

Aragonite ($CaCO_3$) is another carbonate mineral that occurs in a different crystal form and is more rare than either calcite or dolomite. Most of us have seen it as the inner layer of clam shells called "mother-of-pearl." Because both calcite and aragonite are two dif-

FIGURE 4-15 **Comparison of cleavage in halite and calcite.** Halite (left) has perfect cleavage in three directions, forming cubes. Its cubic cleavage differs from the rhombohedral cleavage of calcite (right). The calcite shown here is a particularly transparent variety called Iceland spar.

ferent forms of the identical compound, they are *polymorphs* ("multiple forms") of calcium carbonate.

EVAPORITES Other common nonsilicate minerals are **evaporites**, so-called because they precipitate from bodies of water that are subjected to intense evaporation. Evaporites can occur in thick layers that extend over hundreds of square miles. Three common evaporate minerals are halite (common table salt), gypsum, and anhydrite.

Halite (NaCl) is easily recognized by its salty taste and the fact that it crystallizes and cleaves to form cubes, as shown in the left-hand specimen in Figure 4-15. **Gypsum** is a soft, hydrous calcium sulfate ($CaSO_4 \cdot 2H_2O$) that occurs as several varieties. Satin spar (Fig. 4-16A) has a fine, fibrous structure. Alabaster (Fig. 4-16B) is gypsum's crystalline, massive variety, used in sculpture for its uniform texture and softness. Selenite forms clear crystals that can be cleaved into thin plates.

Evaporite deposits commonly also contain **anhydrite**, a granular, white, anhydrous (water-lacking) calcium sulfate ($CaSO_4$).

▶ **EARTH'S THREE GREAT ROCK FAMILIES AND HOW THEY FORMED**

Geologists divide Earth's rocks into three great families: **igneous, sedimentary,** and **metamorphic.** The basis for this division is the way in which each group forms:

- **Igneous rocks** form when molten rock (magma or lava) cools and crystallizes (solidifies). Igneous rocks are analogous to water freezing to form ice.
- **Sedimentary rocks** form when sediment becomes compressed and/or cemented over time.

A

B

FIGURE 4-16 **(A) The fibrous variety of gypsum called satin spar. (B) Alabaster, the granular variety of gypsum.**

The sediment comes from the weathering of rocks. Sedimentary rocks typically form in widespread layers.

- **Metamorphic rocks** are those that undergo metamorphosis (change in form) due to the action of heat, pressure, or chemical activity. Many are literally deformed or "bent out of shape."

Over Earth's 4.6 billion years, rocks have been recycled many times. A granite does not stay a granite forever. Depending on what happens in its environment, it can be remelted, it can become weathered and reduced to sand and clay minerals to form sedimentary rocks, or it can become metamorphosed. The same is true for any other rock—over time, all are subject to periodic

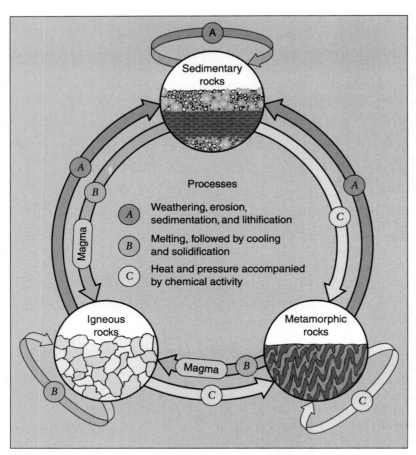

Processes

Ⓐ Weathering, erosion, sedimentation, and lithification

Ⓑ Melting, followed by cooling and solidification

Ⓒ Heat and pressure accompanied by chemical activity

FIGURE 4-17 **The rock cycle.** Geologic processes act continuously on Earth to change one type of rock into another. The process usually is very slow, requiring millions of years. There are exceptions, where the process is compressed into thousands of years. In the case of molten lava from a volcano, lava can solidify into igneous rock in hours, and bits of lava blasted into the air can solidify in seconds.

recycling. This ongoing process is illustrated by the rock cycle in Figure 4-17.

As shown in the rock cycle, any sedimentary or metamorphic rock may become partially or completely melted to produce igneous rocks. Any previously existing rock of any category can be compressed and altered during mountain-building to produce metamorphic rocks. And the weathered and eroded residue of any rock family—igneous, sedimentary, or metamorphic—can be observed today being transported to the sea for deposition and lithification into new sedimentary rocks.

Although geologists classify groups of rocks according to their origin, individual rocks are identified by their discernible characteristics of **texture** (size, shape, and arrangement of particles) and *mineral composition*.

IGNEOUS ROCKS: "FIRE-FORMED"

Igneous rocks constitute over 90% (by volume) of Earth's crust, but you may seldom see them because they are extensively covered by sedimentary rocks. Notable exceptions are in the cores of mountain ranges, where great uplift and deep erosion have laid them bare. They also are exposed to view in deep gorges like the Grand Canyon of the Colorado River; across much

of Canada, where the sediment cover has been stripped away by glacial erosion; or in Hawaii, where you can walk across solidified lavas that flowed from volcanoes only decades ago.

These igneous rocks started out as molten rock beneath the surface. The molten rock, **magma,** is a mixture of molten silicates and gases. Some magma cools and solidifies in place, deep underground. Some penetrates the surface, creating volcanoes (Fig. 4-18), providing an often spectacular reminder of the fiery processes that produce igneous rocks. Magma that erupts through the surface is called **lava** (Fig. 4-19).

The word igneous comes from the Latin *ignis*, meaning "fire." It is a very appropriate name, since these rocks develop from cooling masses of molten material that formed in exceptionally hot parts of Earth's interior—with temperatures in the thousands of degrees.

What Does the *Texture* of Igneous Rocks Tell Us about Their History?

Most igneous rocks form when magma penetrates existing rocks and solidifies before it can reach Earth's surface. Rocks that form this way, beneath Earth's surface, are termed **intrusive igneous rocks** (because they have intruded into older rocks) or **plutonic igneous**

FIGURE 4-19 **Front of a lava flow advancing over an older flow, Kilauea Volcano, Hawaii.** The advancing flow is breaking into a jagged, rough lava that Hawaiians call *aa* (AH-ah), a name adopted by geologists. The underlying older flow, called *pahoehoe* (puh-HOY-HOY) lava, exhibits a ropy texture.

FIGURE 4-18 **A stream of molten lava and a fire fountain pour from Pu'u'O'o vent, Hawaii in 1986.** The lava quickly solidifies to form igneous rock.

rocks (named for Pluto, Roman lord of the underworld). Very large masses of such rocks are called **plutons**. Whenever we see plutonic rocks exposed at Earth's surface (Fig. 4-20), it is because erosion has stripped away the rocks that once covered them.

In contrast, when magma breeches the surface and erupts as lava, we call the resulting rocks **extrusive igneous rocks** (because they are extruded, like you extrude toothpaste from a tube). This group includes rocks that formed from lava, which erupted from volcanoes or flowed from fractures in Earth's crust.

Grain size indicates the cooling history. The grain size of igneous rocks tells us how they cooled, and therefore much about the environment that existed when they cooled (Fig. 4-21). Magmas deep within Earth lose heat very slowly, over millions of years. Interestingly, magmas also retain water, and water inhibits formation of crystal "seeds" or nuclei. With high temperatures and few crystal "seeds" suspended in the molten rock, there is plenty of time and space for the growth of large crystals. In typical intrusive rocks such as granite, diorite, and gabbro, the intergrown crystals are large enough to be seen readily without magnification, as in Figure 4-21A.

FIGURE 4-20 **Part of a granite pluton that once lay deep within Earth, stripped of covering rocks and exposed at Earth's surface, Elephant Rocks State Park, Missouri.** Weathering of the granite along joints has resulted in huge boulders.

A

B

C

FIGURE 4-21 **The textures of igneous rocks tell us how they cooled.** (A) A coarse-grained gabbro. This is an intrusive rock that cooled slowly, allowing its large crystals plenty of time to grow. (B) A porphyry with large phenocrysts of potassium feldspar. It is used here in the ornamental posts outside St. Paul's Cathedral in London. (C) A fine-grained igneous rock, named andesite for its common occurrence in the Andes Mountains. 🔍 *What is the origin of this type of porphyritic texture? Based on color alone, what feldspar mineral is represented by the large phenocrysts?*

In contrast, extrusive igneous rocks have a much finer texture, in which crystals are too small to be seen with the unaided eye. This reflects the rapid cooling (minutes, hours, or days) of molten silicates as they are volcanically ejected at the surface. When a lava is extruded, there is not enough time for the growth of large crystals. Furthermore, water from the melt is quickly lost to the atmosphere. Without water, lots of crystals form—and with a large number of crystals, there is not enough space for them to grow large. A good example is the fine-grained andesite shown in Figure 4-21C.

Basalt is by far the most common extrusive igneous rock. It usually cools quickly and so is composed of fine mineral crystals. The minerals are ferromagnesians (rich in iron and magnesium) and plagioclase feldspar (rich in calcium, sodium, and aluminum). The ocean crust worldwide is mostly basalt, which constitutes about 70% of Earth's surface. Oceanic islands like Samoa, Hawaii, and Tahiti are volcanoes made of basalt. Basalt composes the thousands of kilometers of mid-ocean ridges that run through the oceans. The lava flows of the Columbia and Snake River Plateaus in northwestern United States contain an estimated 75,000 cubic kilometers of basalt. In India, the Deccan Plateau contains over a million cubic kilometers of basalt.

The fine texture of basalt results from rapid cooling of lava. Whether cooled by sudden exposure to air or by water beneath the sea, the sudden quenching of basaltic molten rock leaves little time for the growth of large crystals.

Obsidian is an interesting example of an extrusive rock that cools too fast to form even small crystals. The magma that forms obsidian is nearly pure silica, so it

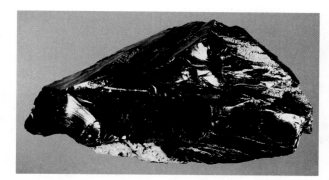

FIGURE 4-22 **The volcanic glass, obsidian.** ◪ *How does obsidian form?*

quickly solidifies to form a natural glass (Fig. 4-22). Obsidian forms from lavas that have lost most of their dissolved gases, so it does not contain bubbles. However, if a lava contains dissolved gases, bubbles will form small cavities called **vesicles**, creating a vesicular glass called **pumice**. It is actually a solidified silicate froth and will float on water. Volcanic rock composed of consolidated ash is called **tuff**.

If coarsely crystalline igneous rock indicates slow cooling and finely crystalline rock indicates rapid cooling, then what would be the cooling history of a rock that displays a mixture of large crystals in a very fine-grained matrix, as in Figure 4-21B? We call this a **porphyritic texture**. The large crystals (phenocrysts) grew

slowly at depth and were then swept upward and incorporated into the lava as it quickly hardened at the surface.

What Does the *Composition* of Igneous Rocks Tell Us about Their History?

Both texture and composition are essential in naming and classifying igneous rocks. For example, granite and rhyolite have an identical composition, yet their textures are completely different, because of the way they formed. Both are composed of feldspar, quartz, and biotite or hornblende. The difference is that **granite** is a coarse-textured intrusive rock that had plenty of time to grow crystals. **Rhyolite** is extrusive and thus has a very fine-grained texture (Fig. 4-23).

Figure 4-24 gives you the full picture of mineral composition, texture, and characteristics of igneous rocks. The minerals in an igneous rock reflect the proportions of whatever elements were available in the parent magma: silicon, oxygen, aluminum, calcium, iron, magnesium, sodium, and potassium. These eight elements combine in various ways to form light-colored feldspars, dark ferromagnesian minerals, micas, and quartz.

If the magma is rich in silica, it is likely that quartz will be present in whatever rock crystallizes from that magma. In general, intrusive igneous rocks are richer in silica than are extrusive rocks. The reason is that

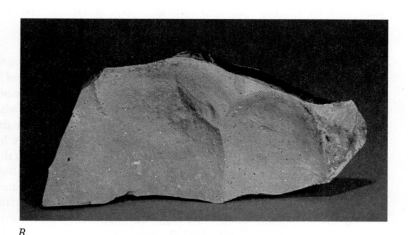

A   *B*

FIGURE 4-23 **Granite and rhyolite.** (A) Granite is a coarse-grained intrusive rock that has the same composition as its extrusive, fine-grained equivalent, rhyolite (B). In the granite, the lighter-colored grains are potassium feldspar and quartz, whereas the black minerals are hornblende and biotite. These same minerals are present in the rhyolite but cannot be seen by the unaided eye.

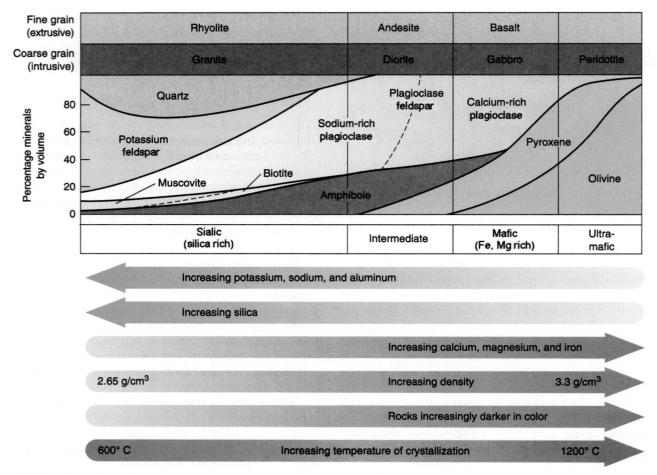

FIGURE 4-24 **Mineral composition, texture, and other properties for common igneous rocks.** You can estimate the abundance of a particular mineral in an igneous rock from the thickness of its colored area beneath the rock name. Use the scale on the left side to estimate the percentage of each mineral. For example, a granite near the midpoint on the figure would be composed of about 26% quartz, 25% potassium feldspar, 31% sodium-rich plagioclase, 12% amphibole, 5% biotite, and 1% muscovite. ◨ *What would be the composition of a mid-point basalt? What minerals might you expect to find in the porphyry shown in Figure 4-21 (B)?*

high silica content makes the magma thick and viscous, so it is more difficult for such a magma to make its way to Earth's surface. This accounts for the intrusive nature of silica-rich rocks like granite. Rhyolite is also rich in silica, but it is often propelled to the surface by high pressures of expanding gases, regardless of the viscosity of the melt.

Representative Igneous Rocks

Among the more familiar igneous rocks are granite, granodiorite, diorite, gabbro, rhyolite, andesite, and basalt. **Granite** (Fig. 4-23A) is a silica-rich, relatively light-colored intrusive rock. It is derived from magma so rich in silica that after all chemical linkages with metallic atoms are satisfied, enough silicon and oxygen remain to form quartz grains. In addition to quartz,

granite contains orthoclase feldspar, sodium plagioclase, and lesser amounts of mica and hornblende.

Granodiorite is another quartz-bearing igneous rock in which plagioclase is the dominant feldspar mineral. Quartz-bearing rocks such as granite and granodiorite are loosely termed "granitic rocks." The granitic rocks of the Sierra Nevada are largely granodiorites. As noted above, the more finely crystalline but compositional equivalent of granite is called rhyolite (Fig. 4-23B).

Diorite is a relatively coarse-grained intrusive rock containing less silica than granite or granodiorite. It is composed primarily of plagioclase feldspar. **Andesite** is the fine-grained equivalent of diorite.

In **gabbro**, the chief mineral is calcium plagioclase, which is darker and more calcium-rich than the plagioclase in diorite. Additional minerals in gabbro include augite and olivine. The fine-grained equivalent of gabbro is basalt (Fig. 4-25). Basaltic lavas have low viscosity and are able to flow for considerable distances across Earth's surface before solidifying. Because of the presence of ferromagnesian minerals and gray calcium plagioclase, low-silica rocks such as basalt and gabbro tend to be black, dark gray, or green.

Some Minerals Form Early, Others Later

You might get the notion that all minerals crystallize simultaneously as a magma cools. But actually, it is not that simple. Just as liquid water has a freezing point of 32°F, each molten mineral has its own freezing point at which it starts to solidify.

A

B

FIGURE 4-25 **Basalt.** (A) A hand specimen of basalt. (B) A thin section of basalt viewed with polarized light. The tabular crystals are plagioclase, and the brightly colored crystals are iron-magnesian silicates. In B, field-of-view width is 2 mm.

For example, if a magma has a basaltic composition, then olivine, pyroxene, and calcium feldspar will be among the earliest minerals to crystallize. Often, these first-to-form crystals are larger and more perfect than those that form later because they have ample space for growth and an abundance of elements to incorporate in their crystal structure. Minerals that crystallize later must grow within the remaining space and thus tend to be smaller and less perfect. Also, minerals enclosed in other minerals must have formed before the enclosing mineral. When rocks are viewed microscopically, you can see these relationships and use them to determine the order in which specific minerals crystallized in a magma.

Sequence of Mineral Crystallization

A century ago, geologist Norman L. Bowen (1887–1956) studied the sequence in which minerals crystallize as a magma slowly cools. Bowen made artificial magmas by melting powdered rock samples in a steel cylinder. He heated it until the powder melted, and then he let it slowly cool to the temperature and pressure selected for the experiment. The same temperature and pressure were maintained for a few months, so that some minerals had time to crystallize within the melt. This created a mixture of crystals and uncrystallized melt. The cylinder was then "quenched" by plunging it into cold water or oil, causing the mixture to quickly solidify as glass, preserving the crystals within the glass.

Bowen repeated this procedure many times, forming crystals at different temperatures before quick-cooling them. By identifying the crystals that formed at each temperature, he was able to determine the order of crystallization for minerals in a cooling magma.

For a melt of basaltic composition, Bowen found that olivine would crystallize first and at the highest temperature. Pyroxene and calcium plagioclase would form next, followed by hornblende, biotite, and sodium-rich feldspars.

The order in which the minerals crystallized came to be called **Bowen's Reaction Series** (Fig. 4-26). It is called a "reaction series" because early-formed crystals react with the melt to yield a new mineral further down in the series. For example, silica in the melt would react with olivine crystals by simultaneously dissolving and precipitating the dissolved components to form pyroxene. The pyroxene crystals would begin to grow in spaces between earlier-formed crystals. At still lower temperatures, the melt would react with pyroxene to form hornblende, and subsequently hornblende would react to form biotite.

To better understand this, on the right branch of Bowen's chart, you can trace the changes that occur as calcium-rich plagioclase reacts with the magma to

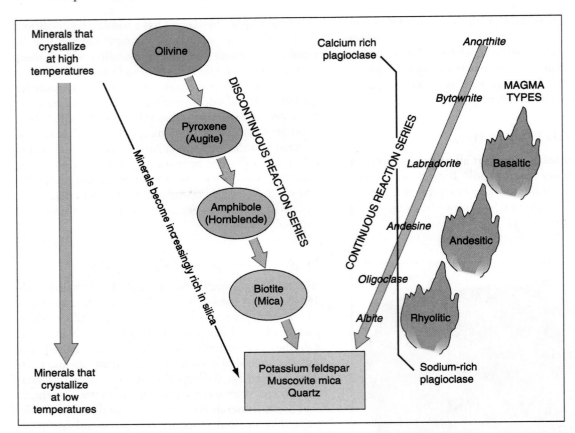

FIGURE 4-26 Bowen's Reaction Series. Note that the earliest minerals to crystallize (at top) are olivine and calcium-rich plagioclase. As crystallization proceeds, each mineral reacts with the melt to form the mineral beneath it. ◀ *What minerals are likely to form when pyroxene crystals react with the remaining liquid magma? Are plagioclase crystals in a granite likely to be of the sodium-rich or calcium-rich variety?*

produce varieties of plagioclase that are successively richer in sodium. Because the plagioclase minerals maintain the same basic crystal structure but change continuously in their content of calcium and sodium, the right side of the diagram is called the **continuous series**.

The left side of the diagram depicts reactions that result in minerals of distinctly different structure. It is therefore called the **discontinuous series**. The three minerals at the base of the chart (in the box) do not react with the melt. By the time they crystallize, little liquid remains. The residue of silicon, aluminum, oxygen, and potassium join to form potassium feldspar, muscovite mica, and quartz.

You can see evidence of the reactions just described when you examine igneous rocks microscopically. You can observe plagioclase crystals in which the innermost zones are more calcium-rich and the outermost zones are more sodium-rich. Similarly, you can see reaction rims of pyroxene around olivine and rims of hornblende around pyroxene.

Because of Bowen's experiments, we can recognize how magmas are able to change through time, and why there are so many mineral variations in igneous rocks. For example, volcanic rocks that erupt early may be rich in iron, magnesium, and calcium. These lavas produce basalts. The parent magma in later eruptions may contain a limited amount of iron, magnesium, and calcium, but may be richer in potassium, sodium, and silica. These lavas produce rocks that are less basaltic and more andesitic in composition.

Volcanic Activity Has Produced Enormous Volumes of Rock

Deciphering the history of a granitic pluton from its composition and texture is essential to discovering Earth's history, but it is unlikely to evoke the feelings of awe and excitement you experience when viewing a volcanic eruption! Volcanoes are simply vents through which pressurized hot gases and molten rock can escape. These extrusion events range from relatively gentle to highly explosive.

Relatively gentle eruptions are exemplified by the Hawaiian volcanoes, which are frequently characterized by nonexplosive eruptions of enormous outpour-

FIGURE 4-27 **A river of low-viscosity lava forms during the 1984 dual eruption of the Mauna Loa and Kilauea volcanoes, Hawaii.**

FIGURE 4-28 **One of many solidified basalt lava flows that form the Columbia Plateau.** The basalt exhibits columnar jointing, formed when lava contracts as it cools.

ings of low-viscosity lava (Fig. 4-27). The lava spreads widely to form the relatively gentle slopes of the volcano. Another example of gentle eruption is the massive, successive outpourings of lava from great fissures that form lava plateaus on the continents. Examples are the Columbia and Snake River Plateaus in the northwestern U.S. (Fig. 4-28) and the Deccan Plateau in India, mentioned earlier.

At the other extreme, explosive eruptions result from the sudden release of molten rock that is driven upward by large pockets of compressed gas. Explosive eruptions can quite literally blow a volcanic cone to bits. One of the most famous examples happened in 1883, when the Indonesian volcanic island of Krakatoa erupted with such violence that it destroyed the island, created a tsunami (giant ocean wave) that killed 36,000 people, and was heard 5000 km away.

Why do we observe such extreme variation in volcano behavior? The answer lies mainly in the viscosity (thickness) of the magma. A magma that is rich in silica is more viscous (thicker) than one with a low silica content. The thick, viscous melt may reach the surface only if driven by enormous pressures, which creates an explosive situation.

The presence of water is another important factor in explosive volcanism. Water's presence accounts for the violence of many Pacific Rim volcanoes, such as Mount St. Helens in Washington State, which erupted violently in 1980. Where does the water come from? It actually is recycled seawater. Here is how it becomes part of the magma: At mid-ocean ridges, extruding basalt reacts with seawater to form hydrous minerals, so-called because they have numerous water molecules trapped in their crystal structure. Over millions of years, these water-bearing minerals are conveyed by seafloor spreading to subduction zones. As the oceanic crust descends into the subduction zone, the trapped water is released from the hydrous minerals. Pressures build in areas containing the trapped water until they release explosively.

Of course, there are degrees of volcanic activity between relatively gentle and violently explosive. In response to changing composition in the magma, some volcanoes actually have shifted from one type of activity to another.

BASALTIC MAGMA The lava that formed Earth's immense volume of basalt originated at depths of about 100 to 350 kilometers from pockets of magma in Earth's upper mantle. How do we know this? We know that volcanic activity today is closely associated with deep earthquakes that occur within the mantle, far beneath the crust. Fractures produced by these earthquakes create passageways for the escape of molten material to the surface.

An interesting question is: How do you get basalt from the upper mantle that is composed of olivine-rich rock called peridotite? The answer is partial melting. As shown in Figure 4-29, **partial melting** occurs when rock that is subjected to high temperature and pressure is only partly melted and the liquid component moves to another location. At the new location, when the separated liquid eventually solidifies, it will form rocks having a different composition (mix of elements) from the parent magma.

The word partial in "partial melting" means that some minerals melt at lower temperatures than others, and so for a time the material being melted resembles a hot slush composed of liquid and still-solid crystals. The molten fraction is usually less dense than the solids from which it was derived and thus tends to separate from the parent mass and work its way toward the surface. In this way, melts of basaltic composition separate from denser rocks of the upper mantle and eventually make their way to the surface to form volcanoes.

ANDESITIC MAGMA Not all lavas found at Earth's surface are basaltic. Volcanoes of the more explosive type that are located at the edge of continents around the Pacific and in the Mediterranean extrude andesitic lavas. **Andesite** contains more silica than does basalt, so its lava is more viscous. The resulting greater resistance to flow contributes to the gas containment that precedes explosive volcanic activity. Andesites are intermediate in silica content between the rocks of the continental crust and those of the oceanic crust.

Andesitic rocks may originate in more than one way. Some result from originally basaltic magmas in which minerals such as olivine and pyroxene form early and settle out, thus leaving the remaining melt relatively richer in silica. This process is called **fractional crystallization**, and it is the reverse of partial melting (Fig. 4-30). Other andesites may result from the mixing and melting of basaltic ocean crust and siliceous marine sediments as they descend into hot zones of the mantle. The water-rich, silica-enriched melts of andesitic composition then rise buoyantly and erupt along volcanic island arcs. We will examine this interesting phenomenon more fully in Chapter 7, which deals with plate tectonics.

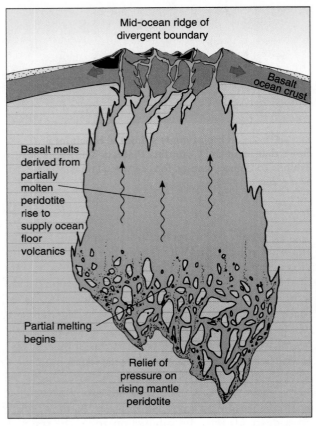

FIGURE 4-29 **How partial melting of peridotite forms basalt.** The entire mid-ocean ridge system worldwide consists of divergent plate boundaries. As the plates move apart, pressure is reduced along the boundary. This allows the rocks below (in the upper mantle) to begin melting and releasing magma. When this magma is extruded as lava along the plate boundary, it cools and solidifies to form basalt. This basalt then moves laterally in the course of seafloor spreading.

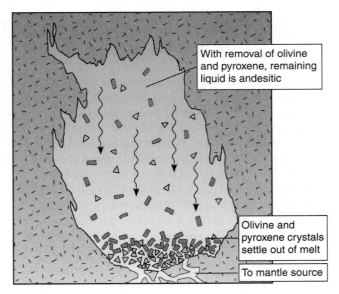

FIGURE 4-30 **An andesitic melt resulting from fractional crystallization of a basaltic magma.**

What Do Igneous Rocks Tell Us about Earth History?

Igneous rocks are primary in Earth history. They were the first—the earliest of rocks to form on Earth. The oldest metamorphic and sedimentary rocks were derived from igneous rocks.

In igneous rocks, we are able to read how mafic crustal rocks like basalt may have separated from an ultramafic (even more mafic) mantle to form an oceanic crust. We are able to understand how partial melting of that basaltic oceanic crust in the descending slabs of subduction zones produced the granitic rocks of the first continents.

Igneous rocks have provided information essential to historical geology. They record the changing positions of Earth's magnetic poles, so that we can know where continents were located in bygone eras. They hold the minerals that contain the radioactive isotopes that permit us to date geological events of the past.

▶ SEDIMENTARY ROCKS: LAYERED PAGES OF HISTORY

For historical geologists, sedimentary rocks can provide more information about Earth's past, and yield it more readily, than can igneous and metamorphic rocks. Layers of sedimentary rocks cover about 75% of the world's land area. Like pages in a history book, these layered rocks inform us about events of the past.

Sedimentary rocks form when loose sediment, such as sand and clay, becomes compressed and/or cemented to form solid rock. This process is called **lithification** (literally, "making into rock"). Lithification in sedimentary rocks may involve only compaction, or the process may also involve cementation. The cement can be mineral matter, like calcium carbonate or silica, which precipitates from water that is held in the spaces among sediment grains. Still other sedimentary rocks are held together because their grains interlock like jigsaw puzzle pieces.

Recall Steno's principle of original horizontality from Chapter 2. It states that "sediment is deposited in layers that are originally horizontal." Thus, a characteristic of sedimentary rocks is their occurrence in horizontal layers called **strata**. The formation of strata usually results from changes in conditions of deposition that cause materials of somewhat different composition or particle size to be deposited for a period of time.

The most abundant sedimentary rocks are sandstone, shale, and carbonate rocks (such as limestone). Sandstones are composed of sand-sized grains of quartz, feldspar, and other particles that are compressed and/or cemented together. Shale is the same, except it consists of much finer particles and includes clay minerals. Carbonate rocks form when carbon dioxide dissolved in water combines with oxides of calcium and magnesium. The material precipitates to the bottom and accumulates to form carbonate rocks.

Where Sediments Come From

Sedimentary rocks originally came from disintegrated, decomposed older rocks. Commonly, the older rocks are igneous; indeed, at first, these were the only rocks on Earth. As an example, let us see how the common components of sedimentary rocks are derived from an abundant igneous rock, granite. Granite is composed of quartz, feldspar, and (commonly) dark mica, called biotite (Fig. 4-31).

Quartz is among the most chemically stable of all the common silicate materials, so it persists almost unchanged during weathering, except for physical breakage. As the parent rock gradually decomposes, quartz grains tend to be washed out and carried away, to be deposited as sand that one day will become sandstone (Fig. 4-32).

The feldspar in granite decays more readily than quartz. Feldspars are primarily aluminum silicates with potassium, sodium, and calcium. In the weathering process, the last three elements are largely dissolved and carried away in solution. The insoluble particles left behind are the chief ingredient of clay. They may lithify to become shale or claystone.

The dissolved elements from the feldspar eventually reach the sea, where they may stay in solution or may precipitate. Limestone may form in this way. If large quantities of the water evaporate, evaporites such as halite or gypsum may form.

Of course, not all the feldspars and micas in the granite necessarily decay. Some may persist as detrital grains that become incorporated into sandstones and other sediments.

Biotite in the granite decomposes in a manner similar to feldspar. Decomposition of biotite, which is a potassium, magnesium, and iron aluminosilicate, yields soluble potassium and magnesium carbonates, small amounts of soluble silica, iron oxides, and clay. The iron oxides add color to many sedimentary rocks, forming shades of brown and red.

Thus, you see that ordinary granite can be the source for all the major components of sedimentary rocks. When rocks other than granite decompose, the sedimentary products will vary. Part of the detective work necessary to deciphering Earth's history is to examine sediment and determine its source.

Classifying Sedimentary Rocks

Sedimentary rocks are classified by their composition and texture. **Texture** refers to the size and shape of individual grains and their arrangement in the rock.

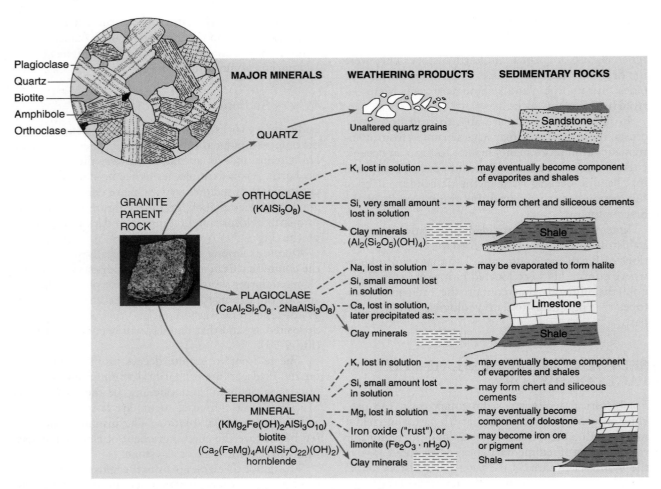

FIGURE 4-31 **When granite weathers, it yields quartz grains that form quartz sandstone, clay particles that form shale, and calcium that forms limestone.** (For simplicity, minor mineral components are excluded.) Some minerals are more durable than others. Quartz is the most durable sandstone, and hence the most common. If weathering is less severe, detrital grains of feldspar may be included in sandstones. ◨ *What is the most abundant, relatively insoluble product of the weathering of orthoclase and plagioclase? Which mineral is most stable (least likely to undergo dissolution during chemical weathering)?*

CLASTIC SEDIMENTARY ROCKS A rock with **clastic** texture is composed of *clasts:* grains and broken fragments of minerals, rocks, and fossils. Other sedimentary rocks are composed of a network of intergrown crystals and therefore have a crystalline texture.

Clastic sedimentary rocks include conglomerates, sandstones, siltstones, and shales, according to the particle size of their component grains.

- Conglomerate (Fig. 4-33) is composed of waterworn, rounded particles larger than 2 mm in diameter. The term breccia (Fig. 4-34) is reserved for clastic rocks composed of fragments that are angular but are otherwise similar in size to those of conglomerates.

- In sandstones, grains range from 2 mm to 1/16 mm in diameter (see Table 5–1 in the next chapter). The varieties of sandstone are then subdivided mainly according to composition.

- In siltstones, particles are finer than in sandstones (1/16 mm to 1/256 mm.)

- Shales may contain abundant clay minerals, which are flaky minerals that align parallel to bedding planes. As a result, shales characteristically split into thin slabs parallel to bedding planes.

The size of the clasts in sedimentary rocks can reveal historical information. Size reflects the processes of erosion and transportation. It helps the historical geologist decide if the sand grains in a sandstone were de-

FIGURE 4-32 **Exposure of weathered granite.** As the parent rock disintegrates, quartz grains and partially decomposed feldspar accumulate. ◨ *In addition to the quartz and feldspar, what other weathering product is present in the debris beneath the granite?*

posited by wind, ocean waves, or an ancient stream. And if it was a stream deposit, did the water flow rapidly or sluggishly? Did deposition occur close to the source area or at a great distance? Such information is needed to reconstruct conditions for particular times and places in Earth history.

FIGURE 4-33 **Conglomerate.** In this sample, you can see pebbles of chert in a carbonate matrix. The largest pebbles are about 2.5 cm in diameter. ◨ *How do conglomerates differ from breccias?*

FIGURE 4-34 **Breccia (the bed of angular fragments across center).** This breccia is composed of angular fragments (clasts) of Precambrian metamorphic quartzite. Mosaic Canyon, Death Valley, California.

CARBONATE SEDIMENTARY ROCKS About 15% of all sedimentary rocks are carbonates. Most of these formed either by precipitation from seawater or by the accumulation of the shelly skeletal parts of sea creatures. Warm shallow seas are the favored environment for the deposition of carbonates. As these waters become warmer, they lose carbon dioxide, and this favors precipitation of calcium carbonate. Warm shallow seas covered vast regions of the continents during the Paleozoic and Mesozoic eras. Carbonate sediment accumulating on the floors of these inland seas entombed millions of ancient marine creatures.

Carbonate rocks contain the minerals calcite and dolomite. Limestone rock is mostly calcite, $CaCO_3$. Dolomite rock is mostly the mineral dolomite, $CaMg(CO_3)_2$, with magnesium content being the difference. Carbonate rocks contain highly variable amounts of impurities, including iron oxide, clay, and particles of sand and silt swept into the environment by currents.

After calcium carbonate particles have accumulated, the particles may grow into a crystalline solid or become otherwise consolidated into rock. Limestones tend to be well stratified (Fig. 4-35), and range in texture from coarsely granular to very fine-grained.

Limestones can be described by textural features (Fig. 4-36) named **micrite, carbonate clasts, oölites,** or **carbonate spar.** Micrite is exceptionally fine-grained carbonate mud. Carbonate clasts are sand- or gravel-sized pieces of carbonate. The most common

FIGURE 4-35 **Roadcuts that expose ancient limestone strata are familiar to Interstate highway travelers throughout the Mississippi Valley region.** The limestone layers in this Missouri road cut formed on the ocean bottom about 470 million years ago. The originally horizontal beds have been gently folded and are now tilted to a dip of about 35°.

clasts are either bioclasts (skeletal fragments of marine invertebrates) or **ooids** (Fig. 4-36C), which are spherical grains formed by the precipitation of carbonate around a nucleus. Carbonate spar is clear, crystalline (sparry) carbonate. It occurs between clasts as a cement. By using these textural terms, we can describe limestones as micritic, clastic, oölitic, or sparry.

One well-known limestone variety is chalk. The chalk we use to write on chalkboards and sidewalks is mixed with clay to give it more durability. When you draw a line with chalk, you are depositing thousands of particles of soft, porous limestone composed of microscopic skeletal remains of marine plankton (tiny floating sea animals and plants). Another variety is lithographic limestone, a dense rock used many decades ago in a printing method called lithography ("stone printing").

The other widespread carbonate rock is dolomite, which contains the element magnesium. (Remember that there is also a *mineral* called dolomite.) Dolomite requires effort to distinguish from limestone. The usual field test is to squirt a drop of cold, dilute hydrochloric acid onto the rock. Limestone effervesces (bubbles) readily, but dolomite does so only slightly, if at all.

The origin of dolomite is different from limestone. The mineral dolomite is not secreted by organisms, nor does it commonly precipitate from seawater. Instead, dolomite forms in the original calcium carbonate sediment when circulating water brings in magnesium atoms that replace some of the calcium atoms.

OTHER SEDIMENTARY ROCKS **Chert** is a form of quartz (SiO_2). It is microcrystalline (crystals too small to see with the unaided eye). Some cherts occur as nodules in limestones (Fig.4-37). The origin of these nodules is still debated, although the majority of geologists think that they form as replacements of carbonate sediment by silica-rich seawater trapped in the sediment.

Other cherts are not nodular. They occur in extensive layers. These so-called bedded cherts are thought to have formed from the accumulation of the skeletal remains of diatoms and radiolarian, which have skeletons made of silica, and from subsequent reorganization of the silica into a microcrystalline quartz. Silica dissolved in seawater from volcanic ash is believed to

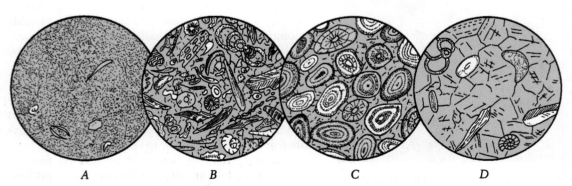

FIGURE 4-36 **Textures of limestones as seen in thin section under a microscope.** (A) Aphanitic limestone or micrite. (B) Bioclastic limestone with fine-grained sparry calcite as cement. (C) Oölitic limestone. (D) Sparry or crystalline limestone.

FIGURE 4-37 **White nodular chert in tan limestone.** Fern Glen Formation west of St. Louis, Missouri.

enhance the process. As evidence of this, many bedded cherts are found in association with ash beds and submarine lava flows.

Evaporites, as indicated by their name, are chemically precipitated rocks that are formed as a result of evaporation of saline water bodies. Only about 3% of all sedimentary rocks are evaporites, but some of these are very important economically. Evaporite sequences of strata are composed chiefly of such minerals as gypsum (used in making drywall), anhydrite, halite (used for table salt and road salt in northern winters), and associated calcite and dolomite.

Coal is the source of roughly half of U.S. electrical power. It is the fuel that power plants commonly burn to produce electricity. Where does coal come from? In swamps and marshes, over thousands of years, millions of plants die and their remains accumulate. Under the right conditions, this plant material becomes lithified through biochemical and physical alteration. For coal to form, the plant matter must accumulate under water or be quickly buried to prevent access to air. Otherwise, oxygen would combine with carbon in the plant remains and would escape as carbon dioxide.

What Do Sedimentary Rocks Tell Us about Earth History?

Sedimentary rocks contain a bonanza of information about the history of our planet. They tell us where ancient seas, deltas, mountains, deserts, and glaciers once existed. They record the climates of eons past and contain the fossil record of the animals and plants that once populated Earth. In sedimentary rocks, we can find clues to great crustal upheavals, global catastrophes, and shifting landmasses. The layers of sandstone, shale, and limestone are like encoded tablets of stone that you are now learning to decipher.

METAMORPHIC ROCKS: CHANGED WITHOUT MELTING

Charles Lyell realized that any rock, if subjected to the right conditions, could become altered to form a quite different rock. He recognized that the rock-altering agents are high temperature, high pressure, the chemical action of solutions and gases, or any combination of these. Lyell used the term **metamorphism** (from the Latin *metamorphosis*, meaning "to change form") to describe the processes that create metamorphic rocks.

Any rock can be converted back to an igneous rock simply by melting it and letting it cool and resolidify. But metamorphic rocks are different, because they do not involve melting.

Conversion of an igneous or sedimentary rock to a metamorphic rock involves actual recrystallization of minerals, but without the rock becoming molten. The recrystallization occurs in response to heat (but not enough to melt), high pressure, and the presence of water.

In recrystallization, the rock texture may change. Also, the existing minerals in the rock may not be able to withstand the new conditions. Their "only way out" is to reorganize their crystals and exchange atoms to form new minerals that are stable under the new pressure and temperature.

New elements need not be introduced. Instead, minerals already present are changed into different and often denser minerals. Variations in heat and pressure may result in different kinds of metamorphic rocks, even from the same parent material.

Contact Metamorphism and Regional Metamorphism

Two major types of metamorphism are contact and regional metamorphism. **Contact metamorphism** refers to alterations of rock that occur immediately adjacent to hot igneous intrusions. The changes that occur in the rock being intruded are largely the result of high temperatures and the action of chemically active fluids given off by the molten rock. Such factors as the size of the magma body, its composition and fluidity, and the nature of the intruded rock also influence the severity of contact metamorphism.

Deposits of various metal ores (gold, silver, lead, zinc, copper, molybdenum, etc.) are frequent economic products of contact metamorphism. The famous mines of the American "Old West" owed their wealth to contact metamorphism.

Regional metamorphism alters rock across much broader areas, under conditions of great confining pressures and heat that result from the deep burial and tectonic processes of mountain-building. In a subsequent chapter, you will see how rocks deposited in troughlike areas of the crust adjacent to continents may be compressed into mountain systems and thus become regionally metamorphosed.

Metamorphic index minerals form under specific temperature and pressure conditions. Geologists use them to determine the growth history of ancient mountain regions, even where only deeply buried "mountain roots" remain. Figure 4-38 shows the temperatures at which certain metamorphic index minerals form in rock that is being subjected to heat and pressure. Note that the mineral **chlorite** forms at relatively low temperatures. The metamorphic index mineral **garnet** forms at higher temperatures and pressures, and **sillimanite** indicates the highest level of temperature and pressure.

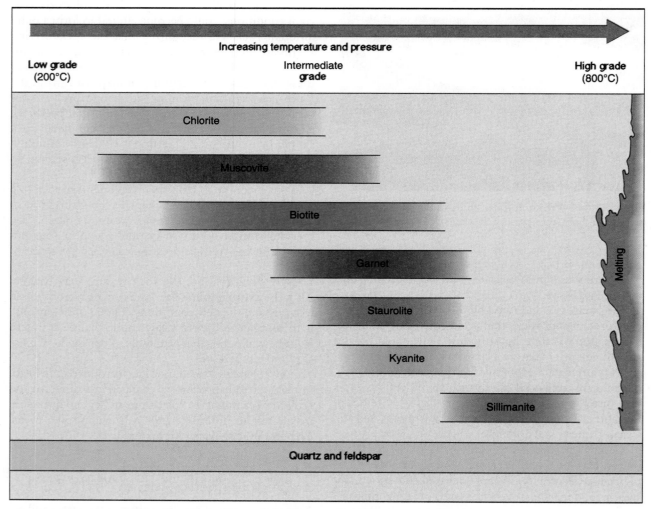

FIGURE 4-38 **How shale progressively metamorphoses to other rock types in response to increasing temperature and pressure.** As the shale (parent rock) is subjected to low-grade metamorphism (relatively modest heat and pressure), the minerals chlorite and muscovite (a mica) form. The shale is metamorphosed to slate, and then to phyllite. With increasing temperature and pressure, the muscovite is joined by new minerals: biotite, garnet, and staurolite. As temperatures approach the level of high-grade metamorphism, the minerals kyanite and sillimanite form. Beyond 800°C, the rock may completely deform by melting.

A geologist working in a region of metamorphic rocks can plot the occurrences of metamorphic index minerals on a map. After many traverses across exposures of metamorphic rocks, he can recognize zones where particular index minerals are highly concentrated. He knows that his travel from outcrops bearing abundant chlorite to areas of abundant garnet and then to rocks with abundant sillimanite indicate a journey from the periphery to the heart of an area subjected to intense tectonic forces.

Kinds of Metamorphic Rocks

Any rock—igneous, sedimentary, or metamorphic—can become metamorphosed in different ways. Consequently, there are hundreds of different kinds of metamorphic rocks. However, you need only know about the most abundant kinds.

Geologists divide metamorphic rocks into two broad groups, based on the presence or absence of foliation. **Foliation** is the parallel alignment of mineral grains in a rock, often quite visible, as in Figure 4-39. Such grain alignment is the result of directional pressures to which the rock was subjected. Platy minerals like mica and chlorite, and long, thin minerals like hornblende, respond to directional forces by orienting themselves at right angles to the force that squeezes the rock.

Where pressures are largely nondirectional, or when heat is the primary metamorphic agent (as in contact metamorphism), foliation is unlikely to develop. Foliation is also unlikely to develop if minerals having platy or elongate form are absent.

FIGURE 4-39 **Coarse foliation in a former granite that has been metamorphosed into a gneiss.** Note the parallel alignment of mineral grains due to directional pressure.

FOLIATED METAMORPHIC ROCKS In order of increasingly coarser foliation, the foliated metamorphic rocks include slate, phyllite, schist, and gneiss. In slate, the foliation is microscopic and is caused by the parallel alignment of minute flakes of silicates such as mica. The planes of foliation are quite smooth, and the rock may be split along these planes of "slaty cleavage." The planes of foliation may lie at any angle to the bedding that existed in the parent rock, and this characteristic helps to differentiate slate from dense shale.

Slate is derived from the regional metamorphism of shale. As heat and pressure are applied to the clay minerals in shale, they are converted to chlorite and mica, two platy minerals that contribute to the rock's slaty cleavage.

Phyllite also is finely textured, although some grains of mica, chlorite, garnet, or quartz may be visible. Fracture surfaces of phyllites often exhibit a wrinkled aspect and are more lustrous than slate. Phyllite represents an intermediate degree of metamorphism between slate and schist. The parent rock is commonly shale or slate.

In **Schist**, platy or needlelike minerals have grown sufficiently large to be visible to the unaided eye. Also, constituent minerals tend to be segregated into distinct layers. Schists are named according to the most conspicuous mineral present. Thus, there are mica schists (Fig. 4-40), hornblende schists, chlorite schists, and many others. Shales are the usual parent rocks for schists, although some are derived from fine-grained volcanic rocks.

Gneiss (pronounced *nice*) is a coarse-grained metamorphic rock (Fig. 4-41). Its often dramatic foliation results when high pressure segregates minerals into bands rich in quartz, feldspar, biotite, or hornblende. The usual parent rocks for gneisses are high-silica igneous rocks (granite) and high-silica sedimentary rocks (sandstones).

NONFOLIATED METAMORPHIC ROCKS Four common **nonfoliated metamorphic rocks** are marble, quartzite, greenstone, and hornfels.

Marble (Fig. 4-42) is limestone that has been metamorphosed. Under pressure, all of the calcite in the original limestone is recrystallized. Any vestige of fossils in the limestone is eradicated. Note that there is no chemical change: Calcite ($CaCO_3$) is the principal mineral in both limestone and marble. Only the texture has changed. Because of marble's uniform texture, softness (easily scratched with steel), and beauty, marble has been favored by sculptors for centuries (Fig. 4-43).

Quartzite is a fine-grained, often sugary-textured rock (Fig. 4-44), composed of intergrown quartz grains. Like the quartz of which it is composed, quartzite is extremely hard. The metamorphism of quartz sandstone yields quartzite.

FIGURE 4-40 **Mica schist.** Cleavage surfaces of the mica sparkle as they reflect light. The schist contains large crystals of a metamorphic mineral, almandine garnet. Such large crystals that grow within a metamorphic rock are called porphyroblasts.

FIGURE 4-42 **Marble.** On close examination, you can discern the lustrous cleavage surfaces of the calcite crystals. Specimen width is about 9.0 cm.

As indicated by its name, **greenstone** is a dark-green metamorphic rock. Except for a relatively few scattered larger crystals, most of the mineral components cannot be seen without magnification. Greenstones are formed during low-grade metamorphism of basaltic rocks.

Hornfels is a very hard, fine-grained metamorphic rock that often contains scattered crystals of mica and garnet that have no preferred orientation. This is because high temperatures emanating from nearby igneous intrusions, rather than directional pressures, cause the metamorphism. Hornfels may form from intensely heated shale or other fine-grained rocks.

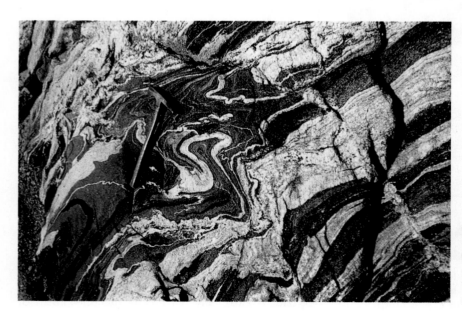

FIGURE 4-41 **The Monson Gneiss of New Hampshire exhibits plastic deformation due to intense pressure and heat.** The dark bands are hornblende, biotite, and plagioclase. Lighter bands are rich in quartz and white feldspar.

What Do Metamorphic Rocks Tell Us about Earth History?

Extensive tracts of metamorphic rocks occur in regions that have been subjected to severe compressional deformation. Such regions of Earth's crust either now have, or once had, great mountain ranges. Thus, where large tracts of low-lying metamorphic rocks are exposed at Earth's surface, geologists conclude that crustal uplift and long periods of erosion leveled the mountains. For example, metamorphic rock exposures at many localities across the eastern half of Canada represent the truncated foundations of ancient mountain systems.

From studies of the minerals in metamorphic rocks, it is possible to reconstruct the conditions under which the rocks were altered. We can learn the directions of compressional forces, how much pressure existed, how much heat was generated, and what kind of parent rock was metamorphosed. Investigators have learned that specific metamorphic minerals form and are stable within finite limits of temperature and pressure. As a result of this knowledge, maps of metamorphic rock zones that formed under specific conditions can be constructed. Commonly, such maps delineate broad bands of metamorphic rocks, each of which formed under progressively more intense conditions.

Here is an example of how we can infer a region's geologic history from the rocks we observe. Imagine a land once underlain by a thick sequence of calcareous shales. Suppose it was subjected to compression that produced mountains, and then slowly lost those mountains by erosion. We could then begin a traverse across the eroded surface on unmetamorphosed shales that were not involved in the mountain-building. These shales would contain only unaltered sedimentary minerals.

Progressing farther toward the area of most intense metamorphism, we might see that the shales had given way to slates bearing the green metamorphic mineral chlorite. Still further along the traverse, we see schists containing intermediate-grade metamorphic minerals. Finally, we might find coarsely foliated schists containing minerals that form only under high temperature and pressure. Where rocks had been heated to melting, hornfels might develop in areas surrounding magmatic bodies.

Although metamorphic rocks reveal the conditions under which they formed, they also contain clues to conditions prior to metamorphism. For example, suppose we find a marble that formed a billion years ago through metamorphism of a limestone. This indicates that the composition of Earth's atmosphere and ocean a billion years ago was similar to that in which carbonates form today.

FIGURE 4-43 **Michaelangelo's famous statue of the biblical David.** He sculpted it from Cararra marble, quarried in Italy's Appenine Mountains.

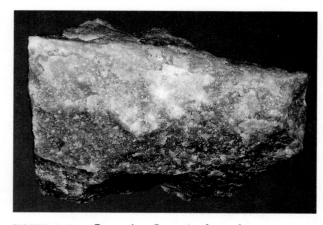

FIGURE 4-44 **Quartzite.** Quartzite forms from sandstone. Metamorphism has transformed the original quartz grains in the sandstone into interlocking crystals of quartz. ▶ *What simple physical test would convince you that this specimen was quartzite and not marble?*

SUMMARY

- Rocks are the materials of which Earth is composed. Rocks are themselves composed of minerals. Minerals, in turn, are constructed of chemical elements. The eight most common elements in rocks are oxygen, silicon, aluminum, iron, calcium, sodium, potassium, and magnesium.
- Silicate minerals (silicon + oxygen, SiO_2) are the most important rock-forming minerals. Silicates include quartz, mica, feldspar, hornblende, augite, and olivine, which are initially crystallized from molten rock (magma). Quartz may also form in voids within sedimentary rocks by precipitation from aqueous solutions.
- Igneous rocks form by cooling and solidifying from a molten silicate body. If solidified beneath Earth's surface, igneous rocks are termed intrusive (granite is an example). Igneous rocks that solidify at Earth's surface, exposed to the atmosphere or ocean, like lava, are termed extrusive (basalt is an example).
- The texture of igneous rocks reflects their cooling history. Coarse-grained varieties cooled slowly, allowing crystals time to grow (these are mainly intrusive), whereas fine-grained varieties cooled rapidly, allowing too little time for crystals to grow (these are mainly extrusive).
- One mechanism that creates a variety of igneous rock types is fractional crystallization, explained by Bowen's Reaction Series. Another is partial melting, in which rock is incompletely melted and molten fractions are removed.
- Clay, calcite, dolomite, gypsum, and halite form through weathering, precipitation, and deposition at Earth's surface.
- Calcite is the main constituent of the carbonate rock limestone. The mineral dolomite forms a carbonate rock that is also called dolomite.
- In water bodies that experience intense evaporation, evaporite minerals may precipitate, such as gypsum, anhydrite, and halite.
- Clastic sedimentary rocks are composed of fragments (clasts) of weathered rock that have been transported some distance from their place of origin. Transported pebbles and cobbles, for example, may be lithified to form the rock called conglomerate. Sand grains may form sandstone. Clay particles may accumulate to form shale.
- Metamorphism ("change of form") includes processes by which any existing rock can undergo change in its mineral makeup and texture. This happens without melting, in the solid state, in response to heat, pressure, and chemically active solutions.
- Two basic kinds of metamorphism are contact and regional. Contact metamorphism occurs relatively locally, around the margins of molten rock. Regional metamorphism occurs on a larger scale and is usually associated with the pressures of mountain-building.
- Metamorphic rocks characteristically occur in distinct metamorphic zones. Each zone is characterized by minerals that form in response to specific conditions of pressure and temperature.

KEY TERMS

andesite, p.61

anhydrite, p.55

augite, p.53

basalt, p.58

Bowen's Reaction Series, p.61

calcite, p.54

carbonate clasts, p.67

carbonate spar, p.67

chert, p.68

chlorite, p.70

clastic sediment, p.66

clay mineral, p.53

cleavage, p.49

color, p.48

coal, p.69

contact metamorphism, p.69

continuous series (of Bowen's Reaction Series), p.62

crystal form, p.49

density, p.49

diorite, p.60

discontinuous series (of Bowen's Reaction Series), p.62

dolomite, p.54

evaporite, p.55

extrusive igneous rock, p.57

feldspar, p.50

ferromagnesian silicate mineral (= mafic silicate mineral), p.53

foliation (in metamorphic rocks), p.71

fractional crystallization, p.64

gabbro, p.60

garnet, p.70

granite, p.59, 60

granodiorite, p.60

greenstone, p.72

gypsum, p.55

halite, p.55

hardness, p.49

hornblende, p.53

hornfels, p.72

igneous rock, p.55

intrusive igneous rock, p.56

lava, p.56

lithification, p.65

luster, p.48

mafic silicate, p.53

magma, p.56

magnetism, p.50

marble, p.71

metamorphic index mineral, p.70

metamorphic rock, p.55

metamorphism, p.69

mica, p.53

micrite, p.67

mineral, p.48

nonsilicate mineral, p.54

nonfoliated metamorphic rocks, p.71

obsidian, p.58

olivine, p.53

oöid, p.69

oölite, p.67

partial melting, p.64

phyllite, p.71

pluton, p.57

porphyritic texture, p.59

pumice, p.59

quartz, p.50

quartzite, p.71

regional metamorphism, p.70

rhyolite, p.59

schist, p.71

sedimentary rock, p.55

sillimanite, p.70

slate, p.71

strata, p.65

streak (of mineral), p.48

texture (of rock), p.56

tuff, p.59

vesicles, p.59

QUESTIONS FOR REVIEW AND DISCUSSION

1. What is a mineral? What characteristics of a true mineral (like quartz, feldspar, calcite, or pyrite) would *not* be present in a piece of glass?

2. What are the eight most abundant elements that make up rocks and minerals?

3. Why are silicate minerals so important in geology? Which silicates would you expect to find in granite? Which silicates would you expect to find in sedimentary rocks?

4. What igneous rock best approximates the composition of continental crust? Oceanic crust?

5. What has happened to an igneous rock that forms through partial melting? Explain how an igneous rock can form by fractional crystallization.

6. What inferences can you draw from the texture of an igneous rock? What is the relation between the viscosity of molten rock and its composition?

7. List the clastic sedimentary rock types in order of increasingly finer grain size.

8. What mineral groups discussed in this chapter are particularly common in sedimentary rocks?

9. If you were a Stone Age (Paleolithic) human and had to choose between limestone and chert as the material for an ax head, which would you select? Why?

10. If you were a sculptor and had to select a metamorphic rock from which to carve a statue, why would you choose marble over schist? What might be your objection to quartzite?

11. List the foliated metamorphic rocks in order of increasingly coarser foliation.

WEB SITES

The Earth Through Time Student Companion Web Site (www.wiley.com/college/levin) has online resources to help you expand your understanding of the topics in this chapter. Visit the Web Site to access the following:

1. Illustrated course notes covering key concepts in each chapter;

2. Online quizzes that provide immediate feedback;

3. Links to chapter-specific topics on the web;

4. Science news updates relating to recent developments in Historical Geology;

5. Web inquiry activities for further exploration;

6. A glossary of terms;

7. A Student Union with links to topics such as study skills, writing and grammar, and citing electronic information.

5

Q
T
K
J
Tr
Pr
P
M
D
S
O
Є
Pre-Є

Sandstones and siltstones of the Uinta Formation in Utah. *The sedimentary layers are floodplain clastics that spread broadly across the Uinta and other intermontane basins as the Rocky Mountains were being eroded, 33 to 53 million years ago.*

CHAPTER 5

The Sedimentary Archives

In the high mountains, I have seen shells. They are sometimes embedded in rocks. The rocks must have been earthy materials in days of old, and the shells must have lived in water. The low places are now elevated high, and the soft material turned into hard stone.

—Chu-Hsi, Chinese philosopher, AD 1200

OUTLINE

▶ TECTONIC SETTING IS THE BIGGEST FACTOR IN SEDIMENT DEPOSITION

▶ ENVIRONMENTS WHERE DEPOSITION OCCURS

▶ WHAT ROCK COLOR TELLS US

▶ WHAT ROCK TEXTURE TELLS US

▶ WHAT SEDIMENTARY STRUCTURES TELL US

▶ WHAT FOUR SANDSTONE TYPES REVEAL ABOUT TECTONIC SETTING

▶ LIMESTONES AND HOW THEY FORM

▶ ORGANIZING STRATA TO SOLVE GEOLOGIC PROBLEMS

▶ SEA-LEVEL CHANGE MEANS DRAMATIC ENVIRONMENTAL CHANGE

▶ STRATIGRAPHY AND THE CORRELATING OF ROCK BODIES

▶ UNCONFORMITIES: SOMETHING IS MISSING

▶ DEPICTING THE PAST

▶ SUMMARY

▶ KEY TERMS

▶ QUESTIONS FOR REVIEW AND DISCUSSION

▶ WEB SITES

▶ BOX 5-1 ENRICHMENT

▶ BOX 5-2 NATIONAL PARK GEOLOGY

Key Chapter Concepts

- The tectonic setting determines the nature of sedimentary deposits. Therefore, by studying a sedimentary rock's composition, texture, and fossil content, we can determine its tectonic setting of formation.

- Environmental conditions at particular locations during the geologic past can be determined by comparing features of modern sediments to analogous features in ancient rocks.

- By examining the rock's textures, structures, colors, and fossils, we can recognize the kind of environment in which the grains of a sedimentary rock were originally deposited.

- Sandstones are particularly useful in determining tectonic setting.

- Thick, widespread limestones imply deposition in clear, warm, shallow ocean water.

- Dolomites form when magnesium-rich solutions react with limestone.

- Facies are lateral changes in sedimentary rocks that reflect different conditions in environments adjacent to the area of deposition.

- Lithostratigraphic correlation of sedimentary rocks involves matching rock bodies by their similarities, without regard to time. Chronostratigraphic correlation matches rock bodies of the same age, without regard to differences in lithology.

- Like a chapter torn from a book, unconformities are gaps in the geologic record where rocks have been removed by erosion or where they have never been deposited.

- After geologic observations have been collected, the data are used to prepare geologic columns, cross-sections, and maps that graphically reveal the geologic history of a region.

TECTONIC SETTING IS THE BIGGEST FACTOR IN SEDIMENT DEPOSITION

As long as Earth has had an atmosphere and hydrosphere (ocean), sediments have been accumulating on its surface. They have accumulated everywhere—on land, sea bottoms, and lake bottoms, and even in ice at the frozen poles. Over time, much of this sediment has been formed into sedimentary rocks. These rocks preserve features that tell us about the environment where they were deposited, and approximately when they were deposited. Thus, by interpreting this evidence in sedimentary rock strata, we can decipher part of Earth's geologic history.

Sedimentary rocks include limestones, sandstones, shales, coal, rock salt, and conglomerate. They range widely in color, texture, and composition. With this remarkable diversity of sedimentary rocks worldwide, what determines the type of sedimentary rock that will form in a specific location? There are several important factors:

- The tectonic setting (active, slowly rising, slowly sinking, or stable)
- The kind of rock being weathered in the sediment-source area (igneous, metamorphic, or other sedimentary rocks)
- The sediment transport medium (water, wind, ice, gravity)
- The active processes in the depositional environment (physical, chemical, and biologic)
- The climate (wet or dry, hot or cold)
- The agents available to change the sediment to solid rock (pressure, cements)
- The time available (years in the thousands, hundred-thousands, millions, ten-millions, etc.)

The first factor—tectonic setting—is the most powerful factor in the big picture. *Tectonics refers to deformation or structural activity of a large area of Earth's lithosphere over a long period of time.* For example, a region may be tectonically quiet and stable, like the U.S. Midwest. Or a region may be tectonically active, like the western margin of North America, where faulting, uplift, volcanism, and compressive forces are producing mountain ranges.

The tectonics of a region powerfully controls the characteristics of the resulting rocks. Where a source area is tectonically active (or recently has been so), crustal compression and uplift form elevated mountains. Erosion of the mountains produces sediment that is transported to adjacent basins.

For example, the layers of sedimentary rock shown in the chapter-opening photograph are alluvial sandstones and siltstones. Their source material eroded from mountains that were tectonically uplifted during the Late Cretaceous and Early Cenozoic (33–53 million years ago). These mountains surround the Uinta Basin of Utah. Thus, the sediment became deposited in this *intermontane* (between mountains) basin.

Some basins of deposition lack surrounding mountains and instead have a huge range along one border. In such basins, regionally extensive blankets of sediment are spread across the basin floor. These products of mountain-range erosion typically feature coarser sediments close to the mountainous source area (conglomerates and coarse sandstones) and finer sediments further from the mountains (sandstones, siltstones, and shales).

In tectonic settings where the source area is stable and topographically flatter, sediment is much finer, so finer clasts and dissolved solids became the most abundant sediment. In these more stable environments, shales and carbonate rocks predominate.

Depositional Basins

Tectonics influences not only the *size* of clastic particles, but also the *thickness* of the accumulating deposit. For example, if a basin is tectonically subsiding (sinking) at the same time that it is receiving an ample supply of sediment, an enormous thickness of sediments—miles thick—can accumulate.

Geologist James Hall (1811–1898) studied sedimentary rocks in the Appalachian Mountains of the eastern United States. These thick rocks accumulated in waters known to have been shallow, so Hall realized that Earth's crust must have been subsiding to permit formation of the tremendous thickness of rock. His reasoning was straightforward: it is easy to visualize filling a 40,000-foot-deep basin with 40,000 feet of sediment. However, where the kinds of fossilized life forms indicate a basin that could not have exceeded several hundred feet in depth, the only way to get several miles thickness of sediment would be for subsidence to occur simultaneously with sedimentation.

In a marine basin that is stable or subsiding very slowly, the area where sedimentation is occurring is likely to remain in the area of wave activity for a long time. Wave action and currents will wear, sort, and distribute the sediment into broad, blanket-like layers. If the sediment supply is small, this type of sedimentation will continue indefinitely. Should the supply of sediment become too great for currents and waves to transport, however, the surface of sedimentation will rise above sea level, causing deltas to form.

Cratons, Shields, Platforms, and Orogenic Belts

Let us zoom out to consider tectonics on the scale of entire continents. We will use North America as a familiar example (Fig. 5-1).

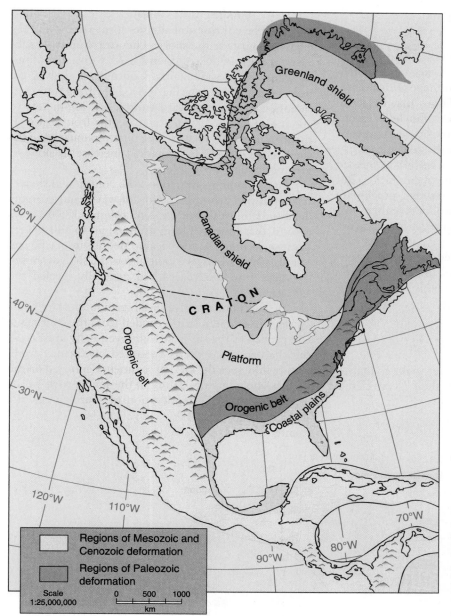

FIGURE 5-1 **The tectonic parts of a continent.** The craton is the stable interior, with its central *shield* of ancient crystalline rock and surrounding *platforms* of accumulated sedimentary rocks. *Orogenic* (mountain-building) *belts* flank the craton. In North America, the shield rocks are exposed throughout eastern Canada. The platform comprises the vast Midwestern area of flat-lying sedimentary layers. And the orogenic belts are the Rocky Mountains and Appalachian Mountains.

Continents have two tectonic components: a *craton* (composed of a *shield* and its surrounding *platforms*) and *orogenic belts.*

The **craton** is the stable interior of a continent, undisturbed by tectonic events since Precambrian time (the past half-billion years). The craton has two components: the shield and platforms. The **shield** is a large area of exposed ancient crystalline rocks, such as in eastern Canada. **Platforms** roughly surround the shield; they are regions where ancient shield rocks are covered by flat-lying or gently warped sedimentary layers.

Orogenic belts are elongated regions that border the craton and that have been deformed by compressional forces. "Orogeny" means mountain-building, so orogenic belts are zones of mountain-building. Young orogenic belts feature frequent earthquakes and volcanic eruptions, whereas older belts display crustal displacement, severely deformed strata, metamorphic rocks, and huge bodies of exposed intrusive igneous rocks.

Earth's history provides many examples of orogenic belts that formerly were depositional basins, receiving great thicknesses of sediment. Similar elongate tracts of sedimentation occur today along the margins of continents. Following long episodes of deposition, the sediments in these basins may be deformed because of an encounter with an oncoming tectonic plate. Mountain ranges then form where once there was only ocean. We will examine these events and their causes in Chapter 7.

ENVIRONMENTS WHERE DEPOSITION OCCURS

Environment of deposition refers to all the environmental factors—physical, chemical, and biologic—under which sediment is deposited. Each type of depositional environment is characterized by conditions that influence the properties of sediment deposited within it. By recognizing those properties in ancient rocks, we can reconstruct depositional environments of the geologic past.

Clastic sediments consist of particles (clasts) formed elsewhere and transported to the site of deposition. But other sediments, such as carbonates that precipitate from sea or lake water, are direct products of their own environment. Their component minerals form and are deposited at the same place.

We distinguish three broad depositional environments: *marine* (ocean), *transitional* (between sea and land), and *continental* (land) (Fig. 5-2). We will now look at all three.

Marine Depositional Environments

We divide the marine realm into shallow marine (which includes the continental shelf), continental slope, and deep marine environments.

SHALLOW MARINE (CONTINENTAL SHELF) ENVIRONMENT The **continental shelves** form a shallow ocean environment near the shore. They are simply the submerged edges of the continents and have clear continuity with the coastal plains. Continental shelves are nearly flat and smooth (Fig. 5-3). They fringe the continents, extending seaward from a few kilometers to about 300 kilometers, at water depths from just centimeters up to 200 meters. The outer boundaries of the shelves are defined by a marked increase in slope to greater depths.

The smoothness of parts of the continental shelves results in part from the action of waves and currents during the most recent Ice Age, roughly 18,000 years ago. At that time, sea level dropped up to 140 meters because so much water became tied up in glacial ice. Waves and currents sweeping across the shelves shifted sediment into low places and generally leveled the surface.

The continental shelves hold great interest for geologists. All sediment that is eroded from the continents and carried to the sea in streams eventually must cross the shelves, and some is deposited on them. Several factors influence the kind of sediment deposited on the shelves, including the nature of the source rock on adjacent landmasses, the elevation of source areas, the

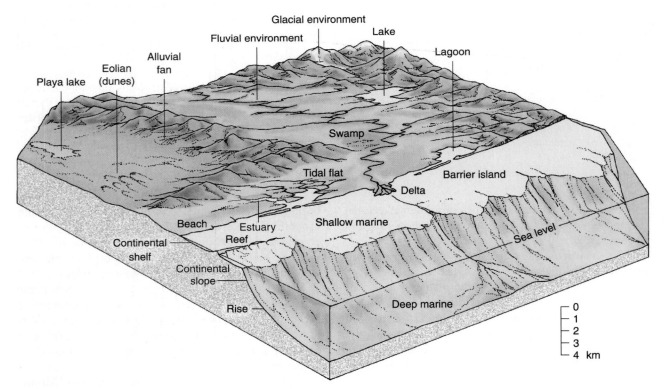

FIGURE 5-2 **Marine, transitional, and continental environments of deposition.** ◗ *What features of a shale formed in an ancient lake might be used to distinguish it from a shale formed in an ancient lagoon?*

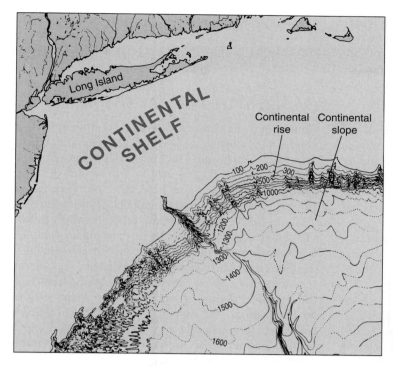

FIGURE 5-3 **Contour map of the continental shelf, rise, and slope east of Long Island, New York.** Continental shelves are nearly flat and smooth.

distance from shore, and the presence of carbonate-secreting organisms.

Because larger grains are heavier than small ones, they tend to drop out sooner and are deposited closer to shore. Also, shallow-water currents and wave action keep finer particles in suspension and carry them farther out to sea. For these reasons, sediment deposited in the shallow marine environment tends to be coarser than material laid down in deeper parts of the ocean. Sand and silt are common.

Where there are few continent-derived sediments and the seas are relatively warm, lime muds of biochemical origin may be the predominant sediment.

Coral reefs are also characteristic of warm, shallow seas. They remind us of the enormous biologic importance of the shallow marine environment.

Over most of this realm, sunlight penetrates all the way to the seafloor, providing both heat energy to warm the water and light energy to drive photosynthesis in plants. Algae and other plants proliferate, creating "sea pastures" that provide food for a multitude of swimming and bottom-dwelling animals (Fig. 5-4).

CONTINENTAL SLOPE ENVIRONMENT Areas of the ocean floor that extend from the seaward edge of the continental shelves down to the ocean depths are called

FIGURE 5-4 **Marine life flourishes in many areas of the well-lit continental shelves.** Marine plants require light to photosynthesize their food and grow, and plants form the base of the food chain that supports marine animals. Members of the Cnidaria, a phylum that includes corals, dominate this scene.

continental slopes (Fig. 5-2). From the sharply defined upper boundary of the continental slope, the surface of the ocean floor drops to depths of 1400 to 3200 meters. At these depths, the slope of the ocean floor becomes gentler. The less-pronounced slopes comprise the continental rises.

TURBIDITY CURRENTS AND TURBIDITES Much of the sediment reaching the continental slope and continental rise is transported there by flowing masses of muddy water termed **turbidity currents** (*turbid* = cloudy, murky, sediment-filled). Water in a turbidity current is denser than surrounding water because it is laden with suspended sediment. It therefore flows down the slope of the ocean floor, beneath the surrounding less-dense clear water. Upon reaching more level areas, the current slows and drops its load of suspended particles.

These deposits, called **turbidites**, often form submarine fans at the base of the continental slope (Fig. 5-5). They are similar to alluvial fans that occur on dry land. In addition to turbidites, slope-and-rise deposits include fine clay that has settled out of the water and large masses of material that have slid or slumped downslope under the influence of gravity.

DEEP MARINE ENVIRONMENT In the deep marine environment far from the continents, most of the sediment that settles to the ocean floor is very fine clay, volcanic ash, and the calcareous or siliceous remains of microscopic organisms. Exceptions include sporadic occurrences of coarser sediments that are carried down continental slopes into the deeper parts of the ocean by turbidity currents. Coarse clastics also may be dropped into deep water from melting icebergs.

Continental Depositional Environments

Continental environments of deposition include river floodplains, alluvial fans, lakes, glaciers, and eolian (windswept) environments. Here is a look at each.

STREAM DEPOSITS The sand, silt, and clay found along the banks, bars, and floodplains of streams are familiar to us. In general, **stream deposits** simply follow the course of the stream. Streams are highly complex systems affected by many interacting variables, including **discharge** (the quantity of water passing a point on the stream in a given time), water velocity, nature of the sediment being transported, and shape of the stream channel. A change in any one of these variables will cause a change in the type and amount of sediment being deposited. Thus, sands, silts, and clays may grade abruptly into one another in stream deposits.

ALLUVIAL FANS Stream-transported materials accumulate quickly when rapidly flowing streams emerge from a mountainous area onto a flat plain. The result of this abrupt deposition is an **alluvial fan** (Fig. 5-6). Alluvial fans are recognized by their lobed form and wedge-shaped cross-section. Ancient alluvial fan deposits lack fossils, except for rare bones of vertebrates, spores, pollen grains, or fragmentary plant remains.

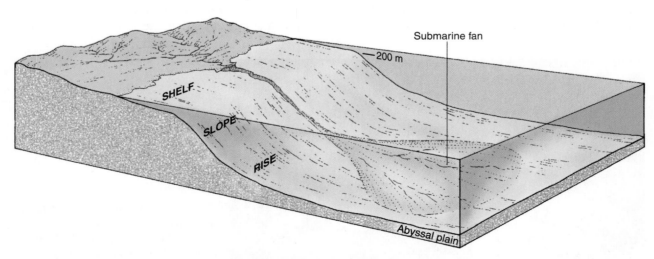

FIGURE 5-5 **Submarine fan built of land-derived sediment emerges from a submarine canyon.** Such fans occur in association with large rivers, such as the Amazon, Congo, Ganges, and Indus. (Vertical exaggeration 200:1.) ◀ *Of the four major kinds of sandstones described in this chapter, which is most commonly associated with such deep-sea fan deposits?*

FIGURE 5-6 **Coalescing alluvial fans to the left of a dry streambed cover part of a floodplain, Gulf of Suez area, Egypt.** The stream is unable to transport the huge amount of debris supplied to it and is dry during part of the year. ◪ *If the mountains on the left side of the photograph are composed of granite, what silicate minerals are likely to be abundant in the sediments of the alluvial fans?*

LAKE DEPOSITS (LACUSTRINE) Lakes are landlocked bodies of water. Lakes are ideal traps for sediment, called **lacustrine deposits.** Silt and clay are common, although a variety of sediment is possible, depending on water depth, climate, and the character of the surrounding land areas. Indicators of a lacustrine depositional environment include fossils of freshwater organisms, tracks made by vertebrates, and mud cracks that form when bottom muds are exposed to air.

GLACIAL DEPOSITS The glacial environment may include a variety of other environments, such as stream, lake, and even shallow marine. Glaciers transport and deposit huge volumes of rock debris, including large fragments. Characteristically, these deposits, called **till,** are unsorted mixtures of boulders, gravel, sand, and clay (Fig. 5-7). Where such materials have been reworked by glacial meltwater, however, they become less chaotic and resemble stream deposits.

WIND DEPOSITS In addition to moving ice and flowing water, wind is a fluid that can erode, transport, and deposit sediment. But wind is very selective in the particle size it can transport. Air has only about 1/1000th the density of water, so it can transport particles only the size of sand or smaller. Environments where wind is an important agent of sediment transport and de-

position are called desert or **eolian environments** (*Aeolus* = Greek god of wind). They are characterized by an abundance of sand and silt, little plant cover, and strong winds. These characteristics are typical of many desert regions (Fig. 5-8).

Transitional Depositional Environments

Between the marine and continental depositional environments lies a transition zone (Fig. 5-2). We call it the shore. Here we see deltas and the familiar shoreline accumulations of sand or gravel that we call beaches. We also see mud-covered tidal flats that are alternately inundated and drained of water by tides, usually twice a day. The endless movement of the daily tides, waves, currents, and storms make the transitional depositional environment one of constant change.

DELTAS **Deltas** are accumulations of sediment that form where a stream flows into relatively quiet water, such as the ocean or a lake. Herodotus proposed the term 25 centuries ago, noting that the Nile River Delta in Egypt has the general shape of the Greek alphabet letter delta (Δ). Not all deltas have this shape, because each stream delta responds differently to the volume of sediment it receives and the erosional removal of that sediment by waves, tides, and currents along a shoreline.

A

B

FIGURE 5-7 **Glacial deposits.** (A) The dark bands in the ice and piles of debris in the foreground are called *moraines*. The bands in the ice are *lateral moraines*, and the heap at the terminus of the glacier is a *terminal moraine*. This glacier is on Baffin Island, Canada. (B) Glacial till (unsorted glacial sediment) exposed on the flanks of Mount Rainier, Rainier National Park, in Washington State.

FIGURE 5-8 **Dunes in sand at Eureka Valley, California.**

In the Mississippi Delta (Fig. 5-9A), sediment supply has historically exceeded the reworking capabilities of ocean waves and currents in the northern Gulf of Mexico. The delta, therefore, has built seaward.

In contrast, the Niger Delta on Africa's west coast (Fig. 5-9B) has a closer balance of sediment supply and sediment removal by marine erosion. The action of waves and currents along the delta shoreline reworks newly deposited sand, silt, and clay almost as fast as it is supplied. Over time, however, sedimentation has somewhat exceeded removal, so the entire front of the delta has grown.

If sediment removal exceeds sediment accumulation, there is no delta at all. This is the case with the Amazon in Brazil, the largest river on Earth, which has no delta. Its sediment load is not great enough to overwhelm the effect of tides, waves, and subsidence near its mouth.

Some of the best exposures of *ancient* deltas can be observed in the eastern United States, in the Allegheny-Cumberland Plateau, which extends from western Pennsylvania and eastern Ohio southwestward to central Tennessee. Deltas in this region developed along the margins of shallow seas that covered western Pennsylvania and Ohio. An ample supply of deltaic sediment was provided by streams flowing from the eroding Appalachian highlands.

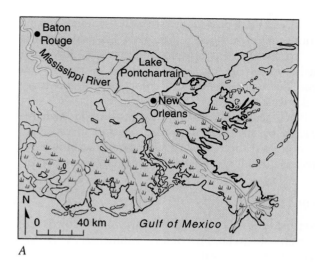

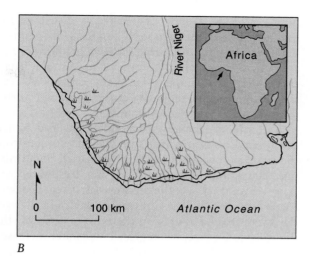

FIGURE 5-9 **Tale of two deltas: the Mississippi River (A) and Niger River (B) deltas.** The Mississippi Delta is a "bird's-foot" delta, named for its shape. It takes that form because the vast sediment supply exceeds the erosion and transport ability of the Gulf of Mexico's wave and current action. Consequently, the sediment slowly spreads seaward. In contrast, Africa's Niger Delta has true delta shape, because the rates of sedimentation and erosion are nearly in balance. ◄ *What changes or events might halt the seaward development of these deltas?*

BARRIER ISLANDS, LAGOONS, TIDAL FLATS, AND ESTUARIES In addition to deltas, the transitional zone includes barrier islands, lagoons, and estuaries that lie between the barrier islands and the mainland, as well as tidal flats (Fig. 5-2). Barrier islands are extensively developed along the U.S. Atlantic Coast (Fig. 5-10) as well as around Florida and the Gulf of Mexico. Several famous barrier islands are familiar to vacationers: Coney Island (New York), Padre Island (Texas), the Outer Banks (North Carolina), and Assateague/Chincoteague (Maryland/Virginia).

Barrier islands may exceed 100 kilometers in length, but most are only a few kilometers or less in width. Sandy sediments predominate in this high-energy environment, where sediment is constantly moved by waves and longshore currents.

Lagoons that lie behind barrier islands are protected from strong ocean waves and currents. As a result, lagoonal deposits usually consist of fine-grained sediments. Lagoonal silts and clays are densely burrowed by mollusks and worms, and if these sediments become lithified, the mollusk shells and borings often are preserved.

On the landward side of lagoons are almost featureless, low-lying plains called **tidal flats.** They are tidal because they are constantly inundated and drained by tides, usually twice daily. Depending on the location, tides rise and fall from several inches to several feet vertically, over just a few hours.

Because lagoons are largely protected from the open ocean, they gradually fill with clay and silt brought by

FIGURE 5-10 **Barrier island and lagoon on the south shore of Long Island.** Waves break along the seaward (right) side of the barrier island, whereas conditions are relatively calm within the lagoon to the left. The areas of brown soil to the left are tidal flats. ◄ *Why are sediments coarser on the seaward side of the island?*

nearby streams. Gentle tidal currents distribute the sediment across the lagoon. The layers of sediment build upward until they reach the level of low tide. The constant rise and fall of water level makes tidal flats a challenging environment for many organisms. Yet, salt-tolerant plants, mollusks, crustaceans, and worms thrive in many tidal flat environments.

Estuaries are the seaward mouths of rivers drowned by the sea. Freshwater from streams and saltwater from the ocean are brought together in estuaries. Some mixing occurs, and typically, the lighter freshwater will flow seaward above the denser saltwater beneath. Many estuaries, like the Chesapeake Bay and Delaware Bay, were created by a rise in sea level that followed the Pleistocene Ice Age, when glacial meltwater increased the volume of ocean water. Other estuaries have formed where coastal areas have subsided tectonically, allowing ocean water to migrate up stream valleys.

Where tidal currents are not too powerful, estuaries trap large volumes of sediment. Sand is common where tides are vigorous, whereas silt and clay predominate in more protected estuaries.

▶ WHAT ROCK COLOR TELLS US

The color of sedimentary rocks tells us much about their environment of deposition. The most significant colors are black (signifying the presence of carbon) and red (signifying the presence of iron compounds).

Shades of Black

Black and dark-gray sedimentary rocks—especially shales—usually result from the presence of carbon. If rocks contain enough carbon to cause black coloration, it implies two things about the depositional environment: an abundance of organic matter in or near the depositional area, and a deficiency of oxygen that kept the remains of organisms from being destroyed by oxidation or bacterial action. These conditions are present today in many ocean, lake, and transitional environments, such as estuaries and tidal flats.

Here is a typical situation: organisms living in or near the depositional basin die, and their remains accumulate on the bottom. The quiet bottom environment typically lacks the dissolved oxygen needed by aerobic bacteria to attack and break down organic matter. There also may be insufficient oxygen for scavenging bottom dwellers that might feed on the debris. Thus, organic decay is limited to the slow and incomplete activity of anaerobic bacteria. Consequently, incompletely decomposed material accumulates, and it is rich in black carbon.

In such an environment, iron combines with sulfur to form finely divided pyrite (iron sulfide, FeS_2), which further contributes to the black color (recall that pyrite produces a black powder in a streak test). Stagnant conditions yield hydrogen sulfide gas (H_2S, which smells like rotten eggs). It poisons organisms, contributing further to the organic debris. You can see that several processes work together to color some sedimentary rocks black.

Black sediments do not always form in such restricted basins. They may also develop in relatively open areas if the rate of organic matter accumulation exceeds the ability of the environment to decompose it.

Hues of Red and Other Colors

Iron oxides may color sedimentary rocks brown, red, or even pale green. Few, if any, sedimentary rocks are free of iron, and less than one-tenth of 1 percent of this metal can color a sediment a deep red.

What actually causes the colors? The answer lies in how iron becomes oxidized. Pure metallic iron has the same number of protons and electrons, so it has no electrical charge. However, at Earth's surface, iron atoms combine with oxygen and lose electrons. With this loss of electrons, the iron atoms become electrically charged ions. (The electrons lost by the iron are gained by the oxygen atoms, and they become oxygen ions).

The combination with oxygen occurs in two ways, depending on the environment. Most commonly, two iron atoms combine with three oxygen atoms to form Fe_2O_3, called **ferric** iron oxide. Ferric minerals such as **hematite** tend to color the rock red, brown, or purple. Strata thus colored by ferric iron are dubbed **red beds** (Fig. 5-11).

In an environment where oxygen is in short supply, such as a stagnant lake or ocean area lacking good circulation, the iron compound formed is likely to be **ferrous** iron oxide, FeO. This differs from ferric in that the iron atoms have lost only two electrons. Ferrous compounds impart hues of gray and green.

The oxidizing conditions required for the development of ferric compounds are more typical of continental than marine environments. Thus, most red beds are floodplain, alluvial fan, or deltaic deposits. Some, however, are originally reddish sediment carried into the open sea. Red beds interspersed with evaporite layers indicate warm and arid conditions.

Although red beds are more likely to represent nonmarine than marine deposition, they are occasionally interbedded with fossiliferous marine limestones. In such cases, the color may be inherited from red soils of nearby continental areas.

▶ WHAT ROCK TEXTURE TELLS US

Just as texture betrays the history of igneous and metamorphic rocks, the texture of a sedimentary rock—its size, shape, and arrangement of grains—tells us much

FIGURE 5-11 **Red beds of the Triassic-Jurassic Navajo Sandstone exposed in an Arizona slot canyon.**

about its origin. In sedimentary rocks, not only the grains themselves, but the textural appearance of a rock, is influenced by materials that hold the particles together.

There are three textural components: clasts, matrix, and cement.

- **Clasts** are the individual particles or fragments derived from the breakdown of existing rock. They are the pebbles in a conglomerate or sand grains in a sandstone.

- **Matrix** is a bonding material, which consists of finer clastic particles (often clay) deposited at the same time as the larger grains, and which fills the spaces between them.

- **Cement** is another bonding material, a chemical precipitate that crystallizes in the voids between grains following deposition. The most common cements are silica (quartz, SiO_2) and calcium carbonate (calcite, $CaCO_3$).

Texture provides clues to a rock's history. For example, in carbonate rocks, extremely fine-grained textures indicate deposition in quiet water, because such tiny particles of calcium carbonate are unlikely to be deposited in turbulent water. The presence of whole, unbroken fossil shells confirms the quiet water interpretation. On the other hand, limestone that contains a mixture of worn and broken shell fragments is likely to be the product of reworking by wave action. Such a limestone is a turbulent shallow-water deposit.

Grain Size and Sorting

A widely accepted scale of particle size was developed in 1922 by American geologist C. K. Wentworth. A simplified version of his scale is shown in Table 5-1. After disaggregation of a rock in the laboratory, the particles are passed through successively finer sieves, and the quantity of particles in each size category is determined.

TABLE 5-1

Particle Size (mm)	Sediment Name	Particle Name	Rock Name
256 128 64 32 16 8 4 — 2 —	Gravel	Boulders Cobbles Pebbles Granules	Conglomerate
1.0 0.5 0.25 0.125 — 0.0625 —	Sand	Coarse sand Medium sand Fine sand	Sandstone
(¹⁄₁₆ mm) — 0.0039 —	Silt	Silt	Siltstone | grading to ↓ — Shale or
(¹⁄₂₅₆ mm)	Clay (as size term)	Clay (as size term)	Claystone

Obviously, a stronger current of water (or wind) is required to move a large particle than to move a small one. Therefore, the size distribution of grains tells us about the turbulence and velocity of currents and the extent of transportation.

Sorting is a geologic term for the degree to which sediment and particles in sedimentary rocks are uniform by particle size. Rocks composed of particles that are all about the same average size are said to be *well sorted* (Fig. 5-12A). A nice beach sand and a uniform-

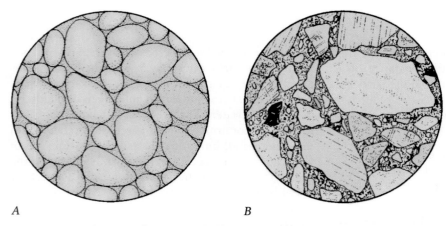

A   *B*

FIGURE 5-12 **Sorting of grains in sandstones as seen under the microscope may range from good sorting (A) to poor sorting (B).** (A) Quartz (light-tan grains) sandstone with carbonate (pink) cement. (B) A sandstone known as greywacke, composed of poorly sorted angular grains of quartz (light tan), feldspar (green), and rock fragments (orange). The graywacke lacks cement; spaces between grains are filled with a matrix of clay and silt. Width of fields is 1.5 mm. ◼ *Which of these sandstones can be considered an immature sandstone? What depositional conditions might account for their differences in composition and texture?*

ENRICHMENT

You Are the Geologist

You are a petroleum geologist working for Paydirt Oil Company. Four parcels of continental shelf in the Gulf of Mexico south of Louisiana are being offered for lease. But your company is willing to lease only one parcel. Your manager asks you to recommend the best parcel to lease.

You know the area is underlain by the Gusher Sandstone. This formation has yielded petroleum in adjacent areas. The formation slopes gently toward the southeast. Cores from a few wells drilled into the Gusher Sandstone reveal grain sizes shown on the map (see Table 5-1).

1. Based on grain size alone, which parcel should your company lease? Why?

2. What additional information do you need to support your recommendation?

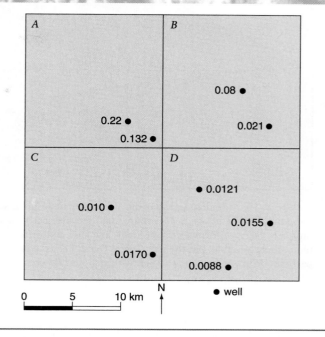

looking sandstone are examples. But if the particles span a broad range of sizes, then the sample is *poorly sorted* (Fig. 5-12B).

Sorting provides clues to conditions of transportation and deposition. Wind winnows the silt and clay particles from sand, leaving grains that are all of about the same size. Wind also sorts the particles that it carries in suspension. Only rarely is wind velocity sufficient to carry sand grains larger than 0.2 millimeter (fine sand). While carrying grains of that size, winds sweep finer particles into the higher regions of the atmosphere. Later, when the wind subsides, well-sorted silt-sized particles drop and accumulate. In general, windblown deposits are better-sorted than deposits formed in an area of water-wave action, and wavewashed sediments are better sorted than stream deposits.

Poor sorting occurs when sediment is rapidly deposited without being selectively separated into sizes by currents. Such conditions occur at the base of mountains, where stream gradient abruptly lessens, water velocity plummets, and sediment thus is quickly deposited. Another example of a poorly sorted rock is glacial till, a heterogeneous mixture of gravel and sand deposited by glaciers (Fig. 5-7B).

If sand, silt, and clay are supplied by streams to a coastline, turbulent near-shore waters sort the particles, creating gradation from near-shore sandy deposits to offshore silty and clayey deposits (Fig. 5-13). Sandstones formed from such near-shore sands can be very porous, providing "void space" between grains for oil and natural gas accumulation. For this reason, geologists make maps showing the grain size of deeply buried ancient beaches and nearshore sandstone formations to determine possible locations where petroleum might be found.

The Shape of Grains

The shape of particles in a clastic sedimentary rock also can be useful in determining its history. We describe shape by the extent of **rounding** of the particle edges and the degree of **sphericity** (how closely the grain approaches the shape of a sphere)—see Figures 5-14 and 5-15.

A particle becomes rounded through wear and tear, by having its sharp corners and edges abraded away through impact and grinding against other particles. Particle mass and water turbulence control the intensity of impact. Heavier impact between water-transported pebbles and granules causes more rapid rounding. Impacts between sand grains in water transport are less severe because grains are more buoyant and because the water actually cushions the impacts.

In conjunction with other evidence, we can infer from the roundness of particles the distance they have traveled, the transporting medium, and the rigor of transport. Very well rounded sand grains also are evidence that a sandstone may have been recycled from older sandstones.

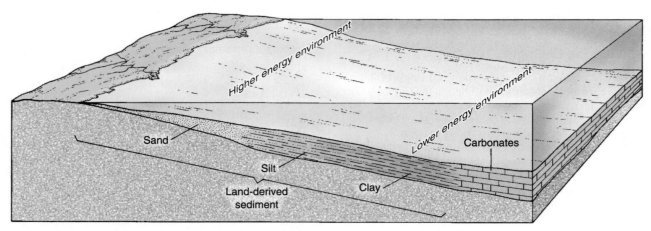

FIGURE 5-13 **Idealized gradation of coarser nearshore sediments to finer offshore deposits.** In this drawing, we assume that the water currents flowing away from the shore are uniform, carry uniform sediment loads, and slow uniformly to deposit the sediments. Reality is much more variable, but the principle is the same: as sediment-laden water slows, the heavier particles drop out first and the lightest drop out last, creating gradation out from shore.

Orientation of Grains

Another element of texture is the arrangement of grains in a clastic rock. Grain orientation is controlled by the transport medium (water, air, or ice), the deposition surface, and the flow direction and velocity.

Grain orientation can indicate the direction of prevailing winds millions of years ago, the flow direction of ancient streams, or the directional trend of shallow ocean currents. In general, sand grains deposited in water currents acquire a preferred orientation in which the long axes of nonspherical grains are aligned parallel to the direction of flow.

If we use a diamond saw to cut a thin section of a sandstone, we can statistically analyze the preferred orientation of sand grains (Fig. 5-16). For coarser sediments, studies of the preferred orientation of glacial and stream-deposited pebbles and cobbles indicate the direction of movement in glaciers and rivers that existed far back into the Precambrian.

▶ WHAT SEDIMENTARY STRUCTURES TELL US

Four structures in sedimentary rocks form during or shortly after deposition, but before lithification. Because these structures result from specific depositional processes, they are extremely useful in reconstructing ancient environments. The structures are mud cracks, cross-bedding, graded bedding, and ripple marks.

FIGURE 5-14 **Well-rounded grains of quartz viewed under the microscope.** From the St. Peter Formation near Pacific, Missouri. (Average grain diameter is 0.25 mm).

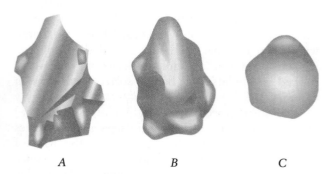

A   *B*   *C*

FIGURE 5-15 **Shape of sediment particles.** (A) This angular particle has undergone little abrasion and therefore little transport. All of its edges are sharp. (B) This rounded grain has traveled a lot, but it has little sphericity. (C) This well-rounded, highly spherical grain is a real veteran, well-traveled and heavily worn. ◧ *Well-rounded, high-sphericity grains of quartz are common, but feldspar grains are less likely to show good rounding and sphericity. What attribute of feldspar accounts for this difference?*

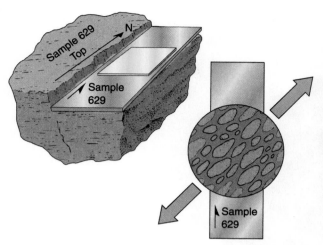

FIGURE 5-16 **Grain orientation study.** One method of studying grain orientation is to record the compass orientation of a rock and then cut an "oriented" thin section of it. Grain orientations are then measured under a microscope, recording the angle between North and each grain's long axis. From multiple measurements, we can determine a mean orientation (in this example, about N 45°E or S 45°W) and evaluate its statistical significance. A thin-section cut perpendicular to bedding might reveal the tilt of the grains and indicate that the transportation medium flowed northeastward rather than southwestward.

Mud cracks indicate drying after deposition. These conditions are common on valley flats, along the muddy margins of lakes, and in tidal zones. Mud cracks develop by shrinkage of mud or clay on drying and are formed on the land surface (Fig. 5-17). Thus, when mud cracks are found in a rock, geologists know that the mud was not deposited in a lake or ocean.

Cross-bedding (cross-stratification) is an arrangement of beds or laminations in which one set of layers is inclined relative to the others (Figs. 5-18A and B). Cross-bedding can form whenever wind or water transports and deposits clastic particles. Wind may heap sand into dunes, or water currents may produce similar features. Grains moved by the currents tumble down the steep downwind (or downstream) side to form layers that are inclined relative to the horizontal layers beneath them.

From cross-bedding, we infer a depositional environment dominated by wind or water currents. The direction in which the beds slope is an indicator of the direction taken by the current. By plotting these directions on maps, geologists have been able to determine the pattern of prevailing winds at various times in the geologic past. Cross-bedding units can form during the advance of a delta (Fig. 5-19) or a dune (Fig. 5-8).

Graded bedding results when flowing water sorts particles by size. This can create beds in which the

A

B

FIGURE 5-17 **Modern and ancient mud cracks.** (A) These contemporary mud cracks formed in soft clay around the margins of an evaporating pond. As water evaporates, the mostly clay sediment contracts, forming the cracks. (B) These mud cracks and wave ripples (caused by wind blowing over a shallow lake) appear in Devonian-age mudstones (about 400 million years old) of the Oneonta Formation near Unadilla in south-central New York. Divisions on the scale are 1.0 cm. ◪ *How might the shape of a mud crack in a cross-section of an ancient mudstone be used to indicate the top of a stratum?*

coarsest grains are at the base and successively finer grains are toward the top (Fig. 5-20). Graded bedding may form simply from faster settling of coarser, heavier grains in a sedimentary mix. But it appears to be particularly characteristic of deposition by turbidity currents.

Turbidity currents can be triggered by submarine earthquakes and landslides along steeply sloping regions of the seafloor. The forward part of a turbidity current contains coarser debris than the tail. As a result,

A

B

FIGURE 5-18 **Cross-bedding (cross-stratification).** (A) Cross-bedded ancient sand dunes of the Navajo Sandstone (Triassic-Jurassic) in Zion National Park and (B) in the Mountain Lakes Formation, Northwest Territories, Canada. The Mountain Lakes Formation is Proterozoic in age, between half a billion and 2.5 billion years old. ◪ *Was the current that produced the cross-bedding in (B) flowing approximately from right to left or left to right?*

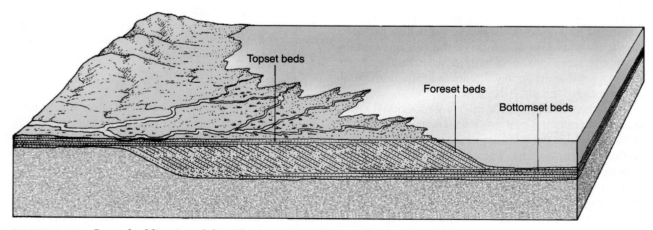

FIGURE 5-19 **Cross-bedding in a delta.** The succession of inclined beds is deposited over bottomset beds that were laid down earlier. Topset beds are deposited by the stream above the foreset beds.

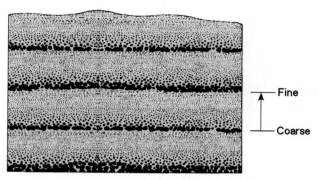

FIGURE 5-20 **Graded bedding results when flowing water sorts particles by size.**

the sediment deposited at a given place on the sea bottom grades from coarse to fine as the "head", and then the "tail" of the current passes over it. The presence of graded beds may indicate former turbidity currents, which frequently characterize unstable, tectonically active environments.

Ripple marks develop in sand and are common along the surfaces of bedding planes (Fig. 5-21). There are two kinds: *symmetric ripple marks* form due to oscillatory motion (back-and-forth) of water beneath waves. *Asymmetric ripple marks* form from air or water currents and are useful in indicating the direction of current movement (Fig. 5-22). For example, ripple marks form at right angles to the current directions;

the steeper side of the asymmetric ripple faces the direction in which the wind or water is moving. Although ripple marks can develop at great depths, they are more common in shallow water areas and in streams.

Which Way Is Up?

The principle of superposition tells us that in undisturbed strata, the oldest bed is at the bottom and higher layers are successively younger. But what if the strata have been completely overturned so that the oldest beds are now at the top, as in Figure 5-23A? If a geologist does not recognize that strata have been overturned, obviously the result can be a serious misinterpretation of geologic events! For this reason, geologists working where strata are deformed must carefully scrutinize rocks for signs of the original tops and bottoms of beds.

Common clues are the structural features just presented above: symmetrical ripple marks, certain types of cross-bedding, graded bedding, mud cracks, and markings that could occur only on the surface of a bed (footprints, trails, raindrop imprints, sole marks, worm holes, and so on).

Symmetrical ripple marks. The sharp crests of the ripples normally identify the tops of beds (Fig. 5-23B). They point toward the younger beds. If fragments of the ancient rippled rock are recycled by erosion and included in overlying strata, the overlying bed must be younger, and the interpretation is confirmed.

A

B

FIGURE 5-21 **Ripple marks.** (A) A crab walks across ripple marks in sand along a modern beach. The current that formed the ripples flowed from left to right. Also note the bird tracks. (B) Cambrian ripple marks (roughly a half-billion years old) preserved on a bedding surface of the Munising Formation, Mosquito Harbor, Pictured Rocks National Shoreline, Michigan.
◄ *What do ripple marks in a marine sandstone indicate about the depth of water in which the sandstone was deposited?*

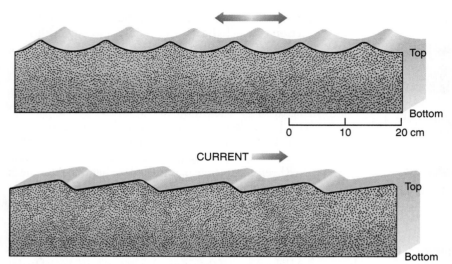

FIGURE 5-22 **Profiles of ripple marks.** (A) Symmetric ripples. (B) Asymmetric ripples.

Cross-bedding. Many cross-beds are concave upward, forming a small angle with beds below and a large angle with beds resting on their truncated upper edges (Fig. 5-23C). But interpretations based on cross-bedding should be confirmed by other top-indicating structures, because some cross-beds do not show the concave upward shape.

Graded bedding. Graded beds (Fig. 5-20) form when fast-moving currents begin to slow, dropping larger particles first (cobbles, then pebbles), followed by progressively finer grains (sand, then silt, then clay). So, if grains are progressively finer toward the top of a bed, you know that the bed is right-side-up.

Mud cracks. When mud dries, shrinks, and cracks, the cracks are wide at the top and narrow at the bottom (Figs. 5-17 and 5-23D). Deposition above mud cracks fills them. This creates a corresponding pattern of ridges that identifies the bottom of the overlying stratum.

Scour marks. As currents flow across beds of sand, they often erode scour marks. An overlying layer of sediment may later fill these depressed markings, forming positive-relief casts in the covering bed (Fig. 5-23E). The casts are termed "sole markings" (like a shoe sole) because they appear on the bottom of the younger stratum.

Fossils of bottom-dwelling organisms such as corals may also be used to determine which way is up, if they were buried in their natural, upright living positions. Some fossils that have been moved by currents may also

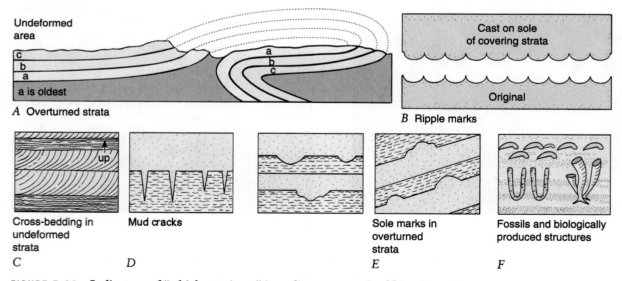

FIGURE 5-23 **Indicators of "which way is up" in sedimentary rocks.** ◧ *In what regions of the United States might one encounter overturned strata?*

be useful. For example, the curved shells of clams washed by currents may come to rest in a convex-upward position, as this is hydrodynamically most stable. Finally, many fossil organisms excavated and lived in burrows, such as the U-shaped burrows shown in Figure 5-23F.

▶ WHAT FOUR SANDSTONE TYPES REVEAL ABOUT TECTONIC SETTING

Sandstones provide an extraordinary amount of information about conditions in and near the site of deposition. The mineral composition of sandstone grains can tell us the source areas and whether the source rocks were metamorphic, igneous, or sedimentary.

Maturity. The minerals let us estimate the "maturity" of the sediments that formed the sandstone. Maturity refers to the amount of transport and erosion experienced by the grains. Rigorous weathering and long transport reduce less-stable feldspars and ferromagnesian minerals to clay and iron compounds and cause rounding and sorting of the remaining quartz grains. So, a sandstone that is rich in angular particles and less-durable feldspars and ferromagnesians is unlikely to have traveled great distances. Such sediments are immature. On the other hand, a sandstone with well-rounded quartz particles and clays is considered mature.

Composition also helps us classify sandstones into four types: quartz sandstone, arkose, graywacke, and lithic sandstone (sometimes termed subgraywacke) (Fig. 5-24). These terms make it easier for geologists to communicate, as each implies particular conditions of deposition.

Quartz sandstone. Quartz sandstones are dominated by quartz, with little or no feldspar, mica, or fine clastic matrix. The quartz grains are well sorted and well rounded (Fig. 5-14). They are most commonly bound together by calcite or silica cement.

Quartz sandstones reflect deposition in stable, shallow-water environments, such as continental shelves or the ancient shallow seas that inundated large parts of low-lying continental regions in the geologic past (Fig. 5-25A). These sandstones, as well as clastic limestones, exhibit sedimentary features, such as cross-bedding and ripple marks that form in shallow water.

Arkose. Sandstones containing 25% or more feldspar (derived from erosion of a granite source area) are called arkoses (Fig 5-26). Although feldspars are an essential component, quartz is the most abundant mineral. Arkoses may originate as basal sandstones derived from the erosion of a granitic coastal area experiencing an advance of the sea. Alternately, they may accumulate in fault troughs or low areas adjacent to granite mountains (Fig. 5-25B).

Graywacke. From the German term *wacken*, meaning "waste" or "barren," graywackes are immature sandstones containing abundant dark, very fine-grained material (Fig. 5-27A). There is little or no cement, and the sand-sized grains are separated by finer matrix particles. The matrix constitutes approximately 30% of the rock, and the remaining coarser grains consist of quartz, feldspar, and rock particles. Graded bedding, interspersed layers of volcanic rocks, and chert (which may indirectly derive its silica from volcanic ash) occur along with graywackes.

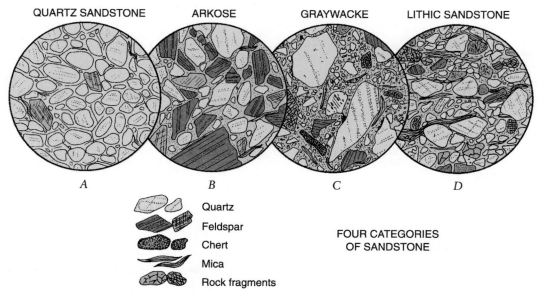

QUARTZ SANDSTONE   ARKOSE   GRAYWACKE   LITHIC SANDSTONE

A   *B*   *C*   *D*

Quartz
Feldspar
Chert
Mica
Rock fragments

FOUR CATEGORIES
OF SANDSTONE

FIGURE 5-24 **Four categories of sandstone as seen in thin section under the microscope.** Diameter of field is about 4 mm.

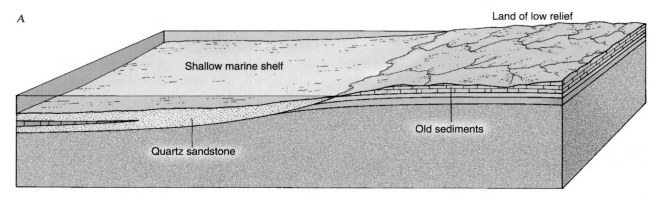

A

Land of low relief

Shallow marine shelf

Old sediments

Quartz sandstone

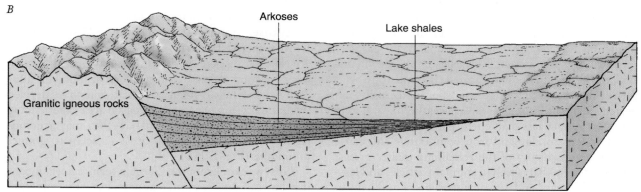

B

Arkoses

Lake shales

Granitic igneous rocks

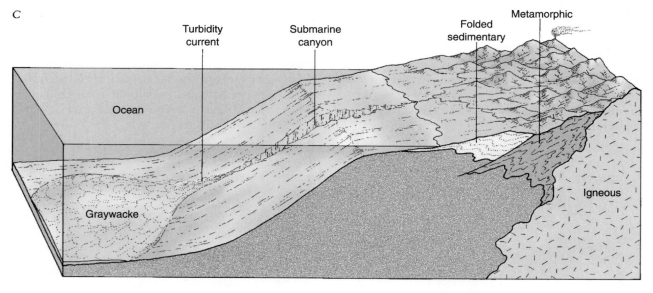

C

Turbidity current

Submarine canyon

Folded sedimentary

Metamorphic

Ocean

Graywacke

Igneous

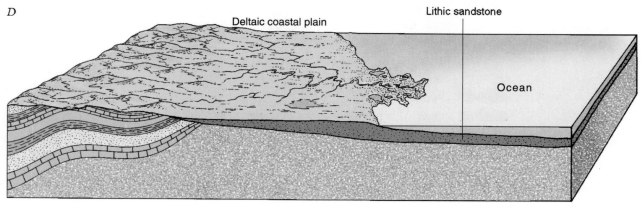

D

Deltaic coastal plain

Lithic sandstone

Ocean

96

FIGURE 5-25   **Idealized geologic conditions under which the four major categories of sandstones are deposited.** (A) *Quartz sandstone.* In this environment there is little tectonic movement. Water depth is shallow, and the basin subsides only slowly. ◨ *What features formed by wave action would you expect to find in these sandstones?* (B) *Arkose.* The high content of feldspar grains in arkose indicates a granitic source area. Arkose (Fig. 5-26) is commonly pink because of the orthoclase grains and an iron oxide cement. (C) *Graywacke.* Graywacke sands are transported by water containing abundant mud, which makes it denser than surrounding clear water. Because of the greater density, the sediment is carried along the sloping seafloor or down submarine canyons. Submarine fans composed of graywacke sands commonly form at the base of the continental slope. Figure 5-27 shows a hand specimen and a thin section of graywacke. In the thin section, note the angularity of grains, poor sorting, and high content of matrix material. (D) *Lithic sandstone.* Lithic sandstones are most commonly associated with deltaic environments of deposition.

FIGURE 5-26   **Arkose.** The pink color commonly seen in arkose resuts from a similar color of the abundant grains of orthoclase feldspar as well as iron oxide cement.

The poor sorting, angularity of grains, graded beds, and volcanics all tell us that graywackes are deposited in a dynamic and unstable tectonic setting. Characteristically, the sediments composing graywackes were deposited offshore of an actively rising mountainous region (Fig. 5-25C).

Lithic sandstone. Quartz sandstone, arkose, and graywacke are distinct kinds of sandstones. A sandstone that has a more transitional composition and texture is termed a **lithic sandstone** (subgraywacke). In lithic sandstones (Fig. 5-25D), feldspars are relatively scarce, whereas more durable minerals like quartz, muscovite, and chert are abundant. Also, quartz grains are more rounded and abundant, sorting is better, and the quantity of matrix is lower in lithic sandstones than in graywackes.

The characteristic environments for lithic sandstones are deltaic coastal plains (Fig. 5-25D), where lithic sandstones may be deposited in nearshore marine environments or swamps and marshes. Layers of coal and micaceous shales are frequently interbedded with lithic sandstones.

A   *B*

FIGURE 5-27   **Graywacke, shown in hand specimen (A) and thin section (B).** The poorly sorted texture of graywacke and the angularity of component grains can be seen in the thin section. Width of field about 1.0 mm.

LIMESTONES AND HOW THEY FORM

Limestones are the most abundant carbonate sedimentary rocks. Most originate in the seas, although some lake deposits exist. Almost all marine limestones appear to have formed from biological processes, either directly or indirectly.

In some limestones, the role of biology is obvious, for the bulk of the rock is composed of shells and calcareous fragments of marine invertebrates. In other limestones, skeletal remains are not present, but nevertheless, the calcium carbonate ($CaCO_3$) that forms the bulk of the deposit was precipitated from seawater because of the life processes of marine organisms.

For example, the relatively warm, clear ocean waters of tropical regions usually are slightly supersaturated with calcium carbonate. In this condition, it takes little change in temperature or seawater chemistry to trigger precipitation of calcium carbonate crystals. A slight temperature increase will do it. Or myriad microscopic marine plants remove carbon dioxide from the water through photosynthesis, and this can trigger precipitation. Also, bacterial decay generates ammonia, making the seawater more alkaline, which also can enhance precipitation. In each case, the calcium carbonate is an inorganic product of an organic process.

Carbonate sedimentation is occurring today in tropical marine areas, such as the Bahama Banks east of Florida (Fig. 5-28). The Bahama Banks carbonates (Fig. 5-29) originate in more than one way. Some are derived from the death and dismemberment of calcareous algae, such as *Penicillus* and *Halimeda*, organisms that secrete tiny, needlelike crystals of calcium carbonate. The microscopic shells of other unicellular organisms also contribute to the carbonate buildup. Some carbonate sediment results from precipitation of tiny calcium carbonate crystals from seawater that has been chemically altered by the biologic processes of marine

FIGURE 5-29 **Carbonate mud accumulating on the sea-floor in the shallow warm waters of the Bahama Banks carbonate platform.** Green algae of the genus *Penicillus* form the tuftlike growths in the background. These algae produce fine, needlelike crystallites of calcium carbonate (the mineral aragonite) that contribute to the production of carbonate sediment. Other algae, such as *Halimeda*, produce similar calcium carbonate particles.

plants. Coarser carbonate particles result from the abrasion of invertebrate shells or are fecal pellets from mollusks and worms.

Where tidal currents flow across the Bahama Banks, oöids accumulate. Oöids (Greek: "egg stone") are tiny spheres composed of concentrically laminated calcium carbonate that form when particles roll back and forth on the seafloor (Fig. 5-30).

Warm, clear, shallow seas are required for calcium carbonate to accumulate. Such conditions once must have existed in western Texas, Alberta, and Michigan, where thick sections of limestones and dolomites have formed. In these areas, optimum conditions for carbonate sedimentation resulted in an accumulation rate that approximately equaled subsidence.

Thick deposits of limestones also accumulated during the Paleozoic Era on shallow submerged carbonate platforms along the eastern margin of North America and in low-lying interior regions that were periodically inundated by seas. During the geologic past, sea levels were typically higher than today, and climates generally were warmer. Thus, carbonate platforms once were more abundant and extensive.

ORGANIZING STRATA TO SOLVE GEOLOGIC PROBLEMS

Rock Units (Formally Called Lithostratigraphic Units)

Pioneering English geologist William Smith demonstrated that distinctive bodies of strata could be traced over many miles and therefore could be mapped. In 1815, he produced a geologic map of England and

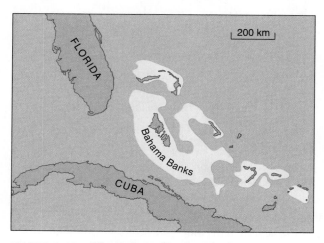

FIGURE 5-28 **The Bahama Banks.** The light blue areas have water depths of less than 180 meters.

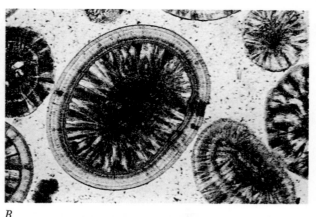

A　　　　　　　　　　　　　　　　　　　　　　　　*B*

FIGURE 5-30　**(A) Ooïds on the weathered surface of limestone (B) thin-section view of an oölitic limestone.** The ooïds are immersed in a cement of sparry (clear, crystalline) calcite. The large ooïd in the center of (B) has a maximum diameter of 0.74 cm.

Wales. Smith's map, the first of such high quality and accuracy, was accompanied by a comprehensive table of rock units he had mapped. Each unit was given a name, such as the "Clunch Clay," "Great Oölyte," or "Cornbrash Limestone."

Thus originated a key concept in geology: the formation. A **formation** is a distinctive rock unit that has recognizable contacts with other distinctive units above and below, and that can be traced across the countryside from exposure to exposure (or from well to well in the subsurface). The formation concept is shown in Figure 5-31.

A formation can consist of a single rock type or a distinctive association of rock types. For example, a formation may be composed entirely of shale beds, or it may be a distinctive sequence of shale interbedded with sandstone, evaporites, or limestone.

Rock units (formally called lithostratigraphic units) are bodies of rock having distinctive features *without regard to time boundaries*. Rock units are mappable and distinctly different from the time-rock unit defined in the previous chapter. Rock units are distinguished and recognized by features such as texture, grain size, whether clastic or crystalline, color, composition, thickness, type of bedding, nature of organic remains, and appearance of the unit in surface exposures (or in the lithologic record of strata penetrated by wells).

Whereas a time-rock (chronostratigraphic) unit represents a body of rock deposited or emplaced during a specific interval of time, a rock unit—such as a formation—may or may not be the same age everywhere it is encountered. For example, the nearshore sands deposited by a sea that is slowly advancing (transgressing) across a low coastal plain may deposit a single blanket of sand (perhaps later to be named the Oriskany Sandstone). However, that sand layer will be older where the sea began its advance and younger where the advance halted (Fig. 5-32).

How We Name Rock Units: Geography + Rock Type

Formations are given a two-part name: geography + rock type. Examples include the Kimmswick Limestone, Tapeats Sandstone, Bright Angel Shale, Pittsburgh Coal, and Pishogue Conglomerate.

The geographic name comes from where the formation is well exposed or was first described. For example, the Kimmswick Limestone was described in 1904 and named for exposures near Kimmswick, Missouri. The Kimmswick Limestone is entirely limestone, so the rock name "limestone" can be used. When formations are composed of multiple rock types, the locality name is simply followed by the term "formation," as in Toroweap Formation.

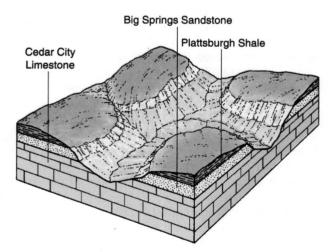

Big Springs Sandstone

Plattsburgh Shale

Cedar City Limestone

FIGURE 5-31　**Formations.** The diagram shows three formations, each named for a geographic location near which it is well-exposed.

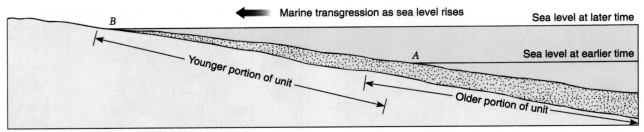

FIGURE 5-32 **Original deposits of a formation may vary in age from place to place.**

There are other rock units in addition to formations. Smaller units within formations may be split out as **members**. For example, in Grand Canyon National Park, the Whitmore Wash, Thunder Springs, Mooney Falls, and Horseshoe Mesa rock units are *members* of the massive Redwall Formation. Or formations may be combined into larger **groups**. In another part of the canyon wall, the Tapeats Sandstone, Bright Angel Shale, and Muav Limestone combine to form a larger rock unit known as the Tonto Group (Table 5-2).

What Are Facies?

*A **facies** (FAY-sheez) is part of a rock body that has characteristics from which we can infer the depositional environment.* For example, a rock body might have a bioclastic limestone along one of its margins and micritic limestone elsewhere. In this case, geologists would delineate two facies: a bioclastic limestone facies, inter-

preted as a nearshore part of the rock body, and a micritic limestone facies, interpreted as a former offshore deposit.

In this case, the distinguishing characteristics are lithologic (rather than biologic), so the facies is designated as a **lithofacies**. In other cases, the rock unit may be lithologically uniform, but the fossil assemblages may differ and permit recognition of different **biofacies**, which also reflect differences in the environment. For example, a limestone unit might contain abundant fossils of shallow-water reef corals along its thinning edge, but elsewhere have the remains of deep-water sea urchins and snails. Thus, there would be two biofacies, one reflecting deeper water than the other.

You can see that facies reflect specific environments of deposition. Today, as we travel across swamp, floodplain, and sea, we traverse different environments of deposition (Fig. 5-33), each characterized by a particular facies.

FACIES SHIFT THROUGH TIME Geologists prepare facies maps for a succession of geologic epochs or ages, and can then see how facies have shifted through time. Consider for a moment an arm of the sea slowly transgressing (advancing over) the land. The sediment deposited on the seafloor may consist of a nearshore sand facies, an offshore mud facies, and a far-offshore carbonate facies. But, as the shoreline advances inland, the boundaries of these facies also shift landward, developing an **onlap sequence** (Fig. 5-34), in which coarser sediments are covered by finer ones.

Should the sea subsequently begin a withdrawal (regression), the facies boundaries will again move in the same direction as the shoreline, creating as they go an **offlap sequence** of beds (Fig. 5-35). In offlap situations, coarser nearshore sediment tends to lie above finer sediments. Also, because offlap units are deposited during marine regressions, recently deposited sediment is exposed to erosion, and part of the sedimentary sequence is lost.

Study of sequential vertical changes in lithology, such as those represented by offlap and onlap relationships, is one method by which geologists recognize ancient advances and retreats of the seas and chart the positions of ancient shorelines.

TABLE 5-2 **Rock Units of the Paleozoic Section in Grand Canyon National Park, Arizona**

SYSTEM	ROCK UNITS		
Permian	Kaibab Fm.		
	Toroweap Fm.		
	Hermit Group	Coconino Fm.	
		Hermit Shale	
Pennsylvanian	Supai Group		
Mississippian	Redwall Fm.	Horseshoe Mesa Member	
		Mooney Falls Member	
		Thunder Springs Member	
		Whitmore Wash Member	
Devonian	Temple Butte Fm.		
Cambrian	Tonto Group	Muav Limestone	
		Bright Angel Shale	
		Tapeats Sandstone	

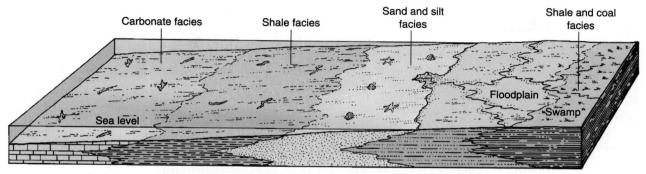

FIGURE 5-33 **Sedimentary facies (lithofacies) developed in the sea adjacent to a land area.** The upper surface shows present-day facies. The side view shows the shifting of facies through time. Notice that bottom-dwelling organisms also differ in environments having different bottom sediment and water depth preferences. (Note: *facies* is both singular and plural.)

WALTHER'S PRINCIPLE Onlap and offlap patterns of sedimentation were recognized as early as 1894 by German geologist Johannes Walther (1860–1937). Walther observed that the lateral succession of facies is also seen in the vertical succession of facies. Thus, to find what facies will be encountered *laterally* from a given locality, we just need to examine the *vertical* sequence of beds at that locality.

For example, Section B in Figure 5-36 shows a typical *fining upward* succession of facies—as you move up-

ward in the section, sediment becomes finer, going from sandstone, to shale, to carbonates. Point X is in the nearshore silt facies and is overlain by finer shale and then limestone. You can see this same sequence of silt, to shale, to limestone as you move westward to Section A. Beneath point X is a coarse beach sand, which can be traced laterally (eastward) in Section C. This relationship, in which the vertical succession of facies corresponds to the lateral succession, has been named **Walther's Principle**.

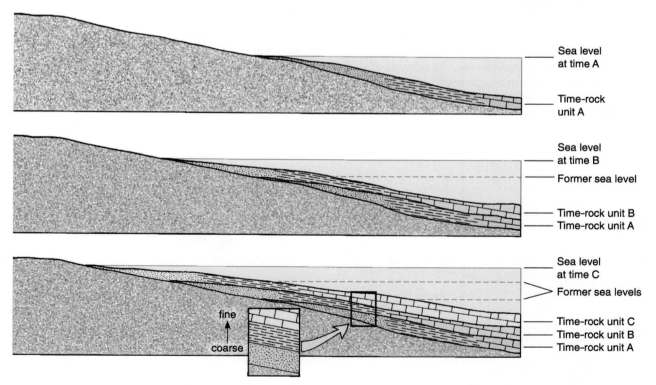

FIGURE 5-34 **Sedimentation during a transgression produces an onlap sequence.** Finer offshore lithofacies overlie coarser nearshore facies (see inset). Nearshore facies are progressively displaced away from a marine point of reference, and older beds are protected from erosion by younger beds.

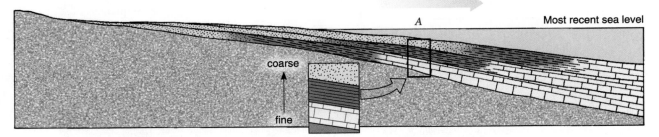

FIGURE 5-35 **Sedimentation during a regression produces an offlap sequence in which coarser nearshore lithofacies overlie finer offshore lithofacies, as shown in A.** The sandy nearshore facies is progressively displaced toward the marine point of reference. Older beds are subjected to erosion as the regression of the sea proceeds. As a result, offlap sequences are less commonly preserved than are onlap sequences.

If the pattern of sediment spread seaward from shorelines always graded from nearshore sands to clays and carbonates, as depicted in Figure 5-13, predicting the locations of ancient rock facies would be easy. In reality, the task is usually more complex. For example, nearshore sandy facies are not present at all along some coastlines. This may be because little sand is being brought to the coast by streams, because vigorous wave and current action has carried the sand grains away, or possibly because sand grains are trapped in submarine canyons farther up the coast.

The nature of sedimentation along a coastline is also controlled by other factors: the direction of longshore currents, the location of major stream mouths that dump their sediment into the sea, the amount of sediment, barriers to sediment dispersal, and whether the sediments are delivered by water, wind, or glacial ice. Any of these factors complicates the study of facies and provides fascinating problems for geologists.

SEA-LEVEL CHANGE MEANS DRAMATIC ENVIRONMENTAL CHANGE

During an "ice age," world temperatures drop, and more of the planet's water becomes frozen in glaciers and polar ice caps. Much of the water that is tied up as ice on the continents comes from the ocean, so sea level drops worldwide. When sea level drops, continental shelves become uncovered, and the amount of exposed continental land increases. Depending on how many meters sea level drops, great areas of former sea bottom can become exposed to the air.

Obviously, this dramatic environmental change from ocean environment to dry land completely alters ecosystems and the distribution of plants and animals. For example, when Pleistocene glaciers had reached their greatest extent and thickness, the Atlantic shoreline moved 100 to 200 kilometers seaward, exposing land that is now beneath over 100 meters of water.

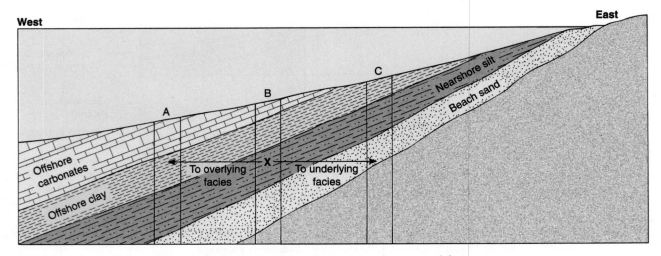

FIGURE 5-36 **An illustration of Walther's Principle, which states that vertical facies changes correspond to lateral facies changes.**

Conversely, during warmer periods, water from melting ice caps and glaciers flows back into the ocean, raising sea level and moving shorelines many kilometers inland. For example, if all the ice in Greenland, the Arctic, and Antarctica were to melt right now, sea level would rise almost 30 meters. All major coastal cities of the world would be inundated.

Precisely how far the sea will advance or retreat during a change in sea level is determined by the amount of change and the land's topography. A low-lying, gently sloping terrain would have a much greater area inundated by a small increase in sea level than would a steeply sloping mountainous tract (Fig. 5-37).

Other events can also change sea level. Upwarping of the floor of an ocean basin (possibly caused by hot ascending convection currents in the mantle, or growth of extensive mid-oceanic ridges) can displace ocean water and result in a rise in sea level.

Changes in sea level profoundly influence the geologic history of continental and shelf areas, for they determine when these areas are inundated and when they are drained. In the geologic past, the continents have experienced many advances of the sea into low-lying regions of the interiors of continents. At times, these marine transgressions covered an astonishing two-thirds of North America. The resulting inland seas are termed **epicontinental**, meaning "a sea over a continent." Each advance of an epicontinental sea was followed by gradual withdrawal back into the major ocean basins. The for-mer seafloor, now land, was then subjected to erosion. Thus, it is possible to recognize sequences of strata having boundaries that are marked by surfaces of erosion.

STRATIGRAPHY AND THE CORRELATING OF ROCK BODIES

Stratigraphy is the study of stratified rocks, including the conditions under which they formed, their relationships, description, identification, and correlation over distance. Because stratified rocks cover approximately 75% of Earth's total land area, and because these strata contain our most readily interpreted clues to past events, stratigraphy is the core discipline of historical geology.

Correlating Rock Bodies

Suppose you see a distinctive limestone in an isolated rock exposure—distinctive by its color, texture, thickness, and prominent fossils. Then you see the same limestone again in an Interstate highway roadcut 20 miles north, and again in a stream bank 50 miles east. You are pretty certain that this limestone is continuous laterally, hidden beneath the cover of soil and trees. And you know that this same limestone, or an equivalent rock body, is likely to continue underground, perhaps even into other states. Recall that this is the *principle of lateral continuity*.

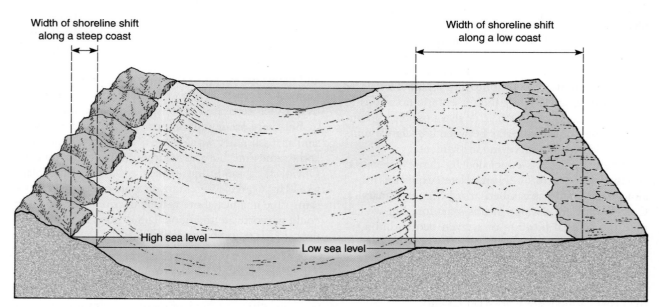

FIGURE 5-37 **Effects of rising or falling sea level on different coasts.** At left, a sea-level rise of 10 meters along a steep shoreline moves the shoreline inland—but not very far, perhaps only a few kilometers, depending on slope. But at right, a 10-meter-rise in sea level along a gradually sloping shoreline would be devastating, moving the shoreline many kilometers inland and perhaps submerging coastal cities.

Determining the equivalence of rock bodies in different localities is called **correlation**. We correlate both rock units and time-rock units. We correlate them from place to place, within continents, between continents, and between boreholes drilled into the seafloor or continents.

Correlation Methods

There are three principal kinds of correlation:

Lithostratigraphic correlation matches up rock bodies by their lithology (composition, texture, color, and so on) and stratigraphic position. This includes recognizing patterns of ocean transgression and regression, distinctive chemical or isotopic characteristics, or electrical or magnetic properties.

Biostratigraphic correlation links rock units by similarity of their fossils.

Chronostratigraphic correlation links rock units by age equivalence, determined by fossils or radioactive dating methods.

Geologists use different types of correlation to solve different problems. If a rock unit's age is not critical, they may use only lithostratigraphic correlation. In other cases, they might employ chronostratigraphic correlation to determine the geologic history of a region. (Chronostratigraphic correlation is the basis for the geologic time scale.)

We will now look at how each type is used.

Using Lithostratigraphic Correlation

If strata are well exposed at Earth's surface—for example, in a desert region with little plant cover—it may be possible to trace distinctive rock units for many kilometers across the countryside simply by walking along the exposed strata. Geologists sketch the contacts between rock units directly onto topographic maps or aerial photographs. They then use these notations to construct a geologic map. It is also possible to construct a map of the contacts between correlative units in the field by using surveying instruments.

Where bedrock is concealed by dense vegetation and thick soil, we must rely on isolated exposures found along valley walls, streambeds, road cuts, and quarries. Correlations are trickier in this situation, but we can note the vertical position of a given bed within a sequence of strata, and we can spot the same pattern in other locations. A formation may change in appearance between two localities, but if it always lies above or below a distinctive stratum of consistent appearance, then correlation is confirmed (Fig. 5-38).

Figure 5-39 illustrates how correlations are used to build a composite picture of the rock record:

1. At Locality 1, a geologist studies the sea cliffs. She recognizes a dense limestone (orange formation F) at the lip of the cliff. The limestone is underlain by formations E (yellow sandstone) and D (gray shale).

2. She continues her survey in the canyon at Locality 2. She recognizes the limestone (formation F) because of its distinctive character and thus correlates it with the limestone at Locality 1.

3. In the canyon, the yellow sandstone E is more clayey than at Locality 1, but she infers it to be the same because it occurs right under the limestone.

4. Working upward toward Locality 3, she maps the sequence of formations G, H, I, J, and K.

5. But she has questions: What lies beneath the lowest formation thus far found (D)? The oil well at Locality 3 provides the answer. Core samples taken from the well reveal that formations C, B, and A lie beneath D. Well logs recorded by petroleum geologists who are monitoring the drilling add to her correlations by matching all the formations penetrated by the drill (B through K) to those found earlier in outcrop.

This is how a network of correlations across an entire region is built, piece by piece.

Using Chronostratigraphic Correlation

To correlate chronostratigraphic units, you cannot depend on similarities in lithology, because similar-looking rocks have formed repeatedly over geologic time (Earth has thousands of different shales, for example, many identical in appearance). Thus, geologists risk correlating two units that look the same but are not, because they were deposited at different times.

Fortunately, fossils help prevent such mismatching, because animals and plants have evolved through geologic time. Distinctive types and ensembles of animals and plants are recognizable in rocks of different ages. Conversely, you can expect rocks of the same age, but from widely separated regions, to contain similar fossils.

But there are complications. For two strata to have similar fossils, they would have to be deposited not only at the same time, but also in rather similar environments. Obviously, a sandstone formed on a river floodplain would have quite different fossils from one formed at the same time in a nearshore marine environment. So, how could we correlate the floodplain deposit to the marine deposit? Here are some possibilities:

- If possible, physically trace the beds along cliffs or valley sides.

- Seek common fossils in both deposits, despite the different environments. For example, wind can deposit the same pollen grains into both environments.

- Look for a distinctive, firmly correlated stratum that lies immediately below, such as a layer of volcanic ash. Ash beds are excellent time markers be-

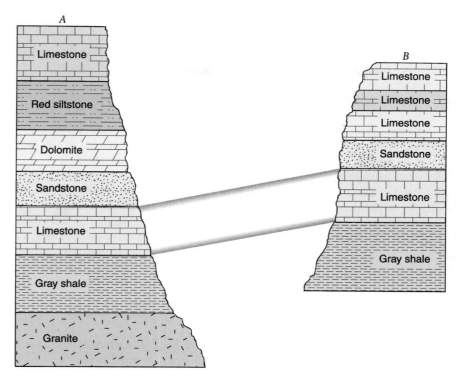

FIGURE 5-38 **When a rock's lithology is not sufficiently distinctive to enable lithostratigraphic correlation between locations A and B, its relation to distinctive rock units above and below may help.** In this example, we can correlate the limestone unit at locality A with the lowest of the four limestone units at locality B because of its position between the distinctive sandstone and gray shale units.

cause they are deposited over a wide area during a relatively brief interval.

- Obtain the actual age of strata, using radioactive dating.

UNCONFORMITIES: SOMETHING IS MISSING

Interpreting geologic history would be much easier if deposition were continuous over time and no erosional losses of sediment ever occurred. Unfortunately, such uninterrupted sequences of strata are rare. Where a gap occurs in the geologic record, there are two possible reasons: either deposition did not occur at all for an interval of time, or the strata were lost to erosion. Either way, gaps in the geologic record may encompass an astonishing number of years of Earth history, in the thousands or millions.

We call these breaks in stratigraphic continuity **unconformities**, because the rocks above and below the gap do not "conform" to one another. Examples of unconformities are abundant on every continent. There are

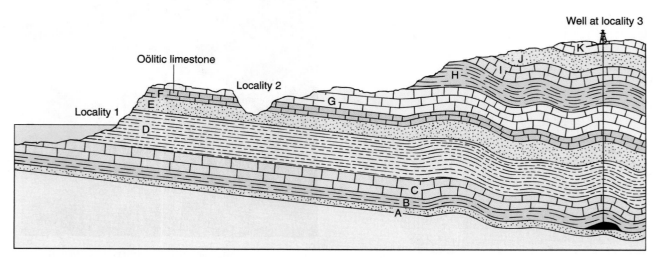

FIGURE 5-39 **Understanding the formation sequence in an area usually begins by examining surface rocks and correlating between isolated exposures.** Study of samples from deep wells permits us to expand the known sequence of formations and to verify the areal extent and thickness of both surface and subsurface formations.

three major types—angular unconformity, disconformity, and nonconformity—as shown in Figure 5-40. The difference among them is the orientation or type of the rocks beneath the erosional surface, as described below.

Angular unconformity. In this type, *nonparallel strata are separated by an erosional surface.* The fact that the lower strata are at an angle is evidence of crustal deformation (Figs. 5-41 and 5-42). As noted in Chapter 2, James Hutton described a now-famous angular unconformity at Siccar Point on the Scottish coast of the North Sea.

Disconformity. In this type, *parallel strata are separated by an erosional surface.* Some unconformities document the simple withdrawal of the sea for a period, meaning the absence of deposition from seawater and the presence of exposure to erosion. This creates a disconformity. What causes the sea to withdraw and re-

turn may be crustal uplift and subsidence, or global sea level changes related to ice ages.

Nonconformity. In this type, *stratified sedimentary rocks rest on older intrusive igneous or metamorphic rocks* (Fig. 5-43). In many nonconformities, crystalline rocks were formed deep within the roots of ancient mountain ranges that subsequently experienced repeated episodes of erosion and uplift. Eventually, the igneous and metamorphic core of the mountains lay exposed as a surface on which the younger strata were deposited.

Although unconformities represent a loss of geologic record, they nevertheless are useful. Like lithostratigraphic units, unconformities can be mapped and correlated. They often record episodes of terrestrial conditions that followed withdrawal of seas. Where regionally extensive unconformities occur, they permit us to recognize distinct sequences of strata of approximately equivalent age.

DEPICTING THE PAST

Geologists create maps, geologic columns, and cross-sections to portray information and illustrate how rock bodies relate to one another.

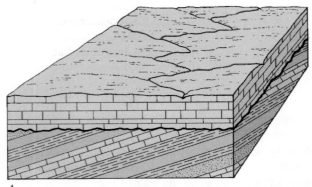

A

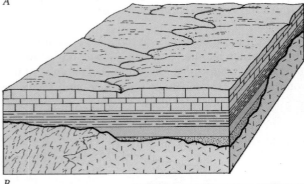

B

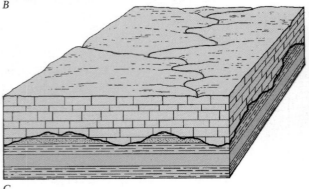

C

FIGURE 5-40 Three types of erosional unconformities. (A) Angular unconformity. (B) Nonconformity. (C) Disconformity.

FIGURE 5-41 Angular unconformity. The ancient erosion surface separates vertical beds of the Precambrian (Proterozoic) Uncompahgre Sandstone from the overlying, nearly horizontal Devonian strata. Elbert Formation at Box Canyon Falls, Ouray, Colorado.

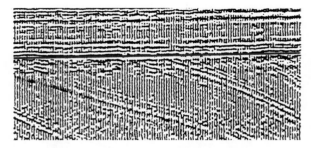

FIGURE 5-42 **Seismic profile reveals an angular unconformity.** The profile depicts tilted (dipping) strata that were beveled by erosion and subsequently covered by horizontal strata. To produce the profile, vibrations are generated in Earth, either using explosives in shallow drill holes or hydraulic vibrators. The vibrations reflect off deep layers of rock and return to Earth's surface, where they are detected by receivers called geophones. Computers then process the data to construct the profile.

Geologic Columns and Cross-Sections

Geologic columns, like the four depicted in Figure 5-44, show the vertical succession of rock units. They are used in correlation and in the preparation of cross-sections, which are cutaway diagrams that show vertical slices through Earth's crust, like a slice of layer cake. There are two basic types of cross-sections, stratigraphic and structural.

STRATIGRAPHIC CROSS-SECTION A stratigraphic cross-section ties together several geologic columns from different locations (Fig. 5-44). Its purpose is to show how rock units change in thickness, lithology, and fossil content across a given area.

To construct a stratigraphic cross-section, we start with a *datum*—some distinctive feature that is observable at all localities used in the cross-section and that we use to tie them together. The datum can be a distinctive rock unit or fossil assemblage. In Figure 5-44, the datum is the trilobite fossil *Glossopleura*.

In drawing the stratigraphic cross-section, we start with a horizontal line that connects points in the geologic columns where *Glossopleura* occurs. The rock units above and below this datum line are then added. Because stratigraphic cross-sections are drawn on such a horizontal datum, they cannot show how strata may be tilted, folded, or faulted (broken). Stratigraphic cross-sections are used primarily to show how formations correlate from one location to another.

STRUCTURAL CROSS-SECTION To add the important dimension of showing the tilting, folding, and faulting of rock units, we draw a **structural cross-section** (Fig. 5-45). The datum for a structural cross-section is not a rock unit or fossil occurrence, but instead is a level line that is parallel to sea level. The tops and bottoms of rock units then are plotted according to their true elevations. Thus, fold and faults are depicted clearly.

Geologic Maps

Suppose you flew over an area where someone stripped away all the grass, trees, buildings, highways, soil, and loose rock to reveal the bare rocks underneath. Also, suppose that they painted each rock formation a different color to make it stand out. This is what a **geologic map** is. It shows the distribution of bedrock of different kinds and ages across the area being mapped.

Unlike road maps, geologic maps show the distribution of rock types and ages, as well as any structural features like faults, anticlines, and synclines.

FIGURE 5-43 **Nonconformity.** The erosional surface is inclined at about 45° and separates Precambrian rhyolitic rock from overlying Upper Cambrian Bonneterre Dolomite. The entire sequence of Lower and Middle Cambrian rocks are missing—about 700 million years of lost history. Taum Sauk Mountain, southeastern Missouri.

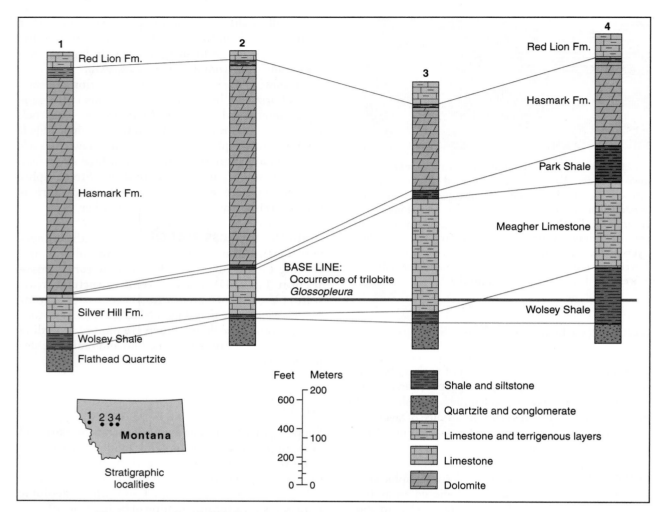

FIGURE 5-44   How we correlate half-billion-year-old rock units (lower Cambrian) in western Montana. Four geologic columns are correlated to form a stratigraphic cross-section. The horizontal datum line (in red) is marked by the occurrence of the trilobite *Glossopleura*. Notice that the Hasmark Formation thins eastward, and the Silver Hill Formation in the west correlates to the Meagher Limestone in the east (Section 4). ◪ *Do the three lowermost formations indicate deposition during a Cambrian marine transgression or regression?*

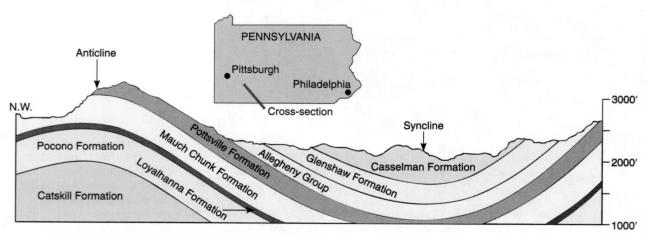

FIGURE 5-45   Geologic structural cross-section. This cross-section is across Paleozoic rocks in the Appalachian Mountains of southwestern Pennsylvania.

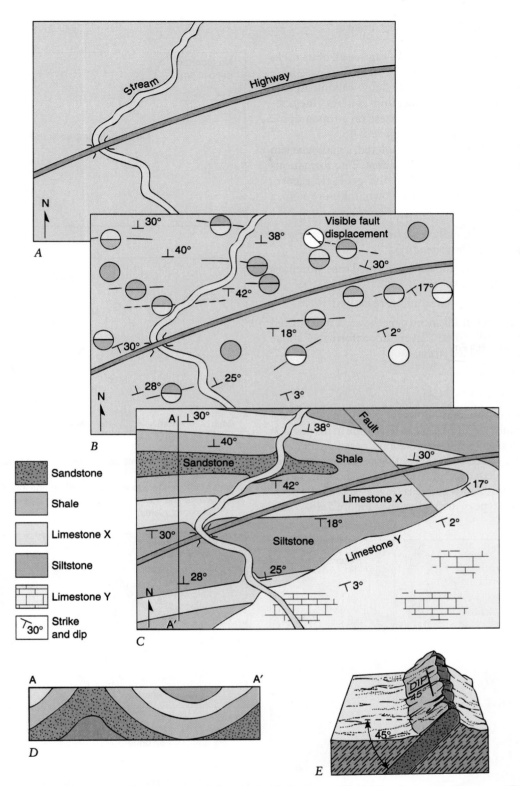

FIGURE 5-46 **How to create a geologic map.** 1. Select a suitable base map (A). 2. Plot data on the base map (B). Working in the field, plot the locations where formations are exposed at the surface in outcrops (circles). Assign a color to each formation and use the colors in the location circles. (Especially useful are outcrops that include contacts between formations. Where contacts can be followed horizontally, trace them onto the base map.) 3. Add the angles at which the rocks lie, called strike and dip. As shown in (E), strike is the direction of a line formed by the intersection of a stratum and a horizontal plane. Dip is the angle the stratum makes with the horizontal. It tells you how steeply the stratum slopes and the direction in which it slopes. 4. Draw the geologic map (C). Draw the formation boundaries to best fit the data. You have to infer a lot when doing this, because you can't actually see most of the rocks. 5. Draw a cross-section. A cross-section is shown along line A–A′. ◀ *What is the oldest rock unit seen at the surface along A–A′? Where is the youngest unit located?*

Actually clearing an area so we can see all of its rocks is neither feasible nor necessary. In reality, we make geologic maps by going into the field, walking an area to observe rock outcrops, locating contact lines between formations, and plotting these on a topographic map. The process is shown in Figure 5-46.

Once the geologic map is completed, a geologist can "read" the geologic history of an area. The formations represent sequential "pages" in the geologic record. From the simple geologic map shown in Figure 5-46, a geologist can deduce:

- The strata originally were horizontal (*Steno's principle of original horizontality*).
- Next, the flat strata were deformed into anticlines and synclines (D).
- Next, the deformed strata were faulted.
- Next, the ocean advanced over the area. It deposited younger sedimentary layers unconformably above the ancient folded strata.

Paleogeographic Maps

A **paleogeographic map** shows the paleogeography (ancient geography) of an area at a specific time in the past. Of course, such maps are interpretations, based on all available evidence collected in the area by geologists and paleontologists. Paleogeographic maps often show the distribution of ancient lands and seas (Fig. 5-47). The process of making this type of map is shown in Figure 5-48.

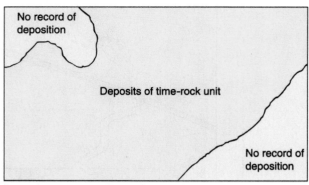

A

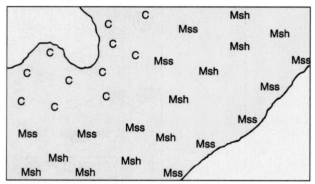

B

C: Continental sediments containing fossils of freshwater clams and land plants
Mss: Marine sandstone with fossils of marine invertebrates
Msh: Marine shale containing abundant fossils of marine microorganisms

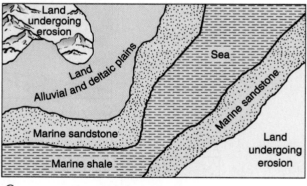

C

☐ Land areas undergoing erosion
▨ Land areas undergoing deposition
▒ Marine areas undergoing deposition

FIGURE 5-48 How to create a paleogeographic map.
1. For the selected area, collect all available data that show the occurrence of the selected time-rock unit. Plot every point where rocks appear for that time period. 2. Plot the rock types observed in that time-rock unit. Determine whether the strata originated on land or in the sea.
3. Draw the paleogeographic map.

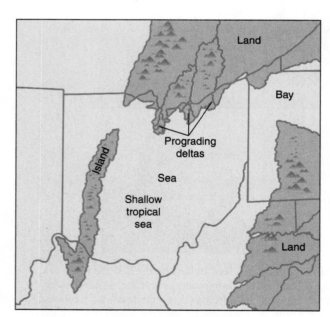

FIGURE 5-47 Paleogeographic map of Ohio and adjoining states over 300 million years ago (early Mississippian Period). At the time, most of this area lay beneath a shallow sea. The data for this map came from outcrops and more than 40,000 records of oil and gas wells.

Isopach (Thickness) Maps

Isopach (EYE-so-pack) means "equal thickness." Isopach maps show changes in the thickness of a formation or time-rock unit. The lines on an isopach map (Fig. 5-49) connect points at which the unit is of the same thickness. Figure 5-50 shows how an isopach map is created.

Lithofacies Maps

Recall that *facies* are lateral changes in sedimentary rocks that reflect different conditions in environments adjacent to the area in which deposition occurred.

Lithofacies maps show rock facies—lateral changes in sedimentary rocks. Such maps add detail and validity to paleogeographic interpretations. Figure 5-51 shows a map of exploratory drilling by an oil company. So far, they have drilled the seven wells shown. Logs for the wells are shown in the correlation above the map. The caption explains how the map was developed.

Figure 5-52 is a lithofacies map of rocks deposited over 400 million years ago in the eastern United States. From this map, we can infer that an ancient mountain system existed along the eastern seaboard, supplying coarse clastics to the basin to the west. The conglomerate in New York probably originated as great alluvial fans built out from the ancient mountain system. Detrital sediments become finer farther from the source, and the section thins westward. In Indiana, the map indicates that all the clastic sediment had been deposited to the east, and only carbonate precipitates were laid down.

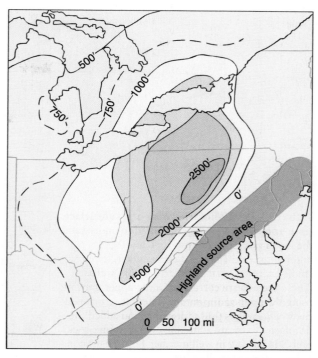

FIGURE 5-49   Isopach map of 450 million-year-old Upper Ordovician formations in Pennsylvania and adjoining states. The pale red areas indicate a center of subsidence in southern New York and Pennsylvania. Over 2000 feet of sediment accumulated here. The isopach pattern further indicates a highland sediment source area to the southeast.

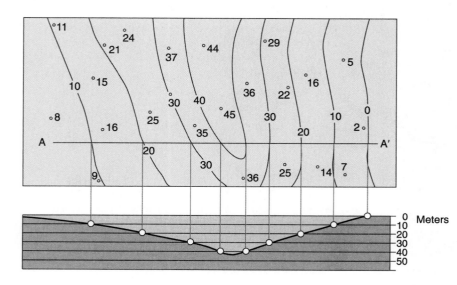

FIGURE 5-50   Isopach maps. Construction of a simple isopach map where strata are undeformed. On a base map, plot the thickness of rock units. The surface thicknesses are from measured exposures; subsurface thicknesses are from data in oil and gas well logs. Then draw isopach (equal-thickness) lines to connect the data points. An isopach map is very useful in determining the size and shape of a depositional basin, the position of shorelines, areas of uplift, and areas of potentially recoverable oil and natural gas.

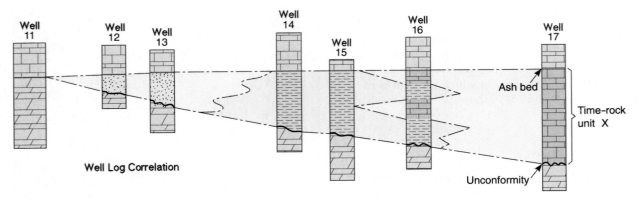

Well Log Correlation

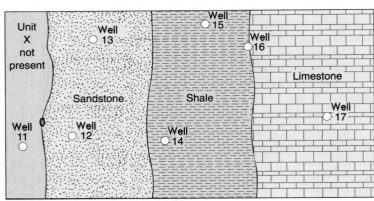

FIGURE 5-51 **Lithofacies map of a subsurface time-rock unit.** Geologists wanted to understand the facies variations of time-rock unit X, highlighted in the upper diagram. First, they studied records from seven exploratory wells. These records are correlated in the upper drawing, using the ash bed as a datum. Then, they mapped the well locations (lower drawing) and added the lithofacies to create a lithofacies map. Time-rock unit X is missing in well number 11, perhaps not being deposited at all or maybe deposited and then eroded away. It is logical that the sandstone facies was deposited adjacent to a north-south trending shoreline. Because there are so few data points (only seven wells over a large area), the lithofacies boundaries on this map are general. ◼ *Why is the ash bed a good datum for the cross-section?*

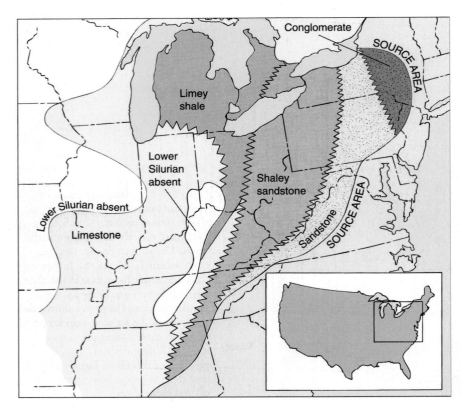

FIGURE 5-52 **Lithofacies map of 430-million-year-old Lower Silurian rocks in the eastern United States.** Data from hundreds of rock outcrops, oil wells, and gas wells were correlated to produce this map.

Grand Canyon National Park, Arizona

In 1540, Hopi Indians directed a band of Spanish conquistadores to the rim of the Grand Canyon of the Colorado River. The great chasm was unlike anything Europeans had witnessed. Today it remains an awesome monument to the erosive force of running water and the gravitational force that pulls erosional debris downward.

The historical geologist sees the Grand Canyon's layers of sandstone, shale, limestones, and lava flows exposed in the canyon walls as a monumental history book.

This "book" reveals geologic events and changing life over a span of 2 billion years, or about 40% of Earth's history, all in one place.

Chapter 1 starts at the canyon's bottom, where former sands and muds record the presence of a shallow water body in the region. We know that volcanoes erupted nearby, for the rocks are interbedded with layers of lava and volcanic ash. About 1.7 billion years ago, mountain-building deformed these sediments and metamorphosed them into the Vishnu Schist (Fig. A), which

rafters on the Colorado River can see close-up.

The Vishnu Schist was then intruded by the Zoroaster Granite. Erosion followed. Then high-silica melts invaded joints and fractures, forming veins and huge masses of igneous rock, which later were metamorphosed to light-colored gneisses. The region was mountainous at this time. Land plants had not yet evolved to retard the forces of erosion. Eventually, the mountains were reduced to lowlands. Only their roots can be seen today in the canyon walls.

Chapter 2 of the Grand Canyon story is written in the 3700-meter thickness of sedimentary rocks and lava flows spread extensively over the Vishnu Schist. (3700 meters = 2.29 miles!) These rocks are called the Grand Canyon Supergroup—see the accompanying geologic column, Figure B.

Following their deposition, tensional (stretching) forces produced mountains caused by multiple normal faults (fault-block mountains). The higher blocks became mountains. Debris that

eroded from these mountains filled the low areas where downfaulted blocks existed. Erosion gradually reduced the entire region to a low-lying terrain, recorded in geologic history by "the great unconformity" that separates Precambrian from Paleozoic strata. Remnants of the Grand Canyon Supergroup are nestled in remaining downfaulted blocks beneath the great unconformity.

Ascending the canyon, we reach Chapter 3, rocks of the Paleozoic Era. Now we are within roughly a half-billion years of the present. The Paleozoic saw repeated inundation and regression by shallow seas. The first inundation deposited sediment that became the nearshore Tapeats Sandstone. As the shoreline shifted eastward, these sands were followed by the Bright Angel Shale and Muav Limestone. The three formations comprise the Tonto Group. Their sequential change in lithology illustrates how rock units may transgress time boundaries.

Chapter 4 is missing! As if pages in our history book had been ripped out, strata of Ordovician and Silurian age

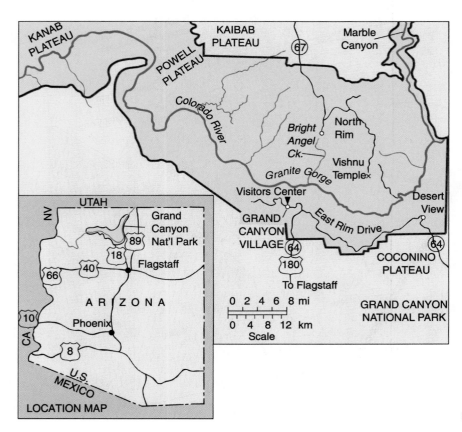

FIGURE A **Grand Canyon National Park.**

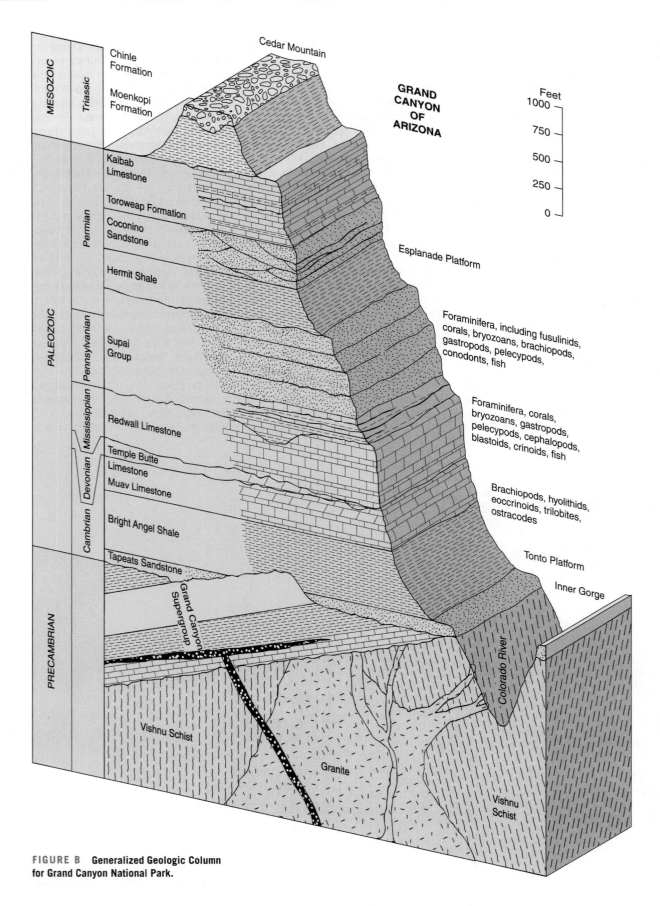

FIGURE B **Generalized Geologic Column for Grand Canyon National Park.**

The following text appears as labels within the figure:

Cedar Mountain

GRAND CANYON OF ARIZONA

Feet
1000
750
500
250
0

MESOZOIC — Triassic
Chinle Formation
Moenkopi Formation

PALEOZOIC
Permian
Kaibab Limestone
Toroweap Formation
Coconino Sandstone
Hermit Shale

Pennsylvanian
Supai Group

Mississippian
Redwall Limestone

Devonian
Temple Butte Limestone

Cambrian
Muav Limestone
Bright Angel Shale
Tapeats Sandstone

PRECAMBRIAN
Grand Canyon Supergroup
Vishnu Schist
Granite
Vishnu Schist

Esplanade Platform

Foraminifera, including fusulinids, corals, bryozoans, brachiopods, gastropods, pelecypods, conodonts, fish

Foraminifera, corals, bryozoans, gastropods, pelecypods, cephalopods, blastoids, crinoids, fish

Brachiopods, hyolithids, eoccrinoids, trilobites, ostracodes

Tonto Platform

Inner Gorge

Colorado River

(400–500 million years ago) do not exist in the Grand Canyon. If ever deposited there, they have been lost to erosion. Thus, an unconformity (a gap in the stratigraphic record) caps the Cambrian rocks.

Chapter 5, above the unconformity, opens with the Temple Butte Limestone of Devonian age (355–410 million years old), laid in a shallow sea. Again, the sea withdrew and then returned during the Mississippian Period (320–355 million years ago) to deposit the cherty carbonates of the Redwall Limestone (Fig. C). The Redwall is well named, for its weathered surface is stained red by iron oxide washed from red beds of the overlying Supai Group and Hermit Shale. The Redwall forms bold cliffs that front many of the canyon's promontories. It is richly fossiliferous with brachiopods, bryozoans, crinoids, and corals.

In Chapter 6, the sea withdrew, creating yet another erosional unconformity, and returned to deposit strata of Pennsylvanian and early Permian age (290–320 million years old). The Supai Group features the park's spectacular temples—The Buddha, Zoroaster, and others. And traces of life appear: nonmarine beds of the

Supai exhibit amphibian tracks, possible reptile tracks, and fern imprints. The overlying Hermit Shale contains evidence of nonmarine deposition in the formation's mud cracks, as well as fossil insects, conifers (evergreens), and ferns.

Chapter 7 is marked by a desert. These floodplains and marshy tracts soon were covered by migrating dunes of the Permian Coconino Sandstone (250–295 million years old). The frosted, well-sorted, well-rounded grains of the white, cross-bedded Coconino sands indicate windblown sediment. Reptiles wandering the dunes left their footprints in these sands of time.

In Chapter 8, the sea returns, and marine limestones and sandstones of the Toroweap Formation rest atop the Coconino. Above the Toroweap are bold vertical cliffs of the Kaibab Limestone, a thick, resistant formation that is the surface of the Kaibab Plateau north of the canyon and the Coconino Plateau on the south side. The Kaibab is the final, youngest Paleozoic rock unit in the Grand Canyon.

Chapter 9 opens the Mesozoic Era about 250 million years ago. Floodplain sands and silts of the Triassic Moenkopi Formation and gravels of

the Shinarump Conglomerate member of the Chinle Formation spread across the region, now mostly swept away by erosion.

Chapter 10 is signaled by uplift and spectacular erosion. Tectonic forces elevate the Colorado Plateau late in the Cenozoic. This uplift causes steeper stream gradients and therefore greater erosion. The Colorado River and its tributaries rapidly deepen their channels. Recent studies of erosion rates indicate that torrential rivers can erode incredibly rapidly. Geologists now propose that the stunningly deep Grand Canyon was eroded in only 1.5 million years (a geologic eyeblink). Although the canyon's rocks are very old, the great chasm itself is geologically young!

The Grand Canyon eroded rapidly, not only from fast-running water but from the impact and abrasion by cobbles and gravels in the rushing currents. Yet, if the river were the only erosion mechanism, the canyon would have perfectly vertical walls. Gravitational mass movements have broadened the canyon's top to its present 17-plus km.

Chapter 11: your own visit to the Grand Canyon. Don't miss it—it is worth the trip!

FIGURE C **View from the Nankoweap Indian Site of the Mississippian Redwall Limestone, Grand Canyon.** The Redwall, a bluish-gray limestone containing chert nodules, takes its colorful name from a coating of red iron oxide stain derived from overlying strata.

SUMMARY

- Sedimentary rocks are the material record of environments that once existed on Earth's surface. For this reason, they are essential for deciphering Earth history.
- Sedimentary rocks are formed by the accumulation and consolidation of weathering products derived from older rock masses, by chemical precipitation, and by the accumulation of organic debris.
- The minerals in clastic sedimentary rocks may provide clues to the kind of parent rock that was weathered and eroded to provide sediment.
- The color of sedimentary rocks may indicate environmental conditions within the area of deposition.
- Fossils in sedimentary rocks are excellent environmental indicators. They tell us whether strata are marine or nonmarine, whether the water was deep or shallow, and whether the climate was cold or warm.
- The texture of sedimentary rocks (size, shape, and arrangement of grains) and the structural features of sedimentary rocks (graded bedding, cross bedding, and ripple marks) provide information about the nature of the transporting and depositional medium and conditions in the depositional environment.
- Quartz sandstones, arkoses, graywackes, and lithic sandstones originate in specific tectonic settings.
- Rock units (lithostratigraphic units) consist of rocks that are sufficiently distinctive in color, texture, or composition to be easily recognized and mapped. Lithostratigraphic units are not necessarily the same age throughout their areal extent. The basic rock unit is the formation.
- A time-rock unit (chronostratigraphic unit) differs from a rock unit in that it is an assemblage of strata deposited within a particular interval of geologic time.
- Within any given time-rock unit, rocks may differ from adjacent rocks in composition, texture, fossil content, and other features. These rock bodies of distinctive appearance or aspect are called facies. Facies reflect deposition in a particular environmental setting.
- Shoreline migrations may be the result of changes in sea level, tectonic movements of continental borderlands, or the seaward advance of the coastline resulting from rapid deposition of sediment brought to the sea by rivers.
- Stratigraphy is the branch of geology that deals with the origin, composition, sequence, and correlation of stratified rocks. Stratigraphic correlation involves determining the equivalence of strata in diverse locations. Lithostratigraphic correlation links units of similar lithology and stratigraphic position. Chronostratigraphic correlation links units of corresponding age.
- Geologists construct columnar sections, cross-sections, paleogeographic maps, and isopach maps to graphically depict the distribution, correlation, and paleoenvironmental significance of rock bodies.

KEY TERMS

angular unconformity, p.106

alluvial fan, p.82

arkose, p.95

barrier island, p.85

biofacies, p.100

biostratigraphic correlation, p.103

cement (in clastic sedimentary rock), p.87

chronostratigraphic correlation, p.103

clast, p.87

continental shelf, p.80

continental slope, p.82

craton, p.79

cross-bedding, p.91

delta, p.83

discharge (of stream), p.82

disconformity, p.106

eolian environment, p.83

epicontinental sea, p.103

estuary, p.86

facies, p.100

ferric, p.86

ferrous, p.86

formation, p.99

geologic map, p.107

graded bedding, p.91

graywacke, p.95

group, p.100

hematite, p.86

isopach map, p.114

lacustrine deposit, p.83

lagoon, p.85

lithic sandstone, p.95

lithofacies, p.100

lithofacies map, p.114

lithostratigrapic correlation, p.103

matrix (in clastic sedimentary rock), p.87

maturity (of sandstone), p.95

member, p.100

mud cracks, p.91

nonconformity, p.106

offlap sequence, p.100

onlap sequence, p.100

orogenic belt, p.79

paleogeographic map, p.110

platform, p.79

quartz sandstone, p.95

red beds, p.86

ripple marks, p.91, 93

rock unit (= lithostratigraphic unit), p.99

rounding (of sedimentary particles), p.89

scour marks, p.94

shield, p.79

sorting, p.88

sphericity (of sedimentary particles), p.89

stratigraphy, p.103

stream deposits, p.82

structural cross-section, p.107

tectonics, p.78

tidal flat, p.85

till, p.83

turbidite, p.82

turbidity current, p.82

unconformity, p.105

Walther's Principle, p.101

QUESTIONS FOR REVIEW AND DISCUSSION

1. What features in a sedimentary rock might indicate that it was deposited in each of the following environments of deposition?

- **a.** Shallow marine environment
- **b.** Deep marine, continental rise environment
- **c.** Transitional, deltaic environment
- **d.** Continental, desert environment

2. Why are sandstones and siltstones of desert environments rarely black or gray in color?

3. How does matrix in a rock differ from cement? What are the most common kinds of cements found in sandstones? Which of these is most durable?

4. What would be the probable origin of a poorly sorted sandstone composed of angular grains in a matrix of mud? What origin might you infer for a well-sorted, ripple-marked quartz sandstone containing fossils of marine clams?

5. What differences in texture and composition serve to distinguish between a mature and an immature sandstone?

6. In a columnar section of sedimentary rocks, a limestone is overlain by a shale, which in turn is overlain by sandstone. What might this *coarsening upward* sequence indicate about the advance or retreat of a shoreline?

7. What conditions in the Bahama Banks result in the high production of calcium carbonate sediment?

8. What features of sedimentary rocks are useful in determining the direction of current of the depositing medium?

9. An isopach map shows an accumulation of 10,000 meters of sediments in a Paleozoic marine basin of deposition, yet the sedimentary rocks contain fossils indicating deposition in water no deeper than 200 meters. What has occurred in the basin of deposition?

WEB SITES

The Earth Through Time Student Companion Web Site (www.wiley.com/college/levin) has online resources to help you expand your understanding of the topics in this chapter. Visit the Web Site to access the following:

1. Illustrated course notes covering key concepts in each chapter;

2. Online quizzes that provide immediate feedback;

3. Links to chapter-specific topics on the web;

4. Science news updates relating to recent developments in Historical Geology;

5. Web inquiry activities for further exploration;

6. A glossary of terms;

7. A Student Union with links to topics such as study skills, writing and grammar, and citing electronic information.

6

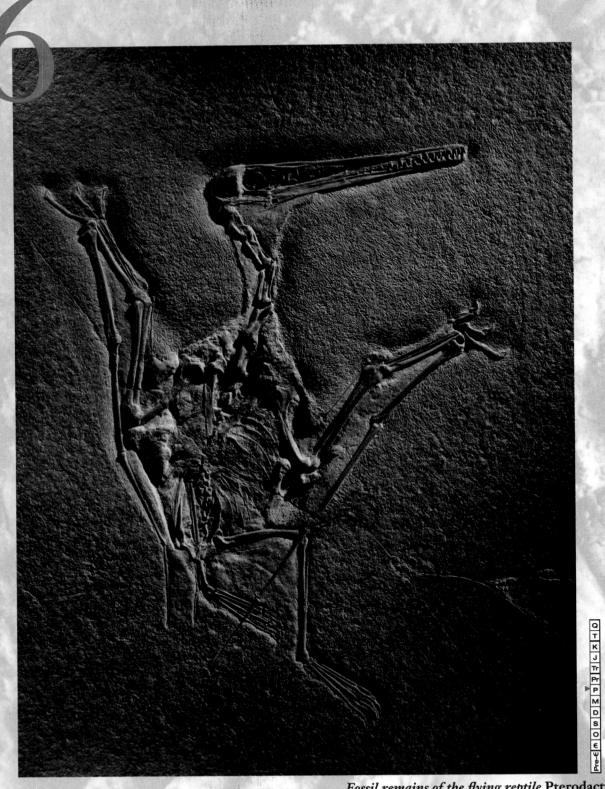

Q
T
K
J
Tr
P
P
M
D
S
O
Є
Pre-Є

Fossil remains of the flying reptile **Pterodactylus**
from the Late Jurassic Solnhofen Limestone of
Germany.

CHAPTER 6

Life on Earth: What Do Fossils Reveal?

Linnaeus thought of himself not as creating an ordered classification of living things, but as discovering and describing an order that already existed in nature.
—U. Lanham, 1968

Key Chapter Concepts

- Fossils are the remains or indications of ancient organisms that are preserved in rock. A fossil can be an actual part of an organism (like a shell or bone), an imprint or cast of the organism, or a trace of the organism's activity (like a footprint or burrow).

- Three common processes by which fossils are preserved are replacement, permineralization, and carbonization.

- Trace fossils are tracks, trails, burrows, and other markings made by ancient organisms in sediment.

- The probability of preserving an ancient organism is greater if the organism was buried rapidly and if it had hard parts.

- In the 1700s, Carolus Linnaeus paved the way for classifying plants and animals.

- To understand and study life, we divide it into three great domains: Archaea, Bacteria, and Eukarya. These are further divided into six kingdoms: Archaebacteria, Eubacteria, Protista, Fungi, Plantae, and Animalia. Each of these in turn is divided into phylum, class, order, family, genus, and species.

- Lamarck's theory of evolution, introduced in the early 1800s, was an important step. However, it was flawed because it stated that organisms can transmit to their offspring traits acquired during their lifespan.

- In the mid-1800s, Charles Darwin demonstrated how natural selection can control the direction of evolutionary change in organisms.

- Natural selection operates through mutation and the rules of inheritance to cause variation in organisms.

- New species can result from shifts to new environments or environmental change in a given location. The new species can then adapt and radiate as new generations exploit new environmental situations or life strategies.

- Evolution may progress in slow and gradual stages or in rapid bursts. The latter, called punctuated evolution, appears to be more prevalent.

OUTLINE

▶ FOSSILS: SURVIVING RECORDS OF PAST LIFE
▶ FIGURING OUT HOW LIFE IS ORGANIZED
▶ EVOLUTION: CONTINUOUS CHANGES IN LIFE
▶ THE CASE FOR EVOLUTION
▶ FOSSILS AND STRATIGRAPHY
▶ FOSSILS INDICATE PAST ENVIRONMENTS
▶ HOW FOSSILS INDICATE PALEOGEOGRAPHY
▶ HOW FOSSILS INDICATE PAST CLIMATES
▶ AN OVERVIEW OF THE HISTORY OF LIFE
▶ LIFE ON OTHER PLANETS: ARE WE ALONE?
▶ SUMMARY
▶ KEY TERMS
▶ QUESTIONS FOR REVIEW AND DISCUSSION
▶ WEB SITES
▶ FIGURES
▶ BOX 6-1 ENRICHMENT
▶ BOX 6-2 ENRICHMENT

- Phylogeny (the evolutionary history of a species or group) seeks to understand the way ancestral organisms evolve into their descendants. Stratophenic phylogeny, inferred from the sequence (stratigraphy) of rock layers, places older ancestral species on the trunk and lower branches of the family tree and the younger species on the higher branches. Cladistic phylogeny arranges species or groups by basic similarities and recent evolutionary innovations.

- The many lines of evidence for evolution include the fossil record, similar body parts in ancient and modern species, vestigial structures (remnants like the tailbone in humans), and the universality of DNA and the genetic code in all species of plants and animals.

- Fossils are important in correlating strata and identifying their relative age, in determining the characteristics of ancient climates and environments, and in reconstructing the geography of ancient times.

- The Universe contains many planets similar to the Earth. Life on one or more of these is a possibility.

It is 100,000 years ago. A weary Neandertal hunter stops to rest on a limestone ledge. As he idly tightens the thongs that hold his hand-ax to its wooden handle, he notices a strange object. It is a fossil, loosened from the limestone by weathering. With the curiosity characteristic of his species, the hunter picks up the fossil, examines it intently, and places it in the skin pouch tied to his fur garment. Perhaps it is charmed, he thinks. Perhaps it will protect him from evil spirits or sickness.

From the time of the Neandertal hunter to the present, humans have been intrigued by the traces of former life we call fossils. Paleontologists carefully scrutinize fossils for clues to how these ancient plants and animals lived and how their descendants fared during their amazing journey through eons of geologic time, down to the present.

► FOSSILS: SURVIVING RECORDS OF PAST LIFE

It is remarkable that fossils are so common, for there are many ways for organisms to be destroyed after death. Scavengers and bacteria attack the deceased. Chemical decay, which we call decomposition, starts immediately. Weathering, erosion, and transport of organism "sediment," especially by running water, make the odds of preserving a dead body very low. However, the chances of preservation are vastly improved if the organism has a hard skeleton of shell or bone, and if it perishes where it can be quickly buried by sediment to protect it from destruction.

How Does Life of the Past Become Preserved?

Conditions on the seafloor often favor preservation. Here, shelled invertebrates flourish and are covered by a continuous rain of sedimentary particles. Not surprisingly, most fossils are found in marine sedimentary rocks, which are former seafloor. However, fossils also occur in terrestrial deposits left by streams and lakes. On occasion, animals and plants have been preserved after becoming immersed in tar, quicksand, or tree sap; trapped in ice or lava flows; or engulfed by rapid falls of volcanic ash (Fig. 6-1).

The term fossil often implies **petrifaction**—literally, "to make into stone." For example, after a shelled organism like a clam or snail dies, scavengers and bacteria normally consume the soft tissue, leaving behind its empty shell. The shell may then be buried by sediment. If it is durable and not easily dissolved, it may remain largely unchanged for a long time. Unaltered shells of marine invertebrates are known from deposits over 100 million years old.

THREE COMMON PRESERVATION PROCESSES Several processes may act upon the shell or bone of a dead organism to enhance its chances for preservation. Most common are permineralization, replacement, and carbonization.

Permineralization means "to spread minerals throughout." It occurs when water that contains dissolved silica, calcium carbonate, or iron circulates through the sediment that encloses a deceased organism. The minerals become deposited in the organism's remains, in bone cavities and canals that once were occupied by blood vessels and nerves or in open spaces vacated by cells. The original composition of the bone or shell remains, but the fossil is made harder and more durable by the minerals (Fig. 6-2).

FIGURE 6-1 **Fossil wasp, victim of a volcano about 30 million years ago.** Sudden showers of ash from a Colorado volcano covered insects, causing them to fall into lakes, where they became buried in bottom sediment. Florissant Fossil Beds National Monument, Florissant, Colorado.

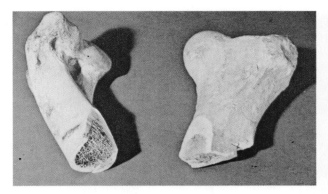

A

B

FIGURE 6-2 **Petrifaction by permineralization.** (A) In these unfossilized fragments from a cow's femur (upper leg bone), note the porous central marrow cavity. (B) Permineralized dinosaur bone in which the porous marrow cavity has been filled with mineral matter.

FIGURE 6-3 **Petrifaction by replacement.** The shell of an extinct marine organism known as an ammonoid. In this 160-million-year-old Jurassic fossil, the original calcium carbonate skeleton has been replaced with the mineral pyrite (commonly called "fool's gold"). ◪ *What other common minerals or compounds frequently occur as replacements? (Answers to questions appearing within figure legends can be found in the Student Study Guide.)*

Replacement is the simultaneous *exchange* of the original substance of a dead plant or animal with mineral matter of a different composition. Solutions dissolve the original material and replace it with an equal volume of the new minerals (Fig. 6-3). Replacement can be marvelously precise, so that details of shell ornamentation, tree rings in wood, and delicate bone structures are accurately preserved.

Carbonization occurs when soft tissues are preserved as thin films of carbon. For example, leaves or tissues of soft-bodied organisms like jellyfish or worms may accumulate, become buried and compressed, and lose their volatile constituents. The carbon often remains behind as a blackened silhouette (Fig. 6-4).

MOLDS AND CASTS Fossils also may be simple molds, imprints, or casts. Any organic structure may leave an impression if it is pressed into a soft material that is capable of retaining the imprint (like pressing your thumb into modeling clay). Among shell-bearing invertebrates, the shell commonly is dissolved after burial and lithification. This leaves a void where the organism was—a vacant **mold** that bears the surface imprint of the original

shell. These features are the reverse of those on the shell (ridges become grooves; knobs become depressions).

If external features of the fossil are visible (like growth lines or ornamentation), the mold is an *external mold*. Conversely, an *internal mold* shows features of

FIGURE 6-4 **Petrifaction by carbonization.** This fossil seed fern from rocks of Pennsylvanian age (about 300 million years ago) has been preserved by carbonization. The frond is approximately 27 cm long.

FIGURE 6-5 **Internal mold (steinkern).** Internal molds form by filling the cavity in a shell, in this case the spiral cavity of an ancient marine snail.

FIGURE 6-7 **Eocene insect remarkably preserved in amber.** This insect is of the Order Diptera, which includes flies, mosquitoes, and gnats. It is approximately 30–50 million years old. Although preservation of insects is uncommon due to their softness, there are indications that they approached modern levels of abundance and diversity early in the Cenozoic Era.

FIGURE 6-6 **Trilobite cast (above) and mold (below).** It formed in a nodule of calcareous shale. The trilobite (*Calliops*) is 4 cm long.

just the inside of a shell, such as muscle scars or supports for internal organs. Many invertebrate shells enclose a hollow space that may be left empty or filled with sediment (Fig. 6-5). The internal filling is called a *steinkern* (German: *stone core*). Steinkerns are also called *internal molds*. Finally, molds subsequently may become filled, forming **casts** that faithfully show the shell's original form (Fig. 6-6).

Although hard parts enhance the chances of preservation, occasionally even soft tissues and organs are preserved:

- Insects and even small invertebrates are found preserved in amber, the hardened resin of conifers (pine tree sap). See Figure 6-7 and Box 6-1, "Amber, the Golden Preservative."

- X-ray images of thin rock slabs sometimes reveal the ghostly outlines of soft tissue of various marine creatures.

- Skin, hair, and viscera of Ice Age mammoths have been preserved in frozen soil (Fig. 6-8) or in tar oozing from natural oil seeps.

ENRICHMENT

Amber, the Golden Preservative

"Fossil" brings to mind remains of ancient organisms that literally have been turned to stone. In this chapter, however, you saw a few rare instances where actual remains have been preserved. For such preservations, nothing beats the organic substance known as amber (Fig. 6-7).

Insects are the usual organisms preserved in amber, because they are readily trapped in the sticky resin. The insects' bodies, except for being dried out, are preserved entirely. By far the most abundant amber-preserved insects are flies, mosquitoes, and gnats of the Order Diptera. Other things found in amber include spiders, crustaceans, snails, mammal hair, feathers, small lizards, and even a frog.

Paleontologists deplore the fact that, during the late 1800s, thousands of tons of raw amber were melted to make varnish, causing the loss of untold numbers of exquisitely preserved organisms.

Amber is a fossil resin produced by conifers. It is initially exuded from cracks and wounds in trees. While it is still soft and sticky, it traps and engulfs insects. An insect's struggle to escape often can be recognized in the swirl patterns around appendages. Later, through evaporation of its more volatile components, the soft resin hardens.

Although amber deposits are known in rocks from the Lower Cretaceous (135 million years ago) to the Holocene (near present), the most famous are those found along the coast of the Baltic Sea near the seaport of Kaliningrad, in Russia. In early Oligocene time (around 30 million years ago), the conifer forests in this area were inundated during a marine transgression. Marine sediment buried the trees and their contained amber. Subsequently, chunks of amber that eroded from the soft clays were distributed across adjacent areas by streams, wave action, and glaciers.

Intrigued by their beauty, neolithic families are known to have gathered the rounded yellow and brown pieces of amber. Such prominent philosophers of antiquity as Aristotle, Pliny, and Tacitus described amber's physical and chemical properties.

As a semiprecious gem, amber was transported along ancient trade routes from the Baltic to the Mediterranean. Traders called amber "the gold from the north." Then as now, it was used in the manufacture of beads for jewelry. Amber beads, amulets, necklaces, and bracelets have been recovered from Etruscan tombs and excavations in Mycenae, Egypt, and Rome. However, for those who wear amber today, note that the person gazing intently at your jewelry could be a paleontologist seeking an embalmed mosquito!

Reference

Poinar, G. O., Jr. 1992. *Life in Amber*. Stanford, CA: Stanford University Press.

• A striking example of soft-tissue preservation was discovered in England, when an excavating machine uncovered the body of a 2000-year-old human. Named Lindow Man (Fig. 6-9), this unfortunate fellow had been ritually slaughtered (possibly by Druids), stripped, garroted, and thrown into a bog. The chemistry of the bog's waters served as a natural pickling agent.

Sometimes Only Traces Remain

Evidence of ancient life does not consist solely of petrifactions, molds, and casts. In the case of animals, sometimes a paleontologist can obtain clues to a creature's appearance and how it lived by examining tracks, burrows, borings, and trails. Such markings are called **trace fossils** (Fig. 6-10).

The trackways of ancient vertebrate animals provide a surprising amount of information (Fig. 6-11). Tracks may indicate whether it was bipedal (walked on two legs), quadrupedal (walked on four), digitigrade (walked up on its toes, like a cat) or plantigrade (walked on the flats of its feet, like us), whether it had a long or short body, whether it was lightly built or ponderous, and sometimes whether it was aquatic (with webbed toes) or possibly a flesh-eating predator (with sharp claws).

FIGURE 6-8 **Baby mammoth dug from permafrost (permanently frozen soil) in northeastern Siberia.** The mammoth stood about 104 cm (3 feet) tall at the shoulders, was covered with reddish hair, and was probably only several months old. Radiocarbon dating indicates it died 44,000 years ago.

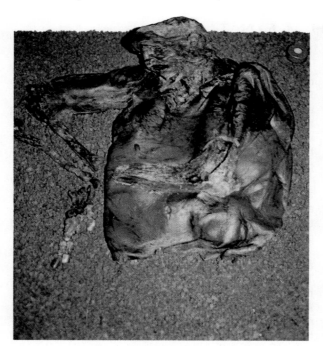

FIGURE 6-9 **Preserved torso, arms, and head of the 2000-year-old Lindow Man.** This example of preservation of soft tissue was found in a peat bog in 1984 at Lindow Moss, England. The lower half of the body was destroyed by an excavating machine. ◨ *The wet sediment of peat bogs is usually deficient in oxygen content. How might this have enhanced preservation of this soft tissue?*

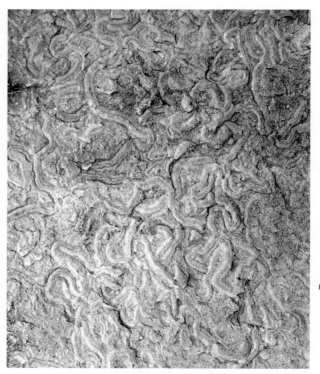

FIGURE 6-10 **Trace fossils of probable annelid worm trails.** (An example of an annelid worm is the familiar earthworm). These trace fossils are in Upper Carboniferous siltstones (about 300 million years old) that form the Cliffs of Moher along Ireland's southwest coast. ◨ *Would this rock be useful for determining grain orientation and the direction of ancient ocean currents?*

Trace fossils of invertebrate animals are more common than vertebrate tracks. They also are useful indicators of the habits of ancient creatures. From traces like those sketched in Figure 6-12, we can sometimes determine if the trace-maker was crawling, resting, grazing, feeding, or living within a relatively permanent dwelling.

Why Are Some Fossils Rare, Yet Others Abundant?

The fossil record of life is very incomplete. If it were a total record, we would have information on every species since life began and for every place on Earth. Clearly, this is unattainable. Only a limited number of animals and plants have been preserved. Many that were preserved remain hidden, not yet exposed by erosion, drilling, or excavation. Among those that are hidden, it is highly probable that some are of species we have not yet discovered.

The fossil record is more complete for marine life that possessed hard external skeletons and for spore and pollen grains, which have highly resistant coverings. The record is less complete for "soft" land life that lacked bone or shell (such as worms and insects). Where burial is rapid and dead organisms are protected from scavengers and agents of decay, the probability of preservation improves.

As we said earlier, favorable places for fossil preservation include the ocean floor and low-lying land areas where deposition predominates. For example, many dinosaur remains have been discovered in the sands and clays deposited by streams that flowed across low plains that lay east of the Rocky Mountains during the Cretaceous Period (145 to 65.5 million years ago). Animals and plants certainly lived in nearby highlands as well, but in these places erosion predominates—not deposition—so fossils of local inhabitants are likely to have been destroyed.

Despite all of the factors that discourage fossilization, or destroy fossils before we find them, the fossil record is sufficient to work with. Based on the ten-thousands of species known from fossils, paleontologists have been able to piece together an amazingly comprehensive history of past life on Earth.

▶ FIGURING OUT HOW LIFE IS ORGANIZED

Linnaeus Leads the Way

Because of the vast number of animal and plant species, both living and fossil, random naming would be confusing. Realizing this, Swedish naturalist Carl von

FIGURE 6-11 **Dinosaur trackways.** The three-toed imprints indicate the passage of a biped, whose tracks are crossed by a quadruped with larger rear feet than front. (This feature is typical of many quadruped dinosaurs.) Claw imprints on the biped suggest that it was a predator. ◄ *What indicates that the quadruped crossed the area after the biped?*

Linné (1707–1778) formulated a systematic method for naming living things. Linné is better known by his Latinized name, Carolus Linnaeus (lin-NEE-us). He proposed what we now call the Linnaean system, which uses observable traits to classify life forms.

In the Linnaean classification, for example, all animals having a vertebral column (which is an observable trait) are placed together as vertebrates. A subgroup of vertebrates that are warm-blooded, have skin with hair, and nourish their young from milk-secreting glands are mammals. Mammals that have an opposable thumb (a thumb that can be touched to other digits on the same hand), flat nails rather than claws, and a well-developed collar bone are primates.

To name species, Linnaeus used "binomial nomenclature," a fancy way of saying that he used two-part names like our own, *Homo sapiens*. In this scheme, the first name is the **genus**, which designates a group of animals or plants that appear to be related because of their general similarity. The second name is the species. Here is an example: *Felis* is the genus for all cat-like animals. There are many kinds of cats, so the species (second name) designates a physically distinct and restricted group under the genus *Felis*. Thus, *Felis domesticus* is the common house cat, *Felis leo* is the African lion, *Felis onca* is the jaguar, and *Felis thomas* is the tomcat.

To avoid confusion, there must be consistency in how organisms are formally named. Thus, no two *genera* (plural of genus) within any of the six kingdoms have the same name. Further, names are in Latin (or are Latinized, like Linné was Latinized to *Linnaeus*), and they are written in *italics*.

What Is a Species?

A **species** is a group of organisms that have structural, functional, and developmental similarities and that are able to interbreed and produce fertile offspring. The species is the fundamental unit in biologic classification.

Although individuals of a species are *generally* similar, they are not identical. They exhibit variation—consider *Homo sapiens*, with our variety of shapes and heights, as well as hair, skin, and eye colors. Because of such variation, it isn't possible for the description of a single individual to include the range of variations present in a species. Much individual variation exists, including differences between sexes or between juveniles and adults

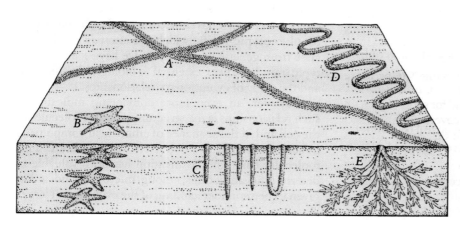

FIGURE 6-12 **Trace fossils reflect animal behavior.** (A) crawling traces, (B) resting traces, (C) dwelling traces, (D) grazing traces, and (E) feeding traces.

(for example, the juvenile monarch caterpillar and the adult monarch butterfly are the same species, just at different growth stages). If "species" did not encompass such variations, we might apply multiple names to different members of the same species.

Do biologists have an advantage over paleontologists when giving a species name to a group of similar organisms? After all, biologists work with living organisms, so they can observe whether different varieties within a group can interbreed, and even examine their DNA. But the breeding habits and DNA of many living animals are unknown. Thus, for biologists and paleontologists alike, the recognition of species relies heavily on physical traits that are constant within the group. They must determine the range of variation and decide how much variation can exist within a proposed species.

What Is Taxonomy?

Taxonomy is the naming and grouping of organisms. The word comes from Greek roots that mean "arrangement method." In a taxonomic classification, species is the basic unit. We arrange the categories of living things in a hierarchy that expresses levels of kinship. Here is the taxonomy widely used in biology, with the broadest category at the top:

- A **domain** is the highest taxonomic level in the classification of life. The three domains are Archaea, Bacteria, and Eukarya.
- A **kingdom** is a large group of related phyla (for example, all the animals). There are six kingdoms.
- A **phylum** (plural: *phyla*) is a group of related classes (like all animals with backbones).
- A **class** is a group of related orders.
- An **order** is a group of related families.
- A **family** is a group of related genera.
- A **genus** (plural: *genera*) is a group of species that have close ancestral relationships.
- A **species** is a group of organisms that have structural, functional, and developmental similarities and that are able to interbreed and produce fertile offspring.

To use yourself as an example, you are a member of:

Domain: Eukarya (organisms having one or more cells that contain visible nuclei)

Kingdom: Animalia (animals)

Phylum: Chordata (animals that have a backbone)

Class: Mammalia (warm-blooded animals that have fur and suckle their young)

Order: Primates (mammals, comprising humans, apes, monkeys)

Family: Hominidae (modern humans and our immediate ancestors)

Genus: *Homo* (human)

Species: *sapiens* (wise)

SIX KINGDOMS Our traditional classification system places all living things in six kingdoms, as shown in Figure 6-13:

Superkingdom Eukarya (single- or multi-celled organisms that have a distinct cell nucleus and cell membrane):

1. **Animalia**—animals: multicellular heterotrophic organisms. Heterotrophs cannot generate their own food and so must consume other animals or plants.
2. **Plantae**—plants: multicellular eukaryotes that typically live on land and undergo embryonic development. They are autotrophs—they generate their own food through photosynthesis.
3. **Fungi**—multicellular eukaryotes, many of which are decomposers (saprophytes) that absorb nutrients from dead organisms or live as parasites on plants.
4. **Protista**—protists, mostly single-celled, animal-like organisms that devour food for energy (heterotrophs), plant-like photosynthesizers (autotrophs), and decomposers (saprophytes).

Superkingdom Prokaryota (single-celled micro-organisms that lack a distinct nucleus and cell membrane):

5. **Archaebacteria**—include the only methane-producing bacteria and bacteria capable of living under extreme conditions of temperature or salinity, called "**extremophiles.**"
6. **Eubacteria**—prokaryotes that live in water or soil or inside larger organisms like us. They include cyanobacteria, formerly called blue-green algae. Because they release oxygen during photosynthesis, cyanobacteria caused dramatic evolutionary change in Earth's early, oxygenless atmosphere.

Although the six-kingdom system is widely used in biology textbooks, it has shortcomings. It is based largely on **phenotypic traits**, which are *observable* traits that arise from genetic processes. As we have seen, phenotypic traits have been essential to biologic classification and the study of evolutionary relationships.

However, an even more convincing indicator of such relationships is the structure of large molecules in the cell and in the sequence of amino acid components in such molecules as ribonucleic acid, or RNA. This molecular basis for classification has been valuable in determining affiliations among microbes, and it has given us better understanding of the evolution of larger

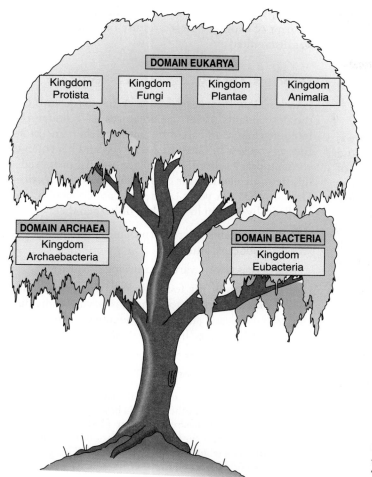

DOMAIN EUKARYA

Kingdom Protista | Kingdom Fungi | Kingdom Plantae | Kingdom Animalia

DOMAIN ARCHAEA
Kingdom Archaebacteria

DOMAIN BACTERIA
Kingdom Eubacteria

FIGURE 6-13 **The two highest levels of our taxonomic system are domains and kingdoms.** There are three domains and six kingdoms.

organisms as well. Molecular studies indicate that superficially dissimilar groups such as plants, protists, animals, and fungi are, in reality, closely related.

To more validly show evolutionary relationships as revealed by molecular sequencing studies, it has been proposed that all life be divided into three great divisions termed **domains** (see Fig. 6-13):

- **Domain Archaea** (equivalent to the kingdom Archaebacteria) includes methane-producing bacteria and an interesting group of heat-loving microbes called thermophiles that populate hydrothermal vent systems on the ocean floor (Fig. 6-14).

- **Domain Bacteria** includes many organisms, such as cyanobacteria, along with purple sulfur bacteria, and many common non-photosynthetic groups of bacteria.

- **Domain Eukarya** includes protists, fungi, plants, and animals, because of their molecular similarities.

From the above, you can see that taxonomy is more than a mere system for cataloguing organisms. Because

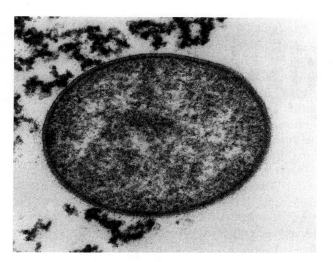

FIGURE 6-14 **Transmission electron micrograph of an organism belonging to the domain Archaea.** This microbe is a thermophile (tolerant of extreme temperatures). It was ejected from a thermal vent on the Pacific Ocean floor along the Gorda oceanic ridge, west of Oregon and Washington. The long axis of the cell is approximately one micron.

it is based on organic structures, embryologic development, and molecular biology, it reflects the broad outlines of evolutionary relationships. In a sense, the classification is a blueprint for constructing the tree of life.

EVOLUTION: CONTINUOUS CHANGES IN LIFE

The Roman poet Ovid (43 BC–17 AD) wrote, "there is nothing constant in the universe, all ebb and flow, and every shape that's born bears in its womb the seeds of change." This thoughtful phrase expresses how life has changed through time, as revealed by the fossil record, with rates of change from startlingly sudden to very gradual. Through untold generations, older forms have evolved into newer forms better able to cope with changes in their environment. This is not to say that the early, simpler forms of life are gone, for many persist along with their often more complex contemporaries.

Lamarck's Flawed Hypothesis

In the early 1800s, the great French naturalist Jean Baptiste de Lamarck (1744–1829) recognized lines of descent from older fossils to more recent ones, and ultimately to currently living forms. From such observations, he stipulated correctly that all species, including humans, are descended from other species. However, Lamarck mistakenly assumed that new structures in an organism appear because of a need or "inner want" of the organism, and that structures once acquired in this way during the lifespan of the organism are inherited by later generations. In a similar but reverse fashion, little-used structures would disappear in succeeding generations.

For example, Lamarck hypothesized that giraffes acquired longer necks by stretching to reach leaves higher in trees than could be reached by other leaf-eaters (Fig. 6-15). He also proposed that snakes evolved from lizards that had a strong preference for slithering on the ground. Because of this "need" to slither, certain lizards developed long, thin bodies, and because legs were less useful in crawling, these structures gradually disappeared.

Lamarck's ideas were challenged almost immediately. There was no way to prove by experimentation that such a thing as "inner want" existed. More importantly, Lamarck's idea that traits acquired during the lifespan of an individual could then be inherited was tested and shown to be invalid. For example, you can acquire a deep suntan, but this *acquired* trait will not result in your children being born with suntans. Circumcision of newborn males has been practiced for over 4000 years, but this *acquired* alteration does not appear in newborn males today.

There is no mechanism through which body cells can forward characteristics to reproductive cells and on to the next generation. Because of this, the Lamarckian concept of evolution based on use or disuse of organs was discredited.

Darwin's Theory of Natural Selection

Charles Darwin praised Lamarck for perceiving that the descendants of former creatures have undergone biologic change and are different from their ancestors. But, as you saw in Chapter 2, Darwin and his younger contemporary Alfred R. Wallace jointly proposed a different mechanism for evolution: **natural selection**. Darwin, through years of careful data collection, and Wallace, by sudden insight, had discerned that:

- More organisms are born than survive to become reproductive adults
- There is always variation among offspring
- Competition for food, shelter, living space, and sexual partners among species having individual variations and surplus reproductive capacity consistently results in survival of the more fit and elimination of the less fit.

In Darwin's own words,

Can we doubt . . . that individuals having any advantage, however slight, over others, would have the best chance of surviving and of procreating their kind? On the other hand, we may feel sure that any variation in the least degree injurious would be rigidly destroyed. This preservation of favorable variations, I call Natural Selection.

Darwin's book, *On the Origin of Species by Means of Natural Selection, or the Preservation of Favoured Races in the Struggle for Life* was published in 1859. It appeared when at least part of the European intellectual atmosphere was more liberal and less satisfied with the theological doctrine that every species was independently created. Although the ideas of Darwin and Wallace continued to disturb the religious feelings of some, they nevertheless increasingly acquired adherents among nineteenth- and twentieth-century scientists.

In fact, Darwin did not consider his views impious. He saw a:

grandeur in this view of life with its several powers, having been originally breathed by the Creator into a few forms or into one; and that . . . from so simple a beginning endless forms most beautiful and most wonderful have been and are being evolved.

Inheritance, Genes, and DNA

Darwin's theory had a weakness. It did not explain the *reason* that offspring exhibit variability. Decades later, the cause was discovered through elegant experiments on garden peas, conducted by a Moravian monk named J. Gregor Mendel (1822–1884).

LAMARCK'S HYPOTHESIS

Short-necked ancestor

Stretching

Necks lengthened by stretching occur in offspring

NATURAL SELECTION

Short-necked ancestors with some neck-length variation

Longer-necked variants obtain food easier, live longer, reproduce more offspring having longer-neck genetic trait

Favorable variation spreads

FIGURE 6-15 **Lamarck's hypothesis versus the natural selection hypothesis.** Lamarck thought that giraffes acquired their long necks by stretching to reach higher foliage, and that they passed this *acquired trait* (stretched necks) on to their offspring. The natural selection hypothesis proposes that longer-necked variants survived better and thus passed along their long-necked genetics, whereas shorter-necked giraffes eventually died out.

Mendel discovered the basic principles of inheritance. His findings, printed in 1865 in an obscure journal, were unknown to Darwin and unheeded by the scientific community until 1900, when his article was rediscovered. Mendel described the mechanism by which traits are transmitted from adults to offspring. In his experiments with garden peas, he demonstrated that heredity in plants is determined by what we now call **genes**. Genes divide in the pollen and ovules and are recombined in specific ways during fertilization. Genes link to form larger units termed **chromosomes**. Mendel's work earned him the posthumous title of "Father of Genetics."

We now know that genes are chemical units or segments of **deoxyribonucleic acid (DNA)**. As indicated by chemical and X-ray studies, the DNA molecule consists of two parallel strands, twisted somewhat like the handrails of a spiral staircase (Fig. 6-16). The twisted strands are phosphate and sugar compounds, linked with cross-members composed of specific nitrogenous

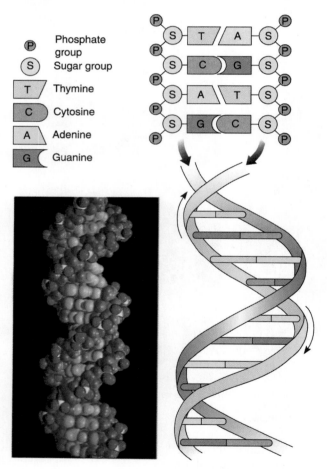

FIGURE 6-16 Portions of the deoxyribonucleic acid (DNA) molecule. At the lower left is a computer reconstruction. The drawing depicts the twisted, double-stranded helix, with side rails composed of alternating molecules of sugar (deoxyribose) and phosphate. Each rung of the twisted ladder is a pair of nitrogenous bases.

bases. Genes are the parts of the DNA molecule that transmit hereditary traits.

The importance of DNA is evident when we realize that it indirectly controls the production of proteins. Proteins are the essential components of many basic structures and organs. Proteins called enzymes even regulate the activities of organisms. Without DNA and its controls, there would be no life as we know it. DNA's ability to replicate itself precisely is the basis for heredity, so organic evolution ultimately depends on the remarkable DNA molecule.

Cell Division and Reproduction: Bringing Variety to Offspring

Reproduction of an organism may be sexual or asexual, or it may alternate between sexual and asexual methods. However, all reproductive methods involve cell division. In sexual reproduction, there is a union of reproductive (sex) cells from separate individuals, whereas in asexual reproduction, cells do not unite. Many single-celled organisms reproduce asexually simply by dividing to form genetically identical daughter organisms. Amoebas reproduce in this way.

Organisms that reproduce asexually have the advantage of increasing their numbers rapidly whenever conditions are favorable. However, one disadvantage is that their offspring develop less genetic variability than sexually reproducing organisms. Because there is no mixing of genes form two parents, they depend largely on mutation for their limited variability. To understand the reasons for this variability, you need some basic facts about chromosomes and their behavior during the development of reproductive cells and during fertilization.

Initially, the kind and number of chromosomes are constant for a single species but differ between different species. Chromosomes are located in the nucleus of the cells and occur in duplicate pairs. Thus, each chromosome has a duplicate mate. In humans, for example, there are 46 chromosomes, or 23 pairs. Cells with paired homologous chromosomes are designated **diploid cells**.

Mitosis. All living things constantly generate new cells to grow and to replace worn-out or damaged cells. In asexual organisms, and in all the body cells of sexual organisms, cell division produces new diploid cells with exact replicas of the chromosomal components of the parent cells. This process is called **mitosis** (Fig. 6-17A).

Meiosis. In most sexual organisms, **meiosis** takes place when **gametes** (egg cells or sperm cells) are formed. Meiosis may occur in unicellular organisms. In multicellular organisms, meiosis occurs only in specialized reproductive organs—testes or ovaries. Meiosis consists of two quickly succeeding divisions, resulting in four daughter cells termed haploid because they *do not* have paired chromosomes. The **haploid cells** are the gametes, or reproductive cells.

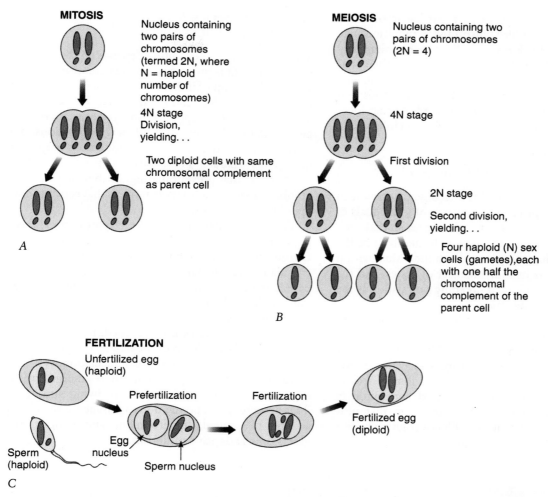

FIGURE 6-17   **Major features of (A) mitosis, (B) meiosis, and (C) fertilization.**

When two gametes meet during sexual reproduction, the sperm enters the egg to form a single cell, which can now be called the fertilized egg. Because two gametes have been combined into one cell, there is now a full complement of chromosomes and genes representing a mix from both parents. This is one source of variety in offspring, of brothers and sisters having different heights, eye color, etc. The fertilized egg now begins a process of growth by mitotic cell division, which eventually leads to a complete organism.

The organism that develops from the union of haploid cells will already vary somewhat from its parents. However, there is yet another aspect of meiosis that produces variation in offspring. During the initial chromosome division, while the chromosomes are still paired, they may break at corresponding places and exchange their severed segments in a process called **crossing over**. The result is an additional mixing of genes.

Sexual reproduction provides variation by means of genetic recombination. In addition, sexually reproducing organisms accumulate considerable "invisible" variation in the form of recessive genes. As environmental changes occur, some of these recessive genes will produce traits in the animal or plant that have survival advantages, and new avenues of evolution will be opened.

Mutation: Source of Variations

Some of the variation we see among individuals of the same species results from gene mixing that occurs during reproduction. However, if genes were never altered, then the number of variations they could produce would be limited. For truly new variation to arise, a process is required that will change the genes themselves. This process is known as **mutation**.

Mutation can be caused naturally by ultraviolet light, by cosmic and gamma rays, or by certain chemicals. It also may occur spontaneously without a specific causative agent. Mutations may occur in any cell, but their evolutionary impact is greater when they occur in the sex cells, because future generations will be affected.

To understand how mutations occur, look again at the configuration of DNA (Fig. 6-16). Note the two-piece "rungs" on the twisted DNA ladder. Genes owe their specific characteristics to a particular order of these rungs. If the order is disrupted, a mutation may result. For example, during cell division, a mishap may occur when the twisted strands separate and then proceed to attract bases to rebuild the DNA molecule. One type of pair may end up occupying the position that another type should have occupied. At that location, the gene is altered and may result in an inheritable change.

Species, Population, and Gene Pool

Evolution is a process of biologic change that occurs in populations. A **population** is a group of individuals of the same species that occupy a given area so that each individual has a chance to mate with members of the opposite sex within the group. Each individual within the population has its individual set of genes, and the sum of all of these within a breeding population is the **gene pool** of that population.

The gene pool is partitioned in each generation and given to new offspring. In each successive generation, new mutations and genetic combinations manifest themselves in the offspring. As natural selection comes into play, some of these traits (such as a longer neck in giraffes) will be passed on to the next generation in greater or lesser numbers than others, so that ultimately, the gene pool is altered. Thus, evolution results from the impact of natural selection on the gene pool.

How New Species Arise and Adaptively Radiate

The entire course of evolution depends on the origin of new species. This process is called **speciation**. A *population* is the group of individuals in a species that occur together at one time and at one place. There is a free flow of genes within a population. However, there is no gene exchange between two *different* populations, because barriers between different populations keep their gene pools separate. These barriers may be reproductive barriers that prevent mating, or they may be geographic barriers, such as ocean tracts that separate animals living on islands.

Wherever reproductive or geographic barriers exist, populations become isolated for many generations. During this period, each isolated species is likely to accumulate enough genetic differences so that interbreeding between the segments eventually becomes impossible. When this occurs, the altered, isolated populations become different species.

Once a new species is established, pioneering segments around the fringes of the habitats may, like their parent species, undergo additional speciation. With innumerable successive speciations, there emerge diverse organisms characterized by diverse living strategies. This branching of a population to produce descendants adapted to particular environments and living strategies is called **adaptive radiation**.

FIGURE 6-18 **The honeycreepers of Hawaii are a fine example of adaptive radiation.** When the ancestors of today's honeycreepers first reached Hawaii, few birds were present. Succeeding generations diversified to occupy various ecologic niches. Their diversity is apparent in the way their beaks have become adapted to different diets. Some have curved bills to extract nectar from tubular flowers; others have short, sturdy beaks for cracking open seeds; still others have pointed bills for seeking insects in tree bark. The bird at the lower right has a relatively unspecialized beak and appears to be similar to the original honeycreeper ancestor.

Hawaiian birds called honeycreepers illustrate adaptive radiation. These birds, comprising many different species, are believed to have descended from a common ancestor. The most striking differences among the species occur in the sizes and shapes of their beaks, which are adaptively related to what they eat (Fig. 6-18). Some have stout beaks for crushing hard seeds, and others have long beaks adapted to probing for insects in crevices or for sucking nectar. These are examples of **adaptations**, a word that means the acquisition of beneficial characteristics that are inheritable.

The catalog of known adaptations is immense. We see them in every creature, including ourselves. An elephant's trunk, a camel's hump, a giraffe's long neck, and a whale's flipper are all adaptations. Here are a few examples of adaptations observed in fossils:

- The broad, spiny shell of the fossil brachiopod *Marginifera ornata* (Fig. 6-19) is an adaptation for providing support and anchorage in a seabed of soft mud.
- The trilobite *Asaphia kowalewskii* (Fig. 6-20) also had an adaptation for living in the soft ooze of the seafloor. The eyes of this ancient arthropod were at the tips of long stalks, like twin periscopes. The animal was able to "see" what existed above while it was tunneling through sediment.
- Reduction of the skeleton to mere spines in the Silurian trilobite *Deiphon* (Fig. 6-21) was probably an adaptation to provide buoyancy for an animal that fed near the surface of the sea.
- Coiling, a trait found in both living and fossil cephalopods (Fig. 12-36), appears to be an adaptation that brings the animal's center of buoyancy higher than its center of gravity. This permits these creatures to swim level without tilting, tipping over, or flipping upside-down.

FIGURE 6-20   **Ordovician-age Trilobite (435–500 million years old).** Its eyes were on long stalks that protruded above the sediment sometimes covering the animal. This specimen, *Asaphia kowalewskii*, is about 3 cm long and was collected near St. Petersburg, Russia.

We have been looking at adaptations in form, but biochemical and physiologic adaptations occur as well. They are more difficult to recognize in fossils.

Above the species level are adaptive radiations among classes, orders, and families. For example, all orders of mammals had a common origin in a single ancestral species from the early Mesozoic (around 250 million years ago). By adapting to different ways of life, the descendant forms eventually diverged more and more from the ancestral stock, thereby providing today's rich

FIGURE 6-19   **Spinose brachiopod *Marginifera ornata*.** The shell has been replaced with silica. The enclosing limestone has been dissolved, providing an excellently preserved fossil with its delicate spines intact. The animal rested on its ventral (bottom) valve, and its spines provided anchorage in the seafloor.

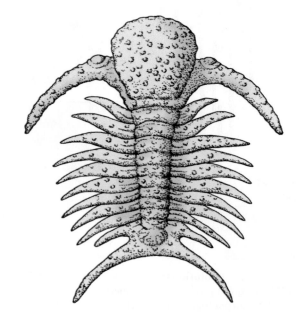

FIGURE 6-21   **Silurian trilobite (410–435 million years old).** *Deiphon*'s extreme spinosity suggests it was a swimmer-floater rather than a bottom-dweller. Spinosity increases surface area without adding weight, thus enhancing buoyancy. This specimen is about 27 mm long.

diversity of mammalian plant eaters, flesh eaters, insect eaters, walkers, climbers, swimmers, and flyers.

Gradual or Sporadic Evolution?

Does evolution proceed gradually in an infinite number of small steps, as Darwin proposed? Or are there sudden sporadic advances? Or both? These are questions widely discussed by paleontologists.

Punctuated equilibrium. Some argue that the fossil record for the past 3 billion years contains many examples of new groups that appear suddenly, and with no disruption in the geologic record. Indeed, they find that the slow and stately advance of evolution is the exception,

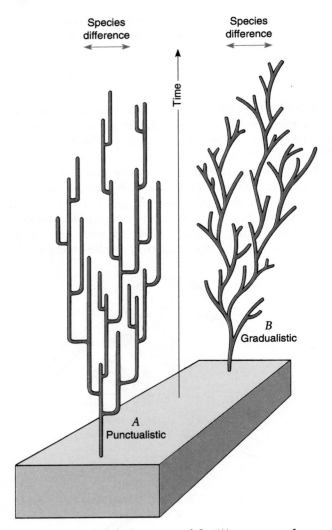

FIGURE 6-22 **Evolutionary models: (A) punctuated equilibrium, (B) phyletic gradualism.** Time is depicted vertically; morphologic change is shown horizontally. The short horizontal side branches of the punctuated equilibrium model depict abrupt change, whereas the inclined branches of the gradualism model suggest slow and uniform change through time.

and that evolutionary progress is more often sporadic. The term punctuated equilibrium was introduced to describe evolution that progresses by sudden advances that "punctuate" long intervals of little change (Fig. 6-22A).

Phyletic gradualism. This is what paleontologists call gradual, progressive change. (*Phyletic* refers to the evolutionary history of an organism.) Like Darwin, supporters of phyletic gradualism think that change occurs by slow degrees along the evolutionary pathway of a lineage (Fig. 6-22B). Where breaks in the progression appear, they result from (1) fossils of intermediate or transitional forms not yet discovered, (2) fossils missing because of erosional loss, or (3) non-deposition of strata able to preserve fossils.

Evidence for phyletic gradualism exists in several invertebrate groups. These include lineages of well-preserved trilobites from the Ordovician in Wales as well as marine planktonic organisms known as foraminifera. Fossil foraminifera provide particularly good evidence for gradualism, for often they occur in complete, well-dated sections of deep-sea cores.

The sudden change (punctuation) that interrupts equilibrium occurs at the periphery of the geographic area occupied by the population. Genetic change is rapid among these individuals, and the gene pool is small. This results in rapid changes in form or physiology, which lead to new species.

Species do not usually originate where their parental stock exists, but rather in boundary zones where variations can be selected against new environmental situations. Should the parent species suffer extinction or severely decline, the new species may either move back into the parental domain or expand into new territory.

When adaptations offer a significant advantage, adaptive radiation and evolution will advance rapidly. This stage of rapid evolution is usually followed by a period of slow evolution and stability that lasts until the population begins to decline. In time, a new episode of rapid speciation will occur.

Phylogeny: Depicting How Ancestors Relate to Their Descendants

Phylogeny refers to the historical development of groups of organisms. Phylogeny can be depicted on a tree or bushlike diagram called a **phylogenetic tree** (Fig. 6-23) or as a **cladogram** (KLAD-o-gram) (Fig. 6-24).

In a phylogenetic tree, the most recently evolved species or groups are on the upper branches and older, ancestral species are on the lower branches and trunk. Thus, the tree depicts change through time.

To construct a cladogram, organisms must be examined objectively for characteristics they share to determine their ancestor-descendant relationships. A cladogram shows closeness of relationship by the arrangement of groups. The shorter the links between groups, the closer the evolutionary relationship.

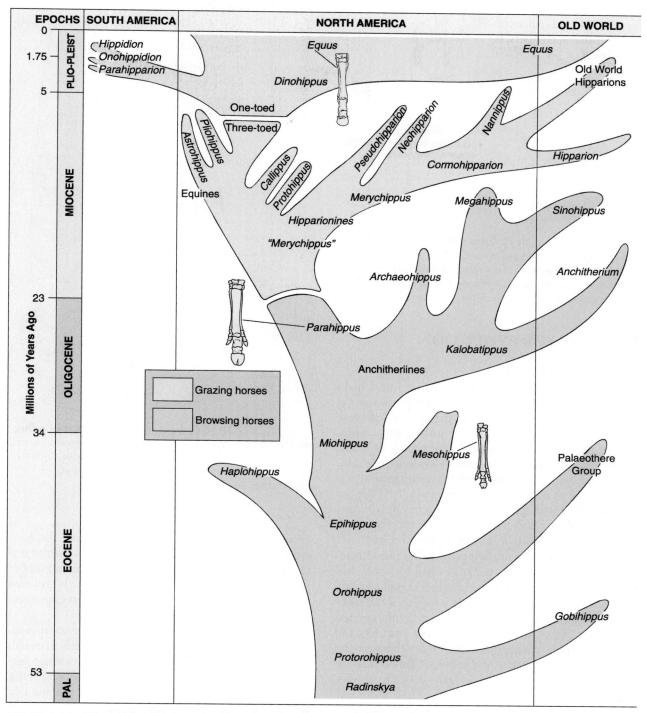

FIGURE 6-23 **The phylogenetic tree of horses.** The evolutionary relationships between genera show the trend from toes to hooves and the transition from browsing to grazing horses.

In the very simple cladogram in Figure 6-24, the camel, tuna, and dolphin share the relatively primitive trait of a spinal column. Although the dolphin and tuna superficially resemble one another, they are not placed close together because only the dolphin and camel share advanced characteristics, such as live birth, off-spring nourished by milk from the female, and endothermy (warm-bloodedness).

Unlike the phylogenetic tree, a cladogram does not directly incorporate information about the time ranges of organisms. Its only purpose is to show how organisms are related.

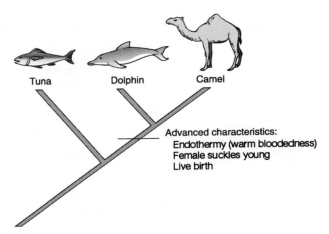

FIGURE 6-24 **A simple cladogram (*KLAD-o-gram*).**
A cladogram shows closeness of relationship by the
arrangement of groups. ◀ *Draw a cladogram depicting reptiles,
mammals, amphibians, and birds. Where along the inclined line do
the advanced characteristics of body hair, lungs, and claws appear?*

THE CASE FOR EVOLUTION

Evidence from Paleontology

During the 1800s, scientists supporting evolution had a
difficult time convincing some of their contemporaries
that Darwin's theory was valid. In part, this was because
our human life span is much too short to witness evolutionary changes across multiple generations of plants
and animals. Fortunately, we can see evidence of evolution by examining the fossil remains of organisms preserved in rocks of successively younger age. If life has
evolved, the fossils preserved in successively younger
rocks should exhibit those changes. Indeed, thousands
of examples are known of such sequential changes.

The most famous example is provided by fossil horses
of the Cenozoic Era (Fig. 6-23). A small browsing animal from the Upper Paleocene (around 55 million years
ago) named *Radinskya* is presently considered the earliest relative of the horses. Unlike modern hoofed species,
the earliest horses had four toes on their front feet and
three on the rear. You can see evolutionary change in the
many branching lineages of the horse "family tree" in
successively higher—hence, younger—rock formations.
The animals grow larger and taller, reduce their side
toes while enlarging the middle toe, develop more complex teeth, and develop deeper, longer skulls.

The change in horse teeth through time exemplifies
the important link between environment and natural
selection. Grass contains silica and is coated with abrasive silica dust because it is close to the ground. Thus, a
diet of grass wears the teeth of plant eaters. Recall the
Mohs hardness scale, on which quartz (silica) = 7, and
apatite (teeth) = 5. Which gets scratched and worn, the
silica or the teeth?

Early horse family members lived in forests. They
browsed on trees and shrubs that were higher above

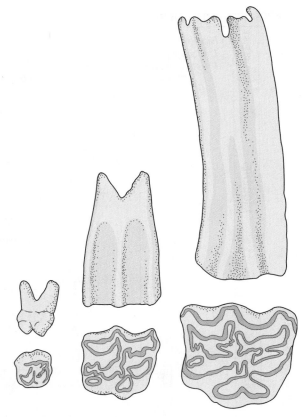

FIGURE 6-25 **Development of high-crowned grinding
molars in horses.** Oral surfaces show complex patterns of
hard enamel (orange) which stand as ridges above the softer
dentine and cement. From left to right, *Radinskya*,
Merychippus, and the modern horse *Equus*.

ground, and less abrasive. Thus, they had the low-
crowned teeth typical of leaf eaters. Low-crowned
teeth do no project very much beyond the gums. But
later horse family members evolved the high-crowned
teeth with complex patterns of enamel that characterize grass-grazing mammals today (Fig. 6-25).

Complicated patterns of hard enamel are set in a matrix of softer dentine and cement. The softer materials
wear away more rapidly than the enamel, leaving a surface with rough ridges that forms a highly effective
grinding surface. This adaptation favored the animals'
survival, because high-crowned teeth with self-renewing
grinding surfaces let them eat grass effectively, live
longer, and thus reproduce more progeny.

In turn, horse evolution probably also affected the
evolution of the predators that pursued them and may
even have contributed to the evolution of grasses that
were more resistant to damage by grazing. As the horse
family evolved, there is paleontologic evidence that
grasslands expanded in North America and Eurasia.
Evolution is an intricate process driven by natural selection, in which every animal and plant interacts with
its neighbors and the physical environment.

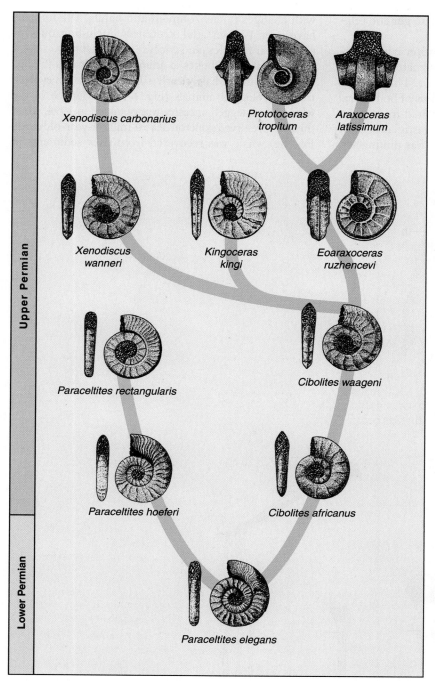

FIGURE 6-26 Evolutionary change in Permian ammonoid cephalopods. Two lineages originated from *Paraceltites elegans*. One terminated in *P. rectangularis*. The other produced *Cibolites waageni*, which became the ancestral stock for three additional lineages.

Although fossil remains of horses provide a fine illustration of paleontologic evidence for evolution, hundreds of other examples representing every major group of animals and plants would serve as well. For instance, the marine invertebrates known as cephalopods provide fine examples of progressive evolutionary changes (Fig. 6-26). Because of this, they are also exceptionally useful in correlation of rock strata.

Evidence from Biology

Biology provides many lines of evidence for evolution. In studies comparing the physiology of organisms, it is common to find body parts of similar origin, structure, and development, even though they may be adapted for different functions. For example, in seed plants, we observe leaves that are not only in the typical form of leaves, but in the form of petals, tendrils, and thorns. In four-limbed vertebrates, limb bones vary in size and shape, but they are fundamentally similar. They even occupy similar relative positions, as we see in birds, horses, whales, bats, amphibians, and humans (Fig. 6-27).

Such basically similar structures in superficially dissimilar organisms are **homologous** (Greek: agreeing). The differences in homologous structures result from variations and adaptations to particular environmental conditions, but their similarities indicate genetic relationship and common ancestry.

Another line of evidence for evolution is the existence of useless, usually size-reduced structures that in other related species are well developed. These **vestigial organs** are the "vestiges" or remains of body parts in earlier ancestral forms. Genes inherited from those ancestors continue to generate these vestigial organs, but their importance to the organism has diminished with changes in environment and habits. We humans have over 100 vestigial structures, including our appendix, ear muscles, and coccyx (tail vertebrae).

Vestigial pelvic bones in animals as diverse as the boa constrictor and whale clearly suggest that they evolved from four-legged animals (Fig. 6-28). Evidence of the whales' four-legged ancestry was obtained in 1994, when the well-preserved skeleton of a 50-million-year-old (early Eocene) whale was recovered from river sediments in

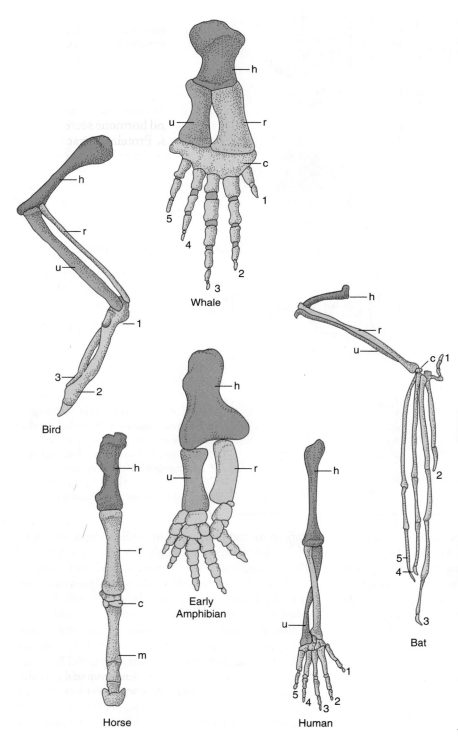

FIGURE 6-27 **Bones of the right forelimb from several vertebrates reveal similarity of structure.** h = humerus, r = radius, u = ulna, c = carpal, m = metacarpal, 1–5 = digits. At the center is the forelimb of the earliest four-legged land animal, an amphibian. Limbs are scaled to similar size for comparison. The color red is used for the humerus, blue for the ulna, and yellow for the radius. ◤ *Suggest a reason for the overlapping arrangement of the ulna and radius in the human forearm.*

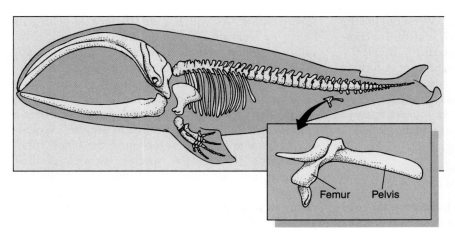

FIGURE 6-28 **The pelvis and femur (upper leg bone) of a whale are vestigial organs.** Vestigial organs exist because animals evolve and adapt to different modes of life. Selective pressure for eliminating such organs is often weak, so the vestige remains for relatively long periods.

Femur   Pelvis

Pakistan. *Ambulocetus* ("walking swimming whale") had front limbs developed as flippers and long hind limbs with elongate toes for the support of webbed feet.

Further evidence of evolution is found in comparative studies of the embryos of vertebrate animals (Fig. 6-29). In their early stages of development, embryos of fish, birds, and mammals are strikingly similar. It appears that all of these animals received a basic set of genes from a common ancestral lineage.

The universality of the laws of inheritance and the DNA molecular structure are other strong pieces of evidence for evolution. The DNA molecule contains a sequence of the nucleotide base pairs (the "steps" on the DNA "ladder"). When the sequences of different animal or plant groups are compared, we can specify the degree to which they are related. If two groups appear to be closely related by form, embryo, or the fossil record, we would predict them to have a greater percentage of DNA sequences in common than less closely related groups. This has proven true in hundreds of analyses.

Also, digestive enzymes and hormone secretions are similar in related organisms. Proteins extracted from corresponding tissues of closely related animals are strikingly similar. The antigenic reactions of blood from various groups of humans are practically identical to such reactions in the blood of anthropoid apes.

FOSSILS AND STRATIGRAPHY

How Do Fossils Reveal the Age of Strata?

In Chapter 2, we learned how William "Strata" Smith used fossils to identify and map stratified sedimentary rocks. Because of evolution, superposition, and the observation that extinct species do not reappear later, we use fossils to approximate the age of a rock unit and its

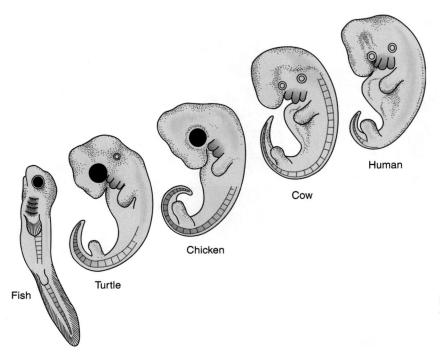

Human

Cow

Chicken

Turtle

Fish

FIGURE 6-29 **Embryos of different vertebrates.** They all share primitive features early in development, such as gills (red) and tails (blue).

position in the stratigraphic column. (We cannot use the inorganic characteristics of strata such as texture or composition, because identical characteristics frequently recur in the geologic column.) Further, rocks formed during the same age in identical environments, but in other locations, often contain similar fauna and flora. This lets geologists correlate different-looking strata of the same age from place to place all around the globe.

The Geologic Range of Fossils: From First Appearance to Last

Before fossils could be used as age indicators, the relative ages of major rock units were determined, using superposition. As you saw in Chapter 2, geologists began by working out superpositional sequences locally and added sections from other localities, assembling the segments into a composite geologic column.

The next step was to determine the fossil assemblage from each time-rock unit and to identify the various genera and species. This work began well over a century ago and is still very much in progress. Gradually, it became possible to recognize the oldest (first) appearance of particular species, as well as their youngest (last) occurrence. The interval between first and last appearance constitutes a species' **geologic range** in time. The geologic range of any ancient organism can be determined with reasonable precision by

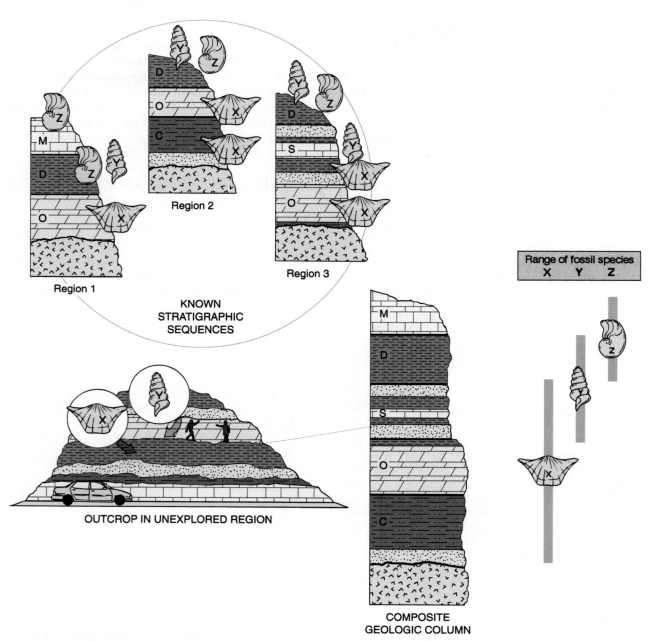

FIGURE 6-30 **Use of geologic ranges of fossils to identify time-rock units.** See text for explanation.

recording its occurrence in numerous stratigraphic sequences from hundreds of locations.

Using Fossils to Correlate Rock Units

The method for identifying time-rock units is illustrated in Figure 6-30:

1. Geologists in Region 1 come upon three time-rock systems of strata, designate them **O, D, M,** and sketch the column you see in the figure.
2. In Region 2, they find units **O** and **D,** but not **M.** They also recognize unit **C,** which is below **O** and hence older. They sketch the column shown.
3. In Region 3, they find new unit **S,** sandwiched between units **O** and **D.** They sketch the column shown.

So, the overall section consists of five time-rock units that decrease in age upward from **C** to **M.**

4. Geologists then draft a composite geologic column (lower right). On it, they plot the ranges of the three fossil species (**X, Y, Z**) in strata **C** to **M.**
5. Geologists next study an unexplored region (lower left). But they have trouble locating the position of this rock sequence in their composite column because the lithologies are different. However, they discover a bed containing species **X,** so at least they can say that the rock sequence might be **C, O,** or **S.**
6. Then they find fossil species **Y** in association with **X.** Now they can say that the outcrop in the unexplored region correlates in time with unit **S** in Region 3. In this way, section by section, rock strata are correlated worldwide.

The preceding illustration is greatly simplified, and future discoveries may extend known fossil ranges. Sometimes key fossils are missing, so it often is safer to use entire assemblages of fossils. Also, certain fossils may be geographically restricted.

For example, two or three million years from now, if we are still here, geologists might have difficulty using fossil evidence to firmly establish that the North American opossum, Australian wallaby, and African aardvark lived during the same range of geologic time. However, if they found *Homo sapiens* fossils with each of these animals, it would indicate their contemporaneity. In this example, we consider *Homo sapiens* to be the *cosmopolitan* species, for it is not restricted to any single geographic location within the terrestrial environment. But the opossum, aardvark, and wallaby are *endemic species,* being confined to specific areas.

In the case of fossilized marine animals, cosmopolitan species have been especially useful in establishing the *contemporaneity* of strata, whereas endemic species are generally good indicators of the *environment* where strata were deposited.

Endemic species may slowly migrate from one locality to another. For example, the peculiar screw-

A

B

FIGURE 6-31 **Two specimens of the guide fossil** *Archimedes.* Specimen (A) shows only the screwlike axis of this bryozoan animal. In specimen (B), you can see the fragile, lacy skeleton attached to the sharp, helical edge of the axis. This forms a netlike colony wound into an erect spiral. Specimens are 8 to 10 cm in height.

shaped marine bryozoan fossil known as *Archimedes* (Fig. 6-31) was endemic to central North America during the Mississippian Period. But it migrated steadily westward, finally reaching Nevada by Pennsylvanian time and Russia by the Permian Period. Some invertebrates that are attached to the seafloor as adults have mobile larvae, which often are widely distributed by ocean currents, facilitating migration of the species.

Pitfalls of Correlating with Fossils

Geologists are cautious when basing correlations on fossils. On one hand, correlation validity increases each

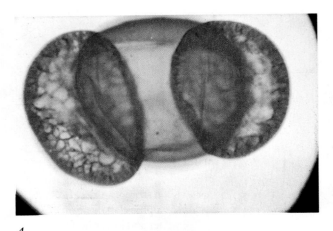

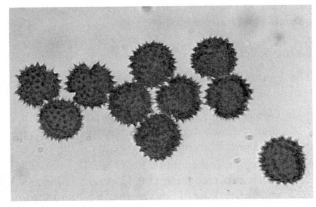

A   *B*

FIGURE 6-32   **Pollen grains.** (A) Pollen grain of a fir tree. The inflated bladders make the pollen grain more buoyant. The grain is about 125 microns in its longest dimension. (B) Pollen grains of ragweed. These grains are about 25 microns in diameter.

time the same sequence of faunal changes is found at different locations worldwide. But on the other hand, not all changes are caused by evolution. Instead, they may indicate that ancient environmental change drove faunal migrations or shifts in flora. A fossil's sudden disappearance need not mean that it became extinct, but rather that it simply moved. The earliest appearance of a fossil in rocks of a region might mean it evolved there, but it also could mean that a species from somewhere else migrated to this locality.

Paleontologists also must consider the possibility of **reworked fossils.** Weathering and erosion can free fossils from their host rock, to become part of new sediment that becomes lithified into a younger bed. Consequently, younger strata might be mistakenly assigned to an older geologic age, based on the fossil. Certain fossil types are particularly resistant to erosion and chemical decay and therefore are susceptible to such reworking. Such resistant fossils include spore and pollen grains (Fig. 6-32) and conodonts (Fig. 6-33). Conodonts are tiny, toothlike structures composed of calcium phosphate that range in age from Late Proterozoic (600 million years ago) to Triassic (200+ million years ago).

Index Fossils: Especially Useful

Index fossils or **guide fossils** are abundant, are widely dispersed, and lived during a relatively short interval of geologic time. Such fossils are helpful in identifying time-rock units and correlating them from area to area.

A guide fossil with a short geologic time range is more useful than one with a long range. For example, a species that lived during the total duration of a geologic *era* (say, all of the Paleozoic, encompassing 290 million years) would not be very useful in identifying rocks of a *period* (say, the Silurian, which lasted only 25 million years). This is why the rate of evolution is im-

portant in the development of guide fossils. Simply stated, the rate of evolution is a measure of how much biologic change has occurred over a given interval of geologic time. Groups that evolve rapidly provide more index fossils than groups that evolve slowly.

Index fossils are of great convenience to geologists. However, correlations and interpretations based on fossil assemblages often are more useful. They are also less susceptible to error or uncertainties caused by undiscovered, reworked, or missing individual species.

Biostratigrapic Zones

A **biozone** is a body of rock deposited during the time when particular fossil plants or animals existed. Biozones vary in thickness and lithology, and they can be either local or global. Three kinds of biozones are range zones, assemblage zones, and concurrent range zones.

A **range zone** is simply the rock body representing the total geologic life span of a distinct group of organ-

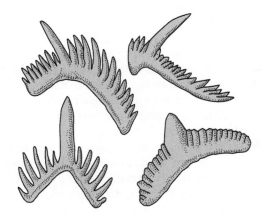

FIGURE 6-33   **Conodont elements.** Image magnified about 50 times.

isms. For example, in Figure 6-34, the *Assilina* range zone is marked by the first (lowest) occurrence of that genus at point A and its extinction at point B.

Geologists also may designate **assemblage zones** based on several species or genera that lived at the same time, and therefore occur together. Assemblage zones are named after an easily recognized, usually common member of the assemblage. If the fossil that lends its name to the assemblage zone is absent, other members permit recognition of the zone.

The **concurrent range zone** is recognized by the overlapping ranges of two or more species or genera. For example, the interval between X and Y in Figure 6-34 might be designated the *Assilina–Heterostegina* concurrent range zone. With concurrent range zones, you can recognize strata that represent smaller increments of time than might be provided by range zones.

Biozones are fundamental in stratigraphy. They are the starting point for all biostratigraphic classification and correlation. They provide the upper and lower boundaries for the time-rock units such as stages, series, and systems.

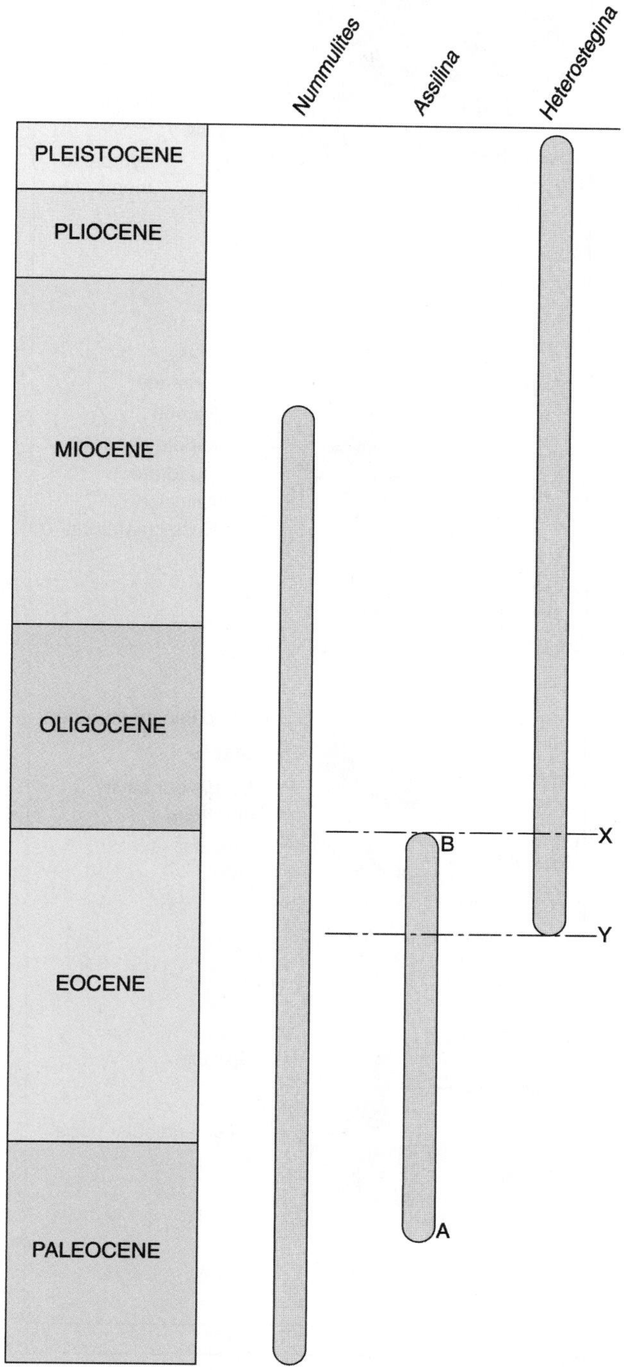

FOSSILS INDICATE PAST ENVIRONMENTS

Ecology of the Past: Paleoecology

Ecology studies relationships between organisms and their environments. Ecologic studies concentrate on the **ecosystem**, which is any selected part of the physical environment, together with the animals and plants in it. Thus, an ecosystem can range from the entire planet or an ocean down to a mountain or valley or lake or even a microscopic area, depending on the scope of the study.

Paleoecology studies how *ancient* organisms interacted with one another and their environments. *Paleoecologists* discover where and how ancient creatures lived

FIGURE 6-34 Geologic ranges of three genera of foraminifera. The interval between A and B is the total range of *Assilina*. The interval between X and Y could be designated the *Assilina–Heterostegina* concurrent range zone.

FIGURE 6-35 Fossil foraminifera. These tiny foraminiferal shells were obtained from a 14-million-year-old limestone.

ENRICHMENT

Earbones Through the Ages

We humans and other terrestrial mammals hear because receptor cells in our inner ear generate electrical impulses in response to vibrations from sound waves. These waves travel through the external ear canal to the tympanic membrane or eardrum (Fig. A).

Sound makes the tympanic membrane vibrate, and the vibrations are transmitted through a chain of three small bones in the middle ear to the sound organ (cochlea) of the inner ear. The three bones (auditory ossicles) are hinged to one another in a manner that allows them to act like an amplifier, increasing the force of sound vibrations.

The evolution of the ear bones and middle ear cavity provides excellent examples of bones that served one function in ancestral animals but are changed to do something different in their descendants.

The first of the three auditory ossicles is the stapes (STAY-peez). You can trace its transformation backward in time to Silurian fish, when their jaws evolved from cartilaginous or bony supports for gills. The bone that would become the stapes was originally an upper gill support. In the evolution of fish with true jaws, the gill support just behind those that evolved into jaws became a supporting prop between the upper jaw and braincase. It is called the hyomandibular.

With the advent of amphibians during the Devonian, the old fish hyomandibular was transformed into the stapes, where it transmitted vibrations from the tympanic membrane to the inner ear. The stapes is the only "ear bone" in amphibians and reptiles.

With the advent of mammals, however, two bones in reptiles that had served as the joint between the upper and lower jaws were changed in function to transmit vibrations. They became the incus (INK-us) and malleus (MAL-ee-us). Thus, over the past 400 million years or so, the bones that were to evolve into our present hearing apparatus were originally gill supports, braincase props, and hinge-bones for jaws.

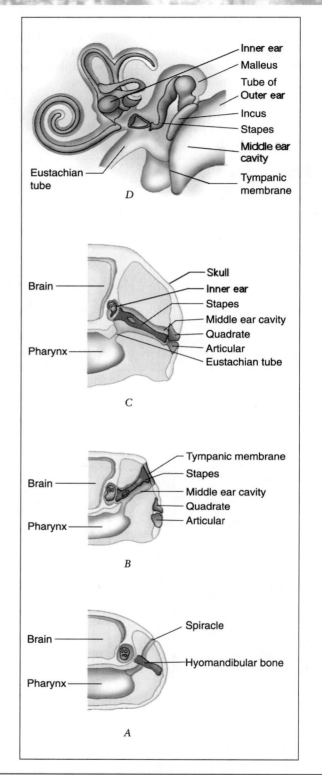

FIGURE A **Evolution of auditory ossicles.** (A) Primitive fish with spiracle and hyomandibular bone, but without middle ear cavity. (B) Primitive amphibian with stapes (also called columella) derived from fish hyomandibular bone located within the former spiracular tunnel, which now functions as a middle ear cavity. (C) A reptile in which the stapes has shifted to a position near the quadrate. (D) Mammal in which the quadrate has been transformed into the incus, and the articular into the malleus.

and what their habits and morphology reveal about the geography and climate of long ago.

The scientific detective work of paleoecology includes comparison of today's organisms with their fossil counterparts. We assume that living and fossil forms had similar needs, habits, and tolerances. So, paleontologists have documented the physical environment for hundreds of living marine foraminifera species. They use this information to deduce the water depth, temperature, and salinity of the ocean in which rocks containing fossil foraminifera (Fig. 6-35) were deposited.

Paleoecologists also examine fossil anatomy to identify features that likely developed in response to biologic or physical conditions in the environment. An example is the hooves and teeth of horses, which indicate a grass-rich prairie environment.

The Marine Ecosystem: Diverse Habitats for Diverse Organisms

To study the ocean ecosystem, ecologists have classified marine environments (Fig. 6-36). The basic division is two realms: the **pelagic** (the water mass lying above the seafloor) and **benthic** (the seafloor). Thus, we have pelagic dwellers above the sea bottom, such as sharks, and benthic dwellers on the sea bottom, such as coral.

PELAGIC REALM (OCEAN WATER)   The pelagic realm is divided into the **neritic zone** above the continental shelf and the **oceanic zone** that extends seaward from the shelves.

Ecologists note two general classes of life in the pelagic realm, "floaters" and "swimmers." Plankton are small

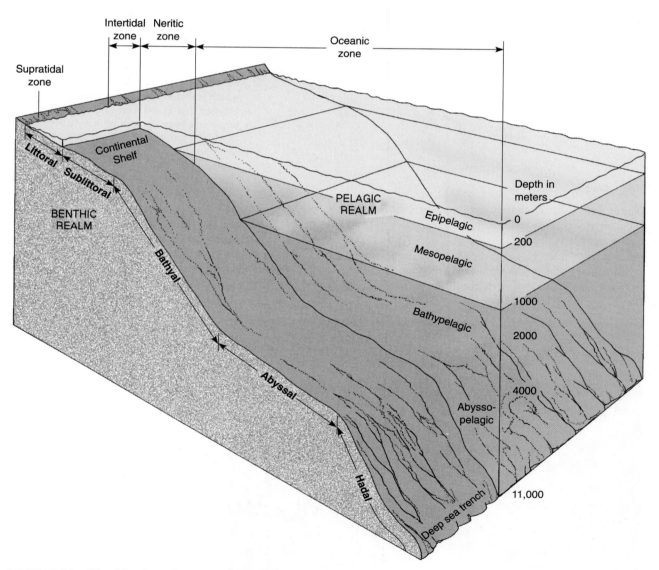

FIGURE 6-36   **Classification of marine environments.**

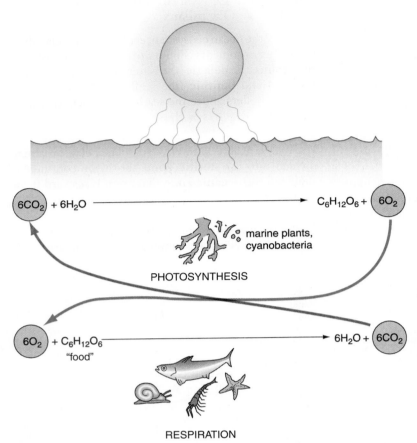

$$6CO_2 + 6H_2O \longrightarrow C_6H_{12}O_6 + 6O_2$$

marine plants, cyanobacteria

PHOTOSYNTHESIS

$$6O_2 + C_6H_{12}O_6 \longrightarrow 6H_2O + 6CO_2$$
"food"

RESPIRATION

FIGURE 6-37 **Interdependence of photosynthesis and respiration.** During photosynthesis, the energy from sunlight is used to make food molecules from carbon dioxide and water. Thus, light energy becomes transformed into chemical energy, which is stored in the food molecules. Oxygen that animals for respiration is an important byproduct of photosynthesis. ▪ *If there was sufficient sunlight to permit photosynthesis to proceed unchecked, what would become depleted?*

plants and animals that float, drift, or feebly swim. Nekton are strongly swimming animals.

Plankton ("floaters") come in two varieties. **Phyto-plankton** are plants and plantlike protistans (including bacteria and algae) that perform photosynthesis. They convert light and carbon dioxide into energy and release oxygen vital for animals (Fig. 6-37). Thus, phyto-plankton enable all other sea life to exist, providing oxygen and being food for everyone else. Phytoplankton are roughly equivalent to plants on land, which do the same thing. Diatoms are among the often-fossilized

FIGURE 6-38 **A living planktonic foraminifera similar to those that became abundant during the Cretaceous and Cenozoic.** The fine rays of cytoplasm are extruded through the pores of the tiny calcium carbonate shell, which is about 0.85 mm in diameter. ▪ *What functions do the pseudopodia (rays of cytoplasm) serve in this tiny animal?*

FIGURE 6-39 **Diorama showing organisms in the sublittoral zone.** See Figure 6-36 for environmental setting. These organisms lived on the seafloor covering the central United States during the Silurian period (410–435 million years ago). Members of this thriving paleocommunity included trilobites, horn and honeycomb corals, algae, branching bryozoan colonies, and small bivalved brachiopods. ◪ *Draw an arrow to a representative of each of these groups and label it.*

types of phytoplankton. **Zooplankton** are animals that include foraminifera (Fig. 6-38), tiny mollusks, small crustaceans, and the mobile larvae of many invertebrate families that live on the seafloor (benthic) as adults.

Nekton ("swimmers") travel under their own power. This is a clear advantage, for a swimming creature can seek its own food and does not depend on food particles in chance currents. Nekton also use their mobility to escape predators and can relocate if the environment grows hostile. Nekton are a diverse group that includes invertebrates (such as shrimp, cephalopods, and certain extinct trilobites) as well as vertebrates (such as fish, whales, and marine turtles).

BENTHIC REALM (SEAFLOOR) The second great division of the ocean ecosystem is the sea bottom or benthic realm. It begins with a narrow zone above high tide called the **supratidal zone** (Fig. 6-36). Relatively few marine plants and animals have adapted to this harsh environment, where ocean spray provides vital moisture but drying poses a constant danger.

Seaward is the **littoral zone**, the area between high and low tide. Organisms in the littoral zone also must be very hardy, able to tolerate alternating wet and dry conditions as the tides move in and out twice daily. Some organisms avoid drying by burrowing into wet sand. Others have adaptations that help them retain body moisture when exposed to air.

Benthic animals and plants are most abundant in the continuously submerged **sublittoral zone**. This extends from low-tide levels to the edge of the continental shelf (about 200 meters deep). Depending on water clarity, light may penetrate to the seafloor in the sublit-

toral zone, although the base of light penetration usually is slightly less than 200 meters.

Algae thrive here, as well as abundant protozoans, sponges, corals, worms, mollusks, crustaceans, and sea urchins (Fig. 6-39). Some of these benthic animals live atop the sediment that carpets the seafloor and are called **epifaunal**, whereas **infaunal** animals burrow into the soft sediment or bore into harder substrates for food and protection (Fig. 6-40).

Burrowers churn and mix sediment, a process known as **bioturbation**. Bioturbation may destroy the original grain orientations in clastic sediments, rendering them

FIGURE 6-40 **Burrows made by infaunal organisms (probably worms) in siltstone of Mississippian age (320–355 million years ago).** They are preserved in the Northview Formation of Missouri.

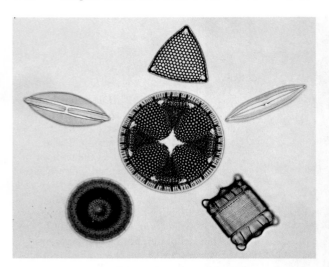

FIGURE 6-41 **Modern marine diatoms.** Diatom shells (called tests) have two perforated structures that overlap like the two parts of a shallow round box for pills. ▮ *In what taxonomic kingdom are diatoms placed?*

FIGURE 6-43 **The coccolithophorid *Coccolithus pelagicus*.** Image magnified 7000 times.

useless for studies of paleocurrent directions. Evidence of cyclic events may be obscured, and magnetic properties of the rock may be altered.

BENTHIC LIFE AND OOZES IN DEEPER WATER Beyond the continental shelves, the benthic environment is one of low temperatures, little or no light, and high pressure. Without light, plants cannot live at these depths. This **bathyal** environment extends from the edge of the shelf to a depth of about 4000 meters. Still deeper levels constitute the **abyssal** environment. The term **hadal** is reserved for the extreme depths found in oceanic trenches.

As you might expect, animals are much less abundant in the abyssal and hadal environments. Most of these deep-water creatures are either scavengers that depend on the slow fall of food from higher levels or predators that eat the scavengers.

Sediments of the bathyal and abyssal zones often consist of microscopic shells and fragments of planktonic organisms. While still wet, the organic sediments have a slippery feel. They are therefore given the delightful name of *oozes*. Depending on the composition of the shells, oozes are either calcareous or siliceous.

Siliceous oozes are composed mainly of tests (shells) of diatoms (Fig. 6-41) and tiny "protozoans" called radiolaria (Fig. 6-42). Siliceous oozes accumulate in colder, deeper oceanic zones where other sediments are lacking, or in regions where an abundance of nutrients promotes high productivity of siliceous organisms.

Calcareous oozes are mainly the calcium carbonate coverings of coccolithophorids (Fig. 6-43) and foraminifera, as well as tiny planktonic snail shells called pteropods (Fig. 6-44). In pteropods, the "foot" seen in ordinary snails is transformed into a pair of fins. If the tiny calcareous shells of coccolithophorids, foraminifera,

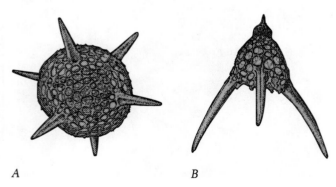

A   *B*

FIGURE 6-42 **Radiolaria.** These protistans build their skeletons of silica. The structures tend to be modifications of either a spherical (A) or helmet shape (B). Image magnified 100 times.

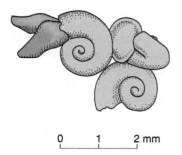

0   1   2 mm

FIGURE 6-44 ***Limacina*, a tiny swimming marine snail, or pteropod.** The foot is modified into a pair of winglike fins, shown at the left. At the right are two empty shells.

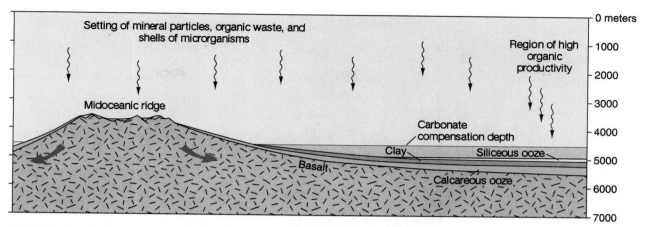

FIGURE 6-45 **Carbonate compensation depth (CCD).** This is the ocean depth below which particles of calcium carbonate from microorganisms are dissolved as fast as they descend through the water column. The depth of the CCD varies at different locations in the ocean. Calcium carbonate accumulates along parts of the midoceanic ridge that are above the CCD. The accumulated layer then is carried away as lithospheric plates diverge from the ridge. When a given region of the seafloor has reached depths below the CCD, calcium carbonate no longer is deposited, but clay particles and siliceous remains of radiolaria and diatoms may accumulate.

and pteropods settle through a column of water no deeper than about 4000 meters, they accumulate to form calcareous ooze (Fig. 6-45).

CARBONATE COMPENSATION DEPTH (CCD) Colder waters below 4000 meters hold more carbon dioxide in solution, which increases their acidity. Such water is corrosive to delicate calcium carbonate shells. At a depth of four to five kilometers, the supply of shells is approximately balanced by the amount being dissolved. As a result, calcareous oozes cannot accumulate.

The water depth at which calcium carbonate is dissolved as fast as it falls from above is termed the **carbonate compensation depth,** or **CCD** (Fig. 6-45). Not all calcareous material dissolves precisely at the CCD, for some shells can survive to greater depths because of their larger size or the presence of protective organic coatings.

HOW FOSSILS INDICATE PALEOGEOGRAPHY

The geographic distribution of today's animals and plants is closely controlled by environmental limitations. Each species has a definite range of conditions for living and breeding, and generally, it is not found outside that range. Ancient organisms had similar restrictions on where they could survive.

Paleogeographic Mapping

If we note the locations of fossil species of the same age on a map and correctly infer the environment in which they lived, we can produce a paleogeographic map for that time interval. We begin with a simple base map and plot locations of marine fossils that lived at a particular time. This provides an idea of which areas were occupied by seas and suggests the locations of ancient coastlines.

For example, Figure 6-46 shows major land and sea regions during the middle Carboniferous Period (around 320 million years ago). The locations of marine protozoans called fusulinids are plotted as red circles. Notice that the Rocky Mountains did not yet exist, and their present location was occupied by a great north-south seaway.

Having obtained a fair idea of where seas and their shorelines existed, we might next look at the evidence for land areas. The fossilized bones or footprints of land animals such as dinosaurs or mastodons suggest a terrestrial paleoenvironment. Fossil remains of land plants, including seeds and pollen, also indicate a terrestrial paleoenvironment.

Through analysis of fossils and their enclosing sediment, it often is possible to recognize deeper or shallower parts of the marine realm or to discern land environments like ancient floodplains, prairies, deserts, and lakes. River deposits may yield the remains of freshwater clams and fossil leaves. A mingling of land and sea fossils might result from a stream entering the sea and perhaps building a delta.

Land Bridges, Isolation, and Migration

The migration and dispersal patterns of land animals, as indicated by fossils, is an important indicator of former land connections, mountain barriers, or ocean barriers that once existed between continents. For ex-

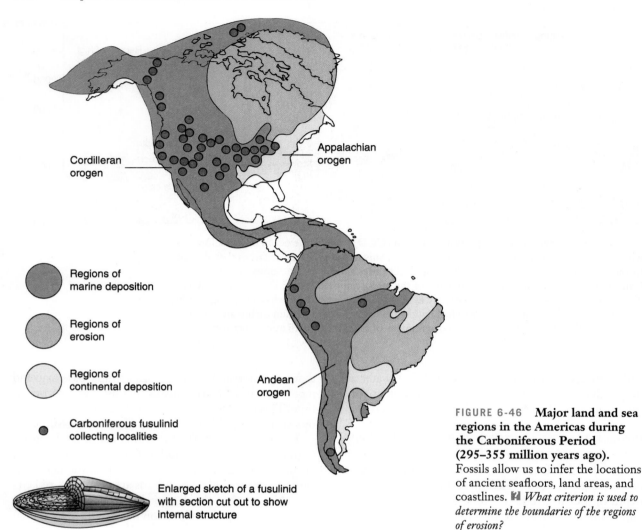

Cordilleran orogen

Appalachian orogen

Regions of marine deposition

Regions of erosion

Regions of continental deposition

Andean orogen

Carboniferous fusulinid collecting localities

Enlarged sketch of a fusulinid with section cut out to show internal structure

FIGURE 6-46 **Major land and sea regions in the Americas during the Carboniferous Period (295–355 million years ago).** Fossils allow us to infer the locations of ancient seafloors, land areas, and coastlines. ▧ *What criterion is used to determine the boundaries of the regions of erosion?*

ample, today the Bering Strait between North America and Asia is about 80 km wide and 30–50 m deep. It prevents migration of land animals between the two continents. However, its shallow depth might lead you to suspect that Asia and North America once were connected by a land bridge.

Indeed, the fossil record shows that a land bridge did connect these two continents on several occasions during the Cenozoic Era (the past 65 million years). The earliest Cenozoic strata on both continents display fossil fauna uncontaminated by foreign species. Somewhat younger rocks contain fossil remains of animals that heretofore were found only on the opposite continent. During the Pleistocene Epoch, camels, horses, mammoths, and a rich variety of other land mammals migrated between North America and Eurasia across the Bering land bridge (Fig. 6-47). The bridge was still in place just 14,000 years ago when Stone Age hunters used the route to enter North America.

Another familiar example of how fossils aid in paleogeographic reconstructions is found in South America. Fossils reveal that South America was isolated from North America in the early Cenozoic Era. Because of this, unique South American fauna evolved over a period of 30 to 40 million years. The establishment of a land connection between the two Americas is recognized in strata only a few million years old (late Pliocene) by the appearance of a mixture of species formerly restricted to either North or South America.

Species Diversity and Geography

Paleontologists also can provide data that help locate Earth's equator, parallels of latitude, and positions of the geographic poles hundreds of millions of years ago. Today, we observe that higher latitudes of the globe have relatively few species but large numbers of individuals (think of northern Canada or Antarctica). In contrast, equatorial regions have a vast number of

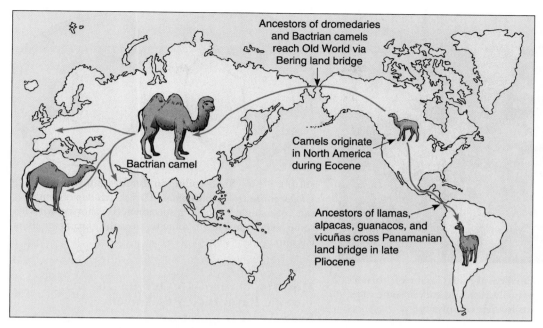

FIGURE 6-47 **Intercontinental migrations of camel family members.** Land bridges enabled the camel, which originated in North America, to spread across four other continents.

species, but with comparatively fewer individuals within each. Stated differently, the *species diversity* for most higher categories of plants and animals increases from the poles toward the equator (Fig. 6-48). This is probably because relatively fewer species can adapt to the rigors of polar climates.

Conversely, the equator offers a stable input of solar energy, less duress from changing seasons, and a more stable food supply. Warmer areas place less stress on organisms and provide opportunity for continuous,

uninterrupted evolution, encouraging more variety. Of course, in specific areas, generalizations about species diversity can be upset by local conditions.

Another way to locate former equatorial regions (and, therefore, the polar regions that lie 90° of latitude to either side) is by plotting the locations of fossil coral reefs of a particular age on a world map. Nearly all living coral reefs (Fig. 6-49) lie within 30° of the equator. It is likely that ancient coral reefs had similar geographic preferences.

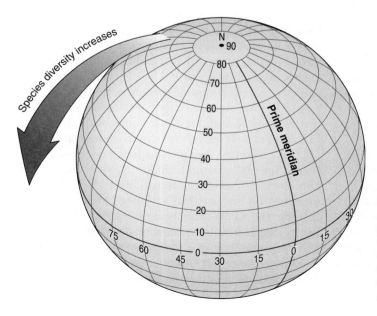

FIGURE 6-48 **Species diversity ranges from low at polar latitudes to high at equatorial latitudes.** With this figure you can review Earth's geographic grid of latitude parallels (east-west lines) and longitude meridians (north-south lines). ◪ *Would a continental location at 70 degrees north latitude and 30 degrees east longitude have a greater or lesser species diversity than a continental location at 20 degrees north latitude and 75 degrees west longitude?*

FIGURE 6-49 **A modern coral reef.** Coral reefs harbor an extraordinary diversity of organisms and, in some ways, rival tropical rain forests in their complexity.

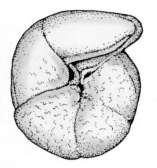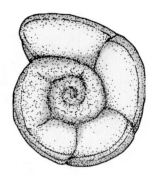

FIGURE 6-50 **Shell coiling in the foraminifer** *Globorotalia truncatulinoides.* This sketch depicts both sides of a single left-coiling specimen. (It is not known why some coil to the left and some to the right.) Diameter about 0.9 mm.

HOW FOSSILS INDICATE PAST CLIMATES

Climate (especially temperature) is one of the most important factors that limit the distribution of organisms. Paleoecologists gain information about ancient climates in several ways. Analyzing fossil spores and pollen grains provides evidence of past climatic conditions. Also, living organisms with specific tolerances can be directly compared with fossil relatives. For example, Corals thrive where water temperatures rarely fall below 18°C, so it is likely that their ancient counterparts were similarly constrained.

However, even if a close living analogue is not known, certain physical features can help determine paleoclimatology. Plants with roots exposed to the air, a lack of annual growth rings, and large wood cells indicate tropical or subtropical climates. Marine mollusks (such as clams, oysters, and snails) with thick shells that often bear prominent spines and ridges are characteristic of warmer ocean regions. In some species of foraminifera, variation in the average size or shell-coiling direction provides clues to cooler or warmer conditions.

The planktonic foraminifera *Globorotalia truncatulinoides* (Fig. 6-50) provides an example of environmentally induced changes in shell-coiling direction. The species coils to the left during periods of relatively cold climates and coils to the right during warmer episodes. Such reversals in coiling may occur quickly and over broad geographic areas. For this reason, they are exceptionally useful in correlation studies. (However, we do not know definitively why they coil in different directions.)

How the Oxygen-16/Oxygen-18 Isotope Ratio Indicates Ancient Seawater Temperature

A widely used method for discovering the water temperature of ancient seas involves analyzing the oxygen isotopes in marine shells. Two isotopes of oxygen are used in these studies. Oxygen-16, the common isotope, has an atomic mass of 16, whereas the rarer oxygen-18 has an atomic mass of 18. The difference in mass is important: As water evaporates from the ocean, the lighter oxygen-16 evaporates faster, whereas the heavier oxygen-18 tends to remain behind.

When the evaporated moisture precipitates back to Earth's surface as rain or snow, water molecules containing the heavier isotope precipitate first, often near coastlines, and flow quickly back into the ocean. Inland, the precipitation from the remaining water vapor is depleted in oxygen-18 relative to its initial quantity. If the interior of a continent is cold and contains growing glaciers, the glacial ice will lock up the lighter isotope, preventing its return to the ocean. This increases the proportion of the heavier oxygen-18 isotope in seawater. As this occurs, the calcium carbonate shells of marine invertebrates will also be enriched in oxygen-18 and thereby reflect episodes of continental glaciation.

Even in the absence of ice ages, the oxygen isotope method may be useful as a paleotemperature indicator. In a famous early study, the oxygen isotope ratio in the calcium carbonate skeleton of a Jurassic (135–203 million years old) belemnite indicated that the average temperature in which the animal lived was 17.6°C, plus or minus 6°C of seasonal variation. Studies on both Jurassic and Cretaceous belemnites (Fig. 6-51) confirmed the inferred positions of the poles during the Mesozoic and indicated that tropical and semitropical conditions were far more widespread during late Mesozoic time than they have been in subsequent geologic time.

A

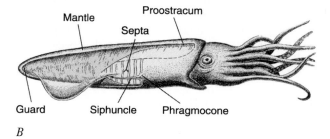

B

FIGURE 6-51 **Belemnites.**
(A) The typical belemnite fossil consists only of the solid part of the internal skeleton, called a guard. The guard here is 9.6 cm long. (B) Belemnites are extinct relatives of squid and cuttlefish. Shown here is an interpretation of a living belemnite.

AN OVERVIEW OF THE HISTORY OF LIFE

The Earliest Traces

As yet, no fossil evidence has been found of life on Earth during the planet's first billion years. The oldest direct indications of ancient life are fossils of primitive prokaryotic microbial organisms more than 3.5 billion years old. They appear at the end of a long period during which living things evolved from nonliving chemical compounds.

We have no direct geologic evidence to tell us how and when the transition from nonliving to living occurred. What we do have are hypotheses supported by observations and experiments. We will examine some of these hypotheses in Chapters 8 and 9, which deal with the Precambrian eons. For this brief overview, we note that life came into existence early in that great 4-billion-year span of time prior to the beginning of the Cambrian Period.

For most of Precambrian time, living things left only occasional traces. Here and there, paleontologists have been rewarded with finds of fossil bacteria, including filamentous cyanobacteria called stromatolites that formed extensive algal mats (Fig. 6-52). These microbial organisms produced large quantities of oxygen by photosynthesis.

However, life was at or below the unicellular level until about 1 billion years ago, when the world's first multicellular organisms left their trails and burrows in rocks of Australia's Torrowangee Group. In rocks deposited about 700 million years ago, fossil metazoans that we recognize as worms, arthropods, and relatives of corals and jellyfish (Fig. 6-53) have been found in scattered spots around the world. Thus, near the end of the Precambrian, the stage was set for the appearance of a wide range of Paleozoic plants and animals. As you read these pages, follow the general progression of life forms through the Paleozoic, Mesozoic, and Cenozoic

FIGURE 6-52 **Stromatolites.** Circular mounds of stromatolites such as these thrived in the shallow waters of Precambrian seas. They are calcareous structures formed by cyanobacteria.

FIGURE 6-53 *Mawsonites*, **a fossil from the Pound Quartzite of Australia.** The formation is late Proterozoic in age. *Mawsonites* is considered to be a jellyfish.

illustrated in "Highlights in the History of Life" inside the front and back covers of this book.

The History of Plants

Long before the first animal appeared on Earth, plants evolved. Plants had their origin among unicellular aquatic protistids of the Precambrian. Among these, the green algae or chlorophytes are the most likely ancestors of land plants. The first invasion from sea to land was made by plants that reproduced with spores (Fig. 6-54). From primitive forms that originated late in the Ordovician (450–435 million years ago), vascular land plants expanded widely during the Devonian and Carboniferous, forming the bulk of vegetation in the great coal swamps.

The largest of these spore-bearers were the scale trees, whose close-set leaves left a pattern of scale-like scars on trunks and branches. The next group of plants to make their debut were pollen and seed producers. These "gymnosperms" appeared in the Devonian and became widely dispersed during the Mesozoic. Plants that bore seeds and flowers, the angiosperms, had evolved by Cretaceous time. They are by far the most abundant plants today and include familiar trees such as oak, maple, sassafras, and birch as well as the grasses that became the primary food source for many Cenozoic mammals.

The History of Animals

We can make some interesting observations from Figure 6-55:

- The principal groups of invertebrates appear either very late in the Proterozoic or early in the Paleozoic (during the Cambrian Period).

- Less advanced members of each phylum characterized the earlier geologic periods, whereas more advanced members came along later.

- Most of the principal phyla that made their appearance during the Paleozoic are still represented by animals today.

- We do not see sudden appearances of bizarre or exotic animals and plants.

- We recognize periods of environmental adversity that caused massive exterminations of life. Such episodes were usually followed by much longer intervals of recovery and more or less orderly evolution.

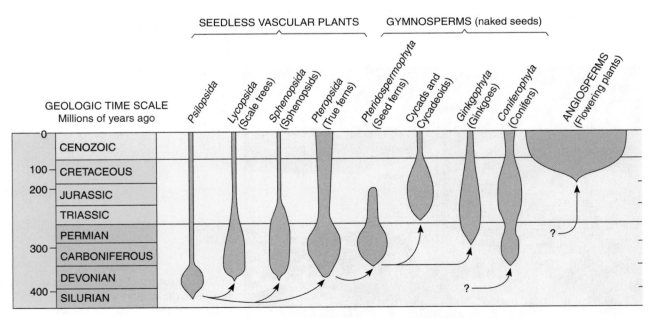

FIGURE 6-54 **Geologic ranges, relative abundances, and evolutionary relationships of vascular land plants.**

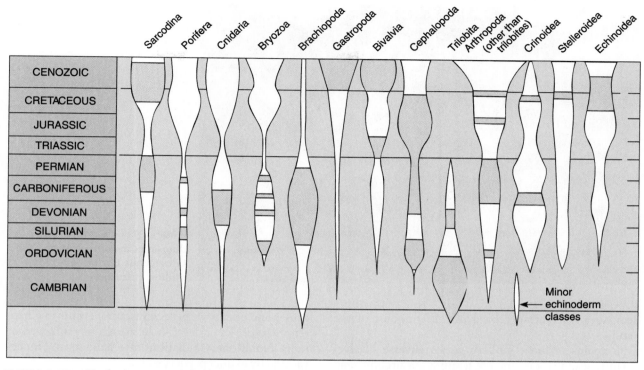

FIGURE 6-55 **Geologic ranges and relative abundances of frequently fossilized categories of invertebrate animals.** Width of range bands indicates relative abundance. Colored areas indicate where fossils of a particular category are widely used in zoning and correlation.

Several generalizations can be made about Paleozoic life. During the Cambrian, the dominant creatures were trilobites, brachiopods that lacked hinged shells, small mollusks, certain soft-bodied worms, and reef organisms called archaeocyathids. Many Cambrian animals fed on microbes and other small particles of food within seafloor sediment or suspended in water. These were deposit and suspension feeders.

Later in the Paleozoic, trilobites were joined by brachiopods, whose shells consisted of two valves hinged along one margin. Ancient species of cephalopods, crinoids, corals, and twig-like bryozoans (moss animals) were abundant in the shallow Paleozoic seas that covered large regions of continents (Fig. 6-56). The Paleozoic was also the era when fishes, amphibians, and reptiles appeared,

FIGURE 6-56 **A large, straight-shelled cephalopod dominates this seafloor scene from the Ordovician Period (470 million years ago).** Also shown are corals, bryozoans, and crinoids.

FIGURE 6-57 **Carnivorous dinosaurs in combat during the Cretaceous Period (65–135 million years ago).**

leaving a fascinating record of the conquest of the land.

Remains of more modern corals, diverse bivalves (clams, scallops, oysters, etc.), spiny sea urchins, and cephalopods called ammonoids characterize the marine strata of the Mesozoic Era. However, the Mesozoic is best known as the era when dinosaurs and their kin dominated the continents (Fig. 6-57). No less important than the dinosaurs, however, were rat-sized primitive mammals that skittered about, perhaps unnoticed by the "thunder beasts." Finally, our planet's first birds made their appearance during the Mesozoic.

The mollusks, ubiquitous in Mesozoic seas, continued in importance during the Cenozoic. However, no ammonoid cephalopods are found in Cenozoic strata. Instead, rocks of the Cenozoic are recognized by distinctive families of protozoans (foraminifera) and a host of modern-looking snails, clams, sea urchins, barnacles, and encrusting bryozoa. Because the Cenozoic saw the expansion of warm-blooded creatures such as ourselves, it is appropriately termed the "age of mammals."

Mass Extinctions

Since that time over a half billion years ago, when most of the principal phyla of animals had become established, the history of life has not been a steady continuum. It has been marked by times of relatively sudden worldwide extinctions so devastating that we call them **mass extinctions**.

Five episodes of mass extinctions have been particularly catastrophic. These occurred at the end of the Ordovician, late in the Devonian, at the end of the Permian, in the late Triassic, and one that killed off the dinosaurs and many less-known animal groups at the end of the Cretaceous Period. We will examine the causes and effects of these devastating events in Chapters 12 and 14 and describe how each was followed by rapid evolutionary radiations into habitats and niches vacated by extinct predecessors.

▶ LIFE ON OTHER PLANETS: ARE WE ALONE?

Is organic evolution unique to our planet? What properties of Earth made it suitable for the origin and evolution of life? Could these conditions exist elsewhere? Or are we truly unique?

Indeed, Earth is special: It was in the right place at the right time, with the right conditions for life as we know it to evolve.

- Earth is massive enough to have sufficient gravity to retain an atmosphere of life-critical gases (oxygen and nitrogen).

- Earth has all of the chemical elements required for life processes, especially the hydrogen-oxygen compound we call water.

- Overall, Earth's surface temperature is just right to provide abundant liquid water. A few degrees cooler, and the whole surface would be ice-covered. A few degrees warmer, and water would all be vapor.

- Overall, Earth's surface temperature is suitable for chemical reactions required for life processes.

- Earth's just-right surface temperature is a consequence of our Sun-star's size, age, and distance from Earth.

- Our Sun-star also has a life span that is sufficiently long, and an energy output that is steady enough, to permit time for the emergence and evolution of life.

Again: Earth's conditions are ideal for life *as we know it*. It is entirely possible that organic (carbon-based) life could exist under different conditions on other worlds. But zebras, stately oaks, bluejays, people, butterflies, slime molds, oysters, viruses, dogs, bacteria, mushrooms, and grasses all thrive under the conditions we know here on planet Earth.

Life in Our Solar System

Has there ever been life beyond Earth in our Solar System? All evidence to date indicates that conditions are currently too harsh on neighboring planets to permit the evolution of *higher* forms of life, either because of their size (too big or too small) or their distance from the Sun (too near or too far). However, several bodies in our Solar System have properties that make them intriguing places to investigate for microbial life: Mars, Jupiter's moons Europa and Io, and Saturn's moon Titan. For example, Europa has water slush or possibly even an ocean beneath its icy surface. Titan has a thick atmosphere of nitrogen molecules, and Io has conditions for life of the kind found on Earth near volcanic vents.

The *Viking* missions to Mars in the 1970s examined the Martian surface for indications of life but found nothing to suggest the existence of present or past organisms. However, many planetary geologists and astronomers suspect that some simple kind of life once existed on the red planet. Of all the planets, Mars has conditions most similar to the Earth, and there is abundant evidence of erosional and depositional features that appear to have been made by water. Exploration intensified in 2004 when twin rovers landed on the Martian surface, photographed layers of water-deposited sediment, and detected minerals that formed by precipitation from water.

Life in the Universe

Earth is indeed biologically special when compared with other planetary bodies in our Solar System. However, it may not be so peculiar in the vast realm of the Universe. Astronomers estimate about 150 billion stars in our galaxy, and the number of galaxies now appears to be nearly limitless. Of those 150 billion stars in our galaxy, a not unreasonable estimate stipulates that at least 1 billion have planets with size and temperature conditions similar to Earth's. These are potentially habitable planets.

For the entire known Universe (of which our galaxy is but a small part), scientists estimate more than 1000 planetary systems similar to our Solar System. Such calculations indicate the probable existence of life out there, somewhere. Indeed, the Universe may be rich in suitable habitats for life.

What might be the nature of extraterrestrial life? If we assume such life formed from the same universal store of atoms and under physical conditions similar to Earth's, then we might recognize extraterrestrial life as "living." But it is highly unlikely that duplicates of humans, cows, or butterflies exist on other planets. There are so many variables in the evolutionary interactions of genetics, environment, and time that making identical species elsewhere is unlikely. For example, to replicate the identical mutations, genetic recombinations, and environmental conditions that evolved a sparrow on Earth seems most improbable.

SUMMARY

- Fossils are the remains or traces of life of the geologic past. Some processes that produce fossils include the precipitation of chemical substances into pore spaces (permineralization), molecular exchange for substances that were once part of the organism with inorganic substances (replacement), or compression of animals and plants to form a film of carbonized remains (carbonization). Molds and casts are additional types of fossils. Some fossils are indirect evidence of life. These include the tracks, trails, and burrows that comprise trace fossils.

- The fossil record of life is not complete. It is biased in favor of animals that possessed resistant parts or lived in a depositional environment where rapid burial occurred.

- Fossils demonstrate that life has changed or evolved through time. Because of this, fossil-bearing rock layers from different periods of geologic time can be recognized and correlated.

- Variation among offspring is essential for evolution. Variation is achieved by mutations and gene recombinations. Mutations are the ultimate source of new genetic material. Recombination (mixing) spreads the new genetic material through the population and mixes the old with the new. Natural selection then sorts out the variations, preserving those that by chance are best fitted to a particular environment.

- Chromosomes are rod-shaped units in the cell nucleus that contain the genes. Genes are discrete units of hereditary information. They are composed of DNA and are responsible for the total characteristics of an organism.

- The evolutionary history of organisms can be depicted on a phylogenetic tree, with descendants on the upper branches and ancestors on the lower branches and trunk. Time is implied in the vertical dimension of the tree. Cladistic phylogeny produces a less treelike diagram called a cladogram, in which related organisms are placed in close proximity on the basis of characteristics they share that indicate a common ancestry.

• The theory and practice of classifying and naming organisms is called taxonomy. Paleontologists, like biologists, use a binomial system of nomenclature in which each type of organism is assigned a two-part name designating genus and species. The taxonomic system used in classifying organisms is hierarchical, with levels including species, genus, family, order, class, phylum, kingdom, and domain. The three domains are the Bacteria, Archaea, and Eukarya. In the six-kingdom system of classification, the kingdoms are the Archaebacteria, Eubacteria, Protista, Fungi, Plantae, and Animalia.

• Adaptation is a process by which populations change so as to become better fitted to their environments and strategy for survival. The changes may be morphological, physiological, or behavioral. When new habitats or other opportunities become available to a species, the new conditions may be rapidly exploited by descendant species. This results in the diversification that we call adaptive radiation.

• Darwin viewed evolution as a gradual change in lineages through time. Evolution may also progress by sudden advances that punctuate periods of relative stability (stasis). The process is termed punctuated equilibrium.

• Paleoecology is the study of the relationships between ancient organisms and their environments. Paleoecology provides information about the distribution of ancient lands and seas, past climates, depth of seas, barriers to migration, and the former location of continents. The paleoecology of marine areas is of particular importance to geologists because of the richness of the marine sedimentary and fossil record.

• Paleontologists are currently considering methods for detecting life on planetary bodies other than the Earth. Within our Solar System, Mars and planetary satellites such as Europa and Titan appear to have physical and chemical conditions essential for the origin of life. There is also a strong statistical probability of past or present life somewhere in the Universe outside the solar system.

KEY TERMS

abyssal marine environment, p. 148

adaptation, p. 133

adaptive radiation, p. 132

Animalia, p. 126

Archaea (domain), p. 127

Archaebacteria, p. 126

assemblage zone, p. 143

Bacteria (domain), p. 127

bathyal marine environment, p. 148

benthic marine realm, p. 145

bioturbation, p. 147

biozone, p. 142

calcareous ooze, p. 148

carbonate compensation depth (CCD), p. 149

carbonization, p. 121

cast, p. 121

chromosome, p. 130

cladogram, p. 135

class (taxonomic) , p. 126

concurrent range zone, p. 143

crossing over (genetic), p. 131

deoxyribonucleic acid (DNA), p. 130

diploid cell, p. 130

domain, p. 126

ecology, p. 143

ecosystem, p. 143

epifaunal, p. 147

Eubacteria, p. 126

Eukarya (domain) , pp. 126, 127

extremophiles, p. 126

family (taxonomic), p. 126

fungi, p. 126

gamete, p. 130

gene, p. 130

gene pool, p. 132

geologic range (e.g., of fossil species), p. 140

genus (taxonomic), p. 125

hadal marine environment, p. 148

haploid cell, p. 130

homologous structures, p. 137

index fossil (=guide fossil), p. 142

infaunal, p. 147

kingdom (taxonomic), p. 126

littoral zone, p. 147

mass extinction, p. 156

meiosis, p. 130

mitosis, p. 130

mold, p. 121

mutation, p. 131

oceanic zone, p. 145

order (taxonomic), p. 126

natural selection, p. 128

nekton, p. 147

neritic zone, p. 145

paleoecology, p. 143

pelagic marine realm, p. 145

permineralization, p. 121

petrifaction, p. 121

phenotypic trait, p. 126

phyletic gradualism, p. 134

phylogenetic tree, p. 135

phylogeny, p. 135

phylum (taxonomic), p. 126

phytoplankton, p. 146

plankton, p. 146

Plantae, p. 126

population (genetic), p. 132

Prokaryota, p. 126

Protista, p. 126

punctuated equilibrium, p. 134

range zone, p. 142

replacement, p. 121

reworked fossils, p. 142

siliceous ooze, p. 148

speciation, p. 132

species, p. 125

sublittoral zone, p. 147

supratidal zone, p. 147

taxonomy, p. 126

trace fossil, p. 123

vestigial organ or structure, p. 138

zooplankton, p. 147

QUESTIONS FOR REVIEW AND DISCUSSION

1. What factors determine whether or not a particular fossil will be valuable as an indicator of the age and correlation of a stratum?

2. Fossil A occurs in rocks of Cambrian and Ordovician age. Fossil B occurs in rocks that range in age from early Ordovician through Permian. Fossil C is found in Mississippian through Permian strata.

   a. What is the maximum possible range of age for a stratum containing only fossil B?

   b. What is the maximum possible range for a stratum containing both A and B?

   c. Which is the better guide fossil, A or C?

3. A chronostratigraphic (time-rock) unit contains a different fossil assemblage at one location than at another location 200 km away. Suggest a possible reason for the dissimilarity.

4. In drilling for oil, geologists recover Devonian conodonts in a stratum known to be Permian in age. Explain how this may have occurred. (Hint: conodonts are very sturdy, durable fossils.)

5. What kinds of vascular land plants existed during the Paleozoic Era?

6. What are the differences between the following?

   a. Mitosis and meiosis

   b. Haploid and diploid

   c. Gymnosperm and angiosperm

   d. Cladogram and phylogenetic tree

7. What is meant by the term adaptation? Cite an example of adaptive radiation.

8. How has the science of paleontology contributed to the following?

   a. Validation of the theory of organic evolution

   b. Recognition of geographic changes in the geologic past

   c. Recognition of climatic changes in the geologic past

9. What are the contributions of Darwin and Mendel to our modern concept of organic evolution?

10. Using fossils for age correlation depends on knowledge of their geologic ranges. How has this knowledge been obtained?

11. Distinguish between the concepts of phyletic gradualism and punctuated equilibrium. Which would be a more appropriate stratigraphic section to study for proof of punctuated equilibrium: (a) a continuous set of deep sea cores, or (b) a section on the continent where there have been repeated episodes of uplift and erosion throughout geologic time?

12. Why is marine life particularly abundant and varied in the sublittoral zone of the ocean?

WEB SITES

The Earth Through Time Student Companion Web Site (www.wiley.com/college/levin) has online resources to help you expand your understanding of the topics in this chapter. Visit the Web Site to access the following:

1. Illustrated course notes covering key concepts in each chapter;

2. Online quizzes that provide immediate feedback;

3. Links to chapter-specific topics on the web;

4. Science news updates relating to recent developments in Historical Geology;

5. Web inquiry activities for further exploration;

6. A glossary of terms;

7. A Student Union with links to topics such as study skills, writing and grammar, and citing electronic information.

7

The Andes Mountains as seen from Pehoe Lake, Torres Del Paine National Park, Chile. *The Andes formed at a convergent tectonic plate boundary as the westward-moving South American plate slowly collided with the Pacific Plate.*

Q
T
K
J
Tr
Pr
P
M
D
S
O
€
Pre-€

Plate Tectonics Underlies All Earth History

The deep interior of the Earth is inaccessible, and no rays of light penetrate to let us see what is below the surface. But rays of another kind penetrate and carry with them their messages from the interior.
—Inge Lehmann, seismologist and discoverer of Earth's inner core, 1959

OUTLINE

▶ EARTHQUAKE WAVES REVEAL EARTH'S MYSTERIOUS INTERIOR

▶ EARTH'S INTERNAL ZONES

▶ EARTH'S TWO TYPES OF CRUST

▶ BROKEN, SQUEEZED, OR STRETCHED ROCKS PRODUCE GEOLOGIC STRUCTURES

▶ PLATE TECTONICS THEORY TIES IT ALL TOGETHER

▶ DRIFTING CONTINENTS

▶ EVIDENCE FOR CONTINENTAL DRIFT

▶ PALEOMAGNETISM: ANCIENT MAGNETISM LOCKED INTO ROCKS

▶ TODAY'S PLATE TECTONICS THEORY

▶ WHAT HAPPENS AT PLATE MARGINS?

▶ WHAT DRIVES PLATE TECTONICS?

▶ VERIFYING PLATE TECTONICS THEORY

▶ THERMAL PLUMES, HOTSPOTS, AND HAWAII

▶ EXOTIC TERRANES

▶ SUMMARY

▶ KEY TERMS

▶ QUESTIONS FOR REVIEW AND DISCUSSION

▶ WEB SITES

▶ BOX 7-1 ENRICHMENT

Key Chapter Concepts

- Seismic waves that pass through the Earth reveal its concentric layers. The uppermost is the crust, followed downward by the mantle, the outer core, and the inner core.

- Seismographs record three types of earthquake wave motion: primary, secondary, and surface waves. Primary and secondary waves are "body waves" capable of traveling through solid rock. Secondary waves do not pass through the outer core, indicating that it is a liquid.

- Earth has two kinds of crust: continental and oceanic. Oceanic crust underlies the ocean basins. The two differ in composition and density. Continental crust is made of granitic rocks, whereas oceanic crust has a basaltic composition.

- Geologic structures like faults and folds provide clues to the nature of forces acting on parts of Earth's Crust.

- Evidence cited by Alfred Wegener in support of his theory of continental drift has been incorporated into the more comprehensive theory of plate tectonics.

- Plate tectonics is a unifying and coherent theory that explains the behavior of Earth's lithosphere. It postulates that the lithosphere is composed of moving lithospheric plates that move apart from spreading boundaries and come together along boundaries that converge or slide past one another.

- Paleomagnetism—magnetism frozen in rocks at the time of their formation—provides evidence of continental displacements and records reversals of Earth's magnetic polarity in the geologic past.

- As evidence of seafloor spreading, magnetic polarity reversals can be mapped in parallel, mirror-image symmetry on either side of spreading boundaries.

- Additional supporting evidence for plate tectonics includes studies of the age of deep-sea sediment; Wadati-Benioff seismic zones where plates converge and form subduction zones; gravity anomalies; volcanoes and volcanic islands formed above thermal plumes; and direct measurements of plate movements from devices aboard artificial satellites.

- One of several ways by which continents increase in size is by incorporation of exotic terranes. Exotic terranes are small fragments of continents or other plate segments that drift and accrete to continental margins.

EARTHQUAKE WAVES REVEAL EARTH'S MYSTERIOUS INTERIOR

Many of the processes operating near Earth's surface are related to conditions deep within the planet. Our deepest wells reach down only about 0.0016 of the way from Earth's surface to its center (10 km out of 6300 km). Thus, our knowledge of Earth's "inner space" relies heavily on interpretation of earthquake waves. These **seismic waves** are the "messages from the interior" mentioned in the epigram at the beginning of this chapter.

Seismic waves permit scientists to determine the location, thickness, and properties of Earth's internal zones. They are generated when rock masses suddenly break or rupture. From the point of breakage, vibrations spread out in all directions, traveling at different speeds through parts of the interior that differ in chemical composition and physical properties. The principal categories of these waves are **primary**, **secondary**, and **surface**. All three are recorded on an instrument called a **seismograph**, which produces a record termed a **seismogram** (Fig. 7-1).

Primary Waves or P-Waves (Compressional)

Primary waves (P-waves) are the speediest of the three wave types, so they are first to arrive at a seismograph station following an earthquake (Fig. 7-1). They typically travel through Earth's crust at about 6 km per second, and even faster in the uppermost mantle, at 8 km per second. Pulses of energy in primary waves are a succession of compressions and expansions that move in the direction of travel of the energy. In other words, P-waves have an accordion-like "push–pull" move-

ment (Fig. 7-2). P-waves can travel through solids, liquids, and gases.

Thus, every particle of rock set in motion during an earthquake is pushed against its neighbor and bounces back. Its neighbor strikes the next particle and rebounds, and subsequent particles continue this motion, "passing it along." This vibrational energy is transmitted through solids, liquids, and gases.

Secondary Waves or S-Waves (Shear)

Secondary waves (S-waves) typically travel 3.5 km per second in the crust and 5 km per second in the upper mantle. They are slower than P-waves, which makes them "secondary" to arrive at seismographs following an earthquake (see Fig. 7-1). Unlike the movement of P-waves, rock segments in secondary waves vibrate at right angles (sideways) to the travel direction of the energy (see Fig. 7-2). It is because of this more complex motion that S-waves travel more slowly than P-waves. Unlike P-waves, secondary waves cannot pass through liquids or gases.

Body Waves

Both P- and S-waves are called **body waves** because they can penetrate deep into the interior or body of our planet. Body waves travel faster in more elastic rocks, so their speeds increase steadily as they move downward into more elastic zones of Earth's interior and decrease as they make their ascent toward Earth's surface. The change in velocity that occurs as body waves pass through rocks of different composition and density results in a bending, or refraction, of the wave. The many small refractions cause the body waves to travel along curved paths through Earth.

Not only are body waves subjected to refraction, but they may also be partially reflected off the surface of a dense rock layer in much the same way as light is reflected off a polished surface. Many factors influence the behavior of body waves. An increase in the temperature of rocks through which body waves are traveling will cause a decrease in velocity, whereas an increase in

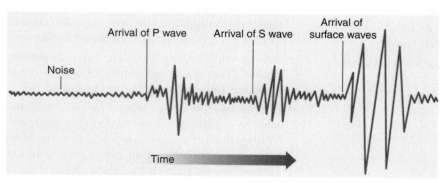

FIGURE 7-1 **Typical seismograph record.** Fast-moving P-waves arrive at the seismograph first, followed by slower-moving S-waves. Surface waves travel even more slowly. *Judging from the size of the vibrations recorded, which of the three kinds of seismic waves is likely to have caused the most damage? (Answers to questions appearing within figure legends can be found in the Student Study Guide.)*

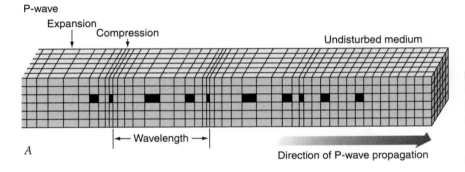

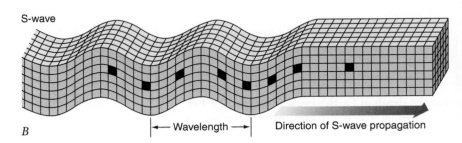

A

← Wavelength →

Direction of P-wave propagation

← Wavelength → Direction of S-wave propagation

B

FIGURE 7-2 **P- and S-type seismic waves.** P-waves have a push-pull motion that causes alternate compressions and expansions. Each particle in a rock will move back and forth *parallel* to the direction of wave motion. Imagine small squares drawn on rock, as shown here. As waves move through the rock, a square will repeatedly change from its square shape to a rectangular shape and back. In S-waves, particle movement is different—it is *perpendicular* to the direction of the wave motion (up-down or side-to-side). In this case, a square will repeatedly change to a parallelogram and then back to a square.

confining pressure will cause a corresponding increase in wave velocity. In a fluid where no rigidity exists, S-waves cannot propagate and P-waves are markedly slowed.

Surface Waves

Surface waves are large-motion waves that travel through the outer crust of Earth. Their pattern of movement resembles the waves caused when a pebble is tossed into the center of a pond. They develop whenever P- or S-waves disturb the surface of Earth as they emerge from the interior. Surface waves are the last to arrive at a seismograph station. They are the primary cause of the destruction that can result from earthquakes affecting densely populated areas (Fig. 7-3). This destruction results because surface waves are channeled through the thin outer region of Earth and their energy is less rapidly dissipated into the large volumes of rock through which body waves travel.

▶ EARTH'S INTERNAL ZONES

From analyses of the countless seismograms, geophysicists have discovered how rocks change with depth. They have been able to recognize the relatively abrupt

FIGURE 7-3 **Destruction from a 2001 earthquake in Bhuj, India.** A woman searches for belongings amidst the rubble of her home.

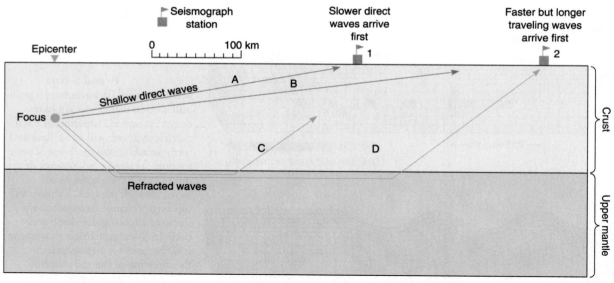

FIGURE 7-4 **How Mohorovičić inferred the depth of Earth's crust from earthquakes.** He based his interpretation on travel paths of seismic body waves: early-arriving P-waves and late-arriving S-waves. *Focus* is the true center of an earthquake and the point at which the disturbance originates. *Epicenter* is a point on Earth's surface directly above the focus.

boundaries between major internal zones. Boundaries where seismic waves experience an abrupt change in velocity or direction are called **discontinuities**. Two widely known breaks of this kind are named after their discoverers, the Mohorovičić and Gutenberg discontinuities.

Mohorovičić Discontinuity (Moho)

Andrija Mohorovičić (*Mo-HOR-o-VITCH-itz*, 1857–1936) was a Yugoslavian seismologist. He based his discontinuity on observations that seismograph stations located about 150 km from an earthquake received earthquake waves sooner than those closer to the focus. Mohorovičić reasoned that below a depth of about 30 km, there must be a zone with physical properties that permit earthquake waves to travel faster—hence, a discontinuity. That layer is the upper **mantle**.

Figure 7-4 illustrates Mohorovičić's discovery. The shallow direct waves (A), although traveling slowly at about 6 km per second, are the first to arrive at seismograph station 1. At the farther station, however, the deeper but faster wave (D) has caught up and actually moved ahead of the shallow direct wave and is therefore the first to arrive at seismograph station 2. The situation is analogous to what many of us do when we take an Interstate beltway around a city to reach a destination, rather than drive a shorter, straighter route through city streets—because faster travel on the Interstate lets us arrive sooner.

The **Mohorovičić discontinuity** (Moho, for short) lies about 30–40 km beneath the surface of continents

and less deep beneath ocean floors. The Moho marks the boundary between the crust and the mantle.

Gutenberg Discontinuity

The **Gutenberg discontinuity** marks the outer boundary of the core. It is located nearly halfway to the center of Earth at a depth of 2900 km (Fig. 7-5). The Gutenberg discontinuity is recognized by an abrupt decrease in P-wave velocity and the disappearance of S-waves.

Earth's Liquid/Solid Core

The outer portion of Earth's core behaves seismically as a liquid. We know this because, below 2900 km depth, S-waves disappear and P-wave velocity slows markedly. Thus, secondary waves generated on one side of Earth fail to appear at seismograph stations on the opposite side. This observation is the principal evidence for an outer core that behaves as a fluid. The outer core barrier to S-waves results in an S-wave **shadow zone** on the side of Earth opposite the earthquake focus (the exact rupture point of an earthquake). Within the shadow zone, which begins 105° from the earthquake's location, S-waves do not appear.

Unlike S-waves, P-waves do pass through liquids. However, they are slowed and sharply refracted downward as they enter the fluid medium of the outer core. The result is a P-wave **shadow zone** that extends from about 105° to 140° from the earthquake focus (Fig. 7-6). Beyond 140°, P-waves are so tardy in their arrival that

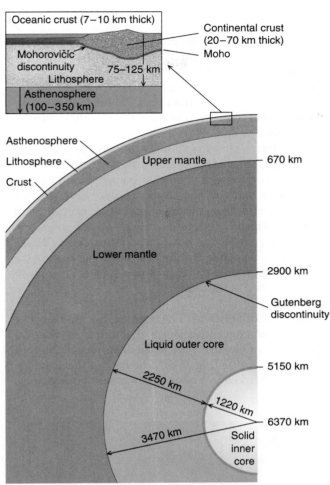

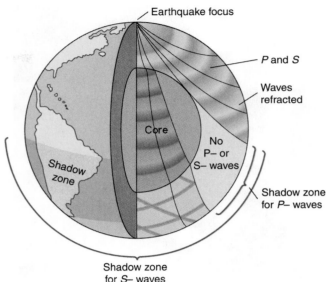

FIGURE 7-6 **Seismic waves refract (bend) as they travel through Earth.** Gradual bends in travel paths occur because velocity increases with depth. Abrupt bending—a *discontinuity*—occurs at the boundaries of major zones. The S-wave shadow zone results from absorption of S-waves in the liquid outer core. P-waves entering the outer core are slowed and bent downward, giving rise to the P-wave shadow zone, within which neither P- nor S-waves are received. ◪ *Name a South American city where seismographs would not record S-waves from an earthquake located at the North Pole.*

FIGURE 7-5 **What's inside Earth.** The interior layers and boundaries are discussed in the text. ◪ *In general, what are the differences in density between the inner core, outer core, mantle, and crust?*

they further validate the inference that they have passed through a liquid medium. At the upper boundary of the core, P-waves are also reflected back toward Earth's surface. Such P-wave echoes are also recorded on seismograms.

The radius of the core is 3470 km. The inner core is solid and has a radius of about 1220 km, which makes it slightly larger than the Moon. Most geologists think the inner core has the same composition as the outer core, but it is solid because the enormous pressure near Earth's center prevents molecules from moving with the freedom of molecules in a fluid.

Evidence for the existence of a solid inner core is derived from hundreds of seismograms. They revealed that weak, late-arriving P-waves somehow were penetrating to stations that were within the P-wave shadow zone. Geophysicists recognized that this penetration could be explained if the inner core behaved seismically as if it were solid.

EVIDENCE FOR A METALLIC CORE We are quite certain that Earth's core has a metallic composition, not a rocky one. Here is the evidence: Earth's overall density is 5.5 g/cm³, yet the average density of surface rocks is much lighter, less than 3.0 g/cm³. Therefore, to achieve the average 5.5 g/cm³ overall density, the deep interior of the planet must compensate by having materials of high density. Calculations indicate that mantle rock density is about 4.5 g/cm³, and that the average core density is very high, about 10.7 g/cm³.

Under the extreme pressure in the core, iron mixed with nickel would have the required high density. However, laboratory experiments suggest that a highly pressurized iron–nickel alloy might be too dense and that minor amounts of elements such as silicon, sulfur, carbon, or oxygen also may be present to "lighten" the core material.

The study of meteorites supports the theory that the core is composed of iron (85%) with lesser amounts of nickel. For example, iron meteorites (Fig. 7-7) consist of metallic iron alloyed with a small percentage of nickel. Geologists suspect that many iron meteorites may be fragments from the core of a shattered planet. The presence of iron meteorites in our solar system suggests that the existence of an iron–nickel core for Earth is plausible.

A

B

FIGURE 7-7 **Meteorites.** (A) A meteorite that landed on the icy surface of Antarctica. (B) An iron meteorite from Mohave County, Arizona that was cut and polished. It shows the interesting pattern of interlocking crystals called the Widmanstatten pattern. ▶ *Why are meteorite-collecting expeditions nearly always more successful in Antarctica than on continents in warmer regions?*

There is further evidence for an iron-rich core. Earth has a magnetic field, and the planet behaves as if a great bar magnet were embedded within it. A magnetic field is generated around any electrical conductor that has electricity flowing through it.

Silicate rocks in the mantle and lithosphere do not conduct electricity very well, but the metals iron and nickel are fine conductors. Heat-driven convection in the outer core coupled with movements induced by Earth's spin may provide the flow of electrons around the inner core that produces the magnetic field. Without a metallic core, this would not be possible.

The Mantle

MATERIALS OF THE MANTLE   As with Earth's core, our understanding of the composition and structure of the mantle is determined from the study of seismograms. These indicate that the mantle has an average density of about 4.5 g/cm^3 and has a stony, rather than metallic, composition. Oxygen and silicon probably predominate, plus iron and magnesium.

Peridotite, a rock rich in iron and magnesium (Fig. 7-8), approximates the kind of material we infer for the mantle. Peridotite is similar in composition to stony meteorites and volcanic rocks that may have reached Earth's surface from the upper part of the mantle. Such suspected mantle rocks are rarely found at Earth's surface. They are rich in olivine and pyroxenes and contain small amounts of minerals, including diamonds, that can form only under pressures greater than those characteristic of the crust.

THE ASTHENOSPHERE   The mantle is not uniform, but has several concentric layers that differ in physical properties. One layer, the **asthenosphere**, is high in the upper mantle (Fig. 7-5). The base of the asthenosphere is marked by a zone of lower seismic velocities,

FIGURE 7-8 **Peridotite.** The green mineral grains are olivine, and the black grains are pyroxene. The specimen was collected at Kilbourne Hole Maar, Dona Ana County, New Mexico. A maar is a crater formed during a violent volcanic explosion. The explosion brought chunks of mantle rock such as this one to Earth's surface.

appropriately named the "low velocity zone." Rock in the asthenosphere is at or near its melting point, and therefore magmas can be generated in this zone. Seismic waves are slowed in the asthenosphere, not because of a decrease in density, but because they enter a region in which rocks are less rigid and more ductile—which means that they can be stretched without breaking.

In this zone, up to 10% of the rock may consist of pockets and droplets of molten silicates. Such material could serve as a slippery layer for the overlying rigid lithosphere, allowing movement. As illustrated in Figure 7-5, the lithosphere includes the crust (oceanic and continental) and the uppermost part of the mantle that is above the asthenosphere. It is the dynamic enveloping shell of Earth and is itself divided into lithospheric plates.

► EARTH'S TWO TYPES OF CRUST

Earth's crust is seismically defined as all of the solid Earth above the Mohorovičić discontinuity. It is the thin, brittle veneer, like an eggshell, that constitutes the continents and the ocean floors. The crust is *not* a uniform shell where low areas are filled with water to make oceans and higher places make continents. Rather, two very distinct kinds of crust determine the existence of separate continents and ocean basins. These two types, oceanic and continental, have very different compositions and physical properties.

Oceanic Crust (More Dense)

Oceanic crust is approximately 5–12 km thick, with an average density of about 3.0 g/cm³ (the average density of basalt and gabbro). Seismograms reveal that oceanic crust has three layers:

- Uppermost is a thin layer of unconsolidated sediment, typically about 200 m thick.
- In the middle are basalts that have been extruded underwater (Fig. 7-9). This layer is typically about 2 km thick.
- The deepest layer consists of gabbro and is about 6 km thick.

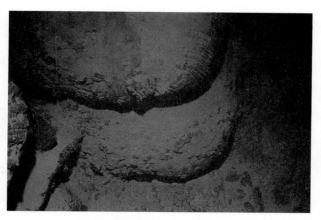

FIGURE 7-9 **Pillow basalt.** It erupted on the seafloor of the Cayman Trough in the Caribbean Sea.

At the boundaries of the ocean basins, the Mohorovičić discontinuity plunges sharply beneath the thicker continental crust (Fig. 7-10). The depth of the Moho beneath the continents varies considerably, but averages about 35 km.

Continental Crust (Less Dense)

Continental crust is thicker and less dense than its oceanic counterpart, averaging about 2.7 g/cm³. As a result, continents "float" higher on the denser mantle than on the adjacent oceanic crustal segments, somewhat like great stony icebergs.

The lower density of continental crust results from its composition. Although it is referred to as granitic, it is really composed of various rocks that approximate granite in bulk composition. Igneous continental rocks are richer in silicon and potassium and poorer in iron, magnesium, and calcium than oceanic rocks. In addition, extensive regions of the continents are blanketed by sedimentary rocks.

Isostasy

Isostasy is a condition of vertical balance—"floating depth," if you will. Objects of different densities float at different heights in water. Similarly, different densities

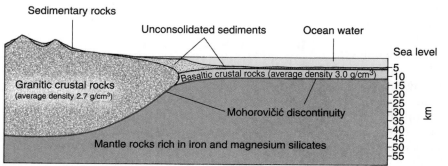

Sedimentary rocks

Unconsolidated sediments

Ocean water

Sea level

Granitic crustal rocks
(average density 2.7 g/cm³)

Basaltic crustal rocks (average density 3.0 g/cm³)

Mohorovičić discontinuity

Mantle rocks rich in iron and magnesium silicates

km

5
10
15
20
25
30
35
40
45
50
55

FIGURE 7-10 **Generalized cross-section showing Mohorovičić discontinuity.**

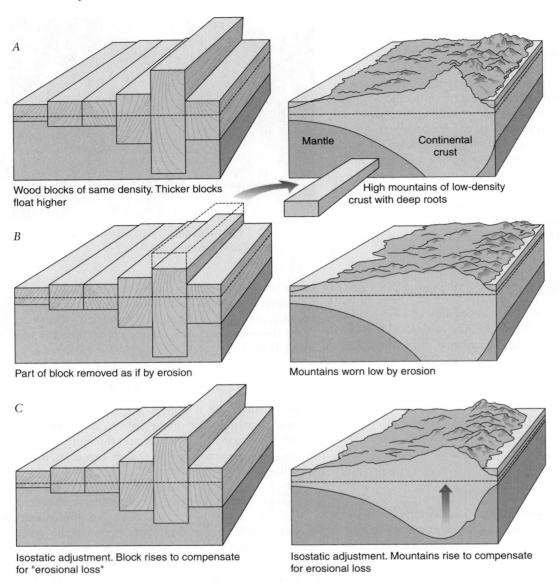

A

Wood blocks of same density. Thicker blocks float higher

High mountains of low-density crust with deep roots

Mantle

Continental crust

B

Part of block removed as if by erosion

Mountains worn low by erosion

C

Isostatic adjustment. Block rises to compensate for "erosional loss"

Isostatic adjustment. Mountains rise to compensate for erosional loss

FIGURE 7-11 **Isostatic adjustments.** Isostasy explains why mountain ranges experience uplifts long after erosion has removed huge volumes of rock. (A) Before isostatic adjustment. (B) Erosional removal by streams, glaciers, and gravity. (C) Uplift to regain isostatic equilibrium.

of Earth's crust float at different heights on the mantle: Denser **oceanic crust** floats lower, and less-dense continental crust floats higher. Isostasy also explains why mountain ranges experience repeated uplifts after erosion removes great volumes of rock and soil.

A little "bathtub geology" (Fig. 7-11) illustrates the concept. Assume you have blocks of low-density wood, such as pine, to represent the continental crust. When placed in a bathtub filled with water, the thicker pile of blocks will float higher. Imagine that the thickest pile of blocks represents a mountain range. Lift the upper block, and notice that the pile of blocks will rise to a new level. Similarly, there would be a rise in a moun-

tain range when erosion removed the "upper block," upset the gravitational balance, and caused an isostatic adjustment.

BROKEN, SQUEEZED, OR STRETCHED ROCKS PRODUCE GEOLOGIC STRUCTURES

Continents, mountains, and deep oceans appear everlasting, but only because our lifespan is too short to let us watch lands in upheaval, continents ponderously splitting and colliding, and ocean basins expanding and contracting. We only witness occasional manifesta-

FIGURE 7-12 **Two faults have broken these sandstone strata.** Note how the light-colored beds labeled "A" have been offset. The strata, exposed near Camberra, Australia, are Silurian in age (410–435 million years old). ◀ *Select one of the beds and color it lightly on both sides of the fault on the left so that the displacement is easier to see.*

tions of these changes, like earthquakes and volcanic eruptions. The results of long-term changes, however, are recorded in rocks as they are deformed—bent, folded, and broken. Among the effects of Earth movements are faults and folds.

Faults

A **fault** is a break in Earth's crustal rocks along which there has been displacement (Fig. 7-12). According to the relative movement of rocks on either side of the plane of breakage (the "fault plane"), faults can be classified as **normal**, **reverse**, or **lateral**. These three types are shown in Figure 7-13.

Lateral faults are further designated as *right lateral* or *left lateral*. To decide which of the two has occurred, you simply look across the fault line to see if the opposite block has moved to your right or left. The San Andreas fault in California is a right-lateral fault (Fig. 7-14). Lateral faults are also called **strike-slip faults**.

The body of rock that lies above the fault plane (the surface along which movement has occurred) is called the **hanging wall** (Fig. 7-15). In normal faults, the hanging wall appears to move *downward* relative to the opposite side or **footwall**. Such faults occur where rocks are subjected to tensional forces—forces that tend to stretch the crust.

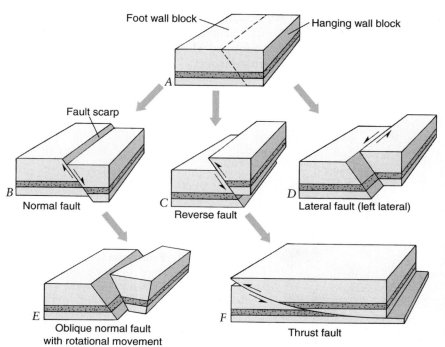

FIGURE 7-13 **Types of faults.** (A) The dashed line shows a potential fault location in unfaulted rock. If a fault occurs, three types of movement are possible, depending on the direction of stress on the rocks: (B) a normal fault, where one side drops down compared to the other; (C) a reverse fault, where one side is pushed up compared to the other; or (D) a lateral fault, where two sides move past one another. Often, faults are not purely of one type but may combine movements, as in (E). A thrust fault (F) is a reverse fault, but the fault plane is inclined at a very low angle. ◀ *Which of these fault types are not the result of compressional forces?*

FIGURE 7-14 **Aerial photograph of the San Andreas fault as it crosses the Carrizo Plain about 100 miles north of Los Angeles.** Note the offset of the stream in the lower part of the photograph. ▌ *What evidence on the photograph indicates this is a right lateral fault?*

On the other hand, in reverse faults, the hanging wall has moved *up* relative to the foot wall. Reverse faults in which the shear zone is inclined only a few degrees from horizontal are termed **thrust faults**. Imagine holding two wood blocks like those in Figure 7-13. They would have to be shoved together, or compressed, to cause reverse faults. Thus, regions of Earth's crust containing numerous reverse faults (and folded strata) are likely to have been compressed at some time in the geologic past. Although examples of all kinds of faults may be found in any great mountain belt, compressional structures are by far the most common.

Folds

Rocks do not always break when stressed; they may bend. Bends in rock strata are termed **folds** (Fig. 7-16). Geologists measure strike and dip to describe the orientation of folded or inclined beds:

- **Strike** is the compass direction of the line produced by the intersection of an inclined stratum (or other feature such as a fault or joint plane) with a horizontal plane (Fig. 7-17).
- **Dip** is the angle of inclination of the tilted layer, fault, or joint plane. It is also measured from the horizontal plane. The dip of the strata being measured by the geologist in Figure 7-18 is 30°.

The principal categories of folds are anticlines, synclines, monoclines, domes, and basins (Fig. 7-19):

- **Anticlines** are uparched layered rocks (think of McDonald's Golden Arches). On *eroded* anticlines, the oldest strata are in the center and the youngest are on the flanks (Figs. 7-19A and 7-20).
- **Synclines** are the opposite—downwardly folded rocks (think "sink-cline"). When eroded, the youngest strata are in the center and oldest strata are on the flanks (Figs. 7-19B and 7-21). These age relationships make it easy to spot anticlines and synclines on geologic maps.
- **Monoclines** consist of a single bend in otherwise horizontal strata (Fig. 7-19C). Monoclines are often formed when there are vertical movements of crustal blocks.

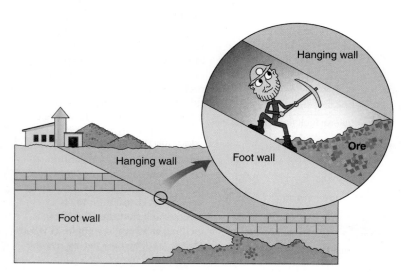

FIGURE 7-15 **Fault terminology.** Miners named the upper wall of a fault (along which ore deposits may occur) the *hanging wall*, because it was the wall "hanging" over their heads. The lower wall, on which miners placed their feet, was termed the *foot wall*. ▌ *What type of fault is depicted?*

FIGURE 7-16 **Severely folded strata.** These are exposed on the north side of the Rhone River valley west of Sierre, Switzerland.

Folds tend to occur together, like a series of frozen wave crests and troughs. The erosion of anticlines and synclines characteristically produces a topographic pattern of ridges and valleys (Fig. 7-22). The ridges develop where resistant rocks project at the surface. Valleys develop along those parts of the fold underlain by more easily eroded rocks like shales.

Domes and **basins** (Fig. 7-19D and 7-19E) are similar to anticlines and synclines except that they have elliptical-to-circular shapes. A dome is uparched strata where the beds dip (are inclined) in all directions away from the center. In contrast, a basin is a downwarp in which beds dip from all sides toward the center of the structure.

The erosionally truncated beds of simple domes and basins may form a circular pattern of ridges and valleys. As is the case with anticlines and synclines, older beds are found in the center of truncated domes and younger strata at the center of basins. Domes and basins vary in size from several meters to hundreds of kilometers in diameter.

PLATE TECTONICS THEORY TIES IT ALL TOGETHER

Perhaps more than other scientists, geologists are accustomed to viewing Earth in its entirety. It is their task to assemble the multitude of observations about the origin of mountains, the growth of continents, and the history of ocean basins into a coherent view of the whole Earth. Exciting and revolutionary discoveries made over the past four decades have provided that kind of integrated understanding of our planet. The current view of Earth's dynamic geology, called **plate tectonics**, is based on the movements of plates of lithosphere driven by convection in the underlying mantle.

As with many scientific breakthroughs, plate tectonics theory was not fully conceptualized until a foundation of relevant data had been assembled. These data arrived from several sources:

- In the years following World War II (1939–1945), research related to naval operations produced submarine detection devices. These also proved useful in measuring the magnetic properties of Earth's oceanic crust.

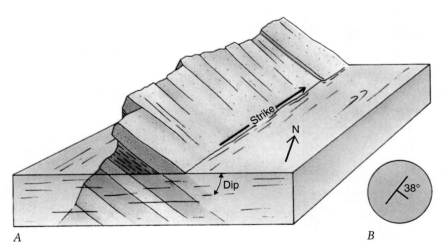

A

B

FIGURE 7-17 **Strike and dip of tilted strata.** (A) *Strike* is the line of intersection of a horizontal plane (the water surface in this example) and the inclined (dipping) bed. *Dip direction* is always perpendicular to the strike. *Dip angle* is the angle formed by the horizontal surface and the tilted strata. In this example, the dip angle is about 38°. (B) The map symbol indicating the strike and angle of the dip.

FIGURE 7-18 **Geologist measuring the dip of strata in a roadcut.**

• With the development of nuclear weapons (which first were exploded in 1945), the need to monitor atomic explosions resulted in the establishment of a worldwide network of seismographs. This network offered an important side-benefit: It provided precise information about the global pattern of earthquakes.

• The magnetic field over large portions of the seafloor was soon to be charted by the use of a newly developed instrument called a magnetometer.

Other technological advances permitted scientists to examine rock that had been dated by isotopic methods and then to determine the nature of Earth's magnetic field at the time those rocks had formed. Geologically recent reversals of the magnetic field were soon detected, correlated, and dated. A massive federally funded program to map the bottom of the oceans was launched. Improved echo-sounding devices not only provided continuous topographic seafloor profiles (Fig. 7-23), but also provided images of the rock and sediment that lay beneath. These innovations in echo-sounding further facilitated the mapping efforts.

A new picture of the little-known ocean floor began to emerge (Fig. 7-24). It was at once awesome, alien, and majestic. Great chasms, flat-topped submerged mountains, boundless abyssal plains, and globe-spanning volcanic ranges appeared on the new maps and begged an explanation. How did the volcanic midoceanic ridges and deep-sea trenches originate? Why were both so prone to earthquake activity? Why was the Mid-Atlantic Ridge so nicely centered and parallel to the coastlines of the continents on either side?

As the topographic, magnetic, and geochronologic data accumulated, the relationship of these questions became apparent. An old theory called *continental drift* was re-examined, and the new, more encompassing theory of plate tectonics was formulated. It was an idea whose time had come.

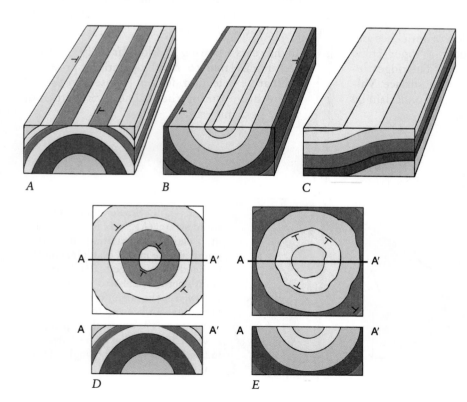

A   *B*   *C*

D   *E*

FIGURE 7-19 **Types of folds.**
(A) Anticline cross-section.
(B) Syncline cross-section.
(C) Monocline cross-section.
(D) Map view and cross-section of a dome. Notice that older strata are in the center of the outcrop pattern. (E) Map view and cross-section of a basin. Younger rocks occur in the central area of the outcrop pattern. Note the dip directions indicated by the strike-dip symbols. ▪ *Find Missouri on the Bedrock Geology map of North America in the Appendices. Is the structure beneath southern Missouri a structural basin or dome? What other clearly domal structure can you discern in nearby states?*

FIGURE 7-20 **Anticline exposed by river erosion, Teton County, Montana.**

FIGURE 7-21 **Syncline illustrating that geologic structures often have no relationship to surface topography.** Looking at this mountain ridge before the road cut was carved, you might expect an anticline to lie beneath it. But it is just the opposite: a syncline, with its rock layers downfolded in the center. The State of Maryland's Geological Survey maintains an interpretive center for visitors to this spectacular Sideling Hill road cut along Interstate 68 in western Maryland. ◪ *As the driver of the "18-wheeler" in the lower right area of the photograph continues toward the axial plane (center) of this structure, will he be passing increasingly younger or older strata?*

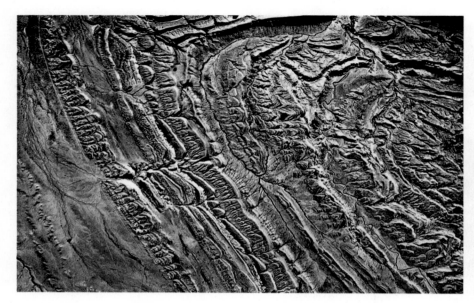

FIGURE 7-22 **Aerial photograph of an anticlinal fold in Wyoming.** Because of the sparseness of vegetation and soil cover, rock layers are clearly exposed. Differences in resistance to erosion cause layers to be etched into sharp relief. Notice how the beds dip away from the center of the anticline. *How has the geologic structure in this area affected drainage patterns?*

DRIFTING CONTINENTS

If you look at a world map, it's hard not to notice that the continental shorelines on either side of the Atlantic Ocean are remarkably parallel. If the continents were pieces of a jigsaw puzzle, you could easily fit the great "nose" of Brazil into the African coastline. Similarly, you could insert Greenland between North America and northwestern Europe. Not surprisingly, earlier generations of map gazers also noticed this fit. It led them to formulate hypotheses for the breakup of an ancient supercontinent.

Early Hypotheses

In 1858, a theorist named A. Snider postulated that before the time of Noah and the biblical flood, there existed a great region of dry land. This antique land developed great cracks encrusted with volcanoes, and during the Great Deluge of the Old Testament, a portion separated at a north–south trending crack and drifted westward, later to be named North America.

Near the close of the nineteenth century, the Austrian scientist Eduard Suess (1831–1914) became particularly intrigued by the many geologic similarities

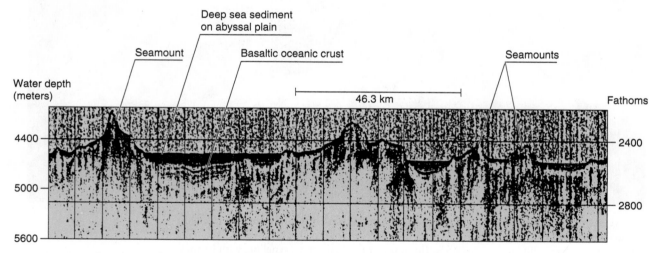

FIGURE 7-23 **Continuously recorded profiles along the edge of the Mid-Atlantic Ridge.** Note the abyssal hills, plains, and seamounts. *The summits of the submarine volcanic mountains are at increasingly greater depth as we proceed from left to right. What might be the reason for this?*

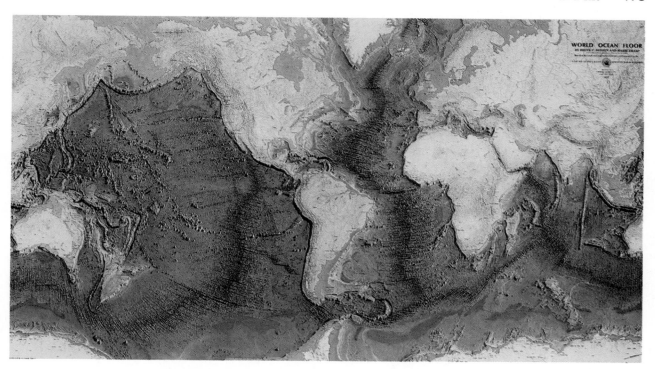

FIGURE 7-24 **Midoceanic ridges, deep-sea trenches, and other features of the ocean floors.**

shared by India, Africa, and South America. He developed a more complete theory involving the breakup of a supercontinent he named Gondwanaland, after a geologic province in east-central India.

Alfred Wegener

The next serious effort to convince the scientific community of the validity of these ideas was made in the early decades of the twentieth century by the German geographer and meteorologist Alfred Wegener (1880–1930). His 1915 book, *Die Entstehung der Kontinente und Ozeane* (*The Origin of the Continents and Oceans*), was a milestone in developing the concept of continental drift.

Wegener's hypothesis was straightforward. Building on the earlier notions of Suess, he argued again for the existence in the past of a supercontinent that he dubbed **Pangea** ("all Earth"). That portion of Pangea that was to separate and form North America and Eurasia came to be known as **Laurasia**, whereas the southern portion retained the earlier designation of Gondwanaland (now shortened to **Gondwana**).

According to Wegener, Pangea was surrounded by a universal ocean named **Panthalassa** ("all seas"), which opened to receive the shifting continents after they had split apart about 200 million years ago (Fig. 7-25). The fragments of Pangea drifted along like great stony rafts on the denser material below. In Wegener's view, the

bulldozing forward edge of the slab would crumple and produce mountain ranges such as the Andes.

Wegener's theory of continental drift was sharply challenged. The chief criticism was against his concept that continents somehow slid through an unyielding oceanic crust. The seafloor was far too rigid to allow passage of continents, no matter what the driving mechanism. But we now know that continents do indeed move—not by themselves, but as passengers on large rafts of lithosphere that glide over a comparatively soft upper layer of Earth's mantle. Nevertheless, much of the evidence for drifting continents that Wegener and others had assembled can be used to substantiate both the older and the newer concepts.

▶ EVIDENCE FOR CONTINENTAL DRIFT

Clues from Global Geography

The most convincing evidence for continental drift is the jigsaw-like geographic fit of continents. The fit is far too good to be coincidental. It remains good even after erosion and deformation modified coastlines following the breakup of the supercontinent Pangea about 200 million years ago. We see an even closer fit at the edges of the continental shelves, which are really submerged portions of the continents. Figure 7-26 is a computerized and error-tested matching of continents.

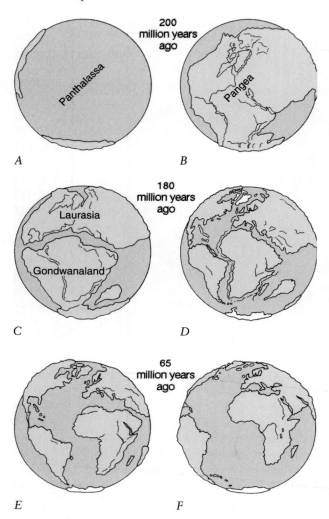

A *B*

C *D*

E *F*

200 million years ago

180 million years ago

65 million years ago

FIGURE 7-25 **The break-up of Pangea.** (A and B) Alfred Wegener's view of Earth 200 million years ago. At this time, there was one ocean (Panthalassa) and one continent (Pangea). (C) 20 million years later, the supercontinent had begun to split into northern Laurasia and southern Gondwanaland. (D) Fragmentation continued and (E) looked this way about 65 million years ago. (F) Further widening of the Atlantic and northward migration of India bring Earth to its present state.

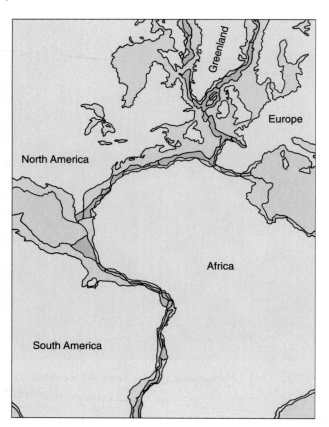

FIGURE 7-26 **Fit of the continents.** The fit was made along the continental slope (green) along the contour line for a depth of 500 fathoms (3000 feet). Overlaps and gaps (orange) probably result from deformations and sedimentation after rifting.

Clues from Paleoclimatology

In widely separated pieces of Gondwana, such as South America, southern Africa, India, Antarctica, Australia, and Tasmania, there are grooved rock surfaces (Fig. 7-27). These were formed by an immense continental glacier that covered part of Gondwana when it was positioned at or near the South Pole. When present-day continents in the Southern Hemisphere are assembled in jigsaw-puzzle fashion to form Gondwana, the glacial grooves match up perfectly (Fig. 7-28). Further evidence includes glacial conglomerates called tillites on each of the continents.

FIGURE 7-27 **Glacial striations on quartzite bedrock.** These striations were scratched into Proterozoic quartzite in Australia during the Permian Period, 250–295 million years ago.

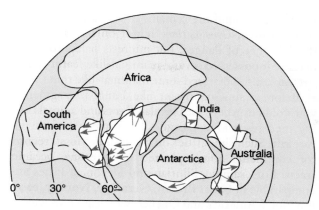

FIGURE 7-28 **Reconstruction of the supercontinent Gondwana near the beginning of Permian time, 250–295 million years ago.** Glacial deposits appear in white. Arrows show the direction of ice movement, as determined from glacial striations on bedrock.

Other clues to paleoclimatology test the concept of moving landmasses. Tropical trees characteristically lack annual rings, which result from seasonal variations in growth. Exceptionally thick coal seams that contain fossil logs lacking growth rings imply that the coal formed in a tropical climate. Yet these coal seams are now located in the United States, Great Britain, Germany, and Russia—hardly tropical areas today. The lands on which the coal seams formed, therefore, must have moved northward from a tropical area.

Rocks deposited in warm arid environments often include deposits of evaporites. From Chapter 4, recall that evaporites are chemical precipitates such as salt and gypsum that characteristically form when a body of water containing dissolved salts is evaporated. Today, evaporites ordinarily form in regions located about 30° north or south of the equator. If you believe that continents have always been where they are now, then it is difficult to explain the great Permian evaporite deposits found in northern Europe, the Urals, and the southwestern United States. The evaporites originated in warmer latitudes before Laurasia migrated northward.

A similar kind of evidence can be obtained by examining the locations of Permian reef deposits. Modern reef corals are restricted to a band around Earth that is within 30° of the equator. Fossil reefs are now found far to the north of the latitudes where they formed.

Clues from Fossils

At least some of the paleontologic support for continental drift was well known to Suess and Wegener. In the Gondwana strata overlying the tillites and glaciated surfaces, there are nonmarine sedimentary rocks and coal beds containing a distinctive assemblage of fossil plants. The plants are the **Glossopteris flora**, named for a prominent member of the assemblage (Fig. 7-29).

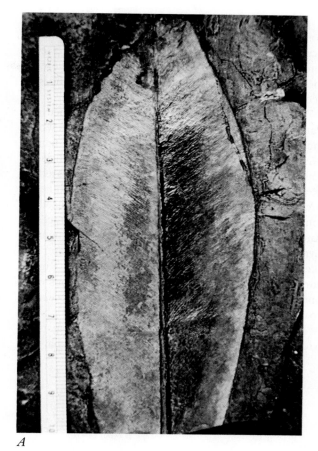

A

B

FIGURE 7-29 **Fossil *Glossopteris* leaf in coal from Antarctica (A) and reconstruction of a *Glossopteris* tree (B).** The leaf is from coal deposits that derived from glossopterid forests of Permian age. This fossil is from Polarstar Peak, Ellsworth Land, Antarctica.

Paleobotanists agree that it would be virtually impossible for this complex flora to have developed identically on the southern continents, if they were separated as they are today. And *Glossopteris* seeds were far too heavy to be blown over such great distances of ocean by the wind.

Vertebrates provide even further evidence for Pangea. A partial list of Gondwana vertebrates would include a small Permian aquatic reptile named *Mesosaurus* (Fig. 7-30), known only from localities in southern Africa and southern Brazil. As indicated by skeletal characteristics and enclosing sediment, *Mesosaurus* inhabited freshwater bodies and used its long sharp teeth to catch fish and small crustaceans. Another vertebrate indicator of adjoined continents is *Lystrosaurus*. This stocky, sheep-sized plant-eater left skeletal remains in Africa, India, Antarctica, Russia, and China, indicating that these separate regions were once a single continent across which *Lystrosaurus* roamed freely.

In somewhat younger strata, paleontologists in widely separated Africa, South America, China, and Russia have found the remains of a small carnivorous reptile named *Cynognathus*. It not only corroborates the existence of Pangea, but interests paleontologists because it possessed several mammal-like features. *Lystrosaurus* also possessed some mammal-like traits.

Before a supercontinent like Laurasia fragments, you would expect to find similar plants and animals living at corresponding latitudes on either side of the future rift. Such similarities do occur in the fossil record for continents now separated by extensive oceanic tracts. For example, Silurian and Devonian fishes are comparable in Great Britain, Germany, Norway, eastern North America, and Quebec, which are now widely separated locations.

In Permian and Mesozoic age rocks, we again see striking similarities in reptilian faunas of Europe and North America. The fossil evidence implies the former existence of continuous land connections, as well as a uniformity of environmental conditions between regions of Laurasia that had been at similar latitudes.

When we turn to Cenozoic mammalian faunas, however, we find a very different situation. The mammals of Australia, South America, and Africa differ. As continents grew separated by ocean barriers, the genetic isolation resulted in morphologic divergence. The modern world's splendid biologic diversity results at least partially from evolutionary processes operating on more-or-less isolated continents. During the periods prior to the break-up of Pangea, the faunas and floras were less diverse.

Clues from Rock Sequences

Regions once close to one another on a former supercontinent should exhibit a general similarity in stratigraphic sequence. Table 7-1 depicts some of these similarities. Near the base of the sections are tillites, indicating glacial conditions. We know that the cold did not persist, because the temperate *Glossopteris* flora lies above the tillite zones. Volcanism, including basaltic lava flows near the top of each section, reflect the opening of fissures as Gondwana began to break apart.

| Pre-Є | Є | O | S | D | M | P | Pr | Tr | J | K | T | Q |

FIGURE 7-30 **Small freshwater reptile *Mesosaurus*.** Fossil remains are from Early Permian strata found in a limited area along the Brazilian and West African coasts.

► PALEOMAGNETISM: ANCIENT MAGNETISM LOCKED INTO ROCKS

Wegener could not explain what breaks continents apart and moves them. But he did assemble convincing evidence that such events had occurred. Discoveries in the late 1950s and 1960s provided further evidence to substantiate his ideas. The new information came from the study of magnetism that is acquired by ancient rocks as they form. To understand this **paleomagnetism**, let us look at Earth's present magnetic field.

TABLE 7-1 **Gondwana Correlations**

System	Southern Brazil	South Africa	Peninsular India
Cretaceous	Volcanics	Marine sediments	Volcanics Marine sediments
Jurassic	(Jurassic rocks not present)	Volcanics	Sandstone and shale Volcanics Sandstone and shale
Triassic	Sandstone and shale with reptiles	Sandstone and shale with Reptiles	Sandstone and shale with reptiles
Permian	Sandstone and Shale with *Glossopteris* flora Shale with *Mesosaurus*	Sandstone and shale with coal and *Glossopteris* Shale with *Mesosaurus*	Shale and sandstone with coal and *Glossopteris* Shale
Carboniferous	Sandstone, shale, and coal with *Glossopteris* Tillite	Sandstone shale and coal with *Glossopteris* Tillite	Tillite

To visualize Earth's magnetic field, think of our planet as having a bar magnet in its interior (Fig. 7-31). The ends correspond to the north and south geomagnetic poles. Today these geomagnetic poles are about 11° of latitude different from Earth's rotational axis, so we know they slowly shift position. However, when averaged over several thousand years, the geomagnetic poles and the geographic poles are located relatively close to each other. If we assume that this relationship has always held true, then we can approximately calculate ancient magnetic pole positions from paleomagnetism in rocks in order to locate Earth's former geographic poles.

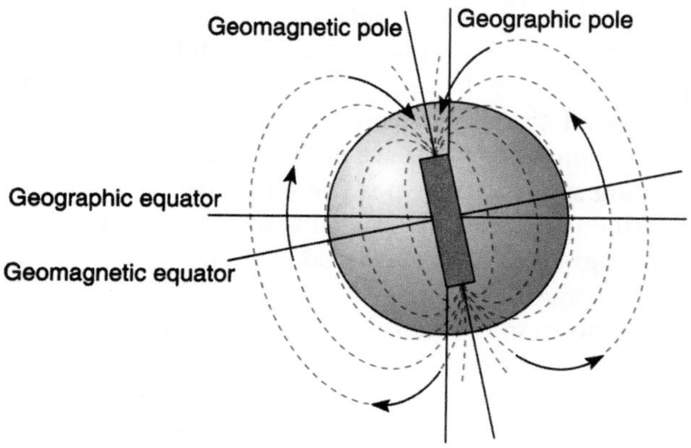

How Is Earth's Ancient Magnetic Field Recorded in Rocks?

Imagine the outpouring of basaltic lava from a volcano. As the lava begins to cool, tiny crystals of magnetite form. When the basalt's temperature falls below 580°C (called the *Curie point*), the crystals become magnetized in Earth's magnetic field. As the lava cools further and solidifies, the magnetite crystals become locked in place and thereby record their geographic position (Fig. 7-32).

Here is a simple analogy: The magnetic orientations of minerals respond as if they were tiny compass needles floating freely in a liquid, aligning parallel to the magnetic lines of force surrounding Earth. They point toward the magnetic poles. Also, the closer they are to the poles, the more they tilt. Thus, at the equator they are horizontal, and at the poles they are vertical (Fig. 7-33). This tilt, or inclination, is used to determine the latitude at which an igneous body containing magnetic minerals cooled and solidified.

FIGURE 7-31 **Dipole model of Earth's magnetic field.** To visualize the magnetic field, think of our planet as having a bar magnet in its interior. The ends correspond to the north and south geomagnetic poles.

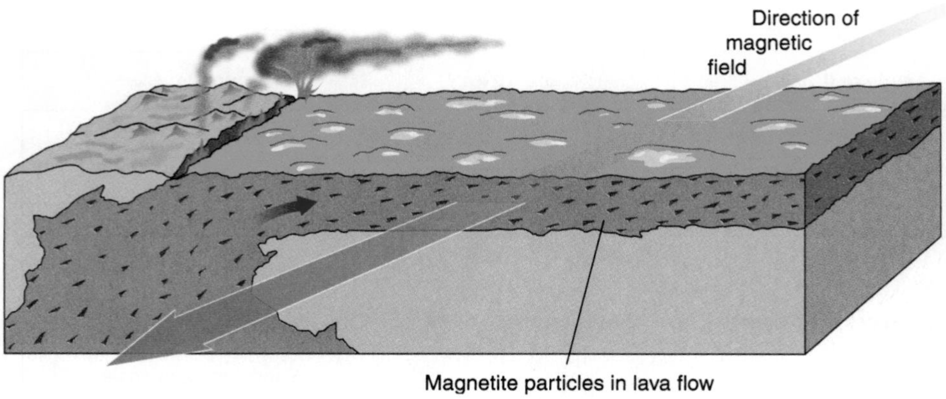

FIGURE 7-32 **Magnetite crystals forming in a lava flow that is being extruded during a fissure eruption.** The magnetite crystals become aligned to Earth's magnetic field in this particular time and place. The crystal orientation becomes preserved as the lava solidifies.

Although igneous rocks are best for paleomagnetic determinations, they are not the only kinds of Earth materials that can acquire magnetism. In lakes and seas that receive sediments eroded from nearby land areas, tiny grains of magnetite settle slowly through the water and rotate so they parallel Earth's magnetic field at the moment. They may continue to move into alignment while the sediment is still wet and uncompacted, but once the sediment becomes cemented or compacted, the paleomagnetism is locked in.

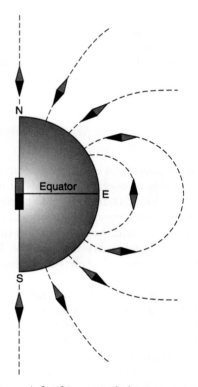

FIGURE 7-33 **A freely suspended compass needle aligns itself in the direction of Earth's local magnetic field.** The inclination (downward dip) of the needle varies from horizontal at the equator to vertical at the poles.

Earth's Wandering Magnetic Poles

Paleomagnetic data confirms that continents have continuously changed position. For example, when ancient pole positions are plotted on maps, we can see that they were in different positions relative to a particular continent at different times in the geologic past. So, either the poles moved relative to stationary continents, or the poles remained fixed while the continents shifted.

If the poles wander and the continents stay put, then a geophysicist working on the paleomagnetism of Ordovician rocks in France, for example, should arrive at the same location for the Ordovician poles as a geophysicist doing similar work on Ordovician rocks from the United States. The paleomagnetically determined pole positions for a particular age would be the same for all continents.

On the other hand, if the continents wander and the poles stay put, then we should find that pole positions for a particular geologic time would be different for different continents. The data suggest this is the correct interpretation.

Another way to view the data of paleomagnetism is to examine what are called **apparent polar wandering paths.** The word *apparent* in the expression is important—if the expression were simply "wandering paths," this would imply that the poles actually do wander. This is highly unlikely. Instead, the *apparent* polar wandering paths are merely lines on a map connecting ancient pole positions relative to a specific continent for various times during the geologic past. They *appear* to have wandered when plotted on a map in this way.

As shown in Figure 7-34, the paths for North America and Europe met in recent time at the present North Pole. This means that the paleomagnetic data from recently formed rocks from both continents indicate the same pole position.

A plot of the more ancient poles shows two similarly shaped but increasingly divergent paths. If this divergence resulted from Europe and North America drift-

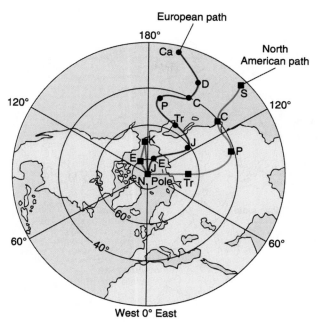

FIGURE 7-34 **Highly mobile locations of Earth's north magnetic pole during the past half-billion years.** To make this general map, its creators averaged multiple data points for each geologic period—**C**ambrian, **S**ilurian, **D**evonian, **C**arboniferous (Pennsylvanian and Mississippian together), **P**ermian, **T**riassic, **J**urassic, **K**(Cretaceous), **E**ocene. The black line shows the path from observations in Europe, and the red line from North America.

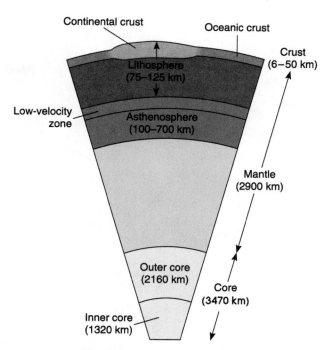

FIGURE 7-35 **Divisions of Earth's interior.** Scale is distorted for easier reading.

ing apart, then you should be able to imagine the movements reversed and see if the paths come together. The Paleozoic portions of the polar wandering path can be brought into close agreement, if North America and its path are slid eastward about 30° toward Europe. This sort of data strongly supports the concept of tectonic plate movements.

◢ TODAY'S PLATE TECTONICS THEORY

Here is a quick plate tectonics review from Chapter 1: Recall that Earth's outer shell, the **lithosphere**, is constructed of thick "plates" or slabs of crust. The larger plates are approximately 75–125 km thick (Fig. 7-35). Of the plates, 7 are huge, with about 20 smaller ones squeezed in between (Fig. 7-36A). Movement of the plates causes them to converge (collide), diverge (spread apart), or grind past one another, resulting in earthquakes along plate margins. When we plot earthquake patterns on a world map, they clearly define tectonic plate boundaries (Fig. 7-36B).

A plate that hosts a continent has the configuration shown in Figure 7-37. Lithospheric plates "float" on the weak, partially molten region of the upper mantle, which is called the **asthenosphere** (Greek *asthenos* = weak).

If a plate is moving away from a **spreading center**, it must simultaneously be colliding with another plate and grinding past still others. Thus, in addition to the divergent plate boundaries that occur along midoceanic ridges, there are convergent and transform boundaries. Let us see what happens at each of the three types of plate boundaries: divergent boundaries, transform boundaries, and convergent boundaries.

Seafloor Spreading (Divergent Boundaries)

Plates move apart at **divergent plate boundaries**, which may manifest as midoceanic ridges complete with tensional ("pull-apart") geologic structures. For example, the Mid-Atlantic Ridge approximates the dividing line between the American plate and the Eurasian–African plates (Fig. 7-38).

As you would expect, such a rending of the crust is accompanied by earthquakes and outpourings of volcanic materials that are piled high to produce the ridge itself. The opening between the separating plates is also filled with this molten rock, which rises from below the lithosphere and solidifies in the fissure. Thus, new oceanic crust (*new seafloor*) is added to the **trailing edge** of each separating plate as it moves slowly away from the midoceanic ridge (Fig. 7-39).

Zones of divergence also may originate beneath continental crust, rupturing the overlying landmass and producing rift valleys like the Red Sea, Gulf of Aden, and East African rift.

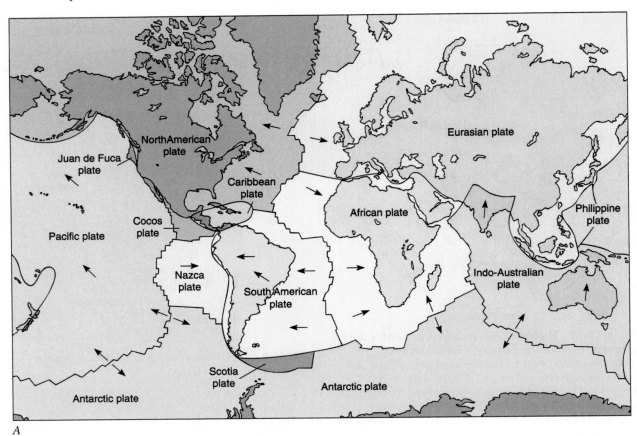

A

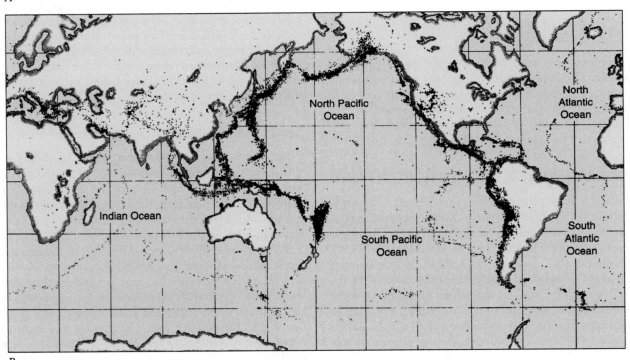

B

FIGURE 7-36 **Earth's major tectonic plates (A) and world distribution of earthquakes (B).** In (A), arrows indicate the general direction of plate movement. For each plate, the pastel color shows the entire plate area, and the deeper color shows the continental mass above sea level on the plate. In (B), note three major earthquake belts: 1. encircling the Pacific Ocean (the "Ring of Fire"), 2. extending from the Mediterranean Sea toward the Himalayas and East Indies, and 3. the midoceanic ridges.

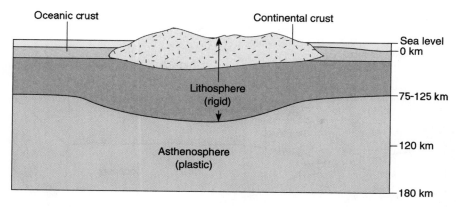

FIGURE 7-37 **Earth's lithosphere.** The lithosphere bears both oceanic and continental crust. It overlies the weak, partially molten rocks of the asthenosphere.

As a continent drifts away from the zone of divergence, the edge of the continent becomes a **passive continental margin**. The eastern margin of North America is a passive continental margin. Such margins are called passive because earthquake and volcanic activity is rare.

This whole process is called **seafloor spreading**. Recognition of seafloor spreading resulted from several observations made by Harry H. Hess (1906–1969), of Princeton University. During World War II, Hess commanded an attack transport in the Pacific. To determine water depth just offshore of enemy-held islands, the ship was equipped with an echo sounder. When not employed for that purpose, however, Hess used the device to produce topographic profiles of the ocean floor.

GUYOTS—FLAT-TOPPED SEAFLOOR VOLCANOES The profiles showed midoceanic ridges, deep-sea trenches, seamounts (submarine volcanoes), and seafloor volcanoes

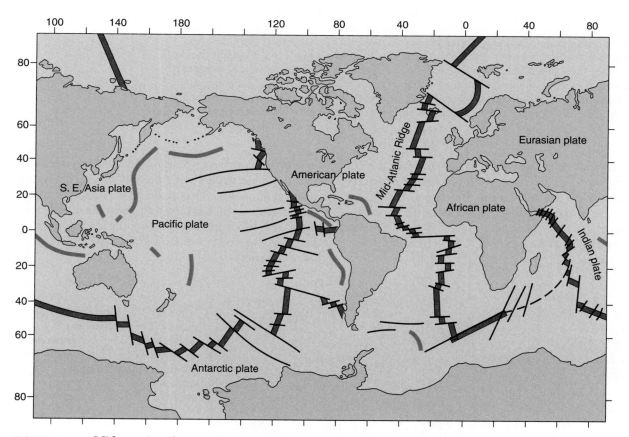

FIGURE 7-38 **Midoceanic ridges (red) and trenches (blue).** Note the many perpendicular transform fault zones that offset each ridge.

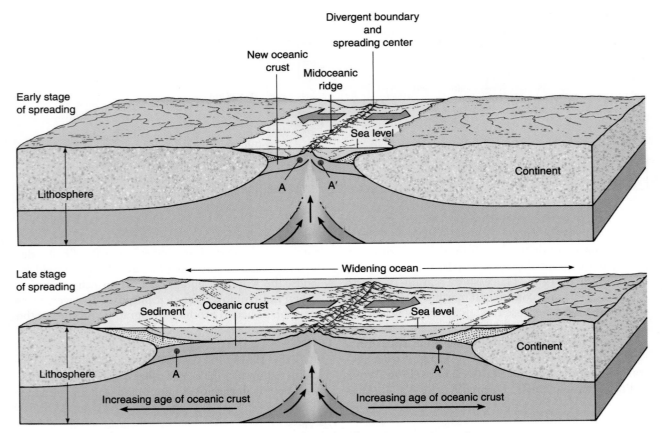

FIGURE 7-39 **Seafloor spreading marks a divergent boundary between two tectonic plates.** *Top:* Early in spreading, the rift widens and newly formed crust of each plate diverges at the ridge axis. In this early stage, reference points **A–A′** are close together. Here the crust stretches and fractures into numerous transform faults (detailed in Fig. 7-41). Lava fills the gap between the diverging plates. As the lavas are cooled by the ocean water and solidify along the ridge, they form the trailing edges of two tectonic plates. *Bottom:* As spreading continues over millions of years, the widening basin becomes an expansive ocean (for example, the Atlantic). The age difference between the leading and trailing edges progressively increases, up to roughly 200 million years. Note that reference points **A** and **A′** have moved far apart.

with flat tops called **guyots** (*GHEE-oh*). Hess noticed that guyots located farther from midoceanic ridges lay at increasingly greater depths. He hypothesized that guyots once were volcanoes that formed at the midoceanic ridges by upwelling lavas. Their summits were originally above sea level, but were truncated by wave erosion.

With time, a guyot was transported on its plate away from the ridge, while simultaneously sinking below the waves (Fig. 7-40). This is because oceanic crust is warmer and at a higher elevation along the midoceanic ridge. As seafloor spreading moves the plate away from the ridge, the crust cools and gradually descends to a lower elevation.

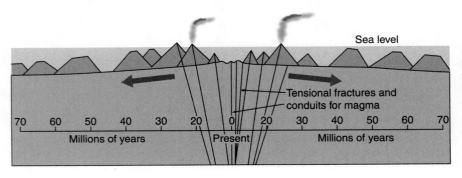

FIGURE 7-40 **The origin of guyots.** The development of guyots begins with the formation of volcanoes along midoceanic ridges. The tops of some of these volcanoes are leveled by wave erosion. These flat-topped volcanoes are then carried by plate movement down the flanks of the midoceanic ridge until submerged.

Hess envisioned ocean floors as continuously moving conveyor belts that moved from midoceanic ridges, where new floor was added by upwelling lavas, to deep-sea trenches, where the crust plunged downward to be consumed in the mantle. This spreading mechanism accounts for the characteristics of guyots, the thinness of the oceanic crust, the absence of oceanic crust much older than about 200 million years, and the lack of sediment older than Jurassic on top of the basalt.

If a plate is moving away from a spreading center, it must simultaneously be colliding with another plate and grinding past still others. Thus, in addition to the divergent plate boundaries that occur along midoceanic ridges, there are transform and convergent boundaries.

Transform Boundaries

The line along a midoceanic ridge where spreading begins is neither straight nor smoothly curving. Rather, it is offset by numerous faults called **transform faults**. They are an expected consequence of horizontal spreading of the seafloor along Earth's curved surface. The relative motions of transform faults are shown in Figure 7-41.

Because the seafloor spreads outward from the ridge, the relative movement between the offset ridge crests is opposite what you would expect from ordinary fault movement. Thus, at first glance, the ridge-to-ridge transform fault (Fig. 7-41A) appears to be a left-lateral fault, but the actual movement along segment X-X' is really right-lateral. Notice that only the X-X' segment shows movement of one side relative to the other. To the west of X and east of X', there is little or no relative movement. Thus, the fault has been "transformed."

Transform faults may occur as the margin of a tectonic plate. Along such a **transform plate boundary**, two plates move sideways past each other.

An example of a transform plate boundary (also called a shear boundary) is California's infamous San Andreas fault. Along this great fault, the Pacific plate moves northward against the American plate (Fig. 7-42). Transform plate boundaries obviously are earthquake-prone (consider the numerous California quakes) but are less likely to develop deep earthquakes or intense igneous activity. No new surface is formed or old surface consumed.

Convergent Boundaries

Convergent plate boundaries develop when two plates move toward one another and slowly collide. As you might guess, these convergent junctions experience frequent earthquakes. They also are zones along which compressional mountain ranges and deep-sea trenches develop (Fig. 7-43).

The structural configuration of the convergent boundary likely will vary according to the velocity of plate movement and whether the plates' leading edges are oceanic or continental crust. When the plates collide, one slab may plunge below the other, producing a **subduction zone**. The sediments and other rocks of this plunging plate are pulled downward (subducted) into the mantle, where they are heated. Water, driven out of the subducted rocks by the extreme heat, penetrates the overlying mantle. This lowers the melting temperature to the point at which partial melting occurs. The melts created in this way supply the lavas for volcanic island arcs, or become incorporated into the upper mantle and crust.

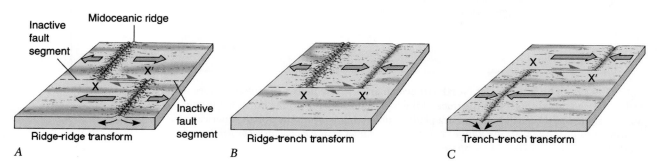

X – X' =
Active fault segment

Inactive fault segment

Midoceanic ridge

X'

X

Inactive fault segment

Ridge-ridge transform

A

Ridge-trench transform

B

X

X'

Trench-trench transform

C

FIGURE 7-41 **Three types of transform faults.** The ridge acts as a spreading center that exists on both sides of the fault. Relative movement on opposite sides along the fault segment is marked with red arrows. The rate of movement depends on the rate of extrusion of new crust at the ridge.

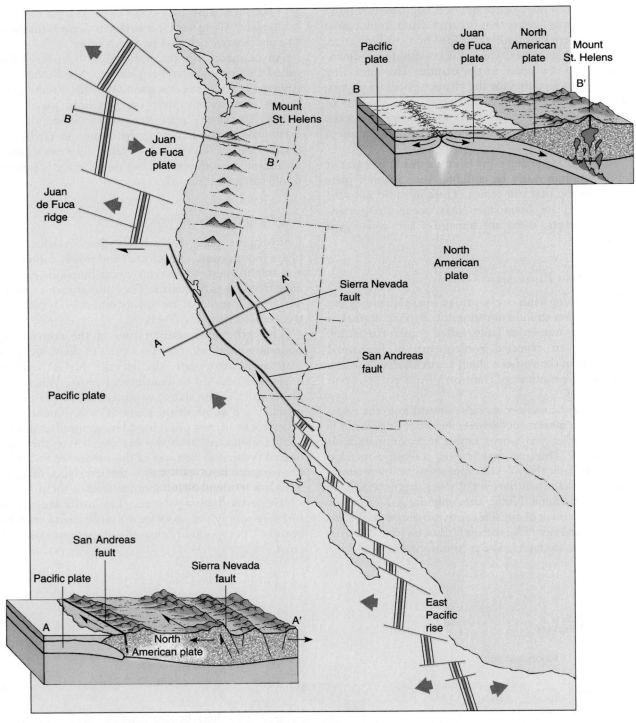

FIGURE 7-42 **The juncture of North American and Pacific plates.** The double lines with red are spreading centers. Note the trace of the San Andreas fault. To the north in Oregon and Washington, the small Juan de Fuca plate is subducting beneath the North American continent to form the Cascade Range. ◄ *Is the San Andreas fault a right-lateral or left-lateral fault? Can it also be considered a transform fault?*

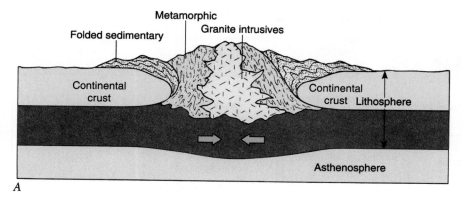

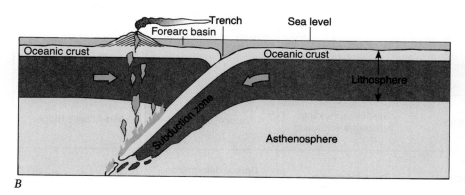

FIGURE 7-43 **Convergence: two types of convergent plate boundaries.** (A) Granitic mountains result when two *continental-crust* plates converge. (B) Volcanic islands result when two *oceanic-crust* plates converge. For convergence of an oceanic plate with a continental plate, see Figure 7-46.

WHAT HAPPENS AT PLATE MARGINS?

Table 7-2 summarizes features of the three plate boundary types—convergent, divergent, and transform. Note the events, topographic features, and examples of each type.

Some of the spectacular landscapes we see—some mountains, volcanoes, and islands—directly result from converging plates. There are three ways for the plates to converge, depending on the kinds of crust involved: (1) continental–continental crust convergence, (2) oceanic–oceanic crust convergence, or (3) continental–oceanic crust convergence.

Continental–Continental Crust Convergence

If the leading edge of a plate is composed of continental crust and it collides with a plate of similar composition, the result is a folded mountain range with a core of granitic igneous rocks (Fig. 7-43A). Because continental plate margins are too light and buoyant to be carried down into the asthenosphere, subduction does not occur in this type of continent-to-continent collision. Instead, the crust at the plate margins is deformed and may detach itself from deeper zones, and slabs of continental crust from one plate may ride up over the other. The zone of convergence between the two plates, rec-

ognized by severe folding, faulting, metamorphism, and intrusive activity, is called the **suture zone**. The typical result is formation of mountains.

A dramatic and ongoing example is the convergence of India with Eurasia, which generated the Himalayan Mountains (Fig. 7-44). Himalayan mountain-building began late in the Mesozoic, over 100 million years ago. At that time, the leading (northern) edge of the Indian tectonic plate was subducting beneath the southern margin of Eurasia (Fig. 7-45A). To the south of the subduction zone, the plate carried the continental mass of India.

As the subduction zone gobbled seafloor, the oceanic tract known as the **Tethys Sea** (which separated Eurasia from northward-moving India) narrowed until the landmasses collided (Fig. 7-45B). Sediments caught in the vise between the two continents were intensely deformed and pushed northward along great faults. Eventually, the northern border of India was firmly sutured against the buttress of Tibet (Fig. 7-45C). Northward movement of the Indian plate continues to this day.

Oceanic–Oceanic Crust Convergence

The second kind of convergent plate boundary involves the meeting of two plates, both of which have oceanic crust at their converging margins (Fig. 7-43B). Such

TABLE 7-2 **Features and Events at the Boundaries of Tectonic Plates**

Boundary	Opposing Plate Markings	Geologic Events	Topographic Features	Modern Examples
CONVERGENT	Continent-Continent	Crustal deformation, mountain building	Mountains	Himalayas
	Continent-Ocean	Subduction deformation, volcanism	Mountains, ocean trenches	Andes
	Ocean-Ocean	Subduction deformation, volcanism	Island arcs, ocean trenches	Western Aleutians
DIVERGENT	Continent-Continent	Continental fragmentation, rising of magma beneath rift, volcanism	Rift valleys	East African Rift
	Ocean-Ocean	Seafloor spreading, submarine volcanism	Midoceanic ridge	Mid-Atlantic Ridge
TRANSFORM	Continent-Continent	Crustal deformation	Crustal deformation along fault	San Andreas Fault
	Ocean-Ocean	Earthquakes	Offset of axes of midoceanic ridges	Offset, East Pacific Rise

(Earthquakes — spanning the Geologic Events column across all rows)

locations develop deep-sea trenches with bordering volcanic arcs, such as those of the southwestern Pacific Ocean.

Continental–Oceanic Crust Convergence

The third type of convergent plate boundary is the collision of plates where one is continental (granitic) and the other is oceanic (basaltic). Because their densities are different, the denser oceanic crust subducts beneath the lighter continental crust. The result is a deep-sea trench located offshore from an associated range of mountains. Volcanic activity accompanying subduction produces a compositional blend of granitic and basaltic lavas. The resulting rock is andesite, named for its prevalence in the Andes Mountains.

Continental–oceanic convergence (Fig. 7-46) produces distinctive rocks and geologic structures. The convergence of the two lithospheric slabs results in subduction of the denser oceanic plate. The less dense continental plate maintains its elevated position, but undergoes intense deformation and melting at depth.

Three terms describe the characteristics of the rocks in this highly dynamic continental–oceanic convergence:

- The contorted and metamorphosed body of rock compressed onto the margin of the continent is called an **accretionary prism**. It grows by accretion as material continues to be scraped off the descending plate and added to the mass already accreted.

- The complexly folded jumble of deformed and transported rocks found in the accretionary prism is called a **mélange**.

- An **ophiolite suite** (Fig. 7-47) is a distinctive assemblage of rocks composed of deep-sea sediment (containing marine microfossils), submarine basalts, and metamorphosed mantle rocks that occur in the accretionary wedge. Ophiolites are splinters of the oceanic plate that were scraped off the upper part of the descending plate and inserted into the accretionary prism at the margin of the continent. They mark the zone of contact between colliding continental and oceanic plates.

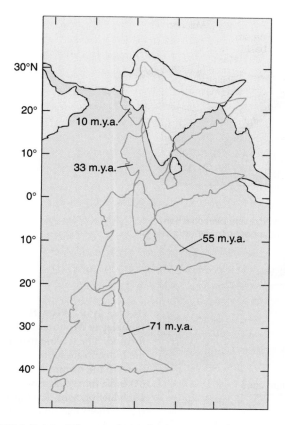

FIGURE 7-44 **The northward migration of India relative to Eurasia.** This shows the subcontinent's positions during four different periods. The shape of India's northern boundary is a matter of conjecture.

Another clue to the presence of such zones is the distinctive metamorphic rock containing blue minerals called **blueschists**. They form at high pressures but relatively low temperatures. This rather unusual temperature/pressure combination occurs in subduction zones where a relatively cool oceanic plate plunges rapidly into deep zones of high pressure.

Wilson Cycles: Closings and Openings of Oceanic Basins

Plate tectonics has been in operation since late Archean times (2.5 billion years ago) and possibly longer. Over this immense interval, there have been many openings and closings of ocean basins. Indeed, plate tectonics controls the birth and death of ocean basins. The Pacific Ocean opened only about 300 million years ago. If rates and directions of seafloor spreading remain the same, it will be closed in another 200 million years.

The opening of a new ocean basin along divergent zones, the expansion of the basin as seafloor spreading continues, and the ultimate closure of the basin as

plates converge is termed a **Wilson Cycle**. The name honors J. Tuzo Wilson (1908–1993) for his contributions to the theory of plate tectonics.

▶ WHAT DRIVES PLATE TECTONICS?

Given the substantial worldwide evidence for the phenomenon of plate tectonics, the next question is: What is the driving force?

Convection Cells in the Mantle

An early hypothesis proposed that the propelling mechanism is huge thermal **convection cells**. They move very slowly, taking millions of years to complete a cycle. They are produced as mantle material heats, expands, consequently becomes less dense, and slowly rises. This displaces cooler material, which sinks. This is the same phenomenon you observe in a pot of heated soup.

The idea goes like this: On rising up against the lithosphere, the flow diverges and drags the overlying slab of lithosphere with it, driving "continental drift." As the material moves horizontally, it loses heat and becomes denser. Ultimately, it encounters an opposing current. Then, both viscous streams descend to be reheated and shunted toward a region of upwelling. Above the descending flow, subduction zones and deep-sea trenches form. Midoceanic ridges mark the location of the ascending flow.

There is much to be discovered about convection cells in the mantle. We do not know if circulation occurs throughout the mantle, or if it is confined to the asthenosphere.

Convection-induced drag at the base of lithospheric plates may account for part of the forces needed to move tectonic plates. However, recent calculations indicate that the force supplied by convection cells to the lithosphere may not be sufficient. Furthermore, the hypothesis does not take into account other forces known to act on the plates.

Ridge-Push and Slab-Pull Model

Many geologists now favor what can be termed the **ridge-push and slab-pull model**. Ridge-push forces arise from the fact that spreading centers, such as midoceanic ridges, stand high on the ocean floor. Their elevation above adjacent regions of the ocean floor results in a tendency for the ridge material to slide downslope, thereby transmitting a *push* to the tectonic plate. At the same time, the mechanism of slab-pull operates at the subduction zones. There, the subducting oceanic plate, being relatively cool and

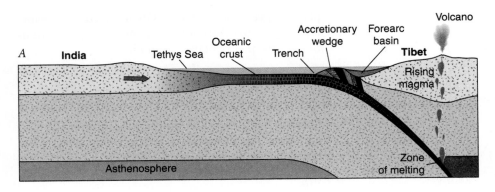

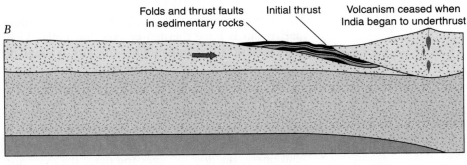

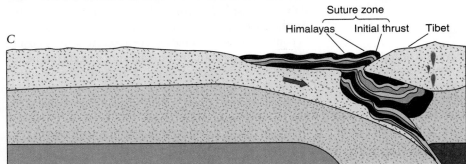

FIGURE 7-45 **How the Himalayan Mountains formed.** The leading margin of the Indian plate subducted beneath the margin of Eurasia. (A) India moved northward and a subduction zone developed at the southern margin of Eurasia. (B) By about 40 million years ago, the plates had collided and the leading edge of India was thrusting beneath southern Tibet. (C) Today, continued underthrusting and compression is crushing Tibet and forming the high Himalayas.

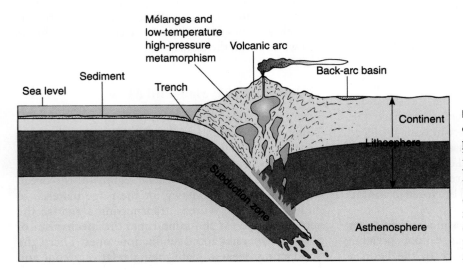

FIGURE 7-46 **Convergence: continental plate and oceanic plate.** The leading edge of the lighter continental plate crumples, whereas the denser oceanic plate buckles downward. This creates an offshore trench. This is a general model for South America's west coast, where the Nazca plate plunges beneath the South American plate (see Fig. 7-36).

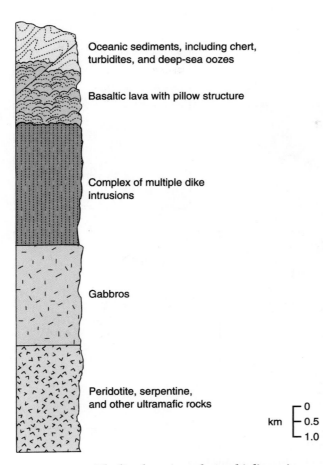

Oceanic sediments, including chert, turbidites, and deep-sea oozes

Basaltic lava with pillow structure

Complex of multiple dike intrusions

Gabbros

Peridotite, serpentine, and other ultramafic rocks

km ⎡ 0
   ⎢ 0.5
   ⎣ 1.0

FIGURE 7-47 **Idealized section of an ophiolite suite.** Ophiolites are thought to be splinters of oceanic crust squeezed into the continental margin during plate convergence.

dense, sinks—pulling the rest of the slab along as it does so (Fig. 7-48).

Thermal Plumes

Mantle material circulates in great rolls. It rises from near the core–mantle boundary in a manner like that of a heat-driven, vertically growing thundercloud. This is termed a **thermal plume**. When a plume nears the lithosphere, it spreads laterally, doming the overlying plate and moving the rifted segments outward from the central area.

As indicated in Figure 7-49, uplift results in a *triple junction* with three radiating fractures. As in the Afar triangle of Ethiopia, two of the fractures open to form narrow oceanic tracts (the Red Sea and Gulf of Aden), whereas the third becomes a fault-bound trough called a *failed arm* that fills with sediment.

► VERIFYING PLATE TECTONICS THEORY

Further Paleomagnetic Evidence

In the 1960s, when Harry Hess began formulating ideas of seafloor spreading, other scientists were puzzling over new data from sensitive magnetometers on research vessels. As they traveled the oceans, these instruments were detecting not only Earth's main geomagnetic field, but also local variations in the paleomagnetism of the oceanic basaltic crust. Maps produced from traverses across the Mid-Atlantic Ridge exhibited a remarkable thing: *mirror-image sets of bands of high- and low-field magnetic intensities* (Fig. 7-50).

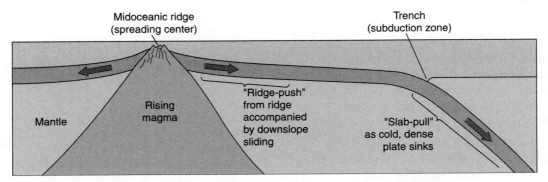

FIGURE 7-48 **The ridge-push/slab-pull mechanism for plate movement.** Near the spreading center, plates ponderously creep down the inclined surface on the asthenosphere, pushing the plates away from the ridge (ridge-push). Simultaneously, at the subduction zones, the cool, dense plates sink, pulling the plates downward (slab-pull).

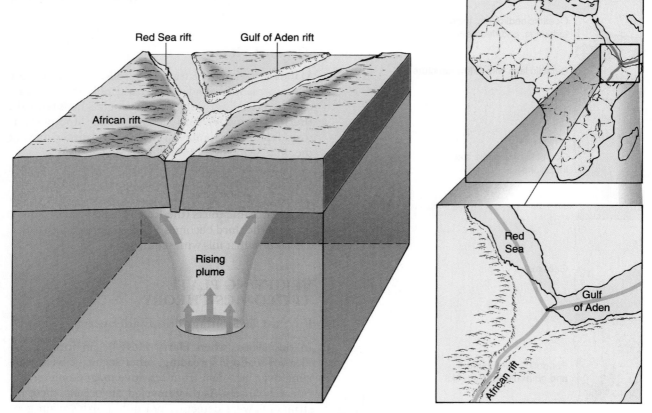

FIGURE 7-49 **Rising plumes of hot mantle may severely rift the crust, often at 120°
angles.** An example is the Afar triangle, shown at the south end of the Red Sea on the map of
Africa. ◀ *Which of the three radiating arms is the failed arm?*

In 1963, F. J. Vine, a research student at Cambridge University, and a senior colleague, Drummond Matthews, suggested that these variations in magnetic intensity were caused by reversals in the polarity of Earth's magnetic field. The magnetometers towed behind the research vessels provided measurements that were the sum of Earth's present magnetic field strength and the paleomagnetism frozen in the crustal rocks of the ocean floor.

- If the paleomagnetic polarity was *opposite* Earth's present magnetic field, the sum would be less than the present magnetic field strength. This would indicate that the crust over which the ship was passing had *reversed* paleomagnetic polarity compared to today's.

- Conversely, where the paleomagnetic polarity of seafloor basalts was the *same* as Earth's present magnetic field, the sum would be greater, so a *normal* paleomagnetic polarity would be indicated (Fig. 7-51).

Since 1963, geophysicists have learned that these irregularly occurring reversals of Earth's magnetic field have occurred frequently over geologic time (Fig. 7-52).

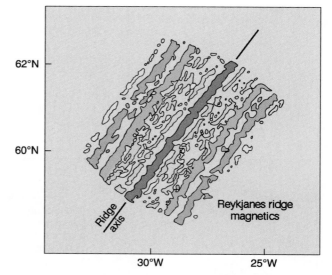

FIGURE 7-50 **Magnetic field of seafloor near Iceland.**
This magnetic field record is from ship traverses made over the Reykjanes midoceanic ridge. The colored bands are magnetically reversed, and the intervening areas have the same magnetic orientation as exists today (termed "normal"). Note the symmetry on either side of the ridge. Age increases away from the ridge. ◀ *By color, which of the magnetically reversed bonds represents the oldest rocks?*

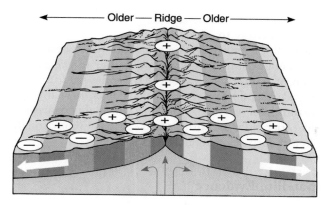

FIGURE 7-51 **Normal (+) and reversed (−) magnetizations of the seafloor.** Note the symmetry on opposite sides of the ridge.

Polarity changes are recorded in the paleomagnetism of lava that cooled into basalts at the midoceanic ridges. The lava acquires the magnetic polarity present at the time of extrusion and then moves out laterally, as previously described.

DETERMINING SEAFLOOR AGE The magnetic stripes discovered by Vine and Matthews verified seafloor spreading. Next, a method was needed to determine the age of each stripe, so geologists would know how fast or slow the seafloor moved. Cores of the basaltic seafloor were not suitable for radiometric dating because when it is hot, basalt is altered by contact with seawater.

By lucky coincidence, a method for determining the age of the stripes emerged in 1963. Geophysicists investigating the magnetic properties of continental lava flows accurately identified periodic reversals of Earth's magnetic field imprinted within layers of basalt (Fig. 7-53). Because they were not altered by seawater, these basalts could be dated by isotopic methods and then correlated to their seafloor counterparts. Using this method, geophysicists determined the time sequence of magnetic reversals for the past 5 million years.

CALCULATING RATES OF SEAFLOOR SPREADING In addition to providing evidence for seafloor spreading, magnetic reversals help construct polarity time scales. Rocks exhibiting a distinctive pattern of reversals characterize particular geologic periods or epochs (Fig. 7-53). Because the reversals are simultaneous worldwide, they are reliable tools for chronostratigraphic correlation.

With knowledge of the age of individual normal or reverse magnetic stripes, it is possible to calculate rates of seafloor spreading and determine the former positions of continents. Figure 7-54 illustrates how. For ex-

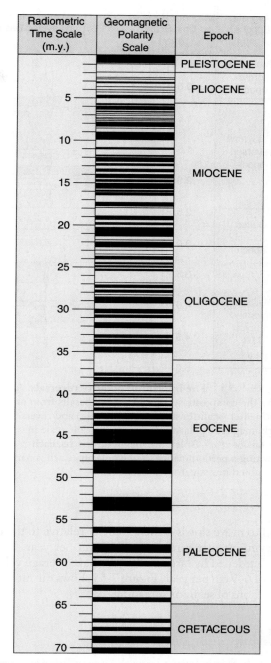

FIGURE 7-52 **Reversals of Earth's magnetic field during the past 70 million years.** Intervals in black indicate when the field was "normal," as today. These changes were measured from seafloor basalt along midoceanic ridges.

ample, if you wish to know the distance between the eastern coast of the United States and the northeastern coast of Africa about 81 million years ago, simply bring together the traces of the two 81-million-year-old magnetic stripes.

The velocity of plate movement is not uniform around the world. Plates that include large continents

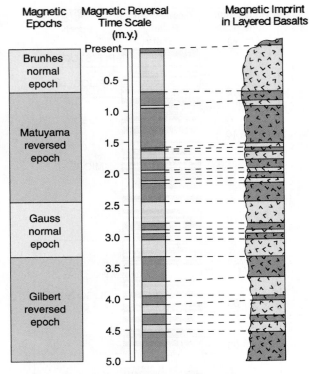

FIGURE 7-53 **The imprint of polarity reversals.** At right, the composite section of layered basalts from many continental localities shows normal (light) and reversed (dark) magnetic polarity. At center is a time scale in millions of years. At left are magnetic epochs, each reflecting a predominance of one polarity. Each is named for a noted investigator.

tend to move slowly—their velocity relative to the underlying mantle rarely exceeds 2 cm per year. Plates not burdened by large continents have average velocities of 6–9 cm per year. Figure 7-55 shows the rates and directions of seafloor spreading.

Oceanic Sediment Evidence

If newly formed crust joins the ocean floor at spreading centers and then moves outward to either side, then the sediments, dated by the fossil planktonic organisms they contain, can be no older than the surface on which they came to rest. Near a midoceanic ridge, the sediments directly over basalt should be relatively young. Samples of the first sedimentary layer above the basalt—but progressively farther away from the spreading center—should be progressively older (Fig. 7-56).

Beginning in 1969, American drilling ships (Fig. 7-57) collected ample evidence to confirm our age distribution of sediments. We know that no seafloor sediments exceed about 200 million years in age, seafloor sediments are relatively thin, and sediments are thinnest closer to midoceanic ridges. This is explained by seafloor spreading. A given area of the ocean floor does not exist long enough to accumulate a thick section of sediments. Seafloor clays and oozes are conveyed to subduction zones, dragged back down into the mantle, and resorbed. Thus, in a continuing conveyor-like operation, the entire world ocean is swept virtually free of deep-sea sediments.

Satellite Evidence

The space program has provided several ways to measure the movement of tectonic plates. During the *Apollo* mission, astronauts placed three clusters of reflectors on the Moon. Laser beams from Earth reflect off these Moon-based mirrors. By recording the time required for the signal to travel to the Moon and back, you can determine the distance between the laser device and the lunar reflector. This is amazingly precise, giving distance measurements correct to within 3 centimeters.

Distances from the Moon to stations on two moving tectonic plates change over time, and such measurements are used to calculate how fast continents are moving toward or away from one another. A similar method employs an artificial satellite to reflect laser pulses back to Earth.

Seismic Evidence

There is abundant evidence for the movement of lithospheric plates, but other evidence is needed to prove that lithospheric plates are dragged back down into the mantle and resorbed. The best indications come from earthquakes generated along a narrow zone that slopes downward under an adjacent continent or island arc at an angle of 45° (Fig. 7-58). The earthquakes are caused by shear between the plunging plate and the mantle, as well as slab-pull by the deeper part of the plate. This sloping zone of earthquake activity is named the **Wadati-Benioff seismic zone** after its two discoverers.

Gravity Evidence

How do gravity measurements confirm the presence of subduction zones? They reveal reduced gravity over areas adjacent to the trenches. Geophysicists refer to such phenomena as negative gravity anomalies. A **gravity anomaly** is the difference between the computed theoretical value of gravity and the actual measured gravity at any point on Earth.

Negative gravity anomalies occur where there are unusual volumes of low-density rock beneath the

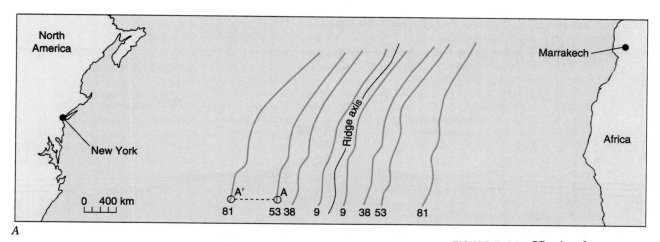

A

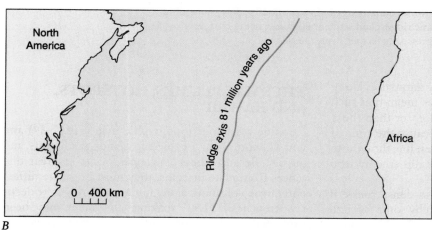

B

FIGURE 7-54 **Viewing the paleogeographic relations of continents today (A) and in the past (B).** (A) Today's map of the North Atlantic, with ages plotted for some of the magnetic stripes. (B) As an example, to see the paleogeographic relation of North America to Africa 81 million years ago, simply move together the two bands marking 81 million years ago, eliminating the younger ones in between. The result is the much narrower North Atlantic that existed 81 million years ago. *In map A, how many kilometers did the ocean floor travel between 53 and 81 million years ago along the dashed line marked A and A'?*

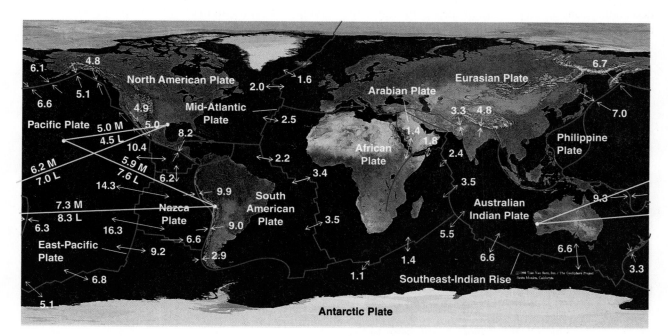

FIGURE 7-55 **Plates move at different rates.** The numbers along midoceanic ridges are rates at which plates are separating in centimeters per year, based on seafloor magnetic reversal patterns. Arrows indicate direction of plate motion. Yellow lines connect stations that are measuring motion rates using satellite laser-ranging methods. Numbers followed by **L** are laser-measured rates; numbers followed by **M** are rates measured by magnetic reversal patterns. *What part of the world appears to have the fastest moving plates at present?*

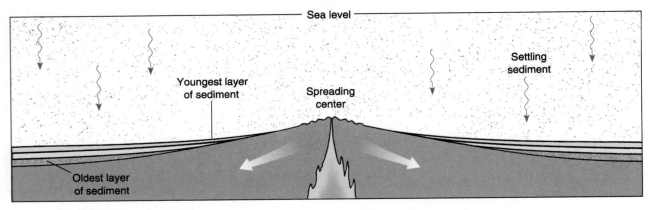

FIGURE 7-56 **Age of sediment along midoceanic ridges.** As a result of seafloor spreading, younger sediments are near the midoceanic ridges, and sequentially older sediments lie directly above the basaltic crust and progressively farther from the ridges.

surface. The strong negative gravity anomalies along the margins of the deep-sea trenches mean that such belts are underlain by rocks much lighter than those at the same depth on either side. Because the zone of lower gravity is narrow in most places, it is thought to mark locations where lighter rocks dip steeply into the denser mantle (Fig. 7-59).

Geophysicists assume that the less-dense rocks of the negative zone must be held down by some force to prevent them from floating upward to a level appropriate to their density. That force might be provided by a descending convection current, or the negative gravity anomaly may reflect stretched and fractured oceanic crust undergoing slab-pull.

THERMAL PLUMES, HOTSPOTS, AND HAWAII

Examine a seafloor topographic map (Fig. 7-24) and you will note chains of volcanic islands, seamounts, and guyots. Because these volcanoes occur at great distances from plate margins, they must originate differently from volcanoes along midoceanic ridges or deep-sea trenches. Their striking alignment has been explained as being a consequence of seafloor spreading.

According to this idea, volcanoes in midocean develop over a **hotspot** in the asthenosphere. The hotspot

FIGURE 7-57 **The JOIDES *Resolution*, successor to the *Glomar Challenger*.** This floating oceanographic research center drills the seafloor and extracts core samples. Onboard facilities include sedimentology, paleontology, geochemisty, and geophysics laboratories for analyzing samples and data. The ship can suspend up to 9000 meters (5.6 miles) of drill pipe to obtain core samples.

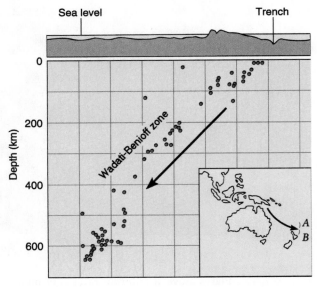

FIGURE 7-58 **Steeply sloping distribution of individual earthquakes (red circles).** They occurred during several months in 1965. Shear caused the earthquakes as the relatively cool edge of the lithosphere plunged into the mantle.

ENRICHMENT

Rates of Plate Movement

The rate of plate movement varies worldwide, ranging roughly between 1 and 10 cm per year. Why does it vary?

If Earth were flat, all locations would move away from a midoceanic ridge at about the same velocity. But Earth is spherical, so the plates are curved. They begin to separate at a location termed a "pole of rotation" and form an ever-widening split. The effect is like pulling away the outer layer of an onion or the peel from an orange. The split in the skin is widest where it is being pulled apart, and the gap narrows toward the pole of rotation (Fig. A).

On Earth, plate movement away from the midoceanic ridge near the pole of rotation is slower because of the lesser distance traversed over a given time interval. Conversely, far from the pole of rotation, plates diverge across a greater distance for the same time interval and hence have greater velocity. You can observe this effect in the North Atlantic, where spreading rates increase from about 1.6 cm per year near Iceland to 3.4 cm per year near the equator. A corollary of this effect is that the width of new oceanic crust also increases with distance from the pole of rotation.

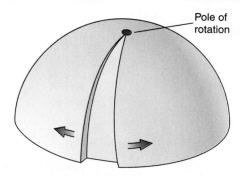

FIGURE A Rates of spreading on a sphere increase with increasing distance from the pole of rotation.

Also, it appears that large expanses of continental crust incorporated into a plate can decrease its velocity. The Pacific and Nazca plates do not carry a heavy load of crust. Thus, these plates have higher velocities than the American and Eurasian plates, which are burdened with large continents.

---

is a manifestation of a plume of upwelling mantle rock. Lava from the plume may work its way to the surface to erupt as a volcano on the seafloor. As the seafloor moves (at rates as high as 10 cm per year in the Pacific), volcanoes formed over the hotspot expire as they are severed from their source of lava and conveyed away. New volcanoes then form at the vacated location. The process may be repeated indefinitely, resulting in a succession of aligned volcanoes that may extend for thousands of kilometers in the direction that the seafloor has moved.

Geologists think the Hawaiian volcanic chain formed in this manner from a single lava source over which the Pacific plate passed on a northwesterly course. Supporting this concept are isotope dates from volcano rocks clearly indicating that those farthest from the source are the oldest (Fig. 7-60). Presently, a new volcano named Lo'ihi is building on the seafloor to the southeast of the

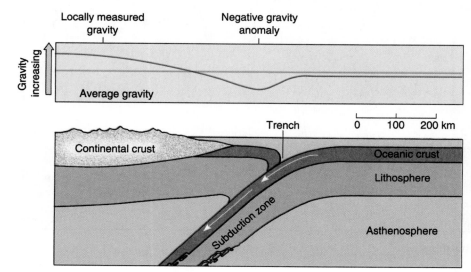

FIGURE 7-59 **Gravity variation over a deep-sea trench.** The trench, subducting sediments, and relatively low-density rocks occupy space that would otherwise be filled with denser rocks. Hence, the force of gravity over the trench and subduction zone is weaker than over Earth generally.

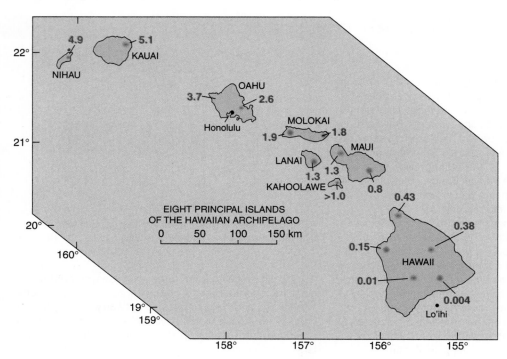

FIGURE 7-60 **The Hawaiian island chain.** There is a hotspot beneath Lo'ihi, the "unborn" Hawaiian Island. Basalts sampled farther to the northwest from this hotspot are successively older. Ages are in millions of years.

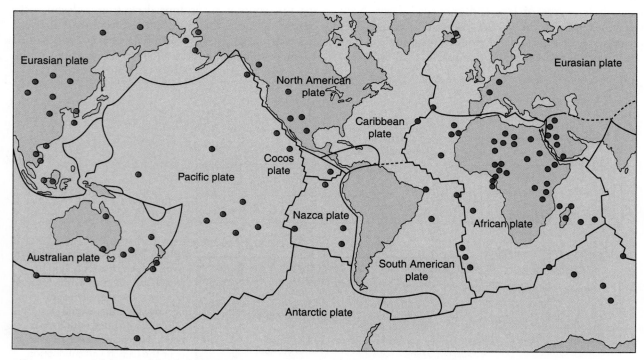

FIGURE 7-61 **Major worldwide hotspots.**

FIGURE 7-62 **Mammoth Geyser, Yellowstone National Park, Wyoming.** Yellowstone is over a hotspot. Surface waters percolating through a system of deep fractures reach the hot rocks below, turn to steam, and erupt in columns of steam and hot water that we call geysers.

island of Hawaii. Described as "an island in the womb," Lo'ihi rises three miles above the seafloor. In 50,000 years, it will be an additional island in the Hawaiian chain.

The ocean around Hawaii is not the only part of the globe that has hotspots. In Figure 7-61, you can see widely dispersed hotspots beneath both continental and oceanic crust. Yellowstone National Park (Fig. 7-62) is over a hotspot that is in the interior of a continent.

EXOTIC TERRANES

We think of continental crust as large landmasses, such as North America or Eurasia. However, many small patches of continental crust are scattered about the globe. As long ago as 1915, Alfred Wegener correctly described the Seychelles Bank (Fig. 7-63) in the Indian Ocean as a small continental fragment that had

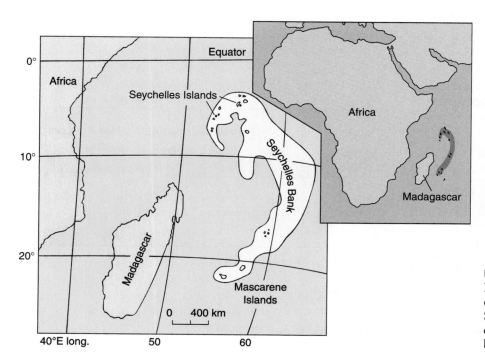

FIGURE 7-63 **Seychelles Bank.** In 1915, Alfred Wegener correctly described the Seychelles Bank as a small continental fragment that had broken away from Africa.

Hawaii Volcanoes National Park

To see geology in action, try Hawaii Volcanoes National Park (Fig. B). The active volcanoes Mauna Loa and Kilauea periodically provide spectacular fiery fountains and rivers of incandescent lava as evidence that the primordial forces of Earth's development are still at work. Mauna Loa attests to the prodigious ability of volcanoes as landbuilders. It is the most massive mountain on Earth, rising more than 9.6 km (6 miles) above the Pacific Ocean floor.

Broad, gently sloping volcanoes such as Mauna Loa are called *shield volcanoes* because their shape resembles that of shields used by early warriors. Their broadly convex shape results from repeated eruptions of highly fluid basaltic lava emerging from vents or along rift zones.

The Kilauea area is ideal for viewing characteristic features of shield volcanoes. You can explore this area along the 11-mile Crater Rim Drive that encircles Kilauea's summit caldera and craters (Fig. C). Along this road, you will see lava flows, steaming vents, and cave-like features called lava tubes. Lava tubes are the vacated conduits through which lava once flowed.

Along Crater Rim Drive you will find good examples of two lava types, pahoehoe (puh-HOY-hoy) and aa (AH-

ah). Pahoehoe has a ropy appearance, produced in the still-plastic surface scum of lava by the drag of more rapidly flowing lava below. In contrast, aa has a blocky, fragmented texture, forming as the thicker upper layer of lava hardens and is carried in conveyor-belt fashion to the front of the flow. There the brittle top layer is broken into jagged chunks that accumulate at the leading edge of the flow. Walking barefoot on this jagged material, you can guess why Hawaiians call it aa.

Basalt is by far the most abundant rock in the park. As is common in many basalts, those of Hawaii contain the mineral olivine. Often you find large crystals of olivine (phenocrysts) in the basalt, and these are cut and polished for jewelry. The gem name for olivine is peridot, and rings and bracelets made from it are popular in Hawaiian tourist shops.

More than 75% of Earth's volcanoes are distributed around the edges of the Pacific plate, forming the Ring of Fire. This includes famous volcanoes like Washington's Mount St. Helens, Alaska's Katmai, Mexico's Popocatepetl, and Japan's Mount Fuji. But Hawaii's volcanoes are *intraplate*, located in the midst of a tectonic plate

rather than along its more dynamic margins. As discussed in this chapter, the intraplate volcanoes of the Hawaiian chain originated as the Pacific plate moved slowly across a hotspot (Fig B).

Over the past 70 million years, the Pacific plate has carried the Hawaiian Islands northwest of the hotspot at an average rate of 10 cm per year. In recent years, the large island of Hawaii has continued to grow, with periodic contributions of lava from Mauna Loa and Kilauea. Hawaii, however, will not be the final island of the Hawaiian Island chain. To the southeast, the seamount Lo'ihi is building its way to the surface of the sea.

For scientists, Mauna Loa and Kilauea are a natural laboratory for studying volcanism. At the Hawaii Volcano Observatory and the University of Hawaii, volcanologists record the periodicity of eruptions, changes in magnetic properties, and variations in the composition and temperature of gases emitted. They record the swelling or tilting of the land surface and carefully consider every tremor associated with subterranean movements of magma. These observations help to forecast eruptions and understand volcano behavior.

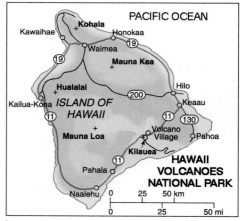

A

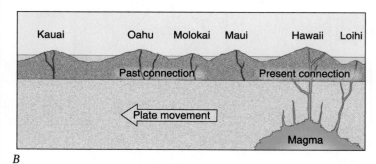

B

FIGURE B **Hawaii Volcanoes National Park and plate movement.** (A) Location map. (B) Origin of Hawaiian Island chain volcanoes as the Pacific plate passed over a hotspot.

FIGURE C **View from the Kilauea caldera rim.** The caldera fills this entire image. In the distance is a pit crater that last erupted in 1967–1968. Ama'Uma' ferns grow in the foreground.

broken away from Africa. The higher parts of the Seychelles Bank project above sea level as about 90 islands that form the nation of Seychelles, but many other such small patches of continental crust are totally submerged.

Geologists use the term **microcontinents** for these bits of continental crust that are surrounded by oceanic crust. We know they are continental crust by their granitic composition, the velocity at which seismic waves traverse them, their general elevation above oceanic crust, and their relative lack of earthquakes.

Microcontinents are small pieces of larger continents that have experienced fragmentation. As these smaller pieces of continental crust are moved along by seafloor spreading, they may ultimately arrive at a subduction zone at the margin of a large continent. But they are composed of lower-density rock and hence are buoyant, so they are a difficult bite for the subduction zone to swallow. Their buoyancy prevents their subduction and assimilation into the mantle.

Exotic Terranes—Continental Crust

These small patches of crust may become incorporated into the crumpled margin of the larger continent as an **exotic terrane**.

Geologists found evidence of exotic terranes long before the theory for their origin was formulated. While mapping Precambrian rocks, they came across fault-bounded areas that were incongruous in structure, age, fossil content, lithology, and paleomagnetic orientation when compared to the surrounding rocks. It was as if these areas were small, self-contained, isolated geologic provinces.

Exotic terranes have been identified on every major landmass, with well-studied examples in the Appalachian Mountains of the eastern U.S., western North America, and Alaska (Fig. 7-64). Geologists have identified an exotic terrane in the Andes that appears to have broken away from the southern margin of the United States and drifted across the ocean to the western margin of South America.

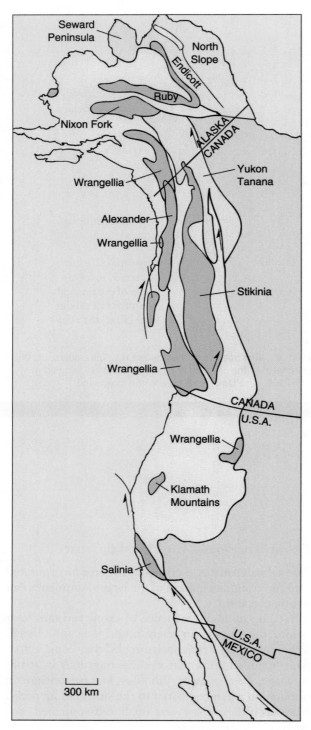

FIGURE 7-64 **The western margin of North America is a jumble of exotic terranes.** Each terrane is bounded by faults and has distinct geological characteristics. The terranes are composed of Paleozoic or older rocks that were accreted during the Mesozoic and Cenozoic eras. The pink terranes may be displaced parts of the North American continent. Green terranes probably originated as parts of other continents. Some, such as Wrangellia, were torn apart during the accretion process and therefore occur in multiple fragments.

Exotic Terranes—Oceanic Crust

If splinters of continents can be transported by spreading seafloor, so can thickened pieces of oceanic crust. Particularly in the Rocky Mountains, we find exotic terranes that were apparently large fragments of oceanic crust, called microplates. The microplates carry volcanoes, seamounts, segments of island arcs, and other features of the ocean floor.

Like passengers on a huge conveyor belt, the microplates were carried to the western margin of North America. As the plate that bore them plunged downward at the subduction zone, volcanoes, seamounts, and the other features were scraped off the descending plate and plastered onto the continental margin as an accretionary prism. The result was a vast collage of accreted oceanic microplates interspersed with similarly transported microcontinents.

It appears that about 50 such exotic terranes exist in the Rockies, all west of the edge of the continent as it existed about 200 million years ago. This process of successive additions of oceanic and continental microplates can significantly increase the rate at which a continent grows.

Exotic Terranes and Earth History

Unraveling the history of mountain ranges containing multiple exotic terranes is complex and requires geologists, geophysicists, and paleontologists. This collaboration can yield dramatic results. For example, paleontologists and geologists working in the Wallowa Mountains of Oregon discovered a massive coral reef of Triassic age. It rested on volcanic rocks with a paleomagnetic orientation indicating that they solidified from lavas at a latitude considerably closer to the equator.

Further investigation indicated that the fossil organisms in the reef were identical to those found in Triassic strata of the Austrian and German Alps. These strata were deposited prior to Alpine mountain building in a seaway called the Tethys, which extended from the present east coast of Japan into the Mediterranean region. The similarity between the fossils of Oregon and those of the Alps strongly suggests that the reefs now in Oregon actually grew around the margins of volcanic islands in the Tethys Sea and—in the subsequent 220 million years or so—were transported thousands of kilometers eastward until they collided with the North American continent.

SUMMARY

- Knowledge of Earth's deep interior is derived from the study of earthquake waves.
- Among the various kinds of earthquake or seismic waves are primary, secondary, and surface waves. Primary and secondary waves (also called body waves) pass deep within Earth's "body" and therefore are the most informative.
- Study of abrupt changes in the characteristics of seismic waves (called discontinuities) at different depths provides the basis for a threefold division of Earth into a central core; an overlying mantle; and a thin, enveloping crust.
- Earth's core is divided into an inner and outer core. The outer core is liquid and is recognized by the disappearance of S-waves at a depth of about 2900 km. This is known because S-waves are not transmitted through liquids.
- The mantle is composed largely of iron and magnesium silicates. The portion of the upper mantle that lies just below the lithosphere is called the asthenosphere. The asthenosphere serves as a weak layer on which the more rigid overlying lithospheric plates move.
- The Mohorovičić discontinuity separates the mantle from the overlying crust. It lies deeper under the continents than under the ocean basins. The continental crust has an overall granitic composition and is less dense than the basaltic oceanic crust.
- Earth's crust floats on the underlying denser mantle. The crust may rise or fall in attempting to maintain its isostatic equilibrium (flotational balance).
- Faults are breaks in crustal rocks along which there has been a displacement of one side relative to the other. According to the directions of that movement, faults are classified as normal, reverse, thrust (low-angle reverse), or lateral (strike-slip).
- Folds are wavelike undulations in strata. The principal types are anticlines, domes, synclines, basins, and monoclines. Both anticlines and domes have uparched strata, but domes are roughly circular. Synclines and basins are composed of down-folded strata, and basins are more or less circular. A monocline resembles a carpet folded over a step.

- Earth's crust and part of the upper mantle constitute a brittle shell called the lithosphere. It is broken into a number of plates that ride over the asthenosphere. Plates move apart and form new seafloor at divergent boundaries, collide at convergent boundaries, or slip past one another at transform boundaries.
- As basaltic lavas are extruded along the fracture zones of midoceanic ridges, they are incorporated into the trailing margin of the plates. They simultaneously take on the magnetism of Earth's field. This process has provided a record of geomagnetic reversals that are used to determine rates of seafloor spreading.
- Thermal convection currents in the mantle and/or ridge-push and slab-pull are the proposed mechanisms for moving lithospheric plates.
- If two plates bearing continental crust at their margins collide, those margins will be compressed to form a mountain range. Where one margin consists of continental crust and the other is oceanic, the oceanic plate will subduct. Where two plates bearing oceanic crust collide, the cooler, denser plate margin will subduct.
- The continental crustal segments of the lithosphere ride as passive "cargo" on the plates. They have a lower density than oceanic crust and do not sink into subduction zones. This explains why continents are older than the ocean floors and why, when continents collide, they produce mountain ranges rather than trenches.
- Where continents straddle zones of divergent movements in the asthenosphere, the landmass may break and produce rift features such as the Red Sea and Gulf of Aden. These fracture zones may then open further, resulting in the formation of widening tracts of ocean floor.
- Aside from the suturing of one large landmass onto another, a continent may grow by addition of metamorphosed and remelted rocks that had accumulated as sediments in marginal depositional basins, or by accretion of exotic terranes.

KEY TERMS

accretionary prism, p. 188

anticline, p. 170

apparent polar wandering path, p. 180

asthenosphere, pp. 166, 181

basin, p. 171

blueschist, p. 189

body seismic wave, p. 162

continental crust, p. 167

convection cell, p. 189

convergent plate boundary, p. 185

dip, p. 170

discontinuity (seismic), p. 164

divergent plate boundary, p. 181

dome, p. 171

exotic terrane, p. 201

fault, p. 169

fold, p. 170

footwall, p. 169

Glossopteris flora, p. 177

Gondwana (=Gondwanaland), p. 175

gravity anomaly, p. 194

Gutenberg discontinuity, p. 164

guyot, p. 184

hanging wall, p. 169

hotspot, p. 196

isostasy, p. 167

lateral fault (strike-slip fault), p. 169

Laurasia, p. 175

lithosphere, p. 181

mantle, p. 164

mélange, p. 188

microcontinent, p. 201

Mohorovičić discontinuity, p. 164

monocline, p. 170

normal fault, p. 169

oceanic crust, p. 167

ophiolite suite, p. 188

paleomagnetism, p. 178

Pangea, p. 175

Panthalassa, p. 175

passive continental margin (trailing edge), p. 183

plate tectonics, p. 171

primary seismic wave (P-wave), p. 162

reverse fault, p. 169

ridge-push, slab-pull model, p. 189

seafloor spreading, p. 183

secondary seismic wave (S-wave), p. 162

seismic wave, p. 162

seismogram, p. 162

seismograph, p. 162

shadow zone, p. 164

spreading center, p. 181

strike, p. 170

strike-slip fault, p. 169

subduction zone, p. 185

surface seismic wave, p. 162

suture zone, p. 187

syncline, p. 170

Tethys Sea, p. 187

thermal plume, p. 191

thrust fault, p. 170

trailing edge, p. 181

transform fault, p. 185

transform plate boundary, p. 185

Wadati-Benioff seismic zone, p. 194

Wilson Cycle, p. 189

QUESTIONS FOR REVIEW AND DISCUSSION

1. If we were able to drill a well from the North Pole to the center of Earth, what internal zones would be penetrated?

2. What are the three major categories of seismic waves? Describe their characteristics.

3. What does the presence of an S-wave shadow zone indicate about the interior of Earth?

4. What is a seismic discontinuity? Where are the Gutenberg and Mohorovičić discontinuities located?

5. How do anticlines (and domes) differ from synclines (and basins) with regard to the age relations of rocks exposed across the erosionally truncated surfaces of these structures?

6. What kind of rock approximates the average composition of the continental crust? The oceanic crust? The mantle?

7. What kind of body waves would be received on a seismograph located 180° from the epicenter of an earthquake?

8. What are the principal categories of faults? What kinds of faults might we find in regions subjected to great compressional forces? What kinds of faults result primarily from tension in Earth's crust?

9. What is a gravity anomaly? What sort of gravity anomaly might you expect over a subduction zone? A midoceanic ridge?

10. What are folds? What are the principal kinds of folds?

11. Compile a list of items that Alfred Wegener might have used to convince a skeptic of the validity of his theory of continental drift.

12. Do midoceanic volcanoes have lavas of granitic or basaltic composition? What is the composition of lavas in mountain ranges adjacent to subduction zones, such as the Andes? Account for the differences in composition.

13. According to plate tectonics, how did the Himalayan Mountains form? The San Andreas fault? The Red Sea and Dead Sea?

14. According to plate tectonics, where is new material added to the seafloor, and where is older material consumed?

15. A moving tectonic plate must logically have a leading edge, a trailing edge, and sides. Where are the leading and trailing edges of the North American tectonic plate?

16. What is paleomagnetism? What is its origin? How is it used in finding ancient pole positions? How has remanent magnetism helped validate the concept of plate tectonics?

17. Why are the most ancient rocks on Earth found only on the continents, whereas only relatively younger rocks are found on the ocean floors?

18. What evidence would you seek to support an interpretation that a particular area in the Appalachians was an exotic terrane? How would an exotic terrane derived from a microcontinent differ from one derived from a volcanic island arc?

19. How has the detection of reversals in Earth's magnetic field been used as support for the concept of seafloor spreading?

20. How do the alignment and age distribution of volcanic islands in the Pacific Ocean provide evidence of seafloor spreading?

WEB SITES

The Earth Through Time Student Companion Web Site (www.wiley.com/college/levin) has online resources to help you expand your understanding of the topics in this chapter. Visit the Web Site to access the following:

1. Illustrated course notes covering key concepts in each chapter;

2. Online quizzes that provide immediate feedback;

3. Links to chapter-specific topics on the web;

4. Science news updates relating to recent developments in Historical Geology;

5. Web inquiry activities for further exploration;

6. A glossary of terms;

7. A Student Union with links to topics such as study skills, writing and grammar, and citing electronic information.

8

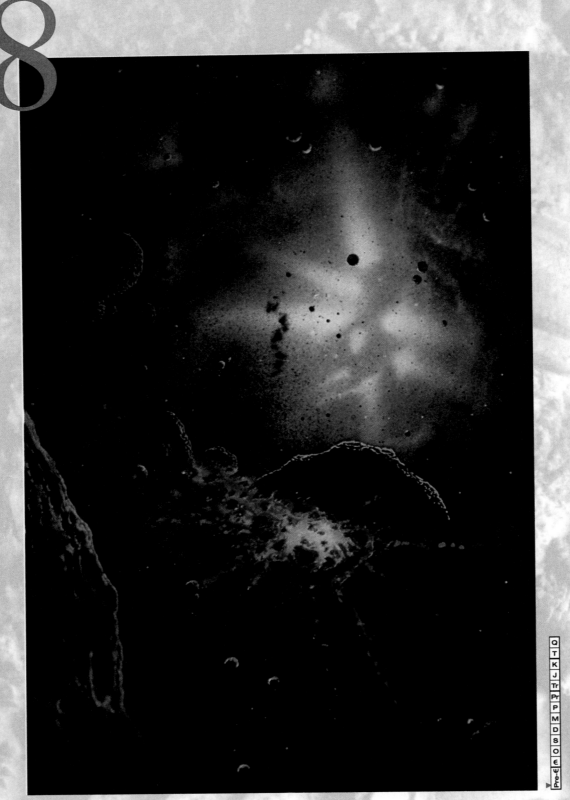

Q
T
K
J
Tr
Pr
P
M
D
S
O
€
Pre

*Collisions between large bodies that had aggregated
in the solar nebula during the formative stages of the
Solar System.*

Earliest Earth: 2,100,000,000 Years of the Archean Eon

What was the Earth like in the very beginning? How does one get from a conglomerate ball of stardust to a planet with concentric structure—inner and outer core, silicate mantle, crust, oceans, and layered atmosphere?

—Preston Cloud, *Oasis in Space*

Key Chapter Concepts

- Earth and other planets of the Solar System formed from a cloud of dust and gas that slowly rotated in space.

- Meteorites contain clues to Earth's origin, age, and source of water.

- "Inner planets" Mercury, Venus, Earth, and Mars are Earthlike planets having relatively small size and high density. "Outer planets" Jupiter, Saturn, Uranus, and Neptune are large and have low densities. Pluto is a distant icy planet that may be an escaped satellite.

- The materials comprising Earth were acquired by accretion of meteorites and other cosmic debris. The internal layering of Earth occurred through differentiation.

- Earth has two kinds of crust. Oceanic crust is mafic in composition. Continental crust is felsic in composition.

- Earth's early atmosphere contained no free oxygen. This anoxic atmosphere was gradually replaced by an atmosphere rich in oxygen that is derived from photosynthetic organisms.

- Water for Earth's hydrosphere was derived from the planet's interior by a process called outgassing.

- Archean rocks occur in regions of continents called cratons, which consist largely of accreted Archean terranes. Granulites and greenstones are the two rock associations occurring in Archean cratons.

- Life evolved from nonliving organic molecules. The first forms of life were anaerobic, prokaryotic, heterotrophic organisms. Photosynthetic autotrophic prokaryotes evolved next, followed by eukaryotic organisms.

- Organelles in eukaryotes represent once independent prokaryotes that established symbiotic relationships with the primary cell.

- Stromatolites, the laminated calcareous structures produced by cyanobacteria, are the most abundant fossils of the Archean.

OUTLINE

- ▶ EARTH IN CONTEXT: A LITTLE ASTRONOMY
- ▶ A SOLAR SYSTEM TOUR, FROM CENTER TO FRINGE
- ▶ FOLLOWING ACCRETION, EARTH DIFFERENTIATES
- ▶ THE PRIMITIVE ATMOSPHERE—VIRTUALLY NO OXYGEN
- ▶ THE PRIMITIVE OCEAN AND THE HYDROLOGIC CYCLE
- ▶ ORIGIN OF PRECAMBRIAN "BASEMENT" ROCKS
- ▶ THE ORIGIN OF LIFE
- ▶ IN RETROSPECT
- ▶ SUMMARY
- ▶ KEY TERMS
- ▶ QUESTIONS FOR REVIEW AND DISCUSSION
- ▶ WEBSITES
- ▶ BOX 8-1 ENRICHMENT
- ▶ BOX 8-2 GEOLOGY OF NATIONAL PARKS AND MONUMENTS

The Archean Eon lasted 30 million of your lifetimes, if you live to be 70. That's a whopping 2.1 billion years. It began about 4.6 billion years ago. We know nothing directly about its earliest years, for no rocks that old survive on Earth, thanks to recycling by plate tectonics. Much of what we know comes indirectly, from beyond Earth: from meteorites that have collided with our planet.

Figure 8-1 shows that the Archean Eon and Proterozoic Eon together span about 87% of Earth's history. Earth's oldest known crustal rocks, found in Canada, are about 4.04 billion years old. However, we now have newer evidence of even older crust, which consists of grains of zircon (Fig. 8-2) in metamorphosed sediments in western Australia. These zircons are 4.4 billion years old.

EARTH IN CONTEXT: A LITTLE ASTRONOMY

When we look at Earth's beginnings, we need the big picture to understand how it came to be. This means looking at Earth in the context of the Universe. Using the metaphor of a zoom lens, we begin fully zoomed out, looking at the Universe, which is vast beyond human imagining. The Universe hosts billions of star galaxies. We rapidly zoom in on just one of those galaxies, our Milky Way (Fig. 8-3). Continuing to zoom in, we see among the Milky Way's billions of stars a small one, our Sun. Around the Sun swirl nine planets, thousands of asteroids and comets, and countless bits of space debris (Fig 8-4).

Let's pause here to explore how the **Solar System** came to be, which explains how Earth came to be. Then we will look at Earth's earliest eon, the Archean.

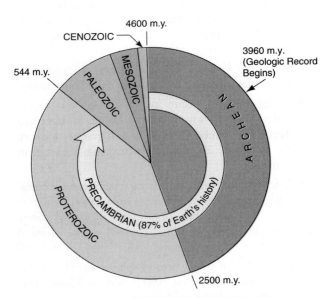

FIGURE 8-1 The Archean and Proterozoic Eons span 87% of Earth's entire history. This long time interval also is called the Precambrian, meaning all time preceding the Cambrian Period, which began 544 million years ago.

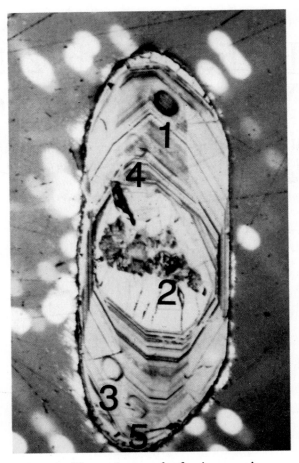

FIGURE 8-2 **Photomicrograph of a zircon grain extracted from gneiss in Canada's Northwest Territories.** This grain is 0.5 millimeters long. Zircon is a zirconium silicate with a hardness of 7.5. Numbers refer to points selected for analysis.

The Solar Nebula Hypothesis

Let us begin with observations, as all scientists do. Looking at the movements of each planet, all are remarkably similar:

- All nine revolve around the Sun in the same direction.

- Their orbits are approximately in the same plane, so the whole Solar System resembles a slowly rotating disk. (Pluto's orbit is an exception, being inclined as shown in Figure 8-4.)

- Each planet rotates in the same direction (except Venus and Uranus).

- As the planets slowly spin like tops, none of their axes are "upright"—all are tilted (which causes the change of seasons on each planet).

- With few exceptions, each moon mimics the movements of its planet, orbiting and rotating in the same direction.

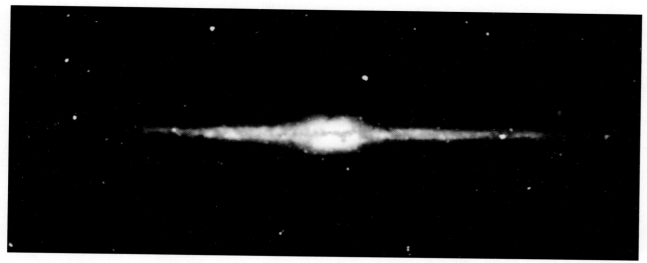

FIGURE 8-3 **Our disk-shaped Milky Way Galaxy.** If you go where no city lights pollute the night sky, you can see our Milky Way Galaxy. Because you are looking edge-on toward the galaxy's center, the many stars in your line of sight create a "milky" band. This image is from the diffuse infrared background experiment on NASA's *Cosmic Background Explorer Satellite* in 1990. ◧ *What would the galaxy look like if viewed from above? (Answers to questions appearing within figure legends can be found in the Student Study Guide.)*

With this degree of alignment and consistency, the Solar System truly is a system. All of its parts must have been set in motion at the same time.

In composition, the planets vary a lot, but in a systematic way. You can compare them in Table 8-1. The inner planets—Mercury, Venus, Earth, and Mars—are small, dense, rocky, and rich in metals. But the outer planets—Jupiter, Saturn, Uranus, and Neptune—are large (Fig. 8-5) and have much lower densities. Like the Sun, Jupiter and Saturn are mostly hydrogen and helium.

The patterns of movement and density lead us to the **nebular hypothesis** for the Solar System's origin. First suggested by German philosopher Immanuel Kant in 1755, this idea proposes that the Solar System distilled from a rotating cloud of dust particles and gases called the **solar nebula**. Knowledge from the space program has strengthened this hypothesis.

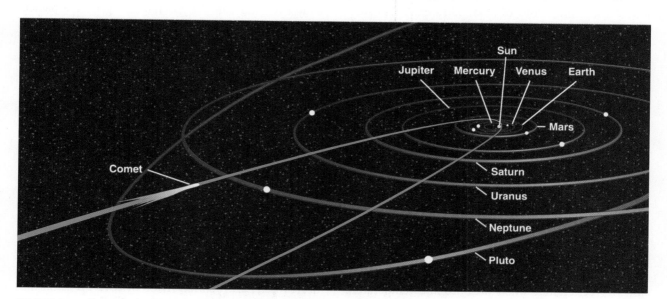

FIGURE 8-4 **Schematic view of the Solar System, showing orbits of the nine planets, the asteroid belt, and a comet.** Moons attend Earth, Mars, Jupiter, Saturn, Uranus, Neptune, and Pluto, although we can't show them at this scale.

TABLE 8-1 Properties of the Planets

Planet	Diameter (km)	Diameter (Earth = 1)	Mass (Earth = 1)	Mean Density (g/cm³)	Rotation Period (days)	Inclination of Equator to Orbit (°)	Surface Gravity (Earth = 1)	Features
Mercury	4,878	0.38	0.055	5.43	58.6	0.0	0.38	Earthlike
Venus	12,104	0.95	0.82	5.24	−243.0	177.4	0.91	Small size
Earth	12,756	1.00	1.00	5.52	0.997	23.4	1.00	Solid
Mars	6,794	0.53	0.107	3.9	1.026	25.2	0.38	Rocky Metallic core
Jupiter	142,800	11.2	317.8	1.3	0.41	3.1	2.53	Gaseous
Saturn	120,540	9.41	94.3	0.7	0.43	26.7	1.07	Large size
Uranus	51,200	4.01	14.6	1.2	−0.72	97.9	0.92	Composed mostly
Neptune	49,500	3.88	17.2	1.6	0.67	29	1.18	of hydrogen and helium
Pluto	2,200	0.17	0.0025	2.0	−6.387	118	0.09	Icy Composed of frozen gases and some rock

(After Fraknoi, A., Morrison, D., and Wolff, S. 2001, *Voyages Through the Universe*. Philadelphia: Saunders College Publishers.

According to the nebular hypothesis, about 5 billion years ago a huge cloud of gas and dust occupying billions of cubic miles existed in space. It was one of many such clouds. The elements forming the gases and particles originally came from nuclear reactions within stars. Then they were scattered here through cataclysmic explosions (called *nova*) that mark the death of stars. Prominent elements in the cloud were hydrogen, helium, oxygen, silicon, and carbon.

The dust cloud began to rotate and contract. It did not shrink into a round ball as you might expect, but into a disk. About 90% of the cloud's mass remained concentrated in the thicker central part. Gravity attracted the particles to one another, and clumps of matter began to form. Smaller particles merged to build clumps up to several meters in diameter, and these larger bodies in turn swept up finer particles within their orbital paths. This process of accumulating bits of matter around an initial mass is called **accretion**.

We call the growing bodies formed from accreting materials **protoplanets** (*proto* = giving rise to). The protoplanets were enormously larger than present-day planets because they had yet to condense. Each rotated as its own miniature dust ball, and each eventually swept up most of the debris in its orbital path and became able to revolve around the central mass (Fig. 8-6). The protoplanets' formation required perhaps 10 million years.

As material near the center coalesced, gravity made it condense, shrink, and become heated to several million degrees. Later, when pressure and temperature within the Sun's core attained critical levels, hydrogen atoms began to fuse together. This hydrogen **fusion** released great energy as it converted hydrogen atoms into helium atoms. At this time the Sun inaugurated its career as a multi-billion-year nuclear reactor. (Today, the Sun converts about 596 million metric tons of hydrogen into 532 million metric tons of helium *each second*. The difference in weight represents matter that has been converted to energy, some of which radiates to Earth to power our surface geologic processes.)

The stream of radiation from the Sun, sometimes called the **solar wind**, drove enormous quantities of the lighter elements and frozen gases outward into space. The solar wind is what causes a comet's tail to stream away from the Sun. As you might expect, the planets nearest the Sun (Mercury, Venus, Earth, and Mars) lost enormous quantities of lighter matter. As a result, they have smaller masses and greater densities than the outer planets (Table 8-1). These "inner planets" are composed of rock and metal that could not be forced away by the solar wind. The outer planets (Jupiter, Saturn, Uranus, and Neptune) were less affected by the solar wind, so they retained their considerable volumes of hydrogen and helium that surround their rocky inner cores.

Meteorites: Samples of the Early Solar System

Meteors are interplanetary bits and chunks of space rock that succumb to Earth's gravity and streak through our atmosphere, superheating and glowing like "falling stars." Most of them fully vaporize in the atmosphere and never reach Earth's surface. **Meteorites** are meteors that survive the heat and reach the ground. Each year, about 500 meteorites that are baseball-sized or larger survive passage through Earth's atmosphere and crash into our

FIGURE 8-5 **Compare the sizes of the planets Mercury through Neptune (Pluto not shown).** Note that the planets are never in the relationship shown; their images are "posed" here to contrast their size and appearance.

planet. Weathering and erosion have erased most of the larger meteorite craters. Only about 70 partially preserved craters or clusters of craters remain. Meteorites and their craters provide clues to the origin and Archean history of the Earth, Moon, and planets.

Meteorites are asteroid fragments, formed when asteroids collide in space and shatter. Such collisions send debris into odd orbits that intersect the orbits of other planets, like Earth's. Many meteorites show compositional and textural evidence of such collisions. Some meteorites are chemically similar to lunar basalts and are believed to have been blasted from our Moon by impact. Even more intriguing are a few meteorites that are similar to Martian rocks analyzed by the *Viking* lander. They arrived on Earth after being blasted from Mars' surface by an impacting asteroid.

Meteorites are of four different compositions, which gives clues to their origin:

- **Ordinary chondrites** are the most abundant and, at 4.6 billion years old, are clearly Archean in age. Chondrites contain spherical bodies called **chondrules** that are solidified molten droplets splashed into space during an impact.

- **Carbonaceous chondrites** (Fig. 8-7) contain about 5% organic compounds. Some are protein-building amino acids, and many contain the basic building blocks of DNA and RNA found in all living things. Thus, carbonaceous chondrites (as well as carbon-bearing asteroids and comets) provided much of Earth's early carbon, and they may have supplied the organic building blocks of life.

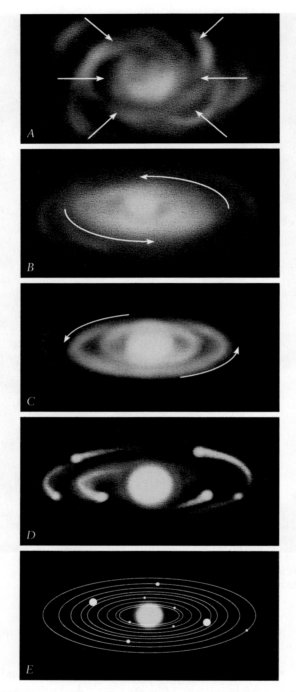

FIGURE 8-6 **The solar nebula hypothesis for the origin of the Solar System.** The solar nebula (A) shrinks and (B) forms a disk as it begins rotation. Much of the material (C) is concentrated centrally, ultimately becoming the Sun. Solid particles in the disk (D) accrete to form the planets, their moons, and the asteroid belt. (E) shows the Solar System today.

FIGURE 8-7 **Small carbonaceous chondrite meteorite.** It fell in western Kentucky.

- **Iron meteorites** (Fig. 8-8) are actually iron-nickel. Asteroids, which are smaller than planets but large enough to have metallic cores, are the probable source for iron meteorites.
- **Stony-iron meteorites** are the least abundant. They are composed of silicate minerals and iron-nickel. They are also derived from shattered asteroids, originating from the area in the asteroid that lies between the iron core and the surrounding rocky shell.

How long ago was the Solar System "born?" It was a protracted process, but based on lunar and meteorite samples and on zircon grains in Earth's oldest known rocks (4.4 billion years old), we place our Solar System's birth at about 4.6 billion years ago.

FIGURE 8-8 **A 15-ton iron meteorite.** Fortunately, most meteorites are much smaller. This one, found in Oregon, is at the American Museum of Natural History in New York City.

The Origin of the Universe

Earth's origin is part of the Solar System's history, and the Solar System's origin is part of the Universe's history. Thus, the story of Earth also includes the story of the Universe. Several hypotheses for the Universe's origin have been offered.

Red Shift

Before we discuss these hypotheses, it's important to know that every hypothesis must conform to an astronomical observation called the **red shift**. To understand it, recall that the "white" light we see really is a blend of all the colors— red, orange, yellow, green, blue, indigo, and violet. The wavelength of each color is different, so when white light passes through a prism, each color is bent (diffracted) at a different angle, spreading out the colors in a display we call the *spectrum*. Falling rain can act as a prism, displaying the spectrum in the sky as a rainbow.

The spectrum for light from any star varies with its composition. But it also varies with the star's movement and speed. For example, if a star is moving away from Earth, wavelengths of the light from the star effectively lengthen and shift toward the red end of the spectrum. (If the star is moving toward Earth, its wavelengths are effectively shortened, shifting toward the blue end.) The greater the velocity of separation, the wider the observed red shift.

Big Bang and the Expanding Universe

By 1914, researchers had found a dozen galaxies that clearly exhibited red shift, rapidly moving away from each other. In 1929, Edwin Hubble discovered that the red shift increased with increasing distances of the galaxies, indicating that the more distant the galaxy, the higher its receding velocity. Thus, red shift indicates that the Universe is expanding.

If we made a video of the galaxies moving apart as indicated by the red shift and played it back in reverse, all the galaxies would merge at a single location, the birthplace of the Universe. The density of this galactic mass was incredible, and it exploded in what astronomers have dubbed the **Big Bang**.

Calculations based on the amount of expansion that has occurred indicate that the Big Bang happened 18–15 billion years ago. It marked the instantaneous creation of all matter in the present Universe. The initial temperature reached about 100 billion degrees Celsius. During the first second, the Universe cooled to about 10 billion degrees (still a thousand times hotter than the center of the Sun today). Atoms could not exist at such high temperatures, so all that could exist was radiant energy, very light particles called neutrinos, and electrons.

The formation of protons and neutrons quickly followed. After about 1.5 minutes, temperatures dropped to about 1 billion degrees Celsius. A few simple atomic nuclei formed by fusion (hydrogen and helium). During the subsequent million years or so, the Universe continued to cool as it expanded. Atoms began to form when the temperature fell to a few thousand degrees. By about a billion years after the Big Bang, stars and galaxies had probably begun to form. In the interior of stars, matter reheated, nuclear reactions started, and synthesis of heavier elements began.

Not So Fast . . .

The Big Bang hypothesis sounds pretty incredible, but it is supported by certain evidence and calculations. However, some astronomers are uneasy about a cosmic evolution that starts with a Big Bang and ends with galaxies hurtling toward the edges of the Universe. They support the **steady-state** cosmology of a Universe that will continue to expand forever. As expansion progresses, new matter, initially hydrogen, forms from combinations of atomic particles in the space between galaxies at about the same rate that older material is receding. This maintains a uniform density of matter in the Universe.

An Oscillating Universe?

Another possibility is to combine the Big Bang and steady-state hypotheses into a single concept that says, "for every big bang, there is a big crunch." According to this **oscillating Universe** cosmology, after the Big Bang, gravitational forces begin to prevail and draw matter back to its place of origin. The expanding Universe thus becomes a contracting one for a while. In the last stages of contraction, matter would hurtle in at enormous speed, compressing the returned particles into an infinitely dense mass for the next Big Bang. They estimate a complete cycle to take 100 billion years.

To summarize, the red shift indicates that the Universe is expanding. Without this knowledge, there would be no basis for the big bang or steady-state hypotheses. Even the oscillating-Universe hypothesis insists on an expanding Universe, at least until a future contraction occurs.

▲ A SOLAR SYSTEM TOUR, FROM CENTER TO FRINGE

The Sun

The Sun is the "CEO" (Chief Energy Officer) of our Solar System, with interior temperatures exceeding 20 million degrees Celsius. Without the Sun's constant radiant energy, all life on Earth would quickly perish. Without the Sun, all nine planets would wander off into space, untethered by solar gravity. Our sun is a small star among billions in the Milky Way Galaxy, yet it is about 1.5 million kilometers in diameter and contains 98.8% of all material in the Solar System. The Sun and Earth are composed of the same elements, but the proportions are vastly different. The Sun is mostly hydrogen and helium, whereas Earth is predominantly iron, oxygen, silicon, and magnesium.

Earth intercepts an enormous amount of solar energy, yet not so much that we are roasted. We enjoy a livable temperature range between −50 and +50°C. Several factors maintain this vital temperature range:

1. **Distance from the Sun.** If we were much closer, Earth would be an inferno. Much farther, and we could rename the entire planet Antarctica.

2. **Earth's rotation.** All sides are heated in a daily cycle, like a roast slowly turning on a spit. As Earth receives solar energy, it also radiates some heat back into space.

3. **Earth's atmosphere.** Some incoming solar radiation is reflected from the atmosphere and clouds back into space, never reaching the surface. Part of the intercepted radiation is absorbed by the atmosphere and radiated back into space without warming Earth's surface.

The Sun is a continuous nuclear fusion reactor. Each second, it transmits energy equivalent to burning 25 billion pounds of coal. This energy is the ultimate force behind many geologic processes that continuously change Earth's surface. For example, the Sun's radiation helps evaporate surface water, which forms clouds and rain, providing the running water required for erosion. The Sun's radiation creates winds and ocean currents. Protracted variations in energy from the Sun may trigger continental glaciation or reduce lush forests to barren wastelands.

The Four Inner Planets

The four inner planets of the Solar System (Mercury, Venus, Earth, and Mars) are all Earthlike: relatively small, rocky, and rich in metals.

MERCURY: HOT, POCKMARKED, AND SWIFT Mercury is small and swift: during an Earth year, it revolves around the Sun over four times (its year lasts only 88 Earth days). The planet has a very thin atmosphere composed of sodium and much smaller amounts of helium, oxygen, potassium, and hydrogen. There is a weak magnetic field, a lightweight, heavily cratered crust (Fig. 8-9), and an iron core.

VENUS: NOTHING WE KNOW COULD SURVIVE HERE Venus (Fig. 8-10) is similar to Earth in density and gravity. However, Venus is hellish—with no liquid water, no oceans, a surface atmospheric pressure roughly equivalent to Earth's ocean pressure 1000 meters down, a 98% carbon dioxide atmosphere, and dense clouds of corrosive sulfuric acid droplets. Needless to say, there is no recognized life on Venus.

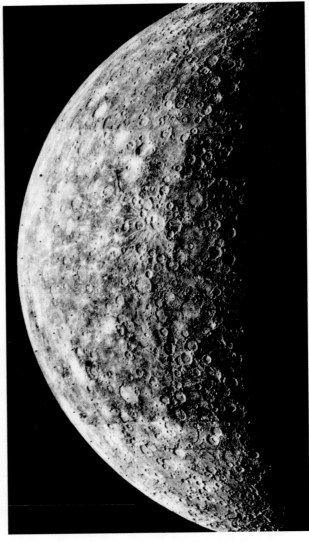

FIGURE 8-9 Mercury, closest to the Sun. Mercury's hot, pockmarked surface is shown in this composite image from *Mariner 10.*

FIGURE 8-10 **Venus, overheated from runaway greenhouse effect.** A view of the surface of Venus as revealed by the imaging radar instrument on the *Magellan* spacecraft. Lava flows extend for hundreds of kilometers across the plains in the foreground. The 5-kilometer-tall volcano Maat Mons is at the center of the image on the horizon.

FIGURE 8-11 **The Moon, showing part of the back side.** The dark regions are basaltic lava flows. The heavily cratered lighter regions show the result of heavy meteor bombardment.

This 98% CO_2 atmosphere exists because there is no ocean in which carbon dioxide can combine with calcium and magnesium to form carbonates. Thus, there is no process to remove CO_2 from the atmosphere. Sunlight heats the surface, and the heat cannot escape through the dense CO_2 atmosphere. This "runaway" greenhouse effect raises the surface temperature to 470°C (hot enough to melt the metal lead). An environmental lesson from Venus is that we must act to avoid increases in the carefully balanced carbon dioxide content of our Earthly atmosphere (which is presently about 0.038%).

EARTH: THE BEST OF ALL POSSIBLE WORLDS   Earth's average density is 5.5 g/cm³, but its surface rocks are 2.5 to 3.0 g/cm³. This disparity indicates that high-density material, such as iron, lies in the interior of Earth. Thus, Earth has zones of different rocks with different densities.

Earth's uniqueness cannot be overstated. It is the only planet known to have bountiful liquid water (instead of mostly ice or mostly vapor), which is essential for life. Earth's vast oceans cover 71% of the surface, leaving only 29% as land surface. Earth's atmosphere is nitrogen (78%) and oxygen (21%), providing these two key ingredients for life. And, of course, it is the only known body in the Universe that is host to life as we know it.

EARTH'S MOON   Earth's Moon is uncommonly large relative to its parent planet—about one-fourth the diameter of Earth. Its mean density of 3.3 g/cm³ is simi-

lar to Earth's upper mantle. The Moon rotates on its axis in exactly the same time (29.5 days) that it takes to orbit Earth (29.5 days), which explains why we always see its same face and never its back side (Fig. 8-11). Images of the far side, from lunar orbiters, reveal a surface more densely cratered than the side facing Earth.

In 1610, Galileo aimed his primitive telescope at the Moon and saw a surface "rough, full of cavities and prominences." Part of what he saw were the **lunar highlands**—the lighter-hued craggy and heavily cratered regions of the Moon. Rocks collected from the highlands by astronauts have been dated to more than 4.2 billion years old (Fig. 8-12A).

The darker areas of the Moon are called **maria** (singular = *mare*). These darker areas form the floors of immense basins that have been flooded with dark basaltic lava. Isotopic dating of mare basalts (Fig. 8-12B) reveals that they were extruded during a period of about 3.8 to 3.2 billion years ago. Unlike the lunar highlands, the maria are not densely cratered. Therefore, meteoric bombardment of the Moon (and Earth as well) diminished sharply after about 3.8 billion years ago.

Why are there craters on the Moon but not on Earth? On Earth, the original crater-riddled crust has been recycled by plate tectonics. Plate tectonics does not operate on the Moon.

Many planetary geologists think that the Moon was originally part of Earth. Our big satellite came to be when another body, somewhat larger than Mars, smashed into Earth about 4.4 billion years ago. The debris orbited around Earth, where it collected to form the Moon.

A

B

FIGURE 8-12 **Moon rocks.** (A) Dark lunar basalt collected by *Apollo 15* astronauts. (B) A light-colored rock from the lunar highlands. It is composed mainly of plagioclase (the white mineral) and olivine (yellow grains). ◀ *How were the spherical holes in the lunar basalt formed? What does this indicate about the rate of cooling of this rock?*

MARS: ONCE WETTER, WARMER, AND HOST TO LIFE?

Earthlings long have speculated that Mars might host some form of life, however humble. Mars is farther from the Sun, but still close enough for some warmth. It has seasonal changes and a richly varied landscape (Fig. 8-13). Spacecraft have revealed rock-strewn surfaces (Fig. 8-14), magnificent craters, colossal volcanic peaks, deep gorges, sinuous channels, dunes, and massive structures of layered rock like those in Earth's Grand Canyon.

We have ample evidence of an early period on Mars when there was abundant water. NASA scientists who have analyzed data collected by orbiting spacecraft and the Mars rover *Opportunity* in 2004 conclude that an ocean once existed on Mars that was at least a half-

FIGURE 8-13 **Heavily eroded canyonlands on Mars.** This *Viking* image shows an area about 60 kilometers across.

kilometer deep and larger than all five U.S. Great Lakes. The evidence includes:

- *Opportunity* images of finely layered bedrock (Fig. 8-15A) with cross-bedding like that formed by sand ripples on Earth. (Sediments such as these may contain traces of past biological activity. In 2009, NASA plans to launch a lander that will be able to detect those biological traces.)

- Polygonal cracks like the mudcracks that form on Earth when water-soaked fine sediment dries and contracts (Fig. 8-15B).

- Mars rocks encrusted with such elements as phosphorus, sulfur, sodium, chlorine, and bromine that are likely to have precipitated from evaporating watery solutions.

Although we have ample evidence of past water on Mars, no liquid water has been found on the planet's surface today. What happened to the water? It is likely that solar wind slowly stripped away Mar's water over about 3.5 billion years. This process, which continues today, operated more powerfully in the distant past when radiation from the sun was greater.

The Martian mass is only 10% of Earth's, so its gravity is much less. Its thin atmosphere is mostly carbon dioxide, but unlike Venus and Earth, Mars has virtually no greenhouse effect. Temperatures at the equator range from +21 to –85°C. Mars' lower density compared to other inner planets, plus its lack of a magnetic field, suggest that the planet has only a small iron-rich core.

FIGURE 8-14 *Viking* **lander photograph of the surface of Mars.** The device on the right is part of the lander that was purposely dropped for scale. It is about 20 cm long. Iron oxide gives the surface its red-orange color.

Like Earth, Mars had its own "Archean Eon." After accreting from countless smaller bodies, Mars became differentiated, with its interior partitioning into zones of different compositions and properties. Differentiation probably involved melting, allowing fluids to migrate from one region of the planet to another. During Mars' first billion years, it also experienced heavy bombardment by meteorites and asteroids that densely cratered its crust.

Early in Martian history, the planet had a dense atmosphere provided by volcanic outgassing (expulsion of interior gases). At that time, the planet also may have had a greenhouse effect that warmed the surface. But eventually, atmospheric water and carbon dioxide became depleted and the planet began to cool. Today, conditions do not appear to be hospitable to life.

THE ASTEROID BELT Beyond the orbit of Mars lies a ring of asteroids, numbering in the thousands. These largest of all interplanetary bodies are composed of rock and metals such as iron and nickel. The largest asteroid is about one-tenth the size of Earth, but the majority are only a few kilometers in diameter. Most

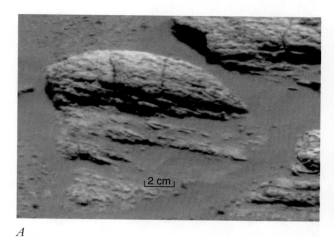

2 cm

A

B

FIGURE 8-15 **Two images of Mars bedrock taken in 2004 by the rover** *Opportunity*. (A) shows evidence of ripple cross-bedding. At the base of the exposure, you can see layers dipping downward toward the right. In (B), fractures divide the rock's surface into polygons. These, like mud cracks on Earth, might have formed when water-soaked sediment dried.

asteroids are objects that never accreted enough space debris to become a full-sized planet. Others are fragments produced by collisions with other asteroids.

The Five Outer Planets

Jupiter, Saturn, Uranus, and Neptune are quite different from the inner group. They are larger, less dense, and are composed of hydrogen and helium. Pluto is a striking exception that does not fit this pattern, and it may actually be an escaped moon of Neptune.

A

B

FIGURE 8-16 **Jupiter.** (A) The planet photographed by *Voyager 1* from a distance of 33,000,000 km. (B) A close-up of Jupiter's Great Red Spot. Note the turbulence to the west (left) of the Great Red Spot.

JUPITER AND SATURN: GIANT GAS BALLS Jupiter, named for the leader of the ancient Roman gods, is the largest planet in our Solar System. Its diameter is 11 times greater than Earth. Its volume exceeds the other planets combined. Its mass is 318 times that of Earth, yet it is only one-fourth as dense because it is a giant gas ball of hydrogen, helium, methane, and ammonia. Jupiter rotates rapidly; its entire day is less than 10 Earth hours. Its rapid spin causes the planet's distinctive atmospheric bands (Fig. 8-16).

With its dozens of satellites, Jupiter resembles a miniature solar system. The four largest moons are especially interesting for trying to discover our Solar System's history (Fig. 8-17). Callisto is riddled with craters and impact basins from meteor and asteroid impacts. Ganymede has sinuous ridges and crisscrossing fractures, suggesting fault movements. Europa is a rocky satellite the size of our Moon. Io's crust is pockmarked by volcanoes that eject plumes of sulfur and sulfur dioxide.

In 2000, magnetic measurements by the *Galileo* spacecraft indicated that Callisto, Ganymede, and Europa have liquid water beneath their frigid crusts. With the presence of water, it is conceivable that life might exist, or once did.

Saturn is the second-largest planet in the Solar System. However, its density is only about 70% that of

FIGURE 8-17 **Jupiter and its four largest satellites.** This image is a spacecraft perspective, reflecting distance from the spacecraft and not actual sizes. From smallest to largest, we see Io, Europa, Ganymede, and Callisto.

FIGURE 8-18 Saturn with its rings, imaged by the Hubble Space Telescope in 2004. Saturn's magnificent rings are shown near their maximum tilt toward Earth. Camera color filters were combined to show the planet as it appears through an Earth telescope.

water, so if you found a big enough swimming pool, Saturn would float in it. Like Jupiter, Saturn is a giant gas ball of hydrogen and helium. There is perhaps no more beautiful sight in the Solar System than the rings of Saturn (Fig. 8-18). They are actually billions of dust particles and larger fragments measuring tens of meters in diameter.

URANUS AND NEPTUNE: THE TWIN PLANETS Neptune and Uranus (Figs. 8-19A and B) are called twins because both are similar in size and density. They are approximately half-scale versions of Jupiter and Saturn. Both have an atmosphere rich in methane and hydrogen. Uranus is noteworthy because its axis of rotation lies nearly in the plane of its orbit—thus, it rotates "on

its side." Collision with another large body may have caused the unusual tilt of Uranus.

PLUTO: AN OUTLIER Pluto is very small and has an odd orbit—at an angle to the rest of the Solar System (Fig. 8-4). These facts suggest that it may not be an original planet, but a satellite of Neptune that escaped. Like Uranus, Pluto orbits on its side, making one rotation every 6.39 days. Its satellite Charon also takes 6.39 days to rotate and revolve, so it always keeps the same face toward Pluto—just like Earth's Moon.

FOLLOWING ACCRETION, EARTH DIFFERENTIATES

About 4.6 billion years ago, Earth was a large ball of accreted materials. It was essentially homogeneous, meaning that it was a random mixture of space debris. It had uniform composition and density throughout, from its center to its surface. It was dramatically different from the planet we know today. Modern Earth is a **differentiated** planet that has three very distinct layers: core, mantle, and crust. Each layer has very different chemical composition and density.

A solid Earth could not have differentiated readily because solid materials are immobile. But liquid materials are very mobile, allowing gravity to sort materials by density. So to become differentiated, Earth must have become at least partially molten. This would allow much of its iron and nickel to move toward the center, to form a dense core (Fig. 8-20). The remaining iron and other metals would combine with silicon

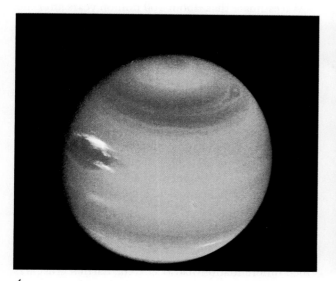

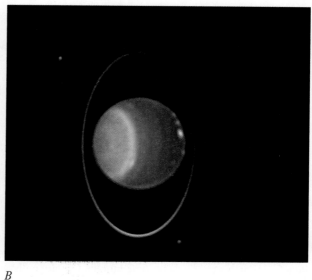

A *B*

FIGURE 8-19 Neptune and Uranus. (A) Neptune, photographed in 1989 by the *Voyager* spacecraft. (B) Uranus and its rings are photographed in near-infrared light by the Hubble Space Telescope. You can see the rings by reflected sunlight. You also can discern some of its moons.

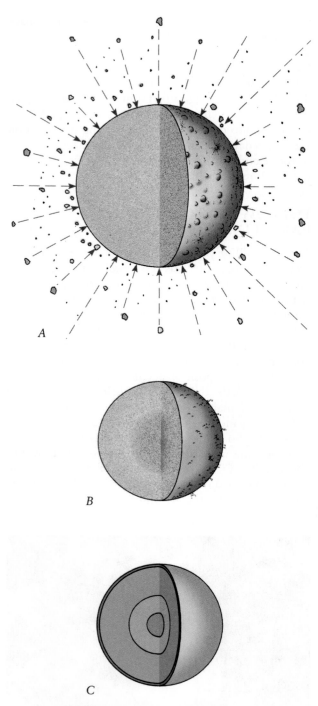

FIGURE 8-20 **First accretion, then differentiation.**
(A) Accretion of particles and bombarding meteorites
make the earliest Earth a homogeneous conglomeration
of space debris. (B) Earth shrinks due to gravitational
compression. This raises its internal temperature high
enough to melt iron and nickel, around 1500°C. Once
liquefied, Earth's materials become mobile. Gravity pulls
denser materials like iron (reddish particles) toward the
center to form Earth's iron core. Lighter silicates float
outward from the center to form the mantle and crust.
(C) Today's Earth is fully differentiated into its core,
mantle, and crust.

and oxygen to form the less-dense mantle that sur-
rounds the core. Still lighter elements would separate
from the mantle to form the lightest, outermost
crustal layer.

You can think of differentiation as analogous to
salad dressing. After shaking a bottle of Italian dressing, the dressing becomes homogeneous, like a planet
just after accretion. But after a few minutes, density differences cause the dressing to separate into layers.

In addition to this process of accretion and differentiation, **partial melting** of Earth was very likely, because complete melting would have allowed too many
volatile gases to escape. For partial melting to occur, an
enormous amount of heat had to come from somewhere. We think the heat came from several processes.

Some heat may have come from the continuing meteorite bombardment as the planet grew. The kinetic
energy of colliding meteorites transforms into heat energy, which is absorbed into the rocks, adding to the
planet's overall heat. Much of the heat generated by
impacting bodies would have remained because subsequent showers of material provided blankets of insulat-
ing debris. Also, as material accumulated at the surface,
the weight on underlying materials would have generated more heat by gravitational compression.

Another extremely important heat source was the
decay of radioactive isotopes. Much of this radioac-
tively generated heat would have accumulated deep
within Earth, where the poor thermal conductivity of
enclosing rocks provided insulation. In Earth's early
days, more heat derived from radioactivity than today
because many isotopes having relatively short half-lives
were still there to generate heat.

We estimate that, about 100 million years after the
solid Earth formed, there would have been sufficient
heat from all these sources to melt iron at depths of 400
to 800 kilometers. Imagine myriad heavy, molten blobs
of iron and nickel gravitating toward the planet's center, displacing lighter materials and forcing them upward toward the surface (Fig. 8-21).

As Earth's iron core developed, it created our
planet's magnetic field. This field is generated by material circulating in the liquid metallic core. As the liquid
metal circulates, it generates an electric current, which
in turn produces a magnetic field. When detected in
other planets, a magnetic field indicates that those bodies have a metallic core.

The Archean Crust

Once differentiation had occurred, Earth's crust was
dominated by iron and magnesium silicates. To understand the origin of this early crust, imagine Earth during its Archean infancy, when it seethed with heat.
There was ample heat to melt the upper mantle. As a
result of the melting, an extensive **magma ocean**

FIGURE 8-21 **Earth as it appeared in the early Archean.** The Moon was closer than today, and slabs of komatiite solidified on the magma ocean.

(Fig. 8-21) may have covered Earth's surface. (A somewhat similar magma ocean existed on the Archean Moon, producing rocks of the lunar highlands.)

The rocks that formed from the cooling magma ocean are called **komatiites** (*koh-MAT-ee-ites*). They take their name from the Komati River in South Africa. Komatiites are ultra-rich in the elements iron and magnesium. They form at temperatures greater than the 1100°C required to melt basalt.

This may have been the earliest time of plate tectonic activity on Earth. Because komatiites are somewhat denser than the partly molten upper mantle, slabs of primordial komatiitic oceanic crust might have sunk as they were conveyed laterally by underlying convection currents in the magma ocean.

Earth has two very distinct crusts today: a denser oceanic crust and a less-dense continental crust (Table 8-2). Could patches of continental crust have formed during the Archean? Recall the 4.4-billion-year-old zircon grains from Australia. They came from a granitic magma that was probably produced above a subduction zone by partial melting of a descending plate. The resulting less-dense granitic magma would then rise to form small islands of continental crust. Apparently, Earth had patches of continental crust 4.4 billion years ago, during the early stages of the Archean.

TABLE 8-2 **Characteristics of Earth's Early Oceanic and Continental Crust**

	Oceanic Crust	Continental Crust
First appearance	About 4.5 billion years ago	About 4.4 billion years ago
Where formed	Oceanic ridges (spreading centers)	Subduction zones
Composition	Komatiite-basalt	Tonalite*-granite
Lateral extent	Widespread	Local
How generated	Partial melting of ultramafic rocks in upper mantle	Partial melting of wet mafic rocks in descending slabs

Source: Condie, K. C. 1989. Origin of the Earth's crust. *Paleography, Paleoclimatology, Paleoecology* 75:57–81, with permission.
*Tonalite is a variety of diorite containing at least 10% quartz.

THE PRIMITIVE ATMOSPHERE— VIRTUALLY NO OXYGEN

Today we are adapted to air that is 78% nitrogen, 21% oxygen, and 1% other gases. But the air we breathe today is far different from the air during the Archean. Our contemporary atmosphere did not just happen. It evolved over the past 4.6 billion years, and it is a product of Earth's highly interactive systems of rock, water, gases, and living things.

Before 3.8 billion years ago, Earth's Archean atmosphere had essentially 0% oxygen, not the 21% we have today. So little free oxygen was discharged into Earth's early atmosphere that it immediately combined with metals such as iron. If we could time-travel back to the Archean, we would quickly suffocate. Archean air consisted mostly of water vapor, carbon dioxide, nitrogen, and lesser amounts of other gases. Significantly, the volatiles (substances easily driven off by heating) released by modern volcanoes approximate the composition of the Archean atmosphere.

Sources of atmospheric gases during Archean days were the accreted particles and pieces of the planet, meteorites, and comets (comets are almost entirely frozen gases, ice, and dust). The volatiles either were scattered throughout the accreting planet or were in the impacting bodies. Today, volatiles and some water are especially abundant in carbonaceous chondrite meteorites.

Outgassing is the process by which water vapor and other gases are released from rocks. The hot environment promoted release of these gaseous elements. Calculations indicate that most of the outgassing of water occurred within the first billion years of Earth history. Evidence for this calculation is the presence of an early ocean, clearly indicated by marine sedimentary rocks dating as far back as 3.8 billion years ago.

Once at Earth's surface, the volatiles underwent a variety of changes (Fig. 8-22). Water vapor condensed and fell as rain, filling low basins to form the seas. Carbon dioxide and other gases that were dissolved in the rain made seawater considerably more acidic than today. This acidity caused rapid chemical weathering, which added calcium, magnesium, and other ions to seawater. Much later, when the seas became less acidic and oxygen more prevalent, these ions would join with carbon dioxide to form limestones and the shells of myriad marine organisms.

Growing an Oxygen-Rich Atmosphere

Earth's early, oxygen-poor atmosphere was followed by an atmosphere that grew increasingly rich in oxygen as life evolved. In fact, the evolution of the atmosphere and living things occurred interdependently, each driving the other to change.

The change from an oxygen-poor to an oxygen-rich atmosphere occurred by Proterozoic time, which began 2.5 billion years ago. Two processes generated oxygen:

1. **Photochemical dissociation.** This is the breakup of water molecules (H_2O) into their hydrogen (H) and oxygen (O) atoms. It occurs in the upper atmosphere when water molecules are split by high-energy ultraviolet light from the

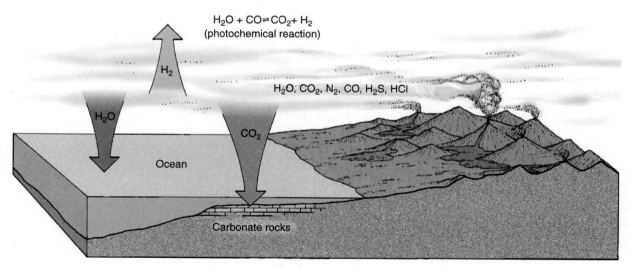

$$H_2O + CO \rightleftharpoons CO_2 + H_2$$
(photochemical reaction)

H_2

$H_2O; CO_2, N_2, CO, H_2S, HCl$

H_2O

CO_2

Ocean

Carbonate rocks

FIGURE 8-22 **The relative amounts of gases in the primordial atmosphere were different from the abundances vented to the exterior of Earth during differentiation.** Nitrogen was retained in the atmosphere, whereas much of the water vapor was lost to the oceans by condensation. Some carbon dioxide was combined with calcium and magnesium, derived from weathering and extracted to form carbonate rocks. Because hydrogen was light, it was lost to space. There was virtually no free oxyen.

Sun. This produced modest oxygen in the atmosphere prior to Earth developing the *ozone layer*. The ozone layer filters ultraviolet radiation, and thus prevents dissociation. Today this process operates too slowly to be significant.

2. **Photosynthesis.** A far more important oxygen-generating mechanism came with the advent of life. Some bacteria evolved the remarkable capability of splitting carbon dioxide into carbon and free oxygen. We know this process as **photosynthesis**. In the section "The Origin of Life" later in this chapter, we will examine the role of photosynthesis in the evolution of early life forms. Photosynthesis has made Earth a truly unique planet in our Solar System.

Geologic Clues to Early Atmospheres

We know Earth's early atmosphere was oxygen-poor from several lines of evidence:

- There are few unoxidized iron minerals from that time. Iron combined with sulfur to form sulfides like pyrite.

- Archean rocks are often dark because they contain free carbon, which would have oxidized in an oxygen-rich atmosphere.

- Rather than oxygen, the Archean atmosphere was rich in carbon dioxide, which combines with water to form carbonic acid. In such an acidic environment, alkaline rocks like limestone and dolomite cannot develop. This may account for the absence of carbonate rocks from this early stage of Earth history.

In rocks younger than 1.8 billion years, evidence of free oxygen in the atmosphere is seen in Proterozoic red shales, siltstones, and sandstones derived from weathering of older rocks on the continents. We also find carbonate rocks (dolomites and limestones) of about the same age. If carbon dioxide were still abundant at this time, these carbonate rocks could not have formed.

BANDED IRON FORMATIONS We infer that oxygen gradually began to appear in Earth's early atmosphere from rocks called **banded iron formations (BIFs)** (Fig. 8-23). BIFs are cherts that exhibit alternating bands of dark rust-red and light gray. The darker red bands are composed mainly of the iron oxide minerals hematite and magnetite, whereas the lighter bands are chemically precipitated quartz.

Banded iron formations formed as precipitates on the floors of shallow seas that occupied parts of the craton. But what was the source of the iron? Although some iron probably came from weathering of iron-bearing rocks on the continents, a larger amount prob-

FIGURE 8-23 **Banded iron formation (BIF).** Bacteria may have played a part in providing the oxygen that oxidized the iron in the red bands. Wadi Kareim, Egypt.

ably came from submarine volcanoes and hydrothermal springs (vents) such as those that exist today along the great midoceanic ridges (Fig. 8-24).

Hot solutions from these vents originate when seawater seeps into fractures and encounters very hot rocks. Once heated, the scalding, chemically corrosive solutions dissolve iron and other elements from the rocks. When these hot solutions rise to the seafloor and encounter cool ocean water, the iron reacts with oxygen to form iron compounds. These become part of the deep-sea sediment.

During the late Archean and early Proterozoic, ample iron may have come from hydrothermal vents. At that time, Earth's interior was hotter, convection in the mantle was more vigorous, and vents were more numerous than today. These primordial sea bottom thermal springs could have provided 25 million tons of iron to the ocean floor each year. This is about 10 times the amount needed to account for all the iron in the banded iron formations.

Banded iron formations have enormous economic importance. They provide most of the iron that is mined worldwide, including the great iron ore deposits of Minnesota, Michigan, the Ukraine, Brazil, Labrador, and Australia. They are the metallic foundation of our industrial civilization.

FIGURE 8-24   **A "black smoker"—a submarine hydrothermal vent—on the Pacific Ocean seafloor along the East Pacific rise.** Seawater is heated as it circulates through hot basalt near a spreading center. The hot water dissolves sulfur, iron, zinc, copper, and other ions from the basalt. When the hot solution contacts cold water, the ions precipitate as "smoke," which actually is tiny mineral grains. The hot, mineral-rich water sustains a diverse plant and animal community.

THE PRIMITIVE OCEAN AND THE HYDROLOGIC CYCLE

The first water bodies that formed at Earth's surface derived from the atmosphere after it cooled sufficiently for water to condense and fall as precipitation. This liquid water probably began to fill low-lying areas as early as 4.4 billion years ago. The early seas formed in this way were acidic, for they contained plentiful carbon dioxide from the early atmosphere. A change to more alkaline water may have occurred rapidly as large amounts of calcium, sodium,

and iron were introduced by submarine volcanoes, neutralizing the acid.

Could the enormous volume of ocean water have come from Earth's interior alone? Yes. Vast amounts of water were locked within minerals in the accreted Earth. This water, "sweated" from Earth's interior and precipitated onto the surface, immediately began to dissolve soluble minerals and carry them to the sea. The initial high acidity of the oceans enhanced the process, so the oceans quickly acquired their saltiness. They have maintained a relatively consistent saltiness by precipitating surplus soluble minerals at about the same rate as they are supplied. The marine fossils suggest that sodium has not varied appreciably in seawater for at least the past 600 million years.

Having outgassed all that water early in its history, Earth has been recycling it ever since. Water is continuously recirculated by evaporation and precipitation—processes powered by the Sun and gravity. This is called the **hydrologic cycle** (Fig. 8-25). Some of the water in the oceans is temporarily sequestered by incorporation into clay minerals on the seafloor. However, these sediments may eventually be melted into magmas that return the water to the surface through volcanic eruption.

ORIGIN OF PRECAMBRIAN "BASEMENT" ROCKS

When early geologists began to describe local rock strata, they sometimes encountered a "basement" of metamorphic and igneous rocks that lay beneath fossil-bearing sedimentary strata. These older rocks were called **Precambrian**, simply meaning that they were older than the Cambrian Period that Sedgwick proposed for overlying younger rocks.

Precambrian rocks clearly represent an enormous expanse of geologic time. They form the very cores of continents. They contain rich deposits of iron and other metals. Notwithstanding their importance, correlating and estimating the age of these apparently unfossiliferous, altered, and deformed Precambrian rocks seemed impossible. Yet a few geologists devoted their lives to successfully deciphering them.

It seemed reasonable to divide Precambrian time into an older **Archean Eon** and a younger **Proterozoic Eon**. The term *Archean* was used to refer to the interval between the birth of the planet and the beginning of the Proterozoic. It was a distinctive eon of high internal heat flow, elevated surface temperatures, limited continental crust, and a non-oxygenic atmosphere. How different it was from the Proterozoic, which had many similarities to today's world.

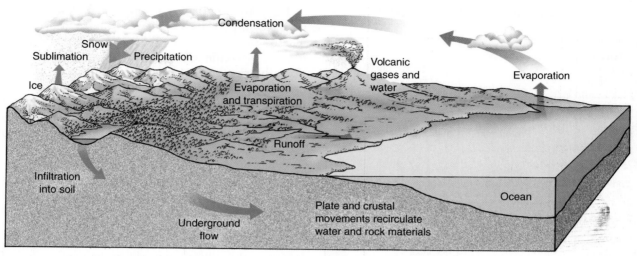

FIGURE 8-25 **The hydrologic cycle.** Water constantly circulates in Earth's atmosphere, oceans, on its surface, and just beneath the surface. To follow the cycle, it is easiest to start at the ocean, where water evaporates. It later condenses and precipitates to the ground (as rain, snow, hail, dew, or frost). It then travels back to the ocean, either overland (streams) or by slowly migrating beneath the surface as groundwater.

Geologists agree that the only reliable basis for correlating fossil-less Archean and Proterozoic rocks is through radiometric dating. Only after the absolute age of a rock is known can it be placed in correct chronological order. Radiometric dating is our basis for mapping Precambrian rocks and deciphering their history. Hundreds of dates thus obtained now permit a tentative calibration of Precambrian geologic time (Table 8-3). The beginning of the Proterozoic is placed at about 2.5 billion years ago. The Archean began 4.6 billion years ago, with Earth's birth.

Where Can We See Precambrian Rocks?

As you would expect from the principle of superposition, most ancient Precambrian rocks worldwide are covered by younger strata. This is demonstrated in Figure 8-26, where the yellow areas are exposed Precambrian rocks. Note the North American area labeled "Canadian Shield."

The most extensive exposures of Precambrian rocks are in geologically stable regions of continents called **shields**. Shields are the oldest parts of continents, and every continent has one or more (Fig. 8-27). In North America, the great **Canadian Shield** extends across 3 million square miles. Continental glaciers of the last Ice Age stripped away the cover of soil and debris, extensively exposing Precambrian rocks of the Canadian Shield.

To the south and west of the Canadian Shield, Precambrian rocks are covered by younger sedimentary sequences of Paleozoic, Mesozoic, and Cenozoic age. Such stable regions where basement shields are blanketed by sedimentary strata are called **platforms**. The platform of a continent, together with its shield, constitutes that continent's **craton**.

We divide the Canadian Shield into **Precambrian provinces** (Fig. 8-28) based on the age of rocks, the directional trends of faults and folds, and the characteristic style of folding. The boundaries of each province are marked by geologic structures that end abruptly and by belts of severely deformed rocks resulting from the collision of crustal elements that came together to form the craton. All the Archean and Proterozoic elements of the Canadian Shield were consolidated by about 1.9 to 1.8 billion years ago.

Continental Crust Appears Worldwide

The continental crust is rich in feldspars, quartz, and muscovite. It is called **felsic**. The first *oceanic* crust formed about 4.5 billion years ago by partial melting of rocks in the upper mantle (Table 8-2). Oceanic crust has less silica and more magnesium and iron than continental crust. It is called **mafic**.

Felsic crust started forming 100 million years later, around 4.4 billion years ago. This occurred in subduction zones, where descending slabs of crust partially melted. The less-dense components in the melt slowly rose buoyantly to the surface, where they cooled to form continental crust.

Ancient patches of continental crust are represented by the 3.8-billion-year-old Amitsoq Gneiss of

TABLE 8-3 Time Divisions of the Precambrian

Time in Billions of Years	Time Divisions		Events
0.5		0.54	
	PROTEROZOIC EON	Neo-Proterozoic	Continental glaciation
1.0		1.0	
		Meso-Proterozoic	Grenville orogeny
1.5		1.6	
2.0		Paleo-Proterozoic	Red beds continental glaciation
2.5		2.5	Kenoran orogeny
3.0	ARCHEAN EON		
3.5			Earliest BIF-
4.0			Earliest surviving Earth's crust
4.0	End of intense meteoritic bombard		
4.1			
4.2	Possible origin of life along midoceanic ridges		
4.3			
4.4	Earliest known Earth material (4.4 b.y. old zircons). Earliest evidence of liquid water and continental crust		
4.5	Asteroid impact ejects material to form Moon(?)		
4.6	Earth accretion		

Greenland (Fig. 8-29) and Canada's 4.04-billion-year-old Acasta Gneiss. These old continental rocks are tonalite gneisses. They were formed by metamorphism of tonalite, a variety of diorite that contains at least 10% quartz.

Proof that Archean continents stood above sea level is apparent in a 3.46-billion year old Australian erosional unconformity that contains a fossil soil. The unconformity and its associated ancient soil represent the oldest land surface known, and they give clear evidence of erosion and weathering on land in the Archean.

Early Archean islands of continental crust probably were small, perhaps less than 300 miles in diameter. These initial patches grew larger by several mechanisms that continue to operate today. Two tectonic plates bearing continental crust might have collided, merging their continental crust. Growth might have occurred as island arcs and microcontinents were added to the host continent along subduction zones. A third phenomenon might have welded sediment onto continental margins.

What did the early protocontinents look like? Geologic evidence indicates that Archean protocontinents were steep-sided, relatively small, narrow landmasses with rugged shorelines. There is no evidence of the wide continental shelves we have today.

As indicated by the distribution of rocks that are 3.0 billion to 2.5 billion years old, some continents had grown large by late Archean time. Once these large landmasses were established, climates differentiated, too. In 2.8-billion-year-old South African sediments, there is evidence of Earth's earliest glaciation. Prior to this time, Earth may have been too warm for such ice sheets to form.

The Earliest Plate Tectonics

When did plate tectonics begin on Earth? Probably not during the first few million years, for the extreme heat would have churned the mantle too rapidly for surface materials to cool and form lithospheric plates. However, by about 4 billion years ago, the surface may have cooled sufficiently to initiate plate formation. We have evidence: Rocks formed at about this time in Canada and Scandinavia display great parallel bands of continental crust that appear to have been moved against one another. We see similar patterns in geologically younger rocks where continents have increased in size by adding microcontinents along subduction zones.

Once plate tectonics was in operation, it supplied the emerging continents with mantle-derived materials that could be partially melted to form additions to continental crust. Archean tectonics would have been more dynamic compared to today because of the greater heat in the mantle. Rates of mantle convection would have been far greater, thermal plumes more numerous, midoceanic ridges longer and more numerous, and volcanism more extensive.

GRANULITES AND GREENSTONES In general, Archean rocks can be grouped into two major rock associations. The first is the **granulite** association, composed largely of gneisses derived from strongly heated and deformed granitic rocks, as well as more gabbroic rocks called **anorthosites**.

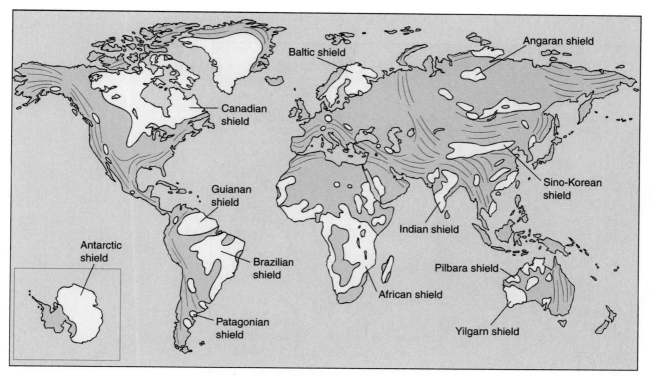

FIGURE 8-26 **Exposed Precambrian rocks.** Yellow areas are exposed Precambrian basement rocks. Green areas are underlain by Precambrian rocks, but the surface consists of younger deposits. Orogenic belts are indicated by the red lines.

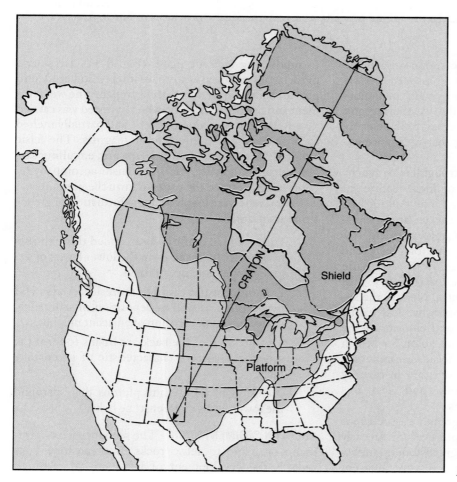

FIGURE 8-27 **North American craton, shield, and platform.**
❓ *What distinguishes a shield from a platform?*

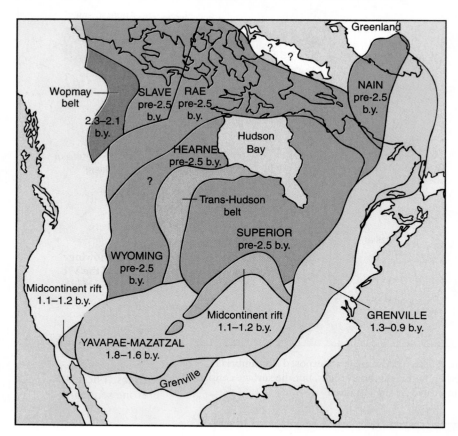

FIGURE 8-28 **Precambrian provinces of North America.**

The second group of Archean cratonic rocks is the **greenstone** association. Greenstone associations are composed of volcanic rocks along with metamorphosed sediments derived from the weathering and erosion of the volcanics. Lavas of greenstone belts exhibit pillow structures, indicating that they were extruded under water (Fig. 8-30).

Greenstones occur in roughly troughlike or synclinal belts (Fig. 8-31). They are prominent features of Archean terranes of all continents. The Abitibi belt (Fig. 8-32) of the Superior Province is the largest uninterrupted greenstone trend in the Canadian shield.

The most intriguing characteristic of greenstone belts is the sequence of rock types, from **ultramafic** near the bottom to felsic near the top (Fig. 8-33). In many, the basal layers are komatiites. Next in the sequence are basalts, which are somewhat less mafic. Green minerals like hornblende and chlorite in the metamorphosed basalts impart the color for which greenstones are named. Next are felsic volcanics (andesites and rhyolites), which are overlain in turn by shales, graywackes, conglomerates, and, less frequently, by banded iron formations.

Where did the greenstone and granulite associations form, and why do they occur in elongate *belts*? An early stage in the development of the greenstone-granulite association was the formation of volcanic arcs adjacent to subducting tectonic plates (Fig. 8-34). Lavas and pyroclastics on the arcs were weathered and eroded to provide sediment to the adjacent trench. The wet sediment and the ocean crust on which it rested was carried downward in the subduction zone and partially melted to form increasingly more felsic magmas. The felsic magmas then worked their way upward, engulfing earlier volcanics. Finally, metamorphism accompanying compression altered the rocks to form the granulites.

What steps were involved in the formation of greenstone associations?

1. Ultramafic to mafic lavas ascended through rifts in back-arc basins to form the lower layers of the greenstone sequence of rocks.

2. Extrusion of increasingly more felsic lavas and pyroclastics, as well as deposition of sediments, derived from the erosion of adjacent uplands.

3. Compression of the back-arc basins to form the synclinal structure characteristic of greenstone belts.

4. Intrusion of the granite plutons that surround the lower levels of greenstone belts.

ARCHEAN SEDIMENTATION The sequence from ultramafic or mafic to felsic rocks in greenstone belts records the development of felsic crustal materials

FIGURE 8-30 **Exposure of greenstone showing well-developed pillow structures, near Lake of the Woods, Ontario, Canada.** The original basalts were extruded underwater during Late Archean time. During the Pleistocene Ice Age, glaciers beveled the surface of the exposures and left behind the telltale scratches (glacial striations).

FIGURE 8-29 **The 3.8-billion-year-old Amitsoq Gneiss of Greenland.** Gneiss is a foliated metamorphic rock. These tonalite gneisses have a felsic composition, indicating that they are continental crustal rocks.

from older, more mafic source rocks. The sedimentary rocks near the top of the greenstone sequence provide clues to the nature of the earliest patches of Archean continental crust. Layers of coarse conglomerates containing pebbles of granite indicate felsic source rocks. Graywackes and dark shales in the greenstone sedimentary sequence bear evidence of deposition in deepwater environments adjacent to mountainous coastlines.

In the greenstone sedimentary association, we do not find blankets of well-sorted, ripple-marked, cross-bedded sandstones. The existence of broad lowlands that could be flooded by shallow interior seas is therefore unlikely. Nor is there any evidence for wide marginal shelf environments.

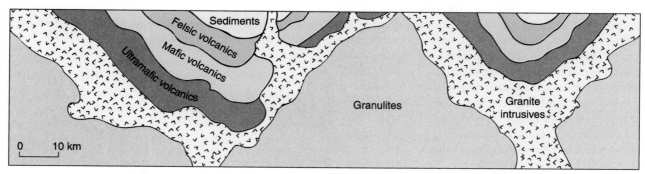

FIGURE 8-31 **Generalized cross-section through two greenstone belts.** Note their synclinal form and the sequence of rock types—from ultrabasic near the bottom to felsic near the top. A late event in the history of the belt is the intrusion of granites. Ultramafic basal layers are particularly characteristic of greenstone belts in Australia and South Africa but do not occur in the Archean of Canada.

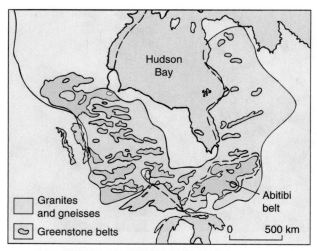

FIGURE 8-32 **Greenstone belts of the Superior Province.**

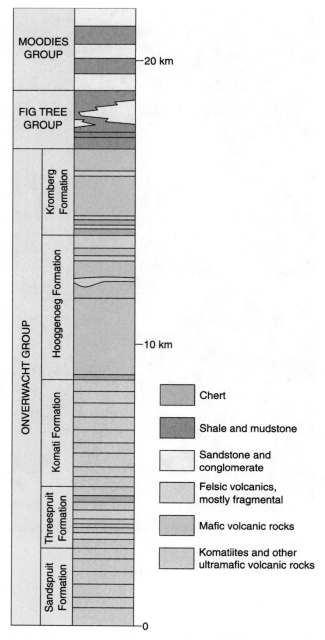

FIGURE 8-33 **Stratigraphic sequence in a greenstone belt in Barberton Mountain Land, South Africa.**

THE ORIGIN OF LIFE

The splendid diversity of animals and plants on Earth today is a consequence of billions of years of chemical and biologic evolution that began during the Precambrian. As described earlier, the basic materials from which microbial organisms could have been assembled initially may have arrived on Earth during the Archean by way of carbonaceous chondrites—meteorites that contain organic compounds.

Unfortunately, we have no unquestioned fossil evidence for the progression from nonliving molecules to living organisms. Our hypotheses for the origin of life are based on present-day biology (again, the present is key to the past), coupled with evidence from Archean rocks, meteorites, and the compositions of planets.

How Molecules Might Combine to Start Life

For the development of life, let us use an analogy: In the nonliving world, neutrons, protons, and electrons combine to form atoms. Atoms in turn combine to form molecules. Molecules join to form complex structures. Perhaps in a similar fashion, the atoms of elements essential to organisms (carbon, oxygen, hydrogen, nitrogen, sulfur, and phosphorus) combine to form organic molecules. At some time in history, these simple molecules might have combined or added components to transform them into the complex organic molecules essential to all living things.

Another analogy might be how small molecules of nitrogen, sugar, and phosphorous compounds combine to form large molecules of DNA, deoxyribonucleic acid (recall Figure 6-16). Some researchers suggest that such large, complex molecules might have organized themselves into **organelles**, bodies capable

of performing specific functions. Organelles might then have combined to form larger entities that grew, metabolized, reproduced, and mutated.

Pulling Together the Pieces of Life

Life requires the elements carbon, hydrogen, oxygen, nitrogen, phosphorus, and sulfur. Each is abundant in the Solar System today and undoubtedly was on primitive Earth. What had to be accomplished was to organize these elements into the systems of an organism and to develop some mechanism for self-replication.

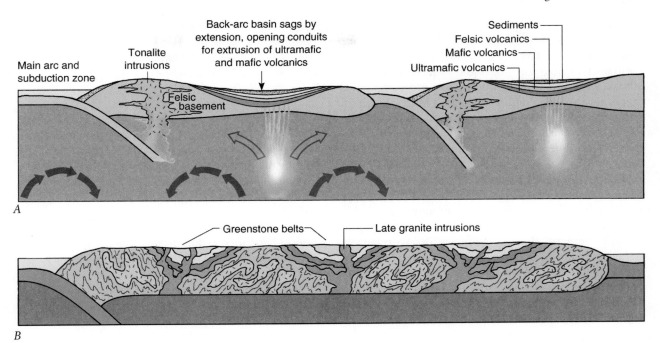

FIGURE 8-34 **A plate tectonics model for the development of greenstone belts and the growth of continental crust.** (A) Plates are in motion, driven by convection cells in the upper mantle. Subduction provides for the emplacement of wedges of oceanic crust and for mixing and melting to provide tonalite intrusions. Tonalite is a variety of the igneous rock diorite containing at least 10% quartz. Behind the main arc, the back arc sags due to stretching, and the greenstone volcanic sequence is extruded. (B) Compression creates the greenstone belts with their synclinal form. Compression also aggregates small continental patches into a larger continental mass. Later, granites are intruded in and around greenstone belts.

There are four essential components of life:

1. **Protein.** Proteins are strings of comparatively simple organic molecules called amino acids. Proteins act as building materials and as catalysts that assist in chemical reactions within the organism.

2. **Nucleic acids.** Nucleic acids are large, complex molecules found in the nuclei of cells. The two classes of nucleic acids are RNA (ribonucleic acid) and DNA (deoxyribonucleic acid). DNA can replicate itself and carries the genetic code that determines the way organisms develop and function.

3. **Organic phosphorous compounds.** These transform light energy or chemical fuel into the energy required for cell activities.

4. **A container.** The cell membrane is an enclosing sack that keeps the cell's components together so they can interact. It relatively isolates the cell's chemistry from outside interference and manages the interaction between the cell's contents and the world outside.

Amino acids are the building blocks of proteins, so they have an important role in developing larger and more complex molecules. Two environmental conditions early in Earth's history may have provided the spark for the natural synthesis of amino acids. First, UV radiation bathed Earth's surface. Experiments demonstrate that UV radiation can separate atoms in water-ammonia-hydrocarbon mixtures and recombine them into amino acids. Second, widespread lightning in Earth's early times could achieve the same thing. Together or separately, UV and lightning may have stimulated the production of amino acids at shallow depths in lakes or oceans. Another possibility is that these important building blocks of life may have arrived on Earth from space. Amino acids have frequently been identified in meteorites.

Then, how did the amino acids aggregate to form proteins? Laboratory experiments suggest that the first proteins could have formed on the surfaces of clay particles. Clay minerals have a layered structure like that of the mineral mica. Clays include metallic ions that could concentrate organic molecules in an orderly array, causing them to align and link themselves into protein-like chain structures.

Simulating the Origin of Life

In 1953, Stanley Miller (1930–), an *exobiologist* (one who studies life beyond Earth), achieved the first laboratory synthesis of amino acids. He simulated Earth's earliest atmosphere (methane, ammonia, hydrogen, and water vapor) in a laboratory flask (Fig. 8-35). He then discharged sparks of electricity (miniature lightning) into the mixture. The liquid yielded a bonanza of amino acids and other complex organic compounds that enter into the composition of all living things. In later experiments by others, similar organic compounds were produced from gases of Earth's pre-oxygen atmosphere (carbon dioxide, nitrogen, and water vapor).

The main requirement for the success of these experiments was the lack (or near absence) of free oxygen. To the experimenters, it seemed almost inevitable that amino acids would have developed in Earth's pre-life environment. Because amino acids are relatively stable, they probably increased gradually to an abundance that would promote their joining together into more complex molecules leading to proteins.

Complete long-chain nucleic acids have not been experimentally produced under pre-life conditions.

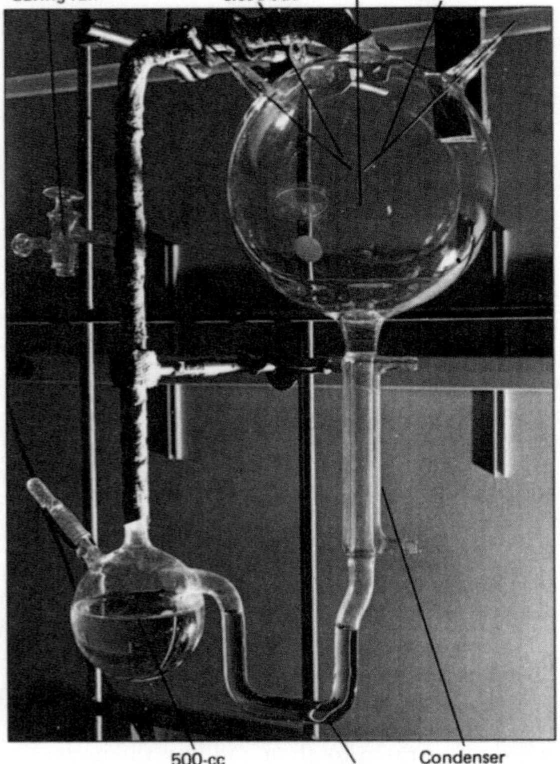

FIGURE 8-35 **An apparatus like this replicated the conditions of early Earth and generated amino acids.** An electric spark in the upper right flask simulated lightning. The gases present in the flask reacted together, forming a number of basic organic compounds.

However, short ordered sequences of nucleic acid components have been produced in the laboratory. Similar sequences could have been formed on the surfaces of clay particles, as described earlier.

Where Did Life Begin?

The idea of the sea as life's birthplace has been with us a long time. Since Darwin's days, biologists have spoken of the primordial ocean as a "rich organic broth" containing all the necessary raw materials for life. However, this venerable hypothesis now is being critically reexamined.

As we have noted, there was very little oxygen in the atmosphere prior to the advent of photosynthesis. However, there was some oxygen being generated by photochemical dissociation. (Recall that this is a process in which ultraviolet light splits water molecules in the atmosphere, liberating oxygen.) Even such small amounts of oxygen released by photochemical dissociation would have destroyed the compounds required for life.

So if life didn't originate in the warm surface waters of the ocean, where did it begin? An answer comes from scientists who cruised midoceanic ridges in diving submersibles. Here they found evidence that life may have originated in total darkness, at great ocean depths, among jets of scalding water rising from hydrothermal vents and volcanic chimneys (Fig. 8-36).

The midoceanic ridges are vast, offering a range of temperatures. Their hot waters dissolve an abundance of elements vital for life, including phosphorus. Clays exist along the ridges, providing the templates needed to construct the organic molecules that preceded life.

Hyperthermophiles and Chemosynthesis

The evidence for such a stygian beginning of life is the presence of microbes called **hyperthermophiles** (literally, high-heat lovers). These thrive in seawater exceeding 100°C (water's boiling point at sea level). These tiny organisms apparently live deep within fissures below the actual vents, for they often are ejected in such great numbers that they cloud the surrounding waters.

In the absence of light for photosynthesis, these organisms derive their energy from **chemosynthesis**. Unlike photosynthesis, by which organisms synthesize their own organic compounds by using carbon dioxide as a source of carbon atoms and sunlight as a source of energy, organisms employing chemosynthesis derive their energy by oxidizing such inorganic substances as hydrogen sulfide or ammonia.

Recent studies strongly suggest that the first organisms were not phototrophs (organisms utilizing photosynthesis), but chemotrophs. The chemosynthetic microbes around vents of the midoceanic ridges form the

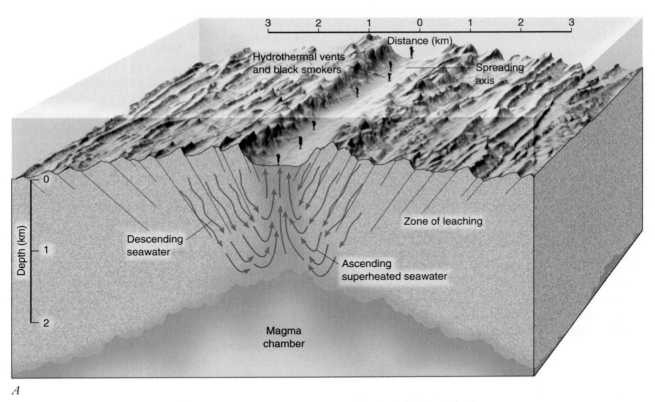

A

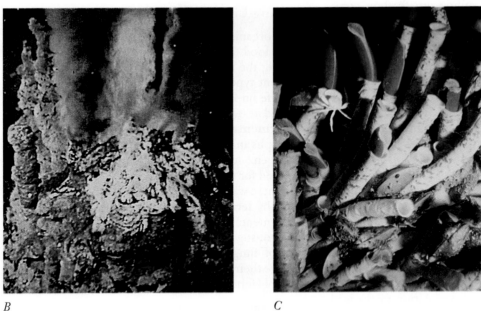

B   *C*

FIGURE 8-36 **Did life begin at vents along the midoceanic ridges?** (A) Section across the central part of a midoceanic ridge (spreading center) shows how hydrothermal vents occur on the ocean floor. Cool water (*blue arrows*) is gradually heated as it descends toward the hot magma chamber. The heated water leaches sulfur, iron, copper, zinc, and other metals from surrounding rocks and returns toward the surface (*red arrows*). It discharges from hydrothermal springs or vents on the seafloor (*small black plumes*), where the load of metallic elements supports a distinctive community of organisms. The hydrothermal vent in (B) is on a midoceanic ridge in the eastern Pacific. Nutrient-rich vent water supports chemosynthetic bacteria that ultimately support a variety of invertebrates, including giant tubeworms (C).

base of a food chain that supports an astonishing variety of invertebrates. These include shrimplike arthropods, crabs, clams, and tubeworms up to 10 feet long (Fig. 8-36C).

Hyperthermophilic microbes have DNA that is sufficiently different from that of bacteria to warrant placing them in a separate branch on the tree of life. In Chapter 6, we noted that all life can be organized into three domains, the Archaea, Bacteria, and Eukarya. The hyperthermophilic and related microbes that feed on hydrogen and produce methane belong in the Archaea. Other organisms, such as plants and animals, whose cell structure includes a nucleus and organelles, are in the Eukarya.

The tubeworms of hydrothermal vents are truly remarkable creatures. They are able to grow more than 33 inches in a year, exceeding the growth rate of any other known invertebrate. As they have no mouth or digestive tract, they depend upon their symbiotic relationship with internal thermophilic microbes for survival. (Symbiotic means that each has some life function that the other needs to survive. If either the tubeworms or microbes disappeared, the other would die out.) Spongy tissue inside the worm is loaded with Archaean organisms. As true chemotrophs, these organisms oxidize hydrogen sulfide into carbon compounds that nourish the tubeworm.

You can envision the steps by which Earth's first hyperthermophiles may have originated. It began with the downward percolation of seawater through fractures and fissures in the seafloor. The water would reach the extremely hot rocks that supply lavas to the midoceanic ridge volcanoes. Here the water would heat to 1000°C, yet not boil due to the tremendous confining pressure. This is like a car's cooling system, which pressurizes the hot "antifreeze" coolant to keep it from boiling.

The superheated water would react with surrounding rocks, extracting elements needed to construct organic molecules. Such molecules would begin to form as the hot waters gradually returned to the surface, cooling as they made their way upward. During that time, amino acids and other organic compounds might be synthesized and combined to form the first protocells. Growth, proliferation, and mutation might lead to primitive chemosynthetic microbes like those seen today in hydrothermal vent environments.

Life in Extremely Hostile Environments

Recent discoveries of organisms living in amazingly harsh environments provide a new view of how inevitable and ubiquitous life on Earth is. The heat-loving hyperthermophiles are an example. We also have detected microscopic life near Earth's frigid poles, where temperatures dip to −113°C. Microscopic life occurs in hot, barren deserts. It also occurs in submicroscopic spaces between grains of solid rock more than 3 kilometers (2 miles) beneath Earth's surface, and in rocks having temperatures exceeding 110°C. One such microbe, found at a depth of 2.8 kilometers (1.7 miles) in the course of drilling for natural gas, was appropriately named *Bacillus infernus*, the "bacterium from hell." These subterranean microbes have been dubbed **lithotrophs** ("rock nourishment").

Most lithotrophs live off energy derived from hydrogen, iron, magnesium, and sulfur. Some thrive on rocks heated by radioactive decay. In the gold mines of South Africa, prodigious numbers of microbes live at roughly 3 kilometers depth. Microbes have been discovered in uncontaminated granites 3600 meters (2.24 miles) below Earth's surface in Sweden.

Did the ancestors of these microbes originate at the surface and work their way downward? Or could life have originated at depth and moved upward as Earth's primordial surface environment became gradually more hospitable? Or both? Or neither? These questions are challenging our ideas about how life originated on Earth and about the probability of life on other planets.

Feeding Life on Earth

Every organism requires nutrients, the substances present in food that can be used as a source of energy for running the machinery to sustain the organism. How different types of organisms "eat" is itself a history of how life on Earth developed. Each organism type has evolved its own strategy for obtaining nutrition.

Fermenters. Food gathered by ancestral heterotrophs might have been externally digested by excreted enzymes before being converted to the energy required for vital functions. In the absence of free oxygen, there was only one way to accomplish this conversion—by fermentation. There are many variations of the fermentation process. The most familiar is fermentation of sugar by yeast. Yeast organisms—which are a type of fungus—disassemble organic molecules, rearrange their parts, and derive energy for life functions. A simple fermentation reaction is:

$$\text{Sugar} + \text{yeast activity} \rightarrow \text{carbon dioxide} + \text{alcohol} + \text{energy}$$

Autotrophs manufacture their own food. Eventually, the original fermentation organisms would have consumed the organic compounds of their environment, creating a food shortage. This scarcity might have caused selective pressures for evolutionary change. At some point prior to depletion of the food supply, organisms evolved the ability to synthesize their nutritional needs from simple inorganic substances. These were the first autotrophic organisms.

The organisms that developed this remarkable ability saved themselves from starvation. Their evolution proceeded in diverse directions. Sulfur bacteria manufactured their food from carbon dioxide and hydrogen sulfide. Nitrifying bacteria used ammonia as a source of energy.

Photoautotrophs, which employ photosynthesis, were more significant to Earth history. They evolved the ability to use sunlight to power the dissociation of carbon dioxide into carbon and free oxygen. The carbon was combined with other elements to enable organisms to grow, and the oxygen was released as a waste gas. This waste gas prepared Earth's surface environment for the next important step in the evolution of primitive organisms. Simplified, here is the reaction for photosynthesis:

$$\text{Carbon dioxide} + \text{water} + \text{sunlight} \rightarrow \text{sugar} + \text{oxygen}$$

Heterotrophs (*troph* = to nourish) can't make their own food, so they scavenge nutrients in their environment. Organisms that do so include animals, such as ourselves. The earliest heterotrophs probably "ate" small aggregates of organic molecules that were present in the surrounding water or cannibalized their developing contemporaries.

Anaerobic organisms flourished during the time when Earth's atmosphere lacked free oxygen. However, as photoautotrophs multiplied, billions upon billions of tiny living oxygen generators began to change the primeval oxygen-poor atmosphere to today's 21% oxygen atmosphere. Fortunately, the change was gradual. Had oxygen accumulated too rapidly, it would have been lethal to early microorganisms. Iron in rocks of the continental crust readily bonded with the emerging oxygen, preventing too rapid a buildup of this chemically aggressive gas. Eventually, organisms evolved oxygen-mediating enzymes that permitted them to thrive in the new atmosphere.

Once the iron on Earth's surface became saturated with oxygen, it could no longer mediate the buildup. At that point, the gas began to accumulate in the atmosphere and to dissolve in fresh water and seawater. Solar radiation converted some oxygen (O_2) to ozone (O_3), forming an absorptive shield against intense and harmful UV from the Sun. This protected still-primitive and vulnerable life forms so they could expand into new environments.

Aerobic organisms. The stage was set for the appearance of aerobic organisms, which rely on oxygen to live. Aerobic organisms are more efficient, because using oxygen to convert food into energy provides far more energy in proportion to food consumed. This energy surplus was critical to the evolution of more complex life forms.

Prokaryotes and Eukaryotes

Lack of fossil evidence means that we cannot date the transition from prebiotic evolution of molecules to the organic evolution of life forms. We can only say that the oldest fossils are prokaryotes.

Prokaryotes. These organisms have genetic material (DNA) that is not packaged into a nucleus (Fig. 8-37A). Prokaryotes reproduce asexually, by cell division (Table 8-4). Cyanobacteria are prokaryotes that are capable of photosynthesis. They first appear in the fossil record late in the Archean. On the family tree of life depicted in Chapter 6, Figure 6-13, prokaryotes are in the Domains Bacteria and Archaebacteria.

Because prokaryotes are asexual, and simply divide to reproduce themselves, they are restricted in genetic variability to whatever was in the original prokaryote. It is probably for this reason that prokaryotes have shown little evolutionary change through more than 2 billion years of Earth history. Nevertheless, such organisms represent an important early step in the history of primordial life.

Eukaryotes. Evolution proceeded from the prokaryotes to eukaryotes. Their cells have a definite nucleus and well-defined chromosomes. Eukaryotes reproduce sexually (Fig. 8-37B). Unlike prokaryotes, eukaryotes contain organelles—bodies that perform specialized functions. Examples are chloroplasts (which convert sunlight into plant sugar) and mitochondria (which metabolize carbohydrates and fatty acids to carbon dioxide and water, releasing energy-rich phosphate compounds in the process). Evidence of eukaryotes extends to at least 2.2 billion years ago and possibly 2.7 billion years ago.

Biologists believe that the organelles in eukaryotic cells were once independent microorganisms that moved into other cells and established symbiotic (mutually beneficial) relationships with the primary cell. For example, one bacterium might engulf, but not digest, another bacterium. If the two survive together, they reproduce and eventually might become organelles (Fig. 8-38). The host cells would have the advantage of having an internal organelle capable of photosynthesis, whereas the other would benefit from a protected environment and availability of nutrients. Eventually, the two would lose their ability to function outside the primary cell.

Eukaryotes reproduce sexually, with a union of egg and sperm to form the nucleus of a single cell. This combines the parental chromosomes, which leads to many possible gene combinations in the next generation. With the development of sexually reproducing eukaryotes, genetic variations could be passed from parent to offspring in a great variety of new combinations. This development was truly momentous, for it led to a dramatic increase in the rate of evolution and

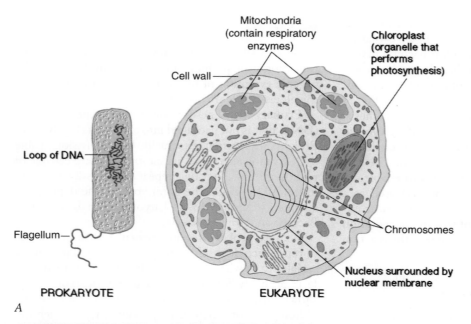

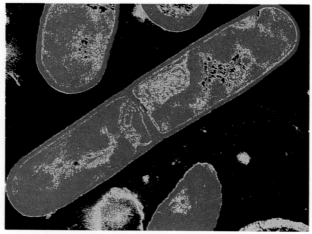

FIGURE 8-37 **A prokaryote cell and eukaryote cell.**
(A) The prokaryotic cell is tiny (0.5 to 1.0 mm) and lacks a
true nucleus. (B) The eukaryotic cell is larger (10 to 100 mm)
and contains a true nucleus and various organelles.

was ultimately responsible for the evolution of complex multicellular animals (Fig. 6-13).

Archean Fossils

Prokaryotic organisms were probably present in the early and middle Archean, but unquestioned fossil remains of delicate and microscopic organisms are lacking. What we do find are **stromatolites**. These are not individual one-celled fossils, but large sedimentary structures formed indirectly as a response to the liberation of carbon dioxide by matlike colonies of photosynthetic cyanobacteria.

Examination of present-day stromatolites, such as those at Shark Bay, Australia (Fig. 8-39), indicates that they develop when fine particles of calcium carbonate settle on the sticky surfaces of cyanobacterial mats to form thin layers (laminae). The bacterial colonies then grow through the layer and form another surface for the collection of more fine sediment. Repetition of this process produces the succession of laminae. In many Archean stromatolites, the calcium carbonate has been replaced by chert.

Stromatolites are more common in Proterozoic rocks than in Archean, perhaps reflecting the more common warm shelf environments during the Proterozoic. However, laminated structures resembling stromatolites (but lacking entrapped fossils of actual organisms) are present in 3.0-billion-year-old rocks of southern Africa and 2.8-billion-year-old rocks of Australia.

Molecular Fossils

In 1999, Australian researchers found evidence of eukaryotic life on Earth 2.7 billion years ago. Prior to this discovery, eukaryotes were thought to have appeared about 1.6 billion years ago. The evidence is indirect—it

TABLE 8-4 **Comparison of Prokaryotic and Eukaryotic Organisms**

	Prokaryotes	**Eukaryotes**
Organisms	Bacteria and cyanobacteria	Protists, fungi, plants, and animals
Cell size	1 to 10 microns*	10 to 100 microns
Genetic organization	Loop of DNA in cytoplasm	DNA in chromomes in membrane-bound nucleus
Organelles	No membrane-bound organelles	Membrane-bound organelles (chloroplasts and mitochondria)
Reproduction	Binary fission, dominantly asexual	Mitosis or meiosis, dominantly sexual

*A micron is ¹⁄₁₀₀₀ of a millimeter.

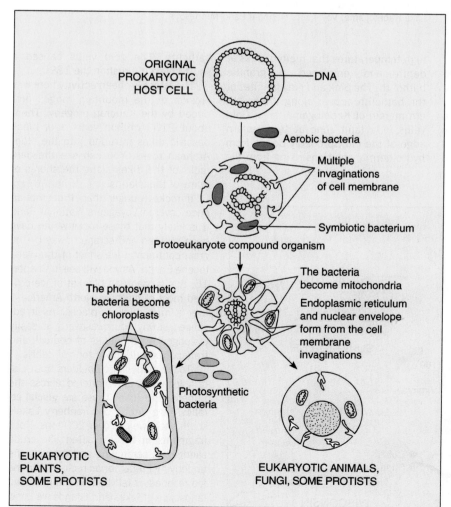

FIGURE 8-38 **The endosymbiotic theory for the origin of eukaryotes.** A prokaryotic organism engulfs other prokaryotes, which function well in the protected environment inside the parent cell. Eventually, the engulfed cells become organelles.

Voyageurs National Park

Voyageurs National Park takes its name from the colorful French-Canadian trappers who traveled the lakes in birchbark canoes. The park lies along the Minnesota-Ontario border, near the southern margin of the Canadian Shield (see map). The park contains more than 30 lakes, which are bordered by rocky shorelines and contain about 900 islands. Visitors leave their automobiles at entry stations and explore by boat.

Like a history book in which all chapters have been destroyed but the first and last, the park displays a record of very early and very recent geologic events. The first chapter is recorded in severely deformed, compressed schists and gneisses of Precambrian age. All exposed bedrock in the park originated during the Precambrian (Fig. A photo), and most formations are Archean. The last chapter is revealed in erosional and depositional features from glaciers.

Archean rocks at Voyageurs National Park include the granulite and greenstone associations described in this chapter. Pillow lava within the greenstone association indicates that much of the Archean volcanic activity here occurred beneath the sea. Metamorphosed graywackes suggest rapid deposition in trenchlike basins.

Voyageurs National Park lies in the Precambrian Superior Province. In the Late Archean, a mountain-building event, the Kenoran Orogeny, created

FIGURE A **Voyageurs National Park.** Exposure of Precambrian gneiss along the shore of Lake Kabetogama, Voyageurs National Park, Minnesota.

high temperatures that melted rocks at depth and emplaced a granitic batholith. The pink and gray granites of the batholith appear along the southern margin of Kabetogama Lake. Gold veins in a fault zone at the western edge of the park probably derive from hydrothermal solutions rising from the batholith. The gold veins caused a short-lived gold rush in the 1890s.

The latest Archean activity here was erosion of the mountain ranges produced by the Kenoran orogeny. Then, about 2100 million years ago, black basaltic dikes intruded into the older Archean rocks. You can see the dark rocks of the dikes along the shores of many of the islands.

If rocks younger than Precambrian once covered Voyageurs National Park, it is likely that these rocks would have been scoured and scraped away by the great continental ice sheet of the Pleistocene Epoch. As you will see in Chapter 15, this gargantuan blanket of ice covered most of northern North America—over a mile thick in places. As it advanced slowly southwestward, it easily dislodged huge chunks of bedrock and transported billions of tons of debris.

Today many large boulders from distant locations are scattered across the Superior Province. These are **glacial erratics**. One erratic on Cranberry Lake's south shore weighs over 200 tons. Rock fragments in the ice acted like crude sandpaper, scratching and grooving the underlying Precambrian rocks. When the ice receded, it left behind the glaciated landscape of lakes and islands we know as Voyageurs National Park.

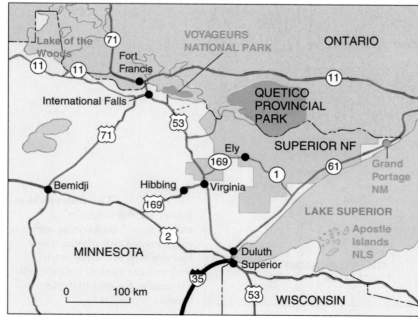

A   *B*

FIGURE 8-39   **Present-day and ancient stromatolites.** (A) Modern stromatolites growing in the intertidal zone of Shark Bay, Australia. Colonial marine cyanobacteria form these structures. Fine particles of calcium carbonate settle between the tiny filaments of the mat-like colonies and are bound with a mesh of organic matter. Successive additional layers result in distinctive laminations. (B) Fossil stromatolites from Precambrian rocks exposed in southern Africa. At top right, note the rock hammer for scale.

does not consist of actual body fossils or trace fossils, but of preserved organic molecules that only eukaryotes can synthesize. These are called **molecular fossils.** They were carefully extracted from black shales of Archean age in northwestern Australia.

Although the molecular evidence indicates that eukaryotes originated in the Archean, they did not begin to diversify until about 1.2 billion years ago, during the Proterozoic. Many believe the expansion of eukaryotic algae is related to the build-up of oxygen in upper layers of the ocean, as well as the production of nourishing quantities of nitrates.

▶ IN RETROSPECT

The Archean is the important first act in the drama of Earth's eventful history. During the Archean, our planet aggregated from a solar nebula; developed its internal zones of core, mantle, and crust; achieved a magnetic field; gave birth to continents; formed a world ocean; and wrapped itself in an atmosphere that became the one we have today. The Archean was the eon when the basic mechanisms governing geologic change were established. It was the time when life appeared.

SUMMARY

- Earth is one of nine planets revolving around the Sun. Mercury, Venus, Earth, and Mars are the inner planets. They are built of rocky and metallic particles and have relatively high densities. Jupiter, Saturn, Uranus, and Neptune are the outer planets. They have lower densities and are larger. Pluto is the smallest planet and has a highly tilted orbit.

- Earth formed from cosmic material in the solar nebula by cold *accretion*. It then underwent *differentiation*, resulting in the formation of the core, mantle, and crust. The heat required for differentiation resulted from radioactive decay, gravitational compression, and intense meteoritic bombardment.

- As Earth was differentiating, gases were released (*outgassed*) from the interior. They accumulated as an initial atmosphere devoid of free oxygen but rich in carbon dioxide, water vapor, and other volcanic gases. An oxygen-rich atmosphere followed the origin and spread of oxygen-generating photosynthetic organisms.

- Archean rocks are most extensively exposed in broadly upwarped, geologically stable Precambrian shields. Within shields, there are Precambrian provinces recognized by their distinctive ages and boundaries marked by ancient orogenic belts.

- The original crust of Earth was mafic and formed by extrusions from the underlying mantle. From the original mafic crust, patches of felsic crust formed by partial melting and recycling of the weathered products of the mafic crust.

- Two main Archean rock complexes are the granulite and greenstone associations. They developed in the course of Archean tectonics when granulite rocks formed along subduction zones adjacent to volcanic arcs and greenstone belts developed in back-arc basins.

- During the Archean, life originated on Earth in an environment deficient in free oxygen and containing all the elements needed to form complex, nonliving organic molecules. The earliest living cells to arise were heterotrophic prokaryotes. These were followed by autotrophic prokaryotes, and eventually eukaryotes.

- Life may have originated not near the surface of the ocean, but at great depths along midoceanic ridges where hydrothermal vents provided warmth and nutrients and where there was no free oxygen to destroy pre-life molecules.

KEY TERMS

accretion, pp. 210, 220

aerobic organism, p. 235

anaerobic organism, p. 235

anorthosite, p. 226

Archean Eon, p. 224

autotroph, p. 234

banded iron formation (BIF), p. 223

Canadian Shield, p. 225

carbonaceous chondrite, p. 211

chemosynthesis, p. 232

chondrules, p. 211

craton, pp. 225, 227

differentiation (planetary), pp. 219, 220

eukaryote, p. 235

felsic, p. 225

fermenter, p. 234

fusion, p. 210

granulite, p. 226

greenstone, p. 228

heterotroph, p. 235

hydrologic cycle, p. 224

hyperthermophile, p. 232

iron meteorite, p. 212

komatiite, p. 221

lithotroph, p. 234

lunar highlands, p. 215

mafic, p. 225

magma ocean, pp. 220–221

maria, p. 215

meteorite, p. 210

molecular fossils, p. 239

nebular hypothesis, p. 209

ordinary chondrites, p. 211

organelles, p. 230

outgassing, p. 222

partial melting, p. 220

photoautotroph, p. 235

photochemical dissociation, p. 222

photosynthesis, p. 223

platform, pp. 225, 227

Precambrian, p. 224

Precambrian provinces, p. 225

prokaryote, p. 235

Proterozoic Eon, p. 224

protoplanet, p. 210

shield (Precambrian), pp. 225, 227

solar nebula, p. 209

Solar System, p. 208

solar wind, p. 210

stony-iron meteorite, p. 212

stromatolite, p. 236

ultramafic, p. 228

QUESTIONS FOR REVIEW AND DISCUSSION

1. At successively greater distances from the Sun, name the planets of our Solar System. In what galaxy are these planets located?

2. Compare Mercury, Venus, Earth, and Mars. What do they have in common and how do they differ?

3. Given that both Earth and the Moon experienced heavy meteoric bombardment during the Archean, why are there so few craters recording that event on Earth?

4. Why are we likely to learn more about the Archean history of Earth by studying the surface and rocks of the Moon?

5. How did the internal layers of Earth develop?

6. What are the principal kinds of meteorites? What is the particular significance of carbonaceous chondrites?

7. Distinguish between the terms *Precambrian shield*, *craton*, and *platform*.

8. Differentiate between the terms *mafic* and *felsic*. Give an example of a felsic extrusive rock and one of a mafic extrusive rock. Name two minerals common to each of these rock types.

9. How might patches of felsic continental crust have been derived from mafic rocks during the Archean?

10. Describe the structural configuration and general vertical sequence of rock types in a greenstone belt. In a plate tectonics scenario, where might greenstone belts form and how would they develop?

11. How do komatiites differ from mafic rocks such as basalt? Where do they occur in greenstone sequences?

12. What geologic evidence suggests that free oxygen was beginning to accumulate in Earth's atmosphere about 3 billion years ago?

13. Why are Archean rocks more difficult to correlate than rocks of the Paleozoic, Mesozoic, and Cenozoic? What is the method used to date and correlate most Archean rocks?

14. Discuss the role of symbiosis in the evolution of eukaryotes. What organelles may have originated by symbiosis?

15. What are hyperthermophiles? What does the presence of hyperthermophiles suggest about the possibility of finding organisms on other planets in the Universe?

WEB SITES

The Earth Through Time Student Companion Web Site (www.wiley.com/college/levin) has online resources to help you expand your understanding of the topics in this chapter. Visit the Web Site to access the following:

1. Illustrated course notes covering key concepts in each chapter;

2. Online quizzes that provide immediate feedback;

3. Links to chapter-specific topics on the web;

4. Science news updates relating to recent developments in Historical Geology;

5. Web inquiry activities for further exploration;

6. A glossary of terms;

7. A Student Union with links to topics such as study skills, writing and grammar, and citing electronic information.

9

Early Proterozoic graywackes (2.5 billion years old), composed of coarse sandstone with firmly cemented particles. Note the graded bedding interstratified with shaley limestone. Deposited east of Great Slave Lake in Canada's Northwest Territories.

The Proterozoic: Dawn of a More Modern World

We crack the rocks and make them ring,
And many a heavy pack we sling;
We run our lines and tie them in,
We measure strata thick and thin,
And Sunday work is never sin,
By thought and dint of hammering.
—A. F. Lawson, *Mente et Malleo* ("*Mind and Hammer*"), 1888

> Lawson proved that Precambrian granites and gneisses of the Canadian Shield resulted from multiple mountain-building episodes.

O U T L I N E

▶ CHAPTER 9—THE PROTEROZOIC: DAWN OF A MORE MODERN WORLD

▶ HIGHLIGHTS OF THE PALEOPROTEROZOIC (2.5 to 1.6 billion years ago)

▶ HIGHLIGHTS OF THE MESOPROTEROZOIC (1.6 to 1.0 billion years ago)

▶ HIGHLIGHTS OF THE NEOPROTEROZOIC (1.0 billion to 540 million years ago)

▶ PROTEROZOIC ROCKS SOUTH OF THE CANADIAN SHIELD

▶ PROTEROZOIC LIFE

▶ SUMMARY

▶ KEY TERMS

▶ QUESTIONS FOR REVIEW AND DISCUSSION

▶ WEB SITES

▶ BOX 9-1 ENRICHMENT

▶ BOX 9-2 ENRICHMENT

▶ BOX 9-3 ENRICHMENT

Key Chapter Concepts

- In contrast to the Archean, the Proterozoic displays a more modern style of plate tectonics, sedimentation, and global climate.

- The Wopmay Orogen, a belt of deformed rocks in Canada's Northwest Territories, developed during a Wilson Cycle involving the opening and closing of an ocean basin.

- Earth's first great ice age occurred during the Paleoproterozoic. In North America, tillites of the Gowganda Formation attest to this frigid episode.

- Most of the world's iron ore deposits occur in Paleoproterozoic banded iron formations (BIFs). The Animikie group in the Lake Superior region and eastern Canada have yielded immense amounts of iron ore.

- During the Paleoproterozoic, Laurasia nearly broke apart. A rift zone (zone of tensional faults) extended from the Lake Superior region to Kansas. Magmas moved to the surface along the fault, flowed onto Earth's surface, and formed a thick sequence of basalt lava flows. The break failed, however, and the continent remained whole.

- Cold climates characterize the Neoproterozoic, as indicated by tillites and glacial features on most of the world's continents—even some that were located near the equator.

The Proterozoic Eon dawned 2.5 billion years ago (Fig. 9-1) with a more modern style of plate tectonics, and ended only 540 million years ago. This enormous period comprises 42% of Earth history. To aid your study of the Proterozoic, we divide the eon into three eras:

- the early **Paleoproterozoic Era** (2.5 to 1.6 billion years ago)

- the middle **Mesoproterozoic Era** (1.6 to 1.0 billion years ago)

- the "new" **Neoproterozoic Era** (1.0 billion years ago to the beginning of the Paleozoic Era, 540 million years ago)

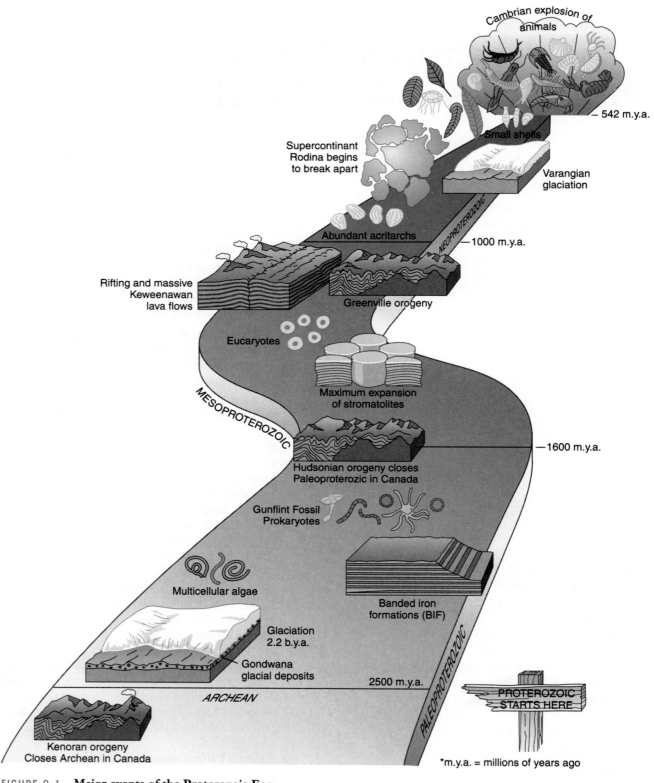

FIGURE 9-1 **Major events of the Proterozoic Eon.**

Because Proterozoic rocks are less altered than Archean rocks, they are easier to interpret. However, difficulties persist, for Proterozoic rocks lack the abundant fossils of more modern Paleozoic-Mesozoic-Cenozoic strata.

A number of large, distinct crustal segments, called **Precambrian provinces**, developed in North America during the Archean Eon. During the Paleoproterozoic, these once-separated segments became sutured together to form Earth's first large continent, **Laurentia**. The suturing occurred along **orogens**—belts of crustal compression, mountain-building, and metamorphism. By about 1.7 billion years ago, the suturing was complete.

For the remainder of the Proterozoic, Laurentia grew as more crustal materials became accreted to continental margins. These additions included sedimentary rocks crushed onto continental margins, as well as microcontinents and island arcs carried to subduction zones by seafloor spreading.

As the continents grew, so did tectonic plate motion, rifting, and seafloor spreading. On and around the growing cratons, tectonic events and sedimentation resembled those of more recent times. By Proterozoic time, wide continental shelves existed, as well as shallow seas that flooded the continental interiors. Such inland seas are termed **epicontinental**. Clean sands and carbonate deposits accumulated. In contrast, such sediments are rare in the Archean. Late in the

Proterozoic, continents became assembled into a new supercontinent named **Rodinia** (Fig. 9-10).

Another contrast between the Proterozoic and the earlier Archean was major glaciation episodes. One such "ice age" occurred during the Paleoproterozoic about 2.4–2.3 billion years ago. The second occurred in the Neoproterozoic 850–600 million years ago.

HIGHLIGHTS OF THE PALEOPROTEROZOIC (2.5 to 1.6 billion years ago)

The 900 million years of the Paleoproterozoic were dramatically eventful. Early plate tectonics was in vigorous operation. There was major mountain-building on all major continents. Earth experienced its first great ice age, and increasing oxygen in the atmosphere indicated that photosynthesis was in progress.

Early Plate Tectonics: Evidence from Canada's Northwest Territories

During the Paleoproterozoic, orogenic belts developed primarily around the margins of the older Archean provinces. One such Paleoproterozoic orogenic belt lies along the western margin of the Slave province in Canada's Northwest Territory (Fig. 9-2). It has been named the Wopmay orogen.

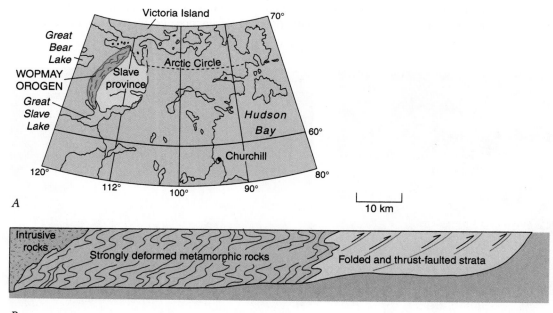

FIGURE 9-2 **Rocks around the Wopmay Orogen in Canada's Northwest Territories show evidence of a Wilson Cycle.** (A) Location on the west margin of the Precambrian Slave Province. (B) Cross-section of the Wopmay orogenic belt. IT depicts a zone of intrusive igneous rocks that pass eastward into metamorphic rocks and then into folded and thrust-faulted stratified rock.

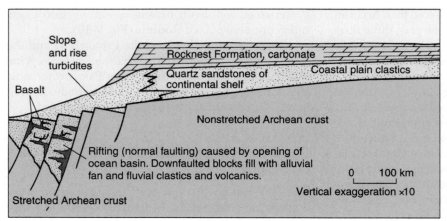

FIGURE 9-3 **Relationship of rock units following the Paleoproterozoic opening of an ocean basin along the western margin of the Slave province.**
🖪 *What type of faults lie beneath the turbidites? Are such faults associated with the leading edges of tectonic plates or with passive margins? (Answers to questions appearing within figure legends can be found in the Student Study Guide.)*

Here we see evidence for a **Wilson Cycle,** named for J. Tuzo Wilson (1908–1993), a pioneer of plate tectonics theory. A Wilson Cycle includes three steps:

1. Opening of an ocean basin. Evidence for the initial opening of an ocean basin consists of numerous normal (tensional) faults caused by rifting, shown in Figure 9-3. Alluvial clastics accumulated in the downfaulted blocks, and lavas flowed upward through the fault planes to become interlayered with sediments.

2. Sedimentation along the margins of separating continents. Continued rifting produced an ocean basin that slowly widened. As it did so, the western edge of the Slave province became a passive shelf that received clastics from the coastal plain and shallow marine environment.

Quartz sandstones (now quartzites) passed upward into massive carbonates of the Rocknest Formation (Fig. 9-4).

3. Closing of the ocean basin through plate tectonics. The leading edge of the Slave province subducted beneath a microplate that approached from the west. The continental shelf buckled downward, forming a deep trough in which only deep-water clastics were deposited. After the trough filled, deltaic and fluvial sands were deposited. All of these sedimentary layers were later to be folded and faulted by plate collisions.

The sequence of events in the Wopmay orogen closely parallels the Paleozoic evolution of the Appalachian Mountains. The history of both areas began with opening of an oceanic tract, progressed to the

A

B

FIGURE 9-4 **Rocknest Formation.** (A) Dolomite and interbedded shales of the Rocknest Formation, Northwest Territories, Canada. The lighter colored layers are dolomite, and the darker, rust-colored beds are shales. (B) Oblique aerial view illustrates dramatic folding due to east-west crustal shortening and accretion during the orogenic phase of the development of the Wopmay orogen.

ENRICHMENT

The 18.2-Hour Proterozoic Day

Earth has not always had its present 24-hour day. It has been slowing at about 2 seconds per day each 100,000 years. The slowdown means that days have been increasing in length through geologic time, and thus the number of days in the year has been decreasing. Why?

Tidal friction causes the slowdown. As Earth rotates, gravitational tug from the Moon and the Sun create two *tidal bulges* that travel around the planet daily, lifting ocean levels from a few inches to several feet. However, the Moon's attraction prevents Earth from dragging the bulges very far. Thus, the two tidal crests tend to act as rather inefficient brakes on either side of the rotating planet, dragging back on the planet's spin as it rotates. The bulges also experience friction along shallower areas of the ocean bottom and are further retarded by continents that stand in their way.

Earth's slowing rotation was calculated by astronomers decades ago. More tangible evidence came in the early 1960s from paleontologist John Wells, who recognized that fine lines (Fig. A) on coral exoskeletons might represent

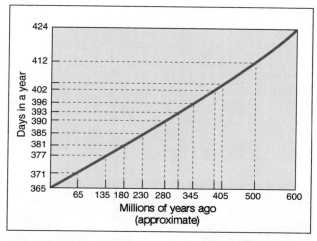

FIGURE B **The changing length of the day through geologic time.**

daily growth increments. A coral might secrete one thin ridge of calcium carbonate each day.

Wells also discerned coarse monthly bands, presumably related to breeding cycles when less calcium carbonate was secreted. He also noted broader annual bands in corals that lived in regions of seasonal change.

Wells counted the growth lines on several species of living corals and found the count "hovers around 360 lines in the space of a year's growth." Next, proceeding to fossil corals successively older in age, he counted correspondingly larger numbers of growth lines in a yearly band (Fig. B). There were, for example, 398 growth lines in the annual band of Devonian corals. This means that the Devonian year had about 38 more days than we have today.

In another method for estimating the days in a paleo-year, Charles Sonett used sedimentary rocks known as *tidal rhythmites*. As seen along shorelines today, they are alternating bands of dark and light silty deposits. Their thickness reflects tidal extremes that mark the lunar month. Variations in the thickness of sets of bands can be related to seasonal change.

Sonett's analysis indicates that about 900 million years ago, during the Neoproterozoic, a day lasted only 18.2 hours. The shorter days would have caused different weathering rates, global temperatures, and weather patterns. We can only speculate about the effect of shorter periods of light and heating upon life 900 million years ago.

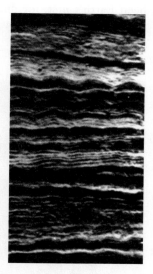

FIGURE A **Growth banding displayed by a specimen of the extinct coral *Heliophyllum halli*.** The finer lines may represent daily growth increments. There are approximately 200 growth lines per centimeter. Together with the annual bands, the growth increments can be used to estimate the number of days in a year at the time the animal lived.

deposition of continental shelf and rise sediments along passive continental margins, and ended with ocean closure and mountain-building.

The Wopmay is just one of many orogens that developed along the margins of Archean provinces during the Proterozoic. Another is the Trans-Hudson orogen, which extends southwestward from Hudson Bay (Fig. 9-5). Rocks of the Trans-Hudson belt record another Wilson Cycle of initial rifting, opening of an oceanic tract, deposition of sediment, and

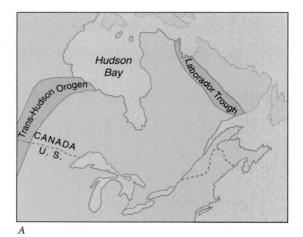

A

B

FIGURE 9-5   The Labrador Trough and Trans-Hudson orogen. (A) Location map. (B) *Landsat* NASA image of folded Proterozoic rocks of the Labrador Trough. The belt was once a chain of mountains, raised by the collision of two continents about 1.8–1.7 billion years ago. Since then, the mountains have eroded, exposing the deeper metamorphic and igneous rocks that were deformed into these folds by the collision. Note the lakes (dark blue) that reflect the trends of the folds. Image taken from a height of about 900 kilometers.

subsequent closure along a subduction zone. This closure, with its accompanying severe folding, metamorphism, and intrusive igneous activity, welded the Superior province to the Hearne and Wyoming provinces that lay to the west (Fig. 8-28). Another piece of continental crust was added to the North American craton.

Evidence of Earth's First Ice Age?

North of Lake Huron lies the Gowganda Formation, which includes conglomerates and laminated mudstones—clear evidence of glaciation.

Laminations in the mudstones represent alternating summer and winter layers of sediment called **varves**. Varved mudstones typically form in glacial meltwater lakes that are adjacent to ice sheets (Fig. 9-6). The varved sediments in the Gowganda Formation alternate with **tillites** (unsorted, lithified glacial debris). This indicates periodic advances of ice into the marginal lakes. Cobbles and boulders in the tillite were scratched and faceted by the abrasive action of an ice mass as it moved across the underlying bedrock.

A basement of 2.6-billion-year-old crystalline rock lies beneath the Gowganda Formation. The formation itself is intruded by 2.1-billion-year-old igneous rocks. Hence, the Paleoproterozoic glaciation occurred some time between these two dates.

Glacial conditions and accompanying cooler climates were widespread—rocks similar to those of the Gowganda are recognized in Finland, southern Africa, and India. This Paleoproterozoic glaciation may have been Earth's first, for during the Archean, the lithosphere was probably too warm for continental ice sheets to form.

FIGURE 9-6   Exposure of varved clays, Ontario, Canada. The average thickness of the lighter colored layers is about 2.5 cm.

Iron, Oxygen, and BIFs

Paleoproterozoic rocks surround the western shores of Lake Superior. They contain rich iron deposits that once were the foundation of the Great Lakes steel industry in Illinois, Indiana, Ohio, and Pennsylvania. Today, mining has sharply declined because the country mostly imports iron ore and steel. The ore minerals are *oxides* of iron, and thus they are evidence of free oxygen in Earth's atmosphere. This implies that photosynthesis, which is an oxygen-generating process, probably was in vigorous operation by this time.

Coarse sandstones and conglomerates deposited in shallow water lie near the base of the **Animikie Group**. These rocks are overlain by cyclic successions of cherts, cherty limestones, shales, and banded iron formations (BIFs). BIFs are also known in Archean sequences, but are not as extensive as those of the Paleoproterozoic (Fig. 9-7). Some Canadian deposits are over 1000 meters thick and extend over 100 km. Within the banded iron sequence is the Gunflint Chert, a formation that contains an interesting assemblage of **cyanobacteria** and other prokaryotic organisms.

End of the Paleoproterozoic

East of the Superior province, along a curving, elongate structural depression called the **Labrador Trough**, are Paleoproterozoic rocks. Like the Wopmay orogen, these rocks were deposited on the continental shelf, slope, and rise. On the western side of the trough, quartz sandstones, dolomites, and iron formations lie above deposits in fault-bound basins. Pillow lavas, basalts, mafic intrusives, and graywackes occupy an adjacent zone to the east. That eastern zone was subjected to intense folding (Fig. 9-5B), metamorphism, and westward thrust faulting. That episode of crustal deformation, called the **Hudsonian orogeny**, marks the close of the Paleoproterozoic.

HIGHLIGHTS OF THE MESOPROTEROZOIC (1.6 to 1.0 billion years ago)

Highlights of the Mesoproterozoic include an aborted ocean rift in the Lake Superior region, massive extrusions of basaltic lava flows, and the addition of copper ores to the already iron-rich Lake Superior region.

An Aborted Rift, Rich in Copper

Figure 9-8 shows a geologic cross-section from a point near Thunder Bay, Ontario (A) southeastward across Lake Superior to the northwest shore of Lake Michigan (A′). Study the cross-section to orient yourself. On the surface, note the little bit of Lake Michigan on the right, and how Lake Superior is interrupted by the land masses of Isle Royale and the Keweenaw Peninsula.

Underlying it all are the crystalline basement rocks of the Archean and Paleoproterozoic Animikian. Above them, the **Keweenawan** rocks extend for hundreds of kilometers. Keweenawan rocks consist of quartz sandstones, arkoses, conglomerates, and basaltic volcanics.

The lava flows are noted for containing pure copper. In the extruding lava, holes (vesicles) formed as gas bubbles, and these provided voids in which the copper was deposited. Copper also filled small joints and pore spaces in associated conglomerates. Long before Europeans

A   *B*

FIGURE 9-7 **Banded Iron Formations (BIFs).** (A) Banded iron formation exposed at Jasper Knob in Michigan's Upper Peninsula. (B) Banded iron formations occur worldwide in Proterozoic rocks, such as these at Wadi Kareim, Egypt. ◄ *What mineral imparts the red color?*

ENRICHMENT

BIF: Civilization's Indispensable Treasure

Iron is the backbone of modern civilization, and much of the iron we use comes from banded iron formations. These formations were deposited in a relatively brief geologic interval between 2600 and 1800 million years ago. They occur in huge bodies that exceed hundreds of meters in thickness and thousands of meters in lateral extent.

The most common type of BIF consists of bands of hematite or magnetite alternating with lighter laminations of chert. Most researchers believe that these deposits were chemically or biochemically precipitated—from somewhere. Fossils of bacteria found in some of the deposits resemble present-day counterparts that precipitate ferric iron hydroxide in oxygen-deficient environments. But there is little agreement on the original source of the iron. Exhaling volcanoes could have been a source, or the iron may have been derived from deep weathering and erosion of iron-rich crustal rocks nearby.

At present, banded iron ores are being mined in North America (in the Labrador Trough and the Lake Superior region), the Ukraine, and in western Australia. Substantial parts of the BIF are usable even as a low-grade ore called *taconite*. Most of the high-grade BIF ores in North America now are depleted, but taconite ore—in which iron ore is concentrated into pellets—is now being shipped to iron smelters. The pellets contain about 60% iron. The most important product made from iron is steel, which is an alloy of iron with a small amount of carbon to strengthen it.

arrived in the late 1600s, the Keweenawan copper deposits were mined by Native Americans. From 1850 to 1950, copper production boomed in this region, but in the mid-1970s, production ceased.

The Keweenawan lavas accumulated to over 25,000 feet thick (nearly 5 miles). The scene of the eruptions was a hell on Earth, as a region the size of Indiana was engulfed in the searing heat of basaltic lava flows and steaming, acrid volcanic fumes.

But even with so great an outpouring, much of the supply of mafic magma was not exhausted. It remained beneath the surface, where it crystallized to form the long mass of black, crystalline Duluth Gabbro—8 miles thick and 100 miles wide.

We have learned that such great outpourings of lava are characteristic of seafloors. When this large quantity of mafic material comes to the surface within the central stable region of a continent, it signals the presence of a *rift zone*, along which a continent may break apart and fill with ocean water. This is the first stage of a Wilson

Cycle. The tract where the break begins typically develops tensional faults, along which mafic magma rises to the surface to form Keweenawan-like accumulations.

Evidence from gravity and magnetic surveys, as well as samples from deep drill holes, indicates that the rift zones associated with Keweenawan volcanism developed about 1.2 to 1.0 billion years ago and extended from Lake Superior southward into Kansas. If these rifts had been extended to the edge of the craton, an ocean tract would have formed within the rift system, and the eastern United States would have drifted away. Rifting ceased, however, before such a separation could occur.

It is important to note that Lake Superior and the other Great Lakes are *not* mini-oceans resulting from a failed continental rift! The Great Lakes, and the thousands of lesser lakes and ponds in the area, result from glaciation. Vast ice sheets spread southward over Canada, gouging great basins that then filled with glacial meltwater when the ice sheets melted back into Canada about 12,000 years ago. They have been lakes ever since.

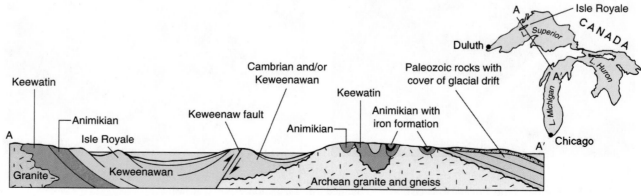

FIGURE 9-8 **Geologic cross-section (A-A') across Lake Superior and Michigan's Upper Peninsula.**

The Grenville Orogeny

The Grenville Province of eastern North America was the last to experience a major orogeny. Exposures of Grenville rocks extend from the Atlantic coast of Labrador to Lake Huron. However, this is only part of their true extent, for they continue beneath Phanerozoic rocks down the eastern side of the United States and westward into Texas (Fig. 8-28).

In the United States, interpretation of Grenville rocks is difficult, because they were heavily altered by building of the Appalachian Mountains during the Paleozoic. Originally, Grenville rocks were carbonates and sandstones (Fig. 9-9), but orogenic forces transformed them into metamorphic rocks containing many igneous intrusions.

Building a New Supercontinent—Rodinia

Deformation of Grenville sediments occurred 1.2 to 1.0 billion years ago during the **Grenville orogeny**. This orogeny was part of the continental collisions involved in the formation of the supercontinent Rodinia (Fig. 9-10). At the time of the Grenville orogeny, the east coast of Laurentia (North America) lay adjacent to a block of western South America termed Amazonia. The west coast of Laurentia lay next to Antarctica and Australia.

About 750 million years ago, Rodinia began to break apart. It was during this continental breakup that the proto-Pacific Ocean—called Panthalassa—formed. A great expanse of ocean now lay west of North America. On the eastern side of North America, Neoproterozoic sediments accumulated in basins

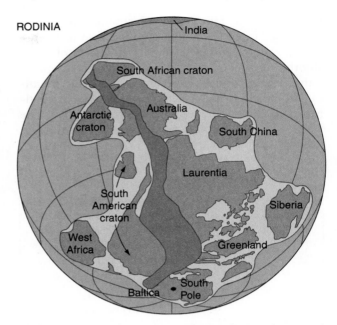

FIGURE 9-10 **The supercontinent Rodinia as it appeared about 1100 million years ago.** The red band shows the location of collisional orogens where continents collided in constructing the supercontinent.

and shelf areas. These sandstones and shales were later deformed during Paleozoic orogenic events. One such event involved rifting and the filling of rift valleys with lava flows and coarse clastics. Ocean water flooded the rift valleys, forming a narrow seaway called Iapetus that widened to form the proto-Atlantic Ocean.

▶ HIGHLIGHTS OF THE NEOPROTEROZOIC (1.0 billion to 542 million years ago)

During the Neoproterozoic, extensive continental glaciations occurred, although they were not always contemporaneous. The evidence: Neoproterozoic rocks on all of Earth's landmasses (except Antarctica) show glacial striations produced by huge ice sheets. We also find tillites, **dropstones** (chunks of rock released from melting icebergs), and varved clays from glacial lakes. The episode of frigid conditions has been named the **Varangian glaciation** after Varangian Fiord in Norway, where Neoproterozoic tillites provided early evidence of this great ice age.

A Big Chill

What conditions during the Neoproterozoic could have caused this widespread glaciation? Not latitude, because paleomagnetic data indicate that by about 850–600 million years ago, the continents rested at low and middle latitudes. Not large polar landmasses, for

FIGURE 9-9 **Folded sandstones (dark) and carbonates of the Grenville Group.** Belmont Township, Ontario, Canada.

there were none. However, the Neoproterozoic equator crossed extensive, highly reflective land surfaces—in contrast to the wide equatorial region of heat-absorbing ocean that exists today. Thus, one hypothesis is that heat lost by reflection alone may have been sufficient to cause cooling. As temperatures dropped, continental glaciers and ice caps formed over the poles. The white surfaces reflected the Sun's energy, resulting in further heat loss.

A second hypothesis proposes that atmospheric carbon dioxide declined, which lowered temperatures. Abundant CO_2 in the atmosphere increases the greenhouse effect and causes warming. Reduced CO_2 decreases the greenhouse effect and causes cooling. Hypothetically, the spread of photosynthetic organisms that extract CO_2 from the atmosphere might have decreased atmospheric CO_2, resulting in cooling.

There are thick layers of limestones between many Varangian tillites. This seems strange, because limestones normally originate in warm seas. How could they be interspersed with glacial deposits? One explanation is that the glacial conditions inhibited photosynthetic uptake of CO_2 by stromatolites and other microbial plants. As a result, sufficient CO_2 may have accumulated periodically and triggered short episodes of greenhouse warming. Thus, we have the paradox of glaciers causing their own destruction!

Earth's Glacial History

Figure 9-11 shows five major glacial episodes in Earth's history. As described earlier, the Gowganda tillites indicate widespread glaciation slightly more than 2 billion years ago. The Neoproterozoic glaciations occurred 700–600 million years ago. Then, glacial ice advanced again over major portions of the continents during the Ordovician and Silurian, during the Carboniferous-Permian, and again very recently, during the Pleistocene Epoch. Each of these major

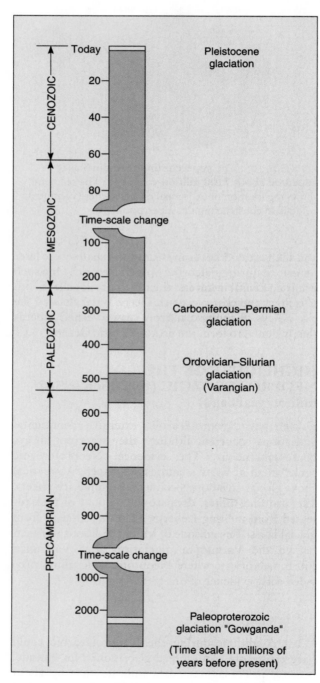

FIGURE 9-11 **Earth has seen several major episodes of widespread continental glaciation (orange).** Each major episode included intervals of glacial advance and retreat.

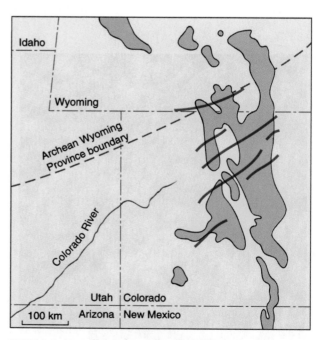

FIGURE 9-12 **Shear zones (red) in Wyoming and Colorado.** These developed when the Archean craton collided with island arc terranes during the Paleoproterozoic. Green areas are Lower Proterozoic outcrops.

FIGURE 9-13 **Chief Mountain near Glacier National Park, Montana.** The upper third is undeformed horizontal limestones of the Neoproterozoic Belt Supergroup. The lower two-thirds are deformed Cretaceous beds. The "Lewis thrust" fault separates the two. It represents the location where the Proterozoic beds were pushed eastward over weaker Cretaceous strata.

glaciations included shorter intervals during which the ice sheets alternately grew (colder climate) and shrank (warmer climate).

PROTEROZOIC ROCKS SOUTH OF THE CANADIAN SHIELD

Precambrian outcrops occur south and west of the Canadian Shield, although they are not as extensive as in Canada. Thick sections are exposed in the Rocky Mountains and Colorado Plateau. These rocks have a complex history that began more than 2.5 billion years ago.

First, an Archean terrane developed, composed of strongly deformed and metamorphosed granitic rocks. From study of remnant volcanic rocks and greenstones in Wyoming, it appears that this old Archean mass collided with one or more island arcs about 1.7 or 1.8 billion years ago. The line of collision is marked by fault zones with severely crushed rocks in southern Wyoming and western Colorado—as shown by the red lines in Figure 9-12.

Following this orogenic event, there was extensive magmatism during the Mesoproterozoic. Magmas that were 1.5–1.4 billion years old intruded across a broad belt of North America from California to Labrador.

The next event south of the Canadian shield was widespread rifting. Large fault-controlled depressions formed and filled with thick sequences of Neoproterozoic shales, siltstones, quartzites, and dolomites of the Belt Supergroup. Exposures in Montana, Idaho, and British Columbia reveal that the Belt Supergroup is over 12 km thick.

Because of their scenic features, Belt rocks are of special interest. Massive carbonates of this unit form towering cliffs in Waterton Lakes and Glacial National Parks (Fig. 9-13). Even though they are exceptionally thick, Belt Supergroup rocks display features such as ripple marks and stromatolites (Fig. 9-14). These indicate that they were deposited in relatively shallow water along the passive western margin of North America.

FIGURE 9-14 **Neoproterozoic stromatolites in carbonates of the Belt Supergroup, Glacier National Park, Montana.**

ENRICHMENT

Heliotropic Stromatolites

Heliotropic means "sun-turning." A common example is the sunflower, which tracks the Sun across the summer sky like a dish antenna, aiming itself for best exposure to solar energy.

FIGURE C 850-million-year-old stromatolites from the Bitter Springs Formation of central Australia. Vertical dimension is 14 cm.

Stromatolites are among the most abundant fossils in Proterozoic rocks. The photosynthetic cyanophyta bacteria that construct stromatolites depend on sunlight for survival and growth. In a study of modern stromatolites, certain species form columnar laminated growths that are inclined toward the Sun (heliotropic) to gather the maximum amount of light on their upper surfaces.

Paleobiologist Stanley Awramic and astronomer James Vanyo have studied 850-million-year-old stromatolites from Australia's Bitter Springs Formation for evidence of heliotropism (Fig. C), and believe they have found it. At the study locality in central Australia, stromatolites curve upward in a distinct sine-wave form. This form appears to have recorded growth that followed the seasonal change in the position of the Sun above the shallow sea in which the stromatolitic cyanobacteria lived.

In their elongate sinuous form, each sine wave represented a year of growth. The stromatolites grew by the addition of a layer or lamina each day. Thus, by counting the laminae in the length of a single sine wave, one can obtain the number of days in the year. The results indicate that there were 435 days in a year 850 million years ago, when the stromatolites of the Bitter Springs Formation were actively responding to Proterozoic sunlight.

Grand Canyon Precambrian Rocks

Another area that features Precambrian rocks is the Grand Canyon region. It has two distinct units, as you can see in Figure 9-15. The lower, older unit is the Vishnu Schist, and the upper unit is tilted strata that we collectively call the Grand Canyon Supergroup.

The Vishnu Schist (Fig. 9-16) is a complex body of metamorphosed sediments and gneisses that have been intensely folded and invaded by granites. These granitic intrusions (and correlative rocks of the southwestern U.S.) were emplaced 1.4–1.3 billion years ago as part of the Mazatzal orogeny.

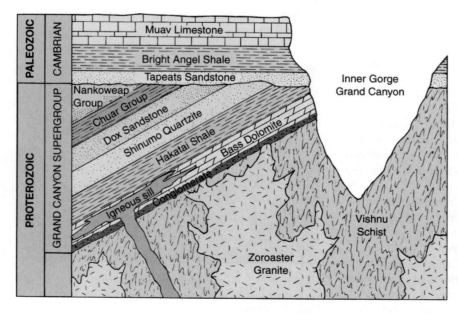

FIGURE 9-15 **Vishnu Schist, Grand Canyon Supergroup, and other rocks in the Grand Canyon of the Colorado River.** ◪ *Indicate by arrows and labels two kinds of unconformities in this cross-section. Is the conglomerate at the base of the Grand Canyon Group an expected lithology for the initial strata above an erosional unconformity?*

FIGURE 9-16 **Vishnu Schist, Grand Canyon National Park.** The schist is exposed along the walls of the Inner Gorge of the Grand Canyon of the Colorado River.

A splendid example of an *unconformity* occurs at the contact between the Vishnu Schist (Fig. 9-15) and the overlying Grand Canyon Supergroup. The latter unit is Neoproterozoic in age and is itself uncomformably overlain by Paleozoic rocks. The Grand Canyon Supergroup correlates with the Belt Supergroup to the north. It consists mostly of clastic rocks (sandstones, siltstones, and shales) that accumulated in a troughlike basin that extended into the craton. The Chuar Group of the Grand Canyon Supergroup contains small, circular carbonaceous structures believed to be fossil algal spheres.

PROTEROZOIC LIFE

Life at the beginning of the Proterozoic was not significantly different from that of the preceding late Archean. Blue-green scums of photosynthetic cyanobacteria (Fig. 9-17) constructed filamentous algal mats around the ocean margins. Myriad prokaryotes floated in the well-illuminated surface waters of seas and lakes. Anaerobic prokaryotes multiplied in environments deficient in oxygen, and these included the thermophiles of deep-sea hydrothermal springs.

Stromatolites, which were relatively sparse during the Archean, proliferated worldwide during the Proterozoic (Fig. 9-18), but declined markedly by the end of the era. Today stromatolites are rare, primarily because the microorganisms that build them are eaten by marine snails and other grazing invertebrates. The decline of Proterozoic stromatolites may be similarly associated with overgrazing by newly evolving groups of marine invertebrates. Even today, stromatolites

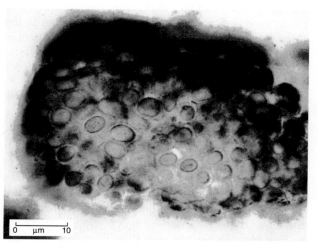

0   μm   10

FIGURE 9-17 **Paleoproterozoic colony of the cyanobacterium *Eoentophysalis*.** The colony appears in a thin section of a Belcher Group stromatolitic chert, Belcher Islands, Canada.

FIGURE 9-18 **Two-billion-year-old columnar stromatolites from the Kona Dolomite near Marquette, Michigan.** Note the rock hammer for scale.

survive only in environments unsuitable for most grazing invertebrates.

Molecular fossils of eukaryotes first appear near the end of the Archean, about 2.7 billion years ago. Although prokaryotes still dominated during the Mesoproterozoic, eukaryotes became more abundant in the fossil record. The latter part of the Neoproterozoic witnessed the evolution of Earth's first multicellular animals (metazoans).

Microfossils of the Gunflint Chert

The prokaryotes, which appeared in the Archean, continued to thrive during the Proterozoic. We find their fossils in nearshore stromatolites and former muds of coastal lagoons and mudflats. Along the northwestern shore of Lake Superior are exposures of the Gunflint Chert, a rock unit 1.9 billion years old which bears these fossils.

You can see various stringlike filaments and ball-shaped cells in thin sections of the chert (Figs. 9-19 and 9-20). Unbranched stringlike forms, some of which are septate, have been given the name *Gunflintia*. More

finely septate forms, such as *Animikiea* closely resemble certain living algae. Other Gunflint fossils, such as *Eoastrion* (literally, the "dawn star"), resemble living iron- and magnesium-reducing bacteria. *Kakabekia* and *Eosphaera* (the "dawn sphere") do not resemble any living microorganism, and their classification is uncertain. These delicate fossils owe their preservation to the glassy chert matrix in which they were hermetically sealed, preventing destruction by oxidation.

What does seem certain is that photosynthetic organisms were abundant in the Paleoproterozoic, actively producing oxygen and thereby altering the composition of the atmosphere. Not only do many of the Gunflint fossils resemble living photosynthetic organisms, but their host rock contains organic compounds regarded as the breakdown products of chlorophyll.

In 1953 when these fossils were found, older fossils were rare and ambiguous. The well-preserved, abundant, and diverse Gunflint fossils alerted scientists that microbial life was abundant about 2 billion years ago and prompted the search for even older fossils.

The Rise of Eukaryotes

The evolution of eukaryotes is a major event in the history of life. Eukaryotes possess the potential for sexual reproduction, providing enormously greater possibilities for evolutionary change. Unfortunately, fossils of the earliest eukaryotes are rare. This is not surprising, considering that the first eukaryotes were microscopic cells whose identifying characteristics (enclosed nucleus, organelles, and so on) are rarely preserved. Size, however, is another important clue to the identification of a fossil as a eukaryote. Living spherical prokaryotic cells rarely exceed 20 microns in diameter, whereas eukaryotic cells are nearly always larger than 60 microns.

Larger (and hence probably eukaryotic) cells begin to appear in the fossil record by about 2.7 to 2.2 billion years ago. This earliest evidence of eukaryotic life is based on biochemical remnants of eukaryotes. These chemical clues are termed **molecular fossils**. Although eukaryotes evolved very early, they did not begin to diversify until about 1.2 to 1.0 billion years ago. It may be that they were unable to expand until the oxygen content of the ocean reached a suitable level, or perhaps they did not diversify until the advent of sexual reproduction.

Acritarchs

Acritarchs (*AK-ri-tarks*) are a group of organisms that are particularly useful in correlating Proterozoic strata. They include diverse unicellular, spherical microfossils with resistant single-layered walls. Their walls may be smooth or variously ornamented with spines, ridges, or papillae (Fig. 9-21). Although their precise nature is uncertain, acritarchs appear to be phytoplankton that

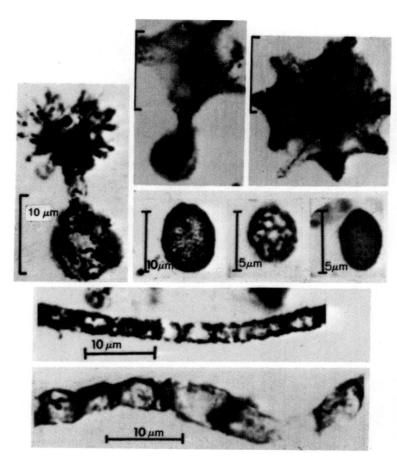

FIGURE 9-19 **Fossil prokaryotes from the Gunflint Chert.** The three specimens across the top with umbrella-like crowns are *Kakabekia umbellata*. The three subspherical fossils are species of *Huroniospora*. The filamentous microorganisms with cells separated by septa are species of *Gunflintia*.

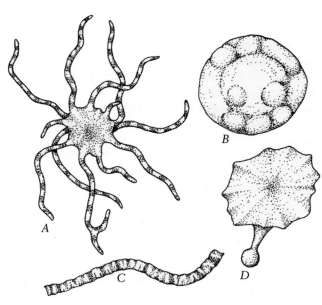

FIGURE 9-20 **(A)** *Eoastrion*, **(B)** *Eosphaera*, **(C)** *Animikiea*, and **(D)** *Kakabekia* from the Gunflint Chert. All specimens are drawn to the same scale. *Eosphaera* is about 30 microns in diameter.

grew thick coverings during a resting stage in their life cycle. Some resemble the resting stage of modern marine algae known as dinoflagellates. (Dinoflagellates are among the organisms that cause "red tides" that periodically poison fish and other marine animals.) Membrane-bounded nuclei can be detected within some acritarchs, and their size is comparable to that of living eukaryotes.

Acritarchs first appear in rocks about 1.6 billion years old. They reached their maximum diversity and abundance 850 million years ago and then steadily declined. By about 675 million years ago, few remained. Their decline coincided with the Varangian glaciation underway near the end of the Proterozoic. A reduction of carbon dioxide and an increase in atmospheric oxygen accompanying glacial conditions may have been responsible for the extinction of all but a few species that managed to survive until Ordovician time.

In addition to acritarchs, protozoan eukaryotes were probably present in the Proterozoic. Protozoans are non-photosynthetic and derive their energy by ingesting other cells (in other words, they are heterotrophs). Foraminifera, amoebas, and ciliates are living examples

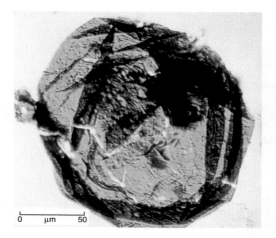

A

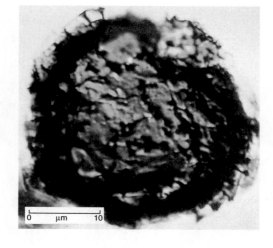

B

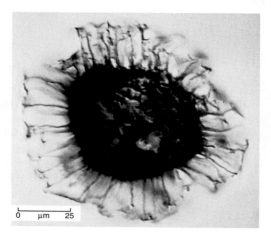

C

FIGURE 9-21 **Acritarchs.** These microfossils represent the resting stage in the reproductive cycle of eukaryotic algae. Their thick organic coverings are resistant to chemical attack.

of protozoans. Protozoan fossils are common in many Phanerozoic rocks, largely because they evolved readily preservable shells.

In contrast, protozoan fossils from the Proterozoic are much less common, because these organisms were naked, shell-less creatures with little chance of preservation. The scarcity of Proterozoic protozoans, however, does not mean that they were not abundant. In fact, it is likely that the algal mats and masses of decaying organic matter in late Proterozoic seas teemed with protozoans exploiting these rich sources of food.

Many-Celled Animals Arrive: The Metazoans

The Neoproterozoic fossil record for larger, multicellular animals is considerably better than that for protozoans. Since the 1960s, important discoveries of large, multicellular Neoproterozoic animals have been made on every continent. Most of the fossils are impressions made in former sediments by animals that clearly were at the metazoan level of evolution (Fig. 9-22).

Metazoans are multicellular animals that possess more than one kind of cell and have their cells organized into tissues and organs. Recently, extraordinary evidence of the presence of metazoans about 570 million years ago was discovered in the uppermost Neoproterozoic Doushantuo Formation of South China. The fossils are eggs and embryos of metazoans. The preserved embryos show precise patterns of cell cleavage at an early stage of embryonic development (Fig. 9-23). Exquisite preservation resulted when phosphate seeped into and coated the cells.

Ediacaran Fauna

The first important discovery of large metazoans in Proterozoic rocks was made in Australia's Ediacara hills in the 1940s. The Australian fossils and remains of similar metazoans from other parts of the world have been named the **Ediacaran fauna**. The oldest members of this fauna have been discovered in China.

Although once thought to have disappeared by Cambrian time, several animals within the fauna sur-

FIGURE 9-22 **Fossil metazoans from the Conception Group, Avalon Peninsula, Newfoundland.** These early metazoans are similar to those of the Ediacaran fauna of Australia. The rocks in which they occur are thought to be Late Proterozoic. The elongate forms may be a form of colonial soft coral. The fossil at the center which resembles a four-leafed-clover has been called a jellyfish, but there are other interpretations.

vived into the early Cambrian. For example, Ediacaran fossils of definite Cambrian age have been reported in 510-million-year-old rocks in Ireland. A few more possible survivors of the Ediacaran fauna occur in the Burgess Shale of the middle Cambrian age. This evidence means that there was no mass extinction of Ediacaran animals near the end of the Neoproterozoic, as once proposed.

Ediacaran organisms can be grouped into three types, based on general appearance: discoidal (flat and circular), frondlike, and elongate. Discoidal forms such as *Cyclomedusa* (Fig. 9-24) are usually interpreted as jellyfish. Another discoidal form, *Tribrachidium* (Fig. 9-25), appears to have no modern counterpart and may be a member of an extinct phylum.

The Ediacaran frondlike fossils resemble living corals informally called sea pens (Fig. 9-26). Sea pens look like fronds of ferns, except that tiny coral polyps are aligned along the branchlets. The polyps capture and consume microscopic organisms that float by. Frond fossils similar to those from Australia are also known from Africa, Russia, and England. In those from England (*Charniodiscus*), the frond is attached to a basal, concentrically ringed disk that apparently held the organisms to the seafloor. The disks are frequently found separated from the fronds, indicating that isolated discs once interpreted as jellyfish are actually the anchoring structures of frond fossils.

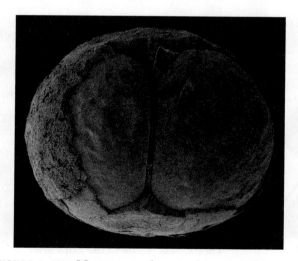

FIGURE 9-23 **Metazoan embryo at the two-cell stage (570 ± 20 million years old).** From the Doushantuo Formation of China, this specimen is about 1mm in diameter.

FIGURE 9-24 **Impression of a soft-bodied discoidal fossil in the Ediacaran Rawnsley Quartzite in southern Australia.** This organism has been interpreted as a jellyfish and named *Cyclomedusa*. However, some paleontologists have concluded that it is unrelated to any living organism.

A

B

C

FIGURE 9-25 **Three members of the Ediacaran fauna from the Edicaran Rawnsley Quartzite in southern Australia.** (A) *Pseudorhizostomites*, a wormlike form of uncertain affinity. (B) *Parvancorina*, possibly an arthropod. (C) *Tribrachidium*, an unusual discoidal form that appears to have no living relatives. Specimen (A) is an imprint in the Rawnsley Quartzite; (B) and (C) are plaster molds made from the original fossils.

FIGURE 9-26 **Diorama of the seafloor in which Ediacaran metazoans lived.** Prominent in this view are silvery jellyfish and large, frondlike organisms interpreted here as soft corals known today as sea pens. ◀ *What characteristics of the frond fossils place doubt on their identification as soft corals (see Vendoza Controversy on page 261)?*

The third group of Ediacaran fossils are ovate to elongate in form. The fossils are regarded as impressions made by large flatworms and annelid worms. Typical examples are *Dickinsonia* (Fig. 9-27), which attained lengths of up to a meter, and *Spriggina* (Fig. 9-28), a more slender animal with a distinctive crescent-shaped structure at its anterior end.

Kimberella (Fig. 9-29) is a particularly significant Ediacaran fossil. Four rather poorly preserved speci-

FIGURE 9-27 ***Dickinsonia costata* in the Ediacaran Rawnsley Quartzite of southern Australia.** This fossil has been interpreted as a segmented worm. Scale in centimeters.

FIGURE 9-28 *Spriggina floundersi*, **interpreted by some as a segmented worm from the Ediacaran Rawnsley Quartzite of southern Australia.** The animal is 3.5 cm long.

mens found at the Ediacaran site resembled jellyfish. Then, in 1993, over 30 specimens of this unusual organism were found on the shores of the White Sea in northern Russia. Clearly, *Kimberella* was no lowly jellyfish, but rather a complex invertebrate ranking much higher on the evolutionary scale. *Kimberella* has evidence of a true **coelum**, or body cavity in which the digestive tract and other internal organs were suspended. At 550 million years old, it is the earliest animal known to possess this important characteristic of all higher animals. It also is bilaterally symmetrical, had a dorsal cover, and possesses a distinctive ruffled border that some interpret as the edge of a mantle (the organ that secretes the shell in mollusks).

Evidently, *Kimberella* crept across the substrate, grazing on algae like some modern snails. Although *Kimberella* appears very mollusk-like, there is no evidence that it had a radula—the rasping tonguelike structure found in most mollusks. But whether or not it is a true mollusk, it is important evidence that advanced, complex invertebrates lived on Earth about 10 million years before the great "Cambrian explosion" of life.

The Vendoza Controversy

There is an interesting controversy about Ediacaran fossils. Many paleontologists have interpreted them as Proterozoic members of existing phyla, such as the Cnidaria (which includes jellyfish and corals) and Annelida (worms). However, this view has been questioned by paleontologist Adolf Seilacher.

Seilacher argues that the resemblance between living sea pens and frond fossils is only superficial. He notes that the branchlets in the frond fossils are fused together and do not have passages through which water currents might pass. Living sea pens have such openings, and this permits the polyps on the branchlets to catch food particles in the passing flow of water.

With regard to the discoidal fossils thought to be jellyfish, Seilacher notes that living jellyfish have radial structures at their centers and concentric structures around the periphery. This arrangement is opposite to that found in the discoidal Ediacaran fossils. Also, the rather superficial resemblance of *Spriggina* and *Dickinsonia* to worms may have little significance, for they exhibit no evidence of organs of modern-day worms (such as a mouth, gut, and anus).

Further, the soft, delicate tissue of Cnidarians and worms rarely is preserved in coarse sandy deposits. Yet the Ediacaran animals have left distinct impressions in the sandstone. For this to occur, they must have had

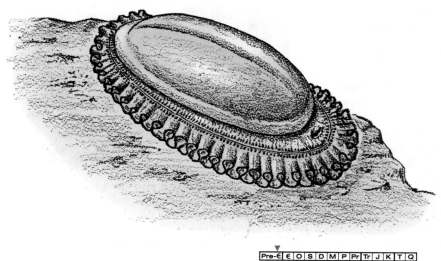

FIGURE 9-29 **Reconstruction of *Kimberella*.** *Kimberella* is a Proterozoic mollusk-like bilaterally symmetrical organism. Specimens range up to 10 cm in length.
🔀 *Name an important characteristic of Kimberella that indicates it is a highly evolved multicellular animal.*

Pre-Є | Є | O | S | D | M | P | Pr | Tr | J | K | T | Q

tough outer coverings. The ribbed and grooved appearance of many Ediacaran impressions suggests that these animals had an exterior construction like that of an air mattress, providing the firmness needed to make an impression in sand.

Seilacher and his American colleague Mark McMenamin postulate that Ediacaran animals lived with symbiotic algae in their tissues, like modern corals. This relationship would have given the animals a way to derive nutrition from the photosynthetic activity of the algae. Also, the broad, thin shapes of many of these animals might have given them sufficient surface area for diffusion respiration. This was essential to animals that had not yet evolved complex circulatory, digestive, and respiratory systems.

The dissimilarities of Ediacaran creatures and animals that exist today suggest that they do not belong in existing phyla, but in a separate taxonomic category. Seilacher proposed the name **Vendoza** (after Vendian, a term used in Russia for the final period of the Neoproterozoic).

What the Edicaran Fossils Tell Us about the Proterozoic

Following their appearance about 630 million years ago, the Ediacaran fauna survived for about 50 million years. The fauna is an important record of Earth's first evolutionary radiation of multicellular animals. Although the record is still too patchy to trace phylogenetic relationships, certain Ediacaran animals may have been ancestors of Paleozoic invertebrates.

The Ediacaran fauna tells us that Neoproterozoic seas were populated largely by soft-bodied organisms. However, some tiny shell-bearing fossils have also been found. One genus, first discovered in Neoproterozoic rocks of Namibia, Africa, is named *Cloudina* after the American geologist Preston Cloud (1912–1991). *Cloudina* secreted a tubular, calcium carbonate shell only a few centimeters long (Fig. 9-30). It has been interpreted as a tube-dwelling annelid worm. Other small shelly fossils that occur worldwide in sediments of latest Proterozoic and earliest Cambrian include possible primitive mollusks, sponge spicules, and tiny tusk-shaped fossils called hyolithids.

Not all Precambrian fossils are "body fossils." Trace fossils of burrowing metazoans exist in Neoproterozoic rocks of Australia (Fig. 9-31), Russia, England, and North America. In every locality, they are in rocks deposited after the late Neoproterozoic Varangian glaciation. The traces consist mostly of relatively simple, shallow burrows, whereas traces in the overlying Cambrian are more complex, diverse, and numerous. There is a similar increase in complexity, diversity, and abundance of metazoan body

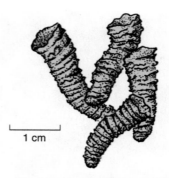

FIGURE 9-30 *Cloudina*, **the earliest known calcium carbonate shell-bearing fossils.** Cloudina was first described from the Proterozoic Nama Group of Namibia. It resembles a tube-dwelling annelid worm.

fossils at the transition from the Proterozoic into the Cambrian.

Why the sudden appearance of Ediacaran fossils near the end of the Proterozoic and the absence of such metazoans in older rocks? It may relate to the accumulation of sufficient free oxygen in the atmosphere to permit oxidative metabolism in organisms. Or, Ediacaran life may have evolved more gradually from earlier small, naked forms that could not leave a fossil record. Perhaps ancestral metazoans lived in "oxygen oases" of marine plants.

After the atmosphere attained about 1% of the present oxygen level, these as-yet-undiscovered animals could have been free to leave their oases and spread widely in the seas. In this view, the evolutionary development of metazoans may not have been abrupt, but their dispersal could have been rapid once suitable conditions became prevalent.

FIGURE 9-31 **Trace fossils made by a possible mollusk as it crawled across soft seafloor sediment.** The host rock occurs at the Proterozoic-Cambrian boundary in British Columbia, Canada. Originally, the traces had the form of elongate depressions. Sediment deposited on top of the original layer then filled the depressions, so the crawling traces are in convex relief.

Oxygen and Climate Changes in the Proterozoic Environment

Earth's early atmosphere and hydrosphere were largely devoid of free oxygen, for good reason: much of the oxygen became bonded to iron and other "oxygen-loving" elements. These elements served as "oxygen sinks" by capturing oxygen that otherwise would join the atmosphere.

But about 2 billion years ago, the oxygen sinks became largely saturated, and free oxygen began to accumulate in the atmosphere. Increased plant photosynthetic activity probably also contributed to this build-up.

(Figure 9-32 correlates this development with others of the time.) As atmospheric oxygen increased, so did oxygen in the sea. It combined with nitrogen in the seawater to form nitrate, an important nutrient for eukaryotic algae. This may help explain the expansion of acritarchs and other eukaryotes during the Paleoproterozoic.

Another result of the oxygen buildup was the extensive accumulation of ferric iron oxide, which stained terrestrial sediments a rusty red. These **red beds** validate the existence of an oxygenic environment. However, the oxygen level probably increased slowly, and likely did not approach 10% of present atmospheric levels of free oxygen until the Cambrian Period.

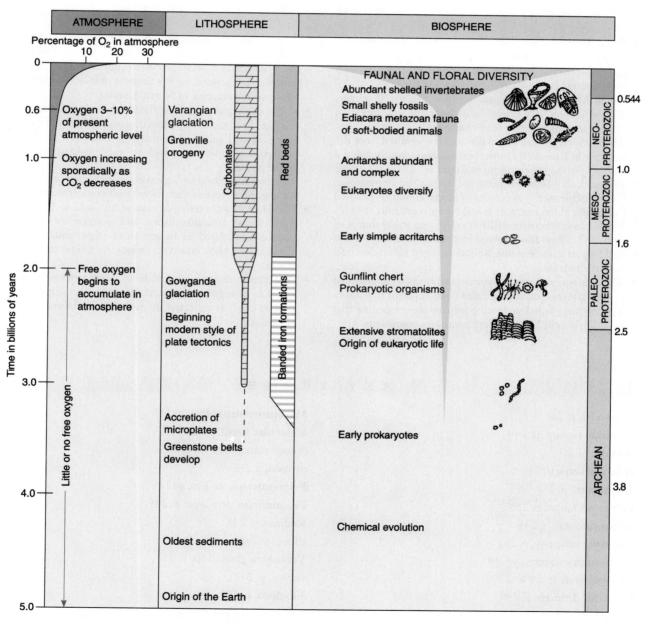

FIGURE 9-32 **Correlation of major events in the biosphere, lithosphere, and atmosphere.**

In general, Proterozoic rocks provide evidence for a wide range of climatic conditions. But we do not think these climates were unique compared to more recent ones. For example:

- Thick limestones and dolomites with reeflike algal colonies were deposited along the equator, where warm, tropical conditions prevailed, much as they do today.
- Proterozoic evaporite deposits in regions now far from the equator indicate warm climates that were likely to have been arid as well.

- In the lower latitudes, climates were more severe, as indicated by consolidated deposits of glacial debris (tillites) and glacially striated basement rocks. As previously discussed, the best known of these poorly sorted, thick, boulderlike deposits is the Paleoproterozoic Gowganda Formation. The ice returned dramatically during the Neoproterozoic, when continental glaciers covered extensive areas of the globe.

SUMMARY

- By the beginning of the Proterozoic, many small cratonic (continental) elements that formed during the Archean had collided and had become sutured together to form large cratons.
- North American Proterozoic orogens like the Wopmay, Trans-Hudson, and Grenville indicate the full operation of plate tectonics during the Proterozoic.
- The Neoproterozoic Grenville orogeny resulted from the suturing of Laurentia to other crustal segments during the assembly of the supercontinent Rodinia.
- Widespread glaciations occurred during the Proterozoic. Particularly extensive episodes of glacial conditions existed during the Paleoproterozoic and Neoproterozoic.
- Banded iron formations (BIFs) are an important source of iron ore. They also provide important evidence of the build-up of sufficient atmospheric oxygen to oxidize iron at Earth's surface.
- The Mesoproterozoic Keweenawan basalts were extruded along a rift zone extending from Lake Superior to Kansas. The rift zone failed to develop into an ocean opening that would have split North America into two continents.

- South of the Canadian Shield, Proterozoic rocks are exposed in the Grand Canyon of the Colorado River and many localities in western North America. The Belt Supergroup is a prominent Proterozoic unit deposited in the rifted western margin of North America.
- Life at the beginning of the Proterozoic resembled that of the Late Archean. It consisted of stromatolites and various stringlike and spherical prokaryotic microbes. By Mesoproterozoic time, stromatolites and prokaryotes were joined by eukaryotes.
- The most significant biological event of the Neoproterozoic was the appearance of metazoans (multicellular animals).
- Fossils of Neoproterozoic metazoans include large discoidal forms, frondlike forms, and elongate forms traditionally considered to be members of presently known phyla. They may, however, belong to a new taxonomic group.
- Proterozoic climates ranged from warm tropical or subtropical, as suggested by extensive stromatolitic carbonates, to cold, as reflected in two episodes of continental glaciation.

KEY TERMS

acritarchs, p. 256

Animikie Group, p. 249

coelum, p. 261

cyanobacteria, p. 249

dropstones, p. 251

Ediacaran fauna, p. 258

epicontinental, p. 245

Grenville orogeny, p. 251

Hudsonian orogeny, p. 249

Keweenawan, p. 249

Labrador Trough, p. 249

Laurentia, p. 245

metazoan, p. 258

Mesoproterozoic Era, p. 243

molecular fossil, p. 256

Neoproterozoic Era, p. 243

orogen, p. 245

Paleoproterozoic Era, p. 243

Precambrian province, p. 245

Rodinia, p. 245

tillite, p. 248

Varangian glaciation, p. 251

varves, p. 248

Vendoza, p. 262

Vendoza controversy, p. 261

Wilson Cycle, p. 246

QUESTIONS FOR REVIEW AND DISCUSSION

1. Describe the sequence of events recorded by the rocks of the Wopmay region of northwestern Canada.

2. With regard to the history of Earth's atmosphere, what is the significance of banded iron formations (BIFs)?

3. When was the supercontinent Rodinia assembled? What orogenic event in eastern North America was the result of the assembly of Rodinia?

4. What kind of tectonic activity was the probable cause of the massive outpourings of Keweenawan lavas in the Lake Superior region? When did this occur?

5. How do eukaryotes differ from prokaryotes (see Chapter 6)? When do eukaryotes appear in the fossil record?

6. Paleontologist Andrew Knoll has stated that "cyanobacteria are the heroes of Earth history." Why do these lowly organisms deserve such praise?

7. Were acritarchs eukaryotic organisms? When did acritarchs reach their maximum diversity, and when did they nearly become extinct? What climatic conditions may have contributed to their decline?

8. What are metazoans? What is the earliest known occurrence of abundant metazoans? With regard to their general appearance, what are the three major groups of Ediacaran metazoans?

9. When rafting through the Inner Gorge of the Grand Canyon of the Colorado River, what Proterozoic rock unit would you see exposed in the walls of the gorge?

10. When did continental glaciation occur during the Proterozoic? What is the evidence that such glaciation occurred? Why is it unlikely that continental glaciers would have formed during the earlier Archean?

11. What characteristics of the Ediacaran discoidal fossils suggest that they may not really be jellyfish, that the frond fossils may not be sea pens (soft corals), and elongate forms such as *Dickinsonia* may not be worms?

12. Stromatolites were exceptionally widespread during the Proterozoic but became relatively sparse thereafter. What other organisms may have contributed to the post-Proterozoic decline of stromatolites?

13. Where was the Belt Supergroup deposited? What evidence indicates that these rocks were deposited in shallow coastal areas?

WEB SITES

The Earth Through Time Student Companion Web Site (www.wiley.com/college/levin) has online resources to help you expand your understanding of the topics in this chapter. Visit the Web Site to access the following:

1. Illustrated course notes covering key concepts in each chapter;

2. Online quizzes that provide immediate feedback;

3. Links to chapter-specific topics on the web;

4. Science news updates relating to recent developments in Historical Geology;

5. Web inquiry activities for further exploration;

6. A glossary of terms;

7. A Student Union with links to topics such as study skills, writing and grammar, and citing electronic information.

10

This vertical cliff of Cambrian sandstone (500–540 million years old) rises about 90 meters (300 feet) above Lake Superior's south shore, near the city of Munising in Michigan's Upper Peninsula.

Early Paleozoic Events

If in late Cambrian time you had followed the present route of Interstate 80, you would have crossed the equator at Kearney, Nebraska.
—John McPhee, *In Suspect Terrain,* 1983

O U T L I N E

▶ DANCE OF THE CONTINENTS
▶ SOME REGIONS TRANQUIL, OTHERS ACTIVE
▶ IDENTIFYING THE BASE OF THE CAMBRIAN
▶ EARLY PALEOZOIC EVENTS
▶ CRATONIC SEQUENCES: THE SEAS COME IN, THE SEAS GO OUT
▶ THE SAUK AND TIPPECANOE SEQUENCES
▶ WAY OUT WEST: EVENTS IN THE CORDILLERA
▶ DEPOSITION IN THE FAR NORTH
▶ DYNAMIC EVENTS IN THE EAST
▶ THE CALEDONIAN OROGENIC BELT
▶ ASPECTS OF EARLY PALEOZOIC CLIMATE
▶ SUMMARY
▶ KEY TERMS
▶ QUESTIONS FOR REVIEW AND DISCUSSION
▶ WEB SITES
▶ BOX 10-1 GEOLOGY OF NATIONAL PARKS
▶ BOX 10-2 A COLOSSAL ORDOVICIAN ASH FALL: WAS IT A KILLER?
▶ BOX 10-3 THE BIG FREEZE IN NORTH AFRICA

Key Chapter Concepts

- The early Paleozoic began 542 million years ago. At that time there were six major continents derived from the breakup of Rodinia: Laurentia, Baltica, Kazakhstania, Siberia, China, and Gondwana.

- The closing of the Iapetus Ocean resulted in the Ordovician Taconic orogeny and the Devonian Acadian orogeny. Erosion of the mountains caused by these orogenies produced immense wedge-shaped deposits of clastic sediments. The Queenston clastic wedge was formed from the erosional detritus of mountains formed during the Taconic orogeny.

- Early Paleozoic continents were composed of quiet stable interior regions and active orogenic belts. Sands and carbonates deposited in shallow epicontinental seas were the dominant sediments of the stable interior. Deep-water deposits interspersed with volcanics characterized marginal orogenic belts.

- Four major cycles of marine transgression and regression dominate the Paleozoic history of the craton. For the early Paleozoic they are the Sauk and Tippecanoe cratonic sequences, followed in the late Paleozoic by the Kaskaskia and Absaroka cratonic sequences.

- The Cordilleran side of North America began as a passive margin, but by Ordovician time a subduction zone formed as the Pacific plate moved against the continent's western margin.

Rocks of the Phanerozoic Eon yield their secrets more readily than Archean and Proterozoic rocks. They are more accessible, less altered, and more fossiliferous. The Phanerozoic Eon includes three eras: Paleozoic ("ancient life"), Mesozoic ("middle life"), and Cenozoic ("recent life"). We now focus on the geologic history of the earliest one, the Paleozoic Era. This chapter looks specifically at its oldest three geologic periods—the Cambrian, Ordovician, and Silurian. These three periods together lasted about 126 million years.

In general, the geologic history of the Paleozoic is characterized by long periods of sedimentation, punctuated by intervals of mountain-building. Just before the Paleozoic began, the supercontinent Rodinia fragmented to form a number of smaller continents. These continents ultimately would converge to form another supercontinent called Pangea. In North America, the mountain-building events are called the **Taconic, Acadian,** and **Allegheny** orogenies.

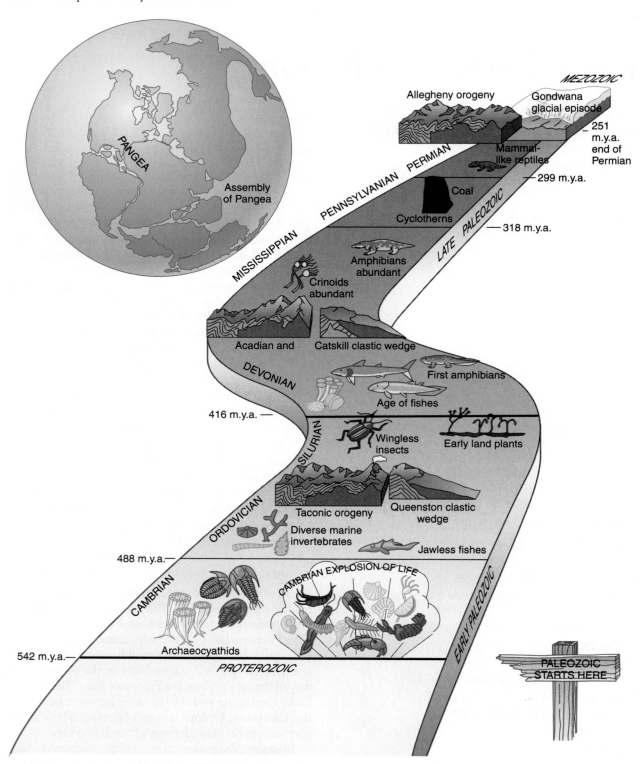

FIGURE 10-1 **Major events of the Paleozoic Era.**

DANCE OF THE CONTINENTS

What was the world like in the early Paleozoic? Where were the continents and what events marked their travels, separations, and collisions? To begin the story, we need to travel back to the late Neoproterozoic. The breakup of the Neoproterozoic supercontinent Rodinia produced six large continents and several smaller microcontinents. The larger continents were:

Laurentia, composed mainly of North America, but including parts of Greenland, northwestern Ireland, and Scotland.

Baltica, composed of Russia west of the Urals and most of northern Europe.

Kazakhstania, the region between the Caspian Sea on the east and China on the west.

Siberia, mostly Russia east of the Urals and north of Mongolia.

China, composed of China, Indochina, and the Malay Peninsula.

Gondwana, composed of South America, Africa, India, Australia, and Antarctica.

After Rodinia broke up about 750 million years ago, Laurentia drifted toward the South Pole and collided with what is today the west coast of Chile (Fig. 10-2). At this time, the South Pole was located across from Laurentia in what is now central Chile. This explains why late Neoproterozoic glacial conditions were widespread in Laurentia, as well as parts of Gondwana, Baltica, and Siberia.

Laurentia had parted company with South America by Cambrian time. The continent drifted northward

until it lay astride the Equator (Fig. 10-3A). Climatic conditions changed from Neoproterozoic cold to Cambrian warm. An ocean tract named **Iapetus** opened east of Laurentia, so that the continent's eastern side became a passive margin.

By Ordovician time (Fig. 10-3B), Gondwana had moved southward so that what is now North Africa was centered near the South Pole. Upper Ordovician tillites in the Sahara Desert reflect glacial conditions resulting from this Gondwana polar location.

The Iapetus Ocean, which had opened during the Cambrian, began to narrow during the Late Ordovician and Silurian (Fig. 10-3C). An active subduction zone formed along the eastern margin of Laurentia. Under the great pressure caused by the subducting plate, the edge of Laurentia folded and broke into multiple thrust faults. It all occurred as part of the **Taconic orogeny.** Closure of the Iapetus Ocean continued into the Silurian and Devonian when the northern end of the sea completely closed, causing Baltica's **Caledonian orogeny.**

SOME REGIONS TRANQUIL, OTHERS ACTIVE

The Stable Interior

In Chapter 8, we noted that continents can be described in terms of cratons and orogenic belts or orogens. A craton is the relatively stable part of the continent consisting of a Precambrian shield and the extension of the shield that is covered by flat-lying or only gently deformed Phanerozoic strata.

Paleozoic strata on the craton were originally wave-washed sands, muds, and carbonates deposited in shallow seas that periodically flooded regions of low relief. These shallow, warm, and well-lighted seas were favorable habitats that allowed for major diversification of marine life. Such extensive inland seas, shown in Figure 10-4, do not exist on Earth today.

Here and there, sedimentary layers deposited in Paleozoic epicontinental seas are warped into broad synclines, basins, domes, and arches, as you can see in Figure 10-5. The resulting tilt to the strata is very slight and is usually expressed in feet per mile rather than degrees (for example, strata might rise in elevation by 2 feet per mile). In the course of geologic history, the arches and domes stood as low islands in the seas or as submarine banks that were barely water-covered.

Domes and arches developed in response to vertically directed forces, unlike those that formed the compressional folds of mountain belts. Possibly, forces associated with plate convergence were transmitted to the craton, causing flexures (domes or arches) in the platform rocks of the craton. Where tensional forces operated, basins developed.

FIGURE 10-2 **Landmasses during the Neoproterozoic, about 750 million years ago.** Note the location of the South Pole and Equator. In those times, most of Earth's landmass was in its southern hemisphere, the opposite of today.

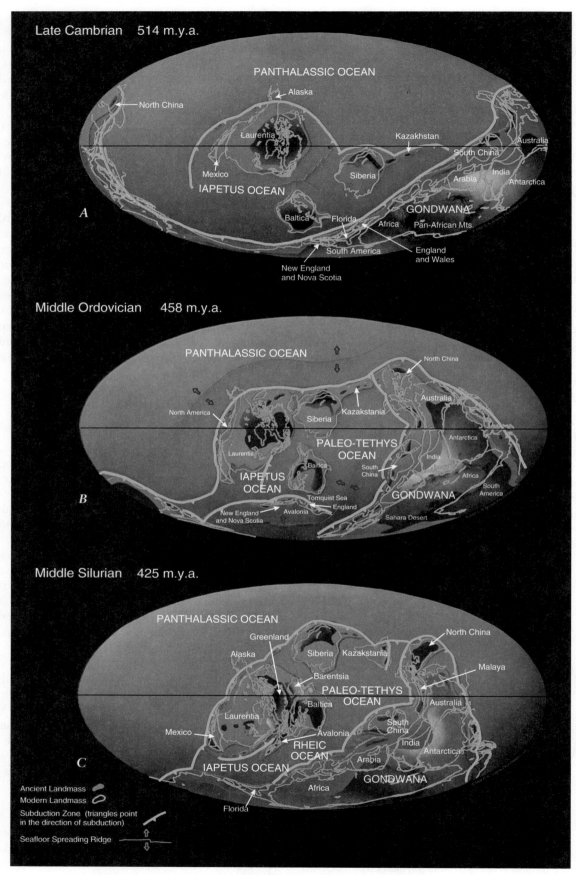

FIGURE 10-3 **Three configurations of Earth's paleogeography, 514–425 million years ago.** (A) Late Cambrian, (B) Middle Ordovician, and (C) Middle Silurian.

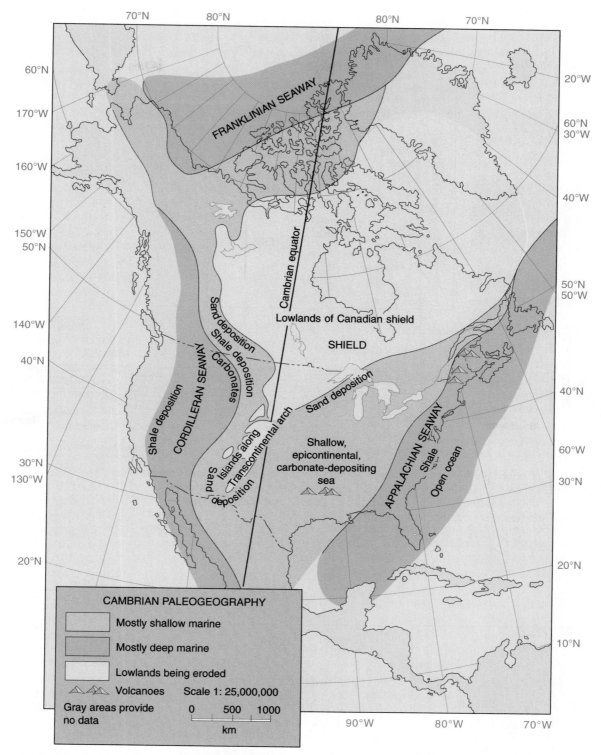

FIGURE 10-4 North America during the Cambrian Period. Note the paleogeographic and tectonic elements. The Cambrian paleoequator runs almost perpendicular to today's equator. (NOTE: On this and other paleogeographic maps, the outlines of today's continents and the Great Lakes are shown for reference only. They did not exist at the time.) ◪ *What were the conditions at the location of your home during the Cambrian period? (Answers to questions appearing within figure legends can be found in the Student Study Guide.)*

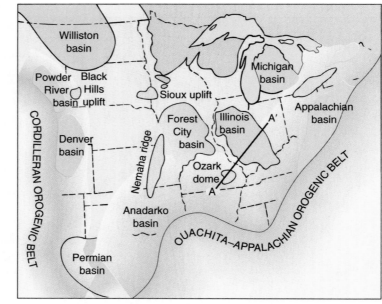

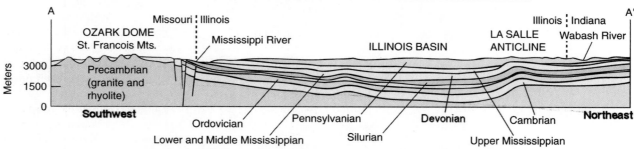

FIGURE 10-5 **Central platform of the United States showing major basins and domes.** (The Great Lakes are shown for reference only.) Structural section A-A9 below the map crosses part of the Ozark dome and the coal-rich Illinois basin. These basins and domes developed at different times during the Phanerozoic.

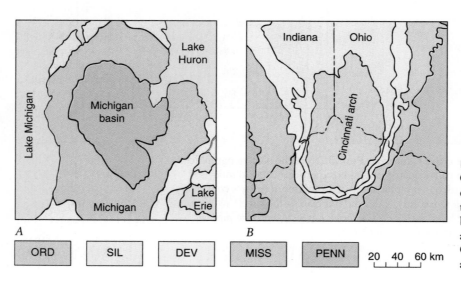

FIGURE 10-6 **Michigan basin and Cincinnati arch.** (A) In an erosionally truncated basin such as the Michigan basin, the youngest beds are centrally located. (B) In a domelike structure such as the Cincinnati arch, the oldest beds are located in the center.

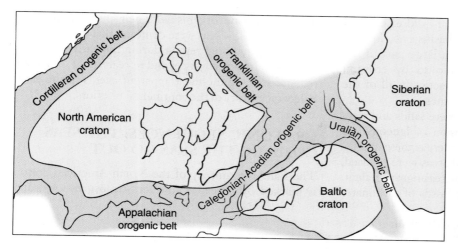

FIGURE 10-7 **Cratons and orogenic belts of North America and Europe.** An orogenic belt is the site of one or more orogenic (mountain-building) events. For example, the Appalachian orogenic belt was the site of the Taconic orogeny during the Ordovician Period, and the Acadian orogeny in the Devonian Period.

We recognize domes and basins by their distinctive pattern of rock outcrops (Fig. 10-6):

- Domes—erosional truncation of domes exposes *older rocks near the centers and younger rocks around the peripheries.* Sequences of strata over arches and domes tend to be thinner. Also, because these structures were periodically above sea level, they have many erosional unconformities.

- Basins—compared to domes, these were more persistently covered by inland seas, and thus have fewer unconformities. They also developed greater thickness of sedimentary rocks. In eroded basins, *younger rocks are near the centers and older rocks are around the edges.*

Orogenic Belts

The North American craton is bounded on four sides by **orogenic belts** that have been the sites of intense deformation, igneous activity, and earthquakes. At least one of these orogenic belts is present on every continent (Fig. 10-7). Most are located along present or past margins of continents, such as the North American Cordilleran belt.

As described in Chapter 7, the passive margin of a continent may experience an early stage in which thick sequences of sediments accumulate along the continental shelf and rise. This stage may be followed by subduction of oceanic lithosphere along continental margins and may terminate in continent-to-continent collisions, as in the classic Wilson Cycle.

IDENTIFYING THE BASE OF THE CAMBRIAN

At one time, it was relatively simple to recognize the boundary between the Cambrian and the underlying Precambrian: The base of the Cambrian System was identified by the first occurrence of shell-bearing multicellular animals. Among these, extinct marine arthropods known as trilobites were used to identify the Precambrian-Cambrian boundary.

But a problem became evident in the 1970s when a distinctive group of shelly fossils was found *beneath* the lowest strata containing the first trilobites. Although these animals had mineralized skeletons, their very small size helped to prevent their earlier discovery. Their discovery, along with detection of new trace fossils, resulted in a new definition of the Precambrian-Cambrian boundary. Today we place the boundary at the lowest (oldest) occurrence of feeding burrows of the trace fossil *Phycodes pedum* (Fig. 10-8). Radioisotopic dates for rocks correlative with this boundary indicate the Cambrian began 542 million years ago.

EARLY PALEOZOIC EVENTS

At the beginning of the Paleozoic Era, North America was relatively stable. We know little of the Paleozoic ocean basins, for those old seafloors exceed 200 million years in age and therefore have long since disappeared

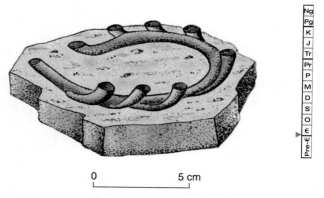

0   5 cm

FIGURE 10-8 **The important trace fossil *Phycodes*.** This sea bottom-dweller mined for food by producing horizontal burrows with upward shafts. The fossil is important because the base of the *Phycodes pedum* biozone marks the Precambrian-Cambrian boundary.

at subduction zones. However, oceans did exist—and in places, they spilled out of their basins onto low regions of the continents. Although a large part of the sedimentary record has been removed by erosion, enough remains to indicate that practically all of the Canadian Shield was inundated at times.

Initially, the dominant deposits were sands and clays derived from weathering and erosion of igneous and metamorphic rocks of the shield. Later, as these quartz-producing and clay-producing land areas were reduced, limestones and dolomites became increasingly prevalent. The limestones contain abundant fossils of carbonate-

secreting marine organisms. Their distribution indicates that shallow seas were common throughout much of Earth's equatorial region during the early Paleozoic (Fig 10-9). In fact, advances and retreats of these epicontinental seas are the most apparent events in the early Paleozoic history of the continental interiors.

CRATONIC SEQUENCES: THE SEAS COME IN, THE SEAS GO OUT

The Paleozoic history of the North American craton is marked by repeated advances (transgressions) and

TABLE 10-1 **Cratonic Sequences of North America**

Geologic Time		Cratonic Sequences		Orogenic Events	Biologic Events	Ice Ages
		Center of craton	Margin of craton			
CENOZOIC			Tejas	Himalayan / Alpine / Laramide	Age of mammals / *Mass extinction*	
MESOZOIC	Cretaceous		Zuni	Sevier / Nevadan	First flowering plants / Climax dinosaurs and ammonites	
	Jurassic				First birds / Abundant dinosaurs and ammonites	
	Triassic			Sonoma	First dinosaurs / First mammals / Abundant cycads / *Mass extinction* (including trilobites) / Mammal-like reptiles	
	Permian	251 m.y.a.	Absaroka			
LATE PALEOZOIC	Pennsylvanian			Alleghenian	Great coal forests / Conifers / First reptiles	
	Mississippian		Kaskaskia	Antler	Abundant amphibians and sharks / Scale trees / Seed ferns	
	Devonian			Acadian–Caledonian	*Extinctions* / First insects / First amphibians / First forests / First sharks	
EARLY PALEOZOIC	Silurian	416 m.y.a.	Tippecanoe		First jawed fishes / First air-breathing arthropods	
	Ordovician			Taconic	*Extinctions* / First land plants / Expansion of marine shelled invertebrates	
	Cambrian		Sauk		First fishes / Abundant shell-bearing marine invertebrates / Trilobites	
NEOPROTEROZOIC		542 m.y.a.			Rise of the metazoans	

retreats (regressions) of epicontinental seas. The regressions exposed old seafloors to erosion, creating extensive unconformities that mark the boundaries of each transgressive-regressive cycle of deposition. We also use these unconformities to correlate particular sequences from one region to another.

Each sequence consists of the sediments deposited as the sea transgressed over an old erosional surface, reached its maximum inundation, and then retreated. These sequences are named **Sauk**, **Tippecanoe**, **Kaskaskia**, and **Absaroka** (Table 10-1).

What caused the transgressions and regressions of the seas? Because the **cratonic sequences** seen in North America also occur on other continents, it is likely that worldwide sea-level changes caused the repeated transgressions and regressions. But what caused global sea-level changes? The favored hypothesis links sea-level changes to the alternate buildup and melting of great ice sheets. Such events would alternately reduce and increase the volume of water in the ocean.

However, glacially linked sea-level change is not the only hypothesis. Many geologists think the sea-level changes resulted from seafloor spreading. Rapid spreading creates high midoceanic ridges. These ridges displace water, causing the sea level to rise globally. When spreading rates slowed, the sea level would be lowered, and epicontinental seas would regress.

THE SAUK AND TIPPECANOE SEQUENCES

The First Major Transgression

During the earliest years of Sauk (Cambrian) deposition, seas were largely confined to the continental margins (continental shelves and rises). Thus, most of the craton was exposed and undergoing erosion. No doubt it was a bleak and barren scene, for vascular land plants had not yet evolved. Uninhibited by protective vegetative growth, erosion gullied and dissected the surface of the land. For at least 50 million years, the Precambrian crystalline rocks underwent deep weathering and must have formed a thick, sandy "soil." Eventually, marine waters spilled out of the marginal basins and flooded the eroded surface of the central craton.

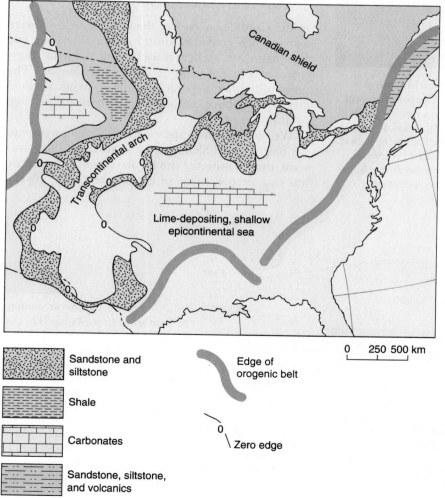

Sandstone and siltstone

Shale

Carbonates

Sandstone, siltstone, and volcanics

Edge of orogenic belt

0 Zero edge

0 250 500 km

FIGURE 10-9 Upper Cambrian lithofacies map. Warm, clear, epicontinental seas covered much of the central United States.

FIGURE 10-10 **The Tapeats Sandstone in vertical walls of Deer Creek Canyon, Grand Canyon, Arizona.** The Tapeats Sandstone was deposited in a nearshore environment of the shallow sea that transgressed a large area of the southwestern U.S. during the Cambrian Period. The sandstone rests on erosionally truncated Proterozoic rocks. It grades upward and eastward into the Bright Angel Shale, which in turn grades into the Muav Limestone.

Islands in the Inland Sea

The craton was not a level, monotonous plain during Sauk time. Instead, it had distinct upland areas composed of Precambrian igneous and metamorphic rocks. During marine transgressions, these uplands became islands in early Paleozoic seas and provided detrital sediments to surrounding areas. The absence of marine sediments over these upland tracts provides evidence of their existence and extent. One of the largest highlands was the Transcontinental Arch (Fig. 10-4), which crossed the craton from Ontario to Mexico.

By Late Cambrian time, seas extended across the southern half of the craton from Montana to New York. An apron of clean sand spread across the seafloor for many miles behind the advancing shoreline. This sandy facies of Cambrian deposition was replaced toward the south by carbonates (Fig. 10-9). Here the waters were warm, clear, and largely uncontaminated by clays and silts from the distant shield. Marine algae flourished and contributed to precipitation of calcium carbonate.

Cambrian Rocks in the Southwest: A Transgressive Succession

In the Cordilleran region, the earliest deposits were sands, which graded westward into finer clastics and carbonates. An excellent place to study this Sauk transgression is in the walls of the Grand Canyon. The Lower Cambrian Tapeats Sandstone is an initial strand line deposit above the old Precambrian surface (Fig. 10-10).

We can trace the Tapeats laterally and upward into the Bright Angel Shale. As the shoreline shifted eastward, the depositional site for the Bright Angel Shale was in a deeper, offshore environment (Fig. 10-11).

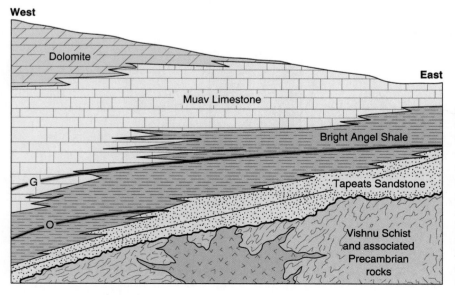

FIGURE 10-11 **East-west section of Cambrian strata exposed in the Grand Canyon.** Line **O** is the top of the Lower Cambrian *Olenellus* trilobite zone. Line **G** is the Middle Cambrian *Glossopleura* trilobite zone. The section is approximately 200 km long, and the portion along the western margin is about 600 m thick. *What does the vertical sequence of rock types indicate about changes in sea level in this region during the Cambrian?*

Next is the Muav Limestone, which originated in a still more seaward environment.

As the sea continued its eastward transgression, the early deposits of the Tapeats were covered by clays of the Bright Angel Formation, and the Bright Angel was in turn covered by the Muav Limestone. Together these formations form a typical transgressive sequence, recognized by coarse deposits near the base of the section and increasingly finer (and more offshore) sediments near the top. Geologists call this a "fining upward sequence."

Cambrian rocks of the Grand Canyon region provide a glimpse of the areal variation in depositional environments, as deduced from changing lithologic patterns. They also illustrate that formations usually do not have the same age everywhere they occur. For example, fossils are reliable evidence that the Bright Angel Formation is Early Cambrian in California but mostly Middle Cambrian in the Grand Canyon. This is an example of *temporal transgression*, which means that sediments deposited by advancing or retreating seas are not necessarily the same geologic age throughout their areal extent.

For a temporal transgression analogy, think of a great snowstorm that begins on Monday in Colorado, reaches Missouri on Tuesday, and New York on Wednesday. A continuous blanket of snow now covers much of the United States, but it is Monday's snow in Colorado, Tuesday's snow in Missouri, and Wednesday's snow in New York.

The carbonate deposition that was characteristic of the southern craton continued into the early Ordovician. Then the seas regressed, leaving a land surface of limestone that experienced deep erosion. The resulting widespread unconformity marks the boundary between the Sauk and the Tippecanoe sequences (Fig. 10-12).

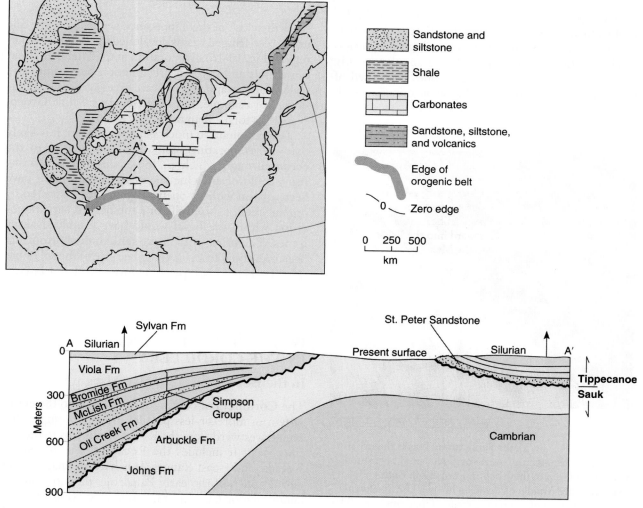

FIGURE 10-12 **Lower Tippecanoe lithofacies map and cross-section along line A-A′.** In the central area, Tippecanoe strata have been removed by erosion. Note the extensive unconformity between the Sauk and Tippecanoe that is depicted on the cross-section.

The Second Transgression

The second major transgression occurred when the Ordovician-Silurian Tippecanoe sea flooded the region that had been vacated as the Sauk Sea regressed. Sediments deposited in the advancing Tippecanoe sea created the initial geologic formations of the Tippecanoe sequence. Again, deposition began with nearshore sands. Among the best known is the Ordovician St. Peter Sandstone (Fig. 10-13). It is nearly pure quartz and thus is valuable for glass manufacturing.

Such exceptionally pure sandstones are usually not developed in a single cycle of erosion, transportation, and deposition. They result from chemical and mechanical processes acting on still older sandstones. In the St. Peter Formation, waves and currents of the transgressing Tippecanoe sea reworked Upper Cambrian and Lower Ordovician sandstones and spread the resulting blanket of clean sand over nearly 7500 square kilometers.

In Chapter 5, you learned that sandstones can be described by their texture and composition as either mature or immature. Sandstones with a high proportion of chemically unstable minerals and angular, poorly sorted grains are immature; sandstones composed of well-rounded, well-sorted, highly stable minerals (such as quartz) are mature. The St. Peter Sandstone is geologically unusual—so pure that we might call it an ultramature sandstone.

The Tippecanoe sandstone phase was followed by deposition of carbonates. Some of these limey deposits were chemical precipitates, many were fossil fragment limestones (bioclastic limestones), and some were great organic reefs. After deposition, the carbonate sediment often underwent chemical substitution—some of its calcium was replaced by magnesium—and in the process became converted from limestone to dolomite.

East of the Mississippi River, the dolomites and limestones were gradually overlain by shales, which form the peripheral sediments of the **Queenston clastic wedge** to the east. The geologic section exposed at Niagara Falls is a classic locality for these rocks (Figs. 10-14 and 10-15).

Evaporites Mark the Sea's "Last Stand"

Near the close of the Tippecanoe sequence in the Silurian, reef-fringed basins developed in the Great Lakes region (Fig. 10-16). Evaporation in the Michigan Basin caused salt and gypsum precipitation on an immense scale. These evaporites indicate excessively arid conditions. For example, in the salt-bearing Salina Group, evaporites total 750 m thick.

To precipitate so much salt and gypsum, it is likely that the evaporating basins had connections to the ocean through which periodic saltwater replenishment could occur. One such connection might have been a basin with its opening to the sea restricted by organic reefs, a raised sill, or a submerged bar. Evaporation within the basin would have produced heavy brines. The brine would have sunk to the bottom and been prevented from escaping because of the sill or bar (Fig. 10-17). This type of feature is called a **barred basin**.

▶ WAY OUT WEST: EVENTS IN THE CORDILLERA

In the Beginning, a Passive Margin

The **Cordillera** of North America refers to the entire system of more-or-less parallel mountain ranges that extend westward from the Rocky Mountains to the Pacific coast. It includes the Rocky Mountains, Sierra Nevada, and Coast Ranges. It is hard to imagine, but during much of the early Paleozoic, this now mountainous region lay near or below sea level (Fig. 10-4).

A coastal plain and shallow marine shelf extended westward from the Transcontinental Arch. This rel-

Pre-Є | Є | O | S | D | M | P | Pr | Tr | J | K | PgNg

FIGURE 10-13 **St. Peter Sandstone.** This is the massive bed directly beneath the overhanging ledge of horizontally bedded dolomite of the Joachim Formation. Both formations are Middle Ordovician in age. The St. Peter is the initial deposit of the transgressing Tippecanoe sea. The brown stains are iron oxide. This exposure is 30 miles southwest of St. Louis, Missouri.

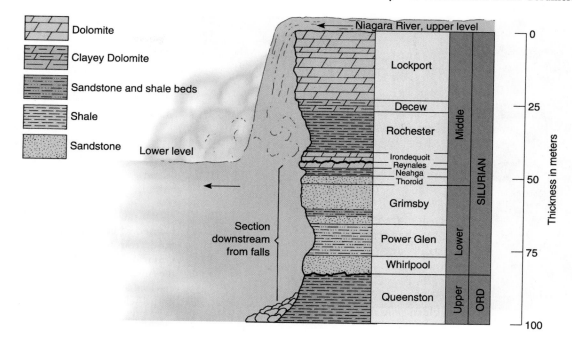

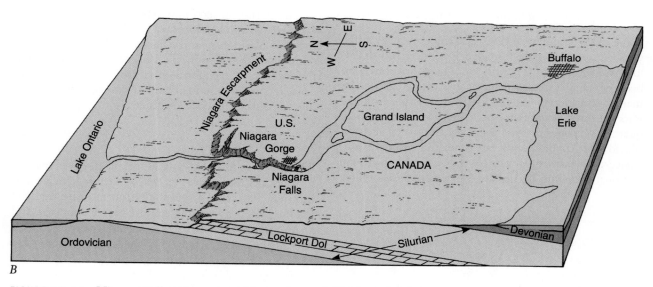

FIGURE 10-14 **Niagara Falls (A) stratigraphic section and (B) block diagram.** The Lockport Dolomite forms the resistant lip of the falls. The rocks dip gently to the south in this area. Where harder dolomite layers such as the Lockport Dolomite intersect the surface, they form a line of bluffs called the Niagaran Escarpment.

atively level expanse was bordered on the west by deep marine basins adjacent to the western continental margin. Many of the marine sedimentary basins of the Cordillera originated during the Neoproterozoic when North America (Laurentia) separated from eastern Antarctica as part of the breakup of Rodinia.

With no evidence of compressional deformation, it is clear that western North America was a passive margin during the early Paleozoic. The Belt Supergroup of Montana, Idaho, and British Columbia, the Uinta Series of Utah, and the Pahrump Series of California were all deposited in basins formed by tensional (normal) faulting during the late Neoproterozoic and early Paleozoic.

FIGURE 10-15 **Niagara Falls formed where the Niagara River flows from Lake Erie into Lake Ontario.** The standard section of the American Silurian System is exposed along the walls of the gorge below the falls.

Early Paleozoic Rocks

Some of the grandest sections of Cambrian rocks in the world are exposed in the Canadian Rockies of British Columbia and Alberta (Fig. 10-18). The formations have been erosionally sculpted into some of Canada's most magnificent scenery (see Box 10-1, about Jasper National Park). Lower Cambrian rocks include ripple-marked quartz sandstones derived primarily from the Canadian Shield. By Middle Cambrian time, the shoreline had transgressed farther eastward. As this occurred, shales and carbonates accumulated in the basin west of the shoreline.

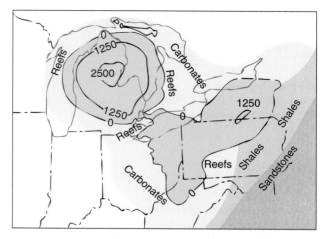

FIGURE 10-16 **Isopach map shows thickness of late Silurian evaporite basins.** Areas of evaporite precipitation were surrounded by carbonate banks and reefs. Gaps in the reefs and banks provided channelways for replenishment of the basins with normal seawater, thus replacing water lost by evaporation.

An interesting section of Middle Cambrian rocks is exposed along the slopes of Kicking Horse Pass in British Columbia, near the border with Alberta. One of the units in this section, the Burgess Shale, has excited the interest of paleontologists around the world because it contains abundant and diverse remains of Middle Cambrian soft-bodied animals.

Orogeny Begins

The Neoproterozoic-to-Cambrian passive margin for the Cordillera changed during the Ordovician and Silurian. Patterns of sedimentation indicate that the Pacific plate was moving against North America, and a subduction zone with an associated volcanic chain had formed (Fig. 10-19). A thick subduction zone complex of graywackes and volcanics was laid down in the trench above the subduction zone. East of the volcanoes, siliceous black shales and bedded cherts accumulated.

The shales are noted for fossils of colonial, planktonic (floating) organisms called **graptolites** (Fig. 10-20). These fossils occur as black shrubby markings in dark shales, where they are preserved as flattened carbonaceous films. Although not very attractive, they are intriguing and valuable fossils—intriguing because they are distant relatives of chordates, and valuable for correlation of Ordovician and Silurian rocks worldwide. Graptolites are so prevalent in lower Paleozoic rocks that the sediments have been called "graptolite facies."

The associated bedded cherts may have derived their silica by dissolution of particles of volcanic ash prevalent at the time. Additional silica was provided by

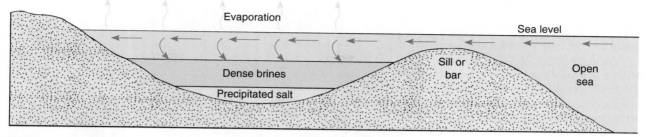

FIGURE 10-17 **Cross-section shows a deposition model for evaporites in a barred basin.** Seawater flows more or less continuously into the basin over a partially submerged barrier (such as a sill or bar). It evaporates to form dense brines. Because of their density, the brines sink and thus are unable to return to the open sea. When the brine becomes sufficiently concentrated, salts precipitate.

dissolution of the delicate shells of siliceous marine microorganisms, such as diatoms and radiolaria.

A thick sequence of fossiliferous carbonates, shales, and sandstones (Fig. 10-21) was spread eastward of the subduction zone complex in the Cordillera. These are the rocks of the back-arc shelf. Because the carbonates contain abundant fossils of marine invertebrates, they are called the "shelly facies."

DEPOSITION IN THE FAR NORTH

Early Paleozoic rocks along the northern margin of North America are sparse, often deformed or metamorphosed, and covered with younger rocks. Nevertheless, there is evidence that the northern edge of the continent formed in the Neoproterozoic when a continental mass split off from the North American craton. The rifting and displacement probably were part of the general dismemberment of Rodinia.

By Ordovician time, conditions had stabilized enough to permit development of a carbonate platform on the continental shelf. To the north, the shelf dropped off abruptly, providing a deep-water environment that trapped turbidities.

The Silurian section in this region is one of the thickest in the world. It includes extensive reefs and thick deposits of evaporites. These rocks imply warm conditions—an interpretation strengthened by paleomagnetic data indicating that this now frigid region lay within 15° of the early Paleozoic equator.

DYNAMIC EVENTS IN THE EAST

At the beginning of the Paleozoic, eastern North America was a passive margin, much like it is today. Relatively shallow-water deposits rich in marine invertebrate fossils were being spread across the continental shelf, while deeper-water sediments accumulated farther seaward along the continental rise.

The initial shelf deposits were quartz sandstones and shales derived from the stable interior, which was still largely emergent in earliest Cambrian time. Gradually, the seas spread inland, and sandy shorelines migrated westward. The initial sandstone deposits of the

FIGURE 10-18 **The Canadian Rocky Mountains in Jasper National Park, Alberta.** The mountains in the distance are part of a thrust sheet composed of lower Paleozoic, horizontal sedimentary rocks. The V-shaped valley in the foreground has been cut into glacial deposits. In the middle distance you can see the many-channeled Athabasca River.

Jasper National Park

For scenic beauty, few places rival Jasper National Park in British Columbia. It extends nearly 200 km along the eastern slope of the Rocky Mountains. Here the Rockies are the product of erosional sculpting of multiple thrust sheets. Strata within the thrust sheets range in age from Neoproterozoic to Jurassic, a roughly half-billion-year span. Like roof shingles, each sheet has been thrust eastward over the one in front of it. As a result, the strata above each of the fault planes are older than those beneath. The bold east-facing cliffs are erosionally resistant Cambrian quartzites and limestones. More easily eroded shaley beds form the gentler slopes.

Both in Jasper National Park and in Banff National Park to the south, the mountains support ice fields and glaciers. The Columbia Ice Field, for exam-

FIGURE A **View toward the terminus of Athabasca Glacier, British Columbia.** The glacier is one of several descending from the Columbia Ice Field.

FIGURE B **Towering wall of massive Cambrian limestone at southeastern end of Maligne Lake, Jasper National Park, Alberta.**

FIGURE C **Cambrian to Devonian strata border Maligne Lake in Jasper National Park, British Columbia.**

ple, sends tongues of glacial ice into valleys on several sides. Notable is the Athabasca Glacier (Fig. A), which is actually on the northernmost tip of Banff National Park.

The rocks of Jasper National Park reveal a fascinating geologic history. Its first chapter records the dismemberment of the supercontinent Rodinia about 750 million years ago. Next, western North America became a passive margin. Initially, sandstones were deposited along this margin as rivers carried sediments eroded from the shield to the coast. Shallow-water limestones (Fig. B), limy shales, and sand-

stones of the Cambrian Gog Group (Fig. C). were deposited next. Some beds contain fossils of extinct trilobites and reef-building archaeocyathids.

The overlying Ordovician and Early Silurian strata contain brachiopods, bryozoans, and graptolites. Late in the Silurian, the warm, shallow sea receded and the former seafloor was eroded. In the Devonian, the sea transgressed over the erosional unconformity, bringing limestone deposition and reefs constructed by extinct corals and calcareous algae.

The general pattern of quiet deposition continued until the Jurassic,

when a microplate that formed in the Panthalassan Ocean docked against the western edge of Canada and welded itself to North America. Later, during the Cretaceous, another mass drifted toward Canada and smashed into the terrane that preceded it. Ponderous compressional forces from these collisions detached the sedimentary sequence from underlying Precambrian granites. These forces ultimately broke it into the thrust sheets that are the hallmark of the Rocky Mountains in Jasper National Park.

---

shelf were soon blanketed by carbonate rocks. The carbonates formed a shallow bank or platform extending from Newfoundland to Alabama. Mud cracks and stromatolites indicate these limy sediments were deposited at or near sea level.

The rather quiet depositional scenario of the Cambrian and Early Ordovician changed dramatically during the Middle Ordovician (Fig. 10-22). At that time, carbonate sedimentation ceased, the carbonate platform was downwarped, and huge volumes of graptolite-bearing black shales and graywackes spread westward over the carbonates.

The Ordovician section in New England contains such coarse detritus, as well as volcanic pyroclastics and interbedded lava flows. The rocks record the development of a subduction zone with an accompanying vol-

canic chain along the eastern margin of North America. The Iapetus Ocean, which opened during the Neoproterozoic and Cambrian, was beginning to close. That closure produced a subduction-volcanic arc complex and ultimately brought together the continents that were to compose Pangea.

Taconic Orogeny

Pulses of orogenic activity that began in the Early Ordovician were followed by several more intense deformational events in Middle and Late Ordovician. These constitute the **Taconic orogeny**. It was caused by the partial closure of the Iapetus Ocean. During that closure, an island arc, and possibly other terranes, collided with the formerly passive margin of North

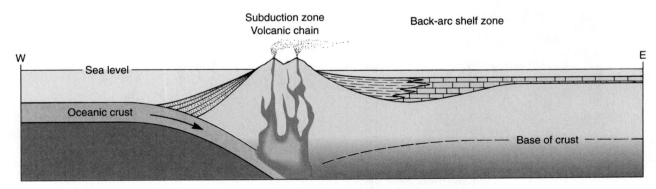

FIGURE 10-19 Interpretive cross-section of conditions across the Cordilleran region during early Paleozoic time. Sedimentation patterns indicate that during the Ordovician and Silurian, the Pacific plate was moving against North America. This movement formed a subduction zone with an associated volcanic chain.

ENRICHMENT

A Colossal Ordovician Ash Fall: Was it a Killer?

Great volcanic eruptions of ash are said to have caused massive extinctions of species. Well, maybe.

Ordovician sedimentary rocks in eastern North America and northwestern Europe contain many widespread ash beds that have been altered to bentonites (clays). These beds were deposited in a geologic "instant," so they are excellent time planes for correlating sedimentary rocks of identical age.

The most extraordinary Ordovician ash fall is the Millburg bed (in Europe, it is called the Big Bentonite). This bed is equivalent in volume to an estimated 1140 km^3 of dense rock pulverized to ash-sized particles. The bentonite layer, now compacted to 1–2 meters thick, spans several million square kilometers. The ash mineralogy indicates a single sustained eruption over 10–15 days. Its thickness distribution suggests the parent volcano was between east-

ern North America and Sweden. Uranium-lead ages from zircon crystals in the bentonite confirm that the beds now on opposite sides of the Atlantic are of the same age, and that the eruption occurred 454 million years ago.

This was the largest ash-producing eruption in the past 590 million years. Yet, it seems to have had little effect on Ordovician marine organisms. This observation casts doubt on hypotheses that invoke atmospheric dust and ash as a primary cause of global extinctions.

Reference

Huff, W. D., Bergstrom, S. M., and D. R. Kolata. 1992. Gigantic Ordovician volcanic ash fall in North America and Europe: Biological, tectonomagmatic, and event-stratigraphy significance. *Geology* 20: 875–78.

America. Later orogenic episodes resulted from further compression of opposing plates. In this scenario, Europe and Africa lay opposite and east of North America.

NORTHERN APPALACHIAN REGION The effects of the Taconic orogeny are most apparent in the northern Appalachians. Caught in the vise between closing lithospheric plates, sediments were crushed (Fig. 10-23), metamorphosed, and thrust northwestward along a

great thrust fault. Along the fault, continental rise sediments were shoved over and across some 48 km of shelf and shield rocks. Today, the remnants of this activity are exposed in the Taconic Mountains of New York.

Ash beds, now weathered to a clay called **bentonite**, attest to the violence of volcanism. Masses of granite now exposed in the eastern Appalachian Piedmont Province (see the map in Appendix B-1) help record the great pressures and heat to which deposits in the subduction zone complex were subjected. But even if

A

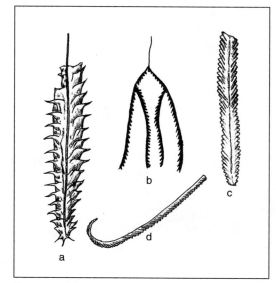

B

FIGURE 10-20 **Graptolites.** (A) Branches, called stipes, of the graptolite *Diplograptus*. It is common in dark shales of Ordovician age in Europe and North America. (B) Four Ordovician graptolite colonies. Each individual (zooid) lived in one of the tiny cone-like tubes.

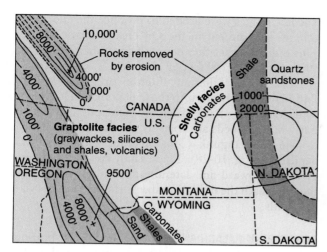

FIGURE 10-21 Ordovician sedimentation in the Williston basin region. Isopach lines indicate thickness of the Ordovician time-rock unit. Deep-water sediments reach thicknesses of 2900 meters (9500 feet) in Idaho and 3080 meters (10,000 feet) in British Columbia.

the igneous rocks had never been found, geologists would know mountains had formed, for the great apron of sandstones and shales that outcrop across Pennsylvania, Ohio, New York, and West Virginia must have had their source in the rising Taconic ranges (Fig. 10-24).

From a feather edge in Ohio, this barren wedge of rust-red terrestrial clastics called the **Queenston clastic wedge** (Fig. 10-25) became increasingly coarser and thicker toward ancient source areas to the east (Figs. 10-26 and 10-27). Streams flowing from those mountains were laden with sand, silt, and clay. As they emerged from the ranges, they deposited their load of detritus as huge deltaic systems that grew and spread westward into the basin, often covering earlier shallow marine deposits.

The deltaic aspect of many of these deposits accounts for the alternate name of Queenston delta for the Queenston clastic wedge. It has been estimated that over 600,000 cubic kilometers of rock in the

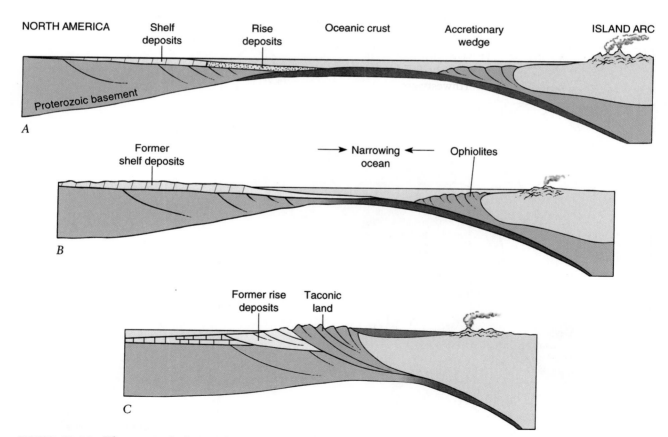

FIGURE 10-22 Plate tectonic forces that caused the Taconic orogeny (A) Following the Neoproterozoic breakup of Rodinia, eastern North America had a passive margin. (B and C) Subsequently, a large island arc converged on the passive margin and converted it to an orogenic belt with growing mountain ranges.

FIGURE 10-23 **Ordovician shale beds on Newfoundland's west coast.** They are intensely folded as a result of the Taconic orogeny of late Ordovician time. The location is Black Point, Port au Port.

Taconic ranges were eroded to produce the enormous volume of sediment in the Queenston clastic wedge. This indicates that the mountains may have exceeded 4000 meters (13,100 feet) in elevation.

During the Silurian, intense orogenic activity shifted northeastward into the Caledonian belt. Meanwhile, erosion of the Taconic highlands continued. Early Silurian beds are coarsely clastic, as represented by the Shawangunk Conglomerate and Tuscarora Sandstone (Fig. 10-28). These sandy and pebbly strata give way upward and laterally to sandstones such as those beneath the dolomites that form the "lip" of Niagara Falls.

SOUTHERN APPALACHIAN REGION Silurian iron-bearing sedimentary deposits (Fig. 10-29) accumulated in the southern Appalachian region. The greatest development of this sedimentary iron ore is in central Alabama, where there are also coal deposits of Pennsylvanian age and extensive limestone deposits. The coal is used to manufacture coke—which, along with limestone, is

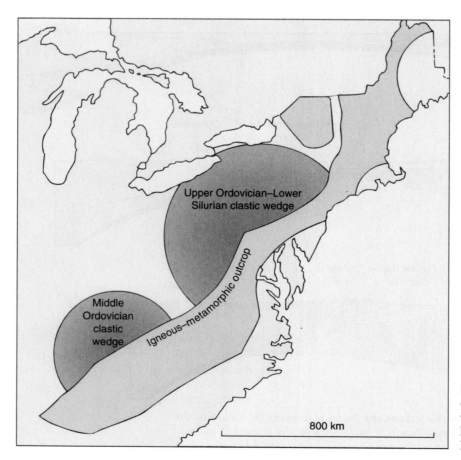

FIGURE 10-24 **Great wedges of clastic sediments spread westward as mountain belts that had developed during the early Paleozoic eroded.**

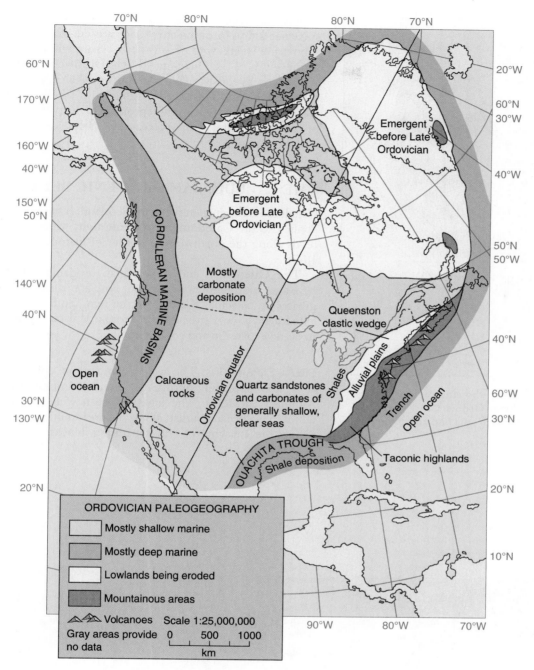

FIGURE 10-25 **Paleogeography of Ordovician North America.**

used to manufacture steel. The fortunate occurrence of iron ore, coal, and limestone in the same area accounts for the once-thriving steel industry of Birmingham, Alabama.

Extending across the southern margin of the North American craton is the Ouachita-Marathon depositional trough (Fig. 10-28). Although over 1500 km long, only about 300 km of its folded strata

are exposed. Additional information about the distribution of early Paleozoic rocks within the belt is derived mostly from oil wells and geophysical surveys.

Overall, nearly 10,000 meters of Paleozoic sediment filled the **Ouachita-Marathon trough**. However, of this thickness, only about 1600 meters are early Paleozoic in age. Cambrian sediments are

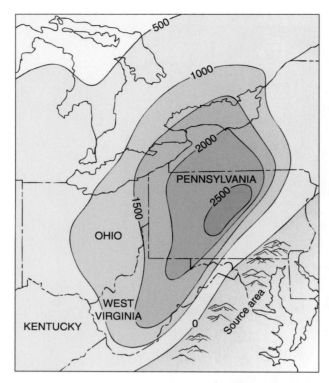

FIGURE 10-26 Isopach map showing regional thickness variation of Upper Ordovician sedimentary rocks in Pennsylvania and adjoining states.

largely graywackes and bedded cherts. The abundant silica needed to form the chert beds was derived from submarine weathering of ash spewed from volcanoes to the south of the Ouachita trough. Ordovician and Silurian rocks consist of black graptolitic shales (graptolite facies) and fossiliferous limestones (shelly facies). The former represent deposition in deep waters of the continental rise, whereas the carbonates are shelf deposits.

▶ THE CALEDONIAN OROGENIC BELT

Caledonia is the ancient name for Scotland. It has been applied to the **Caledonian orogenic belt**, which extends along the northwestern border of Europe (Fig. 10-7).

The Caledonian and Appalachian orogenic belts have a generally similar history, for both are part of the greater Appalachian-Caledonian system. Both elongated depositional sites evolved from a Wilson Cycle of ocean expansion and contraction.

OCEAN EXPANSION The cycle began with an episode of seafloor spreading during the Late Precambrian to Middle Ordovician. The Iapetus Ocean widened to admit new oceanic crust along a spreading center. Along the margins of the separating blocks, huge

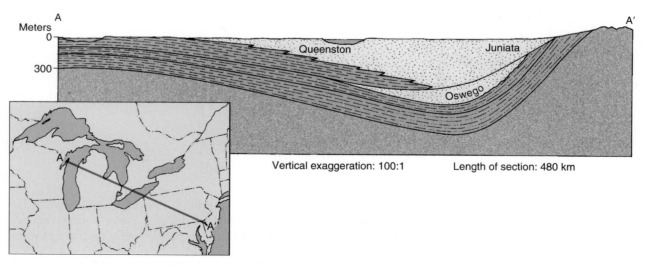

Vertical exaggeration: 100:1   Length of section: 480 km

FIGURE 10-27 **Restored section of Upper Ordovician rocks from Delaware to Wisconsin.** We interpret the cross-section as a rising highland area to the east that supplied clastic sediments to the basin until it filled. Continued sedimentation forced the retreat of the sea westward and extended a clastic wedge toward the west.

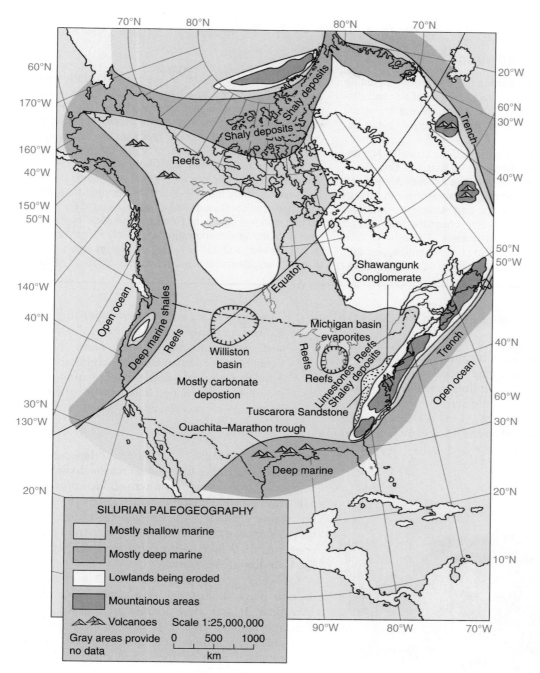

FIGURE 10-28 **Silurian North America Paleogeography.**

ENRICHMENT

The Big Freeze in North Africa

We are causing many environmental problems—including air and water pollution, acid rain, deforestation, and disruption of water systems. Yet environmental and ecological crises occurred many times in the geologic past, long before humans.

Consider the effects on life of the Ordovician Ice Age described in this chapter. Evidence is that the Ordovician South Pole was in the now-barren area between Algeria and Mauritania. Evidence for an enormous ice sheet in this region includes glacial grooves and striations, outwash plains, moraines, and meltwater channels that extend across thousands of square kilometers. There must have been a truly enormous buildup of continental ice, and that accumulation would surely have resulted in cooler tempera-

tures in ocean regions far to the north. Eustatic and climatic changes would have had a pervasive effect on life.

Indeed, paleontologists have long noted Late Ordovician extinctions of many marine invertebrates, including entire families of bryozoans, corals, brachiopods, sponges, nautiloid cephalopods, trilobites, crinoids, and graptolites. Of all known animal families, 22% were wiped out, one of the greatest mass extinctions in the geologic record.

It seems likely that the extinction resulted from general cooling of tropical seas, as well as the draining of epicontinental and marginal seas as sea level was lowered. The loss of the shallow-water environments in which early Paleozoic invertebrates thrived, and the crowding of species into remaining habitats, would explain this Ordovician ecological disaster.

volumes of sediment accumulated, with two distinctly different facies:

1. The graptolite facies includes more than 6000 meters of volcanics, graywackes, and graptolitic shales. Because graptolites are floating creatures, they drifted freely across the ocean, providing our most exact means for determining the equivalence of strata in widely separated regions.

2. The shelly facies is clean sandstones and fossiliferous limestones that accumulated along the shallower continental shelves. Here and there along the Caledonian trough margins are Upper Silurian freshwater shales that contain fossil remains of early fishes and strange arthropods known as eurypterids (Fig. 10-30).

OCEAN CONTRACTION Closure of the Iapetus Ocean and crumpling of the Caledonian marine basins began in Middle Ordovician time, when subduction zones developed along the margins of the formerly separate continents. Little by little, the Iapetus Ocean continued to narrow until the opposing continental margins converged in a culminating mountain-building event: the **Caledonian orogeny**.

Orogenic activity was more or less continuous at one place or another during the Silurian and Devonian. It was most intense in Norway, where Precambrian and lower Paleozoic sedimentary rocks were metamorphosed and thrust-faulted. Today, the truncated folds of the Caledonians end at the western coast of Ireland. Nonetheless, they can be traced to northeastern Greenland and Spitzbergen and then across the Atlantic in Nova Scotia and Newfoundland.

FIGURE 10-29 Clinton iron ore from the Silurian Clinton Group near Birmingham, Alabama. The ore is an oölite of the iron oxide mineral hematite. It includes shells of marine fossils that have been replaced by iron oxide. Nearby coal beds provide the fuel needed to produce steel from the iron ore.

ASPECTS OF EARLY PALEOZOIC CLIMATE

During the early Paleozoic, Earth had latitudinal and topographic variations in climate similar to today. As you have seen, climates probably were milder and more uniform during great marine transgressions, and more

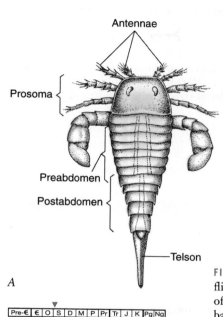

B

FIGURE 10-30 **Two genera of eurypterids.** *Eurypterus* (A) is noted for its broad, flipperlike paddles and blunt frontal margin. *Pterygotus* (B) is distinguished by a pair of formidable-looking frontal pincers. The animal swimming in the center background is a primitive jawless fish.

Labels on image A: Antennae, Prosoma, Preabdomen, Postabdomen, Telson

extreme and diverse when continents stood high and great mountain ranges diverted atmospheric circulation patterns. But several other factors contributed to the early Paleozoic climate:

- Earth spun faster and had shorter days.
- Tidal effects were stronger.
- No vascular plants existed to absorb the Sun's radiation until the Late Ordovician.

During the Cambrian and Ordovician, the paleoequator crossed North America from Mexico to the Arctic (Figs. 10-4 and 10-25). Extensive limestones and coral reefs indicate generally warm conditions for Laurentia during the early Paleozoic. Periods of aridity are recorded by thick deposits of salt and gypsum.

There also were severe climates during the early Paleozoic. As noted, extensive Late Ordovician glacial deposits attest to frigid conditions in the region of today's Sahara Desert. These widespread deposits suggest emplacement by continental glaciers and certainly were accompanied by lower annual temperatures at middle and high latitudes. The distribution of glacial deposits, direction of glacial striations, and paleomagnetic data confirm a North African location for the Ordovician South Pole.

SUMMARY

- Six major continents and several small microcontinents existed at the beginning of the Paleozoic, now placed at 542 million years ago. In addition, an oceanic tract called the Iapetus Ocean had opened and separated Laurentia from Baltica and Siberia.
- As plates converged during the Ordovician and Silurian, closure of the Iapetus Ocean caused the Taconic and Caledonian orogenies. Orogenic activity in the Caledonian orogenic belt continued to Middle Devonian time.

- The Taconic orogeny raised mountains along the eastern margin of Laurentia (North America). Rapid erosion of these mountains produced huge volumes of detritus, which was spread westward by streams and deposited as a great blanket of sediment called the Queenston clastic wedge.
- Erosion of mountains raised during the Caledonian orogeny also resulted in a Devonian clastic wedge across much of northwestern Europe.
- Deep-water deposits with interspersed volcanics characterize marginal orogenic belts. Shallow-water sands and

fossil-bearing carbonates typify the cratonic or stable interior regions.

- Large-scale structural domes and basins characterize the stable interiors of continents. Domes have an overall thinner section of rocks with many unconformities. Basins have a thicker sequence of sediments and fewer unconformities.

- Western North America was tectonically quiet during the Cambrian, but by Ordovician time the Pacific plate had moved against the western margin of the continent.

This produced a subduction zone and a chain of volcanic islands.

- The range of climatic conditions during the early Paleozoic was not significantly different from those of the more recent geologic past. Fossil-rich carbonate deposits and evaporites confirm the warm conditions of continents. These continents are known by paleomagnetic evidence to have been close to the equator. Cold climates are indicated by glacial striations and tillites in Gondwana continents.

KEY TERMS

Absaroka sequence, p. 275

Acadian orogeny, p. 267

Allegheny orogeny, p. 267

barred basin, p. 278

bentonite, p. 284

Caledonian orogenic belt, p. 288

Caledonian orogeny, pp. 269, 290

Cordillera, p. 278

cratonic sequences, p. 275

graptolites, p. 280

Iapetus Ocean, p. 269

Kaskaskia sequence, p. 275

orogenic belt, p. 273

Ouachita-Marathon trough, p. 287

Queenston clastic wedge, p. 285

Sauk sequence, p. 275

Taconic orogeny, p. 269

Tippecanoe sequence, p. 275

QUESTIONS FOR REVIEW AND DISCUSSION

1. If you wanted to go cross-country skiing during the Silurian, which of the six major Silurian continents would you visit?

2. What geologic evidence indicates that the Transcontinental Arch was above sea level during most of the early Paleozoic?

3. What is a barred basin? Why is a barred basin a particularly effective place for precipitation of evaporites? Where and when did such basins develop during the early Paleozoic?

4. What is a clastic wedge? Are clastic wedges primarily marine or nonmarine? Under what circumstances do they develop? What clastic wedge is associated with the Taconic orogeny?

5. Why are the unconformities that form the boundaries of cratonic sequences considered the result of eustatic lowering of sea level rather than local tectonic uplift of parts of the craton?

6. Why is the unconformity that separates the Sauk from the Tippecanoe not the same age at widely separated locations?

7. Domes and basins are characteristic structural features of the central stable regions of North America. Define each and explain how they differ. Why are there more erosional unconformities in domes? How might one distinguish between domes and basins on a geologic map?

8. The guide fossil for the base of the Cambrian is a trace fossil (*Phycodes*). What are trace fossils?

9. What evidence might you seek in the field to confirm the following:

   a. The location of the paleoequator

   b. Former conditions of extreme aridity

   c. Former extensive episodes of glaciation

   d. Mountain-building associated with tectonic plate collision

10. What is there about graptolites that makes them so useful in global correlation of Ordovician and Silurian rocks?

WEB SITES

The Earth Through Time Student Companion Web Site (www.wiley.com/college/levin) has online resources to help you expand your understanding of the topics in this chapter. Visit the Web Site to access the following:

1. Illustrated course notes covering key concepts in each chapter;

2. Online quizzes that provide immediate feedback;

3. Links to chapter-specific topics on the web;

4. Science news updates relating to recent developments in Historical Geology;

5. Web inquiry activities for further exploration;

6. A glossary of terms;

7. A Student Union with links to topics such as study skills, writing and grammar, and citing electronic information.

11

Ng
Pg
K
J
Tr
▶ P
P
M
D
S
O
€
Pre

Red sandstone in Garden of the Gods, Colorado Springs, Colorado. *Erosion continues to sculpt these nearly vertical Lyons Formation beds, which formed during the Permian Period (295–250 million years ago) into dramatic pinnacles and spires, some standing 15 meters high. The sand that formed these beds came from the eroding ancestral Rocky Mountains to the west. The beds were tilted during the mountain-building orogeny that formed today's Rocky Mountains.*

Late Paleozoic Events

We need not be surprised if we learn from geology that the continents and oceans were not always placed where they now are, although the imagination may well be overpowered when it endeavors to contemplate the quantity of time required for such revolutions.
—Sir Charles Lyell, *The Student's Elements of Geology,* 1871

Key Chapter Concepts

- As in early Paleozoic rocks of the craton, the sedimentary rocks of the late Paleozoic can be divided into "packages" that have unconformities at both their upper and lower boundaries. Each "package" is called a cratonic sequence.

- The lower unconformity in a cratonic sequence marks where sediment from a transgressing sea was deposited on an old erosional surface. When the sea regressed, a second erosional surface developed on the now-exposed seafloor. Then, when covered with sediment deposited in another transgression, that surface became the unconformity at the top of the cratonic sequence.

- The late Paleozoic-Triassic has two cratonic sequences, the Kaskaskia (Devonian-Mississippian age) and the Absaroka (Pennsylvanian-Triassic age). Each began with a marine transgression over an older erosional unconformity and ended with a marine regression followed by erosion to produce another unconformity.

- In the North American craton, the Mississippian Period is characterized by limestones deposited in warm epicontinental seas. Shales, sandstones, and coal beds are more characteristic of the Pennsylvanian Period.

- In North America and Europe, Pennsylvanian coal deposits occur in repetitive layers of nonmarine and marine sediments called cyclothems. The transgressions and regressions of inland seas that caused cyclothems may be related to advances and retreats of continental glaciers in Gondwana.

- The Acadian orogeny of the Devonian Period was caused by two collisions: (1) At the northern end of Laurentia's eastern margin, Laurentia and Baltica collided. (2) To the south, Laurentia collided with a microcontinent named Avalonia. The Acadian Mountains were formed during the two collisions. Sediments derived from the erosion of these mountains were spread westward to form the vast Catskill clastic wedge.

- The Alleghenian orogeny and Ouachita orogeny were events in the final assembly of the supercontinent Pangea. The part of Gondwana that is today the western bulge of Africa was sutured to the southern part of Laurentia.

O U T L I N E

▶ EPICONTINENTAL SEAS ON THE CRATON
▶ UNREST IN THE WEST
▶ TO THE EAST, A CLASH OF CONTINENTS
▶ SEDIMENTATION AND OROGENY IN THE WEST
▶ EUROPE DURING THE LATE PROTEROZOIC
▶ GONDWANA DURING THE LATE PALEOZOIC
▶ CLIMATES OF THE LATE PALEOZOIC
▶ MINERAL PRODUCTS OF THE LATE PALEOZOIC
▶ SUMMARY
▶ QUESTIONS FOR REVIEW AND DISCUSSION
▶ WEB SITES
▶ BOX 11–1 ENRICHMENT
▶ BOX 11–2 GEOLOGY OF NATIONAL PARKS AND MONUMENTS

• During the Devonian, a subduction zone developed along the formerly passive western margin of Laurentia. As a result, mountain-building occurred within the western part of the North American craton: the Antler orogeny during the Devonian Period and the Sonoma orogeny during the Permian-Triassic.

• Great ice sheets spread across a vast region of Gondwana during the Carboniferous (Pennsylvanian and Mississippian Periods) when the Gondwana portion of Pangea was located above the South Pole.

The late Paleozoic comprises 160 million years (410–250 million years ago). It is divided into four periods: the

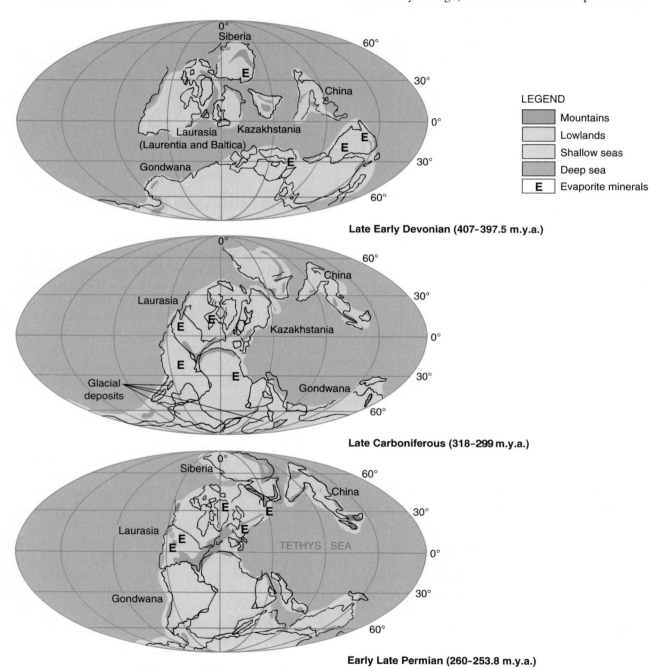

FIGURE 11-1 **Paleogeography of the Late Paleozoic.** Compare how much continental movement occurred from map to map. The time interval between the upper and middle maps is about 100 million years, during which tectonic forces slowly merged Gondwana and Laurasia. The time interval between the middle and lower maps is about 50 million years, during which Siberia collided with Europe to form the Ural Mountains. ◧ *During the Pennsylvanian Period, what continent would you cross into if you traveled south from the present day location of Louisiana? (Answers to questions appearing within figure legends can be found in the Student Study Guide.)*

Devonian, Mississippian, Pennsylvanian, and Permian. By Permian time, most of the separate landmasses of earlier periods were assembled into the great supercontinent of **Pangea**. Before the joining of shorelines, the margins of larger continents (such as eastern North America) grew by accretion of island arcs and microcontinents. Eventually, the larger continents collided.

The **Caledonian** and **Acadian** orogenies spanned the interval between middle Silurian and middle Devonian. The two orogenic events brought Laurentia (North America) and Europe together as a combined landmass named **Laurasia** (Fig. 11-1).

Subsequently, the plate bearing Gondwana began to close on Laurasia, causing the Late Carboniferous **Hercynian orogeny** of central Europe and the **Alleghenian orogeny** of eastern North America. Mountains in the southern Appalachian region resulted from a collision with northwestern Africa. Similarly, the now deeply eroded and largely buried mountains of the Ouachita orogenic belt were produced by collision with Gondwana. By the Late Permian, the clustering of once-separate continents was nearly complete.

In addition to continental collisions and recurring orogenic activity, the late Paleozoic was a time of diverse sedimentation, diverse climatic conditions, and progress in organic evolution. During this time, there was widespread colonization of the land by both large land plants and vertebrates. Amphibians and reptiles, along with spore-bearing trees and seed ferns, were prominent in the landscapes of the late Paleozoic. Near the end of the era, conifers and reptiles became abundant, able to tolerate drier, cooler climates. Marine invertebrates thrived in shallow epicontinental seas. In the stable U.S. interior, Mississippian limestones deposited in these inland seas are over 700 meters thick.

Adjacent to the orogenic belts, sediment-laden streams deposited their sand and mud. Plant material accumulated in swampy areas, forming the great coal deposits that have enabled the manufacture of steel and generation of roughly half of U.S. electric power.

Arid conditions existed at many locations around the globe, as evidenced by aeolian cross-bedded sandstones (lithified desert dunes) and layers of evaporites (Fig. 11-2).

The more southerly parts of Gondwanaland were located near the pole, and glaciation occurred. Late Carboniferous or Permian ice sheets are indicated by tillites and glacial striations (Fig. 11-3) on the now-separated landmasses of South America, Africa, Australia, Antarctica, and India. Decades before geologists had an understanding of plate tectonics, ancient tillite deposits on these continents were cited as evidence of continental drift.

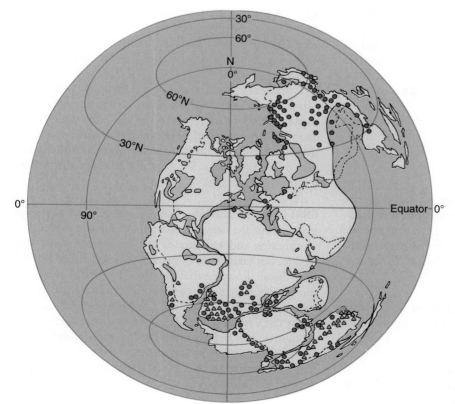

FIGURE 11-2 **Evidence for ancient climates and movement of ancient continents during the Permian Period, about 250 million years ago.** Blue triangles show the distribution of glacial tillites; red circles represent coal; irregular green areas represent evaporites.

FIGURE 11-3 **Glacial striations.** Rocks embedded in the base of a moving glacier gouged the grooves or striations into underlying bedrock.

EPICONTINENTAL SEAS ON THE CRATON

The last event of the early Paleozoic cratonic region was the withdrawal of inland seas. Following this withdrawal, many early Paleozoic formations were subjected to erosion that exposed older and deeper formations in uparched areas.

Gradual flooding of that early Paleozoic erosional surface was the first event of the late Paleozoic, creating a new inland epicontinental sea, the Kaskaskia Sea (Fig. 11-4). In it were deposited marine rocks of the **Kaskaskia cratonic sequence** (see Table 10-1). The Kaskaskia Sea did not regress until the end of the Mississippian Period. Then an interval of erosion prevailed before advance of a second epicontinental sea and the deposition of the **Absaroka cratonic sequence**.

We will now look at the Kaskaskia and Absaroka sequences.

The Kaskaskia Sequence

The initial deposits of the transgressing Kaskaskia Sea were clean quartz sands that became lithified into sandstones. The best known is the Oriskany Sandstone of New York, Pennsylvania, Virginia, and West Virginia (Fig. 11-5). Because of its purity, Oriskany Sandstone is extensively used in making glass.

MINERAL ASSEMBLAGES Mineral grains in the Oriskany Sandstone illustrate how geologists determine where grains in a rock came from and how much of a beating they took before being deposited. Although quartz is by far the most abundant mineral in the Oriskany, the formation also contains a small percentage of other sil-

icate materials, which, like quartz, are resistant to weathering and erosion but are heavier than quartz. Because they are heavier than quartz, they are called **heavy minerals**.

Two distinct assemblages of heavy minerals occur in the Oriskany Sandstone. South of New York, the formation contains only the most erosion-resistant heavy minerals (tourmaline, zircon, and rutile). The grains in these sandstones are exceptionally well-rounded and worn, and less-stable heavy minerals are missing. Geologists reason that this area of the Oriskany was derived from older clastic sedimentary units to the east and north.

In contrast, the Oriskany within New York contains a heavy mineral assemblage of more easily weathered heavy minerals (pyroxene, hornblende, and biotite). These minerals decompose more readily because they contain iron, which combines with oxygen to form iron oxides. They also have good cleavage, allowing corrosive solutions to penetrate into fresh mineral surfaces. Pyroxene, hornblende, and biotite occur in a variety of igneous and metamorphic rocks. In addition, the grains show very little wear. Therefore, we may infer that the sands composing the Oriskany of New York may have had an igneous or metamorphic source area and were not recycled from older sedimentary rocks.

SEDIMENTATION Oriskany sands spread across the erosional unconformity that remained following the regression of the early Paleozoic Tippecanoe Sea. The deposition of nearshore sands was followed by carbonate deposition. During the carbonate phase, corals built extensive reefs on the limy seafloor. Where water circulation was restricted, salt and gypsum were deposited.

During the Middle and Late Devonian east of the Mississippi Valley, carbonate sedimentation gave way to clay-sized particles, which lithified into shales. The change from chemical precipitation to clastic deposition was caused by the erosion of mountains that were raised during the Acadian orogeny along North America's eastern margin.

CHATTANOOGA SHALE The Catskill clastic wedge grew along the western side of the mountains. Conglomerates and sandstones are characteristic of this wedge, but farther west, suspended muds were carried into the shallow sea that covered the platform. These muds were deposited as a thin but extraordinarily widespread formation called the **Chattanooga Shale**. Because the Chattanooga shale is widespread and easy to recognize, it is an important marker for regional correlation. Coincidentally, it contains some uranium which lets us determine its age by radioactivity measurements: 350 million years old.

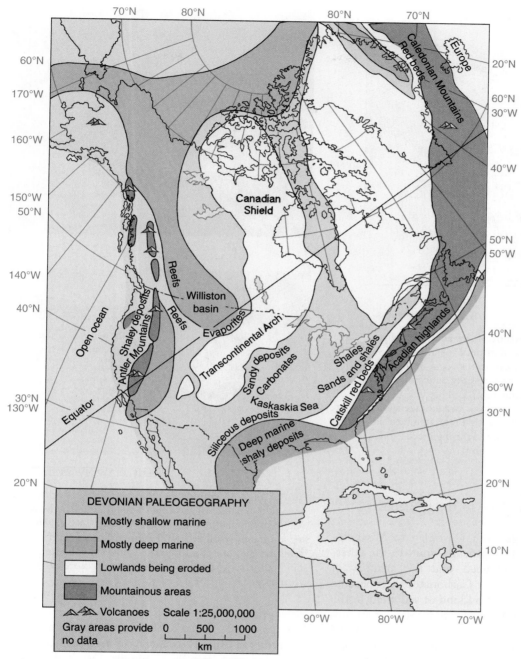

FIGURE 11-4 **Devonian Period paleography.** This is what North America looked like during the 55-million-year interval from 410 to 355 million years ago. (Note: On paleogeographic maps in this book, we show the Great Lakes for visual reference only. They are modern, formed late in the Pleistocene Epoch when glacier-carved basins filled with meltwater from the retreating ice sheet around 11,000 years ago.)

In addition to its uranium content, the Chattanooga is rich in finely disseminated pyrite (an iron sulfide mineral) and organic matter. It is black or dark gray in color and lacks fossils of bottom-dwelling invertebrates. These qualities indicate that the Chattanooga accumulated in stagnant, oxygen-deficient water.

It is difficult to conceive how a sea occupying a depositional area as vast as the Chattanooga's could sustain such oxygen-deficient conditions. One interesting suggestion is that phytoplankton in the Chattanooga Sea proliferated catastrophically, consuming and thereby depleting the oxygen supply. Without the oxygen required for decay of organic matter, carbon would

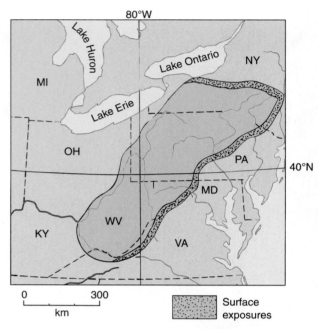

FIGURE 11-5 **Areal extent of the economically important Oriskany Sandstone.** Most of this formation is subsurface; the surface exposures occur in the speckled areas. Because of its purity, Oriskany Sandstone is extensively used in making glass. And because the rock is porous and permeable, it contains vast storage space among the sand grains for natural gas that migrated into the formation. This fuel is extracted through wells and pipelined for heating and cooking.

A

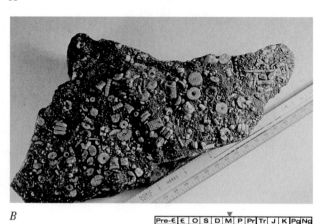

B

Pre-Є | Є | O | S | D | M | P | Pr | Tr | J | K | Pg | Ng

FIGURE 11-6 **Crinoids.** (A) Reconstruction of crinoids (invertebrate animals) growing on the floor of the Kaskaskia Sea (Mississippian). Note their long stems (stalks), which are commonly fossilized. (B) Fragments of fossil crinoid stems in limestone (Burlington Formation). The stems stand out in relief because the limestone is more easily weathered.

accumulate and contribute to the formation's black color.

During the passage from Devonian to Mississippian, the mountains that provided mud to the Chattanooga Shale were reduced, and the quantity of muddy sediment decreased. Carbonates then became the abundant and widespread kind of sediment in the epicontinental seas of the platform. Cherty limestones (see Fig. 4-37), shelly limestones, limestones composed of the remains of billions of crinoids (Fig. 11-6) and other invertebrates, and limestones containing myriads of oöids (Fig. 11-7) formed extensive beds across the central and western parts of the craton.

The sea in which these crinoidal, oölitic, and other types of limestones were deposited was the most extensive that North America had experienced since Ordovician time (Fig. 11-8). The widespread blanket of carbonates deposited in this sea is often referred to as the "great Mississippian lime bank." It records the last great Paleozoic flooding of the North American craton.

Sands, clays, and thin layers of carbonates were deposited in Late Mississippian time as the Kaskaskia

Sea regressed. Today, these rocks form petroleum reservoirs in Illinois, so they have been studied extensively for economic development. Detailed maps of some of the sandstone units show that they are thickest along branching, sinuous trends that suggest old stream valleys developed on the former seafloor. Studies of grain size, cross-bedding (Fig. 11-9), and current-produced sedimentary structures indicate that the detritus was derived from the northern Appalachians and was transported southwestward across the central interior.

The far western part of the craton was too far from the Acadian Mountains to receive clastic particles. Middle and Upper Devonian rocks are largely limestones. In a depression called the Williston basin (extending

A

B

FIGURE 11-7 **Oölitic limestone used architecturally and viewed microscopically.** (A) Photomicrograph of Salem Limestone reveals fossil debris and oöids cemented with clear calcite. Width of section is 2.5 mm. (B) This sculpture of a *Uintatherium,* an Eocene mammal, is one of many carved in the Mississippian Salem Limestone that decorates Wilson Hall at Washington University in St. Louis. The Salem Limestone is extensively quarried in southern Indiana and used as trim and building stone in public buildings nationwide.

from South Dakota and Montana northward into Canada), extensive reefs developed. Arid conditions and restricted circulation in reef-enclosed barred basins resulted in the deposition of great thicknesses of gypsum and salt. Petroleum seeped into these highly permeable reefs, creating some of Canada's richest oil fields.

The Absaroka Sequence

MISSISSIPPIAN + PENNSYLVANIAN = CARBONIFEROUS When the sea finally left the craton at the end of Mississippian time, the exposed terrain eroded to form one of the most widespread regional unconformities in the world. Erosion removed entire systems of older rocks over arches and domes.

The resulting unconformity separates strata into the Mississippian and Pennsylvanian systems, which are equivalent to the Carboniferous of Europe. We use these two names in North America instead of the combined term Carboniferous because of the unconformity and because the rocks above the unconformity differ markedly from those below. The overlying Pennsylvanian strata were deposited under very different conditions.

The Absaroka cratonic sequence includes rocks ranging in age from Pennsylvanian through Triassic. In general, the Pennsylvanian rocks near the eastern highlands are thicker, and virtually all are continental sandstones, shales, and coal beds (Fig. 11-10). This eastern section of Pennsylvanian rocks gradually thins away from the Appalachian belt and changes from predominantly terrestrial to about half marine and half nonmarine rocks. Still farther west, the Pennsylvanian outcrops are largely marine limestones, sandstones, and shales.

CYCLOTHEMS A notable aspect of Pennsylvanian sedimentation in the middle and eastern United States is the **cyclothem,** a cyclic repetition of groups of marine and nonmarine strata. For example, a typical Pennsylvanian cyclothem in Illinois contains ten units, idealized in Figure 11-11A. In an actual example, shown in Figure 11-11B, beds 1 through 5 are continental deposits, the uppermost of which is a coal bed. Strata above the coal represent an advance of the sea over an old vegetated area.

Cyclothems are valuable because they can be correlated over great distances. Those of the Appalachian region, for example, can be correlated with cyclothems in western Kansas. In Missouri and Kansas, geologists recognize at least 50 cyclothems within a section only 750 meters thick, and some of these extend across thousands of square kilometers.

Cyclothems result from repetitive advance and retreat of seas. But what caused these oscillations in the

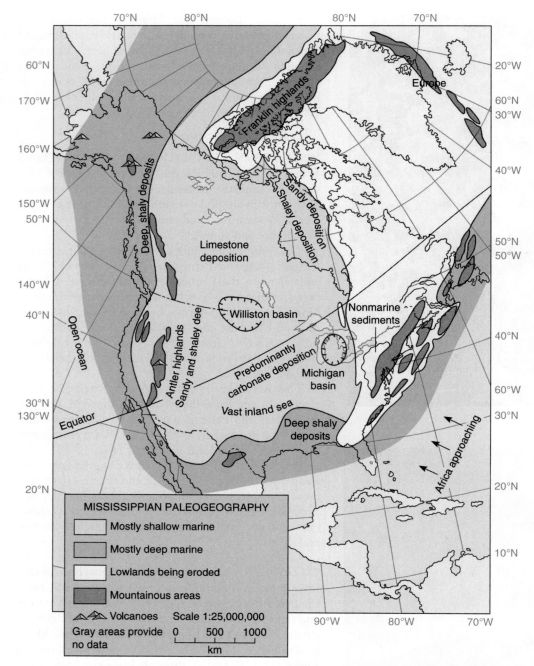

FIGURE 11-8 **Mississippian Period paleogeography.** This is what North America looked like during the 35-million-year interval from 355 to 320 million years ago. Note how most of the United States was covered by water. Was your hometown location under water?

first place? The favored hypothesis is worldwide change in sea level related to glacial and interglacial climatic conditions in Gondwana.

Large regions of Gondwanaland were covered with glacial ice from Late Mississippian through Permian time. When the ice sheet grew in size, sea level lowered because of water removed from the ocean and precipi-

tated as snow on Gondwana. During warmer episodes, meltwater returning to the ocean raised sea level. These sea-level oscillations caused the shorelines to shift back and forth, producing cyclothems. Additional support for this idea comes from the presence of cyclothems on continents other than North America, indicating a global cause.

The column on the left margin shows abbreviated geologic periods: Q, T, K, J, Tr, P, M, D, S, O, €, Pre-€.

FIGURE 11-9 Cross-bedding in sandstones, southern Illinois. Studies of grain size, cross-bedding, and sedimentary structures caused by water currents indicate that the sediment was derived from the northern Appalachian Mountains and transported southwestward across the interior U.S. This image shows the Upper Mississippian Tar Springs Formation.

UNREST IN THE WEST

Colorado's Uncompahgre Mountains

Ordinarily, cratonic areas are stable. But the southwestern part of the North American craton is an exception, for during the Pennsylvanian this was a region of mountain-building.

What caused this mountain-building? Very likely it was the collision of Gondwanaland with North America along the southern margin of Laurentia. This would have generated stress in the area to the north, and crustal adjustments to relieve the stress produced the highlands and associated basins.

The resulting highlands formed the ancient Uncompahgre Mountains in southwestern Colorado and the Oklahoma Mountains of western Oklahoma. (*Uncompahgre* is a Ute Indian word meaning "hot water spring.") These mountains and related uplifts resulted from nearly vertical movements of crustal blocks along large faults.

The Colorado Mountains included a range that extended north-south across central Colorado (the Front Range-Pedernal uplifts) and a segment curving from Colorado into eastern Utah (the Uncompahgre uplift). The central Colorado basin lay between these highland areas (Fig. 11-12). A separate range, the Kaibab-Defiance-Zuni uplift, extended across northeastern Arizona on the southwestern perimeter of the Paradox basin.

To the east of the Colorado Mountains lay the southeastward-trending Oklahoma Mountains. Eroded stumps of this once rugged range form today's greatly reduced Arbuckle and Wichita Mountains. Remains of the Amarillo Mountains are now buried beneath younger rocks and are known from rock samples encountered when drilling for oil.

Judging from the tremendous sediment volume that was eroded from the Uncompahgre Mountains, their height probably exceeded 1000 meters. Also, they probably experienced repeated uplift. Erosion of these highlands eventually exposed their Precambrian igneous and metamorphic cores.

As erosion and weathering continued, sediment spread onto adjacent basins (Fig. 11-13). A small part of this massive accumulation of red clastic sediment is exposed in Colorado—at the Garden of the Gods near Colorado Springs, in the Red Rocks Amphitheatre near Denver, and at the "flatirons" near Boulder (Fig. 11-14).

PARADOX BASIN The basins near the ancient Uncompahgre Mountains are filled with sediments. They provide clues to the geologic history of the southwestern region of the craton. The Paradox basin is especially interesting. It was flooded by the Absaroka Sea in Early Pennsylvanian time. Initially, shales were deposited over Mississippian limestone bedrock that is now riddled with solution features (like caverns and sinkholes).

By Middle Pennsylvanian time, the western access of the Absaroka Sea to the Paradox basin became partially blocked, and thick beds of salt, gypsum, and anhydrite were deposited (Fig. 11-15). Fossiliferous and oölitic limestones developed around the periphery of the basin, and patch reefs grew along the western side. The association of porous reefs (oil "reservoir rocks") and lagoonal deposits (oil "source rocks") was suitable for the later entrapment of petroleum.

Near the end of the Pennsylvanian, the Paradox basin was filled to above sea level by arkosic sediments shed from the recently uplifted Uncompahgre highlands. Also at this time, the Absaroka Sea began a slow and irregular regression. The withdrawal was still incomplete in Early Permian time, and marine sediments continued to be deposited in a narrow zone from Nebraska to western Texas. Fossiliferous limestones characterized these inland seas, although near highlands, in Colorado, Texas, and Oklahoma, coarse clastics accumulated thickly enough to bury surrounding uplands.

EVAPORITES The deposits along what was the eastern edge of the seaway have been eroded away. But at several places along the western side, exposures show a change from richly fossiliferous beds below to barren shales, red beds, and evaporites above. The thick and

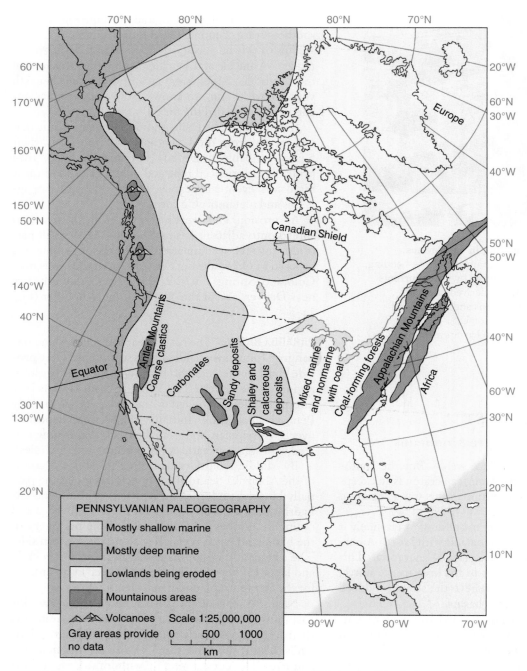

FIGURE 11-10 **Pennsylvanian Period paleogeography.** This is what North America looked like during the 25-million-year interval from 320 to 295 million years ago.

extensive salt beds of Kansas provide testimony to the gradual restriction and evaporation of Permian seas in the central United States.

The Permian Basin of West Texas and New Mexico

The last, but most important, Permian record in North America is found in western Texas and southeastern New Mexico, where 14,000 feet of lagoon, reef, and open-basin sediments were deposited (Figs. 11-16 and 11-17). These rocks are wonderfully exposed in the Guadalupe Mountains of West Texas. For geologists, they provide a splendid natural laboratory for studying facies relationships and the influence of submarine topography on the control of carbonate and evaporite deposition.

In this region, irregularly subsiding basins developed between shallow submerged platforms. Dark-colored

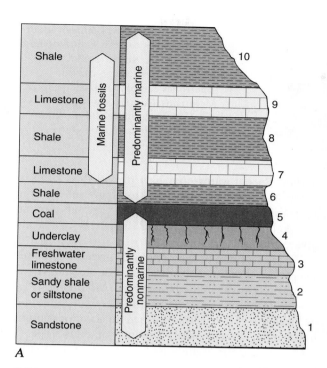

A

B

FIGURE 11-11 **Coal-bearing cyclothems.** (A) Idealized coal-bearing cyclothem, showing a
typical sequence of layers. Many actual cyclothems lack some of the units shown. The units may not
have been deposited because changes from marine to nonmarine conditions may have been abrupt,
or because units may have eroded away following marine regressions. Bed number 8 (shale) usually
represents maximum inundation. If identified elsewhere, it provides an important correlative
stratigraphic horizon. (B) Part of an Illinois cyclothem. At bottom is the coal seam (bed 5), followed
upward by shale (6) near the geologist's hand, limestone (7), shale again (8), another limestone (9),
and the upper shale (10). Part of another sequence caps the exposure. This cyclothem is part of the
coal-rich Carbondale Formation of the southern Illinois coal fields. ◳ *If you came across a limestone
that was part of a cyclothem, how could you tell that it was a marine limestone and not a freshwater limestone?*
◳ *Would rocks deposited above bed 10 be predominantly marine or nonmarine?*

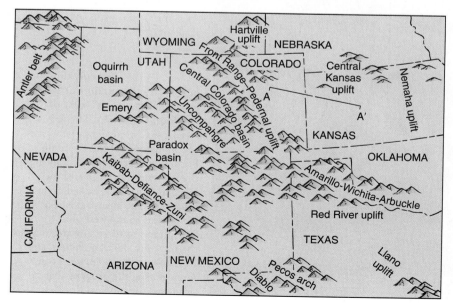

FIGURE 11-12 **Principal highland
areas, southwestern part of the
craton, Pennsylvanian time.**

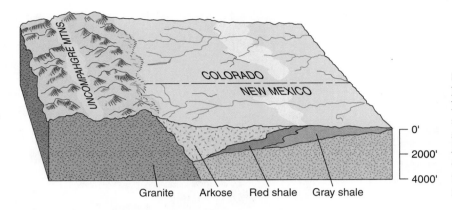

FIGURE 11-13 **The Pennsylvanian Period across eastern Colorado and New Mexico.** Note the great accumulation of coarse arkosic sandstones east of the Uncompahgre Mountains. (Arkosic sandstones contain 25% or more feldspar grains that derived from granitic source rocks.)

limestones, shales, and sandstones were deposited in the deep basins, whereas massive reefs formed along the basin edges. In lagoons behind the reefs, thin limestones, evaporites, and red beds were deposited. Late in the Permian, the connections of these basins to the south became so severely restricted that the waters gradually evaporated, leaving behind great thicknesses of gypsum and salt.

Much paleoenvironmental information has been obtained from study of the West Texas Permian rocks. The lack of medium-grained and coarse-grained clastics indicates that surrounding regions were low-lying. The gypsum and salt suggest a warm, arid climate in which seawater evaporated from basins was periodically replenished with more seawater.

Mapping of the rock units has helped to establish an estimated depth of about 500 meters for the deeper basins and only a few meters for the intrabasinal platforms. The basin deposits were dark in color and rich in organic carbon as a consequence of accumulation under stagnant, oxygen-poor conditions. Upwelling

of these deeper waters may have provided an influx of nutrients on which phytoplankton and reef-forming algae thrived. Along with the algae, the reefs contain the skeletal remains of over 250 species of marine invertebrates.

Today, because of their relatively greater resistance to erosion, these ancient reefs form the steep El Capitan promontory in the Guadalupe Mountains (Fig. 11-18). Here, one can examine the fore-reef composed of broken reef debris that formed an accumulation of loose angular rocks called a submarine talus slope. It was caused by the pounding of waves along the southeastern side of the reef.

▶ TO THE EAST, A CLASH OF CONTINENTS

During the latter half of the Paleozoic Era, the Appalachian and Ouachita belts experienced their final mountain-building. The crumpling of these former depositional tracts into highlands resulted from the

FIGURE 11-14 **Steeply dipping beds of the Fountain Formation form the "flatirons" west of Boulder, Colorado.** The red arkosic sandstones, conglomerates, and mudstones of the Fountain Formation were deposited during the late Pennsylvanian and early Permian Periods. The sediment source was the ancestral Rocky Mountains to the west. The formation was tilted upward to form the flatirons, dramatic ridges, and hogbacks of the Colorado Front range during the orogeny that produced the modern Rockies.

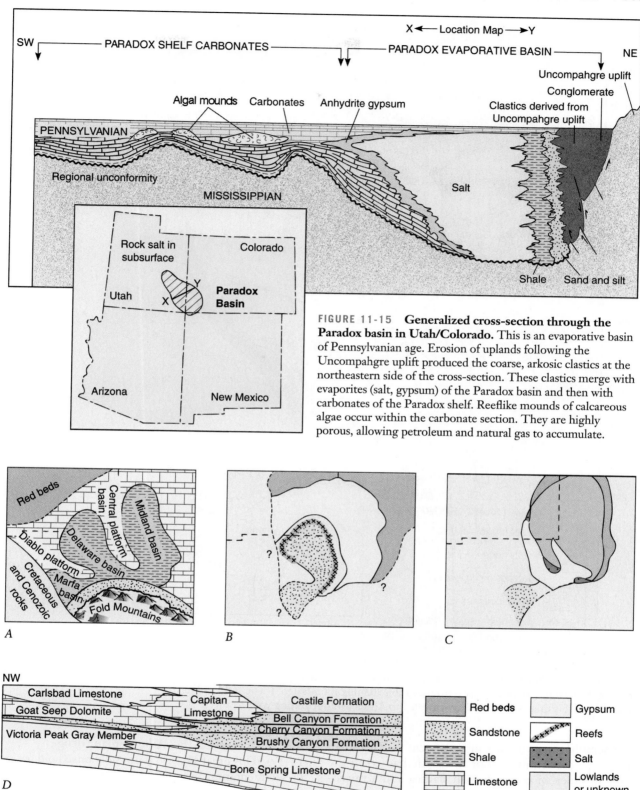

FIGURE 11-15 **Generalized cross-section through the Paradox basin in Utah/Colorado.** This is an evaporative basin of Pennsylvanian age. Erosion of uplands following the Uncompahgre uplift produced the coarse, arkosic clastics at the northeastern side of the cross-section. These clastics merge with evaporites (salt, gypsum) of the Paradox basin and then with carbonates of the Paradox shelf. Reeflike mounds of calcareous algae occur within the carbonate section. They are highly porous, allowing petroleum and natural gas to accumulate.

FIGURE 11-16 **Division of the Permian Period into four stages in North America.** From oldest, these stages are the Wolfcampian, Leonardian, Guadalupian, and Ochoan. The paleogeography and sedimentation in West Texas are shown during the (A) Wolfcampian, (B) Guadalupian, and (C) Ochoan. D is a simplified cross-section of Leonardian and Guadalupian sediments of the Guadalupe Mountains, indicating the relationship of the reef to the other facies.

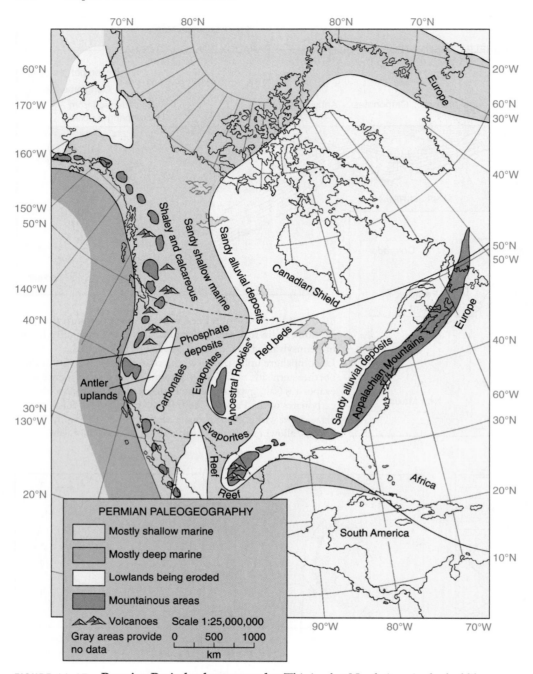

FIGURE 11-17 Permian Period paleogeography. This is what North America looked like during the 45-million-year interval from 295 to 250 million years ago. If you lived in North Carolina at this time, you could have traveled to Africa without getting your feet wet.

reassembly of continents into Pangea. By Devonian time, the eastern margin of Laurentia had taken the double shock of collision with Baltica and a microcontinent named the **Avalon terrane.** Together, these two collisions comprise the **Acadian orogeny** (Fig. 11-19).

Another collision occurred in late Carboniferous time, when northwestern Africa collided with the southern part of the Appalachian belt. That encounter

was the cause of the **Alleghenian orogeny** (Fig. 11-19). This great smash-up not only raised old basins of sedimentation but also transmitted compressional forces into the interior, causing deep-seated deformations such as those that raised the Colorado and Oklahoma Mountains.

The effects of the Acadian orogeny are seen in a belt extending from Newfoundland to West Virginia, where thick, folded sequences of turbidites are inter-

ENRICHMENT

The Wealth of Reefs

Reefs are of great interest to geologists for many reasons. They comprise immense masses of carbonate material generated entirely by marine invertebrates and algae. Reefs often are well-preserved and therefore provide intact representations of complex biologic communities, both ancient and recent, from which a wealth of ecologic information can be obtained. Petroleum geologists, aware that ancient reefs serve as excellent traps for petroleum, carefully map reef locations and trends.

Reefs are shallow-water, wave-resistant buildups of calcium carbonate formed around a rigid framework of skeletal material. For most of the Phanerozoic, that framework was built by corals, although the families of corals that construct reefs have varied through time as various groups evolved or became extinct.

Before there were abundant reef-building corals, calcareous algae and bryozoa were the principal frame-builders. The calcareous skeletons of other reef-dwelling organisms such as crinoids, brachiopods, mollusks, sponges, and algae add mass to the reefs and help to trap sand and lime mud.

In most ancient reefs, such as the spectacular Permian Capitan Reef of western Texas (Figs. 11-16 and 11-18), you can recognize three major reef components or facies:

1. The **reef core** itself, formed of organisms that have built the reef upward from the shallow seafloor to sea level. Waves constantly crash along the front (windward) side of the reef core. In response to this vigorous wave action, organisms living along the reef front form low-growing or encrusting skeletal structures.
2. The **fore-reef facies** resembles a submarine talus deposit. It is a steeply dipping apron of rubble that forms when waves continuously break off pieces of the reef.
3. A more sheltered **back-reef facies** lies landward of the reef core. Here carbonate sands, muds, and oöids accumulate among more erect-growing frame-builders. The back-reef is known for its extraordinary biodiversity.

---

spersed with rhyolitic volcanic rocks and granitic intrusions. The intensity of the compression that affected these rocks is reflected in their metamorphic minerals, which indicate temperatures exceeding 500°C and pressures equivalent to burial under 15 km of rock.

The Acadian orogeny demolished the marine depositional basin along the east side of Laurentia and established mountain ranges in which erosion prevailed. Consequently, in isolated basins among the mountains, Devonian nonmarine sediments were deposited. However, the greatest volume of erosional detritus spread outward from the highlands as a great wedge of terrigenous (continental) sediment, the **Catskill clastic wedge** (Fig. 11-20A and B).

The Catskill Clastic Wedge

Devonian rocks of the Catskill clastic wedge (also called the Catskill delta) have long interested geologists. The Catskill sediments present varied facies of marine and nonmarine depositional environments.

Catskill sediments typically exhibit rapid lateral changes from sandstones to shales. Such relationships form traps for petroleum, which is why thousands of gas wells have been drilled into Catskill strata. Also, flat slabs of red sandstones have provided flagstones for buildings in eastern cities.

Why study the Catskill rocks? Historical geologists seek the clues these rocks provide about the time and location for each phase of the Acadian orogeny. The

rocks reveal that the Acadian orogeny was caused by the convergence of the Avalon terrane, and possibly others, against the irregular eastern margin of the North American craton. Orogenic pulses occurred as the westward-moving Avalon terrane encountered parts of the cratonic margin that projected eastward as promontories.

The first orogenic pulse occurred early in the Middle Devonian, when the Avalon terrane encountered promontories in the vicinity of today's St. Lawrence River valley in Canada. Orogenic pulses followed in succession as the Avalon terrane converged on more southerly promontories of the central and southern Appalachians. Study of the directional properties of the fluvial sandstones indicates deposition from many small streams, all flowing westward out of the Acadian highlands (Fig. 11-21).

Unlike the earlier Queenston rocks described in Chapter 10, the Catskill sediments were laid down when land plants were abundant (Fig. 11-22) and provided a green mantle for the alluvial plains and hills. The vegetation indicates a tropical climate in this part of the Devonian world.

Nonmarine Catskill sedimentary rocks are dominated by sandstones and shales with red iron-oxide (hematite) coloration. The majority of these "red beds" represent deposits of braided or meandering streams (Fig. 11-23).

There are many similarities between the nonmarine rocks of the Catskill clastic wedge and those that spread

FIGURE 11-18 **Facies change at the southern end of the Guadalupe Mountains near El Paso, Texas.** The prominent cliff, called El Capitan, is a gigantic Permian reef. It is rich in fossils of sponges and other marine invertebrates. Evaporites, dolomites, and unfossiliferous limestones indicate abnormally high salinity. Behind the reef were backreef lagoons and in the foreground was a forereef basin in which normal marine sediments were deposited. This area is notable for the detail the rocks provide in reconstructng the depositional environment. ▨ *What kind of sedimentary rock probably underlies the prominent slope beneath El Capitan?*

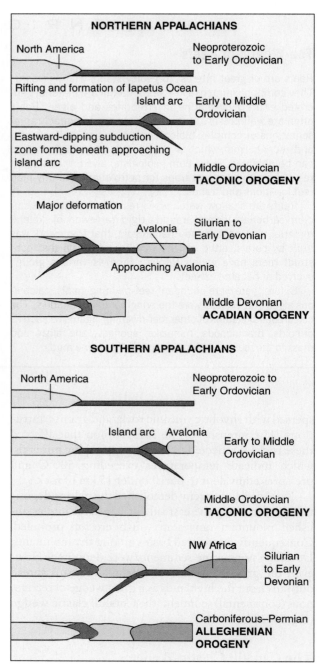

FIGURE 11-19 **Interpretation of plate tectonic events in the evolution of the northern and southern Applachians.**

across Europe south of the Caledonian orogeny. This region of largely nonmarine deposition is named the Old Red Continent after its most famous formation, the Old Red Sandstone (Fig. 11-24).

After the Devonian

Mississippian strata crop out in the Appalachian region from Pennsylvania to Alabama. Nonmarine shales and sandstones predominate, evidence of sediment from erosion of the Acadian Mountains. Some of the finer clastics spread westward onto the craton to form a widespread blanket of Lower Mississippian black shales. Sandstones and conglomerates were deposited as part of the **Pocono Group**.

Pocono sandstones form some of the resistant ridges of the Appalachian Mountains in Maryland (Fig. 11-25). Westward, the Pocono section thins and changes imperceptibly into marine siltstones and shales. Thus, there developed another great clastic wedge of alluvial deposits that sloped westward and merged into deltas along the coast of the epicontinental

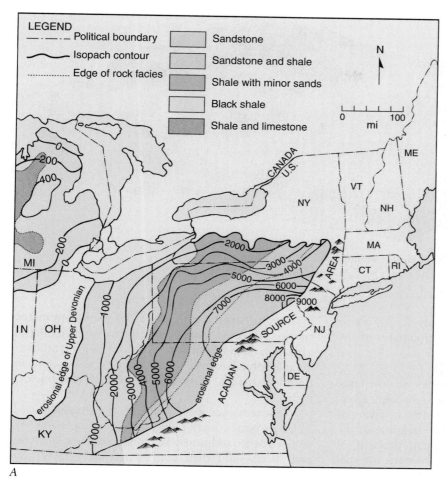

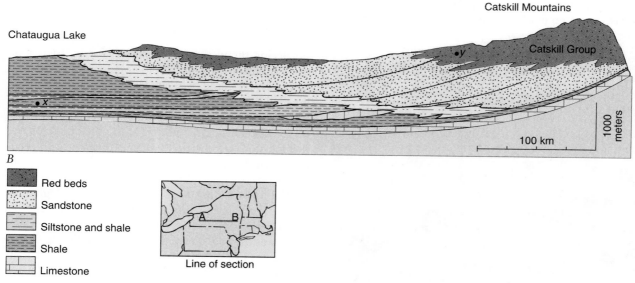

FIGURE 11-20 **Upper Devonian sedimentary rocks in the northeastern U.S.; Catskill clastic wedge in New York.** (A) The Acadian orogeny established mountain ranges where erosion prevailed. The greatest sediment volume spread outward to form the Catskill clastic wedge. (B) East-west section across the Catskill clastic wedge. Note how continental red beds interfinger with nearshore marine sandstones. These in turn grade toward the west into offshore siltstones and shales. Continental deposits prograded upon the sea, pushing the shoreline progressively westward.
Are the shale beds at point **x** *older, younger, or about the same age as the sandstone beds at point* **y**?

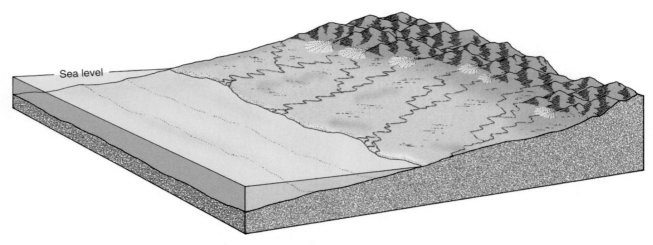

FIGURE 11-21 **Panorama of the Catskill clastic wedge viewed from a point above south-central Pennsylvania.** The shoreline trends northeastward for about 300 km.

sea. The plains and deltas, standing only slightly above sea level, were backed by the rising mountains of the Appalachian fold belt.

COAL COUNTRY Pennsylvanian rocks of the Appalachians are characterized by cross-bedded sandstones and gray shales that were deposited by rivers or within lakes and swamps. Numerous coal seams of the Pennsylvanian System reflect luxuriant growths of mangrove-like forests, an ideal environment for coal formation.

Plants that grew in the poorly drained swampy areas were frequently inundated and died. Immersed

FIGURE 11-22 **Fossil plant root traces in mudstones of the Catskill clastic wedge.** Oneonta Formation near Unadilla, New York. Catskill sediments were deposited when land plants were abundant, indicating a tropical climate in this part of the Devonian world.

in water or covered with muck, the dead plant material was protected from being destroyed by oxidation to carbon dioxide and water. It was, however, attacked by anaerobic bacteria. These organisms broke down the plant tissues, extracted the oxygen, and released hydrogen.

What remained was a peat-like, fibrous sludge with a high carbon content. Later, such peat-like layers were covered with additional sediment—usually siltstones and shales—and then compressed and slowly converted to coal.

The Alleghenian Orogeny

Bordering the interior platform of North America on the east from Newfoundland to Georgia is the Appalachian orogenic belt. (Its southwestward extension is called the Ouachita orogenic belt.) The Appalachian belt records three major orogenic events, plus many minor disturbances. The first of the major events was the Taconic orogeny that occurred during the Ordovician period. The second was the Acadian orogeny of the Devonian.

The culminating third event was the Alleghenian orogeny, which began in the Mississippian and continued throughout the remainder of the Paleozoic. Its cause was the collision of the African and South American regions of Gondwana with the underside of Laurentia (Fig. 11-26). The Alleghenian orogeny affected a belt extending across 1600 km, from southern New York to central Alabama. Study of the rocks deformed during these events has been the basis for classic theories of mountain-building and for the concept of the Appalachian belt as a collage of microcontinents and other terranes accreted to the eastern edge of North America by plate convergence.

A
Pre-Є | Є | O | S | D | M | P | Pr | Tr | J | K | PgNg
B

FIGURE 11-23 **Sedimentary rocks of the Catskill clastic wedge.** (A) Catskill Formation sandstone filling an abandoned stream channel (scale divisions are 10 cm). (B) Tidal flat deposits of the Catskill clastic wedge near Altoona, Pennsylvania. The red sandstones, siltstones, and shales were carried westward by streams with headwaters in the Acadian Mountains.

The Appalachian depositional basin can be divided into two belts:

- A western belt of shallow-water shales, limestones, and sandstones and an eastern belt that is largely deep-water graywackes, volcanics, and siliceous shales. Today, these rocks underlie the Valley and Ridge Province and the Appalachian Plateau (Fig. 11-27).

- An eastern belt that has been highly metamorphosed and intruded by granite plutons. Today, these rocks underlie the Blue Ridge and Piedmont Provinces.

FIGURE 11-24 **Cross-bedded Devonian Old Red Sandstone in Scotland.** The Caledonian Orogeny created a land area on which red beds of the Old Red Sandstone were deposited, and also produced a mountainous source area for the sediment. Height of image is 60 cm.

All three orogenic episodes that molded the Appalachian provinces occurred during the Paleozoic. As described in the previous chapter, the first was the Taconic orogeny, which occurred during the Ordovician Period. The second was the Acadian orogeny of the Devonian Period. The culminating third event, the Alleghenian orogeny, occurred from Mississippian to Permian time as Gondwana (Africa and South America) converged on North America and Europe.

The effects of the Alleghenian orogeny were profound. They included the compression of the early continental shelf and rise sediments, as well as sediments that had been deposited along the bordering tract of the craton. The great folds, visible in the Valley and Ridge Province, were developed during this orogeny (Fig. 11-25).

Less visible at the surface, but no less impressive, are thrust faults formed along the east side of the southern Appalachians. Many of these folds are asymmetrically overturned toward the northwest, and the surfaces of the thrust faults are inclined southeastward. This suggests that the entire region was pushed forcibly across the edge of the central craton (Fig. 11-27B).

This kind of deformation, in which basement rocks are largely unaffected and the overlying "skin" of weaker sedimentary rocks breaks into multiple thrust faults, is known as **thin-skinned tectonics**.

Erosion of the mountains produced during the Alleghenian event produced another great blanket of nonmarine sediments. These mostly continental red sandstones and gritty shales compose the **Dunkard Group** (Fig. 11-28) and **Monongahela Group** of Pennsylvanian-Permian age.

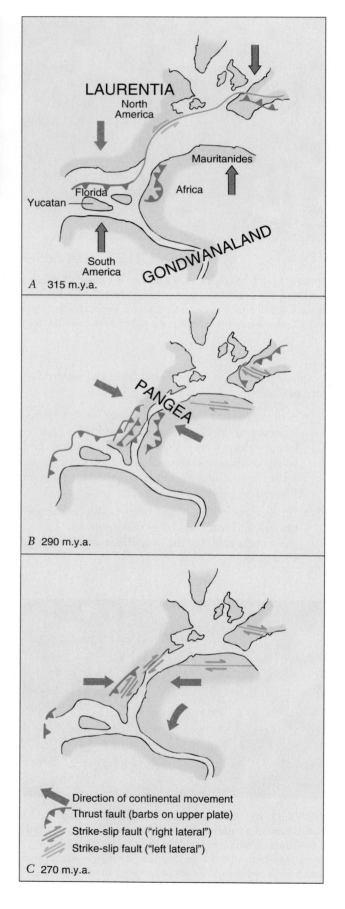

FIGURE 11-25 **Lower Mississippian Purslane and Rockwell formations (Pocono Group) in a road cut through Sideling Hill, western Maryland.** The section includes thick, cross-bedded fluvial sandstones as well as a few dark, coaly shales. ☘ *What kind of geologic structure is seen here?*

The Ouachita Deformation

The Ouachita orogenic belt is a continuation to the west of the Alleghenian orogenic belt (Fig. 11-26). Ouachita deformation (Fig. 11-29) began rather late in the Paleozoic, for the rock record indicates that the region experienced slow deposition interrupted only by minor disturbances from Early Devonian to Late Mississippian. Carbonates predominated in the more northerly shelf zone, whereas cherty rocks known as novaculites accumulated in the deeper marine areas.

Novaculites are hard, even-textured, siliceous rocks composed of microcrystalline quartz. They are formed from bedded cherts that have been subjected to heat and pressure. Arkansas novaculite is used as a whetstone in sharpening steel tools (Fig. 11-29B and C).

Deposition rates increased near the end of the Mississippian, when more than 8000 meters of graywackes and shales spread into the depositional basin. This clastic flood continued into the Pennsylvanian, forming a great debris wedge that thickened and coarsened toward the south, where growing mountain ranges were rapidly eroding (Fig. 11-30).

Radiometric dating of now deeply buried basement rocks from the Gulf Coast states indicates that these rocks were metamorphosed during the late Paleozoic and were the source for Pennsylvanian clastics. The coarse sediments document pulses of mountain-building

FIGURE 11-26 **Plate tectonic model for late Paleozoic continental collisions.** Views show the relationships of North America, South America, and Africa during (A) Early Pennsylvanian, (B) Late Pennsylvanian, and (C) Permian. The Alleghenian orogeny began in the Mississippian and continued through the remainder of the Paleozoic. It was caused by the African and South American regions of Gondwana colliding with the underside of Laurentia.

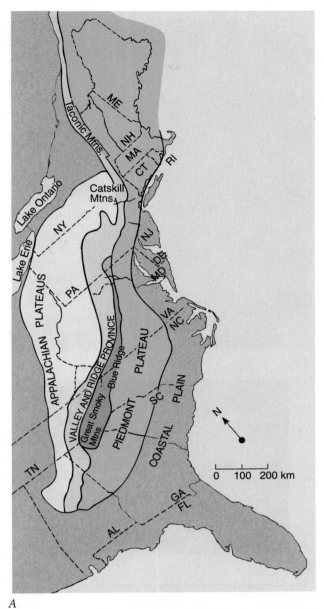

A

FIGURE 11-27 **Eastern U.S. physiographic provinces.**
(A) Each province is characterized by distinctive surface expressions: Piedmont rocks are mainly slates, schists, gneisses, and intrusive igneous rocks that form a region of low relief. Strongly deformed and metamophosed Precambrian and lower Cambrian metamorphosed rocks underlie the Blue Ridge, which widens to the south as the Smokey Mountains. As suggested by its name, the Valley and Ridge Province is composed of long ridges of erosionally resistant sandstones and valleys cut by streams into less resistant shale and limestone. Rocks of the Appalachian Plateau are mostly broadly folded late Paleozoic sedimentary strata. (B) Cross-section A-B is based on deep seismic-reflection profiles.

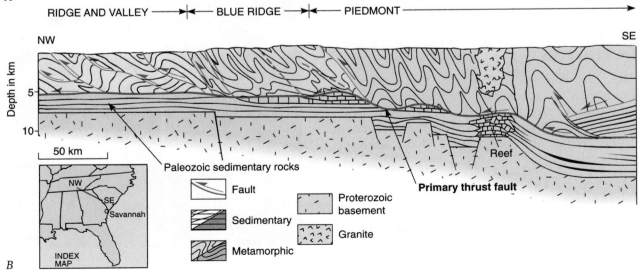

B

FIGURE 11-28 **Waynesburg Sandstone, Dunkard Group, along the Ohio River in Meigs County, Ohio.** Rocks of the Dunkard Series are the only Permian sediments preserved in the Appalachian region.

that produced mountains along the entire southern border of North America.

Since Permian time, erosion has leveled the once lofty mountains. The only remnants are the Ouachita Mountains of Arkansas and Oklahoma and the Marathon Mountains of southwest Texas. Although the Ouachita belt has been traced over a distance of nearly 2000 km, only about 400 km are exposed. The actual trend of the belt has been determined by well samples and subsurface data obtained during petroleum drilling activities.

▶ **SEDIMENTATION AND OROGENY IN THE WEST**

The late Paleozoic history of the western or Cordilleran belt was almost as lively as that of the Appalachian. During the early Paleozoic, a passive margin existed along the western side of the North American craton. Continental shelf, slope, and rise sediments

A

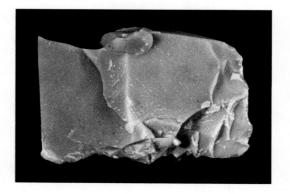

B

C

FIGURE 11-29 **The Ouachita orogenic episode.** (A) Late Paleozoic strata in the Ouachita Mountains of Arkansas showing the effects of compressional deformation from the orogeny. (B) Hand specimen of Arkansas novaculite. (C) Close-up of deformed novaculite; the prominent bed is 8 cm thick. ◾ *What is novaculite used for?*

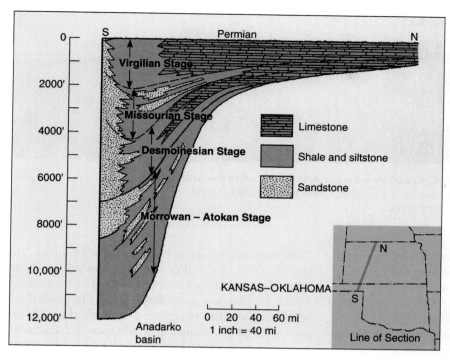

FIGURE 11-30 **Geologic section of Pennsylvanian rocks across Kansas and Oklahoma.** The thick wedge of sediment was shed from mountain ranges. ◀ *What is the predominant Pennsylvanian rock in the Oklahoma panhandle? What type of sedimentary rock predominates in northern Kansas? Why?*

were deposited along that quiet passive margin. More dynamic conditions began in the Devonian, when subduction of oceanic lithosphere beneath the western margin of the continent began. This was the beginning of a disturbance known as the **Antler orogeny.**

During the Antler orogeny, a volcanic island arc converged on the western margin of North America, crushing sediment in the intervening basin. The convergence was accompanied by massive thrust-faulting. Its effects are visible today in the Roberts Mountains thrust fault of Nevada (Fig. 11-31). Continental rise and slope deposits have been thrust as much as 80 km over shallow-water sediments of the former continental shelf.

The Antler orogeny began in the Late Devonian and continued actively into the Mississippian and Pennsylvanian. Erosion of the Antler Mountains provided detrital sediment that was transported into adjacent basins (Fig. 11-32). Thick sequences of Pennsylvanian and Permian shelf sediments accumulated in the area now occupied by the Wasatch and Oquirrh Mountains (Fig. 11-33).

Mississippian and Pennsylvanian deposits west of the Antler highlands include volumes of coarse detritus and volcanic rocks. These materials were swept eastward from a volcanic arc that lay along the western side of North America. More than 2000 meters of sandstones, shales, lavas, and ash beds are found in the Klamath Mountains of northern California (Fig. 11-34). Volcanic rocks in western Idaho and British Columbia indicate continuous volcanism from the Mississippian through the Permian.

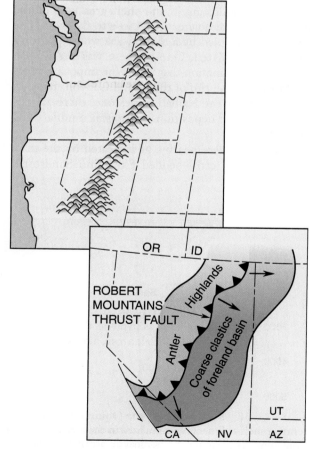

FIGURE 11-31 **Highland areas associated with the Antler orogeny; Roberts Mountains thrust fault in Nevada.**

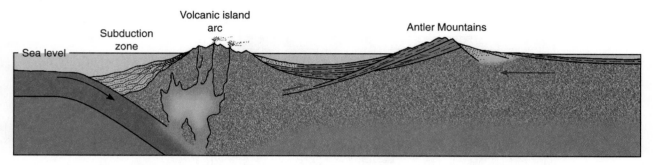

FIGURE 11-32 **Interpretation of the Cordilleran orogenic belt shortly after the Antler orogeny (Early Mississippian time).** Erosion of the Antler Mountains provided detrital sediment that was transported into adjacent basins.

The Permian-Triassic disturbances of the Cordilleran region have been named the **Cassiar orogeny** in British Columbia and the **Sonoma orogeny** in the southwestern United States. Like the earlier Antler Orogeny, the Sonoma event was caused by the collision of an eastward-moving island arc against the North American continental margin. Oceanic rocks and remnants of the arc were forced into the edge of the continent and became part of North America.

Permian conditions in the shelf area east of the Antler uplift were quieter than those to the west. The region was occupied by a shallow sea where platform deposits accumulated. One of these was the Kaibab Limestone, a formation that forms the imposing vertical cliffs along the rim of the Grand Canyon. Beneath and eastward of the Kaibab Limestone, there are red beds that reflect deposition on coastal mudflats and floodplains.

Sand dunes were nearby, as indicated by the massive, extensively cross-bedded Coconino Sandstone.

Driven by northeasterly paleowinds, sand dunes also spread across the region that now lies north of the Grand Canyon of the Colorado River. The dunes, now frozen in time, are part of the DeChelly Sandstone, which displays the sweeping cross-beds (Fig. 11-35) and pitted quartz grains ("frosted grains") characteristic of eolian (wind-transported) deposits.

While the Kaibab was being deposited, a relatively deep marine basin developed to the north in the region now occupied by Wyoming, Montana, and Idaho. In this basin, sediments of the Phosphoria Formation were deposited. Although the Phosphoria includes beds of cherts, sandstones, and mudstones, it takes its name from its many layers of dark phosphatic shales, phosphatic limestones, and phosphorites (Fig. 11-36).

This unusual phosphate concentration may have resulted from upwelling of phosphorus-rich seawater from deep parts of the basin. Metabolic activities of microorganisms may have assisted in precipitation of the phosphate salts.

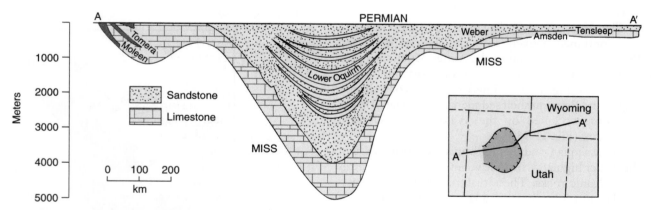

FIGURE 11-33 **Section across the Oquirrh basin.** Thick sequences of Pennsylvanian and Permian shelf sediments accumulated in the area now occupied by the Wasatch and Oquirrh Mountains. The Oquirrh basin received about 5000 meters of sediment eroded from the Uncompahgre Mountains, just during Pennsylvanian time. The basin must have subsided rapidly to accommodate this huge thickness of sediment. ◪ *What mountain-building episode provided a highland source area for this great accumulation of sandstone?*

FIGURE 11-34 **The Mississippian Bragdon Formation, eastern Klamath Mountains of California.** The Bragdon is composed of siltstones, sandstones, and felsic ash shed from a volcanic island arc that formed next to North America during late Paleozoic time. ◗ *What is the approximate angle of dip in these beds?*

EUROPE DURING THE LATE PROTEROZOIC

South of the Caledonian orogenic belt lay a region of dry land that during the Devonian received sands and clays eroded from the growing Caledonan Mountains. Today this great accumulation of sediment would be called the "Caledonian clastic wedge," but early British geologists dubbed it the Old Red Continent because sediments of the clastic wedge are rust-red in color.

Fossil remains of plants and fresh-water fish are abundant in these sediments. They indicate that Europe's climate was tropical and sometimes semiarid during the Devonian. The literary geologist Hugh Miller (1802–1856) was fascinated by these fossils when he was a poor laborer in a quarry near Cromarty, Scotland. His book, *The Old Red Sandstone and Footprints of the Creator*, was a best-seller in nineteenth-century Britain.

During the late Paleozoic, Europe was bordered on the east by a sea called the Uralian Sea. Closure of this sea at the end of the Paleozoic produced the Ural Mountains. To the south of the Old Red Continent, another ocean existed called the Hercynian. Like the Uralian Sea, the Hercynian was transformed into mountain ranges during the assembly of Pangea. The

FIGURE 11-35 **Permian cross-bedded DeChelly Sandstone, Canyon de Chelly National Monument, northeastern Arizona.** The dunes display sweeping cross-beds and pitted quartz grains ("frosted grains") characteristic of wind-borne deposits.

late Paleozoic orogenic event has been named the Hercynian orogeny.

The Hercynian orogeny occurred at approximately the same time as the Alleghenian orogeny in North America. For the most part, the eroded stumps of the Hercynian mountains are now buried, but here and there patches of the younger covering rocks have been eroded away, revealing the intensely folded, faulted, and intruded older rocks that lie beneath.

From the Hercynian uplands, gravel, sand, and mud were carried down into basins and coastal environments. The clastic deposits were quickly clothed in dense tropical forests. Burial and slow alteration of vegetative debris from these forests provided the material for coal formation in the great European coal basins. Plant fossils found in these coal seams are tropical; they differ from the more temperate Gondwana flora of the Southern Hemisphere.

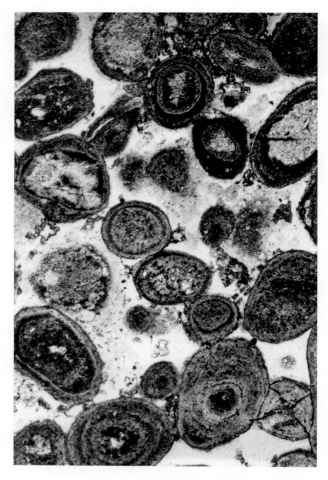

FIGURE 11-36 **Phosphorite.** Composed of pellets and oöids with microcrystalline cement, phosphorite is mined from the Phosphoria Formation in Montana for the manufacture of fertilizers and other chemical products. Oöid average diameter is 0.8 mm. ◖ *As world populations increase, why do phosphate deposits become increasingly important?*

FIGURE 11-37 **Upper Carboniferous tillite, evidence of great continental glaciers at Kimberley, South Africa.** The tillite, at the base of the slope beneath the hammer, is composed of Precambrian basalt. ◖ *How does tillite differ from consolidated stratified drift?*

GONDWANA DURING THE LATE PALEOZOIC

During the late Paleozoic, the great landmass of Gondwana remained fairly intact. As you refer to Figure 11-1, note that Gondwana moved over the South Pole and more fully entered the side of Earth where Laurasia lay. In its northward migration, Gondwana closed the ocean that separated it from Laurasia, causing the Hercynian orogeny of Europe and the Alleghenian orogeny of North America.

Orogenic activity associated with subduction zones also was evident in the late Paleozoic history of Gondwana, particularly along the Andean belt of South America and the Tasman belt of eastern Australia.

The most dramatic paleoclimatologic event of Gondwana's late Paleozoic history was the growth of

vast continental glaciers. The scour marks of glaciers (Fig. 11-3) and extensive layers of tillite (Fig. 11-37) exist at hundreds of locations in South America, South Africa, Antarctica, and India. These indicate at least four glacial advances, suggesting a pattern of cyclic glaciation not unlike that experienced by North America and Europe less than 100,000 years ago.

The orientation of striations chiseled into bedrock by the moving ice suggests that the glaciers moved northward from centers of accumulation in southwestern Africa and eastern Antarctica. During the warmer interglacial stages and in outlying less frigid areas, *Glossopteris* and other plants tolerant of the cool, damp climates grew in profusion and provided the materials for thick seams of coal.

In time the ice receded, and Permian nonmarine red beds and shales were deposited on the Gondwana craton. Some of these sediments contain the fossil remains of the ancestors of the Earth's first mammals.

CLIMATES OF THE LATE PALEOZOIC

The main climatic zones of the late Paleozoic paralleled latitudinal lines, just as they do today. Of course, the continents were located very differently. The South Pole was in South Africa and the North Pole was in the open ocean. The paleoequator extended northeastward across Canada and eastward across Europe (Fig. 11-38). Within 30° of the paleoequator, coal beds of tropical plants, evaporite deposits, coral reefs, and desert red

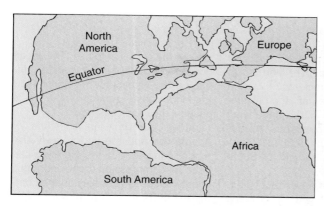

FIGURE 11-38 **Approximate relationship of continents to the equator during the Carboniferous (Mississippian and Pennsylvanian Periods).**

A

B

FIGURE 11-39 **Mining coal underground and at the surface.** (A) A remote-controlled mining machine in a West Virginia underground mine. This "continuous mining" machine can work almost nonstop, clawing out the coal and passing it through the machine to a conveyor, which carries the coal out of the mine for cleaning and shipping to power plants. (B) Overburden is removed to reveal the coal in an Illinois surface mine.

beds developed. Warmth-loving amphibians and reptiles lived in forests of dense ferns and scale trees.

Such balmy conditions were not characteristic of the Gondwana continents. South Africa's position over the South Pole certainly was the primary contributor to the cold that gripped southern continents near the end of the Paleozoic.

Today, carbon dioxide keeps our planet warm by trapping heat radiated from Earth's surface. This is the phenomenon we call the *greenhouse effect*. The more carbon dioxide, the warmer the atmosphere. The less carbon dioxide, the cooler the atmosphere.

If cooling of the atmosphere during the Carboniferous was caused by a decrease in carbon dioxide, what could have caused that decrease? The Carboniferous is notable for prodigious amounts of plant matter that became buried to form huge volumes of coal. We know that the conversion of dead vegetation to coal requires *burial*, for if organic matter is left exposed to air, carbon in that material will be combined with atmospheric oxygen and removed as carbon dioxide. Burial of organic matter thus prevents addition of carbon dioxide to the atmosphere. With less carbon dioxide, perhaps the greenhouse effect was reduced sufficiently to contribute to the late Paleozoic Gondwana ice age.

MINERAL PRODUCTS OF THE LATE PALEOZOIC

Although a great variety of mineral deposits formed during the late Paleozoic, particularly significant are the fossil fuels (coal, oil, and gas).

Coal

Coal occurs in all post-Devonian systems. In Northern Hemisphere continents, it is particularly characteristic

of the Late Carboniferous (Pennsylvanian). Thick deposits of Pennsylvanian-age coal occur in the Appalachians, the Illinois basin (Fig. 11-39), and the industrial heartland of Europe. Some sequences of Pennsylvanian strata include multiple coal beds, as in West Virginia, where 117 different coal layers have been named.

In the anthracite ("hard coal") district of eastern Pennsylvania, orogenic compression has partially metamorphosed coal into an exceptionally high-carbon, low-emission variety that is prized for industrial use. Permian coal seams are found in China, Russia, India, South Africa, Antarctica, and Australia.

Petroleum

Commercial quantities of oil and gas frequently are found in Upper Paleozoic strata. Devonian reefs

Acadia National Park

Maine's Acadia National Park is a delightful place to observe geologic features and processes. Thanks to the work of Pleistocene glaciers in removing soil and debris, you can walk directly on bedrock in many parts of the park. Along the shoreline, granite, diorite, and schist are revealed in wave-eroded cliffs and platforms.

At Acadia, you can witness the effects of igneous activity, explore glacial moraines, examine exotic rocks on cobble beaches, and observe powerful wave action sculpting a rugged coast. All of this is easily reached by traveling southeast from Bangor through Ellsworth to Bar Harbor (Fig. A).

Ice sheets covered Acadia and the rest of New England repeatedly during the Pleistocene Epoch. Each advance of the ice destroyed features formed by the previous advance, leaving features of the final glaciation clearly in evidence. The most apparent are glacial striations, glacial polish, and scars made in bedrock as rock debris carried along in the base of the glacier chipped out pieces of the underlying surface. There are numerous *erratics* (Fig. B), including one called "Mammoth Rock" on the summit of South Bubble Mountain.

Valleys in Acadia have the characteristic U-shape gouged by glaciers. This distinctive shape results from the ability of glacial ice to erode the valley walls, unlike streams, which erode only the bottom and create a characteristic V-shape. As the glaciers moved down valleys, some areas were excavated deeper than others. Later, the deeper excavations filled with water to form lakes (such as Sea Cove Pond, Eagle Lake, and Jordan Pond).

When the great Ice Age came to a close, meltwater returning to the ocean caused the sea level to rise. The ocean flooded into valleys and low areas, producing the remarkably irregular New England coast. Appropriately, it is called a drowned coastline.

Along the drowned coastline of Acadia you find the park's most dramatic scenic attractions. Here you can get a vivid impression of the enormous energy released by waves as they batter sea cliffs and erupt in spray (Fig. C). Thunder Hole, south of Bar Harbor, provides a spectacular view of wave erosion in action. Thunder Hole is a deep gorge eroded along joints by pounding waves. When seas are heavy, water surges into the narrow chasm, compressing the trapped air, which on release, resounds like the clap of thunder. Surging waves along the coast lift and carry gravel which is hurled against cliffs, undercutting them and causing collapse. These processes ultimately produce sea caves, sea arches, and wave-cut platforms. In quieter areas, the waves deposit their load, forming cobble and shingle beaches.

A particularly intriguing aspect of Acadia's geology relates to the derivation of its most ancient rocks. Acadia is a part of an exotic or alien terrane, other components of which have been traced from Nova Scotia down into the southern Appalachians. As noted in Chapter 7, alien terranes are pieces of continental crust or island arc moved by seafloor spreading from a distant

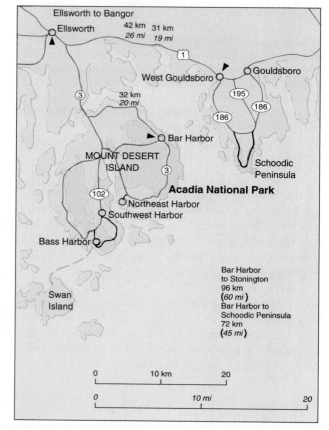

FIGURE A   Location map of Acadia National Park.

FIGURE B   **Glacial erratic boulder on Cadillac Mountain, Acadia National Park.** An erratic is a rock fragment that has been carried by glacial ice and deposited at some distance from the outcrop from which it was derived. Usually, erratics rest on rock of different lithology and age.

FIGURE C **Granite exposed along the rugged coast of Mount Desert Island, Acadia National Park.** The impact of waves exerts pressure of as much as 6000 pounds per square foot, readily fragmenting and eroding huge masses of rock.

FIGURE D **Basalt dike cutting across lighter-colored Cadillac Granite.** This is on Acadia's Schoodic Peninsula, the only part of the park on the mainland.

place of origin to the margin of a continent, and then sutured to the host continent at its subduction zone. They can be recognized as alien terranes because they differ from their host continent in rock type, age, fossils, and paleomagnetic orientation.

The alien terrane that includes Acadia National Park has been named **Avalonia**. It contains early Paleozoic rocks with trilobite and graptolite faunas distinctly different from those of adjacent areas of North America. In fact, these faunas more closely resemble those of Europe (Baltica).

When an alien terrane arrives at a continent, it is said to have docked (rather like ships that arrive at a port one by one). Alien terranes older than Avalonia had already docked against North America during the Ordovician Taconic orogeny. Avalonia came along somewhat later during the Devonian.

Rocks of the terrane indicate it is a vestige of an island arc that arrived at the subduction zone during the Acadian orogeny. After Avalonia docked, at least three additional terranes were incorporated into the Appalachian orogen.

The Ordovician and Silurian rocks of Acadia are mostly metamorphosed sediments and volcanic ash formed long before Avalonia had moved to North America. From oldest to youngest, these rocks consist of the Ellsworth Schist, Cranberry Series, and Bar Harbor Series. An erosional unconformity marks the upper boundary of these three units.

Intruded into these units (and hence younger) are rocks of Devonian age. The earliest intrusives are diorites that form dikes and massive sills. Some of the dioritic magma penetrated to the surface and erupted as fiery lava flows. After most of the dioritic magma had been emplaced and had crystallized, granitic magmas penetrated the diorites and meta-sediments. The initial fine-grained granites gave way to coarser-grained, and finally medium-grained, granites, reflecting variations in the cooling history of the intruded bodies.

After these granitic rocks had been emplaced, basaltic lavas rose into fracture systems. Their dark color and sharp contacts make them clearly visible against the lighter granites. The best area in which to view the basalt dikes is in the part of the park called the Schoodic Peninsula (Fig. D). All of this igneous activity was accompanied by compressional deformation of the eastern margin of North America. Ultimately, however, compression gave way to tensional forces as the break-up of Pangea began following the Permian Period. Erosion and intermittent uplifts were the principal post-Paleozoic events until the coming of the glaciers during the Pleistocene.

FIGURE 11-40 **Drilling for oil in upper Paleozoic rocks of eastern Kansas.** *What purpose does the derrick serve when drilling for oil?*

within the Williston basin of Alberta and Montana have been exceptionally productive reservoir rocks for petroleum. Devonian petroleum also has been produced in the Appalachians. In fact, in 1859, the first U.S. oil well was drilled into a Devonian sandstone in Pennsylvania—and oil was struck at a depth of only 20 meters!

Mississippian and Pennsylvanian formations in the Rocky Mountains, the midcontinent (Fig. 11-40), and the Appalachians also contain oil reservoirs. But the greatest amount of oil has come from wells drilled into Upper Paleozoic reefs and sandstones of the Permian basin of West Texas. Elsewhere, oil trapped in Upper Paleozoic strata beneath the North Sea is being produced for use in Europe.

Other Economic Minerals

Arid, warm climatic conditions, particularly in northern continents during the late Paleozoic, provided a suitable environment for deposition of sodium and potassium salts. In Late Permian time, enormous amounts of phosphates also were deposited as part of the Phosphoria Formation, exposed in Montana, Idaho, Utah, and Wyoming. The Phosphoria shales are extensively quarried for phosphate as an important plant food (fertilizer).

Mountain-building, with its attendant igneous activity, nearly always emplaces metallic ores. The Hercynian and Alleghenian orogenies of the late Paleozoic generated ores of tin, copper, silver, gold, zinc, lead, and platinum. Deposits of all the precious metals, as well as copper, zinc, and lead, are found in the Urals of Russia and in China, Japan, Burma, and Malaysia. Tin, tungsten, bismuth, and gold are mined in Australia and New Zealand.

It is obvious that late Paleozoic rock sequences, with their stores of both metallic and nonmetallic economic minerals and their content of fossil fuels, are vitally important to the welfare of modern civilizations.

SUMMARY

- The assembling of continents that began in the early Paleozoic continued throughout the remainder of the era. As plates converged, they carried island arcs and microcontinents to the margins of Europe and eastern North America, and these land areas became welded to their larger host continents.
- The collisions produced the Taconic, Caledonian, Acadian, and Alleghenian orogenies. The Alleghenian orogeny occurred near the end of the era as northwestern Africa moved against North America. To the south, the plate bearing South America converged on the underside of North America and caused the Ouachita orogeny.
- While the process of bringing continents into contact resulted in mountain-building along converging margins, the stable interiors were affected primarily by broader, gentler changes, causing advances and withdrawals of epicontinental seas. In North America, these shifts in shorelines are reflected in the marine and nonmarine sediments that compose the Kaskaskia and Absaroka cratonic sequences.

- Kaskaskia rocks are predominantly fossiliferous limestones, with sandstones and shales increasing in volume toward the eastern and southern depositional areas. As the Kaskaskia Seas withdrew, deltaic and fluvial deposits spread across the old seafloor. The regression resulted in a great regional unconformity that marks the boundary between Mississippian and Pennsylvanian systems in North America.
- The most distinctive feature of Pennsylvanian sediments is their cyclic nature (cyclothems). In the midcontinent, they consist of alternate marine and nonmarine groupings. To the east, the Pennsylvanian sediments were largely terrestrial, whereas marine deposition prevailed in the western part of the craton. Near the end of the Paleozoic, epicontinental seas regressed. Evaporite and red bed sequences were deposited in the Permian basins of New Mexico and West Texas.
- During the Devonian a subduction zone developed along the craton's formerly passive western margin. Stresses re-

sulting from the converging oceanic plate caused disturbances that raised mountains from Arizona to Idaho.
- The Old Red Continent, which extended across Ireland, Wales, England, Scotland, and Scandinavia, was the dominant feature of Europe at the beginning of the Devonian. For a time, relative stability prevailed. The quiet period

ended as the continent collided with Gondwana, causing the Hercynian orogeny.
- During the late Paleozoic, Gondwana developed an immense ice sheet as it moved over the South Pole. Tillites and glacial striations resulting from this major episode of glaciation are found on all Southern Hemisphere continents.

KEY TERMS

Absaroka cratonic sequence, p. 298
Acadian orogeny, p. 308
Alleghenian orogeny, pp. 308, 312–313
Antler orogeny, p. 317
Avalon terrane, p. 308
Caledonian orogeny, p. 297
Cassiar orogeny, p. 318

Catskill clastic wedge, p. 309
Chattanooga Shale, p. 298
Cyclothem, p. 301
Dunkard Group, p. 313
heavy minerals, p. 298
Hercynian orogeny, p. 319
Kaskaskia cratonic sequence, p. 298

Laurasia, p. 297
Monongahela Group, p. 313
novaculite, p. 314
Pangea, p. 297
Pocono Group, p. 310
Sonoma orogeny, p. 318
thin-skinned tectonics, p. 313

QUESTIONS FOR REVIEW AND DISCUSSION

1. Describe the relation between the orogenies that produced the Appalachian and Ouachita Mountains and the movement and position of tectonic plates.

2. In terms of plate tectonics, what is the reason for the approximate time equivalence between the Alleghenian and Hercynian orogenies?

3. In the study of an ancient mountain range, how might a geologist recognize a displaced terrane? Having recognized a terrane as a displaced (alien or exotic), how might he or she determine that the terrane was from an island arc? from a microcontinent?

4. Why is it logical to divide the late Paleozoic into two cratonic sequences? What are the names and durations of those sequences?

5. When did the Old Red Continent develop and where was it located? What kind of sediment was being deposited across the Old Red Continent?

6. How do the deposits of the Old Red Continent compare with those of the Catskill clastic wedge?

7. What is the geologic evidence for the occurrence of a geographically extensive episode of continental glaciation in the Southern Hemisphere during the late Paleozoic?

8. What is the paleoenvironmental significance of an association of red sandstone, salt, and gypsum deposits? Where and when during the late Paleozoic did deposition of this association of rocks occur?

9. What is a cyclothem? Prepare a list and discuss each of the eustatic and tectonic conditions that might account for the cyclicity for which cyclothems are named.

10. Describe the environmental conditions under which the Chattanooga Shale was deposited. What problems are associated with hypotheses for the origin of this far-ranging blanket of dark shales?

11. Approximately where was the South Pole located during the Permian Period? How have geologists ascertained that position? What confirmation for the pole position can be obtained from the global distribution of corals?

12. Name three metallic and three nonmetallic mineral resources extracted from Upper Paleozoic rocks.

13. Explain the occurrence of oil and gas in the calcareous algal mounds found in the shelf carbonates of the Paradox basin.

WEB SITES

The Earth Through Time Student Companion Web Site (www.wiley.com/college/levin) has online resources to help you expand your understanding of the topics in this chapter. Visit the Web Site to access the following:

1. Illustrated course notes covering key concepts in each chapter;

2. Online quizzes that provide immediate feedback;

3. Links to chapter-specific topics on the web;

4. Science news updates relating to recent developments in Historical Geology;

5. Web inquiry activities for further exploration;

6. A glossary of terms;

7. A Student Union with links to topics such as study skills, writing and grammar, and citing electronic information.

12

Q
T
K
J
Tr
Pr
P
M
D
S
O
€
Pre-€

***Reconstruction of some arthropods preserved in the
Cambrian Burgess Shale of British Columbia.*** *The
organisms provide a view of marine life 530 million years
ago.*

Life of the Paleozoic

When you were a tadpole and I was a fish
In the Paleozoic Time,
And side by side on the ebbing tide
We sprawled through the ooze and slime,
Or skittered with many a caudal flip
Through the depths of the Cambrian fen,
My heart was rife with the joy of life,
For I loved you even then.
—Langdon Smith, "Evolution"

Key Chapter Concepts

- Because of the proliferation of animals with shells, the fossil record improves dramatically at the beginning of the Paleozoic Era.
- Vertebrates (including fishes, amphibians, reptiles, and mammal-like reptiles) evolved during the Paleozoic Era.
- Representatives of most major invertebrate phyla were present during the Paleozoic, including protozoans, sponges, corals, bryozoans, brachiopods, mollusks, arthropods, and echinoderms.
- The evolution of land animals (tetrapods) from fishes depended on the evolution of the amniotic egg.
- From primitive land plants that appear near the end of the Ordovician, vascular plants expanded across the continents and formed great Devonian forests.
- Plant evolution during the Paleozoic progressed from seedless, spore-bearing plants to plants with seeds but lacking flowers (gymnosperms). Flowering plants (angiosperms) did not appear until the Cretaceous Period of the Mesozoic Era.
- Of Earth's four major episodes of mass extinctions, three occurred during the Paleozoic Era. The cause of the extinctions may have been terrestrial (climate and atmospheric changes) or extraterrestrial (asteroid impact).

The pace of evolution appears to have gone into double time at the start of the Phanerozoic Eon, 540 million years ago. The rich fossil record of the past half-billion years contrasts startlingly to the skimpy record of the Precambrian. Life during most of the Precambrian consisted of single-celled microbes, with metazoans appearing only in uppermost Proterozoic strata, as in the Ediacara Hills in Australia. Except for a few tiny tubular creatures, Precambrian animals had not acquired the ability to create their own shells.

Hard shells were the key improvement that produced the superior fossil record of the Phanerozoic. During the Cambrian, shell-bearing trilobites and brachiopods were particularly abundant. An even greater expansion of shelly and coralline animals followed in the early Ordovician. Epicontinental seas that spread across the cratons of the world provided a multitude of habitats and opportunities for diversification and expansion of invertebrates. Almost every phylum of common shell-bearing invertebrates living today had originated by Ordovician time, 488 million

OUTLINE

▶ ANIMALS WITH SHELLS PROLIFERATE— AND SO DOES PRESERVATION

▶ THE CAMBRIAN EXPLOSION OF LIFE: AMAZING FOSSIL SITES IN CANADA AND CHINA

▶ CONTINUING DIVERSIFICATION: EACH CREATURE FOUND ITS ECOLOGICAL NICHE

▶ PROTISTANS: CREATURES OF A SINGLE CELL

▶ MARINE INVERTEBRATES POPULATE THE SEAS

▶ ADVENT OF THE VERTEBRATES

▶ THE RISE OF FISHES

▶ CONODONTS: VALUABLE BUT ENIGMATIC FOSSILS

▶ ADVENT OF TETRAPODS

▶ PLANTS OF THE PALEOZOIC

▶ MASS EXTINCTIONS

▶ SUMMARY

▶ KEY TERMS

▶ QUESTIONS FOR REVIEW AND DISCUSSION

▶ WEB SITES

▶ BOX 12–1 ENRICHMENT

TABLE 12-1 Summary of Invertebrate Phyla with a Good Fossil Record

PHYLUM	BRIEF DESCRIPTION	EXAMPLES
SARCODINA	Single-celled eukaryotes with pseudopodia, including foraminifera and radiolana	
PORIFERA	Simple, multicellular animals forming colonies and with bodies preforated by many pores. The sponges.	
ARCHAEOCYATHA	Extinct, double-walled, vase- or cup-shaped animals with pores in walls.	
CNIDARIA	Radially symmetrical animals with stinging cells, including corals, jellyfish, and sea anemones	
BRYOZOA	Tiny, colonial animals with U-shaped row of tentacles, often building branching colonies.	
BRACHHIOPODA	Marine invertebrates with shell composed of two parts (valves), one dorsal and the other ventral.	
ARTHROPODA	Animals with jointed appendages, segmented body, and armor-like exoskeleton.	
MOLLUSCA	Unsegmented, mostly shell-bearing invertebrates, including bivalves (clams, oysters) snails, chambered nautilus, and octopods.	
ECHINODERMATA	Spiny-skinned invertebrates with radially symmetrical adult bodies and water vascular system. Starfishes, sea urchins, crinoids.	

years ago (Table 12-1). Shells probably evolved for protection and to support soft tissue organs.

Plant evolution also progressed as the transition from water to land was accomplished. Plants having woody tissue capable of conducting fluids (vascular plants) spread across the continents, ultimately producing densely forested regions.

Animals with backbones—vertebrates—appeared during the Paleozoic. The earliest vertebrates were jawless fishes now known from rocks in China as old as Cambrian. Subsequently, fishes with articulated lower jaws developed. This triggered a Devonian burst of evolution that produced the major groups of fishes, many of which are still with us today. Also in the Devonian, an advanced lineage of fishes with stout fins and primitive lungs gave rise to four-legged animals called **tetrapods**. Paleontologists call these earliest of four-legged animals **basal tetrapods**. Although capable of walking on land, the first tetrapods were linked to swamps, lakes, and streams by eggs that survived only in water. Thus, we call them amphibians.

Another evolutionary innovation, the **amniotic egg**, freed tetrapods from their reproductive dependency on water bodies. The amniotic egg has adaptations to retain water, and provide for gas exchange, support, and protection. Vertebrates with such an egg became totally independent of the ocean, lakes, and streams for their reproduction. They were the early *amniotes*. By Permian time, amniotes called **therapsids** possessed some morphologic features of mammals. However, the arrival of true mammals would be a Mesozoic event.

Life during the Paleozoic had its calamities. There were intervals when the global environment became inhospitable or when catastrophic events occurred. During such times, entire families of animals were exterminated. We call such biological disasters **mass extinctions**. During the Paleozoic Era, mass extinctions occurred late in the Ordovician, Devonian, and Permian periods.

ANIMALS WITH SHELLS PROLIFERATE—AND SO DOES PRESERVATION

Multicellular animals already were present before the Paleozoic Era, as indicated by the Neoproterozoic Ediacaran fauna. Fossil discoveries in Namibia (Africa) show that Ediacaran animals, although characteristic of the late Precambrian, ranged into the Cambrian. However, Ediacaran creatures were soft-bodied, so they were infrequently preserved. When animals began to develop hard parts, their probability of preservation improved immensely. The first animals to achieve this milestone are called **small shelly fossils** (Fig. 12-1).

Although usually found in strata at the base of the Cambrian System, small shelly fossils date back to the late Neoproterozoic (Fig. 12-2). They were widely

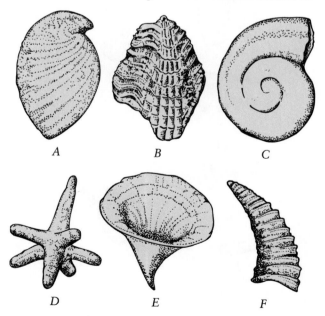

FIGURE 12-1 **Tiny shell-bearing fossils from the Late Precambrian and Early Cambrian in Siberia.**
(A) *Anabarella*, a gastropod; (B) *Camenella*, affinity uncertain; (C) *Aldanella*, a gastropod; (D) sponge spicule, (E) *Fomitchella*, affinity uncertain; and (F) *Lapworthella*.

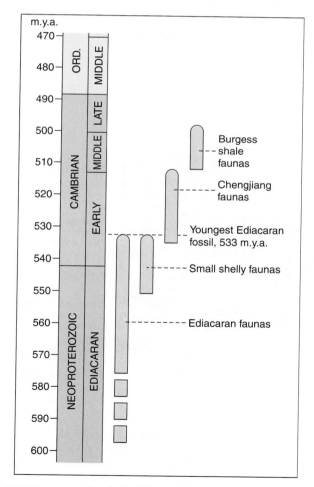

FIGURE 12-2 **Geologic time scale across the Proterozoic–Cambrian boundary.** It shows the positions of the Ediacaran, Chengjiang, and Burgess Shale faunas.

FIGURE 12-3 *Anabarites*, **an early Cambrian shelled fossil.** The shell is only 5 mm long, smaller than an M&M.

distributed around the globe. The tiny fossils rarely exceed a few millimeters in size and include shells and skeletal elements of mollusks and sponges. The fossils also include small invertebrates of uncertain classification that secreted tubular or cap-shaped shells:

- *Cloudina* (Fig. 9-30), an organism that lived from Neoproterozoic time into the Cambrian. It secreted a ringed calcareous tube.

- *Anabarites* (Fig. 12-3) secreted a phosphate shell that resembles three "tubes" open to one another along the length of the shell. Each tube is surmounted by a narrow keel that may have given the animal stability in soft seafloor mud.

- The coiled shell of *Aldanella* was secreted by a gastropod (snail).

- Another form, *Lapworthella* (Fig. 12-1), has the shape of a curved cone ornamented with grooves and ridges. It probably was not the single shell of an animal, but one of many similar elements that covered the body. This interpretation is supported by the discovery of Lapworthella elements fused together in a side-by-side arrangement.

Most of the small shelly fauna had disappeared before the end of the Early Cambrian. Immediately thereafter, large-shelled invertebrates became abundant and diverse. Familiar fossils such as trilobites and brachiopods (Table 12-1) dominate the fossil record, but there were soft-bodied creatures as well. Some are known from extraordinary preservation at only a few localities, such as the Burgess Shale in British Columbia and Chengjiang in China.

► THE CAMBRIAN EXPLOSION OF LIFE: AMAZING FOSSIL SITES IN CANADA AND CHINA

Two remarkable fossil localities—in British Columbia and China—provide a panoramic view of life at the dawn of the Paleozoic. They reveal Cambrian seas teeming with a splendid diversity of animals already in an advanced stage of evolution (see the chapter-opening

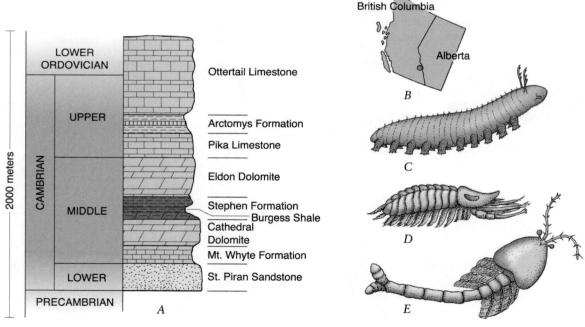

FIGURE 12-4 **The Burgess Shale Fauna.** (A) Cambrian geologic column at (B) Kicking Horse Pass, British Columbia. Here Charles D. Walcott discovered the fauna in 1909.
(C) *Aysheaia*, an invertebrate onycophoran (velvet worm), is of particular interest because it appears to be intermediate in evolution between segmented worms and arthropods.
(D) *Leanchoila* and (E) *Waptia* are among the many kinds of arthropods found at this locality. Length ranges from about 1 to 3 cm.

FIGURE 12-5 *Haplophrentis*. This photograph illustrates the nature of preservation of Burgess Shale fossils. *Haplophrentis* had a tapering shell surmounted by a lid (operculum). The lid could be closed for protection. The lateral blades on either side may have served as props. Shell length is 2 cm, or about 3/4 of an inch.

illustration). The abrupt appearance of so many varied animals about 535 million years ago, and the speed with which they radiated, is called the *Cambrian explosion*. During that explosion—also dubbed "evolution's big bang" in the popular press—all principal invertebrate phyla evolved, except the Bryozoa.

The Burgess Shale Fauna

High on a ridge near Mount Wapta, British Columbia, is an exposure of Middle Cambrian **Burgess Shale**. It contains one of the most important faunas in the fossil record (Fig. 12-4). The fossils are shiny, black impressions on the shale bedding planes (Fig. 12-5). Many are the remains of animals that lacked hard parts. Altogether, they form an extraordinary assemblage (Fig. 12-6). The assemblage includes four major groups of arthropods (trilobites, crustaceans, and members of the taxonomic groups that include scorpions and insects) as well as sponges, onycophorans (Fig. 12-4C), crinoids, mollusks, three phyla of worms, corals, chordates, and many species that defy placement in any known phylum.

Fossils like those in the Burgess Shale were unknown until 1909, when they were discovered by Charles D. Walcott (1850–1927). The initial discovery consisted of fossil-bearing slabs of rock that had been carried

FIGURE 12-6 **Burgess Shale diorama at the U.S. National Museum of Natural History.** This reconstruction, based on actual fossil remains, depicts a Middle Cambrian benthic marine community. Slumping along the steep submarine escarpment in the background contributed to preservation of the Burgess Shale fauna by burying the organisms. Along the wall, you can see green and pink vertical growths of two types of algae. The large purplish creatures that resemble stacks of tires are sponges (*Vauxia*). The blue-colored animals are trilobites (*Olenoides*). The brown arthropods with distinct lateral eyes are named *Sidneyia*. Climbing out of the hollow on the seafloor are crustaceans called *Canadaspis*. The yellow animals swimming toward the right above the seafloor are *Waptia*. *Opabinia* has crawled out of the left side of the hollow. Burrowing worms are visible in the vertical cut at the bottom.

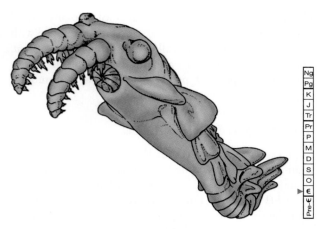

Ng
Pg
K
J
Tr
Pr
P
M
D
S
O
€
Pre-€

FIGURE 12-7 *Anomalocaris,* **"invertebrate equivalent of the dinosaurs."** Thus described by paleontologist Desmond Collins, this fierce creature was the largest predator of its time (often 60 cm, or 2 feet in length). On its head were two stalked eyes and a pair of feeding appendages that captured prey and conveyed victims back to the circular mouth, ringed with sharp outer and inner circles of teeth that match the pattern of wounds on fossils of other community members. The side flaps were used in swimming, like underwater wings.

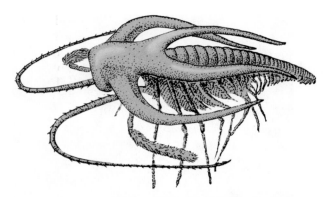

FIGURE 12-9 *Marrella,* **the most elegant and common arthropod in the Burgess Shale fauna.** It is readily recognized by the four spines extending backward from its head. Walcott nicknamed these animals "lace crabs." Length is about 2 cm.

downslope by a snow slide. They were found at the end of the field season, when rock exposures higher up the slope were already blanketed with snow. As a result, Walcott and his crew of fossil collectors (including Mrs. Walcott and two sons) were forced to leave the site and return the following year. They then traced the fossiliferous

slabs up the mountain to their source in the Burgess Shale. Quarrying was begun immediately and continued during the field seasons of 1912, 1913, and 1917.

Altogether, about 60,000 specimens were collected and stored in the U.S. National Museum of Natural History, where Walcott served as secretary of the Smithsonian Institution. In the 1960s, these fossils were re-examined by British paleontologist Harry B. Whittington, who then joined paleontologists from the Geological Survey of Canada in reopening the quarry and assembling a second collection. Whittington and

FIGURE 12-8 *Opabinia,* **the strangest predator of its day.** This Burgess Shale animal had five eyes and a flexible nozzle equipped with nippers used to capture victims. Each side was covered with overlapping lobes with narrow bands, interpreted to be gills. Length is 7 cm (about finger-length).

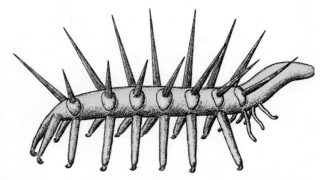

FIGURE 12-10 **The early Cambrian Burgess Shale fossil** *Hallucigenia.* *Hallucigenia* was misinterpreted from scant original information in its compressed and carbonized remains. It was thought to walk on stilt-like leg pairs, like some sea urchins walk on movable spines, with a single row of "tentacles" extending along its back. But then, Swedish paleontologist Lars Ramskold flaked away some shale covering a specimen and observed a claw at the tip of a "tentacle." Claws normally occur on feet, not on tentacles. So, *Hallucigenia* was flipped over. "Tentacles" turned out to be legs, and spines turned out to be protective dorsal spines. *Hallucigenia* is an onycophoran, sharing many characteristics of annelids and arthropods, so it is considered intermediate in evolution between the two groups.

his colleagues devoted the next 15 years to study of the Burgess Shale fauna.

The Burgess Shale fauna awakened paleontologists to the remarkable diversity of life at the beginning of the Paleozoic. That diversity includes such incomparable creatures as *Anomalocaris* (Fig. 12-7), *Opabinia* (Fig. 12-8), *Marrella* (Fig. 12-9), and *Hallucigenia* (Fig. 12-10).

Predators like *Anomalocaris* and *Opabinia* would have caused selective pressures in prey. The need to avoid being eaten probably encouraged the evolution of hard protective shells. It would also cause an increase in the diversity of prey animals, as they evolved into new species better able to survive attacks from predators.

Among the unique creatures of the Burgess Shale fauna were small, elongate animals interpreted as chordates. **Chordates** are animals that, at some stage in their development, have a **notochord** (an internal supportive rod) and a nerve cord that extends along the dorsal (upper) side of the notochord. Chordates like ourselves, in which the notochord is replaced with a series of vertebrae, are called **vertebrates**. The early

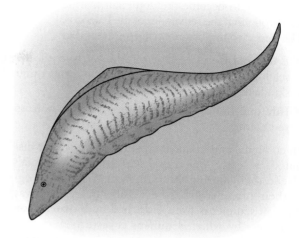

FIGURE 12-12 **The world's oldest fish, the Early Cambrian *Myllokunmingia*.** It was discovered in 1999 in richly fossiliferous beds of the Maotianshan Shale near the town of Chengjiang in China.

chordate found in the Burgess Shale, *Pikaia*, is shown in Figure 12-11.

The Chengjiang Fauna

For a time, *Pikaia* was the oldest known chordate. But in 1984, another extraordinary Cambrian fossil site was discovered. It is in the Yunnan Province of China (which borders Vietnam and Laos), near the town of **Chengjiang**. In general, the fauna resembles that of the Burgess Shale, but the Chengjiang fossils are older and even better preserved. More than 100 species have been recovered. The fossils exhibit extraordinary preservation, and include many soft-bodied creatures that are usually not preserved. For example, jellyfish show the detailed structure of tentacles, radial canals, and muscles. Even on soft-bodied worms, eyes, segmentation, digestive organs, and patterns on the outer skin are readily recognized.

Two close relatives of *Pikaia*, named *Cathaymyrus* and *Yunnanozoon*, were found in the Chengjiang strata. They are 535 million years old. The Chengjiang fossils also include the world's oldest known fish (Fig. 12-12).

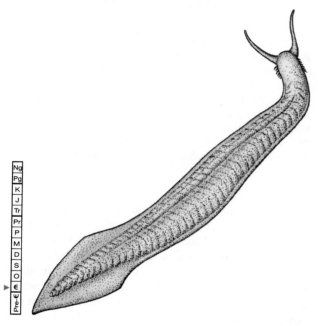

FIGURE 12-11 **Reconstruction of *Pikaia*, the earliest known member of our own phylum, Chordata.** *Pikaia* fossils exhibit two distinctly chordate features: a notochord (the rod along the animal's back) and a series of V-shaped muscles along the sides of the body, as is characteristic of fish musculature. These muscles, working against the flexible notochord, provide the sinuous, fishlike body motion required for swimming. *Pikaia* is named for Mount Pika near the Burgess Shale discovery site. Length is about 4 cm. ◪ *Name another chordate feature seen in Pikaia fossils.(Answers to questions appearing within figure legends can be found at the back of the text.)*

◤ CONTINUING DIVERSIFICATION: EACH CREATURE FOUND ITS ECOLOGICAL NICHE

The Cambrian biologic radiation was followed in Ordovician time (500–435 million years ago) by a *tripling* of global biodiversity. This included trilobites, brachiopods, bivalve mollusks, gastropods, and coralline animals. Overall, the families of known marine organisms increased from about 160 to 530 during the Ordovician Period. The new groups spread widely to become the dominant marine invertebrates throughout the remainder of the Paleozoic.

Here are two important terms to remember:

- **Epifaunal** animals live on the surface of the seafloor (epi = *outer*; as in epidermis, your outer skin layer; fauna = *animals*).
- **Infaunal** animals burrow into the soft bottom sediment.

Both epifaunal and infaunal creatures were widely distributed in early Paleozoic seas. Infaunal animals evolved rapidly during the Cambrian. You can recognize their presence by trace fossils, and in the way Cambrian sediments were disturbed by **bioturbation** (the process by which sediment is disturbed and reworked by organisms). In the preceding Neoproterozoic, fine layers of sediment were usually undisturbed. A covering of cyanobacterial mats probably protected large tracts of the ocean floor. Extensive bioturbation began in the Cambrian, and marked the evolution of infaunal burrowers. Some refer to the dramatic change in the fabric of seafloor sediments as the "Cambrian substrate revolution."

All animals must eat, and each early Paleozoic creature found its niche, adapting to an environment and extracting nutrients from it. There were borers that drilled into hard substrates and burrowers that tunneled through soft sediment. There were stationary attached life forms. There were mobile crawlers, swimmers, and floaters.

Some were **filter feeders** that strained tiny bits of organic matter or microorganisms from the water. Others were **sediment feeders**, which passed the mud of the seafloor through their digestive tracts to extract nutrients, as earthworms do today. Some animals grazed on algae that covered parts of the ocean bottom. And some carnivores ate the grazers. A host of scavengers processed organic debris and thereby aided in keeping the seas clean and suitable for life.

▶ PROTISTANS: CREATURES OF A SINGLE CELL

The principal groups of Paleozoic unicellular animals that have a significant fossil record are foraminifera and radiolaria.

Foraminifera

Foraminifera first appear in the Cambrian and survive to the present. "Forams" build their tiny shells (called *tests*) by adding chambers singly, in rows, coils, or spirals. Chambers connect to one another by a series of openings, or *foramina*, from which the animals derive their name. Some species construct their tests from tiny silt particles. Others secrete tests made of calcium

A

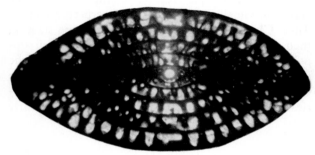

B

FIGURE 12-13 **Fusulinids.** (A) Fusulinid limestone of Permian age from the Sierra Madre of West Texas. (B) Fusulinids are identified according to the characteristics of their complex internal structure that is seen in thin sections. Because they evolved rapidly and are so widespread and abundant, fusulinids are among the most important invertebrate guide fossils of the Pennsylvanian and Permian Periods. ◼ *To what phylum of invertebrates do these organisms belong? During what geologic period did fusulinids become extinct?*

carbonate. Early Paleozoic foraminifera were mostly simple, saclike, tubular, or loosely coiled bottom species with tests built from sediment. They became more numerous and varied by Carboniferous time.

The increase in diversity is striking among **fusulinids**, shell-bearing foraminifera that resemble wheat grains (Fig. 12-13). Although they look alike on the outside, fusulinids have complex internal features that are different for groups living at different stages of geologic time. Fusulinids had global distribution during the Pennsylvanian and Permian. They were sometimes so abundant that they constituted a high percentage of the bulk volume of entire strata. Because they evolved rapidly and are so widespread and abundant, fusulinids are among the most important invertebrate guide fossils of the Pennsylvanian and Permian Periods.

In modern oceans, the tests of planktonic (floating) calcareous foraminifera rain continuously on the seafloor like a limy snowfall. The accumulated debris

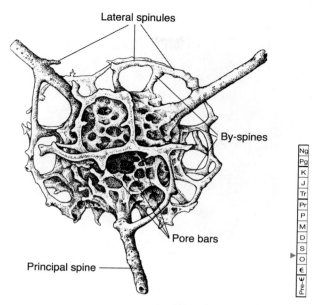

FIGURE 12-14 **Late Ordovician radiolarian, Hanson Creek Formation, Nevada.** Because of the fragile and delicate structure of the radiolarian skeleton, it is unusual to find well-preserved radiolarians in rocks of the early Paleozoic.

forms a deep-sea sediment called *globigerina ooze* (*Globigerina* and related species are prevalent).

Radiolaria

Radiolarians are single-celled planktonic (floating) organisms that have existed at least since the beginning of the Paleozoic Era. Their ornate, lattice-like skeletons are composed of opal-like silica. In some ocean regions today, radiolarian skeletons accumulate to form deposits of siliceous ooze. Although they do occur in Lower Paleozoic rocks (Fig. 12-14), radiolarians are rare and not yet useful for Paleozoic stratigraphic correlation. They are more abundant in Mesozoic and Cenozoic rocks and are widely used for correlation.

MARINE INVERTEBRATES POPULATE THE SEAS

Shell-bearing animals that lived in shallow seas are by far the most abundantly preserved fossils of the Paleozoic. There were cup-like fossils of archaeocyathids, sponges, corals, and colonies of bryozoan "moss animals." Trilobites spread far and wide during the Cambrian, but by Ordovician time they were outnumbered by brachiopods. Clams, snails, and cephalopods populated most epicontinental sea environments. Among the "spiny-skinned" echinoderms, crinoids were especially abundant. Some Mississippian limestones are composed largely of the skeletal parts of crinoids.

Not all Paleozoic invertebrates were bottom dwellers. Planktonic graptolites were carried by currents around the ocean, providing index fossils for global stratigraphic correlation.

Cup Animals: Archaeocyathids

Archaeocyatha (Greek: *ancient cups*) constructed conical, cuplike or vase-shaped skeletons from calcium carbonate (Fig. 12-15). Their features suggest sponges and corals but are not distinctive enough to warrant placement in either group. **Archaeocyathids** grew on the shallow seafloor, often among stromatolite algal mats.

Although abundant during the Early Cambrian, archaeocyathids declined and became extinct by the end of the Middle Cambrian. They are exceptionally useful in the stratigraphic correlation of Lower and Middle Cambrian strata.

Sponges: Phylum Porifera

Among the many animals to colonize the early Paleozoic seafloor were the sponges. They belong to the phylum **Porifera** (meaning *porous*—water continually flows through them). Sponges evolved from single-celled animals that grew in colonies Thus, they offer insight into how the transition may have occurred from unicellular to multicellular animals.

Sponges have a long history. Cambrian fossils include all but one of today's classes of Porifera. Fossil

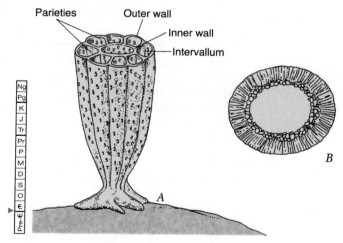

FIGURE 12-15 **Archaeocyathid skeleton.**
(A) Longitudinally fluted cup of an archaeocyathid, about 6 cm tall. The parieties are vertical plates like the septa in corals, and the space between two plates is called the intervallum. (B) Slice of an archaeocyathid with closely spaced parieties. Where archaeocyathids grew together in clusters, they formed reefs. Some Australian archaeocyathid reefs are over 60 meters thick and extend horizontally in narrow bands for over 200 km. ◪ *During what period of the Paleozoic were archaeocyathids particularly prevalent? In what kind of environment did they flourish?*

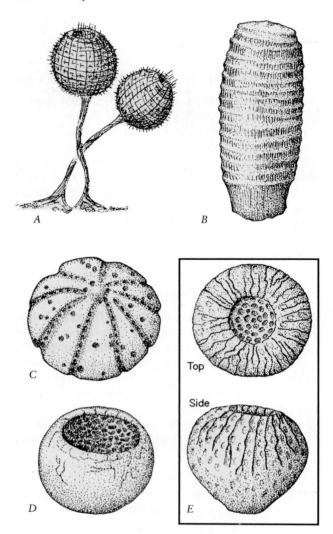

FIGURE 12-16 **Early Paleozoic sponges.**
(A) *Protospongea* (Cambrian), (B) *Nevadocoelia* (Ordovician),
(C) *Caryospongea* (Silurian), (D) *Palaeomanon* (Silurian),
(E) top and side views of *Astylospongea* (Silurian).

FIGURE 12-17 **Silurian sponge *Astraeospongea*.** This
fossil sponge takes its name from its star-shaped, six rayed
spicules, some of which are visible in the photograph.

FIGURE 12-18 **Devonian siliceous sponge *Hydnoceras*.**
The specimen is 15 cm tall. 🔲 *Describe the pattern made by the
spicules on the surface of the fossil.*

sponges like those shown in Figures 12-16, 12-17, and
12-18 are used as markers in stratigraphic correlation.
Fortunately, many sponges are preserved as fossils because their skeletons are durable. They are composed
of tiny, variously shaped skeletal elements called
spicules—composed of calcium carbonate, silica, or
spongin (an elastic flexible material, seen in natural
bath sponges).

Sponges vary greatly in size and shape, but their
basic structure is a highly perforated vase modified by
folds and canals (Fig. 12-19). Sponges lack true organs.
Water currents moving through the sponge are created
by the beat of *flagella*. These currents bring in suspended food particles, which are then digested.

A group of early Paleozoic mound-building sponges
are the **stromatoporoids** (Fig. 12-20). Stromato-

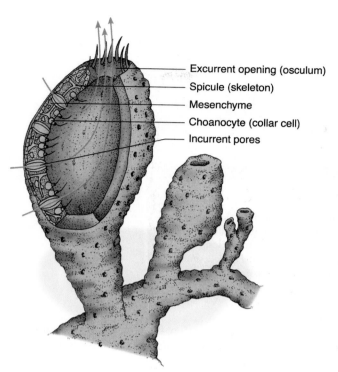

FIGURE 12-19 **Schematic diagram of a sponge with the simplest type of canal system.** A sponge's basic structure is a highly perforated vase modified by folds and canals. The body is attached to the seafloor at the base. Arrows show the path of water currents.

poroids constructed calcareous colonies consisting of pillars and thin laminations. Especially during the Ordovician and Silurian, they grew profusely in association with corals, brachiopods, and other reef-dwellers.

Corals and Other Cnidaria

The phylum **Cnidaria** is noted for its great diversity and beauty (Fig. 12-21). Cnidaria include sea anemones, sea fans, jellyfish, the tiny *Hydra*, and reef-forming corals.

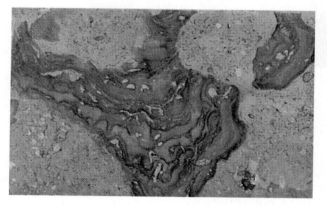

FIGURE 12-20 **Stromatoporoid from Devonian rocks in Ohio.** This polished limestone slab contains *Stromatoporella*.

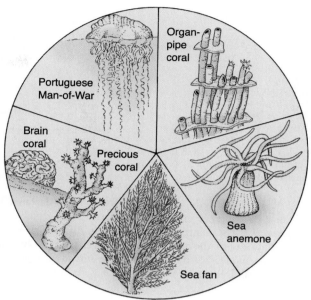

FIGURE 12-21 **Common cnidarians.** Pronounced *nid-AIRY-ans*, this remarkably diverse group features stinging cells. They were formerly called Coelenterata.

Their outer body wall (endoderm) contains primitive sensory cells, gland cells that secrete digestive enzymes, flagellated cells, and cells to absorb nutrients. A unique characteristic of all Cnidaria is stinging cells (**cnidocytes**). When activated, they can inject a paralyzing poison. Body form in Cnidaria may be either a *medusa* or a *polyp* (Fig. 12-22).

Corals are classified according to the nature of their septa, tabulae, and other skeletal characteristics. *Septa* are vertical plates inside the cup, or *theca*, of the coral. The septa distinguish two major groups of corals:

- In the *rugose* or *tetracorals*, septa are added to the inside of the theca in multiples of four.

- In the *scleractinid* or *hexacorals*, septa are added to the inside of the theca in multiples of six.

In the Paleozoic corals called *tabulates*, septa are absent or only poorly developed. Instead, horizontal plates called *tabulae* are their most important feature. Tabulate and rugose corals lived only during the Paleozoic. Scleractinids are the important corals of the Mesozoic and Cenozoic Eras.

The first corals were the tabulates, recognized by their simple, often clustered or aligned tubes that are divided horizontally by tabulae (Fig. 12-23) as in the honeycomb coral *Favosites*. Tabulate corals were the principal Silurian reef formers. They declined after the Silurian, and their reef-building role was assumed by the rugose corals. Rugose corals (Fig. 12-24) were abundant in the Devonian and Carboniferous, but became extinct during the Late Permian.

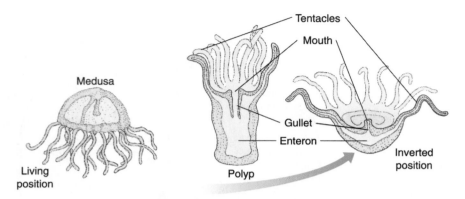

FIGURE 12-22 **Medusa and polyp forms in cnidarians.** The medusa form (left), seen in the jellyfish, resembles an umbrella. They swim by rhythmic contractions of the umbrella. The undersurface has a centrally located mouth. All jellyfish have cnidocyte-laden tentacles, commonly around the margin of the umbrella. The polyp form (right), seen in *Hydra*, corals, and sea anemones, has the mouth on top, surrounded by tentacles. In stony corals, the polyp secretes a calcareous cup, in which it lives. The animal may be solitary or may combine with other individuals to form large colonies.

Immense coral reefs developed in the warm Paleozoic epicontinental seas. Many were discovered during oil drilling. They are important for two very different reasons:

- Because of their porosity, buried Paleozoic reefs provided reservoirs that accumulate petroleum.

- They provide a way to approximately locate Earth's paleoequator. Nearly all living coral reefs

lie within about 30° of the equator. It is reasonable to assume that ancient reef-building corals lived at similar latitudes.

Moss Animals: Bryozoa

Bryozoans are tiny, bilaterally symmetric animals. They grow in colonies resembling crusts or twiglike branches. The individual animal is a *zooid*. They live in

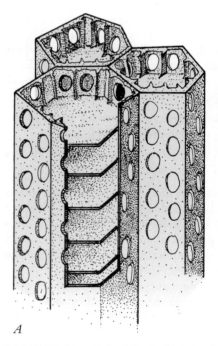

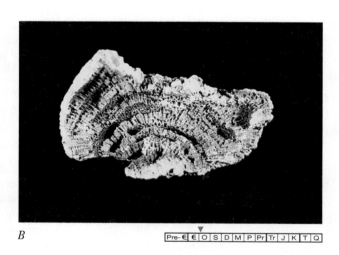

B

Pre-€ | € | O | S | D | M | P | Pr | Tr | J | K | T | Q

A

FIGURE 12-23 **Colonial tabulate coral.** (A) Skeletal structure of honeycomb coral, *Favosites*. The cup (*theca*) may be divided by vertical plates (*septae*) that separate tissue layers and provide support. As the animal grows, it also secretes horizontal plates (*tabulae*), which give the coral its name. An individual polyp resided in each of the tubes, which are about 3 mm wide. (B) A colony of *Favosites*

FIGURE 12-24 **Devonian rugose corals.** (A) The solitary, horn-shaped rugose coral *Zaphrenthis* with radiating septa. (B) The colonial rugose coral *Lithostrotionella*. (C) Polished slab of colonial rugose coral *Hexagonaria*. Water-worn fragments of this coral are found along Lake Michigan's shore at Petoskey, Michigan, giving it the name Petoskey stone, the State Rock of Michigan (although it is not a rock!). (D) Reconstruction of colonial and solitary rugose corals on a Devonian epicontinental seafloor. ◪ *What was the purpose or function of the septa in rugose corals?*

a calcified capsule (zooecium), which often is preserved. They appear as pinpoint depressions on the outside of the colony (zoarium). The zooid has a complete, U-shaped digestive tract with a mouth surrounded by a tentacled feeding organ (lophophore).

More than 4000 species of bryozoans live today, but nearly four times that number are known as fossils. Their earliest unquestioned occurrence is from Lower Ordovician strata. But they did not become abundant until Middle Ordovician to Silurian time, when their remains sometimes are much of the bulk of entire formations. Like corals, bryozoa contributed to reef frameworks. Some Paleozoic bryozoans are shown in Figure 12-25.

Brachiopods

Brachiopods are probably the most abundant, diverse, and useful fossils in Paleozoic rocks.

These marine animals have two enclosing half-shells (*valves*) that together constitute the shell of the animal (Fig. 12-26A). They resemble clams in having two-part shells, but their symmetry and soft-part anatomy are very different. Brachiopod valves almost always are symmetrical, but the two valves differ from each other in size and shape. One is dorsal (top) and one is ventral (bottom) The valves in most clams are similar in appearance, and are right and left.

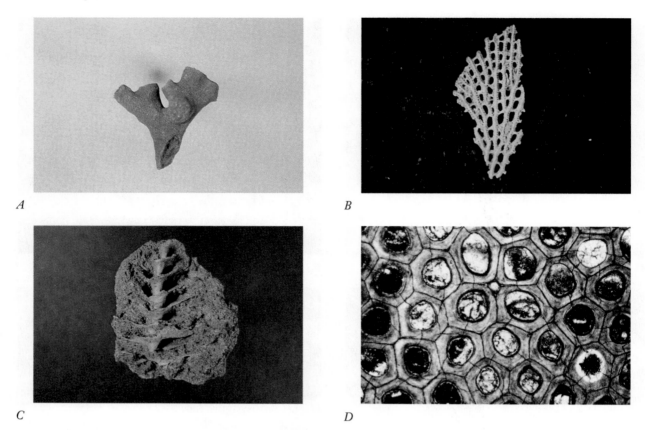

A

B

C

D

FIGURE 12-25 **Paleozoic bryozoans.** (A) *Hallopora* is one of many branching twig bryozoans; different species develop massive, incrusting, or branching colonies (from Ordovician rocks of Kentucky). (B) Some Late Paleozoic bryozoans constructed lacy, delicate, fanlike colonies like this *Fenestella* (from Devonian limestones at Falls of the Ohio River). (C) Bizarre Mississippian corkscrew bryozoan known generically as *Archimedes*, with part of the spirally encircling frond of a lacy bryozoan colony attached. The name derives from the broad-threaded screw that the Greek mathematician Archimedes invented to raise water from a well. (D) Section of *Batostoma*. Individual zooids lived in the circular areas.

In *articulate* brachiopods, the valves are hinged (*articulated*) along the back margin and are prevented from slipping sideways by teeth and sockets. *Inarticulate* brachiopods lack this definite hinge, and their valves are held together by muscles.

Brachiopod valves may be ornamented with radial ridges or grooves, spines, nodes, and growth lines. Calcium carbonate usually forms their valves, although some are mixtures of a protein material called chitin and calcium phosphate. This is particularly true of the inarticulates (Figs. 12-26C and 12-27).

Inarticulate brachiopods characteristically have simple spoon-shaped or circular valves. Although brachiopod larvae swim freely, the adults are anchored or cemented to objects on the seafloor by a fleshy stalk (pedicle) or by spines. Some simply rest on the seafloor.

Brachiopods still live in the seas today, although in far fewer numbers. The earliest brachiopods were almost entirely the chitinous inarticulate types. These inarticulates increased in diversity during the Ordovician but then declined. Very few species remain today.

The articulates also first appeared in the Cambrian Period but became truly abundant during the succeeding Ordovician Period. Their shells dominate many limestone formations and provide stratigraphers with valuable markers for correlation.

Abundant early Paleozoic articulate brachiopods were strophomenids, orthids, rhynchonellids, and pentamerids (Fig. 12-28). During the Late Paleozoic Devonian Period, spiriferid brachiopods became abundant. (Fig. 12-29C).

Most characteristic of the Carboniferous and Permian were large, spiny brachiopods called productids (Fig. 12-30). They were so numerous during the Carboniferous that the period might well be dubbed the age of productids.

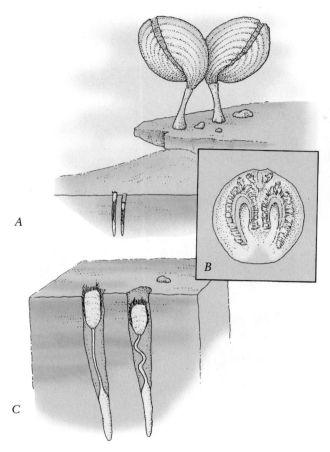

FIGURE 12-26 **Dwelling positions of articulate and inarticulate brachiopods.** (A) The articulate brachiopod *Magellania* attaches to the seafloor by a pedicle. (B) Interior of the dorsal valve showing ciliated *lophophore*. Brachiopods take their name from the lophophore which resembles two arms (Latin brachium = *arm*). The lophophore consists of two ciliated, coiled tentacles that circulate water between the valves, distribute oxygen, and remove carbon dioxide. Water currents generated by cilia on the lophophore move food particles toward the mouth and short digestive tract. (C) The inarticulate brachiopod *Lingula* excavates a tube in bottom sediment and lives within it. The pedicle secretes mucus that glues the animal to the tube.

Mollusks: Clams, Snails, Squid, and Kin

Stroll along almost any seashore, and you will see that mollusks are today's most familiar marine invertebrates. The phylum Mollusca includes snails, clams, chitons, tooth shells, octopods, and squid (Fig. 12-31). Although most mollusks possess shells, some, such as slugs and octopods, do not.

The various classes of Mollusca look different, but have similarities in their internal structure, which is why they are classified in the same phylum. In all Mollusca, the sensory, digestive, and circulatory organs in-

Pre-€ | € | O | S | D | M | P | Pr | Tr | J | K | PgNg

FIGURE 12-27 **Cambrian inarticulate brachiopod *Dicellomus*.** Maximum diameter is 2.2 centimeters.

dicate advanced evolution. Their muscular *foot* is for locomotion. In cephalopods, the foot is modified into tentacles. A fleshy fold called the *mantle* secretes the shell. In aquatic mollusks, respiration is with gills.

Placophorans are primitive mollusks that have multiple paired gills and, in shelled forms, a creeping foot like a snail's. Most familiar are the **polyplacophorans**, represented by chitons (Fig. 12-32). Their fossil record is scant but lengthy. It begins in the Cambrian and extends to the present day.

Monoplacophorans have a single shell resembling a flattened or short cone. They were once thought to have lived only from the Cambrian to the Devonian, until living specimens were dredged from a deep sea trench near Costa Rica. The specimens, later named *Neopilina* because of their similarity to the Silurian monoplacophoran *Pilina* (Fig. 12-33), displayed a segmental arrangement of gills, muscles, and other organs. This indicates that mollusks are derived from a segmented annelid worm ancestor.

Bivalvia or **Pelecypoda** (Fig. 12-34) are a class of mollusks that includes clams, mussels, scallops, and oysters. They originated in the Cambrian, but did not become abundant until the Carboniferous and Permian.

Bivalves have two calcareous valves joined on the dorsal side by a hinge and tough elastic ligament. Unlike brachiopods, which use muscles to open their valves, these mollusks open their valves by the hinge ligament. When muscles inside the shell close the valves, the elastic ligament is stretched. When these muscles relax, the ligament causes the valves to open. Most bivalves (oysters are an exception) have valves that mirror one another.

Gastropods (Latin: *stomach-foot*) first appear in the Lower Cambrian. Earliest forms had small conical

FIGURE 12-28 **Some early Paleozoic brachiopods.** Ordovician strophomenid brachiopods: (A) *Rafinesquina*, (B) *Strophomena*, and (C) *Leptaena*. (D) *Hebertella*, an Ordovician orthid brachiopod. (E) *Lepidocyclus*, an Ordovician rhynchonellid. (F) an internal mold of *Pentamerus*, a Silurian pentamerid.

shells; coiled shells became common later in the Cambrian and Ordovician (Fig. 12-35). The oldest known air-breathing gastropods (pulmonates) appeared as fossils in cavities within trees of a Devonian forest in New York State. Gastropods had become abundant and diverse by Pennsylvanian time. The succeeding Per-

mian Period ended with widespread extinctions among gastropods and all other marine invertebrates.

Cephalopods (Latin: *head-foot*) may be the most complex of all invertebrates. Today, this marine group is represented by the squid, cuttlefish, octopus, and the attractive chambered nautilus. The nautilus provides us

C

FIGURE 12-29 **Devonian spiriferid brachiopods.**
(A) *Mucrospirifer*. (B) *Platyrachella*. (C) Spiriferid brachiopod
with shell broken to reveal calcified internal spiral supports
for the lophophore, from which they take their name. All
are natural size.

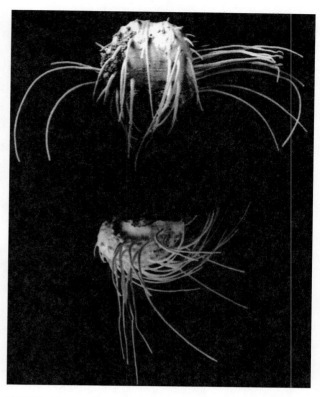

FIGURE 12-30 **Permian spinose productid brachiopod,
Marginifera ornata.** Ventral view (top); side view
(bottom); from the Salt Range of western Pakistan. Valves
(excluding spines) about 2 cm wide. **❓** *What was the probable
purpose or function of the spines?*

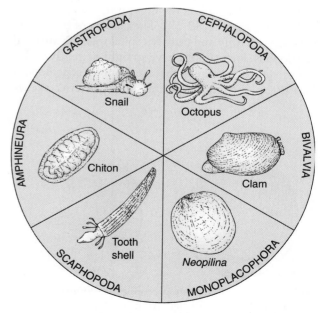

FIGURE 12-31 **Common members of the phylum
Mollusca.**

with important information about the soft anatomy and
habits of a vast array of cephalopods known only by
their preserved shells. The living genus *Nautilus* (Fig.
12-36), has a bilaterally symmetrical body, a prominent
head with paired image-forming eyes, and tentacles. It

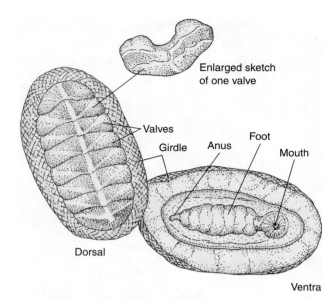

FIGURE 12-32 **A common placophoran, the Atlantic Coast chiton.** It has eight overlapping calcareous plates covering an ovoid, flattened body. Chitons are highly adapted for adhering to and grazing on algae that grows on rocks and shells.

forcefully ejects water through its tubular "funnel" to provide swift, jet-propelled movement.

Cephalopod shells superficially resemble gastropods. However, most gastropod shells coil in a conical spiral, whereas the cephalopod shell forms a straight cone or (most commonly) coils in a plane. Gastropods lack chambered shells, whereas cephalopod shells are divided into chambers by partitions called *septae* (Fig. 12-36A). The bulk of a cephalopod's soft organs reside in the large outer chamber.

Where the septa join the inner wall of the shell, suture lines form (Fig. 12-37). The pattern of these lines is used to identify and classify cephalopods. **Nautiloid** cepalopods, for example, have straight or gently undulating sutures, whereas the **ammonoid** cephalopods have complex sutures. Specific suture patterns charac-

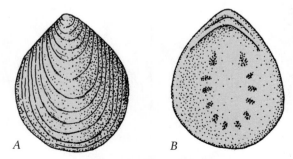

FIGURE 12-33 **The monoplacophoran *Pilina*.** (A) External view of the shell. (B) Internal view showing muscle scars. Maximum diameter is 3.5 cm.

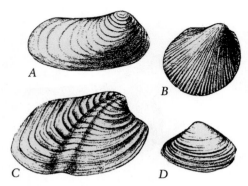

FIGURE 12-34 **Paleozoic bivalves.** (A) *Allorisma* (Mississippian-Pennsylvanian), (B) *Cardiopsis* (Mississippian-Pennsylvanian), (C) *Grammysia* (Silurian-Mississippian), and (D) *Ctenodonta* (Ordovician-Silurian). ◄ *How does the symmetry of these bivalves differ from that typically seen in articulate brachiopods?*

terize each ammonoid species and can be used to precisely correlate cephalopod-bearing strata.

The oldest fossils classified as cephalopods have small, conical shells. They occur in Lower and Middle Cambrian rocks. The class gradually grew in number and diversity and became common in Ordovician and Silurian seas. By Silurian time, a great variety of shell forms—from straight to tightly coiled—had developed (Fig. 12-38). Some of the

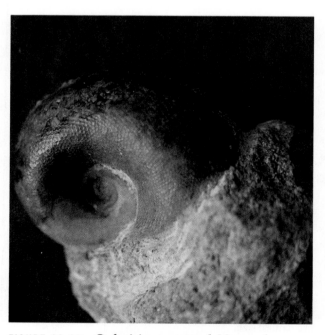

FIGURE 12-35 **Ordovician gastropod.** Many early gastropods coiled in a plane, but in later forms spiral coiling was more common. Gastropods of the Ordovician and Silurian had shapes similar to those of living species. Average diameter is 4 cm. ◄ *How would the interior of this gastropod shell differ from that of a similarly coiled cephalopod shell?*

A   *B*

FIGURE 12-36 *Nautilus*, **a modern nautiloid cephalopod.** (A) Shell of a nautiloid sawed in half to show large living chamber, septae, and septal necks through which passed a porous tube called the siphuncle. (B) Living animal photographed at a depth of about 300 m.

straight-cone nautiloids exceeded 4 meters in length. The first signs of a decline in nautiloid populations can be detected in Silurian strata. After Silurian time, the group continued to dwindle until today only a single genus, *Nautilus*, survives.

During the Devonian, the first ammonoid cephalopods appeared. These were the goniatites, characterized by angular and generally zigzag sutures without any additional wrinkles (Fig. 12-39). The **goniatites** persisted throughout the late Paleozoic and gave rise to the **ceratites** and ammonites of the Mesozoic.

Arthropods: Jointed Bodies and Limbs

Arthropods (Latin: arthro = *joint*; pod = *foot*) are a phylum of invertebrates includes living animals like insects, spiders, lobsters, and a host of others that possess *exterior skeletons* composed of chitin, obviously segmented bodies, paired and jointed appendages, and highly developed nervous systems and sensory organs. Arthropoda that left a particularly significant fossil record include trilobites, ostracods, and eurypterids.

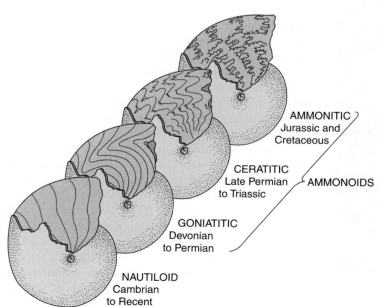

AMMONITIC
Jurassic and
Cretaceous

CERATITIC
Late Permian
to Triassic

AMMONOIDS

GONIATITIC
Devonian
to Permian

NAUTILOID
Cambrian
to Recent

FIGURE 12-37 **Cephalopod suture patterns.** Nautiloidea cephalopods have straight or gently undulating sutures. Ammonoidea have more complex sutures.

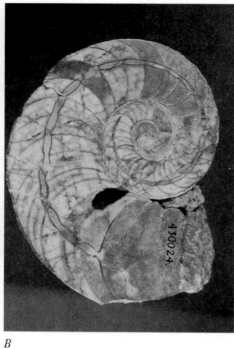

FIGURE 12-38 **Variation in shell shape among Silurian nautiloid cephalopods.** (A) A sawed and polished section of the straight conch of *Orthoceras potens*. Length is 22.5 cm. (B) Sawed and polished section of *Barrandeoceras*, which exhibits a coiled form. Diameter is 18 cm.

Trilobites were swimming or crawling arthropods (Fig. 12-40). They take their name from the division of their upper surface into three longitudinal *lobes*. Their skeleton was of chitin, strengthened by calcium carbonate in parts not requiring flexibility.

In some forms, the body segmentation was so flexible that the animal could roll into a tight ball to protect soft parts from predators (Fig. 12-41). As in many other arthropods, growth was accomplished by molting. Although some trilobites were sightless, the majority had either single-lens eyes or compound eyes (like a fly's) composed of numerous lenses.

Following their appearance in the Early Cambrian, trilobites expanded rapidly. The exoskeletons of most early species have numerous thoracic segments and small or missing pygidia ("tail" segment). This indicates that trilobites evolved from the many-segmented annelid worms.

Near the end of the Cambrian, trilobites declined in number and diversity. Contributing factors may have included predation (by anomalocarids and cephalopods) or adverse environmental conditions associated with the restriction of epicontinental seas.

Survivors of the late Cambrian extinction provided the stock for a second radiation of trilobites that began in the Ordovician and continued until Middle and Late Devonian, when they experienced another extinction event. Only a few groups survived to carry on into the Carboniferous. The final blow for trilobites came during the middle Permian, when all remaining families disappeared. Although they were unable to survive, do not regard trilobites as biologic failures! They were important animals on our planet for more than 300 million years.

Most trilobites were bottom dwellers (benthic), living in shallow shelf areas. Many were restricted to particular regions, which diminishes their usefulness in worldwide stratigraphic correlations. However, they do provide evidence for plate tectonics. During the Paleo-

FIGURE 12-39 **Goniatite ammonoid cephalopod exhibiting zigzag sutures.** Diameter is 4.6 cm.

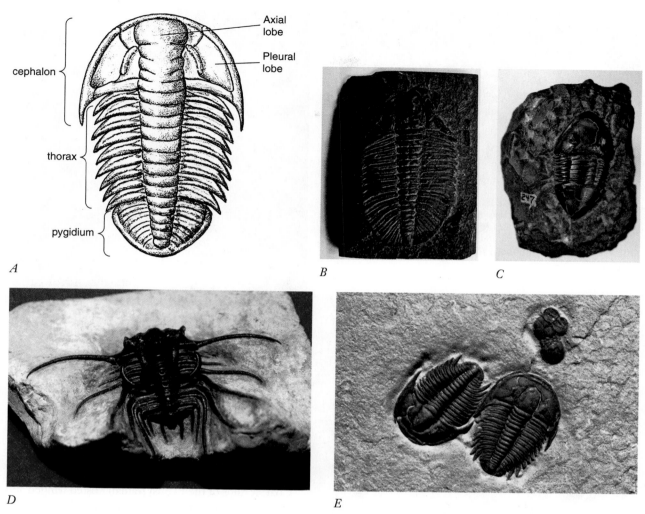

FIGURE 12-40 **Trilobites.** (A) Drawing of the Cambrian trilobite *Bathyuriscus* with cephalon, thorax, and pygidium labeled. There are three lobes: the central raised area, called the axial lobe, which is bordered by two pleural lobes—hence the *"tri"* in trilobite. (B)The Middle Cambrian guide fossil *Ogygopsis klotzi* from Mount Stephens, British Columbia, near the site of the famous Burgess Shale quarry. Length is 6.0 cm. (C) *Isotelus* from Ordovician limestones in New York State. Length is 3.8 cm. (D) *Dicranurus* from the Devonian rocks of Morroco. Length is 3.0 cm. (E) Cambrian trilobites *Modicia* (the larger two specimens), and *Agnostis* (the smaller specimens).

zoic, Europe and North America were separated by the Iapetus Ocean. Trilobite fauna living in shelf areas on either side of the ocean were distinctly different. When the Iapetus Ocean began to close, marginal seas of both continents converged. This allowed the formerly separated trilobite faunas to intermingle and to lose their distinctive differences. The resulting cosmopolitan fauna thus provides evidence of plate convergence.

Ostracods were arthropod companions to the trilobites in Paleozoic seas (Fig. 12-42). They first appeared early in the Cambrian, and many species still exist today. They occur in both marine and freshwater sediments. Because of their small size, they are brought to

the surface in wells drilled for oil. Along with foraminifera and radiolaria, they are used to correlate strata in oil fields.

Eurypterids (Fig. 12-43) are an impressive group of early Paleozoic aquatic invertebrates. These predators had five pairs of appendages and fearful-looking pincers (chelicerae). Some also had a spine that projected from their posterior segment. Although many were of modest size, some were nearly ten feet long. Had these giants survived, they would be suitable subjects for a Hollywood horror film.

Eurypterids lived in brackish estuaries from Ordovician until Permian time, but were especially abundant

FIGURE 12-41 **A rolled-up *Flexicalymene*, Ordovician limestones, Ohio.**

FIGURE 12-43 **Fossil eurypterid.** *Eurypterus lacustris* is from the Bertie Waterlime of Silurian age, Erie County, Pennsylvania. This well-preserved specimen is about 28 cm long.

during the Silurian and Devonian. Because of their rarity, Eurypterids are less useful in stratigraphic studies than trilobites or ostracods.

Not all Paleozoic arthropods lived in the sea. With their sturdy legs and protective exoskeletons, members

FIGURE 12-42 **Ostracods (*Leperditia fabulites*) in limestone.** These ostracods are unusually large, averaging about 1 cm in length. Two even larger strophomenid brachiopods are also present. Ostracods were small, often bean-shaped, and had a bivalved shell vaguely suggestive of a clam. However, their bivalved shell enclosed a segmented body from which extended seven pairs of jointed appendages. Most adults are about 0.5–4 mm in length. The valves are composed of chitin and calcium carbonate and are hinged along the dorsal margin. Specimens shown are in the Ordovician Plattin Limestone, near St. Louis in Jefferson County, Missouri.

of this phylum were *preadapted* (already well-equipped) for coming ashore. Fossils of the many-legged centipedes and millipedes are found in Silurian nonmarine deposits of Britain. Devonian rocks have yielded fossil remains of mites, primitive wingless arthropods called archaeognaths, centipedes, and the earliest spiders.

Coal swamps of the Pennsylvanian hosted winged and wingless insects, including dragonflies (Fig. 12-44), scorpions, and giant cockroaches four inches long. The Pennsylvanian insect fauna includes the largest insect known, a giant dragonfly with a wingspread of over 2 feet.

Spiny-Skinned Animals: Echinoderms

Echinoderms (*ih-KINE-oh-derms*, Fig. 12-45) are animals with mostly five-way symmetry (like a starfish) that masks an underlying primitive bilateral symmetry. Their endoskeleton consists of calcium carbonate plates, and in many groups it also bears spines. A unique characteristic of this phylum is a system of soft-tissue tubes—the **water vascular system**—which functions in respiration and locomotion (Fig. 12-46). Members of the phylum are exclusively marine, typically bottom-dwelling, and either attached to the seafloor or able to move about slowly.

Among the many classes of the phylum Echinodermata, several are abundant and useful in geologic studies: **Asteroidea** (starfish), **Ophiuroidea** (brittle stars), **Echinoidea** (sea urchins), **Edrioasteroidea**, **Crinoidea** (crinoids), **Blastoidea** (blastoids), and **Cystoidea** (cystoids). The crinoids, blastoids, and cystoids that attached themselves to the substrate were particularly common in the Paleozoic.

ENRICHMENT

The Eyes of Trilobites

"The eyes of trilobites eternal be in stone, and seem to stare about in mild surprise at changes greater than they have yet known," wrote American paleontologist T. A. Conrad. Indeed, trilobites may have been the first animals to look out upon the world, for they possessed the most ancient visual systems yet discovered.

The eyes of trilobites include both simple and compound types. A simple eye appears as a single, tiny lens, resembling a small node. A few receptor cells probably lie beneath each simple lens. Simple eyes are rare among trilobites. Far more abundant are compound eyes composed of a large number of individual visual bodies, each with its own lens.

Each lens is composed of a single crystal of calcite. One of the properties of a clear crystal of calcite is that in certain orientations, objects viewed through the crystal produce a double image. Trilobites, however, were not troubled by double vision. This is because the calcite lenses were oriented with the principal optic axis (the light path along which the double image does not occur) perpendicular to the surface of the eye.

In some compound trilobite eyes—called *holochroal*—the many tiny lenses are covered by a continuous, thin, transparent cornea. Beneath this smooth cover there may be as many as 15,000 lenses. In other trilobites, such as *Phacops* (Fig. A) and *Calliops*, there are discrete individual lenses, each covered by its own separate cornea and sepa-

FIGURE A **Side view of the trilobite *Phacops rana*.** Note the discrete visual bodies of the large compound eye. Length is 3.2 cm.

rated from its neighbors by a cribwork of exoskeletal tissue. Such compound eyes are termed schizochroal.

Trilobite eyes can sometimes provide clues useful in determining the habits of certain species. Eyes along the anterior margin of the cephalon, for example, may indicate active swimmers. Eyes located on the ventral side of the cephalon ("head") may indicate that the trilobite was a surface-dweller. In the majority of trilobites, the eyes are located about midway on the cephalon, in a position appropriate for an animal that crawled on the sea bottom or occasionally swam above it. Trilobites that either lacked eyes or were secondarily blind appear to have been adapted for burrowing in the soft sediment on the ocean floor.

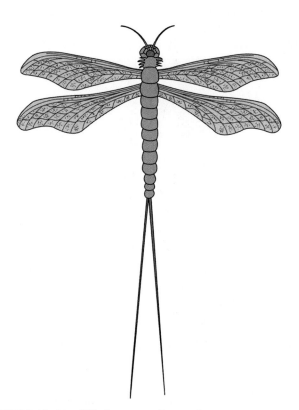

FIGURE 12-44 ***Mischoptera*, a Pennsylvanian-age dragonfly.** Length is 8.4 cm.

Echinoderms probably evolved late in the Proterozoic, although fossil remains from this time are few and often enigmatic. The Ediacaran fauna, for example, includes a globular fossil named *Arkarua* that has five rays on its surface and may be related to edrioasteroids.

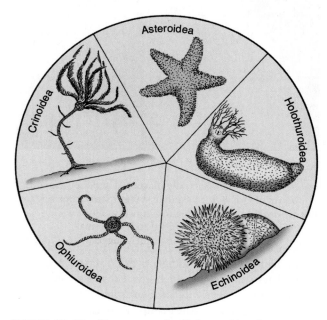

FIGURE 12-45 **Representative living echinoderms.**

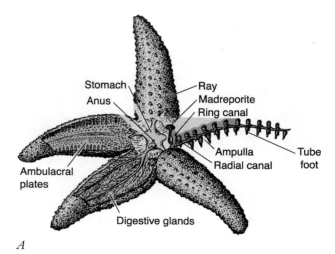

A

B

FIGURE 12-46 **(A) Partially dissected starfish showing elements of the water vascular system and other organs. (B) Underside of starfish showing tube feet.**

Helicoplacus (Fig. 12-47), a bizarre Cambrian echinoderm, had plates and food grooves arranged in a spiral around its spindle-shaped body. Edrioasteroids (Fig. 12-48) appeared early in the Cambrian, but by the end of the Carboniferous, the group had become extinct.

Pre-Є | Є | O | S | D | M | P | Pr | Tr | J | K | T | Q

FIGURE 12-47 **Spiraled, spindle-shaped Early Cambrian echinoderm *Helicoplacus*.** This unusual Cambrian echinoderm had plates and food grooves arranged in a spiral around its spindle-shaped body. Length is about 2.6 cm.

Stemmed or stalked echinoderms first occur in Middle Cambrian strata but do not become abundant until Ordovician and Silurian time. Stalked forms called cystoids (Fig. 12-49) are the most primitive among this group. Although cystoids range from Cambrian to Late Devonian, they are chiefly found in Ordovician and Silurian rocks.

Stalked echinoderms known as blastoids (Fig. 12-50) have an orderly arrangement of plates and a

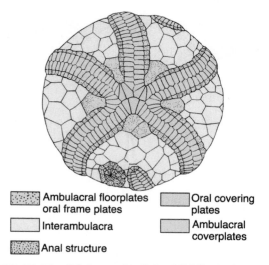

░ Ambulacral floorplates oral frame plates

▓ Anal structure

☐ Interambulacra

▒ Oral covering plates

▒ Ambulacral coverplates

FIGURE 12-48 ***Edrioaster bigsbyi*, a Middle Ordovician edrioasteroid.** Many paleontologists consider edrioasteroids to be ancestral to starfish and sea urchins. They constructed globular, discoidal, or cylindrical tests (shells), many of which have concave lower surfaces to facilitate attachment to hard substrates. This specimen is 45 mm in diameter.

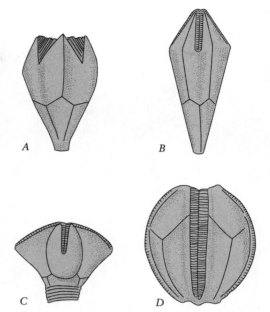

FIGURE 12-50 **Some common Paleozoic blastoids.** The five radial areas (ambulacral areas) are prominent and have slender branches or brachioles along their margins. Unfortunately, the delicate brachioles are rarely preserved in fossil specimens. (A) *Codaster* (Dev.), (B) *Troosticrinus* (Sil.), (C) *Orophocrinus* (Miss.), and (D) *Cryptoblastus*. (Miss.). All are approximately natural size.

FIGURE 12-49 **Well-preserved specimen of Silurian cystoid *Caryocrinites ornatus*, Lockport Shale, New York.** The five-part symmetry is less well developed in cystoids. Beginning students of paleontology often recognize cystoids by their distinctive patterns of pores.

The Echinoderm-Backbone Connection

Echinoderms have a close relationship to chordates (creatures with backbones). You can see the evidence in the early development of a chordate and echinoderm embryo when a small sphere of cells forms, called the **blastula**. A group of cells move inward on the blastula to form an opening called the **blastopore**. In echinoderms and chordates, the blastopore develops into the *anus*. The mouth develops later from another opening.

Animals like echinoderms and chordates that form the anus this way are termed **deuterostomes**. Along with starfish and sea urchins, we are deuterostomes. The companion term **protostomes** refers to animals like arthropods, mollusks, and annelid worms, in which the blastopore develops into a *mouth*.

Protostomes and deuterostomes also differ in their patterns of early cell division. In chordates and echinoderms, early cells cleave one above the other in a *radial* arrangement (Fig. 12-54). In protostomes, cell divisions are *spiral*, so successive rows nestle neatly within depressions between earlier formed cells. Also, in embryologic development, the mesodermal layer of cells as well as certain other elements of the body arise in the same way in both echinoderms and chordates.

Even the larvae of echinoderms resemble those of primitive chordates. Biochemistry provides a final line

well-developed water vascular system. Blastoids first appeared in Silurian time, expanded in the Mississippian, and declined to extinction in the Permian.

Crinoids (Fig. 12-51) are the most abundant stalked echinoderm. They range from the Ordovician to the present. Some Ordovician, Silurian, and Carboniferous rocks contain such great quantities of crinoid skeletal plates that they are called crinoidal limestones (Fig. 11–6B). The four main parts of a crinoid are shown in Figure 12-52.

Although one group of crinoids survived the wave of extinctions at the end of the Paleozoic and gave rise to those living today, most died out before the end of the Permian. Free-living echinoderms such as echinoids (Fig. 12-53) and starfish were locally abundant during the late Paleozoic, but less numerous than they would become in the Mesozoic and Cenozoic Eras.

A

B

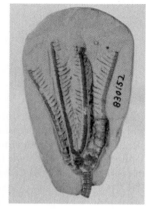

C

FIGURE 12-51 **Crinoids.** (A) This diorama depicts crinoids on a Mississippian epicontinental seafloor. (B) Calyx and arms of the Mississippian crinoid *Taxocrinus*. Height is 5.1 cm. (C) *Scytalocrinus*. Height is 4.5 cm. Both fossil crinoids are from the Keokuk Formation of Mississippian age, near Crawfordsville, Indiana.

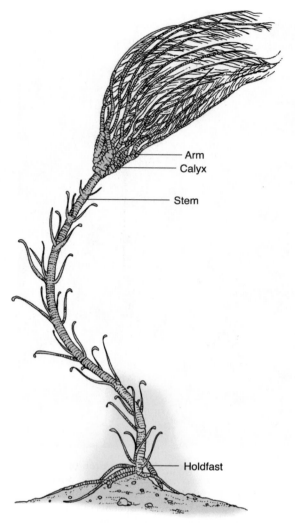

FIGURE 12-52 **Crinoid in living position on the seafloor.** Its four main parts are (1) the cup-shaped calyx (which contains the vital organs), (2) the stem with its anchoring holdfast, (3) the arms, which bear ciliated food grooves and, like the brachioles of blastoids, move food particles toward the mouth, (4) the holdfast, which anchors the crinoid to the bottom.

of evidence for a relationship between echinoderms and chordates by revealing chemical similarities associated with muscle activity and the chemistry of oxygen-carrying pigments in the blood.

Graptolites

Graptolites are extinct, planktonic creatures with preservable chitinous skeletons that housed colonies of tiny individual organisms (Fig. 12-55). Fossils occur as flattened and carbonized impressions in dark shales. Rarely, uncompressed specimens are found. These reveal an unexpected relationship to primitive living chordates called **pterobranchs**. Both groups secrete tiny enclosed tubes, and both have a very similar structure of the thecae. These pieces of evidence mean that a close relationship is virtually certain.

Graptolites appeared at the end of the Cambrian, but they did not become abundant until the Ordovi-

cian. They disappear from the fossil record by the end of the Mississippian. In 1989, however, a possible surviving species was recovered from South Pacific waters, and named *Cephalodiscus graptoloides*. More specimens were later found in waters surrounding Bermuda. The protein in the skeletal material and the general morphology of these living forms closely resemble those of Paleozoic graptolites.

During the Ordovician and Silurian, graptolites were so abundant that they may have formed large floating masses. They were apparently carried about by ocean currents and thus achieved the worldwide distribution that accounts for their value as Ordovician and Silurian guide fossils.

FIGURE 12-53 **Large Mississippian echinoids** *Melonechinus*, **St. Louis Limestone, St. Louis, Missouri.** Average diameter of specimens is 12 cm.

ADVENT OF THE VERTEBRATES

Vertebrates are animals having a segmented backbone, or dorsal **vertebral column**. The vertebral column is composed of individual vertebrae. Arches of the vertebrae encircle and protect a hollow spinal cord (nerve cord). In addition, vertebrates have a cranium or skull that houses a brain.

Vertebrates are diverse and include both water-dwelling and land-dwelling tetrapods (Latin: *four feet*). Tetrapods include vertebrates that walk on four legs (quadrupedal), those that walk only on their hind legs, like us (bipedal), those whose forelimbs that have been modified into wings, and those whose limbs have become flippers for swimming (like dolphins).

An important difference in reproduction separates vertebrates into two major groups:

- **Amniotic vertebrates** (amniotes)—all higher vertebrates have evolved internal fertilization and an amniotic egg (enclosed egg).

- **Non-amniotic vertebrates**—fish and amphibians must be in water (or at least wet) to reproduce. Their eggs are naked, in that they lack a covering. Such eggs must be fertilized externally.

The embryo in an amniotic egg is enveloped in an **amniotic membrane**, which allows oxygen to enter but which retains water (Fig. 12-56). The membrane encloses the embryo in a cushioning watery environment. There is also a yolk sac containing nutrients to nourish the embryo. Another membrane, the allantois, holds the embryo's waste products. Finally, the chorion regulates the exchange of oxygen and carbon dioxide.

Most amniotes have a shell or leathery covering to protect the embryo. However, in mammals, the embryo grows within an amnion that develops within the mother's womb. For all amniotes, the remarkable amniotic egg provided freedom from dependency on water bodies. It facilitated the exploitation of diverse terrestrial environments and is thus an extraordinary milestone in the evolution of vertebrates.

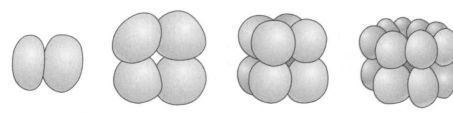

Radial cleavage, as seen in the development of the deuterostone embryo

 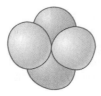 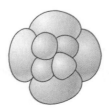 

FIGURE 12-54 **Patterns of embryologic cleavage.** Upper row, radial cleavage in deuterostomes (like starfish and us). Lower row, spiral cleavage in protostomes (like arthropods).

FIGURE 12-55 **Part of a stem (stipe) of Ordovician graptolite *Orthograptus quadrimucronatus*.** The tiny cups (*thecae*) in which the individuals lived are clearly visible on each side. They aligned along stems (*stipes*). There may be a single stipe or several formed into a system of two or more branches. A tiny polyp-like animal lived in each cup. In general, evolution progressed from multibranched forms to those having only a single stipe. The entire colony is a *rhabdosome*. (Image is magnified 15 times.)

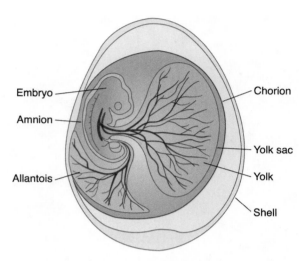

FIGURE 12-56 **Amniotic egg.** The *amnion*, from which the egg takes its name, encloses the embryo in water. The *allantois* serves as a reservoir for waste, the *chorion* regulates the exchange of gases, and the *yolk* (also present in fish and amphibian eggs) serves as a storage area for fats, proteins, and other nutrients needed by the developing embryo.

THE RISE OF FISHES

All vertebrates, including ourselves, are members of a larger group called chordates. At least at some stage in their life history, all chordates have:

- A stiff, elongate supporting structure
- A dorsal, central nerve cord
- Gill slits
- Blood that circulates forward in a main ventral vessel and backward in the dorsal

In primitive chordates, the supportive structure is a notochord. It has been studied by generations of biology students in such animals as the fishlike lancelet named *Branchiostoma* (Fig. 12-57). In taxonomic hierarchy, vertebrates are chordates in which the notochord is supplemented or replaced by a series of cartilaginous or bony vertebrae.

The vertebrates that we informally call fishes include five taxonomic classes (Fig. 12-58):

1. **Agnathans**: the jawless fishes
2. Acanthodians: spiny fishes with jaws
3. Placoderms: plate-skinned fishes with jaws
4. Chondrichthyes: fishes with cartilaginous skeletons, like sharks, rays, and skates
5. Osteichthyes: fishes with bony skeletons

The first three of these—Agnatha, acanthodians, and placoderms—are most frequently found in rocks of the early Paleozoic.

Agnathids (Jawless Fish)

The oldest known agnathids were discovered in the Early Cambrian fossil beds near Chengjiang, China. These fossils, *Myllokunmingia* (Fig. 12-12) and *Haikouichthys*, have several features that favor their placement in the Vertebrata: V-shaped musculature needed for swimming, a relatively complex skull, gill supports, and fin supports. Both of these fishes were covered by soft tissue. The tiny fishes resemble the larvae of the lamprey, a living agnathid.

Armored agnathids were also present during the early Paleozoic. The clumsy-looking *Astraspis* (Fig. 12-59), with its jawless slit-like mouth and row of gill openings, is one of the earliest. Early Paleozoic agnathids are collectively termed **ostracoderms** ("shell skins"), even though some lacked heavy bony armor. The diverse ostracoderms included unarmored forms (*Theolodus* and *Jamoytius*) and armored forms (*Pteraspis* and *Hemicylaspis*) (Fig. 12-60).

Why the bony armor on many ostracoderms? An obvious view is protection against predators, but another hypothesis is that the dermal armor stored seasonally available phosphorus. Phosphates could be accumulated as calcium phosphate in the armor during times of greater availability and then used during periods when supply was deficient. This cache of phospho-

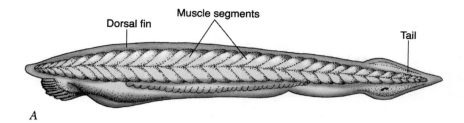

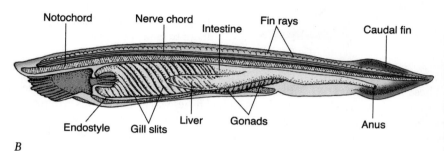

FIGURE 12-57 *Branchiostoma*, a fishlike member of the subphylum Cephalochordata. The tiny animal has a dorsal nerve cord lying above a notochord, V-shaped muscles for swimming, and a prominent gill structure—all chordate characteristics. (A) External view. (B) longitudinal section. Length is 3–5 cm.

rus may have been needed to maintain a suitable level of muscular activity.

Hemicyclaspis (Fig. 12-60D) is one of the most widely known of the ostracoderms. It is recognized by its large, semicircular head shield, with four openings on top: two for the upward-looking eyes, a single nostril and a pineal opening that may have housed a light-sensitive third eye. The mouth was located in a position suitable for taking in food from the surface layer of soft sediment. Depressed areas along the margin of the head shield covered a system of nerves and may have had a sensory function.

The ostracoderms continued into the Devonian but ended there. They were mostly small, sluggish animals restricted to mud-straining or filter-feeding. They would be replaced gradually by fishes having bone-supported, movable jaws.

Evolution of the Jaw

Evolution of the jaw was no small accomplishment. It enormously expanded the adaptive range of vertebrates.

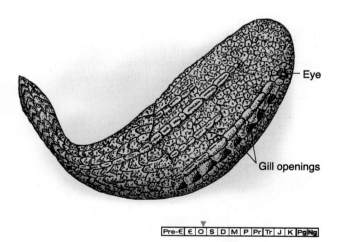

FIGURE 12-59 **Ordovician agnathan *Astraspis*, from Harding Sandstone, Colorado.** Image is about ⅔ its natural size.

FIGURE 12-58 **Evolution of the five major categories of fishes.** The width of the vertical red areas indicates the approximate relative abundance of each group.

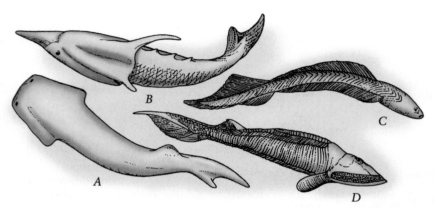

FIGURE 12-60 **Early Paleozoic ostracoderms.** (A) *Thelodus*, (B) *Pteraspis*, (C) *Jamoytius*, and (D) *Hemicyclaspis*, drawn to same scale. These animals ranged 18–40 cm in length.

Jawed fishes could bite and grasp. These new abilities led to more varied and active lifeways and to new sources of food unavailable to the jawless agnathids.

Currently, we have two hypotheses for the origin of the jaw. The older idea is that jaws formed by modification of a forward pair of gill arches—the thin strips of bone or cartilage that support the soft tissue of the gills. More recently it has been proposed that jaws arose from the velum, a structure that functions in respiration and feeding in larval lampreys. Both ideas are based on the anatomy and embryology of living fishes, because the first jaws in the fossil record give little evidence of their origin.

Acanthodians and Placoderms (Fish with Jaws)

The oldest fossil remains of non-armored jawed fishes occur in nonmarine rocks of the Late Silurian. These fishes, called **acanthodians** (Fig. 12-61), became most numerous during the Devonian and then declined to extinction in the Permian. Although Acanthodians had jaws, they were primitive and distinctly different from modern fishes.

Another group of fishes includes the **placoderms**, or "plate-skinned" fishes. They too arose in the late Silurian, expanded rapidly during the Devonian, and then in the latter part of that period began to decline and be replaced by the ascending sharks and bony fishes.

Placoderms varied considerably. Most formidable of these plate-skinned fishes was a carnivorous group called **arthrodires** (Latin: *jointed neck*). The name refers to a ball-and-socket joint between shoulder and head that allowed the head to be rotated backward. Thus the jaws, lined with knife-sharp plates of bone, could be opened widely to ensnare even large victims. One gigantic Devonian arthrodire, *Dunkleosteus* (Fig. 12-62), exceeded 10 meters (33 feet) in length. Other placoderms, called **antiarchs** (Fig. 12-63), had the heavily armored form and mud-grubbing habits of their ostracoderm predecessors.

Chondrichthyes (Fish with Cartilaginous Skeletons)

Among the early Paleozoic acanthodians and placoderms were ancestors to cartilaginous and bony fishes. When we look at rocks of Late Silurian and Devonian

FIGURE 12-62 **Gigantic armored skull and thoracic shield of formidable late Devonian placoderm fish,** *Dunkleosteus*. *Dunkleosteus* was over 10 m (about 30 feet) long. The skull shown here is about 1 m tall and has large bony cutting plates that functioned as teeth. Each eye socket was protected by a ring of four plates. A joint at the rear of the skull permitted the fish to raise its head, making possible an extra large bite. *Dunkleosteus* ruled the seas 350 million years ago.

FIGURE 12-61 **Early Devonian acanthodian fish** *Climatius*. This small fish was only a few inches long.

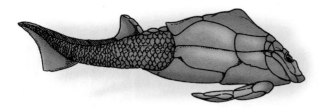

FIGURE 12-63 **Devonian antiarch fish *Pterichthyodes.*** It had the heavily armored form and mud-grubbing habits of their ostracoderm predecessors.

age, we find entire fossils from many localities around the world of **chondrichthyans** (cartilaginous fishes) and **osteichthyans** (bony fishes).

Chondrichthyans that populate modern seas include sharks, rays, and skates. Among the better-known late Paleozoic sharks were species of *Cladoselache* (Fig. 12-64A). Remains of this shark are frequently encountered in the Devonian shales that crop out on the southern shore of Lake Erie. Modern sharks arose from cladoselachian ancestors. During the Late Carboniferous, a chondrichthyan named *Xenacanthus* (Fig. 12-64B) lived in lakes and streams.

Osteichthyes (Fish with Bony Skeletons)

Bony fishes played a key role in the evolution of tetrapods (four-legged animals). They are also the most numerous, varied, and successful of all aquatic vertebrates. For these reasons, their evolution is especially important. We divide bony fishes into two categories: **actinopterygians**, the familiar ray-fin, and **sarcopterygians**, the lobe-fin.

Actinopterygians lack a muscular base to their paired fins, which are thin structures supported by radiating rods or rays. Unlike the Sarcopterygians, they lack paired nasal passages that open into the throat.

The ray-fins began their evolution in Devonian freshwater lakes and streams and quickly expanded into the marine realm. They became the dominant fishes of the modern world, where they include salmon, perch, bass, carp, tuna, herring, and minnows. The more primitive Devonian ray-finned bony fishes are well represented by the genus *Cheirolepis* (Fig. 12-65). The more advanced bony fishes evolved during the Mesozoic and Cenozoic from such fishes as these.

Sarcopterygians had sturdy, muscular lobe-fins and a pair of openings in the roof of the mouth that led to

A

Pre-Є | Є | O | S | D | M | P | Pr | Tr | J | K | PgNg

B

Pre-Є | Є | O | S | D | M | P | Pr | Tr | J | K | PgNg

FIGURE 12-64 **Shark models.** (A) Devonian marine shark *Cladoselache.* (B) Pennsylvanian freshwater shark *Xenacanthus.*

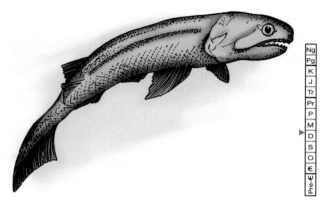

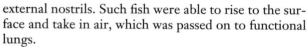

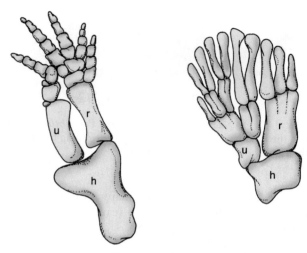

FIGURE 12-65 *Cheirolepis*, the ancestral bony fish that lived during the Devonian Period.

FIGURE 12-67 **Limb bones of an early amphibian (left) and crossopterygian fish (right).** Some early amphibians may have had more than five digits.

external nostrils. Such fish were able to rise to the surface and take in air, which was passed on to functional lungs.

Lungs and gills do not occur together in most modern fishes, but in late Paleozoic fishes the combination was more common. Studies of living sarcopterygians indicate that lungs probably began their evolution as sac-like bodies developed on the underside of the esophagus and then became enlarged and improved for the extraction of oxygen. (In most modern fishes, the respiratory function of the lung has been lost. It is converted to a swim bladder, which fish use to maintain their depth without rising or sinking.)

We divide Devonian sarcopterygians into two major groups: dipnoans and crossopterygians. **Dipnoans**, represented in the Devonian by *Dipterus* (Fig. 12-66), were not on the evolutionary track that would lead to tetrapods. Nonetheless, they are an interesting group that includes living freshwater lungfishes of Australia, Africa, and South America. Their restricted presence south of the equator suggests that Gondwanaland was the probable dispersal center for dipnoans.

The taxonomic term Dipnoi means "double breather." The name was suggested by living species that are able to breathe by means of lungs during dry seasons. At such difficult times, they burrow into the mud before the water is gone. When the lake or stream is dry, they survive by using their accessory lungs, and

when the waters return, they switch to gill respiration. The dipnoans are an enduring group. One species living today in Queensland, Australia, has a fossil record that extends back 100 million years.

Crossopterygian fossils are considered the ancestors of the earliest land invaders—the non-amniotic tetrapods. This is because of the bone arrangement in their muscular fins (Fig. 12-67), the skull element pattern (Fig. 12-68), and their teeth structure (Fig. 12-69). A rather advanced example of Devonian crossopterygians is *Eusthenopteron* (Fig. 12-70).

In this genus, the paired fins were short and muscular. Internally, a single basal limb bone called the humerus (equivalent to the femur in the hind limbs) articulated with the girdle bones. It was followed by the ulna and radius for the pectoral fins, and the tibia and fibula for the posterior fins.

Why did the robust skeleton and sturdy limbs of crossopterygians evolve? Not because fishes developed a desire to live on land. The adaptations simply improved survival odds—by assisting movement in shallow water, or by enabling the animal to abandon a water body that was drying out or stagnating and move to another that offered a better chance of survival and reproduction.

FIGURE 12-66 *Dipterus*, a Devonian lungfish.

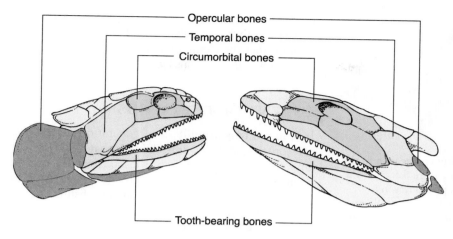

Opercular bones
Temporal bones
Circumorbital bones
Tooth-bearing bones

FIGURE 12-68 **Skulls and lower jaws of a crossopterygian (left) and Devonian amphibian *Ichthyostega* (right).** ☑ *Which senses (sight, hearing, smell, taste, and touch) would have been improved by the longer distance from nostril to eye orbit developed in* Ichthyostega?

During the Devonian, two distinct branches of crossopterygians had evolved. The rhipidistians led ultimately to non-amniotic tetrapods, and the other branch led to fishes called coelacanths. Coelacanths (Fig. 12-71) were thought to have become extinct during the Late Cretaceous, until several were caught in 1938 off the coast of the Comoro Islands (in the Indian Ocean west of Madagascar).

CONODONTS: VALUABLE BUT ENIGMATIC FOSSILS

For 150 years, paleontologists have used tiny fossils called conodont elements as guide fossils for rock strata over the 700–800-million-year span from Neoproterozoic to Triassic. Conodont elements consist of cones, bars, and blades, each bearing tiny denticles, cusps, or ridges (Fig. 12-72, upper). They are called "elements" because the complete animal was unknown until recently.

Since the discovery of conodont elements by Russian zoologist Christian Pander in 1856, paleontologists have proposed that the elements were structures variously suited to grasping, filtration, or support for diverse animals like worms, fish, snails, cephalopods, and arthropods. However, recent evidence indicates that conodont elements are from one specific chordate animal.

To determine which elements occurred together in the **conodont**, paleontologists had to find them in undisturbed natural associations. A few such assemblages have been discovered, allowing reconstruction of how the elements were arranged in the living animal.

Several conodont animal fossils have been reported over recent decades. The most informative was found in 1995 in the Ordovician Soom Shale of South Africa. It is the remains of a creature with an elongate, eel-like body, about 40 mm long (Fig. 12-72, lower). The creature had distinctive eye musculature similar to that in primitive jawless fish.

Like the living cephalochordate *Branchiostoma* (Fig. 12-57), muscles along each side of the conodont animal were V-shaped units lying sidewise, with the point forward. Fossilized fibrous muscle tissue is similar to

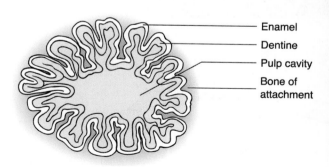

Enamel
Dentine
Pulp cavity
Bone of attachment

FIGURE 12-69 **Cross-section of crossopterygian tooth.** Note the distinctive pattern of infolded enamel, also characteristic of teeth in amphibian descendants of crossopterygian fishes.

FIGURE 12-70 **Devonian crossopterygian lungfish *Eusthenopteron*.** Its sturdy fin structure foreshadowed that of its four-footed land descendants.

FIGURE 12-71 *Latimeria,* **a surviving coelacanth in the ocean near Madagascar.** *Latimeria* is a large fish, nearly 2 meters long. It swims slowly, using its paired fins in a manner resembling the way four-legged animals walk on land. Sharp movement of the tail fin provides speed bursts needed to catch prey.

muscle tissue in fish. The Soom Shale specimen, named *Promissum,* had a longitudinal band that can be interpreted as a notochord and conodont elements at its anterior end.

This evidence strongly favors chordate stratus for the conodont. But what function was served by individual conodont elements? Were they supports for food-gathering organs? Did they act as sieves, filter-feeding particles from water? Or were they used for chewing?

If conodont elements functioned as teeth, they should show wear from grasping food, just as our teeth do. Scanning electron microscope images of some conodont elements reveal fine scratches, pitting, and wear—all suggesting the elements processed food. However, other elements have a laminar structure, indicating that the elements were covered by soft tissue that secreted the laminae.

▶ ADVENT OF TETRAPODS

It required tens of millions of generations to convert crossopterygian fishes into animals that could live on land (Fig. 12-73). Even so, the conversion was not complete, for the early tetrapods we call amphibians continued to return to water to lay their fishlike naked eggs. From these eggs came fishlike tadpoles which, like fish, used gills for respiration.

Coming Ashore: First Four-Legged Vertebrates

Important changes accompanied the shift to dwelling on land. A three-chambered heart developed to route blood more efficiently to and from the lungs. Limb bones and the skeletal structures that supported them were modified to overcome the constant tug of gravity and to better hold the body above the ground. The spinal column, a simple structure in fishes, was transformed into a sturdy, flexible bridge of interlocking elements.

The first organs for hearing evolved underwater. Thus, as creatures moved onto land, these organs had to adapt to hearing in the air. The old hyomandibular

3 mm

FIGURE 12-72 **Conodont elements (upper) and animal restoration (lower).** The elements shown (magnified 50 times) are typical, occurring as disassociated fragments. They usually are smaller than a millimeter but are durable, composed of a calcium phosphate variety of the mineral apatite. The restoration of the conodont animal (below) is interpreted from fossil remains in the Lower Carboniferous Granton Shrimp Bed of Scotland.

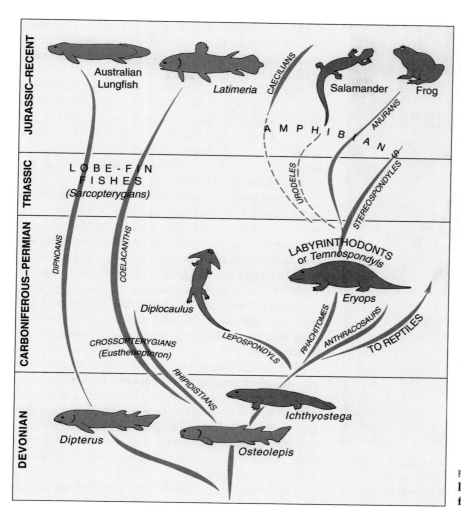

FIGURE 12-73 **Evolution of lobe-fin fishes and amphibians from Devonian fishes.**

bone, used in fishes to prop the braincase and upper jaw together, became transformed into an ear ossicle—the stapes. (See "Earbones Through the Ages" in Chapter 6.) The fish spiracle (a vestigial gill slit) became the eustachian tube and middle ear. To complete the auditory apparatus, a tympanic membrane (eardrum) developed across a prominent notch (otic notch) in the rear of the skull.

The fossil record for amphibians begins in the Late Devonian with a group of fish called **ichthyostegids**

(Fig. 12-74). As suggested by their name, these creatures retained many features of their fish ancestors, including the tail fin, bony gill covers, fishlike vertebrae, and skull bones that closely resemble the bones of crossopterygians.

The amphibians that followed the ichthyostegids fall mostly within a group informally called **labyrinthodonts**. The name refers to their teeth—like those of the crossopterygians (Fig. 12-69), they were characterized by labyrinthic folding of enamel. During

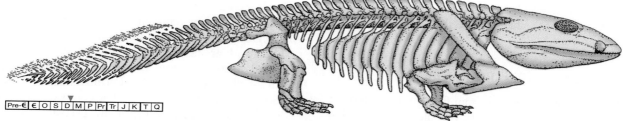

FIGURE 12-74 **The *Ichthyostega* skeleton retains the fishlike form of its crossopterygian ancestors.**

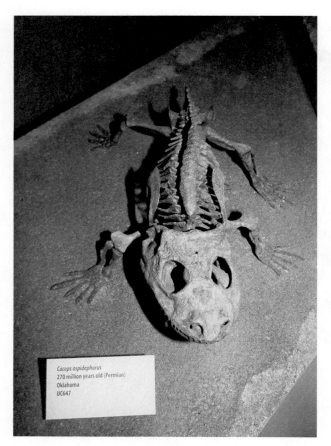

FIGURE 12-75 *Cacops*, **a small labyrinthodontic amphibian from the Lower Permian.** Like many labyrinthodonts, *Cacops* had a heavy skull pierced by five openings: two eye orbits, two nostril openings, and a pineal opening that contained a median light receptor. A prominent notch (otic notch) at either side of the back of the skull accommodated a large eardrum.

the Carboniferous, large numbers of labyrinthodonts wallowed in swamps and streams eating insects, fish, and one another.

Except for the need to return to water to reproduce, many Carboniferous and Permian labyrinthodonts lived out most of their lives on solid ground. *Cacops* (Fig. 12-75) was one such land-dweller. Typical of labyrinthodonts, *Cacops* had a heavy skull pierced by five openings: two eyes, two nostrils, and a pineal opening that served as a light receptor. A prominent notch on each side of the back of the skull accommodated a large eardrum.

Amniotes: Reptiles, Birds, and Mammals

The evolutionary advance from fish to tetrapods was a great biologic achievement, yet there was another equally significant event. Among the evolving land ver-

tebrates were some that developed a way to reproduce *without returning to the water*. This new manner of reproduction came with the development of the *amniotic egg* (Fig. 12-56).

The oldest fossil amniotes have been recovered from 300-million-year-old swamp deposits in Nova Scotia. Named *Hylonomus*, they were slender reptiles about 24 cm long. *Paleothyris* was a similar reptile found in beds of the same age. Both resembled modern lizards. Their skeletal remains are often found in the hollowed stumps of fossil trees.

Judging from their small, sharp teeth, they ate insects and millipedes. We identify these lightly built tetrapods as true reptiles because the postcranial skeleton and bones at the rear of the skull are distinctly reptilian, and they lack the otic notch seen in amphibian precursors.

Other early reptiles have been found in somewhat younger Carboniferous strata in Kansas. The most common is *Petrolacosaurus*, which has a longer neck and legs than *Hylonomus*. Its skull details suggest it was an early member of the lineage that would eventually lead to dinosaurs.

By Late Carboniferous, amniotes had diverged into two branches: reptiles and **synapsids**. Both groups first appear in the Pennsylvanian. The traditional view of synapsids, still favored by some paleontologists, is that they are a subclass of Reptilia. However, cladistic analyses indicate otherwise, because they diverged from ancestors completely different from *Hylonomus* and other true reptiles.

The small and relatively primitive groups of the Pennsylvanian provided the stock from which a succession of synapsids evolved. The most spectacular were **pelycosaurs**, several species of which sported erect "sails" supported by rodlike extensions of the vertebrae. The pelycosaurs were varied, including plant-eaters like *Edaphosaurus* and flesh-eaters like *Dimetrodon* (Figs. 12-76 and 12-77). Flesh-eaters are identified by their arched skulls, great jaws, and sharp recurved teeth.

Many researchers have attempted to explain the function of the pelycosaurian sail. Most logically, it was a "cooling fin," regulating the animal's body temperature by alternating as a solar heat collector when the animal felt cold and as a radiator of surplus heat when it felt hot. Because pelycosaurs had several mammalian skeletal characteristics, they are interpreted as early representatives of the reptilian group from which an advanced group of synapsids arose, known as therapsids.

Therapsids are fascinating fossil vertebrates, widely dispersed during the Permian and Triassic. They have left an excellent fossil record in Permian beds of, Russia, South America, India, Antarctica, and the Karoo basin of South Africa, as well as in Triassic beds in Madagascar.

FIGURE 12-76 **Permian sailback reptiles.** The carnivorous sailback *Dimetrodon* is in the foreground. *Edaphosaurus*, a plant-eater, is in the distance.

FIGURE 12-78 **Mammal-like reptiles.** Three carnivorous forms (*Cynognathus*) are about to attack a plant-eating therapsid reptile (*Kannemeyeria*). Mammal-like reptiles had fewer skull bones than generally found in reptiles, and a mammal-like enlargement of the lower jaw bone at the expense of more posterior elements of the jaw. A double ball-and-socket articulation evolved between the skull and neck. Ribs were shorter in the neck and lumbar region for greater overall flexibility. Teeth in *Cynognathus* show a primitive, crude differentiation into incisors, canines, and cheek teeth.

Therapsids were predominantly small to moderate-sized animals having several mammalian skeletal traits. These mammal-like features are well developed in the therapsid *Cynognathus* (Fig. 12-78).

One group of therapsids became particularly common during the Triassic. These were the cynodonts (Latin: *dog-toothed*). Mammal-like traits were even further refined in this group. For example:

- Cheek teeth changed to more efficiently slice food for easier digestion—important because efficient food processing allows faster energy from food.

- A bony palate permitted breathing while chewing, an important adaptation for an animal evolving toward mammalian warm-bloodedness. (Efficient breathing produces ample oxygen needed to derive heat energy from food.)

- A bone on the snout portion of the skull had probable whisker pits, suggesting a covering of hair to slow loss of body heat.

FIGURE 12-77 **Mounted skeleton of the Permian "sail-reptile" *Dimetrodon gigas*.** The sail may have been a cooling fin, regulating the animal's body temperature by alternating as a solar heat collector and a radiator of surplus heat.

► PLANTS OF THE PALEOZOIC

The history of plants is very, very long. Stromatolites, those fossil masses formed of countless tiny blue-green algae, are a good indicator. They existed in the late Archean, expanded during the Proterozoic, and are

FIGURE 12-79 **Fossil cyanobacteria.** Hummocky stromatolites of Cambrian age along the banks of Missouri's Black River.

FIGURE 12-80 **The receptaculid *Fisherites*, from the Ordovician Kimmswick Formation of Missouri.** Sometimes called sunflower corals, receptaculids are the fossil remains of a type of green algae—they are not actually related to corals. About half of a complete specimen is shown here. Note the pattern of intersecting spiral lines that define small rhombic openings. (About 10 × 16 cm.)

present in limestones of the Phanerozoic. Cambrian stromatolite reefs (Fig. 12-79) were widespread but became less common during the Ordovician. Then as now, they may have flourished only where grazing marine invertebrates were few. Many grazing invertebrates had appeared by late Cambrian time.

Cyanobacteria were responsible for stromatolite growth. Their evolution was an important first step toward colonizing continental environments. Most Precambrian stromatolites grew in shallow marine and intertidal environments, but some grew in freshwater bodies as well. Thus, stromatolitic cyanobacteria were hardy organisms that were able to migrate into brackish bays and estuaries and eventually into streams and lakes.

Chlorophytes, a green algae, probably was the next step in the evolutionary journey toward land plants. A close relationship between chlorophytes and land plants is suggested by the adaptation of some species to freshwater bodies and even moist soil. Both green algae and land plants possess the same kind of green pigment and produce the same kind of carbohydrate during photosynthesis.

Receptaculids are another relatively common group of marine algal fossils in lower Paleozoic rocks (Fig. 12-80). Receptaculids are produced by organisms of uncertain classification, but investigators interpret them as lime-secreting algae. They resemble the seed-bearing part of a sunflower, with its criss-crossing spirals.

Land Plants

Today's land plants include **bryophytes** (mosses, liverworts, and hornworts) and **tracheophytes** (trees, ferns, and flowering plants). Tracheophytes have vascular tissues that transport water and nutrients from one part of the plant to another.

The importance of the vascular system is obvious when you recall that above-ground moisture is undependable in most land areas. There is nearly always a greater persistence of moisture and plant nutrients within the pore spaces of soil. Beneath the surface, however, there is no light for photosynthesis. The vascular system is a solution: In tracheophytes, it permits part of the plant to function underground, where there is water but no light, and another part to grow where water supplies are uncertain but there is sunlight for photosynthesis.

The earliest fossil evidence for land plants consists of spores, which are microscopic plant reproductive structures. Four spore bodies grouped into a *tetrad* (tetrahedral arrangement) occur in rocks as old as Ordovician. These tetrads are the reproductive and dispersal elements of ancestors for both bryophytes and tracheophytes. By Silurian time (about 425 million years ago), spores with a distinctive three-rayed scar appeared. These *trilete* spores are characteristic reproductive bodies of primitive tracheophytes.

Three major advances occurred in land plant history, each involving development of increasingly more effective reproductive systems:

1. Seedless, spore-bearing plants, such as those common in the great coal-forming swamps of the Carboniferous.

2. Late Paleozoic evolution of seed-producing, pollinating, but nonflowering plants (gymnosperms).

3. Late Mesozoic evolution of plants with both seeds and flowers (angiosperms).

Early Paleozoic land plants profoundly altered the environment. Their roots gripped the soil and slowed erosion. Decaying vegetation promoted formation of soils. Even more important, plants provided food and shelter for emerging land animals.

The transition from aquatic plants to land plants apparently was difficult, for it was late in coming. The first unquestioned vascular plant remains occur in Silurian rocks (435–410 million years ago). An example is *Cooksonia* (Fig. 12-81). By Devonian time, other species appeared—*Aglaophyton* (Fig. 12-82) and *Rhynia* (Fig. 12-83). True woody vascular tissue can be discerned in fossils of *Rhynia*.

Although small, early plants paved the way for the evolution of large trees. Because of the evolution of wood, plants were able to stand against the vertical pull of gravity and horizontal force of strong winds. By Devonian time, lofty, well-rooted, leafy trees had made their appearance (Fig. 12-84).

Among Carboniferous plants were so-called scale trees, or **lycopsids**. Today they are represented by their smaller descendents, the club mosses, including *Lycopodium*. The forked branches of *Lepidodendron* reached 30 meters toward the sky (Fig. 12-85). Side-by-side with the Carboniferous lycopsids grew the sphenopsids (Figs. 12-86 and 12-87) and true ferns (Fig. 12-88).

Seed plants evolved during the late Paleozoic. The earliest are called "seed ferns," because they had fern-like leaves, but unlike true ferns, they reproduced by means of seeds. They were the dominant plant group of the Late Paleozoic.

In Gondwana continents, *Glossopteris* is an important Carboniferous and Permian seed plant (Fig. 7-29). Most species of *Glossopteris* had thick, tongue-shaped leaves.

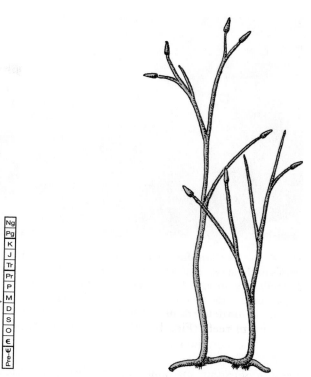

FIGURE 12-82 **Early Devonian non-vascular land plant, *Aglaophyton*.** *Aglaophyton* grew to heights of 20–25 cm and had a simple conducting strand for moving water up the stem. This restoration is from Scotland.

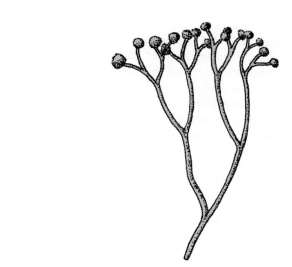

FIGURE 12-81 **Late Silurian/Early Devonian vascular plant, *Cooksonia*.** These were small, leafless plants with thin, evenly branching stems surmounted by the spore-bearing bodies called sporangia. Height is about 4 cm.

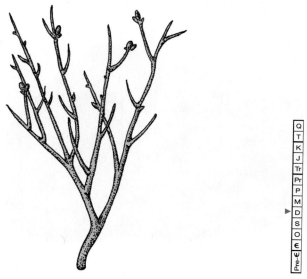

FIGURE 12-83 **Middle Devonian vascular land plant, *Rhynia*.** Xylem, the vascular tissue that conducts water and dissolved minerals, has been unmistakably identified in fossils of *Rhynia*. Because of its thick-walled cells, xylem also gives strength to a plant. A major group of early land plants, the rhynophytes, takes its name from *Rhynia*.

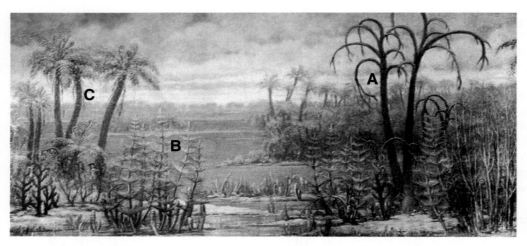

FIGURE 12-84 Middle Devonian forest in eastern U.S. One such forest containing thick stands of the spore-bearing tree *Archaeopteris* covered the present area of the Sahara Desert 370 million years ago. Another Devonian forest stood near the present site of Gilboa, New York. Some of the Gilboa trees were over 7 m tall, but were to be dwarfed by their Carboniferous descendants. (A) An early lycopod, *Protolepidodendron*. (B) *Calamophyton*, an early form of the horsetail rush. (C) Early tree fern, *Esopermatopteris*.

Because of certain anatomic traits and because of their association with glacial deposits, *Glossopteris* and associated plants were probably adapted to cool climates.

Fossils of seed plants include the cordaites (Fig. 12-89). These primitive conifers included very large trees, some 50 meters tall. Cordaites and more modern cone-bearing conifers spread widely during the Permian, perhaps as a consequence of drier climatic conditions. The first ginkgoes appeared during the Permian

and became abundant during the Jurassic. Today, only a single species, *Ginkgo biloba*, survives from this once-flourishing group. The tree is readily identified by its fanlike leaves.

Extinction overtook many plant groups near the end of the Permian Period. Many species of lycopsids, seed ferns, and conifers disappeared. However, small ferns that grow in damp areas were not profoundly affected by the crisis.

◀ MASS EXTINCTIONS

For most of the Paleozoic, Earth was populated by a rich diversity of life. However, there were times when the planet was less hospitable, and large groups of organisms suffered extinction (Fig. 12-90). Early geologists saw evidence of these mass extinctions in the fossil record and used the extinctions to define the boundaries between geologic systems, as you can see in the figure.

Of the many episodes of mass extinction, five were particularly catastrophic: one at the end of the Ordovician, one late in the Devonian, one near the end of the Permian, one late in the Triassic, and one that marks the end of the Cretaceous.

Late Ordovician Extinctions

Extinctions near the end of the Ordovician Period occurred in two phases. During the first phase, the principal victims were planktonic (floating) and nektonic (swimming) organisms such as graptolites, acritarchs, many nautloids, and conodonts, as well as such benthic (sea bottom) creatures as trilobites, bryozoa, corals,

Pre-Є | Є | O | S | D | M | P | Pr | Tr | J | K | Pg Ng

FIGURE 12-85 Fossilized bark of Carboniferous tree, Lepidodendron. The elongate leaves of these scale trees emerged directly from the trunks and branches, leaving a regular pattern of leaf scars upon being released. In *Lepidodendron*, the scars are arranged in diagonal spirals, whereas species of *Sigillaria* have leaf scars in vertical rows. Leaf scars are about 1 cm wide.

FIGURE 12-87 *Annularia*, an abundant Pennsylvanian-age sphenopsid. Living sphenopsids include the scouring rushes or horsetails.

FIGURE 12-86 *Calamites*, a sphenopsid. Plants shown are about 3 to 5 m tall. Fossil sphenopsids possessed slender, unbranching, longitudinally ribbed stems with a thick core of pith and whorls of leaves at each transverse joint. At the top, a cone bore the spores that would be scattered in the wind.

and brachiopods. In the second phase, several trilobite groups perished, and corals, conodonts, and bryozoans were severely reduced in numbers and diversity.

Both phases of extinction appear to be related to global cooling associated with the growth of Gondwana ice caps. The fossil record reveals that habitable zones of many plant and animal groups were shifted and compressed toward the equator in response to cooling at higher latitudes. With cooler conditions, tropical organisms were hit the hardest.

As ice accumulated in the continental glaciers, sea level was lowered. This caused a loss of epicontinental seas and shallow marginal shelf environments. These seas had been optimum areas for the proliferation of many groups of marine invertebrates.

The second phase of Ordovician mass extinction accompanied the global rise in sea level caused by melting of ice sheets and warmer conditions. Organisms

tolerant of cooler conditions became distressed and many became extinct.

Late Devonian Extinctions

Following the late Ordovician extinction, life expanded again (Fig. 12-90), building slowly in the early Silurian but attaining rich levels of diversity during the Devonian. By the end of the Devonian, however, marine invertebrates again were confronted with environmental stress, as indicated by the decimation of once extensive Devonian reef communities.

Reef-building tabulate corals and stromatoporoids are rarely seen in rocks deposited after the Devonian. Of rugose coral species numbering in the scores, only a few species survived. Brachiopods, goniatites, trilobites,

FIGURE 12-88 *Pecopteris*, a true fern from the Pennsylvanian strata in Illinois. (The penny shows scale.) True ferns lived in the coal forests, and many were tall enough to be classified as trees. Like lycopsids and sphenopsids, they reproduced by spores carried in regular patterns on the undersides of their leaves.

FIGURE 12-89 *Cordaites.* The end of the branch shows straplike leaves, some of which were as long as 1 m. The clustered bodies produced the plant's male gametes.

conodonts, and placoderms were severely reduced in numbers and variety. Altogether, a remarkable 70% of the ocean's families of invertebrates disappeared in the Late Devonian.

DEVONIAN EXTINCTIONS—ASTEROID, OR SOMETHING SLOWER?

You probably have heard the many hypotheses that a sudden event, such as the impact of a large asteroid, caused the extinctions. But Late Devonian extinctions occurred over a span of 20 million years. So, it is unlikely that any momentary event caused the Devonian extinctions. Nor have we found heavy metal concentrations that might result from a shattered asteroid.

More likely, the cause was something much slower and more complex: an ecological crisis. For example, the extensive forests of huge trees that spread across the continents for the first time would have produced organic acids that might have accelerated rates of weathering. Rapidly decomposing bedrock could have released huge volumes of phosphorus and other minerals that, on reaching the ocean, might have produced explosive growths of algae. Bacteria decomposing the remains of dead algae would have robbed the ocean of oxygen. The process, called eutrophication, causes fish kills today in lakes subjected to similarly excessive algal growth.

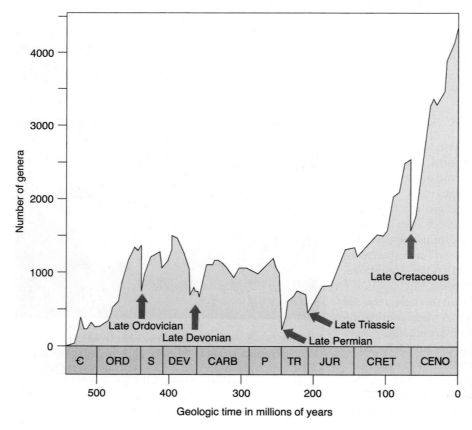

FIGURE 12-90 **Five major mass extinction episodes.** Mass extinctions are worldwide extinctions over a short span of geologic time. This graph shows the diversity of marine animals, compiled from a database of first and last occurrences of more than 34,000 genera.

Evidence of anoxic conditions is present in the extensive tracts of Devonian black shales, a rock type that develops in oxygen-deficient waters. The Chattanooga Shales mentioned in the previous chapter are an example. Also, many late Devonian limestones are rich in the isotope carbon-13 and relatively poor in carbon-12. An explosive growth of algae would explain this, because during photosynthesis, algae use more carbon-12 than carbon-13. Thus, the percentage of the heavier isotope would increase in ocean water, and become incorporated into limestones and the shells of marine invertebrates. The eutrophication hypothesis may also explain why corals were so severely decimated. Corals live in warm regions adjacent to land areas densely covered by tropical forests.

Some geologists prefer a hypothesis that blames the Late Devonian crisis on continental glaciation. By Late Devonian time, South America had drifted to a position over the South Pole, and glaciation occurred. Global cooling and reduction of shallow-water environments may have contributed to the extinctions.

Late Permian Extinctions— Terrestrial Causes?

Decimation of life at the end of the Permian Period 251 million years ago has been described by Smithsonian paleontologist Douglas Erwin as "the mother of mass extinctions." The loss of biologic diversity at this time exceeded that of all other extinctions, including the one involving the demise of the dinosaurs 65.5 million years ago. By some estimates, more than 90% of all existing marine species and 70% of animals living on land disappeared or were severely reduced. The land animals that disappeared included entire families of amphibians, primitive reptiles, and synapsids.

Of all groups, however, tropical marine invertebrates experienced the most extensive losses. Fusulinids that had once populated large areas of the seafloor disappeared. So did rugose corals, many families of crinoids, productid brachiopods, lacy bryozoa, and many groups of ammonoids. Trilobites were not involved in this extinction, for they already had disappeared in an earlier stage of the Permian.

Many factors may have contributed to the late Permian mass extinction:

- At the time, the supercontinent Pangea had completed its development. More rigorous climatic conditions existed across the great continent, as is characteristic of the interiors of large continents today. Long-term climatic change is reflected in the way spore-bearing plants gave way to conifers, ginkgoes, and other plants tolerant of cooler conditions.
- Frigid polar regions existed at both the north and south ends of Pangaea. Organisms accustomed to warm waters shifted toward the lower latitudes, but many did not survive. Many tropical organisms could not survive the cooling, which also prevented construction of organic reefs and thereby impeded the formation of limestones.
- Epicontinental seas had drained or were very limited, reducing favorable sites for shallow marine invertebrates.
- Pangea blocked equatorial circulation in the ocean, disrupting the life zones of many organisms.
- The late Permian also saw extraordinary volcanic activity, with one of the Earth's greatest episodes of flood basalt volcanism occurring in Siberia. Carbon dioxide and dust released into the atmosphere during that volcanic episode may have increased the greenhouse effect. With the onslaught of warm conditions, large stores of methane gas frozen in sediments on the ocean floors may have been released, driving global warming to even more disastrous levels.

Late Permian Extinctions— Extraterrestrial Cause?

Could the end-Permian mass extinction be due to an extraterrestrial body striking Earth? Recent evidence from analysis of sedimentary rocks at the Permian-Triassic boundary indicate the possibility of such a cataclysmic event. The sediments contain a high concentration of soccer-ball shaped carbon molecules called *fullerenes* (also dubbed "buckyballs" because they resemble the geodesic dome invented by Buckminster Fuller). Atoms of helium and argon are trapped within the fullerenes, and the ratio of the two is similar to that found in meteorites.

Also, the isotope of helium found in the fullerenes is extraterrestrial helium-3, whereas the terrestrial isotope is helium-4. Significantly, similar fullerenes have been found in two well-known carbonaceous chondrite meteorites: the Allende and Murchison meteorites. This evidence supports an impact by an extraterrestrial body.

If an extraterrestrial body did strike Earth at the end of the Permian, the event would be only the first domino to fall in a cascade of disastrous events:

- The impact would have generated tsunamis much greater than the 2004 Indian Ocean tsunami (which killed over 200,000 people). These giant tsunamis would have devastated low-lying continental margins.
- The impact would throw so much water vapor and dust into the atmosphere as to block much of the sun's radiation.
- Disturbance of the crust likely would trigger rampant volcanism, which in turn would change the composition of the atmosphere, harmfully affect climate, and cause ocean anoxia. All living things would be in danger of extermination.

SUMMARY

- The fossil record of the Paleozoic is immensely better than that of the preceding Proterozoic. Shell-building is one important reason for the improvement.
- Archaeocyathids, brachiopods, and trilobites were abundant during the Cambrian, and by Ordovician time, all major invertebrate phyla were well established.
- The best fossils of soft-bodied invertebrates are those of the Early Cambrian Chengjiang fossil site and the Middle Cambrian Burgess Shale fauna. They include many previously unknown arthropods, echinoderms, sponges, and cnidarians as well as animals believed to be the earliest known chordates and jawless fishes.
- Among the other important invertebrate groups of the Paleozoic were foraminifera (fusulinids expanded during the Pennsylvanian and Permian), sponges, corals (tabulate and rugose), bryozoa, mollusks, echinoderms (especially blastoids and crinoids), and graptolites.
- The first fossils of vertebrate animals are found in rocks of Cambrian age. They are remains of a group of jawless fishes known as agnathids. Ostracoderm agnathids expanded during the Ordovician and Silurian.
- Archaic jawed fishes appeared late in the Silurian but did not begin to dominate the marine realm until Devonian time. This was also the time when cartilaginous (chondrichthyan) and bony (osteichthyan) fishes began their rise to dominance.
- A special group of bony fishes called crossopterygians arose from osteichthyan stock. These fishes could breathe air by means of accessory lungs and possessed muscular fins that provided short-distance overland locomotion.
- The bones within the fins of crossopterygians resembled those of primitive amphibians. Other skeletal traits clearly suggest that these Devonian fishes were the ancestors of the first tetrapods—the ichthyostegids.
- From a start provided by ichthyostegids, tetrapods called labyrinthodonts underwent a successful adaptive radiation that lasted until the close of the Triassic. Long before their demise, however, they provided a lineage from which synapsids and reptiles evolved.
- The transition from non-amniotic tetrapods like amphibians to amniotic reptiles involved the development of the amniotic egg, an evolutionary advancement that liberated land vertebrates from the need to return to water bodies to reproduce.
- During very late Pennsylvanian and Permian, amniotes underwent an elaborate evolutionary radiation that produced the fin-back reptiles and therapsids as well as several other major reptilian groups.
- The earliest Paleozoic precursors to plants were mainly unicellular and microbial, with the most obvious larger fossils being stromatolites.
- Fossil spores indicate that by late Ordovician, spore-bearing vascular land plants were present. It is probable that they evolved from the green algae (chlorophytes). These primitive plants were followed in the Devonian and Carboniferous by lycopsids, sphenopsids, and seed ferns. In temperate regions of Gondwana, a flora dominated by the *Glossopteris* developed. Permian forests included plants more tolerant of drier and cooler conditions, such as conifers and ginkgoes.
- Life did not expand steadily throughout the Paleozoic Era. There were difficult times when major groups of organisms became extinct. These mass extinctions occurred at the end of the Ordovician, Devonian, and Permian periods. During the extinction at the end of the Permian, nearly half of the known families of animals disappeared.
- The decimation of marine animals was particularly dramatic. It included the loss of fusulinids, spiny productid brachiopods, rugose corals, two orders of bryozoans, and many taxa of stemmed echinoderms, including blastoids.
- Among the vertebrates, over 70% of vertebrate families perished. Fortunately, a few groups survived the Permian crisis and were able to continue their evolution during the Triassic.
- The cause of the Permian biologic crisis is not definitely known. Many believe it was the result of terrestrial causes such as global warming resulting from massive and prolonged volcanic activity. Other scientists put the blame on the catastrophic impact of a large meteorite or asteroid.

KEY TERMS

acanthodian, pp. 354, 356

actinopterygian fish, p. 357

Agnathans, p. 354

ammonoid, p. 344

amniotic egg, pp. 329, 353–354

amniotic membrane, p. 353

amniotic vertebrate, p. 353

antiarch, p. 356

archaeocyathid, p. 335

arthrodire, p. 356

arthropod, p. 345

Asteroidea, p. 348

basal tetrapod, p. 329

bioturbation, p. 334

Bivalvia (or **Pelecypoda**), p. 341

Blastoidea, pp. 348, 350

blastopore, p. 351

blastula, p. 351

brachiopod, p. 339

bryophyte, p. 364

bryozoan, p. 338

Burgess Shale fauna, p. 331

cephalopod, p. 342

ceratite, p. 345

Chengjiang fauna, p. 333

chlorophytes, p. 364

chondrichthyan fish, p. 354, 357

chordate, pp. 333, 354

Cnidaria, p. 337

cnidocyte, p. 337

conodont, pp. 359–360

crinoid, pp. 348, 351

crossopterygian fish, p. 358

Cystoidea, pp. 348, 350

deuterostome, p. 351

dipnoan fish, p. 358

echinoderm, p. 348

Echinoidea, p. 348

Edrioasteroidea, p. 348

epifaunal animal, p. 334

eurypterid, p. 347

filter feeders, p. 334

foraminifera, p. 334
fusulinid, p. 334
gastropod, p. 341
goniatite, p. 345
graptolite, p. 352
ichthyostegid, p. 361
infaunal animal, p. 334
labyrinthodont, p. 361
lycopsid, p. 365
mass extinctions, p. 329
mollusk, p. 341
monoplacophoran, p. 341
nautiloid, p. 344

notochord, p. 333
Ophiuroidea, p. 348
osteichthyan fish, pp. 354, 357
ostracod, p. 347
ostracoderm, p. 354
pelycosaur, p. 362
placoderm, p. 354, 356
placophoran, p. 341
polyplacophorans, p. 341
Porifera, p. 335
protostome, p. 351
pterobranchs, p. 352
radiolaria, p. 335

recaptaculid, p. 364
sarcopterygian fish, p. 357
sediment feeders, p. 334
small shelly fossils, p. 329
stromatoporoid, p. 336
synapsid, p. 362
tetrapod, pp. 329, 353
therapsid, p. 363
tracheophyte, p. 364
trilobite, p. 346
vertebrate, pp. 333, 353
water vascular system, p. 348

QUESTIONS FOR REVIEW AND DISCUSSION

1. Which of the Chengjiang and Burgess Shale fossils can be chordates? What is the basis for this assignment?

2. To what geologic system(s) of the Paleozoic would rocks containing the following fossils be assigned?

   a. Fusulinids

   b. Archaeocyathids

   c. *Archimedes*

3. The phylum Cnidaria was formerly termed Coelenterata. Why is Cnidaria an appropriate name for these animals? What classes of Cnidaria lived only during the Paleozoic?

4. How do the sutures of nautiloids differ from those of ammonoid goniatites?

5. What are conodont elements? What other animal group has hard tissue (tooth or bone) of the same composition as conodont elements?

6. What groups of Paleozoic invertebrates had become extinct by the end of the Paleozoic?

7. How are shallow shelf and epicontinental sea environments affected by episodes of continental glaciation?

8. Define and discuss the principal characteristics of the Echinodermata, Trilobita, Mollusca, Brachiopoda, Bryozoa, and Porifera.

9. Discuss the advantages that accrued to invertebrates that evolved shells.

10. Distinguish between the following:

   a. Ostracoderms and placoderms

   b. Osteichthyes and chondrichthyes

   c. Crossopterygians and labyrinthodonts

   d. Infaunal and epifaunal invertebrates

   e. Synapsids and diapsids

11. Discuss those characteristics of therapsid reptiles that indicate that they were on the main line of evolution toward mammals.

12. What group of algae is considered the probable ancestor of the first land plants (tracheophytes)? For what reasons is this group considered ancestral to land plants?

13. Why was the evolution of a vascular system critical to the invasion of the lands by plants? What were the first plants to make the transition, and what were their characteristics?

14. Describe in general terms the appearance of synapsid reptiles.

WEB SITES

The Earth Through Time Student Companion Web Site (www.wiley.com/college/levin) has online resources to help you expand your understanding of the topics in this chapter. Visit the Web Site to access the following:

1. Illustrated course notes covering key concepts in each chapter;

2. Online quizzes that provide immediate feedback;

3. Links to chapter-specific topics on the web;

4. Science news updates relating to recent developments in Historical Geology;

5. Web inquiry activities for further exploration;

6. A glossary of terms;

7. A Student Union with links to topics such as study skills, writing and grammar, and citing electronic information.

13

Dipping Jurassic and Cretaceous sandstones along Waterpocket Monocline, Capitol Reef National Park, Utah.

Q
T
K
Jr
Tr
▶ P
M
D
S
O
€
Pre-€

Mesozoic Events

The Earth, from the time of the chalk to the present day, has been the theater of a series of changes as vast in their amount as they were slow in their progress. The area on which we stand has been first sea and then land for at least four alternations and has remained in each of these conditions for a period of great length.

—Thomas Huxley, "On a Piece of Chalk"

O U T L I N E

▶ THE BREAKUP OF PANGEA
▶ THE MESOZOIC IN EASTERN NORTH AMERICA
▶ THE MESOZOIC IN WESTERN NORTH AMERICA
▶ THE TETHYS SEA IN EUROPE
▶ GONDWANA EVENTS
▶ SUMMARY
▶ KEY TERMS
▶ QUESTIONS FOR REVIEW AND DISCUSSION
▶ WEB SITES
▶ BOX 13–1 GEOLOGY OF NATIONAL PARKS AND MONUMENTS
▶ BOX 13–2 GEOLOGY OF NATIONAL PARKS AND MONUMENTS
▶ BOX 13–3 ENRICHMENT
▶ BOX 13–4 ENRICHMENT

Key Chapter Concepts

- The supercontinent Pangea began to break apart during the early Mesozoic. Along the rift zones created by the breakup, magma rose through faults and fractures, forming multiple basaltic lava flows.

- During the Triassic Period along the eastern margin of North America, nonmarine red beds, arkosic sandstones, and lake shales filled basins that had formed by downfaulted crustal blocks during the initial stages of rifting.

- As the fragmentation of Pangea progressed, rift zones lengthened to form narrow ocean tracts. These widened progressively to form Earth's present major ocean basins.

- The Gulf of Mexico began to form in the Late Triassic and Jurassic as North America separated from South America. Restricted circulation and warm climates resulted in evaporite deposition. Salt domes, which formed due to confining pressure on deep salt beds, have provided traps for oil and gas.

- During the Cretaceous, epicontinental seas spread widely across North America and Europe. Chalk was a particularly common sediment type deposited in these seas.

- Accretionary tectonics occurred in the Cordilleran orogenic belt during the Mesozoic, as old island arcs, deep sea ridges, and microcontinents accreted to the western margin of North America.

- The Triassic sedimentary record of the western United States consists largely of sands, silts, and clays spread by streams. These largely fluvial deposits are represented by the Moenkopi and Shinarump formations. Late Triassic sedimentation is dominated by aeolian sands such as those of the Navajo Formation.

- The Sundance Formation consists of deposits formed in a Middle Jurassic epicontinental sea that occupied much of the North American Cordillera. As the sea regressed, its former basin was filled with nonmarine sands and silts of the Morrison Formation, noted for its dinosaur remains.

- During the Mesozoic, western North America was subjected to four orogenies: Sonoma, Nevadan, Sevier, and Laramide.

- Outside of North America during the Mesozoic:

1. The Tethys Sea between Eurasia and Africa began to narrow as a result of compression from the northwardly moving African plate.

2. South America and Africa separated. A subduction zone developed along the western margin of South America.

3. During the Late Cretaceous, volcanism in India was accompanied by extrusion of basaltic lavas in immense volumes, forming the Deccan Traps.

The extinction of animals and plants at the end of the Permian Period provides a natural boundary to mark the end of the Paleozoic Era ("ancient life") and the commencement of the Mesozoic Era ("middle life"). The overlying younger strata contain distinctly different populations of animals and plants that form the middle chapter in the history of life. Like the Paleozoic, the Mesozoic ended in a mass extinction that separates it from the youngest era—the Cenozoic ("recent life").

The Mesozoic Era lasted about 185 million years, ending approximately 65.5 million years ago. We divide the Mesozoic into three unequal periods. The Triassic lasted about 51 million years, the Jurassic 55 million years, and the Cretaceous about 80 million years.

During the Mesozoic, many new groups of plants and animals evolved and experienced often spectacular radiations. This was the era in which two new vertebrate classes appeared: birds and mammals. It was also

a time in which the supercontinent Pangea began to rift apart. As the continents drifted apart, the seafloor between them spread. A process of fragmentation and drift had begun that would ultimately lead to the present physical geography of the planet.

THE BREAKUP OF PANGEA

Any history of the Mesozoic Era must begin with Pangea. The same tectonic forces that had drawn the plate segments together to form this great supercontinent now operated to break it up. The dismemberment of Pangea occurred in four stages (Fig. 13-1).

Stage 1 in the breakup of Pangea occurred during the Triassic, with rifting and volcanism along normal fault systems. As North America separated from Gondwana, tensional forces pulled the crust apart. This caused normal faulting along North America's eastern margin, accompanied by emplacement of dikes along fault planes and fractures, by lava flows, and by vent eruptions.

As rifting progressed, Mexico was separated from South America, and the eastern border of North America broke away from the Moroccan bulge of Africa. Oceanic basalts were added to the seafloor of the newly formed and gradually widening Atlantic Ocean. There are many striking similarities between the eastern United States and Morocco in their largely tensional geologic structures and their Triassic volcanic and clastic sedimentary rocks. These now-widely separated regions were on opposite sides of the same axis of spreading and therefore were affected by similar forces and events.

Recently, geologists used seismic techniques to locate the ancient (300-million-year-old) suture zone

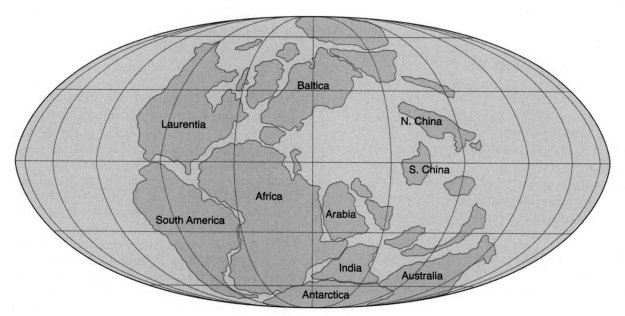

FIGURE 13-1 **The breakup of Pangea begins.** Paleogeographic reconstruction of Earth's surface about 180 million years ago.

formed when Africa and North America originally converged. The suture zone extends east-southeast across southern Georgia. It was left behind when the two continents separated later, about 190 million years ago. This was particularly interesting to paleontologists, who had fossil evidence of the convergence and later separation of the two continents.

Throughout Triassic time, Laurasia north of Nova Scotia remained intact. South America and Africa also remained connected.

Stage 2 in the breakup of Pangea involved the opening of narrow oceanic tracts between southern Africa and Antarctica. This rift extended northeastward between Africa and India and developed a branch that separated northward-bound India from the yet unseparated Antarctica-Australia landmass. The rifting was accompanied by great outpourings of basaltic lavas.

Stage 3 of the breakup of Pangea saw the Atlantic rift begin to extend northward. Clockwise rotation of Eurasia closed the eastern end of the Tethys Sea, a forerunner of today's Mediterranean Sea. By the end of Jurassic time, South America began to split away from Africa. The rift worked its way from the south, creating a long seaway like today's Red Sea between Egypt and Saudi Arabia.

Australia and Antarctica remained connected, but India had moved well along on its journey to Asia. By Late Cretaceous time, about 70 million years ago, South America had completely separated from Africa, and Greenland began to pull away from Europe. However, northeasternmost North America remained attached to Greenland and Baltica.

Stage 4 in the breakup of Pangea occurred early in the Cenozoic Era. The North Atlantic rift separated Laurentia from Baltica, and Australia separated from Antarctica. This final separation of continents occurred only about 45 million years ago. The total time required for the fragmentation of Pangea was about 150 million years.

◤ THE MESOZOIC IN EASTERN NORTH AMERICA

Triassic and Jurassic Periods

The rugged Appalachian Mountain ranges were raised during the Allegheny orogeny at the end of the Paleozoic, and at the beginning of the Mesozoic, they were eroding. Coarse clastics from the uplands filled intermountain basins and other low areas during the Early and Middle Triassic. Then, during the Late Triassic and Early Jurassic, North America began to experience the tensional forces and rifting that preceded actual separation of the continents. Fault-bounded rift basins developed along the east coast (Fig. 13-2).

Downfaulted blocks provided traps that accumulated great thicknesses of poorly sorted arkosic red sandstones and shales (green areas in Figure 13-3).

These nonmarine rocks indicate that rifting had not proceeded to the point where ocean water could enter the rift zone. Sediments deposited in the downfaulted basins constitute the Newark Supergroup of Late Triassic and Early Jurassic age.

The poor sorting of the coarser Newark clastics and their high content of relatively unweathered feldspar grains indicate transportation and deposition by streams flowing swiftly from granitic highlands that bordered the fault basins (Fig. 13-4).

Mudcracks and raindrop impressions provide further evidence of nonmarine deposition. In some places, drainage in the basins became impounded and lakes formed. Remains of freshwater crustaceans, fish, and even the footprints of early dinosaurs are present in Newark rocks (Fig. 13-5).

As the Newark sediments were being deposited, tremendous volumes of lava flowed from fissures, and volcanoes spewed ash onto surrounding hills and plains. All of this volcanic activity was associated with the initial breakup of Pangea. Globally, the outpourings of black basaltic lavas covered a vast 7,000,000 square km, an area the size of the Sahara Desert. And it may have released enough carbon dioxide to cause global warming, which in turn may have contributed to Late Triassic mass extinctions.

Three particularly extensive lava flows and an imposing sill are included within the Newark Supergroup in the New Jersey-New York area. The exposed edge of one sill with columnar jointing forms the well-known Palisades of the Hudson River (Fig. 13-6).

By Late Triassic time, the fault-block mountains produced by rifting were reduced by erosion. The highlands were further worn during the Jurassic and Early Cretaceous until only a broad, low-lying landscape remained. At the contact of resistant igneous and metamorphic rocks of the Appalachian Piedmont and more easily eroded sedimentary rocks of the Coastal Plain, waterfalls developed. The contact is called the "fall line." In colonial times, water wheels were built where streams made their sudden descent, providing water power for textile mills and grist mills for grinding grain. This source of power helped determine the location of major cities like Philadelphia, Baltimore, Washington, Richmond, and Augusta.

South of the old Appalachian-Ouachita orogenic belt, a new depositional basin began to form: the Gulf of Mexico. Beginning in the late Triassic and continuing in the Jurassic, evaporites were deposited in this region (Fig. 13-7). The Gulf of Mexico formed a great evaporating basin, concentrating seawater and precipitating gypsum and salt to thicknesses exceeding 1000 meters (Fig. 13-8).

One of the thick salt beds, the Louann Salt of Jurassic age, is the source of the Gulf Coast's famous **salt domes**. These huge, cylindrical domes are economically

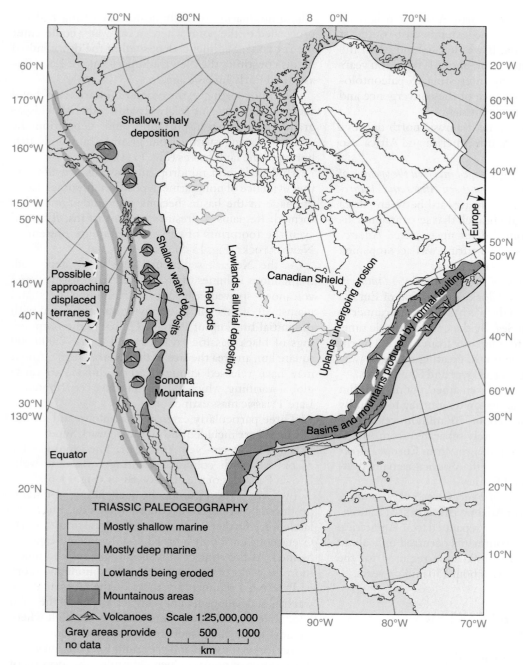

FIGURE 13-2 **Generalized paleogeographic map for the Triassic Period (250–203 million years ago).** ◀ *What caused the faulting along the continent's eastern margin?*

important structures that help entrap oil and natural gas. When compressed by the weight of thousands of meters of overlying sediment, salt flows plastically. As it moves upward in the direction of lower pressure, it produces folds and faults in overlying strata. These structures become oil and gas traps (Fig. 13-9).

Evaporite conditions ended later in the Jurassic. Several hundred meters of normal marine limestones,

limy muds, and sandstones accumulated in the alternately transgressing and regressing seas of the Gulf embayment. Although there are surface exposures of these rocks in Mexico, in the United States they are deeply buried beneath a cover of Cretaceous and Cenozoic sediments. Were it not for the drilling activities of oil companies, little would be known about these rocks.

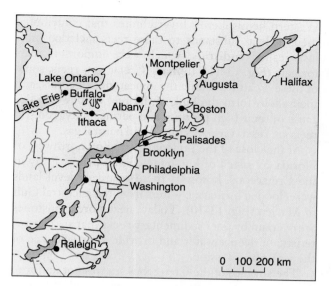

FIGURE 13-3 **Troughlike deposits of Late Triassic rocks in eastern North America.**

Cretaceous (the "Chalk" Period)

The Cretaceous was a time of high sea level and vast epicontinental seas. The Atlantic and Gulf coastal regions received their full share of flooding (Fig. 13-10). The present locations of cities from Boston to Miami lay beneath many meters of seawater, and Minnesota, Iowa, and Missouri would have had shorelines along the interior epicontinental sea.

The Atlantic Coastal Plain, which had undergone erosion for 100 million years since the beginning of the Mesozoic, began to subside early in Cretaceous time. On the subsiding coastal plains, alternating layers of marine and deltaic deposits accumulated, gradually building a great sediment wedge that thickened seaward (Fig. 13-11). Today, the thinner eastern border of the wedge is exposed in New Jersey, Maryland, Virginia, and the Carolinas. The greatest volume of Cretaceous sediment was deposited farther east along the present continental shelf.

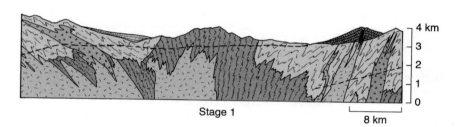

Stage 1

8 km

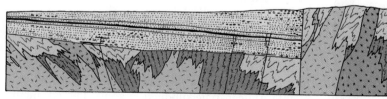

Stage 2

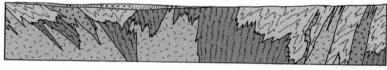

Stage 3

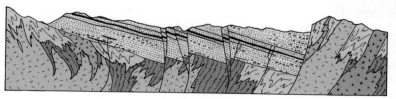

Stage 4

FIGURE 13-4 **Four stages in the Triassic history of the Connecticut Valley.**
(1) Erosion of complex structures develop during the Allegheny orogeny. (2) Mountains have been eroded to a low plain; Triassic sedimentation has begun. (3) Newark sediments and basaltic sills, flows, and dikes accumulate in trough-like fault basins. (4) In Early Jurassic time, the Palisades orogeny breaks the area into a complex of normal faults.

NgPgKJTrPrPMDSO€Pre-€

FIGURE 13-5 **Three-toed footprint cast of a Triassic reptile in Newark Group sandstone.** The footprint is roughly the size of your hand, about 20 cm long. Linear ridges in the slab are casts of mud cracks.

To the south, Florida was a shallow submarine bank during the Cretaceous. The oldest strata are limestones, but later in the period, streams from the southern Appalachians transported clays, silts, and sands into the area. After the Appalachian source areas were worn down, carbonates were deposited above the older sandstones and shales.

During the Cretaceous, carbonate reefs were extensive in the warmer regions of both the Eastern and

FIGURE 13-6 **Palisades of the Hudson River.** "Palisade" (French: *a fence of stakes*) alludes to the vertical *columnar jointing* commonly present in basaltic sills like this. Radiometric dates indicate the Palisades basalt solidified about 190 million years ago. Although most Newark beds are Triassic, spore and pollen studies suggest that in some areas, sediments and lava flows were still accumulating at the beginning of the Jurassic.

Western hemispheres. Invertebrates that contributed their skeletal substance to these reefs included mollusks called **rudistids**, which were important reef-formers (Fig. 13-11). In many Cretaceous reefs, some of these creatures mimic the appearance of corals, and their shells form the bulk of the reef framework. Because of their high porosity and permeability, rudistid reefs are reservoir rocks for oil and natural gas.

At the close of the Jurassic, the region northwest of Florida consisted of low-lying coastal plains. Early in the Cretaceous, however, these relatively level lands were flooded by marine waters from the ancestral Gulf of Mexico (Fig. 13-10). Today, nearshore sandstones are overlain by finer sedimentary rocks that are characteristic of deeper water and provide clear evidence of the sea's transgression.

The advance of the Cretaceous sea was not uniform. An extensive regression occurred near the end of the first half of the period. However, flooding resumed in Late Cretaceous time, when a wide seaway extended from the Gulf of Mexico to the Arctic Ocean (Fig. 13-12).

Chalk (Fig. 13-13) is particularly prevalent worldwide among Upper Cretaceous sediments. Chalk is a white, fine-grained, soft variety of limestone composed mainly of the microscopic calcareous platelets (coccoliths) of golden-brown algae. In fact, the Cretaceous Period takes its name from *creta*, the Latin name for chalk.

▶ THE MESOZOIC IN WESTERN NORTH AMERICA

As eastern North America was separating from Europe and Africa and experiencing crustal extension, the opposite situation prevailed in the west: compressional forces. As the newly formed Atlantic Ocean widened, North America moved westward, overriding the Pacific plate. Thus, deformation in western North America was related to events in the east. It has been shown that the pace of tectonic activity in the North American Cordillera was most intense when seafloor spreading was most rapid in the Atlantic.

Accretionary Tectonics: A Way to Grow

During the Mesozoic, a subduction zone was active along North America's western margin. This was not a simple Andean-type subduction zone. The advancing Pacific plate carried not just ocean basalts and seafloor sediments, but transported entire sections of volcanic arcs, fragments of distant continents, and pieces of oceanic plateaus to North America's western margin. You can see them approaching in the Triassic map (Fig. 13-2).

These fragments, now incorporated into the Cordilleran belt, constitute displaced, alien, or **exotic terranes**. The better documented of these can be iden-

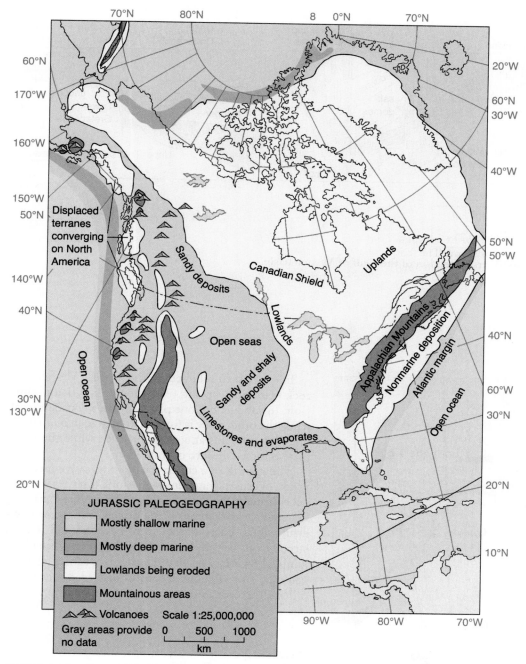

JURASSIC PALEOGEOGRAPHY

- Mostly shallow marine
- Mostly deep marine
- Lowlands being eroded
- Mountainous areas
- Volcanoes   Scale 1:25,000,000
- Gray areas provide no data

0   500   1000
km

FIGURE 13-7 **Generalized paleogeographic map.** Jurassic Period (199.6–145.5 million years ago). ◪ *Describe the conditions at the site of your school during the Jurassic Period.*

tified by distinctive age, rock assemblages, and mineral resources, which indicate whence each fragment came. Many fragments have one or more fault boundaries.

More than 50 exotic terranes have been recognized in the Cordillera. They may constitute as much as 70% of the total Cordilleran region. They indicate that North America grew not only by accretion of materials along subduction zones but also by incorporation of huge crustal fragments formed elsewhere and conveyed to North America by seafloor spreading.

Some of these fragments were so large (and difficult for the subduction zone to "swallow") that they changed the shape of the subduction zone. Others, composed of low-density rocks that were too buoyant for subduction, were scraped off the subducting plate, splintered, and thrust against the continental margin.

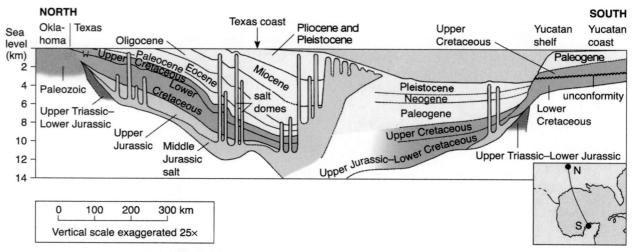

FIGURE 13-8　**North-south cross-section of the Gulf of Mexico basin.**

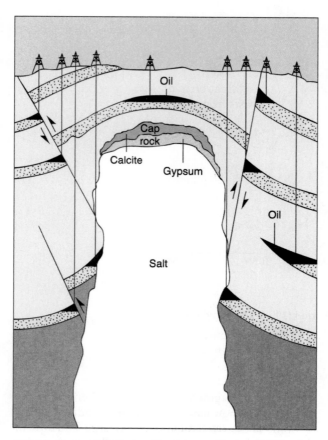

FIGURE 13-9　**Salt dome.** You can see the possibilities for oil entrapment in this structure, due to uparching of strata, faults, and blockage of oil-bearing strata against impermeable rocks.

In contrast to subduction, the process whereby one rock mass rides up and over another is called **obduction**. In subduction, one plate stays at a constant level, while the other plate is forced beneath it; in obduction, one plate remains at a constant level, while the other plate is forced upward over it.

The growth of a continent by progressive incorporation of crustal fragments and exotic terranes is called **accretionary tectonics**. In an oversimplified analogy, imagine ocean icebergs being driven onshore by powerful winds. Each crashes into the coast and is then struck from behind by subsequent arrivals. Finally, a jumbled region of displaced icebergs is constructed. In geology, an analogous region of displaced terranes is called a **tectonic collage**.

Triassic Period: Volcanism, Orogenies, and Arizona's Petrified Logs

We divide the Cordilleran region during the Mesozoic into a western belt of thick volcanic and siliceous deposits and a wide eastern tract adjacent to the more stable interior of the continent (Fig. 13-2). Clastic sediments and volcanics nearly 800 meters (half a mile) thick are exposed in southwestern Nevada and southeastern California—attesting to the instability of the western zone. Geologists speculate that this belt may have resembled the modern Indonesian island arc, with its abundant volcanic and earthquake activity.

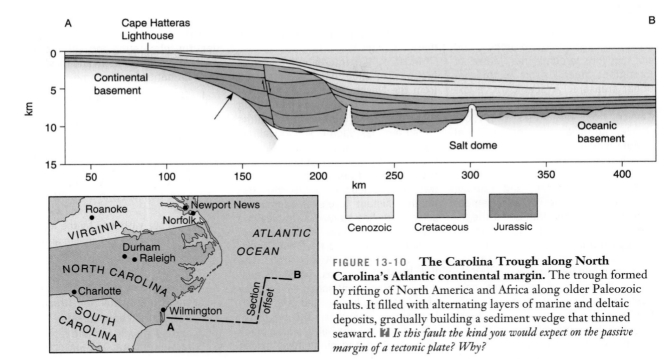

FIGURE 13-10 **The Carolina Trough along North Carolina's Atlantic continental margin.** The trough formed by rifting of North America and Africa along older Paleozoic faults. It filled with alternating layers of marine and deltaic deposits, gradually building a sediment wedge that thinned seaward. ▪ *Is this fault the kind you would expect on the passive margin of a tectonic plate? Why?*

FIGURE 13-11 **Cretaceous rudistid bivalve mollusk from Texas (*Eoradiolites davidsoni*).** Rudistids appeared in the Jurassic and proliferated in the shallow tropical seas of the Cretaceous, becoming important reef-building organisms. Rudistids cemented themselves onto the substrate and became so highly modified that they do not even look like bivalves. Height is 8.6 cm.

SONOMA OROGENY The initial orogenic event of the Cordilleran region is difficult to date precisely, but evidence for deformation at or near the Permian-Triassic boundary exists from Alaska to Nevada. The **Sonoma orogeny** (as it is called in the United States) occurred when an eastward-moving volcanic arc collided with North America's Pacific margin (Fig. 13-14).

At that time, North America was bordered by a westward-dipping subduction zone. The collision welded the island arc onto North America's western edge, adding up to 300 km to the continent's east-west dimension (that's about the present width of California). Oceanic rocks were thrust eastward onto the eroded structures of the Late Devonian Antler orogen. The larger thrust sheets moved scores of kilometers eastward and were themselves broken into many smaller internal faults.

Triassic rocks of the far western Cordillera include thick volcanics and graywackes derived from the island arc. It is uncertain whether these rocks were deposited where they are today or whether they were part of the displaced **Sonoma terrane** that originated in the Pacific Ocean.

Lower Triassic rocks of the eastern Cordillera are shallow marine sandstones and limestones. The thickest section is in southeastern Idaho, where nearly 1000 meters of Lower Triassic sediments accumulated. Eastward from the Cordillera, these marine beds interfinger with continental red beds.

Zion National Park

Towering "temples" of Navajo Sandstone, colorful cliffs and arches, and slot canyons eroded into Mesozoic formations greet visitors to this national treasure. A scenic drive extends through spectacular Zion Canyon from the park's south entrance, northward to the erosional feature called Temple of Sinawava (Fig. A).

Zion Canyon's history is one of relentless mass wasting and running water. Mass wasting is the movement of rock and soil downhill under the influence of gravity. In Zion, this includes large blocks sliding downslope when resistant caprock is undermined by erosion and weathering. Rockslides and rockfalls are common. But mass wasting alone cannot carve a canyon; this demands the erosive and transportational power of streams. The stream is an abrasive conveyor belt, eroding its channel and carrying away the material, plus debris supplied by mass wasting.

Most of the interesting erosional features in Zion have been carved from the Navajo Sandstone, part Triassic and part Jurassic in age. Among these features are towering monoliths called temples. Zion is an old Hebrew name for sanctuary, and refers to these cathedral-like features, reminiscent of the citadel in Palestine that was the nucleus of Jerusalem. Beneath resistant layers of sandstone, softer rock has spalled away, leaving inset arches (blind arches). Freestanding arches also have developed, the largest being Kolob Arch. The beauty of the arches, canyon walls, and temples is enhanced by an array of colors: red, brown, and green from iron oxides, gray-green and bright orange from lichens, and shiny black from the iron and manganese oxides of desert varnish.

With clearly exposed rocks, Zion is ideal for interpreting geologic history. Let us start at the bottom and move upward, formation by formation:

1. The lowermost (hence oldest) rock unit is the Permian (295–250 million-year-old) Kaibab Limestone, a yellow-gray unit exposed in Hurricane Cliffs north of the Kolob Visitor Center. The Kaibab sediments were deposited in a shallow sea.

2. That sea withdrew by the end of the Permian, and shales and sandstone of the Triassic Moenkopi Formation were deposited on the old sea bed.

3. Uplift rejuvenated the streams, and their steep gradients carried gravels into the region to form the Shinarump Conglomerate, a basal member of the Chinle Formation.

4. Most of the Chinle Formation is sands, shales, and volcanic ash beds. Tree branches, logs, and stumps in the Chinle have become petrified, identical to the tree remnants in Petrified Forest National Park. Apparently, much of the silica required for silicification of the wood was derived from dissolution of volcanic ash in the Chinle.

5. Above the Chinle beds are ripple-marked and cross-bedded siltstones of the Moenave Formation.

6. Next up are siltstones of the Kayenta Formation, with dinosaur footprints along its bedding surfaces.

7. Above the Kayenta is the massive Navajo Sandstone, over 220 meters (670 feet) thick in places and the park's principal cliff-former. Lower beds of the Navajo show bedding features, indicating shallow-water deposition. But the greater part of the unit displays curved and wedge-shaped cross-bedding, characteristic of wind-

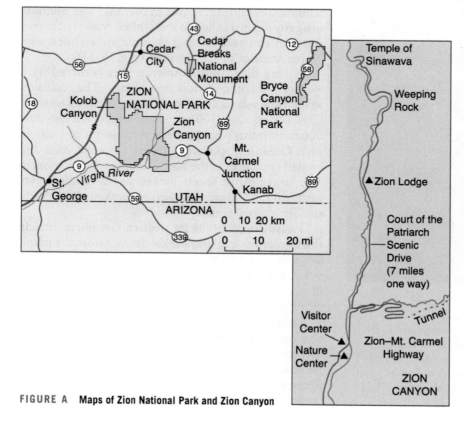

FIGURE A Maps of Zion National Park and Zion Canyon

transported sediment. This interpretation is strengthened by the prevalence of frosted quartz grains, resulting from grain-to-grain impact during wind transport. Cross-bedding is magnificently displayed at Checkerboard Mesa (Fig. 13-25). You can envision the depositional area as a great coastal desert like the Sahara, where dunes migrated along the coastal plain of a shallow sea.

8. Streams laden with red mud provided the material for the fluvial shales of the Temple Gap Formation, which rests atop the Navajo Formation. And finally, these beds were covered by shallow-water limestones of the Jurassic Carmel Formation.

FIGURE B Navajo Sandstone, North Fork of the Virgin River, Zion Canyon

Although seas remained in Canada during the Middle Triassic, in the United States they regressed westward, leaving vast areas of the former seafloor exposed to erosion. Upper Triassic formations rest on the unconformity resulting from this erosion. The Upper Triassic here is continental deposits, delivered by rivers flowing westward across an immense alluvial plain. Starting at the bottom, the strata are:

- The shaley Moenkopi Formation.
- The pebbly Shinarump Conglomerate (Fig. 13-15), derived from uplifted neighboring areas in Arizona, western Colorado, and Idaho.
- The vivid shales and sandstones of the Chinle and Kayenta formations. Their sediments were deposited in stream valleys and lakes.
- The Chinle and Kayenta formations alternate with dune accumulations of the Navajo Sandstone and the Lower Jurassic Wingate Sandstone (Fig. 13-16). Exposed in the walls of Utah's Zion Canyon, they display sweeping cross-bedding like that formed in windblown sandy deposits.

Arizona's Painted Desert is developed mostly in Chinle rocks (Fig. 13-17). The Chinle Formation is known for its petrified logs of conifers. Each year, thousands of people visit Petrified Forest National Park to examine these ancient trees, now turned to colorful agate (Fig. 13-18).

Jurassic Period Orogenies: Nevadan, Sevier, and Laramide

NEVADAN OROGENY Most Mesozoic orogenic activity in the Cordillera resulted from the continuing subduction of oceanic lithosphere beneath the North American plate's western margin. That subduction varied in rate, inclination, and to a small degree in direction. It resulted in eastward-shifting phases of deformation, initially affecting the far western Cordillera and then proceeding eastward to the craton's margin. This is called the **Nevadan orogeny**.

The Triassic, and increasingly the Jurassic and Cretaceous, saw folding, faulting, and metamorphosis of the graywackes, mudstones, cherts, and volcanics

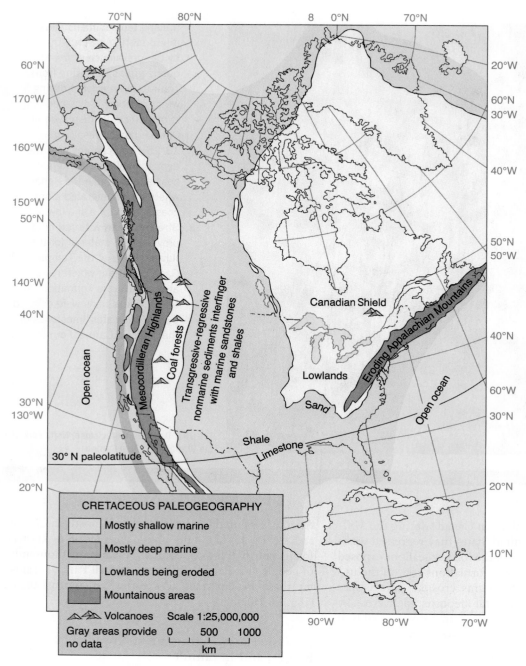

FIGURE 13-12 **Generalized paleogeographic map.** Cretaceous Period (135–65 million years ago).

that had been swept into the subduction zone. The crumpled, altered rock sequences that are trapped between converging plates are called a **mélange** (may-LAWNJ), French for "jumble." The Franciscan fold belt of California is a good example (Fig. 13-19).

In addition to this sedimentary deformation, the subduction zone also generated great volumes of

magma that intruded overlying rocks repeatedly, cooling to form granite batholiths during the Jurassic and Cretaceous. The Sierra Nevada batholith (Fig. 13-20), Idaho batholith, and Coast Range batholith (Fig. 13-21) illustrate the vast scale of this magmatic activity, both during and after the Nevadan orogeny.

A

B

FIGURE 13-13 CHALK. (A) Sea cliffs of chalk along the Dorset coast of England. (B) Close-up view of the chalk with dark nodules of flint, a gray or black variety of chert.

SEVIER OROGENY A second tectonic phase called the **Sevier orogeny** occurred east of the present-day Sierra Nevada Mountains. It affected shallow-water carbonates and terrigenous clastics deposited from Middle Jurassic to earliest Cenozoic. During the Sevier orogeny, strata were sheared from underlying Precambrian rocks and broken along parallel planes of weakness to form multiple low-angle thrust faults, like overlapping roof tiles (Fig. 13-22).

Thrust faults during the Sevier orogeny shortened Earth's crust by over 100 km (about 60 miles) in the Nevada-Utah region. Several large folds and intricately folded strata also are present.

The term "Sevier orogeny" usually is applied to the Nevada-Utah region, but "Sevier-type" deformation occurred to the north in Montana, British Columbia, and Alberta (Fig. 13-23). The major ranges in this region are fault blocks of Paleozoic strata that were thrust eastward.

LARAMIDE OROGENY Magmatic activity along the far western edge of the North American plate diminished near the end of the Cretaceous. By Paleogene time, starting about 65 million years ago, much of the major thrusting along the marginal shelf region had ended. Once again, deformational events shifted eastward to the cratonic region, where today's Rocky Mountains of New Mexico, Colorado, and Wyoming are located. These more-eastward disturbances constitute the **Laramide orogeny**.

High-angle reverse faults (which become thrust faults at depth) developed during this orogeny, but more characteristic structures are broadly arched domes, basins, monoclines, and anticlines (Fig. 13-24).

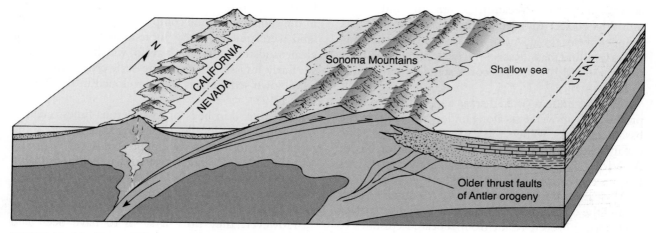

FIGURE 13-14 **Part of the western United States during Late Permian/Early Triassic times.**

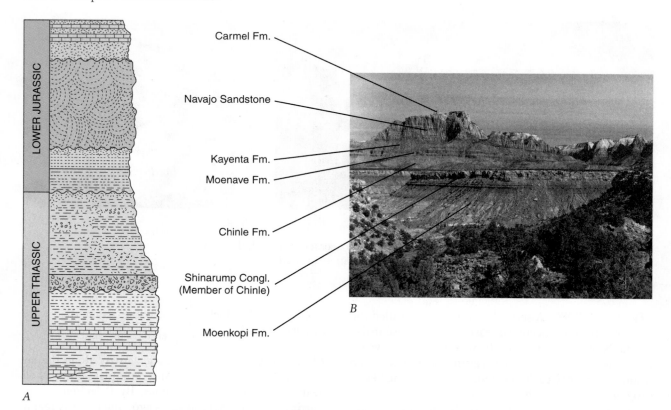

A

FIGURE 13-15 **Geologic column (A) and Panoramic view (B) of Triassic and Jurassic formations in Zion National Park, Utah.** The erosion of shaley beds characteristically forms slopes, and in this scene, the lowermost slope is formed by easily eroded shales of the Moenkopi Formation. Above that slope is a prominent ledge of harder Shinarump sandstones and conglomerates. The next slope up is in the softer Chinle Formation. The Chinle in Zion contains silicified logs, as in Petrified Forest National Park. Cliffs at the top are eroded from hard sandstones of the Navajo and Carmel formations.

Many of the larger faults represent renewed movement along ancient Precambrian faults. Strata composing the domes and anticlines are draped over central masses of Precambrian igneous and metamorphic rocks. In several instances, erosion has stripped away the cover of strata, exposing the central crystalline mass. Resistant layers of inclined beds that surround the central cores stand as mountains today.

Many of these uparched areas appear to have been produced by movements along underlying faults in the basement rocks. When the eastward-moving Pacific oceanic lithosphere scraped the underside of North America's cratonic margin, these uparched structures may have developed by drag.

Most *structures* of today's Rocky Mountains result from the Laramide orogeny. Today's surface *landscapes*, however, result from repeated episodes of Cenozoic erosion and uplift. Erosion acted upon the already existing geologic structures to sculpt the final scenic design.

Jurassic: Habitats for Marine Reptiles and Dinosaurs

During the Jurassic, when the mélange was forming in the subduction zone complex, depositional sites inland were quieter. Early Jurassic deposits consist of clean sandstones like the Navajo, which is Late Triassic/Early Jurassic in age. Large-scale cross-bedding of windblown sediment is well-developed in the Navajo (Fig. 13-25).

However, thin beds of fossiliferous limestone and evaporites occur locally, indicating that at least some parts of the formation were deposited in water, not by wind. The Navajo and associated sand bodies probably were deposited in a nearshore environment. It is likely that some were part of a coastal dune environment (Fig. 13-26).

However, they do not seem to have been laid down on the floor of a vast interior desert, as often has been speculated. Judging from studies of cross-bedding orientations, the source area for these clean

FIGURE 13-16 **Colorado National Monument near Grand Junction, Colorado.** Foreground erosional features are sculpted from the Lower Jurassic Wingate Sandstone, beneath which lies the red Chinle Formation. The cliffs in the far background are also Wingate Sandstone.

FIGURE 13-18 **Petrified logs, Petrified Forest National Park, Arizona.** Triassic floods left trees on sandbars or trapped in log jams, later to be covered by sediment. Minerals dissolved in percolating groundwater subsequently replaced the wood with silica, "petrifying" the wood ("turning it to stone," in popular language). The petrified logs and wood fragments are of the Arizona state fossil, *Araucarioxylon arizonicum.* This conifer grew to 45–60 meters in height and 1.0–1.5 meters at the base. The logs have been weathered from the Triassic Chinle Formation.

FIGURE 13-17 **Typical exposure of the Triassic Chinle Formation, Painted Desert, Arizona.** Some red bands are ancient soils (paleosols).

sandstones probably was the Montana-Alberta region of the craton. Somewhat older but similar quartz sandstones were recycled and spread southward into Wyoming and Utah, and then westward across Nevada.

Marine conditions became more widespread in the Middle Jurassic at that time. The entire west-central part of the continent was flooded by a broad seaway that extended into Utah (Fig. 13-27). This great embayment has been dubbed the **Sundance Sea.**

Deposited within the Sundance Sea were sands and silts of the **Sundance Formation,** famous for its Jurassic marine reptile fossils. Sediment for the Sundance Formation and overlying younger Jurassic rock units was derived from the Cordilleran highlands to the west. These highlands continued to grow throughout the Jurassic, ultimately extending from Mexico to Alaska.

Eventually, the Sundance Sea regressed, leaving behind a vast, swampy plain across which meandering rivers built floodplains. These deposits compose the **Morrison Formation** (Fig. 13-28), which extends

ENRICHMENT

Did Seafloor Spreading Cause Cretaceous Epicontinental Seas?

We know that during the Cretaceous the continents were extensively inundated by epicontinental seas. In North America, Europe, Australia, Africa, and South America, the Cretaceous marine sedimentary record indicates that about a third of today's land area was underwater. Because all continents experienced marine transgressions at about the same time, the cause must have been a worldwide rise in sea level.

However, there is little evidence that the cause of a global rise in sea level was the melting of continental glaciers. The preferred explanation is that rapid growth of ocean ridges caused the flooding.

Geologists have estimated the volume of new ocean crust produced during 10-million-year intervals as far back as the beginning of the late Cretaceous. The rate at which ocean crust was being produced along mid-ocean ridges during the Cretaceous was exceptionally rapid. Newly formed ocean crust is hotter than older crust and therefore occupies more volume. This is evident on the present seafloor, where the surface of newly formed crust stands 2.8 km below sea level, as contrasted to an average of 5 km below sea level for older, cooler oceanic crust.

Also, when spreading rates are high (producing high-volume crust), the corresponding subduction of cooler, low-volume crust is also greater. Taken together, these seafloor spreading processes decrease the space for water in the ocean basins, causing sea level to rise and resulting in marine transgressions across the continents.

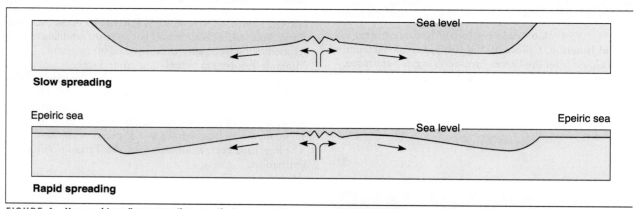

FIGURE A   How rapid seafloor spreading can displace seawater onto continental margins.

FIGURE 13-19 Strongly deformed beds of the Franciscan Formation, San Francisco. These crumpled, altered rock sequences, called a *mélange*, were formed from abyssal plain and continental shelf graywackes, pillow basalts, and cherts trapped in a subduction zone as the westward-moving North American plate overrode the Pacific Ocean lithosphere. Horizontal distance is 12 meters.

FIGURE 13-20 **Granodiorite of the Sierra Nevada batholith in Yosemite Valley, California.** The dramatic vertical cliffs result from vertical joints that form in the granodiorite.

across millions of square kilometers of the American West. The bones of more than 70 dinosaur species have been recovered from the floodplain deposits of the Morrison Formation.

The Great Cretaceous Epicontinental Sea

During the Cretaceous, North America's Pacific border was a land of lofty mountains (see Fig. 13-12). Erosion of these ranges brought sediments into adjoining basins that were rapidly subsiding, and many were open to the Pacific Ocean. In some places, volcanics and clastics accumulated more than 15,000 meters thick (over 9 miles). Cycles of folding and intrusion occurred repeatedly, with exceptional deformations during Middle and Late Cretaceous time.

In the more eastward regions of the Cordillera, the Cretaceous began with the advance of marine waters, both from the ancestral Gulf of Mexico and the Arctic. These Early Cretaceous epicontinental seas did not meet, so an area of dry land existed in Utah and Colorado. Their advance was interrupted by a regression that produced the unconformity that marks Early and Late divisions of the Cretaceous.

Flooding that followed in the Late Cretaceous was the greatest of the entire Mesozoic Era. This time, the embayment from the north joined with the southern seaway and effectively separated eastern and western North America.

Sedimentation in Cretaceous epicontinental seas was controlled largely by local conditions. Along the Gulf Coast, plentiful sediment accumulated—limestones, chalk, and marls (clayey limestones) (Fig. 13-29). North of Texas, however, the great bulk of sediments consisted of terrigenous clastics supplied by streams from the Cordilleran highlands.

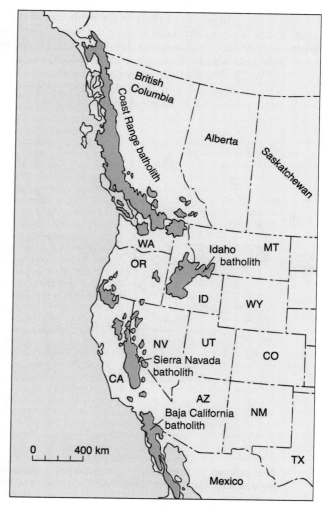

FIGURE 13-21 **Mesozoic batholiths of western North America.** These massive emplacements of granitic rock developed during the subduction that caused the Nevadan orogeny.

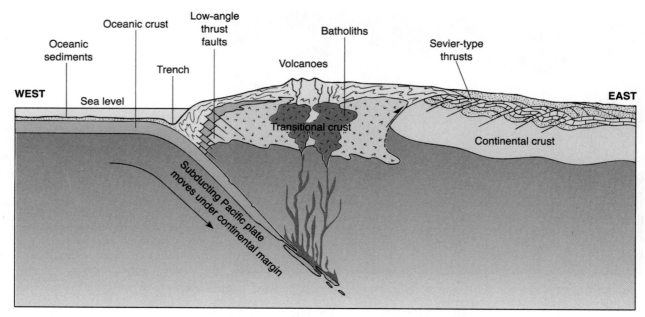

FIGURE 13-22 **An advanced stage in the evolution of the North American Cordillera.**
As the Pacific oceanic plate thrust beneath the North American continental plate, these
structures resulted: the ocean trench, low-angle thrust faults, melting of subducted rock that
results in batholiths and volcanoes, and overlapping low-angle thrust (Sevier-type) faults to the
east. The French word *décollement* (*day-COAL-mont*)—which means "unsticking"—describes
the Sevier-type thrust-faults, in which older rocks are thrust atop younger ones in multiple,
nearly parallel slabs. ◪ *Where along the cross-section would one find an ophiolite suite of rocks?*

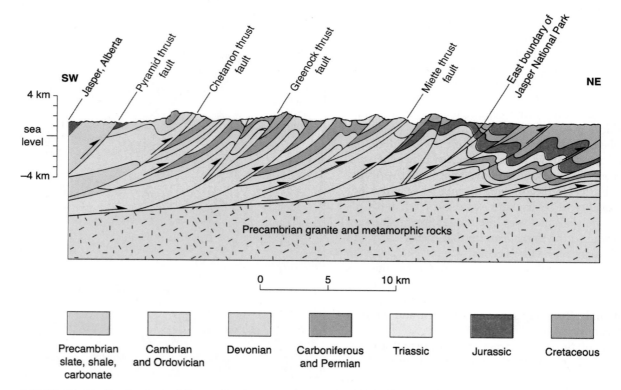

FIGURE 13-23 **Sevier-type deformation in western Canada.** This section extends 40 km
northeastward from the town of Jasper in Alberta. The section shows Sevier-type deformation
with multiple overlapping thrust faults. The faults flatten at depth, merging into a single plane
of detachment along the surface of the Precambrian igneous-metamorphic basement.

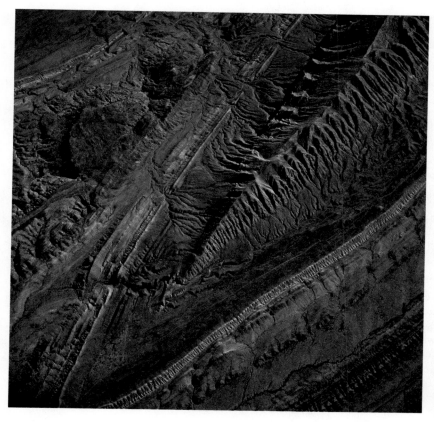

FIGURE 13-24 **Arial view of Sheep Mountain anticline, Wyoming.** This anticline is one of many folds formed during the Laramide orogeny.

FORELAND BASIN AND ECONOMIC RESOURCES In plate tectonic terms, the depositional basin that received these sediments is called a **foreland basin** (Fig. 13-30). The foreland basin of North America's western interior was immense, extending from the Arctic to the Gulf of California. In places it was over 1600 km (1000 miles) wide. It was bounded on the west mostly by fold and thrust belts, and on the east by the craton. Today, many rock formations in the basin yield commercial quantities of oil, natural gas, and coal.

Decades of exploration for these resources—drilling thousands of test wells—have provided the information needed to map the Cretaceous facies. These maps show coarse, terrigenous clastics immediately

FIGURE 13-25 **Cross-stratification in Navajo Sandstone, Checkerboard Mesa, Zion National Park, Utah.**

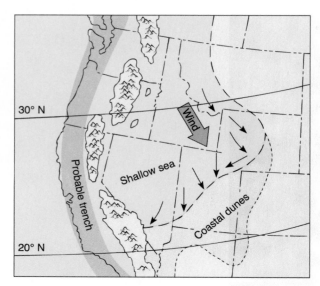

FIGURE 13-26 **Paleogeographic map, early Jurassic in the western United States.** Note the extent of the sea and coastal dunes. Also note the paleolatitudes—today, 20° and 30° North latitude are way to the south, with 20° near Mexico City and 30° crossing southern Texas.

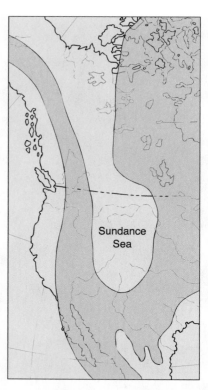

FIGURE 13-27 **Region inundated by the Middle Jurassic Sundance Sea.**

FIGURE 13-28 **Jurassic Morrison Formation near Grand Junction, Colorado.**

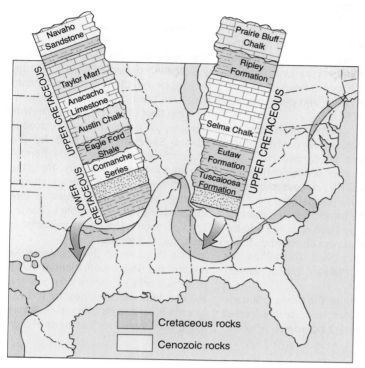

FIGURE 13-29 **Outcrop Area of Cretaceous limestone and marl rocks in the Atlantic and Gulf coastal plains.** Generalized columnar sections show the Cretaceous in Texas and Alabama. Here, plentiful sediment accumulated—limestones, chalk, and marls (clayey limestones).

adjacent to the Cordilleran highlands. The terrigenous sands and clays interfinger with coal-bearing continental deposits that similarly interfinger with marine rocks farther to the east (Fig. 13-31). Marine and nonmarine facies shift repeatedly with alternate transgressions and regressions of the Cretaceous sea.

As is common, the transgressive phase is marked by sandstone beds, such as those of the Cretaceous **Dakota Group** (Fig. 13-32). Rocks of the Dakota are exposed in many places along the eastern front of the Rocky Mountains, where the inclined beds form prominent ridges called *hogbacks* (Fig. 13-33). In parts of the Great Plains, the Dakota sandstones are an important source of the underground water essential to farmers and ranchers.

Cretaceous rocks of Wyoming and Colorado also include extensive beds of **bentonite**, a soft, plastic, light-colored clayey rock. Bentonite is composed of clay minerals that formed by alteration of volcanic ash. During the Late Jurassic and Cretaceous, volcanoes in the Idaho region explosively ejected tremendous volumes of ash that westerly winds carried into adjacent states. The resulting ash beds, subsequently converted to bentonites, represent single, isotopically datable geologic events that can be traced for great distances, even across changing facies. Hence, bentonite beds are of great value in chronostratigraphic correlation.

Cretaceous carbonate formations include chalk and chalky shales of the **Niobrara Formation** (Fig. 13-34).

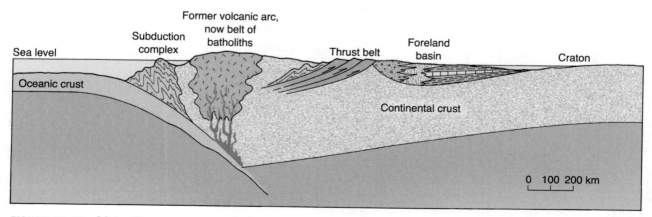

FIGURE 13-30 **Major Cretaceous tectonic features in the western United States.**

Grand Staircase–Escalante National Monument

President Clinton's 1996 proclamation established Grand Staircase–Escalante National Monument in southern Utah about 290 miles south of Salt Lake City. It is 1.9 million acres of breathtaking buttes and mesas, labyrinthine canyons, and multihued rock formations piled one upon the other like layers in a giant stone staircase.

The monument includes three distinct sections: the *Grand Staircase*, the *Kaiparowits Plateau*, and the *Canyons of the Escalante* (Fig. A). The Grand Staircase is the western third, consisting of level "steps" or benches interrupted by cliffs (termed "risers," just as in a staircase). The first riser you encounter is Chocolate Cliffs, composed of the Triassic Moenkopi Formation (Figure B). The Moenkopi is capped unconformably by a hard, conglomeratic Shinarump member of the Chinle Formation. This resistant layer forms the surface of the bench leading to the next riser, the Vermillion Cliffs. Stream and lake sediments of the Jurassic Moenave and Kayenta Formations provide the gray, beige, and pink colors of Vermillion Cliffs.

The imposing White Cliffs form the next riser. Bedrock here consists of the Navajo Sandstone, capped by a thin limestone member of the Carmel Formation. Gray Cliffs, composed of resistant Cretaceous sandstones, are the next riser. The limestones and marls of Pink Cliffs have been sculpted into the erosional features of nearby Bryce Canyon. Above Pink Cliffs lies the Paunsaugunt Plateau, the uppermost step.

To the north of the Grand Staircase lie buttes, mesas, and steeply incised gorges of the Kaiparowits Plateau.

Kaiparowits is a Paiute Indian name meaning, "Big Mountain's Little Brother." Although this section is topographically a dissected plateau, its underlying architecture is that of a broad structural basin. As you climb upward from the floor of one of the Kaiparowits canyons, you encounter the Cretaceous Dakota Sandstone, followed by the Tropic Shale, Straight Cliffs Formation, and Waheap Formation. Along the western margin of the Plateau, steeply dipping beds of a monocline are sculpted into a line of hogbacks named The Cockscomb (for its resemblance to the comb of a rooster).

The Canyons of the Escalante lie north of the Kaiparowits Plateau. It is a wondrous area of narrow canyons and colorful formations eroded by the Escalante River and its tributaries. A hike along Calf Creek provides views of high

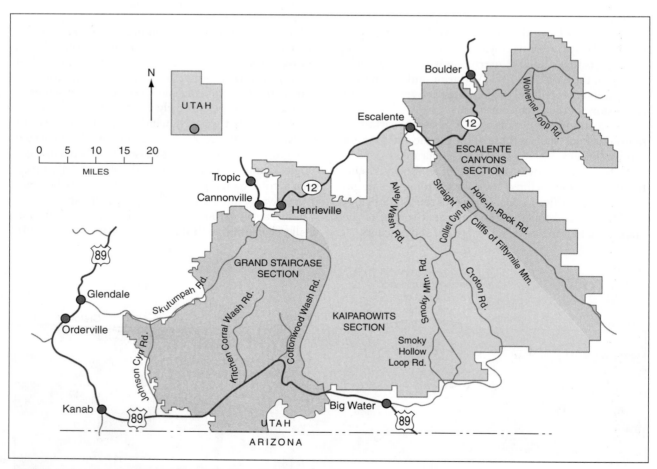

FIGURE A Location map: Grand Staircase–Escalante National Monument.

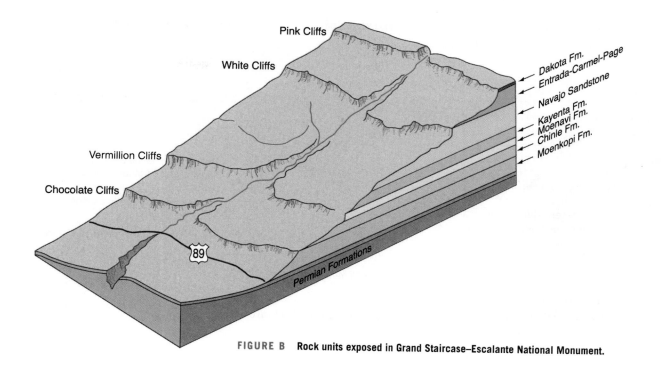

FIGURE B   Rock units exposed in Grand Staircase–Escalante National Monument.

waterfalls and excellent exposures of the Navajo and Kayenta Formations. Hole-in-the-Rock Road takes intrepid hikers to Devil's Garden, where erosion of the Jurassic Entrada Formation has produced features named goblins, rock babies, and hoodoos. Dinosaur tracks occur on the bedding surfaces of the Entrada.

Nearly 270 million years of Earth's history are recorded in the rocks and fossils of Grand Staircase–Escalante National Monument. During the Per-mian, the region was a low coastal area occasionally inundated by a shallow sea. The Permian units include the Her-mit Shale, Coconino Sandstone, Toroweap Formation, White Rim Sand-stone, and Kaibab Limestone. The Tri-assic Moenkopi Formation begins the Mesozoic section. Fossils in the Moenkopi include footprints and bones of reptiles and labyrinthodonts. The bones of labyrinthodonts and crocodile-like reptiles called phytosaurs also occur in the overlying Chinle Formation.

For more, see http://www.ut.blm.gov/ monument/visitor.

Reference

Doelling, H.H., Blackett, R.E., Hamlin, A.H., Powell, J.D., and Pollock, G.L. *Geology of Grand Staircase-Escalante National Monument, Utah* in Sprinkle, D.A., Chidsey, T.C. Jr., and Anderson, P.B. eds. 2000. *Geology of Utah's Parks and Monuments*. Utah Geologic Association Publication 28. Salt Lake City: Publishers Press.

FIGURE C   Navajo Sandstone exposed in the White Cliffs of the Grand Staircase section. The Navajo is capped by a thin limestone member of the Carmel Formation.

FIGURE D   Erosion-etched cross-beds in the Navajo Sandstone, Escalante Canyons section.

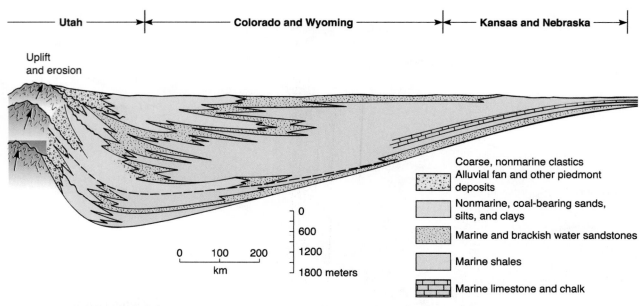

Uplift
and erosion

Coarse, nonmarine clastics
Alluvial fan and other piedmont
deposits

Nonmarine, coal-bearing sands,
silts, and clays

Marine and brackish water sandstones

Marine shales

Marine limestone and chalk

0
600
1200
1800 meters

0   100   200
km

FIGURE 13-31 **Generalized cross-section, Cretaceous rocks of the back-arc basin, U.S. western interior.**

FIGURE 13-32 **Impressive I-70 road cut (north side) west of Denver.** Here, Dakota Group rocks dip steeply toward the east (right). Sandstones of the Dakota Group are an important artesian groundwater source beneath the Great Plains. They also have yielded oil and gas in eastern Colorado, Nebraska, and Wyoming.

The Niobrara has yielded the remains of a variety of marine creatures, including enormous numbers of oysters, a large Cretaceous diving bird (Fig. 13-35), marine reptiles, and the large flying reptile *Pteranodon*.

Toward the end of the Cretaceous, seas that supported these creatures regressed from central North America. At the same time, the Laramide deformation was occurring, which produced the folds and faults of the Rocky Mountains. As the seas withdrew, deltaic and other continental sediments suitable for forming coal spread across the old seafloor.

THE TETHYS SEA IN EUROPE

The great Alpine-Himalayan mountain belt stretches from Gibraltar at the western end of the Mediterranean Sea, across southern Europe and Asia to the Pacific Ocean. Most of the rocks within the belt were laid down in the ancient **Tethys Sea**. They were deformed into mountain ranges when northward-moving segments of Gondwana collided with Eurasia.

The geologic history of the Tethys Sea traces back to the Paleozoic. But unlike the Appalachians, for which most deposition and tectonic activity occurred during the Paleozoic, the Tethys experienced its most eventful geologic history later, in the Mesozoic and Cenozoic.

FIGURE 13-33 **Hogbacks near Golden, Colorado.** These north-south trending ridges were formed by the eastward-dipping Dakota Sandstone. The Dakota at this location is a beach deposit of the Cretaceous sea, so well-developed ripple marks and occasional dinosaur footprints are preserved.

During the Triassic Period, the Tethys was a limy sea (Fig. 13-36) containing rich populations of bivalves, cephalopods, crinoids, and corals. North of the sea, a highland tract called the Vindelician Arch separated marine sedimentation to the south from very different sedimentation in north-central Europe. In the north, the Triassic record begins with continental red and brown clastic rocks that resemble those deposited under arid conditions during the preceding Permian Period. These continental deposits are overlain by shoreline sandstones, marls, evaporites, and limestones deposited during a temporary marine invasion in central Europe.

The sea did not remain for long. Nor was it able to reach as far north as Great Britain, where a nonmarine sequence known as the New Red Sandstone was being deposited.

Marine conditions grew more widespread during the Jurassic. Epicontinental seas spread northward from the Tethys across Russia and into the Arctic Ocean. But this invasion was short-lived, and shallow seas regressed back into the Tethys Sea by the end of the Jurassic.

The relative quiet of the Jurassic changed during the Cretaceous, when the African plate drifted northward, narrowed the Tethys Sea, and powerfully compressed Tethys sediments. The record of this event is complex

FIGURE 13-34 **Cretaceous Niobrara Formation, west-central Kansas.** The Niobrara has yielded the remains of varied marine creatures, including enormous numbers of oysters, the large Cretaceous diving bird in Figure 13-35, marine reptiles, and the large flying reptile *Pteranodon.*

FIGURE 13-35 **The Cretaceous diving bird *Hesperornis*.** This bird was over a meter tall when standing. Propulsion was provided by its large, webbed feet. A notable primitive trait of this bird was its teeth.

overturned folds and thrust faults that trend east-west across southern Europe.

Like North America, Europe experienced extensive marine transgressions during the Late Cretaceous (Fig. 13-37). A great embayment from the Tethys worked its way northward, ultimately joining a similar southward encroachment from the North Atlantic. One result was a rapid interchange of marine organisms from the two formerly separated regions.

In the Tethys-Alpine region, block-faulting that had begun during Late Jurassic time continued until about the Middle Cretaceous. Then, further compression from the south folded the thick sequence of marine Tethys sedimentary rocks. The axes of these folds parallels the east-west trend of the Tethys.

Most of the great anticlines and synclines remained submerged, but the tops of some folds were above sea level. They existed as elongated islands along the north side of the narrowing Tethys. Erosion of these anticlinal islands produced clastic sediments that were periodically swept downward by turbidity currents into the adjoining basins.

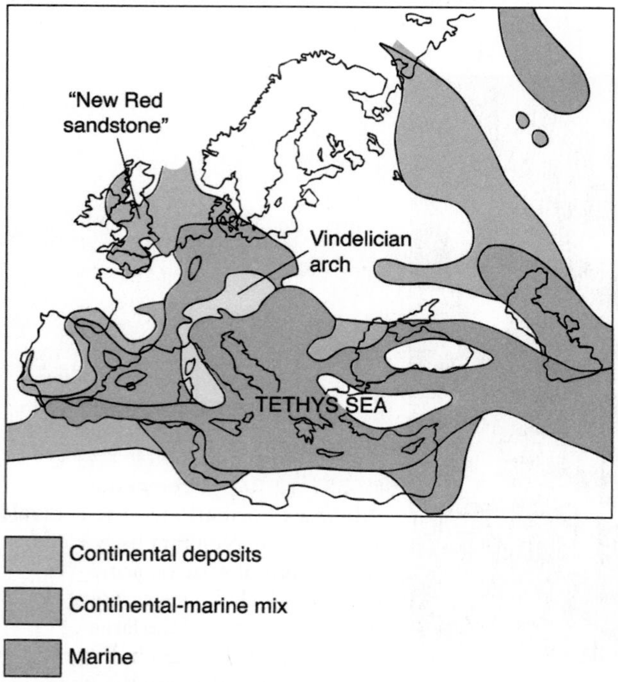

FIGURE 13-36 **Paleogeographic map of Europe.** Triassic Period (250–203 million years ago).

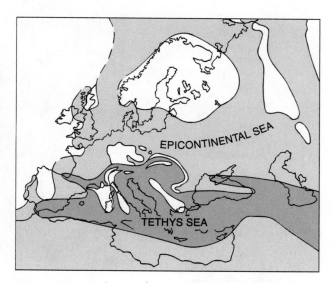

FIGURE 13-37 **Areas of western Europe covered by Cretaceous seas.**

ENRICHMENT

Chunneling Through the Cretaceous

Visionaries had long dreamed of a 50-kilometer (31-mile) tunnel beneath the English Channel to connect England and France. That dream became reality in October 1990, when tunnelers from England and tunnelers from France shook hands deep beneath the floor of the English Channel. They had excavated and installed the lining for part one of a three-part tunnel system, which was completed in the summer of 1993. The Channel Tunnel ("Chunnel") is the longest undersea tunnel on Earth. It has two traffic tunnels flanking a central service tunnel. High-speed trains carry passengers through the traffic tunnels from Folkestone, England (southwest of Dover) to France, in only 30 minutes.

About 85% of the Chunnel was excavated in a Lower Cretaceous formation known as the Chalk Marl. The term *marl* indicates that the chalk contains clay, and the mixture of chalk and clay results in a rock that is relatively strong, stable, and impermeable. It is an excellent natural material in which to excavate a tunnel.

In addition, throughout most of the tunnel route, the Chalk Marl is over 25 meters thick and thus easily accommodates the 8-meter diameter of the traffic tunnels. The Chalk Marl is overlain by the Upper Cretaceous Gray and White Chalk Formations. Beneath the Chalk Marl is the Gault Clay, a unit unsuitable for tunneling because it is plastic and inherently weak.

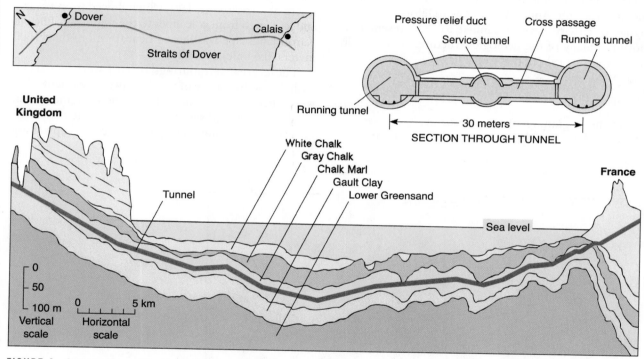

FIGURE A Cross-sections of the Chunnel. The cross-section showing the channel route has large vertical exaggeration.

GONDWANA EVENTS

South America

Along the western margin of South America, two parallel deposition zones existed at the beginning of the Mesozoic:

1. The more dynamic western tract included turbidites, conglomerates, and siliceous sediments, deposited in the deeper waters of the continental rise and slope.

2. A quieter eastern tract composed of carbonates and shales, deposited in shallow shelf zones.

We can trace the approximate boundary between these two suites of rocks along the central ranges of the Andes. In addition to these marine tracts, broad basins in the interior accumulated continental deposits. Particularly remarkable are the eolian and fluvial sandstones and siltstones that spread across southeastern South America during the Triassic. The beds contain a rich fauna of vertebrates similar to those of the same age in southern Africa.

By Jurassic time, a rift zone between South America and Africa widened into a configuration resembling today's Red Sea. New seafloor formed along the spreading center of this developing South Atlantic.

Meanwhile, the western side of the continent was at the leading edge of the South American tectonic plate. It was opposed by the Pacific plate, the forward margin of which plunged beneath South America as an eastward-inclined subduction zone.

In the subduction zone, oceanic crust and its cover of seafloor sediment were carried down to deep zones and melted. The magmas migrated toward the surface, cooling to form batholithic intrusions and immense outpourings of andesite lava. (The rock name *andesite* comes from the Andes Mountains.) Deformation and igneous activity have continued to the present day, forming many of the geologic structures now seen in the towering peaks of the Andes.

Africa

Near the beginning of the Mesozoic, northwestern Africa and eastern North America had begun to separate. However, no rift existed between Africa and South America. Thus, there was no South Atlantic Ocean. Most of Africa was relatively stable throughout the Mesozoic, with only relatively minor transgressions along its northern and eastern borders.

Although the Mesozoic rock record in the more interior parts of Africa is unexceptional, some regions yield particularly interesting geologic information. One such region is the Karoo basin at the southern end of Africa (Fig. 13-38). The Karoo was a continental basin that was formed late in the Carboniferous and received swamp, lake, and river deposits until late in the Triassic Period. Rocks of the Karoo sequence are known to paleontologists around the world for their fossils of mammal-like vertebrates.

As the Triassic drew to a close, terrestrial siltstones and sandstones of the Karoo sequence were covered by layers of low-viscosity lava, which flowed from fissures and less often from volcanoes to the southeast. The old Karoo landscape was buried beneath more than 1000 meters (0.6 mile) of basalt.

These great outpourings of lava were associated with the pulling away of the Gondwana segments that were once part of South Africa. Such fragmentation would have caused severe fracturing of the continental crust, providing multiple avenues for upwelling

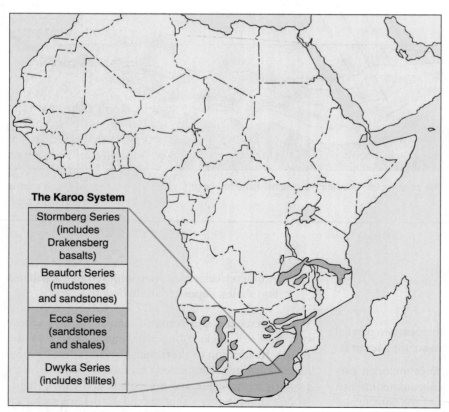

The Karoo System

| Stormberg Series (includes Drakensberg basalts) |
| Beaufort Series (mudstones and sandstones) |
| Ecca Series (sandstones and shales) |
| Dwyka Series (includes tillites) |

FIGURE 13-38 **Karoo System outcrops of South Africa.** Rocks of the Karoo sequence are known to paleontologists around the world for their fossil mammal-like vertebrates.

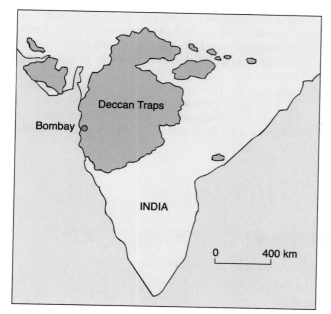

FIGURE 13-39 **Present-day distribution of Cretaceous Deccan Trap basalts.** They form the greatest volume of continental basalt on Earth's surface. (*Trap* in Dutch means "stair steps," and is used here because of the step-like appearance of some lava flows.) Much of the original lava cover has been lost to erosion, as indicated by isolated remnants east of the large area.

molten rock. The extrusions continued well into the Jurassic Period. Contemporaneous lava floods and volcanism also occurred on the separating landmasses of South America, Australia, and—somewhat later—Antarctica.

India

During the Mesozoic, India moved steadily northward on its remarkable voyage to Laurasia. The Indian continent itself remained tectonically stable, even though it was moving at a rapid rate toward its ultimate collision with Laurasia in Cenozoic time. The landmass experienced only relatively minor marginal incursions of the sea.

Across the interior, erosion of upland areas spread terrigenous clastic sediments into the lowlands and plains. Paleontologists have recovered dinosaur bones and superb plant fossils from these sandstones and shales.

By Cretaceous time, India was nearing the Tethyan region. Here and there, the floor of the Tethyan trough buckled into elongate ridges—which were bellwethers of the coming tectonic collision with Eurasia. As these ridges developed, the northwestern half of India flooded with immense quantities of low-viscosity basaltic lava. This formed the flood basalts of the **Deccan Traps**. It is likely that the Deccan Traps record the passage of India across a fixed hot spot in the mantle (Fig. 13-39).

The Deccan Traps cover about 500,000 square km of western and central India with an aggregate volume exceeding 1,000,000 cubic km. They form the greatest volume of continental basalt on Earth's surface. The extensive fiery extrusions and huge volumes of carbon dioxide released along the web of fissures may have triggered climatic warming and chemical changes that adversely affected life worldwide, both on land and sea. In fact, Deccan volcanism may have contributed to the episode of mass extinction that closed the Cretaceous Period.

SUMMARY

- Within the 185 million years of the Mesozoic Era, global geography changed as large continental blocks of crust separated from Pangea. During the Triassic, North America broke away from Gondwana. By the Late Jurassic, rifts had developed among all Gondwana segments except Australia and Antarctica.
- In eastern North America, tensional forces associated with the fragmentation of Pangea resulted in block faulting and volcanic activity. In downfaulted lowlands, arkosic red beds and lake shales were deposited.
- The ancestral Gulf of Mexico began to form in the Late Triassic. During the Jurassic and Cretaceous, it acquired sequences of limestones, evaporites, sandstones, and calcareous shales.
- Continental deposits, particularly red beds, characterized the Triassic of the Rocky Mountain regions of North America. Marine conditions existed along the Pacific margin, where volcanic rocks and graywackes accumulated to great thicknesses.
- During the Jurassic, the western margin of North America began to override the subduction zone at the leading edge of the eastward-moving Pacific plate. The advancing Pacific plate carried large fragments of volcanic arcs, midoceanic ridges, and microcontinents to the western margin of North America. These crustal fragments became incorporated into the Cordillera as a tectonic collage of displaced terranes. Growth of a continental margin in this way is referred to as accretionary tectonics.
- Deformation of the North American Cordillera began along the Pacific coast and moved progressively eastward. By Late Cretaceous time, inland seas that had occupied the western interior of North America were displaced. Majestic mountain ranges stood in their place.

- The focus of Eurasian Mesozoic events was the great east-west-trending Tethys Sea. The Tethys region was affected by powerful compressional forces as the African plate moved northward.
- The Mesozoic history of the interior regions of Africa is read primarily from continental deposits. Among these are beds of the Karoo basin of South Africa, which contain fossil remains of mammal-like vertebrates.
- During the Mesozoic, India rifted away from Gondwana and drifted toward its Eurasian destination. While moving northward, India passed over a hot spot, resulting in massive extrusions of lava flows known as the Deccan Traps. By Cretaceous time, India had approached close enough to Eurasia to cause initial crustal buckling in the eastern Tethys.
- South America separated from Africa during the Jurassic. The separation continued during the remainder of the Mesozoic, providing space for the South Atlantic Ocean. A subduction zone developed along the continent's leading western margin.

KEY TERMS

accretionary tectonics, p. 380

bentonite, p. 395

chalk, p. 378

Dakota Group, p. 395

Deccan Traps, p. 401

exotic terranes, p. 378

foreland basin, p. 393

Laramide orogeny, p. 385

mélange, p. 384

Morrison Formation, p. 387

Nevadan orogeny, p. 383

Niobrara Formation, pp. 395–396

obduction, p. 380

rudistid, p. 378

salt dome, p. 375

Sonoma orogeny, p. 381

Sonoma terrane, p. 381

Sevier orogeny, p. 385

Sundance Formation, p. 387

Sundance Sea, p. 387

tectonic collage, p. 380

Tethys Sea, pp. 396–398

QUESTIONS FOR REVIEW AND DISCUSSION

1. What geologic events precede the separation of a continent into two or more parts? Cite examples from Mesozoic history.

2. Describe the general structural and environmental conditions associated with the deposition of the Newark Group. Why did normal faulting predominate, rather than reverse faulting? Did older structures influence the location of Triassic faults? Explain.

3. Discuss the characteristics of structures produced by the Sevier orogeny.

4. What environmental conditions account for the presence of Jurassic evaporites in the Gulf Coast region? How are these evaporites related to petroleum traps in overlying Cretaceous and Cenozoic strata?

5. Explain the mechanisms by which North America increased in size during the Mesozoic.

6. Discuss the possible effects of Deccan volcanisms and extensive ocean floor volcanism on global climate near the end of the Mesozoic Era.

7. Describe and contrast obduction with subduction. Cite examples of each.

8. What is chalk? In what Mesozoic geologic period was it particularly abundant? What is the relationship between chalk and marine plankton?

9. What is the total duration in years of the Mesozoic Era? Which period was the longest?

10. What was the probable source area for the sediments of the Morrison Formation of the Rocky Mountain region? What evidence indicates that the Morrison Formation was continental rather than marine?

11. At what time during the Mesozoic were epicontinental seas most extensive? During which period were such marine incursions most limited?

12. What was the Tethys Sea? What water bodies today are remnants of this seaway? Where might one go to examine Mesozoic rocks deposited within the Tethys Sea?

The Earth Through Time Student Companion Web Site (www.wiley.com/college/levin) has online resources to help you expand your understanding of the topics in this chapter. Visit the Web Site to access the following:

1. Illustrated course notes covering key concepts in each chapter;

2. Online quizzes that provide immediate feedback;

3. Links to chapter-specific topics on the web;

4. Science news updates relating to recent developments in Historical Geology;

5. Web inquiry activities for further exploration;

6. A glossary of terms;

7. A Student Union with links to topics such as study skills, writing and grammar, and citing electronic information.

*The Cretaceous Period dinosaur, **Tyrannosaurus rex**, shown as a fearsome predator—yet considerable evidence indicates that it scavenged much of its food. T. Rex was about 12 meters in length, had 15-cm teeth, and could run an estimated 40 km per hour.*

Life of the Mesozoic

Tyrannosaurs, enormous bipedal caricatures of men,
would stalk mindlessly across the sites of future cities
and go their way down into the dark of geologic time.
—Loren Eiseley, *The Immense Journey,* 1959

Key Chapter Concepts

- Climate strongly influenced the diversity and distribution of animals and plants during the Mesozoic Era. Fossils and oxygen isotope studies indicate that a warm and uniform climate prevailed during most of Mesozoic time, except in polar regions.

- Corals of modern families built great reefs in epicontinental seas of the Jurassic and Cretaceous Periods. Bivalve mollusks—such as rudistids and oysters, sea urchins, crinoids, and belemnites—were abundant. Ammonite cephalopods, known for the intricacy and variety of their septal fluting, were particularly successful during the Mesozoic. However, they did not survive beyond that era.

- The most significant event in the evolution of plant life during the Cretaceous Period was the appearance and expansion of angiosperm plants. Angiosperms dominated during all of later geologic time.

- Dinosaurs appeared during the Triassic Period and held dominion over all land vertebrates until the close of the Cretaceous Period.

- Important predators in Mesozoic seas were marine reptiles, such as the plesiosaurs and ichthyosaurs, and the giant Cretaceous marine lizards known as mosasaurs.

- Reptiles achieved gliding flight during the Triassic. By the Jurassic, aerial reptiles such as the pterosaurs excelled in the wing-flapping ability to fly efficiently. The earliest known true bird is *Archaeopteryx* of the Jurassic.

- During the Triassic, mammals evolved from mammal-like reptiles. The Triassic species were mostly mouse-sized animals whose classification is based mainly on differences in their teeth. However, by Cretaceous time, mammals the size of dogs had evolved.

- The Mesozoic closed with a major extinction event. Among the animals that became extinct were the dinosaurs, pterosaurs, ichthyosaurs, plesiosaurs, mosasaurs, rudistid mollusks, and ammonites. Many other groups were decimated.

- Mass extinctions at the end of the Cretaceous may have been caused by impact of a large meteorite or comet, or by massive volcanic eruptions.

OUTLINE

▶ CLIMATE CONTROLS IT ALL
▶ MESOZOIC INVERTEBRATES
▶ MESOZOIC VERTEBRATES
▶ DINOSAURS: "TERRIFYING LIZARDS"
▶ DINOSAURS: COLD-BLOODED, WARM-BLOODED, OR BOTH?
▶ DINOSAUR PARENTING
▶ THE FLYING ARCHOSAURS
▶ A RETURN TO THE SEA
▶ THE RISE OF BIRDS
▶ THE MAMMALIAN VANGUARD
▶ SEA PLANTS AND PHYTOPLANKTON
▶ LAND PLANTS
▶ LATE CRETACEOUS CATASTROPHE
▶ SUMMARY
▶ KEY TERMS
▶ QUESTIONS FOR REVIEW AND DISCUSSION
▶ WEB SITES
▶ BOX 14–1 ENRICHMENT
▶ BOX 14–2 ENRICHMENT
▶ BOX 14–3 GEOLOGY OF NATIONAL PARKS AND MONUMENTS

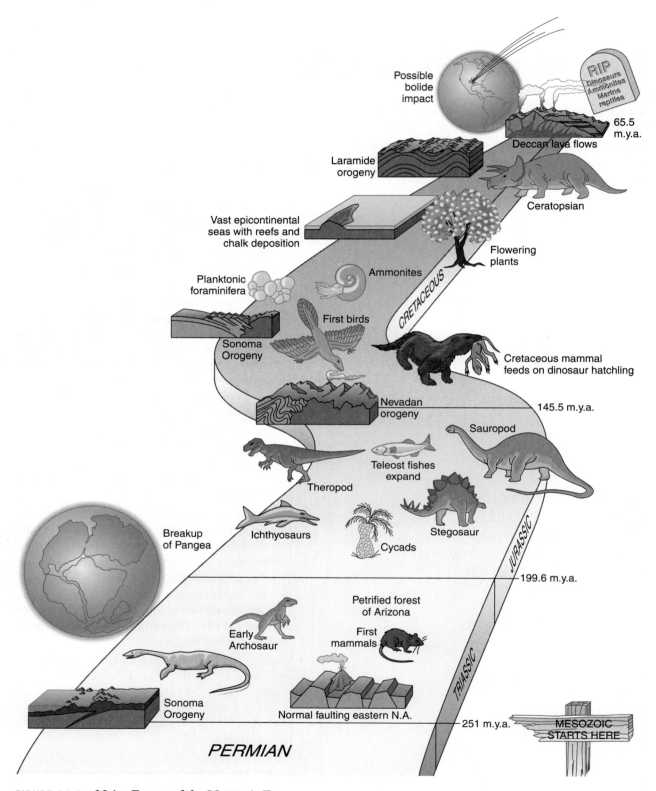

FIGURE 14-1 **Major Events of the Mesozoic Era.**

In this chapter, we examine Earth's inhabitants during the Mesozoic Era. Among these, none have captured public interest more than the dinosaurs. Yet, for all our fascination with them, we should not forget that it was during this era that birds and mammals also made their appearance (Fig. 14-1).

Many new families of marine invertebrates, reptiles, and fishes evolved during the Mesozoic. Planktonic foraminifera appeared in the Cretaceous and expanded dramatically. Their calcareous skeletons (called tests) along with the calcium carbonate plates of coccolith-ophores (coccoliths) accumulated by the trillions in limy ocean sediments.

Plants also evolved dramatically during the Mesozoic. Flowering plants (angiosperms) arrived at the Jurassic-Cretaceous boundary. With this event, modern trees with seeds, nuts, and fruits expanded across the continents.

▶ CLIMATE CONTROLS IT ALL

In today's world we see far more different kinds of animals and plants living in climatic zones near the equator than near the poles. There was a similar diversity gradient during the Mesozoic. Other factors also influenced climate, including the splitting apart of Pangea which caused changes in oceanic and atmospheric circulation that directly affected the evolution, distribution, and diversity of life.

Influences on Climate

Climate is influenced by many factors. However, the primary global control of climate is the *balance between incoming and outgoing radiation from the sun.* That balance is affected by the following factors:

1. The configuration and dimensions of the oceans and continents.

2. The development and location of mountain systems and land bridges between continents.

3. Changes in snow, cloud, or vegetative cover—each affects the amount of solar radiation reflected back into space from Earth.

4. The carbon dioxide content of the atmosphere—which can trigger greenhouse warming or global cooling.

5. The location of the poles and equator.

6. The amount of radiation-blocking aerosols thrown into the atmosphere by volcanoes.

7. Astronomic factors, such as changes in Earth's axial tilt and orbit.

Water bodies retain heat far better than land areas. Furthermore, ocean waters constantly circulate, distributing their warmth around the globe. Hence, when the pro-portion of ocean to land increases, warmer climates may result. The reverse holds true as well. Thus, the reduced ocean area at the end of the Paleozoic probably was an important cause of cooler climates.

Continents and Currents

During the waning days of the Paleozoic Era, cool climates prevailed on many continents. Contributing causes were the vastness of the Pangea supercontinent, the upheaval of mountains, overall increases in land elevation, and withdrawal of inland seas.

Then gradually, the world climate warmed. Glaciers began to melt in Africa, Australia, Argentina, and India as these continents drifted away from the South Pole. During most of the Jurassic and Cretaceous Periods, climates for most regions were warm and lacked extreme seasonal temperature differences. At the same time, extensive regions were covered by epicontinental seas, which contributed to warmer conditions.

Although rifting of Pangea began in the Triassic, the continents still were tightly clustered. The paleo-equator extended from central Mexico across the northern bulge of Africa (Fig. 14-2). As noted in Chapter 13, the Triassic was a time of general emergence of the continents. Mountains, thrust upward at the end of the Paleozoic, inhibited the flow of moist air into the more centrally located regions, causing widespread aridity. Evaporites, dune sandstones, and red beds accumulated at both high and low latitudes. They attest to relatively dry and warm conditions.

Reconstructions based on paleomagnetic studies suggest that during the Jurassic, the continents were at the approximate latitudinal positions they occupy today. Marine waters extended northward into the space formed by the opening of the Atlantic Ocean. In many places, shallow inland seas spilled from deep basins onto the continents. An arm of the Proto-Pacific Ocean extended westward as the Tethys Sea. Warm, westward-flowing equatorial currents penetrated far into the Tethys. These same equatorial currents, deflected by the east coast of Pangea, also were diverted to the north along coastal Asia and to the south along northeastern Africa and India.

To complete the cycle, the cooled currents returned to the equator along the west side of the Americas. The presumed ocean and wind currents and the rather extensive coverage of seas, both upon and adjacent to continents, brought mild climates to many regions during the Jurassic. There is also evidence, however, of aridity, as in the Gulf of Mexico where thick beds of salt were deposited. There are no glacial deposits of Jurassic age, and coal beds indicating warm or temperate conditions occur in Antarctica, India, China, and Canada.

During most of the Cretaceous, climates were generally warm and stable as they had been during the

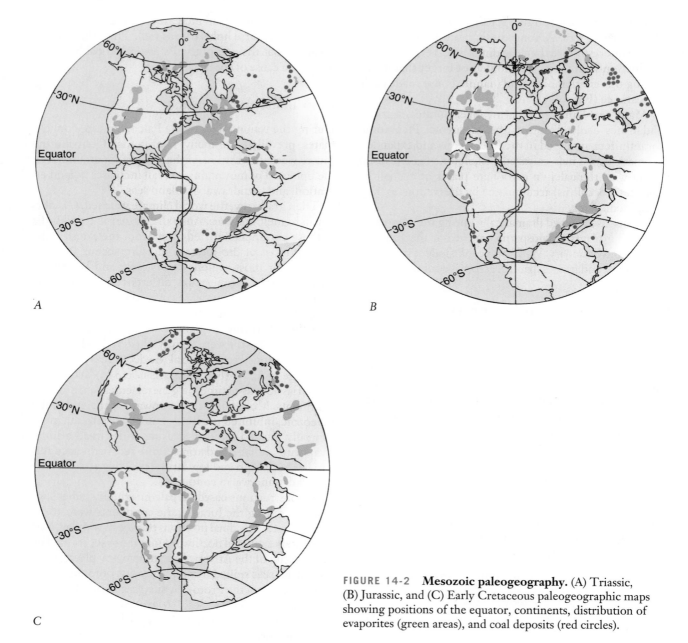

A

B

C

FIGURE 14-2 **Mesozoic paleogeography.** (A) Triassic, (B) Jurassic, and (C) Early Cretaceous paleogeographic maps showing positions of the equator, continents, distribution of evaporites (green areas), and coal deposits (red circles).

Jurassic. A remarkably homogeneous flora spread around the world, with subtropical plants thriving in latitudes 70° from the equator. Coal beds formed on nearly every continent and even at high latitudes.

Cretaceous Cooling

However, such warm conditions did not persist, and toward the end of the Cretaceous, climates began a slow backward swing toward cooler conditions. The Late Cretaceous South Pole was centrally located in Antarctica, and the North Pole was located at the north edge of Ellesmere Island, Canada. Europe and North America had moved somewhat farther north, and the

widespread inland seas had begun to recede. Worldwide regressions were accompanied in some regions by episodes of mountain-building.

Several lines of evidence indicate terminal Cretaceous cooling. Tropical cycads diminished sharply, and ferns declined in North America and Eurasia. Hardier plants, such as conifers and angiosperms, extended their realms. Oxygen isotope ratios (see Chapter 6) from open ocean planktonic calcareous organisms indicate declining ocean temperatures, beginning about 80 million years ago. Such a dip in worldwide mean annual temperatures would have had a damaging effect on plant and animal life. It may have contributed to the mass extinction at the end of the Mesozoic.

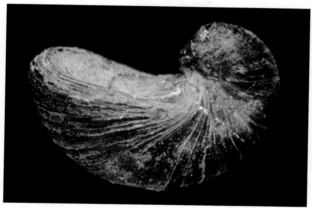

A

B

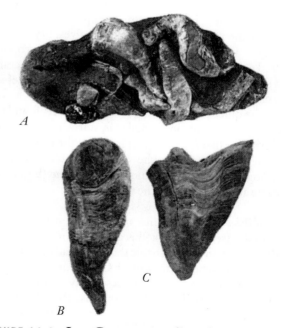

C

FIGURE 14-3 **Mesozoic oysters.** (A) *Exogyra arietina* from Cretaceous rocks in Texas. Note how the beak of the larger left valve is twisted to the side so as not to overhang the small right valve. (B) *Gryphaea* is found in shallow marine Jurassic and Cretaceous strata around the world. This oyster family member has a large valve that is arched up and over the smaller valve. (C) Shells of *Gryphaea* weathering from the Cretaceous Mancos Shale. ⊞ *How does the shape of these oyster shells differ from clam or mussel shells? (Answers to questions appearing within figure legends can be found in the Student Study Guide.)*

MESOZOIC INVERTEBRATES

Marine Invertebrates—Back from the Brink

At the end of the Paleozoic, about 95% of marine invertebrate species came perilously close to extinction. But it was not the end of the line. In the Mesozoic, marine invertebrates expanded, if somewhat tardily, as indicated by limited Early Triassic faunas. However, once a few groups became established, marine life expanded quickly.

BIVALVES Bivalves became increasingly common from the Middle Triassic on and eventually surpassed the brachiopods in colonizing the seafloor. Among the most successful bivalves were the oysters, represented by such genera as *Gryphaea* and *Exogyra* (Fig. 14-3). Some members of the oyster family grew odd conical shells that resembled corals and snails. These so-called **rudistids** (Fig. 14-4) became important components of Jurassic and Cretaceous reefs.

CORALS Corals were abundant in warm, shallow seas of the Mesozoic. For example, during the Late Jurassic, prodigious numbers of reef-building corals lived in the Tethys Sea. As is true today, reef-building corals thrive only in clear water of normal salinity. They prefer depths no more than about 50 meters, and water temperatures no lower than about 20°C. One reason corals require shallow water is for access to sunlight, because

A

C

B

FIGURE 14-4 **Late Cretaceous rudist pelecypod *Coralliochama orcutti* from Baja California.** (x 0.5) (A) Cluster of specimens from the reef deposit. (B, C) Individual specimens.

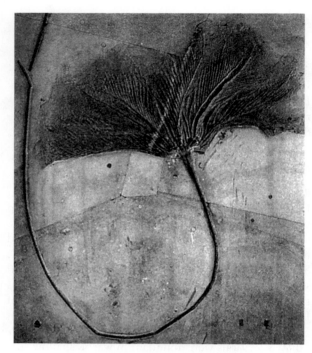

FIGURE 14-5 **Large, well-preserved crinoid from Lower Jurassic strata, Holzmaden, Germany.** The extended arms of this giant *Pentacrinus subangularis* crinoid span about 1.2 meters.

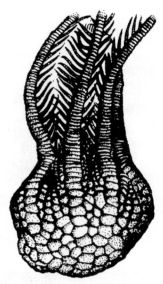

FIGURE 14-6 **The Late Cretaceous crinoid** *Uintacrinus.* This unusual crinoid possessed a large calyx with ten long arms on top. Because it lacked an attachment stem, it was formerly thought to be a free swimmer or floater. But recent studies indicate *Uintacrinus* lived on the seafloor with its massive calyx embedded in sediment. Some of its arms extended upward to intercept food particles, whereas others stretched along the seafloor for support.

they have a symbiotic relationship with zooxanthellae algae. The algae live within the coral polyp and depend on sunlight for photosynthesis.

The great reefs of corals offered food and shelter to a host of oceanic life. Surviving groups of brachiopods clung to the reef structures, as did bivalves, bryozoans, sponges, and other sedentary creatures. Gastropods grazed ceaselessly along the reef structures, while crabs and shrimp scuttled about, seeking food in the recesses and cavernous hollows of the reefs.

In the quieter lagoonal areas behind the reefs and on the floors of the epicontinental seas, starfish, sea urchins, and crinoids like *Pentacrinus* (Fig. 14-5) thrived. Not all crinoids were attached to the seafloor by stems. Bottom-dwelling stemless varieties such as *Uintacrinus* (Fig. 14-6) became abundant during the Cretaceous.

ECHINOIDS Those mobile relatives of crinoids, the echinoids (sea urchins), became far more diverse and abundant in the Mesozoic than they had ever been in the preceding era. Many Lower Cretaceous formations contain prodigious remains of these spiny creatures.

There are two kinds of echinoids. **Regular echinoids** have five-fold symmetry and a shell that is more or less spherical (Fig. 14-7A). **Irregular echinoids** have a flattened shell (Fig. 14-7B) and are bilaterally symmetrical. Regular echinoids were especially abundant during the Jurassic but were overtaken by irregu-

lar echinoids during the Cretaceous. Other members of the Phylum Echinodermata were common in Mesozoic seas, such as starfish and ophiuroids (Fig. 14-8).

"Age of the Ammonoids"

A paleontologist specializing in invertebrates might call the Mesozoic "the Age of Ammonoids." Not only were these mollusks abundant; they were so varied that they are exceptionally useful in worldwide stratigraphic correlation. As pelagic swimming animals, ammonoids attained global distribution. Even after death, their gas-filled shells were widely dispersed by currents. Using ammonoid guide fossils to develop zones permits correlation of Mesozoic time-rock units with a level of precision often surpassing that of isotopic techniques.

CEPHALOPODS You may recall that two orders of cephalopods arose during the Paleozoic Era: the **Nautiloidea**, having relatively straight sutures, and the **Ammonoidea**, having wrinkled sutures. **Sutures** of cephalopods are lines formed on the inside of the shell where the edge of each chamber's partition, or **septum**, meets the inner wall. Wrinkled sutures are a reflection of septa that, like the edges of a pie crust, are fluted. Knowledge of the exact suture pattern of ammonoid cephalopods is necessary for their identification and hence their use in correlation (Fig. 14-9).

A

B

FIGURE 14-7 **Mesozoic echinoids.** (A) Regular echinoid *Cidaris* from marine Jurassic strata. (B) The Cretaceous irregular echinoid *Hemiaster*.

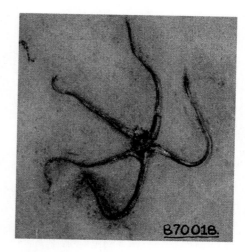

A

B

FIGURE 14-8 **Two members of the echinoderm class Stelleroidea.** (A) An ophiuroid (brittle star or serpent star) from Triassic rocks in England. Slab is about 10 cm wide. (B) An imprint of an asteroid (starfish) in a California Cretaceous sandstone. The starfish is about 15 cm in diameter.

Based on the complexity of the suture patterns, we divide **ammonoids** into three groups:

- **Goniatites** (see Chapter 12, Fig. 12-39), which lived from Devonian to Permian time
- **Ceratites**, abundant in Permian and Triassic marine areas (Fig. 14-10)
- **Ammonites** (Fig. 14-11)

Ammonites, although represented in all three Mesozoic periods, were most abundant during the Jurassic and Cretaceous.

WRINKLED SEPTA AND SUTURES Why were the edges of ammonoid septa wrinkled? Perhaps, like the corrugated steel panels used in buildings, fluted septa provided greater strength. Comparison studies of living cephalopod Nautilus have revealed that the gas-filled chambers exert only a slight outward pressure, whereas the water pressure on the outside of the conch wall is considerable. A shell crushed by water pressure would kill a cephalopod, so septal fluting may have helped the animal withstand pressure differences. Proponents of this concept call attention to the fact that ammonoid shells (unlike nautiloid shells) tend to thin toward the large final living chamber.

Not all investigators agree that the fluted margins of ammonoid septa evolved to resist hydrostatic pressure.

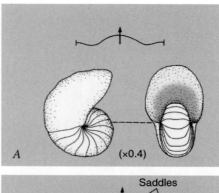

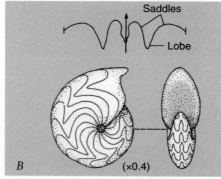

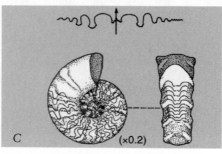

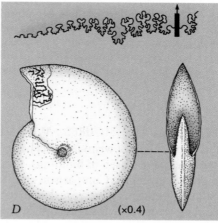

FIGURE 14-9 Sutures of cephalopods. (A) Nautiloid (arrow at mid-ventral line points toward conch opening). (B) Ammonoid with goniatitic sutures. (C) Ammonoid with ceratitic sutures. (D) Ammonoid with ammonitic sutures. *The wrinkled patterns of sutures trace the similarly wrinkled edges of septa. What may have been the purpose of the wrinkled septal margins?*

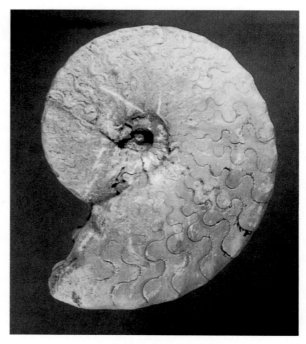

FIGURE 14-10 Triassic ammonoid cephalopod. Note the ceratitic sutures with tiny serrations on the lobes. This specimen is from the Upper Muschelkalk strata of Germany.

Another proposal is that the septal folds served as places that could be filled with soft tissue to provide multiple attachment sites. In this way, the light, buoyant conch could be held firmly to the heavy body of the cephalopods and prevent it from being torn away.

The great variety of Mesozoic ammonoids is an indication of their success in adapting to a variety of marine environments. They seem to have expanded not only within the shallow epicontinental seas but also in the open oceans. During the Cretaceous many departed from the normal pattern of coiling in a plane and evolved straight shells or shells that coiled in a helix fashion, like the shells of many snails. However, near the end of the Cretaceous, the entire diverse assemblage began to decline and, rather mysteriously, became extinct by the end of the era. Only their close relatives, the nautiloids, survived.

BELEMNITES Another group of cephalopods that became particularly common during the Mesozoic were the squidlike **belemnites** (see Fig. 6-51). The belemnite conch was inside the animal. Its pointed end was at the rear, and the forward part was chambered. A few remarkable specimens from Germany are preserved as thin films of carbon; they show the 10 tentacles and body form. Much like the modern-day squid, the belemnites were probably able to make rapid reverse dashes by jetting water out of a funnel at their anterior end.

Snails and Crustaceans

Marine **gastropods** also were abundant during the Mesozoic. Many are found in sediments that represent old beach deposits. Then, as now, cap-shaped limpets grazed across wave-washed boulders, while a variety of snails with tall, helicoid conchs crawled about on the surfaces of shallow reefs. For the most part, the gastropod fauna had a modern appearance and included many colorful and often beautiful forms that have present-day relatives.

Modern types of marine **crustaceans** were abundant by Jurassic time, such as crayfish, lobsters, crabs, shrimps, and ostracods. At some localities, barnacles grew in profusion on reefs and wave-washed rocks.

Protistans

Among the single-celled protozoans that crawled or floated in Mesozoic seas were the **radiolarians** (Fig. 14-12) and **foraminifera** (Figs. 14-13). As noted in Chapter 12, both of these groups appeared in the Paleozoic. Radiolarians make their latticelike skeletons from opaline silica. In some regions today, radiolarian and diatom skeletal remains accumulate on the seafloor to form extensive deposits of siliceous ooze. In the past, these have contributed to the formation of chert beds and the porous siliceous rock called diatomite.

The tests of foraminifers are more readily preserved than those of radiolarians. "Forams," as they are called, left an imposing Mesozoic fossil record that is very important in stratigraphic correlation. Forams are especially important in petroleum exploration. Because of their small size and strong tests, large numbers of foraminifers can be obtained unbroken from the small pieces of rock recovered while drilling for oil. They are then used in tracing stratigraphic units from well to well. Forams also are sensitive indicators of water temperature and salinity. They therefore provide data useful in reconstructing ancient environmental conditions.

Foraminifers were only meagerly represented in the Triassic but began to proliferate thereafter. By Cretaceous time, they attained their greatest number of species. Planktonic species colonized the upper levels of the ocean in prodigious numbers. Their empty tests, accompanied by myriads of coccoliths, blanketed the seafloor and became part of thick beds of chalk and marl that characterize marine Cretaceous sections in many parts of the world.

FIGURE 14-11 **Cretaceous ammonoid cephalopods.** These are from the Mancos Shale of New Mexico. (A) *Hoplitoides sandovalensis.* Vertical diameter is 9.3 cm. (B) Apertural view of another *Hoplitoides sandovalensis.* (C) Part of outer whorl of *Tragodesmoceras socorroense.* Maximum height is 16 cm.

Belemnites were highly successful during the Jurassic and Cretaceous. Triassic belemnites may well have been the ancestors to the squids, which also were numerous during the Jurassic and Cretaceous. Octopods, lacking a shell, have a poorer fossil record consisting only of occasional imprints in once-soft sediment.

Terrestrial Invertebrates

Information is abundant about Mesozoic marine invertebrates. However, less is known about continental groups, simply because the marine species were more likely to be preserved. Rare fossils of air-breathing snails have been found. Somewhat more common are freshwater clams

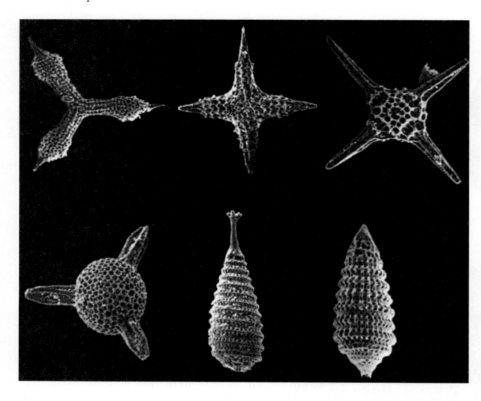

FIGURE 14-12 **Scanning electron photomicrographs of Jurassic radiolarians.** These are from the Coast Ranges of California. Top row: *Paronaella elegans*, *Crucella sanfilippoae*, and *Emiluvia antiqua*. Bottom row: *Tripocyclia blakei*, *Parvicingula santabarbarensis*, and *Parvicingula hsui*. ◀ *In what taxonomic kingdom do we place radiolarians?*

and snails. Ostracods are common in Mesozoic lake bed deposits and are used in correlating these sediments.

It is likely that many varieties of worms existed, but their soft bodies left few traces. Spiders, millipedes, scorpions, and centipedes that had been abundant in Carboniferous forests undoubtedly were present in the Mesozoic too, but fossils are rare. Most major insect groups appeared before the end of the Jurassic. Many of the best insect fossils are from the Solnhofen Limestone, a fossil-rich Jurassic formation in Bavaria. Cretaceous fossils of insects also are found preserved in amber. Interestingly, the list of Mesozoic insects does not include fleas. They appeared in the early Cenozoic—at the same time that their hosts, the mammals, were evolving.

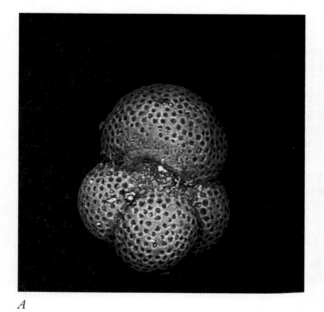

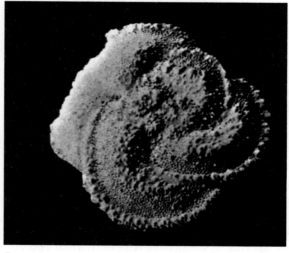

A

B

FIGURE 14-13 **Electron micrographs of Cretaceous planktonic foraminifera.**
(A) *Globigerinoides*. (B) *Globotruncana*. Magnified 100 times.

MESOZOIC VERTEBRATES

The Ray-Finned Bony Fishes

Ray-finned bony fishes were abundant members of marine faunas since Devonian time. In the Cretaceous, however, earlier forms were replaced by the most advanced group of fishes: the teleosts. The teleosts produced an astonishing variety of species, and became the dominant fishes on Earth. Today they are more abundant and varied than any other group of living vertebrates.

The Rise of Modern Amphibians

Among the tetrapods of the late Paleozoic, a group of amphibians known as **temnospondyls** was able to survive the wave of extinctions at the end of the Permian. About 17 families of temnospondyls lived during the Triassic. Thereafter, they declined, with only two lineages surviving into the Jurassic and one into the Early Cretaceous. Their successors include frogs, newts, salamanders, and limbless amphibians called caecilians.

The oldest known frog is *Triadobatrachus* (Fig. 14-14A) from the Early Triassic rocks of Madagascar. The skull of this little amphibian is very froglike in appearance, but the skeleton behind the skull is not as highly modified for jumping as its Jurassic descendants (Fig. 14-14B). Salamanders and newts are less specialized than frogs. They retained four walking legs and evolved a flattened tail for swimming. The oldest known salamander is *Karaurus* from Late Jurassic beds in Kazakhstan.

The limbless caecilians should take the prize for the oddest of modern amphibians. Externally they resemble earthworms, whereas internally, they are supported by as many as 200 vertebrae. They live in tropical regions, where they find their food in ponds or by burrowing through leaf litter. The oldest caecilian thus far discovered is *Eocaecilia* from the Early Jurassic of Arizona. This creature still retained very reduced limbs.

The Triassic Transition

The general crustal unrest and regression of epicontinental seas near the end of the Paleozoic resulted in a variety of continental environments. This provided the environmental stimulus needed to maintain the spread and diversification of land vertebrates.

Although marine faunas changed abruptly in passing from the Paleozoic to the Mesozoic, there was more continuity among land animals. The main Carboniferous amphibians continued into the Early Cretaceous before becoming extinct. The mammal-like therapsids were also able to cross the era boundary. The most progressive therapsids succeeded their Permian precursors to become contemporaries of primitive Triassic mammals.

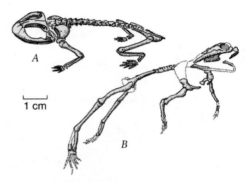

FIGURE 14-14 **Frog skeleton restorations.** (A) Ancestral Triassic frog *Triadobatrachus*. (B) *Prosalirus*, an early modern frog from the Lower Jurassic Kayenta Formation of Arizona.

Many new reptile groups appeared in the Triassic. Among these were the first turtles. Although they were similar to modern turtles, many retained teeth in their jaws. The Triassic also was the time during which many lineages of marine reptiles appeared. **Rhynchocephalians**, represented today by the tuatara of New Zealand, were abundant during the Triassic. Most interesting of all, however, were an important group of reptiles known as archosaurs.

Basal Archosaurs

Archosaurs are what biologists term diapsid reptiles. The diapsids have two skull openings behind the eye orbits (*di* = two; *apsis* = openings). They are divided into two groups, the **lepidosaurs** (lizards, snakes, and their ancestors) and the **archosaurs**. Archosaurs include the dinosaurs, as well as pterosaurs (flying reptiles), and crocodilians. Cladistic analyses indicate that birds are archosaurs as well.

Several groups of archosaurs were present during the Triassic. They are termed "basal archosaurs" because they are at the starting point of archosaurian evolution. As exemplified by *Hesperosuchus* (Fig. 14-15), the basal archosaurs were small, agile, lightly built animals with long tails and short forelimbs. They had already developed the unique habit of walking upright on their hind legs.

Although many of the basal archosaurs were bipedal sprinters, some reverted to a four-footed stance and evolved into either armored land carnivores or large crocodile-like aquatic reptiles called **phytosaurs** (Fig. 14-16). Occasionally in the history of life, initially dissimilar organisms from separate lineages gradually become more and more similar in form. The once-distinct groups change over many generations so that they are better adapted to a particular environment. We call this evolutionary process of producing similar forms in unrelated organisms **convergence**.

Phytosaurs and crocodiles are good examples of evolutionary convergence. The most visible distinction

FIGURE 14-15 *Hesperosuchus,* **a basal archosaur.** These small (adults are about 4 feet long), agile, lightly built animals had long tails and short forelimbs. They developed the unique habit of walking up on their hind legs, an important evolutionary innovation. It permitted basal archosaurs to move about more speedily than their sprawling ancestors. Because their forelimbs were not used for support, they could be employed for catching prey. This Triassic example is from the southwestern United States.

FIGURE 14-16 *Rutiodon,* **a large crocodile-like aquatic phytosaur.** Phytosaurs and crocodiles are good examples of evolutionary convergence, where dissimilar organisms evolve similar forms in response to their environment. Phytosaurs were among the largest land animals of the Triassic, some attaining lengths of 11 meters (about 35 feet). ◀ *What living reptile is an example of convergent evolution with Rutiodon?*

between the two groups is the position of the nostrils, which are at the end of the snout in crocodiles but were just in front of the eyes in phytosaurs. Phytosaurs were among the largest land animals of the Triassic. Some attained lengths of 11 meters (about 35 feet).

DINOSAURS: "TERRIFYING LIZARDS"

Dinosaur derives from Greek: *deinos* = terrifying and *sauros* = lizard. Dinosaurs are the most awesome and familiar of prehistoric beasts (Fig. 14-17). We divide them into two groups, based in part on the arrangement of bones in the hip region: **Saurischia** (lizard-hipped) and **Ornithischia** (bird-hipped).

- In Saurischia, the pelvis had three bones on each side (Fig. 14-18). The uppermost **ilium** bone was firmly clamped to the spinal column. The bone ex-

tending downward and slightly backward is called the **ischium**. Forward of the ischium is the **pubis**. So, in the saurischians, the three pelvic bones were in an arrangement called triradiate (having three radiating bones). This is the same as in their basal archosaurian ancestors.

- Ornithischians had the same three bones, but in a different arrangement. The pubis was swung downward and backward so that it was parallel to the ischium, as in modern birds (Fig. 14-18).

Another key difference between ornithischians and saurischians is their teeth. In saurischians, teeth either extended around the entire margins of the jaws or were limited to the frontal area. Ornithischians lacked teeth in the front of both upper and lower jaws (Fig. 14-19).

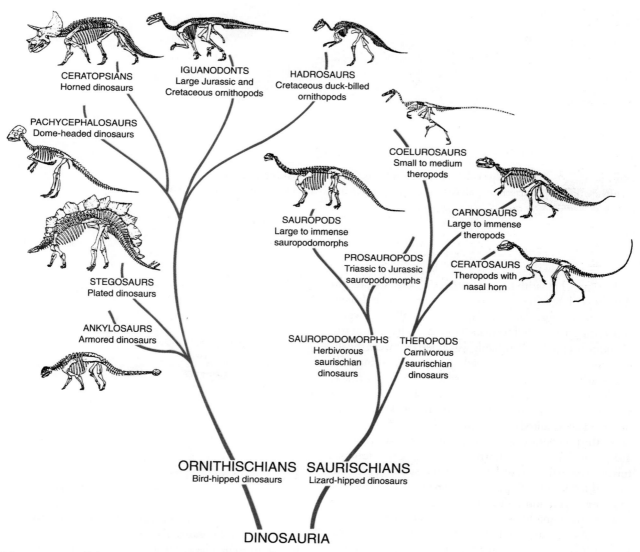

CERATOPSIANS
Horned dinosaurs

IGUANODONTS
Large Jurassic and Cretaceous ornithopods

HADROSAURS
Cretaceous duck-billed ornithopods

PACHYCEPHALOSAURS
Dome-headed dinosaurs

COELUROSAURS
Small to medium theropods

SAUROPODS
Large to immense sauropodomorphs

CARNOSAURS
Large to immense theropods

PROSAUROPODS
Triassic to Jurassic sauropodomorphs

CERATOSAURS
Theropods with nasal horn

STEGOSAURS
Plated dinosaurs

ANKYLOSAURS
Armored dinosaurs

SAUROPODOMORPHS
Herbivorous saurischian dinosaurs

THEROPODS
Carnivorous saurischian dinosaurs

ORNITHISCHIANS
Bird-hipped dinosaurs

SAURISCHIANS
Lizard-hipped dinosaurs

DINOSAURIA

FIGURE 14-17 **Dinosaur family tree.** *Which of the groups shown were predators?*

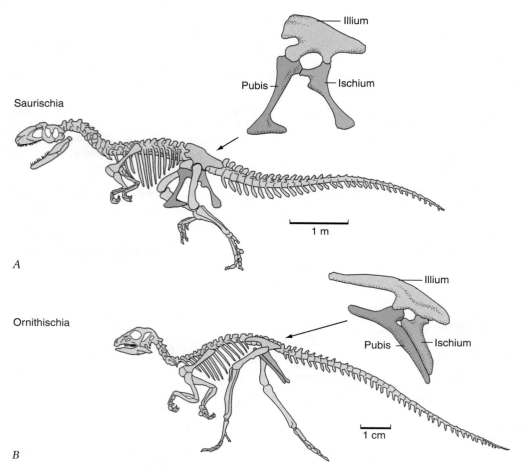

A

B

FIGURE 14-18 **Dinosaur pelvic structures: a basis for distinguishing Saurischia from Ornithischia.** (A) Saurischian pelvis of *Allosaurus*. (B) Ornithischian pelvis of a small dinosaur, *Heterodontosaurus*.

In saurischians, the teeth and jaws were adapted to cutting and tearing—but not chewing—loose leaves, shoots, and twigs. Parts of plants were swallowed whole and reduced to a pulpy mass later by a huge gizzard. Grinding of the food probably was aided by stones the dinosaurs swallowed, which acted as a sort of grinding mill. Bacteria in the stomach helped break down woody tissue.

Ornithischians lacked teeth in the front of both upper and lower jaws. In many, the front part of the skull and jaws became a beak, suited to cropping vegetation. A single new bone, the predentary, was added to the lower jaw (Fig. 14-19B). It may have had a horny covering like the beak of a turtle. Teeth that were limited to the sides of the jaws formed a "pavement" perfectly shaped for crushing and grinding vegetation. With this efficient battery of teeth, ornithischians might have eaten even coarse vegetation.

In the dinosaur family tree (Fig. 14-17), you can see that we divide the Saurischia into two groups: **theropoda** (typically bipedal meat-eaters) and **sauropodomorpha** (typically large, quadrupedal herbivores).

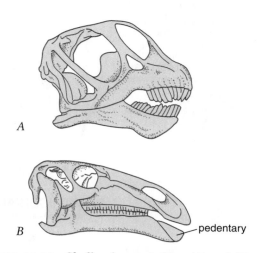

A

B

FIGURE 14-19 **Skulls of a saurischian (A) and (B) ornithischian dinosaur.** The ornithischian skull is distinguished by a lack of teeth in the forward part of the jaws, and a predentary bone at the forward end of the lower jaw.

FIGURE 14-20 *Herrerasaurus,* **one of the oldest known true dinosaurs.** This bipedal flesh-eating theropod, about 3 meters (10 feet) long, is from Triassic rocks of Argentina.

Lizard-Hipped Bipedal Meat-Eaters (Saurischian Theropods)

The earliest dinosaurs were theropods, represented by *Eoraptor* and *Herrerasaurus* (Fig. 14-20). Potassium-argon dating of volcanic ash beds associated with these predators indicates they roamed Earth about 225 million years ago, during the late Triassic. The early theropods were smaller; *Eoraptor* was only about a meter long and *Herrerasaurus* around 3 meters. *Eoraptor's* carnivorous lifestyle is indicated by its clawed three-fingered hands and curved, serrated teeth.

Coelophysis (Fig. 14-21) was a theropod from the Triassic Chinle Formation. One New Mexico site yielded skeletons from over 100 individuals, apparently killed during a local catastrophe such as a flash flood. In this large collection, males could be distinguished from females by skeletal differences, providing evidence for sexual dimorphism among dinosaurs.

During the Cretaceous Period a distinctive group of theropods called ornithomimosaurs ("bird mimics") appeared. A representative is the ostrich-like *Ornithomimus* (Fig. 14-22), with its long neck, slender hind limbs, and toothless, beak-like jaws.

In part because of movies like *Jurassic Park* and *The Lost World*, the great meat-eaters of the Jurassic and Cretaceous have become as familiar to us as modern zoo animals. Among the favorites are the Jurassic *Allosaurus* (Fig. 14-23), Cretaceous *Deinonychus* (Fig. 14-24), Cretaceous *Tyrannosaurus* (Fig. 14-25), and *Velociraptor*.

For sheer size and frightening appearance, *Tyrannosaurus* has always been the most famous of the theropods. It attained lengths of 13 meters (42 feet) and weighed over 4 metric tons (around 9000 pounds, about the weight of a Hummer).

In addition to its size, *Tyrannosaurus* had many other intimidating features:

- Its teeth were large, curved, laterally compressed, and serrated for slicing through meat.
- Its powerful jaws could exert over 3000 pounds of biting force. (An adult lion can exert only a little more than 900 pounds of biting force.)
- Its mobile skull-to-neck attachment provided rapid movement of the head.
- Its eyes were large and stereoscopic, suggesting that the animal located its prey visually. (There is no evidence for the *Jurassic Park* premise that they could not detect prey unless it moved.)
- Its nostrils were far forward, and coupled with an enlarged olfactory lobe of the brain, probably provided an excellent ability to smell.

FIGURE 14-21 **Small, agile theropod** *Coelophysis.* Living about 220 million years ago during the Late Triassic, these fast, agile, bipedal predators may have pursued their prey in packs. There is evidence that they occasionally even ate juveniles of their own species. Length is about 3 meters (10 feet).

FIGURE 14-22 **Ostrich-like *Ornithomimus* ("bird mimic").** This Cretaceous dinosaur resembled a featherless ostrich and probably lived in much the same way. Note the saurischian tri-radiate pelvic bones. ▐ *Label the pubis, ischium, and ilium on the drawing.*

- As *Tyrannosaurus rex* hunted (or possibly scavenged for carcasses), it held its body nearly horizontal, pivoting its weight on the pelvic region. It held its tail horizontally to counterbalance the forward part of its body.

Tyrannosaurus was the largest known terrestrial carnivore until 1995, when the more massive *Gigantosaurus* was discovered in Argentina. Giant theropods also roamed Africa, as indicated by the bones of *Carcharodontosaurus saharicus* (the shark-toothed reptile from the Sahara).

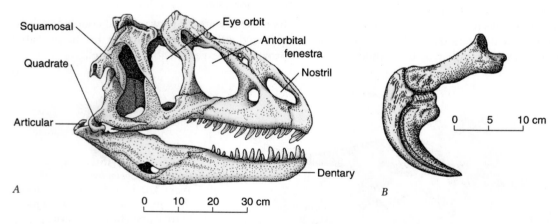

FIGURE 14-23 **Teeth and claw of *Allosaurus*, a serious predator.** (A) The long jaws and dagger-like teeth were effective in capturing and shredding prey. The skull is a framework of sturdy arches with powerful muscles attached. (B) One of the three powerful curved claws on the hand of *Allosaurus*. Such claws doubtless were effective at tearing flesh from prey.

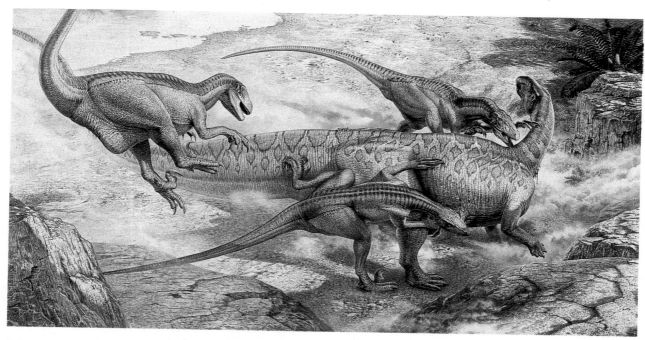

FIGURE 14-24 *Deinonychus* **("terrible claw"), another serious predator.** Its large, serrated teeth and a greatly enlarged "terrible claw" on the second digit of its hind feet made this creature a killer. The size of its claw compelled the animal to run on only two toes. Shown here are three *Deinonychus* attacking the large ornithischian *Tenontosaurus*.

FIGURE 14-25 **A Cretaceous street gang.** From left, these giant predatory dinosaurs are *Daspletosaurus*, *Tyrannosaurus*, and *Tarbosaurus*. *Tyrannosaurus* attained lengths of 13 meters (42 feet) and weighed over 4 metric tons (the weight of two small pickup trucks).

Lizard-Hipped Quadrupedal Plant-Eaters (Saurischian Sauropodomorphs)

We divide the herbivorous sauropodomorpha into **prosauropods** and **sauropods**. The prosauropods were the likely ancestors of the sauropods and lived from Late Triassic to Early Jurassic. Their remains have been found on all continents except Antarctica.

Prosauropods were quadrupeds, but their forelimbs were shorter than the hindlimbs (Fig. 14-26). This indicates that they very likely could rear up on their hind legs to reach food on higher tree branches. In this upright posture, sauropods also could survey the horizon for menacing predators.

During the Early Jurassic, the prosauropods were replaced by the famous giant sauropods. The most impressive were gargantuan, long-necked, long-tailed animals that required a sturdy four-legged stance to support their tremendous bulk. *Apatosaurus* (formerly *Brontosaurus*) is a sauropod favorite of schoolchildren, weighing 30 tons (about the weight of four African bull elephants). Another favorite is *Brachiosaurus* (Fig. 14-27). Although awesome in size, *Apatosaurus* and *Brachiosaurus* were relative lightweights when compared to the 80–100 ton *Supersaurus* (Fig. 14-28).

Despite their frightening size, sauropods were plant eaters, mostly "high browsers" whose long necks permitted munching the upper foliage of trees. The longer forelimbs of *Brachiosaurus* allowed an even higher reach into the treetops. To avoid a burden of weight on the neck, the heads of sauropods were relatively small.

Could these giants have been water dwellers? For many years, paleontologists speculated that they could not have supported their own weight continuously, and might have lived in the buoyant waters of lakes and streams. But little evidence supports this. Sauropod footprints and foot structure indicate that they walked on the tips of the toes on the front feet, with the heels of their hind feet resting on large pads like those of elephants. Clearly, they were land-dwellers whose massive limbs provided adequate support on dry land.

Large size afforded certain advantages to the sauropods. Predators often avoid encounters with huge animals. In addition, great size in cold-blooded animals provides better temperature regulation. A large pot of

FIGURE 14-26 *Plateosaurus*, a Late Triassic prosauropod. Very likely it could rear up on its hind legs to reach food and to spot predators.

FIGURE 14-27 **Huge Jurassic sauropod *Brachiosaurus*.** Displayed at the Field Museum in Chicago.

hot water loses heat more slowly than a small pot, and a large animal loses heat more slowly than a small animal. (The ratio of surface area to volume for an animal decreases as size increases. Consequently, a large animal has a proportionately smaller surface over which heat can be lost.)

Sauropods began their major expansion during the Early Jurassic, and by the end of that period they peaked in diversity, abundance, and size. In North America, they declined somewhat after the Jurassic, but continued in diminished numbers until the close of the Cretaceous. A reduction in sauropods during the Cretaceous did not occur in Southern Hemisphere continents, where they continued as dominant land herbivores.

Bird-Hipped Plant-Eaters (Ornithischians)

The other major dinosaur line, the Ornithischia, evolved near the end of the Triassic and thrived throughout the following Jurassic and Cretaceous. Ornithischians were herbivores. The group included both quadrupedal and bipedal varieties, with the bipedals considered more primitive. Even the most advanced quadruped ornithischians had short forelimbs, indicating their descent from bipedal forms.

ORNITHOPODS (BIPEDAL ORNITHISCHIANS) The bipedal ornithischians are known as **ornithopods**. Their evolutionary history began in the Triassic, with small, lightly

FIGURE 14-28 **A colossal trio: Sauropods, the largest land animals to evolve.** *Seismosaurus* (left) with two *Supersaurus*. The *Supersaurus* at center was formerly named *Ultrasaurus* but subsequently was identified as a *Supersaurus*.

ENRICHMENT

Can We Bring Back the Dinosaurs?

Imagine visiting a zoo or theme park and safely viewing a living *Tyrannosaurus* or *Triceratops*. Michael Crichton described such a place in his 1990 science fiction novel *Jurassic Park*, which appeared three years later as a Hollywood movie. The novel describes a dinosaur-populated theme park constructed on an isolated tropical island. Dinosaurs dwelling in the park are copies of real animals, produced by cloning of their Mesozoic counterparts.

Cloning has been accomplished in several kinds of living animals. The process involves transplanting DNA from a somatic cell (a *body cell*, not involved in reproduction) to an egg that has been stripped of its nucleus. With DNA from the animal to be cloned, the egg completes development without fertilization. The result is a precise copy of the animal that contributed the DNA.

In the *Jurassic Park* tale, the DNA was obtained from dinosaur blood extracted from Mesozoic mosquitoes preserved in amber (see Fig. 4-6). Missing DNA segments were added from the DNA of frogs.

The premise of *Jurassic Park* is clever, but we have no way of knowing whether Mesozoic mosquitoes sucked dinosaur blood. Perhaps they preferred the blood of the small furry mammals that were scurrying about. But assuming that they did bite dinosaurs, would blood extracted from the belly of an amber-encased mosquito come from one species of dinosaur, or from two or more? Some spectacular DNA forensics would be required to make the necessary identifications.

Another problem is the extreme improbability that DNA could survive intact for tens of millions of years. Lengthy sections would be decomposed or destroyed. It would be necessary to reconstruct missing parts, filling the gaps by specifically and repetitively copying segments between defined nucleotide sequences.

If we make the extravagant assumption that we would be able to produce a dinosaur gene segment, or even a few hundred segments, these would still represent only a tiny fraction of the billions of segments that once were present in complete dinosaur DNA. Further, there are specific proteins that coat chromosomes and bind to key locations. These proteins govern how genes are expressed. Without them, the chromosomes are ineffective.

But assume that all of the above problems are solved and we have well-preserved DNA with correctly linked proteins. When the DNA is implanted into the egg of another species, will it develop into a dinosaur embryo? Eggs are not just passive containers waiting to receive DNA from any provider. They contain specific directions about how cell division is to proceed and how the embryo is to be positioned within the confines of a shell of particular size and shape. It is improbable that the egg of a living reptile or bird would so closely resemble that of a *Tyrannosaurus rex* to allow complete development.

We have witnessed some truly astonishing advances in molecular biology over the past several decades. Maybe someday some of the seemingly insurmountable obstacles to dinosaur cloning may find solutions. For the foreseeable future, however, don't expect to see dinosaurs on your next visit to the zoo.

FIGURE 14-29 *Iguanodon* and *Baryonyx*. *Iguanodon* (rear) was a herbivorous ornithischian of the Early Cretaceous. Approaching from the right is the carnivorous *Baryonyx*. The youth in the painting is shown for scale.

were large ornithischians in which the forward part of the skull was broad, flat, and toothless. It resembled a duck's bill, so the nickname "duck-billed dinosaurs" applies to members of this group (Fig. 14-30). Behind the toothless forward part of the jaw were rows of interlocking lozenge-shaped teeth cemented together to form rasplike grinding surfaces well-adapted to chewing coarse vegetation.

An interesting peculiarity of some hadrosaurs was a bony skull crest, often containing tubular extensions of the nasal passages (Figs. 14-31 and 14-32). The cranial crests could have been used to catch the eye of sexual partners and could be further employed as vocal resonators used to call a breeding partner.

FIGURE 14-30 **Cretaceous duck-billed dinosaurs (hadrosaurs).** Displayed at the American Museum of Natural History in New York.

built species that lived primarily on dry land. A representative large Jurassic ornithopod is *Camptosaurus*, a medium-sized bipedal dinosaur with a heavy tail, short forelimbs, and long hind legs. The articulation of its jaw brought the teeth together all at the same time, a design frequently seen in herbivores down through the ages. *Camptosaurus* cropped leaves and stems with the forward, beaklike part of the jaws and passed the vegetation back to the cheek teeth for chopping and chewing.

The larger Cretaceous ornithopods developed from camptosaurid ancestors. Among these was *Iguanodon* (Fig. 14-29), one of the first dinosaurs to be scientifically described. *Iguanodon* sometimes is called the "thumbs-up dinosaur" for the horny spike that substituted for a thumb.

Many ornithopods lived and traveled in enormous herds. A 200-acre Cretaceous fossil bed in Montana contains the bones of about 10,000 ornithopods, all of the same species. A layer of volcanic ash above the bone bed suggests that the huge herd was suffocated by ash, lethal gases, or possibly mud flows associated with a nearby volcanic eruption. Such volcanic events were commonplace in the Rockies during the Cretaceous.

By Cretaceous time, ornithopods had moved into a variety of terrestrial environments. A particularly successful group was the trachodonts, or hadrosaurids. Hadrosaurids

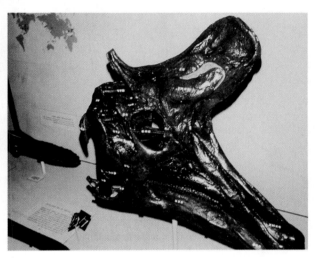

A

B

FIGURE 14-31 **Bellowing resonators? Skulls of Cretaceous crested hadrosaurs ("duck-billed" dinosaurs).** (A) Hadrosaur *Lambeosaurus* skull with a peculiar hatchet-shaped crest. (B) *Corythosaurus* with a helmet-shaped crest. These crests may have functioned as vocal resonators for bellowing. Skulls are approximately 0.75 meter (30 inches) long.

Dinosaur National Monument

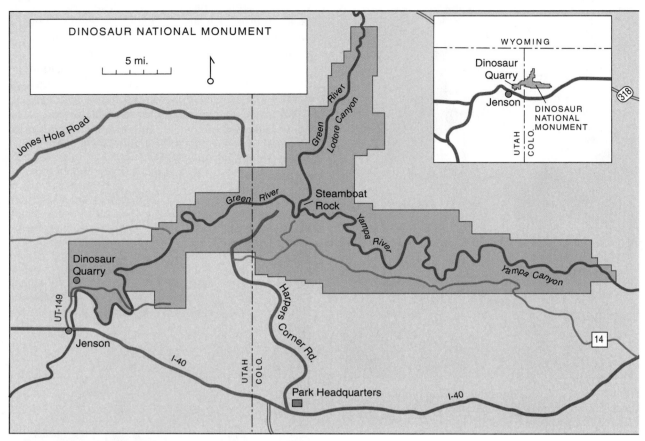

FIGURE A **Dinosaur National Monument and Dinosaur Quarry.**

The story of Dinosaur National Monument begins with industrialist Andrew Carnegie (1835–1919), who decided he wanted something really big for the new exhibit hall of his Carnegie Museum in Pittsburgh. A huge dinosaur would fit the "big" requirement very well, so Carnegie dispatched paleontologist Earl Douglass (1862–1931) to Utah's Uinta Mountains to find an impressive specimen.

In the summer of 1909, as Douglass scrutinized stream banks and canyon walls for exposed bone, he sighted eight intact vertebrae of the huge sauropod *Apatosaurus*. The bones were embedded in a steeply dipping bed of the Jurassic Morrison Formation. He had found a bonanza fossil site, "dinosaur ledge," and supervised excavation there from 1909 to 1924. Over 350 tons of specimens were shipped east by rail.

FIGURE B **Dinosaur National Monument entrance.** This building at the west entrance to Dinosaur National Monument encloses the quarry face.

FIGURE C **Dinosaur Quarry.** A museum technician examines exposed dinosaur bones.

The bipedal ornithischians also include a group called **pachycephalosaurs**, exemplified by the "bone-head" dinosaur *Pachycephalosaurus* (Fig. 14-33).

Stegosaurs ("Plated" Quadrupedal Ornithischians)

The best known of the quadrupedal ornithischians are the stegosaurs (Fig. 14-34). *Stegosaurus*, the classic stegosaur, had two pairs of heavy spikes mounted on its tail, used for defense. However, a more distinctive feature was the plates that stood upright along their backs. The plates were not attached to the spinal column, but were held in place by ligaments and muscles.

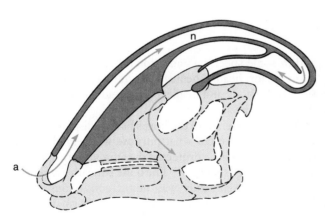

FIGURE 14-32 **Internal structure, skull crest of *Parasaurolophus cyrtocristatus*.** The bone of the left side has been removed to expose the left nasal passage (**n**). Air enters the nostrils at **a**, moves up and around the partition in the crest, and from there moves down and back to internal openings in the palate. When models of the passages are constructed, they can be used to generate sounds like those of a trombone. Perhaps *Parasaurolophus* made such sounds to attract a mate or intimidate a rival.

FIGURE 14-33 **The "bone-head" dinosaur, *Pachycephalosaurus*.** Its skull is mostly solid bone, with a tiny brain space. Perhaps they used their skulls as battering rams against one another during competition for territory or mates. Abrasions on skulls and tendons that braced the neck vertebrae support this interpretation. Similar head-butting is seen today in bighorn sheep.

427

FIGURE 14-34 "PLATED DINOSAURS" OF THE JURASSIC. Best known of this group is *Stegosaurus* (upper left), weighing 1–2 tons. Its relatively small head terminated in a toothless, narrow beak suitable for cropping plants. Chewing plant food was the task of leaf-shaped teeth in its cheek regions. Although a large animal, its brain weighed about 2.6 ounces (the size of a small potato). Its distinctive diamond-shaped plates were covered with grooves and canals that mark the location of blood vessels. These plates may have functioned in temperature regulation, somewhat like solar panels and radiators. Other stegosaurs shown are *Tuojiangosaurus* (upper right), *Dacentrurus* (lower right), *Lexovisaurus* (lower center), and *Kentrosaurus* (lower left).

The plates may have functioned in the regulation of body temperature by serving as body heat dissipaters. Indeed, their arrangement, size, and shape plus the presence of branching surface grooves for blood vessels all favor a thermal-regulation use. The plates also may have served as sexual display, camouflage, or protection.

Stegosaurs were the dominant quadrupedal ornithischians of the Jurassic. Their contemporaries in the Jurassic were the **nodosaurs** and **ankylosaurs** (Fig. 14-35), both of which surpassed the stegosaurs during the Cretaceous.

Ceratopsians (Horn-Faced Dinosaurs)

The last major group of dinosaurs to evolve were the **ceratopsians**, first appearing in the Middle Cretaceous. They take their name from horns that grew on the face of all but the earliest forms. Typically, ceratopsians possessed a median horn just above the nostrils, and in some species an additional horn pair projected from the forehead. The head was quite large in proportion to the body and displayed a shieldlike bony frill at the back of the skull roof.

You can understand the horns having display and defensive functions, but what purpose was served by the shield? Might it have protected the animal from a frontal attack by a large theropod? The bone beneath the skin of many ceratopsian frills was not a continuous plate. Rather, it contained a pattern of large and small holes. The larger holes would have reduced the shield's weight and hence pressure on the beast's neck. Perhaps the smaller holes accommodated blood vessels whose function was to help the animal radiate excess body heat, or, if needed, to absorb heat from the environment.

In addition, the frill provided attachment sites for the massive ceratopsian jaw muscles. All ceratopsians possessed jaws shaped like a parrot's beak (Fig. 14-36). Judging from scars that mark the shield bones of some

FIGURE 14-35 **Ankylosaurs, the heaviest-armored Cretaceous dinosaurs.** These bulky, squat ornithischians wore closely fitted bony plates that protected their entire 6-meter backsides. Their small heads sometimes were covered with armor. *Euoplocephalus* (top) had a bony tail club that may have been defensive, perhaps by damaging the legs of an attacking theropod. Nodosaurids like *Edmontonia* (foreground) were somewhat less ponderous and could move about more quickly.

ceratopsians, they were often attacked by predatory theropods. No doubt they frequently emerged the victor, for they would not have been easy for a large predator to bring down. Many reached lengths of 8 meters (28 feet) and weighed more than 9 tons.

DINOSAURS: COLD-BLOODED, WARM-BLOODED, OR BOTH?

Since the late 1700s, when dinosaurs were first studied scientifically, they have been regarded as reptiles.

Hence, they also were presumed to be **ectothermic**, or cold-blooded (*ecto-* means external, meaning that the creature took on the temperature of its environment). Ectothermal animals have little or no ability to maintain uniform body temperature through their own physiologic processes. However, they may regulate their body temperature by seeking either sun or shade in response to temperature needs. In living reptiles, the pineal gland may play a role in directing this behavior.

In extinct reptiles, certain anatomic features, such as the sail in *Dimetrodon*, the plates on the back of

FIGURE 14-36

Ceratopsians: horns, huge heads, parrotlike beaks. These beasts take their name from horns that grew on the face of all but the earliest forms. From left: the well-known *Triceratops*, *Pachyrhinosaurus*, and the multihorned *Styracosaurus*.

Stegosaurus, or the head frill on *Triceratops* may have served two opposite temperature-regulating purposes: to absorb heat from the sun's rays when the body needed warming and to radiate excessive body heat when cooling was needed.

In contrast, **endothermic** (warm-blooded) animals, such as mammals and birds, maintain a constant body temperature by internal production of heat and the radiation of excess heat away from the body. Mammals produce heat by oxidizing food. However, when body temperature rises, the hypothalamus (a part of the brain) triggers internal mechanisms to dissipate the heat, including expansion of blood vessels in the skin, perspiring, or (in furry animals) panting. When temperatures fall, other mechanisms minimize heat loss, such as restriction of blood vessels in the skin and shivering.

In the 1860s, biologist Thomas Huxley (1825–1895) suggested that dinosaurs might have been warm-blooded (endothermic). This idea was revived a century later, and paleontologist Robert Bakker became an enthusiastic supporter of the warm-blooded dinosaur concept. Bakker knew that birds and dinosaurs have many similarities. If dinosaurs were truly endothermic, it would be logical to reclassify vertebrates by removing dinosaurs from the reptile category and building a new classification that would include dinosaurs and birds under Archosauria. In this reclassification, dinosaurs still "live" today—they are birds.

Bakker supported his hypothesis of warm-blooded dinosaurs with several other lines of evidence. One relates to the way dinosaurs stood and walked. Today's lizards and salamanders are cold-blooded, and most have a sprawling stance, with their limbs directed more or less to the side. In contrast, the limbs of warm-blooded mammals and birds are held directly beneath the body. Dinosaur stance resembles that of mammals and birds. Does this indicate that dinosaurs were endothermic? Possibly, but dinosaur posture might be only an evolutionary solution for supporting their enormous weight.

Bakker and others also point to the microscopic structure of dinosaur bones. Some species are richly vascular, like the bones of mammals. But most reptiles have less vascular bones, indicating a poorer supply of blood to the bone tissue. However, this correlation is not absolute— some bones in living crocodiles have considerable vascularity, yet crocodiles are primarily cold-blooded. Perhaps the vascular bone in dinosaurs reflects growth rate or size and was not related to warm-bloodedness.

Isotope analysis of bone also provides evidence. Cold-blooded animals exhibit large differences in the oxygen isotope content of bones of the extremities versus bones of the body core. Warm-blooded vertebrates do not exhibit this. Analysis of bone from Cretaceous theropods, ceratopsians, and hadrosaurs shows isotope variability similar to that in warm-blooded vertebrates.

Yet another argument for dinosaur warm-bloodedness is correlation between the proportions of predators to prey in mammals (endotherms) as opposed to living reptiles (ectotherms). Today's warm-blooded communities are about 3% predators and 97% prey (plant-eaters). Clearly, it takes a lot of food to fuel the energy requirements of an endothermic predator such as a lion or wolf. In the cold-blooded community, 33% of the animals are predators and 66% are prey. Determining the proportion of predators to prey among dinosaurs is tricky because the fossil record cannot be as precise as data obtained from living communities. At present, the evidence is inconclusive.

Feathered Theropods

Birds have feathers. So, if dinosaurs and birds are closely related, should we not find evidence of feathers on some dinosaurs? Unfortunately, feathers are delicate and have a low probability of being preserved. Nevertheless, in 1996, Chinese paleontologists discovered the remains of a small carnivorous theropod with a covering of hollow fibers that resemble simple feathers. They named their fossil *Sinosauropteryx*. Their report suggests the fibers were simple feathers that insulated the animal. After this discovery, many other feathered theropods have been found, including *Caudipteryx* and *Protarchaeopteryx* from China.

▶ DINOSAUR PARENTING

Dinosaurs reproduced by laying eggs. Clutches of dinosaur eggs have been found at dozens of localities around the world. But did dinosaurs care for the eggs after they were laid? Did they nurture the hatchlings? Discoveries made during the last few decades in Montana and Mongolia show that at least some dinosaurs cared for their young.

Dinosaur eggs from Montana occur in the Cretaceous Two Medicine Formation in the western part of the state. Dubbed "Egg Mountain" by its discoverer John (Jack) R. Horner, the fossil site includes an entire hatchery of hadrosaurian dinosaurs complete with nests, clutches of eggs, embryos, and nestlings. There is evidence that hadrosaur babies were nurtured by their parents, lived within the social structure of large herds, were warm-blooded, and in general behaved more like birds than like today's reptiles.

The hadrosaurs hollowed out bowl-shaped nests in soft soil and laid about 20 eggs in neatly arranged circles within each nest. Plant impressions in the sediment suggest that the eggs were covered with decaying vegetation so that its fermentation would provide warmth. Horner observed that many of the nests contained the bones of juveniles that were about a meter long. Thus, the babies (which were only about 30 cm long when hatched) stayed in their nests where food was brought to them until they had grown sufficiently to fend for themselves. In addition, the teeth of these juveniles exhibited distinct signs of wear, suggesting that they had been feeding for some time

FIGURE 14-37 **Fossil dinosaur eggs from the Upper Cretaceous of Mongolia.** Discovery of eggs containing fully formed embryo skeletons indicates that the eggs belonged to the theropod dinosaur *Oviraptor.*

while still in the nest. Horner named these dinosaurs *Maiasaura* from the Greek for "good mother lizard."

The gentle plant-eating hadrosaurs were not the only dinosaurs to "sit" their nests like birds. Predators appear to have had the parenting instinct as well. In 1993, Mark A. Norell discovered a nest of dinosaur eggs in Cretaceous rocks of Mongolia's Gobi Desert. Within the nest was the nearly complete skeleton of an embryonic theropod. The tiny dinosaur was about to hatch and resembled a miniature adult. It was readily identified as the embryo of the predatory dinosaur *Oviraptor.*

The fossil eggs found by Norell are identical to those found in 1922 in the Gobi Desert in Mongolia by famous fossil hunter Roy Chapman Andrews (Fig. 14-37). At the time, the eggs were thought to belong to the ceratopsian dinosaur *Protoceratops*. However, on top of the nest that Norell found were the bones of *Oviraptor*, so-named because that theropod was assumed to have died while attempting to steal or eat the eggs. Actually, *Oviraptor* was falsely accused, for the eggs were its own. It died while protecting or incubating them. As evidence of this, several *Oviraptor* skeletons were recently found squatting over their nests in precisely the manner of birds.

THE FLYING ARCHOSAURS

Triassic Gliders

This same pattern repeats throughout the history of life: a small group of animals initially adapts to a narrow range of ecologic conditions, and then their descendants, through evolutionary processes, spread into peripheral environments. As the descendants diverge from their ancestral lineage, they change in ways that make them fit their new surroundings.

This process, called **adaptive radiation**, is well demonstrated by Mesozoic vertebrates. Their adaptive radiation produced a rich diversity of carnivores, herbivores, omnivores, land dwellers, water dwellers, swimmers, walkers, and even vertebrates that could fly.

The first reptiles to attempt flight probably were gliders, similar to present-day "flying lizards." For such animals, gliding was an easy way to move from treetops to the ground or from branch to branch. The earliest gliding reptile was *Coelurosauravus*. The skin membranes that served as its wings were supported along the sides of the body by at least 22 long, slender bones extending outward from each side of the chest. At one time these elongate bones were thought to be ribs, but a fossil discovered in 1996 revealed that the bones did not attach to the rest of the skeleton. When not used in gliding, they could be closed like a Japanese fan.

The Triassic lizard *Icarosaurus* had a similar wing for gliding (named for Icarus of Greek mythology, who used homemade wings to fly). One of the more unusual aerial animals of the Triassic was *Sharovipteryx* (Fig. 14-38), found in Kyrgyzstan in 1969.

The Flying Pterosaurs

The pterosaurs were highly successful aerial reptiles, dominating the skies for over 100 million years—from

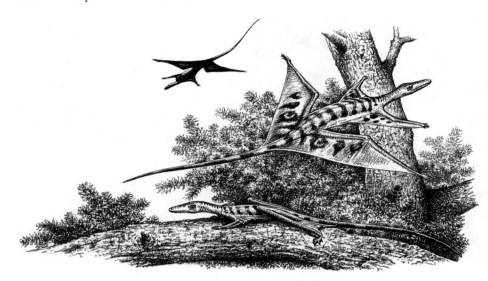

FIGURE 14-38
Sharovipteryx, **a primitive, gliding, diapsid reptile from the Triassic of central Asia.** The flight surface in this small, lightly built archosaur was a thin skin extending from hind legs to tail. *Sharovipteryx* could manuever while gliding by changing the position of its hind limbs. From snout to tail tip, this early glider was only about 24 cm long.

Late Triassic until Late Cretaceous. The most familiar Jurassic and Cretaceous pterosaurs typically were long-jawed, with large heads and eyes. In most forms, the jaws were lined with thin, slanted teeth. The bones of the fourth finger were lengthened to help support the wing, whereas the next three fingers were of ordinary length and terminated in claws. The wing was a sail of skin, stretched among the elongate fourth digit, sides of the body, and the hind limbs.

There were two general groups of pterosaurs. The rhamphorhyncoids evolved first, with long tails that terminated in a diamond-shaped vane (Fig. 14-39). The more advanced pterodactyloids were tail-less, as seen in *Pteranodon* (Fig. 14-40). Much like large sea

birds today, *Pteranodon* probably soared above the waves, snapping up sea creatures in their toothless jaws. Relative to body size, pterosaurs had somewhat larger brains than their land-dwelling relatives. Perhaps this was a result of the higher level of nervous system control and coordination needed for flight.

First prize for size among pterosaurs certainly goes to *Quetzalcoatlus northropi* (Fig. 14-41) from Upper Cretaceous beds of western Texas. And if *Quetzalcoatlus* was the largest, *Pterodausto*, with its strainer teeth, was the most peculiar (Fig. 14-42).

Some pterosaur fossils had a covering of soft hair, prompting the hypothesis that they were warm-blooded. This would be an important adaptation for flying verte-

FIGURE 14-39
Pterosaurs, most famous of the flying reptiles. *Eudimorphodon* (right foreground) was a rhamphorhyncoid with long, sharp teeth to catch and hold slippery fish. The creature was about 60 cm long. Also shown is *Peteinosaurus* (left), with a rudderlike membrane or vane at the end of its tail.

brates. Without a regulated body temperature, cold air would limit their power of exertion, and they might have difficulty staying aloft. In one of these flyers, *Sordes pilosus* ("hairy devil"), the fur is longest on the animal's underside, prompting speculation that it also served to incubate eggs or insulate hatchlings.

▶ A RETURN TO THE SEA

Nothosaurs and Placodonts

The marine habitat is one in which the archosaurs were not notably successful. Only one archosaurian group—the sea crocodiles—was able to invade the oceanic environment (Fig. 14-43). However, other groups adapted well to life in the sea: ichthyosaurs, plesiosaurs, mosasaurs, and sea turtles (Fig. 14-44). Many adaptations were needed to change a land animal into an ocean dweller, including the evolution of streamlined

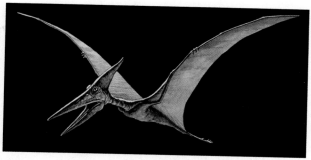

FIGURE 14-40 **The Cretaceous pterodactyl *Pteranodon.*** *Pteranodon* had a large wingspan over 7 meters (23 feet), but its body was only goose-sized. Its skeleton was lightly constructed, as required of any aerial vertebrate. The crested point at the back of the skull had branching channels thought to have contained blood vessels. Blood flowing through the crest could have been cooled as air rushed over the crest or warmed in the morning sun. Thus, the odd-looking pterosaur crest may have functioned in temperature control.

FIGURE 14-41 **The gigantic Cretaceous pterosaur *Quetzalcoatlus.*** With an astonishing wingspan of about 12 meters (40 feet), *Quetzalcoatlus* probably flew much like a modern condor, using thermal air currents and winds to help keep it aloft. It was named for an Aztec god who took the form of a feathered serpent.

FIGURE 14-42 **The distinctive pterosaur *Pterodaustro*.** *Pterodaustro* had long, curved jaws. Its upper jaw held rounded teeth for crushing the shells of invertebrate prey. The lower jaw had more than 400 flexible, wire-thin teeth for straining tiny organisms from water. While feeding, *Pterodaustro* probably folded back its wings, dipped its curved snout into the water, and swept the lake bottom for mollusks and crustaceans. On lifting its head, water would drain through its mesh of teeth, straining out the food behind to be swallowed. It was found in Cretaceous lake sediments in Argentina.

bodies for efficient movement through water, paddle-shaped limbs to replace feet, and highly efficient lungs to hold air while submerged.

Marine reptiles with paddle-shaped limbs for locomotion already were present during the Triassic Period. One group, the **nothosaurs**, were just beginning to take on adaptations that would be perfected in their descendants, the plesiosaurs.

Nothosaurs were joined in the Triassic by a group of mollusk-eating, flippered reptiles known as **placodonts** (Fig. 14-45). These bulky animals had distinctive pavement-like teeth in the jaws and palate. They formed an effective tool for crushing the shells of the marine invertebrates upon which they fed.

Plesiosaurs

Plesiosaurs were by far the best known of the paddle swimmers, with their large, many-boned flippers. Slender curved teeth lined their jaws, suitable for ensnaring fish. They had short, broad bodies, and in some species the extraordinarily long neck terminated in a smallish head. *Elasmosaurus*, a well-known long-necked Cretaceous plesiosaur (Fig. 14-46), attained an overall length exceeding 12 meters (40 feet).

Conversely, some plesiosaurs had short necks that supported large heads. It is likely that the short-necked models were aggressive divers. *Kronosaurus*, a giant, short-necked form from the Lower Cretaceous

of Australia, had a 3-meter-long skull that probably holds the record for any known reptile.

Ichthyosaurs

The most fishlike of marine reptiles were the Triassic to Early Cretaceous **ichthyosaurs** (Fig. 14-47). In many ways, they were the reptilian counterparts of present day toothed whales. Ichthyosaurs had fishlike tails, boneless dorsal fins to help prevent sideslip and roll, and paddle limbs for steering and braking.

Ichthyosaurs were active predators with highly adapted vision and high-speed swimming ability. Their large eyes (some larger than a dinner plate) let them see in the vanishing light at great depths. A ring of bony plates surrounding the eyes protected them against high water pressure. Their heads were a pointed wedge, suitable for cutting rapidly through water.

FIGURE 14-43 **Toothy skull of Jurassic marine crocodile *Geosaurus*.** The sea crocodiles were the only archosaurian group able to invade the oceanic environment. Length of this skull is about 45 cm (1.5 feet).

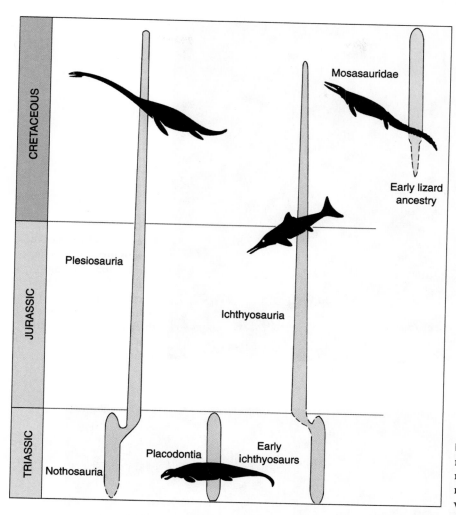

FIGURE 14-44 **Stratigraphic ranges of Mesozoic marine reptiles.** Ichthyosaurs, plesiosaurs, mosasaurs, and sea turtles adapted well to life in the sea.

Mosasaurs

The mosasaurs were a highly successful group of Cretaceous giant marine lizards. They had large, sharp teeth, elongate bodies, and porpoiselike flippers. The lower jaw had an extra hinge at mid-length (Fig. 14-48), allowing the animal to open its mouth widely to grasp large prey.

Sea Turtles

Less spectacular than the mosasaurs, but far more persevering, were the sea turtles. This group tended toward giganticism; for example, the Cretaceous turtle *Archelon* was nearly 4 meters long (13 feet). As an adaptation to its aquatic habitat, the marine turtle's carapace

FIGURE 14-45 **Triassic nothosaurs and placodonts.** Nothosaurs such as *Ceresiosaurus* (top left) and *Nothosaurus* (right foreground) were medium-sized marine reptiles with long necks and jaws set with sharp teeth for snaring fish. Placodonts, such as *Placodus* (bottom left), were massively constructed reptiles adapted for plucking mollusks from the seafloor. *Placodus* was about 3.5 meters long.

FIGURE 14-46
Cretaceous long-necked plesiosaurs. On viewing the skeleton of a long-necked plesiosaur, Thomas Huxley said the animal reminded him of "a snake threaded through a turtle." Their earliest remains lie in Jurassic strata.

FIGURE 14-47 **Ichthyosaurs were the supreme marine reptiles of the Mesozoic.**
Readily recognized by their dolphin-like bodies, Ichthyosaurs were the reptilian counterparts of present-day toothed whales. Exceptional vision and swimming ability made them powerful hunters. Shown here is the large Jurassic ichthyosaur *Grendelius*, about 4 meters (13 feet) long. There were giants among the ichthyosaurs, including one from British Columbia measuring 23 meters (more than 75 feet) long.

FIGURE 14-48 **Cretaceous mosasaur.** These giant marine lizards attained lengths over 9 meters (30 feet). They were primarily fish-eaters, but some frequently dined on large ammonites. Puncture wounds on the shells of ammonite fossils precisely match the dental pattern of mosasaurs.

(back shell) was greatly reduced, and its limbs were modified into broad paddles.

THE RISE OF BIRDS

From the time of Darwin, naturalists have recognized the structural similarities between birds and reptiles. They prompted Thomas Huxley to remark that birds are "glorified reptiles" that evolved wings and feathers and lost their teeth. But Huxley's comment depreciates the marvelous attainments of birds, including their superior power of flight and high level of endothermy.

Both attributes are related to the evolution of feathers from reptilian scales. In their earliest development, feathers may not have functioned in flight, but instead as insulation, camouflage, or display.

Although birds may have evolved from basal archosaurs, there is considerable evidence that small theropods were their ancestors. Both birds and early theropods already were similar in their bipedal stance, as well as in the structure of their limbs, shoulder girdle, and skull. Several theropods had feathers, hollow bones, and keeled breastbones.

Archaeopteryx, the first undisputed fossil bird, is a close evolutionary link between small, bipedal theropods and modern birds (Fig. 14-49). Except for its distinctly fossilized feathers (arranged on the forelimbs as in modern flying birds), its skeleton resembles that of a small theropod. Its jaws bore teeth and it had a long, feathered, but otherwise lizardlike tail.

On the wings of modern birds, the bones of the digits coalesce for greater strength. However, the primitive wings of *Archaeopteryx* retained claw-bearing, free

fingers for climbing and grasping. The sternum lacked a keel, indicating that the sturdy muscles needed for sustained flight were lacking.

Small, delicate, hollow-boned animals with tiny bones are not readily preserved, and thus the fossil record for early birds is not good. Fossils of larger birds, like the Cretaceous aquatic bird *Hesperornus* (see Fig. 13–35), are more common. Nevertheless, the fossil record is sufficient to indicate that many different kinds of birds lived during the Cretaceous Period.

THE MAMMALIAN VANGUARD

While Mesozoic reptiles had their heyday, small, furry animals were scurrying about in the undergrowth and unwittingly awaiting their day of supremacy. These shrewlike descendants of the mammal-like reptiles were the first primitive mammals. Among the earliest were *Megazostrodon*, *Eozostrodon*, and the more widely known *Morganucodon* (Fig. 14-50) from Upper Triassic rocks of southern Wales.

We know these early mammals from all three systems of the Mesozoic, based on rare finds of tiny bone and teeth remains. Here is some of the evidence we have that these creatures were mammals:

- Differentiation of teeth is a mammalian trait. Like us, these tiny creatures had incisors, canines, premolars, and molars. Further, the teeth grew from "baby teeth" like we have, suggesting that the young were suckled.
- Reptiles have a single ear bone (the stapes). But more efficient hearing was achieved in these early

A   *B*

FIGURE 14-49 **The Jurassic bird *Archaeopteryx*.** (A) *Archaeopteryx* in Germany's Solnhofen Limestone, which formed from lime mud deposited on the floor of a tropical lagoon. This extraordinarily preserved specimen resides in the Berlin Museum of Natural History. (B) Restoration of crow-sized *Archaeopteryx*.

mammals by two additional ear bones, the malleus and incus.

- Whisker pits on the upper jaw bones indicate a covering of hair.
- The articulation of the jaw to the skull was mammalian, and the lower jaw was functionally a single bone, the dentary, as in mammals.

Tooth morphology is of particular importance in identifying early mammals (Fig. 14-51). The group called **docodonts**, for example, had multicusped molar teeth, which suggests they may have been the stock from which evolved present-day **monotremes** (egg-laying mammals). Very likely, the docodonts fed on insects, as did many of these primitive mammals.

Symmetrodonts had molars constructed on a more or less triangular ("symmetrical") plan. **Multituberculates** had teeth with many tubercles or cusps on their molars. Their chisel-like incisors and a gap between the incisors and molars gave them a rodentlike appearance (Fig. 14-52). The multituberculates were a persevering lineage. They appeared during the Late Jurassic and were present until the Eocene epoch.

Triconodonts are recognized by their cheek teeth, which have three cusps aligned in a row. The braincase, vertebrae, and pelvis in these early mammals are distinctly mammalian in form.

In 2005, exceptionally large triconodonts were discovered in the fossil-rich Yixian Formation of China's Liaoning province. The mammals show excellent preservation resulting from sudden burial in a fall of volcanic ash. The discovery included two species, a larger animal named *Repenomamus giganticus*, which rather resembled a large badger, and a smaller species named *Repenomamus robustus* (Fig. 14-53). The large pointed incisors, pointed canines, cheek teeth, and evidence of strong jaw musculature indicate *Repenomamus* was well-adapted for catching, holding, and tearing prey.

These animals lived about 130 million years ago during the Early Cretaceous. You might surmise that, in the com-

FIGURE 14-50 *Morganucodon*, **an early mammal from the Late Triassic of Wales.** This shrewlike descendant of mammal-like reptiles was among the earliest of the primitive mammals.

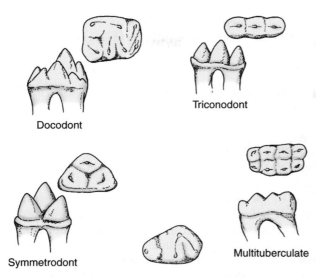

FIGURE 14-51 **Molars of Mesozoic mammals.** Side views of lower molars (as viewed from inside the mouth) and top views of the oral surfaces.

FIGURE 14-53 **Large Cretaceous mammals dined on small Cretaceous dinosaurs.** About 130 million years ago, triconodont mammals preyed on juvenile dinosaurs. In the foreground is *Repenomamus robustus*, a mammal about the size of an opossum. The larger species, named *R. giganticus* was about 3 feet long. The fossils were found in northeastern China and reported by Yaoming Hu of the Institute of Vertebrate Paleontology and Paleoanthropology in Beijing and Jin Meng of the American Museum of Natural History in New York.

petition between dinosaurs and early mammals, it was always the mammals that were eaten. But fossilized stomach contents of *R. robustus* contain the bones of a young dinosaur, *Psittacosaurus*, about 6 inches long. Small dinosaurs beware: Early mammals were on the prowl.

Mammal Types

Taxonomists divide mammals into two groups. The first are called **prototherians** and include the triconodonts, multituberculates, and the monotremes. Monotremes are represented today by the platypus and spiny anteater of Australia.

The second group consists of **therians**. Marsupials (also called euherians) and placental mammals (like us) are therians. Both prototherians and therians were present by Middle Jurassic time. The earliest known placental mammal was recently discovered in Lower Cretaceous beds in China. Named *Eomaia* ("dawn mother" in Greek), the tiny fur-covered animal had feet well

adapted for climbing among the branches of trees 125 million years ago.

For the mammals, the Mesozoic was a time of evolutionary experimentation. They lived among the great archosaurs while steadily improving their nervous, circulatory, and reproductive systems. With their ability to control body temperature, they could thrive in both cold and warm climates. As the dinosaur population declined near the end of the era, mammals quickly expanded into vacated habitats.

► SEA PLANTS AND PHYTOPLANKTON

Animal life on Earth ultimately depends on plant life. This generalization was as valid during the Mesozoic as it is today. Then, as now, plants made up the broad base of the food pyramid. Their nutritious starches, oils, and sugars made possible the evolution and continuing existence of animals.

Plants are a fundamental part of Earth's essentially self-sustaining ecologic system. The operation of the system depends on oxygen and carbon dioxide. Animal respiration provides the carbon dioxide needed for plant photosynthesis, whereas plants, by means of photosynthesis, supply the oxygen needed by animals.

In the geologic past, variations in plant productivity may have caused corresponding changes in the amount

FIGURE 14-52 **The rodentlike multituberculate *Taeniolabis*.** Chisel-like incisors and the gap between the incisors and the cheek teeth gave them a rodentlike appearance and indicate that they may have been plant-eaters and probably had rodentlike gnawing habits as well.

of carbon dioxide and oxygen in the atmosphere. Such variations may have favored the evolution of some animals over others and may have been responsible for the demise of particular groups.

Marine Phytoplankton

Photosynthetic organisms that live suspended in water do not require the vascular and supportive systems that characterize land plants. Most of these organisms are unicellular, although they may grow together in impressive colonies and aggregates. They are part of that vast realm of floating organisms termed **plankton**.

Those having internal organelles called **chloroplasts** can photosynthesize their food and are generally called **phytoplankton**. The more familiar of these are members of the Protista. The geologic record of the most important fossil phytoplankton groups is shown in Figure 14-54.

Dinoflagellates

Fossil dinoflagellates are important aids in Mesozoic and Cenozoic stratigraphy. From the Jurassic forward, they were among the primary producers in the marine food chain. For propulsion, they used two flagella: one longitudinal and whiplike, the other transverse and ribbonlike.

During their life cycle, dinoflagellates develop a mobile planktonic form and a cyst phase that is formed within the mobile organism. Only dinoflagellate cysts, which have an organic covering that is extremely resistant to decay, are known as fossils (Fig. 14-55).

Coccolithophorids

The coccolithophorids also began their expansion during the Early Jurassic. These calcium carbonate-secreting organisms have a splendid fossil record. Their abundant remains formed many of the extensive chalk deposits of the Mesozoic and early Cenozoic. Today, they are frequently present in the deep-sea sediment known as *calcareous ooze*.

The coccolithophorid organism is one of several varieties of unicellular golden-brown algae. These algae deposit calcium carbonate internally on an organic matrix and construct tiny, shieldlike structures called **coccoliths**. Once formed, the coccoliths move to the surface of the cell and form a calcareous armor (Fig. 14-56).

Coccoliths have the right traits to be extremely useful in stratigraphic correlation of Cretaceous to Holocene rocks: They are abundant fossils, have undergone frequent evolutionary changes through time, and are widely dispersed by oceanic currents.

Silicoflagellates and Diatoms

Silicoflagellates are flagella-bearing organisms that secrete delicate siliceous skeletons in simple latticelike frameworks. Radiating spines characterize most genera (Fig. 14-57).

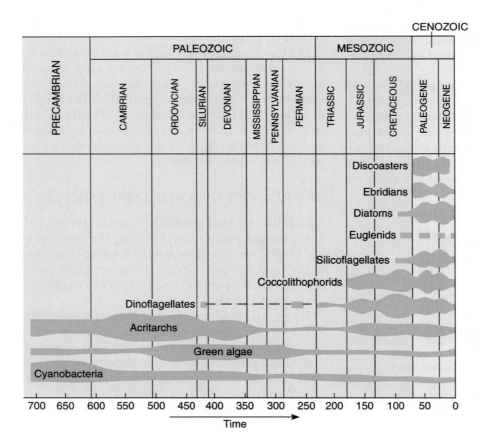

FIGURE 14-54 **Geologic distribution and abundance of phytoplankton.** The most abundant Mesozoic phytoplankton groups were the coccolithophorids and dinoflagellates.

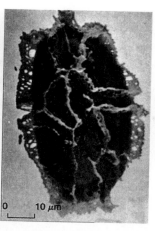

FIGURE 14-55 Fossil dinoflagellate cyst. *Prionodinium alveolatum*, from Cretaceous rocks of Alaska.

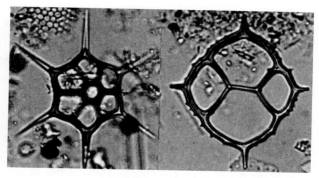

FIGURE 14-57 Silicoflagellates from the Mid-Atlantic Ridge. They range in size from 0.02 to 0.1 mm. Silicoflagellates and diatoms, along with the coccolithophorids, are members of the phylum Chrysophyta. Image is magnified 3800 times.

Diatoms also secrete siliceous coverings (Fig. 14-58). The covering is called a **frustule**. In the past, a proliferation of diatoms was often associated with volcanic activity. Apparently, silica supplied to seawater as fine volcanic ash stimulated diatom productivity.

The earliest known silicoflagellates and diatoms appeared in the Cretaceous. Along with other phytoplankton, they experienced a decline at the end of the Cretaceous, and then all groups expanded again into the early epochs of the Cenozoic.

LAND PLANTS

The Mesozoic is sometimes called the Age of Reptiles. With equal validity, the era could be dubbed the Age of Cycads. These are seed plants lacking true flowers (Fig. 14-59). Jurassic cycads included tall trees with rough branches marked by the leaf bases of earlier growths and by crowns of leathery pinnate (feather-like) leaves. Cycads experienced a marked decline in the Late Cretaceous, and only a few have survived to the present time, such as the sago palm, a common house plant.

There were three important episodes in the evolution of land plants:

1. Development of the spore-bearing plants, like ferns.

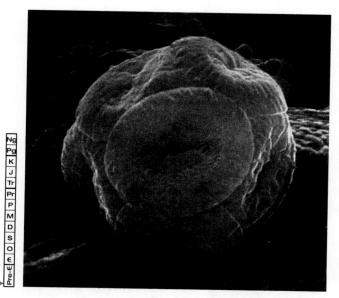

FIGURE 14-56 Scanning electron micrograph of Late Cretaceous coccolith. Although coccoliths measure only 0.002 to 0.01 mm in diameter, an electron microscope lets us discern their intricate construction. They consist of one or two elliptical or round plates that are concave on one side to fit snugly around the surface of a spherical cell. Each disc or plate is itself composed of smaller elements uniformly arranged, in a circular, radial, or spiral plan. Image is magnified 6000 times. ◀ *What are the individual plates that compose the coccosphere called?*

FIGURE 14-58 Modern marine diatoms. Diatom shells (tests) are composed of an upper part (the epitheca) and a lower part (the hypotheca) that fit together like a lid and a box. The covering frustule may be circular, cylindric, triangular, or a variety of other often beautiful shapes. ◀ *In what taxonomic kingdom are diatoms placed?*

2. Evolution of the **gymnosperms**—nonflowering, pollinating seed plants like the cycads, seed ferns, conifers, and ginkgoes (Fig. 14-60).

3. The Early Cretaceous appearance of angiosperms—species having enclosed seeds and flowers. Pollen grains produced by angiosperms provide the earliest evidence of these flowering plants.

Gymnosperms

Among the nonflowering, pollinating seed plants, all but seed ferns have living representatives. Six groups of conifers were present during the Jurassic and Cretaceous, including large numbers of pines. In 1994, the remains of 39 huge pines were discovered in Wollemi National Park, Australia. Dubbed the "Wollemi pines," many of the trees are species known only from this locality.

Angiosperms

By the Middle Cretaceous, angiosperms had become widespread. Forested areas included stands of birch, sycamore, magnolia, holly, palm, maple, walnut, beech, poplar, willow, and sassafras (Fig. 14-61). Before the period came to a close, **angiosperms** had surpassed the nonflowering plants in both abundance and diversity. Flowering trees, shrubs, and vines expanded across the

FIGURE 14-60 *Ginkgo biloba*, the ginkgo or maidenhair tree. So-named because the leaves resemble those of the maidenhair fern. Note the naked, fleshy seeds that grow on female trees. Fossils of these plants that are over 200 million years old are nearly identical to living forms.

FIGURE 14-59 **Cycad growing in a South African forest.** Cycads are seed plants that lack true flowers. Pineapple-like structures at the top are pollen cones.

FIGURE 14-61 **Fossil sassafras leaf, Cretaceous Dakota Sandstone, Ellsworth, Kansas.** The slab is about 12.5 cm (5 inches) wide. By Middle Cretaceous time, such angiosperms had become widespread.

lands and, except for the absence of grasses, gave the landscape a modern appearance.

Angiosperms provide many examples of coevolution with Mesozoic insects, dinosaurs, mammals, and birds. **Coevolution** occurs when two or more different organisms develop a close, reciprocal relationship in which the evolution of one organism is partially dependent on the evolution of the other.

The coevolution of insects and flowering plants is a classic example. Angiosperms, by encouraging insect visits, use insects as delivery agents for their pollen. This provides a far more efficient means of pollen dispersal than random wind pollination. The selective competition for efficient insect pollinators has induced evolution of constantly changing variations in both plants and insects. In the angiosperms, the need for each plant to be recognizably different results in a spectacular floral variety (color, shape, scent, height) that has persisted from the Cretaceous to the present.

In addition to their interdependence with insects, early angiosperms likely developed coevolutionary relationships with birds, mammals, and even dinosaurs. Many of the great sauropods subsisted largely on ferns, horsetails, tree-sized club mosses, and conifers. But such plants are slow to grow and regenerate. At times, regeneration may not have kept pace with dinosaur consumption, leaving over-browsed areas that could be invaded quickly by the evolving lineages of shrublike angiosperms.

Unlike gymnosperms, angiosperms reproduce, disperse, and grow rapidly. Thus, they are highly resistant to over-browsing. The presence of this expanding angiospermal food supply undoubtedly influenced the evolution of the ornithischians that were so numerous during the Cretaceous.

The duckbills took on habits not unlike today's antelope and bison, whereas ceratopsians may have lived rather like rhinoceroses. The new plants provided nutritious fruits and nuts for these dinosaurs, and the dinosaurs in turn aided in the dispersal of angiosperm seeds by passing them unharmed through their digestive tracts.

◀ LATE CRETACEOUS CATASTROPHE

Just as the end of the Paleozoic was a crisis for animal life, so also was the end of the Mesozoic—specifically, at the end of the Cretaceous. Primarily on land, but also at sea, extinction overtook many seemingly secure groups of vertebrates and invertebrates. Altogether, the Late Cretaceous catastrophe eliminated about one-fourth of all known animal families.

In the seas, the plesiosaurs, mosasaurs, and the once highly successful ammonites perished. Entire families of echinoids, bryozoans, planktonic foraminifers, and calcareous phytoplankton became extinct.

On land, the most noticeable loss was among the great clans of reptiles. Gone forever were the magnificent di-

nosaurs and soaring pterosaurs. The only reptiles to survive the great biologic crash were turtles, snakes, lizards, crocodiles, and New Zealand's reptile tuatara (*Sphenodon*).

What caused this spectacular decimation of animal life at the end of the Mesozoic? Scores of hypotheses, some scientific but many preposterous, have been offered. The most credible hypotheses attempt to explain the simultaneous extinctions of marine and terrestrial animals and seek a single or related sequence of events as a cause.

The hypotheses fall into two broad categories:

- An external or extraterrestrial event, such as an encounter with an asteroid or comet. Such events are considered catastrophic because their effects are concentrated within a relatively short span of time.
- An event or trend right here on Earth, such as extreme volcanism.

Did a Bolide Impact Cause the Mass Extinctions?

Since geologists first became aware of the extinctions at the end of the Cretaceous, they have speculated about collisions with asteroids or comets, or lethal cosmic radiation. But tangible evidence was lacking until 1977. Geologist Walter Alvarez discovered a thin clay layer near Gubbio, Italy at the boundary between the Mesozoic's Cretaceous Period and the Cenozoic's Paleogene Period (Fig. 14-62). Alvarez sent samples of the clay to his physicist father Luis, who had the clay analyzed, with startling results. The samples contained approximately 30 times more of the metallic element *iridium* than is normal for Earth's crustal rocks.

Evidence: Iridium and Shocked Quartz

Where could this high concentration of iridium have come from? Iridium probably is present in Earth's core and perhaps the mantle, but how could the metal from so deep a source find its way into a clay layer at the boundary between Cretaceous and Paleogene beds? Volcanism is an obvious possibility.

But iridium also occurs in extraterrestrial objects such as asteroids and meteorites. For this reason, the father-and-son Alvarez team favored an extraterrestrial hypothesis for the iridium in the clay layer. They proposed that an iridium-bearing asteroid crashed into Earth at the end of the Cretaceous Period (Fig. 14-63).

The explosive, shattering blow from the huge body (presumed to be over 10 km in diameter) would have thrown dense clouds of iridium-bearing dust and other impact ejecta into the atmosphere. Transported by atmospheric circulation, the dust might have formed a lethal shroud around the planet, blocking the Sun's rays and killing marine and land plants on which all other forms of life ultimately depend. As the dust settled, it would have formed the iridium-rich clay layers

FIGURE 14-62 **Coin marks iridium-rich clay layer separating Cretaceous and Paleogene rocks near Gubbio, Italy.** The gray Cretaceous limestone below the coin contains abundant fossil coccolithophores, but few of these phytoplankton exist in the Paleogene beds above the iridium-rich layer.

found at Gubbio and subsequently at many other places around the world (Fig. 14-64).

In addition to the iridium-rich clay layer found at many sites around the world, there is other evidence for the impact of a large extraterrestrial body at the end of the Cretaceous. There is a widespread occurrence of **shocked quartz** in the boundary layer (Fig. 14-65). These mineral grains are recognized by distinctive parallel sets of microscopic planes (called *shock lamellae*) that are produced when high-pressure shock waves, such as those emanating from the impact of a large meteorite, travel through quartz-bearing rocks.

In the stratum containing shocked quartz grains, there also are tiny glassy spherules thought to represent droplets of molten rock thrown into the atmosphere dur-

ing the impact event. These are **tektites**. A rare, dense, high-pressure silicate mineral known as **stishovite**, found at Meteor Crater in Arizona and other known impact structures, also is found in the boundary clay. It is taken as evidence of sudden extremely high pressures, such as those that would be associated with the impact of an asteroid. Finally, sediment in the boundary layer often includes a carbon soot that may be the residue of forests burned during the firestorm caused by the impact.

Any large extraterrestrial object that explodes upon striking Earth is called a **bolide**. Many bolides have collided with Earth in the geologic past, but the scars they left on continents have mostly been obliterated by weathering and erosion. Nevertheless, a few can be discerned on photographs taken from spacecraft or dur-

FIGURE 14-63 **Artist's conception of an asteroid colliding with Earth 65.5 million yeas ago.**

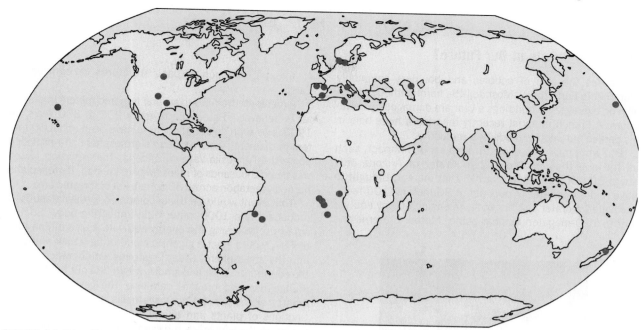

FIGURE 14-64 **Occurrences of iridium-rich sediment layer at Cretaceous-Paleogene boundary.**

ing geologic investigations. For example, geologists have discovered a large, Jurassic-age bolide impact crater on the seafloor north of Norway. Named the Mjølnir Crater, it is about 40 km (25 miles) in diameter. Sediment within and around the crater contains shocked quartz and high concentrations of iridium.

The Chicxulub Structure

But the Mjølnir Crater is of Jurassic age. Where is the crater produced by the bolide that allegedly caused mass extinctions at the close of the Cretaceous?

Currently, the best candidate is a buried crater in the Gulf of Mexico just offshore from the Yucatán Peninsula, named for the nearby town of Chicxulub (Fig. 14-66). At this location is a buried circular structure about 180 km (112 miles) in diameter, revealed by magnetic surveys, gravity surveys, and oil well cores and logs. The southern rim of the crater is visible on radar images taken by the space shuttle *Endeavor*. Andesitic rock in the central core of the structure has an isotopic composition similar to that of tektites that are abundant in the Cretaceous-Cenozoic boundary layer at many locations in the Caribbean region.

Further evidence of the bolide impact is in core samples of rocks penetrated during oil drilling in and

FIGURE 14-65 **Shocked quartz grain.** It was found in Haiti and thought to have been ejected during the Chicxulub asteroid impact. Note the parallel microscopic planes, called shock lamellae (a lamella is a thin, flat layer). Maximum grain diameter is 0.3 mm.

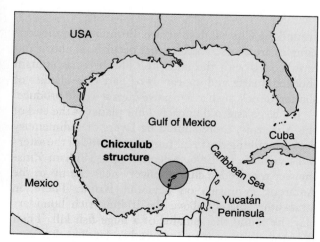

FIGURE 14-66 **Location map for the Chicxulub structure.**

ENRICHMENT

Is There a Bolide in Our Future?

Meteorites the size of cobbles and boulders commonly plummet to Earth today. Most do little harm, although globally, on average, 16 buildings a year are damaged by meteorites. A French motorist recently reported a huge hole in his parked automobile made by a meteorite.

But what might we expect from a serious impact, such as the one that created Meteor Crater of Arizona only about 50,000 years ago (Fig.A)? That meteor, thought to have had a diameter of only about 30 meters (100 feet), excavated a crater 1.2 km wide, releasing energy equivalent to a 20-megaton nuclear bomb. Meteor Crater is one

FIGURE A   **Meteor Crater in Arizona.** It was formed by a large meteorite about 50,000 years ago.

of about 130 such impact structures recognized on Earth.

A meteorite from 50 to several hundred meters in diameter is believed to strike Earth every 200 to 300 years. In 1908, one such object, thought to have been about 60 meters in diameter, entered Earth's atmosphere and exploded above the Tunguska Valley in Siberia. The explosion, which was heard thousands of miles away in London, flattened and burned vegetation across 50 square km of forested area.

That event would be trivial compared to the catastrophe produced by a 1000-meter-wide impacting body. Such a smashup, releasing the energy equivalent of a billion tons of TNT, would eject a gigantic cloud of fiery rock and dust into the atmosphere. Great firestorms would sweep across the continents. Dust and smoke would blot out the Sun and plunge the world into total darkness. The loss of sunlight for months would prevent photosynthesis and cause mass extinctions of plants and animals both on land and in the ocean. At the very least, the event would have the potential to wipe out over a quarter of the planet's human population.

As an indication that the possibility of future bolide impacts should be taken seriously, space scientists are currently examining ways to detect, track, and intercept bodies headed our way. The program has been dubbed "Spacewatch" and includes a global network of telescopes for early detection of approaching large meteorites and asteroids. If the incoming body is spotted sufficiently early, it may be possible to push it into a nonthreatening trajectory by exploding a nuclear device off to its side.

On the side nearest the explosion, heat from radiation would vaporize the surface of the meteorite, and the resulting jet of vapor might act like a rocket engine and nudge it off course. (It would be important that the bomb not strike the body directly, for that could shatter it, causing a barrage of rock shrapnel on Earth.) Other plans are still being formulated as scientists grapple with possible ways to safeguard civilization against such a catastrophe.

---

around the Chicxulub structure. Prominent in the core samples are coarse breccias that occur both above and interbedded with the andesite. The breccias contain shocked quartz and appear to be part of a blanket of shattered rock, such as a massive impact would produce.

If a large asteroid impacted our planet at the end of the Mesozoic, there should be layers of sedimentary rock containing the fossil remains of some of the asteroid's victims. In 1996, paleontologist William Zinsmeister reported such a discovery near the tip of the Antarctic Peninsula on Seymour Island. There, in strata immediately above the iridium-rich boundary clay, he found the remains of a huge fish kill. Their doom would appear to be directly related to the event that produced the iridium-rich boundary clay.

Not only fish, but ocean plankton would have been affected by the impact. In 1997, a core sample was taken from beneath the ocean floor about 320 km (200 miles) east of Jacksonville, Florida. The core represents the entire time span of the bolide event. At its base, there is a layer of white sediment containing the shells of millions of microfossils (mostly foraminifera). They represent life just before the impact. Above the fossiliferous white layer is a green clay thought to represent the dust and ash from the impact explosion. The green layer is capped by a red clay that contains no fossils at all, suggesting that plankton had been exterminated. At the top of the core, fossils again appear, indicating that several thousand years later, recovery was beginning to take place.

Does the timing of the proposed impact correlate with extinctions that were widespread at the end of the Cretaceous? Rocks that had been melted by the event were found by isotopic dating to be 65.2 ± 0.4 million years old. In addition, while these rocks had been melted by the heat of impact, they acquired remanent magnetism indicating that they solidified during the episode of reverse geomagnetic polarity known to exist at the time of deposition of the boundary layer.

If the Chicxulub structure is indeed an impact crater, it is the largest on Earth. The bolide that produced it had an estimated diameter of about 10 km (6 miles). As it splashed down, it would have produced a tsunami-like ocean wave over 34 meters (112 feet) high that would flood land areas around the Caribbean and Gulf of Mexico. Rocks would have been vaporized, and the atmosphere would have filled with dust, water vapor, and carbon dioxide from melted limestones. Could life on Earth sustain such an event?

Did Global Volcanism Cause the Mass Extinctions?

On first examination, the bolide-impact hypothesis seems a tidy way to account for the extinction of dinosaurs and many of their animal and plant contemporaries. Like all hypotheses, however, it must stand the rigorous test of scientific scrutiny. Were the Alvarezes correct in assuming that the iridium was derived from an impacting asteroid? Or could there be another explanation?

Geologists have found evidence that iridium in clays such as that at Gubbio might have its source in Earth's mantle, from which it can move to the surface by way of conduits and blast into the atmosphere as iridium-rich volcanic ash and dust. Volcanism was prevalent during the late stages of the Cretaceous. Especially significant were the tremendous outpourings of lava at the end of the Cretaceous that formed the Deccan Traps in India. Intensive volcanic activity at the end of the Cretaceous also was vigorous in western North America, Greenland, Great Britain, Hawaii, and the western Pacific.

Volcanoes produce dust and aerosols, such as sulfuric acid, that block solar radiation and thus cause temperatures to decline. Sulfuric acid in the atmosphere generates acid rain, and such precipitation could change the alkalinity of the oceans, placing lethal stress on plankton, other invertebrates, and life higher in the food chain. Thus, volcanism can be considered harmful to animals both on land and in the sea.

Proponents for a volcanic source for the iridium in the boundary clay note that the element is often distributed across sedimentary thicknesses of 30 to 40 cm. This suggests deposition of ash-derived iridium over a span of several thousand years, an interval considered reasonable for global volcanism. If the iridium were of bolide origin, one would expect to find it confined to a thin layer of sediment.

Finally, those favoring the volcanic hypothesis also note the presence of antimony and arsenic in some of the beds containing the iridium. Although common in volcanic ash and lava, these elements are rare in meteorites.

Did Environmental Change Cause the Mass Extinctions?

Several hypotheses that seek to explain the demise of plants and animals near the end of the Cretaceous propose that events on Earth upset the ecologic balance between organisms and the environment to which they had become adapted. Recall that continents during the Cretaceous were extensively covered by shallow, warm, epeiric seas. Many geologists believe these transgressions were caused by displacement of ocean water when midoceanic ridges were raised as a consequence of accelerated seafloor spreading. Whatever the cause, the warm inland seas that resulted were very favorable for marine life and helped to moderate and stabilize climates on the continents as well. Under such conditions, plants and animals experienced remarkable increases in variety and abundance. However, these favorable conditions were soon to change.

Studies of the stratigraphic sequence across the Cretaceous-Paleogene boundary indicate a global lowering of sea level at the end of the Mesozoic. Perhaps this was caused by a slowing of seafloor spreading rates. Whatever its cause, the change in sea level spelled disaster to animals and plants of formerly extensive shallow coastal areas, especially to the phytoplankton adapted to the shallow sea environment.

Without the moderating effects of vast epicontinental seas, landmasses also would have experienced harsher climatic conditions and more extreme seasonal changes. Many families of organisms that were adjusted to the previous environment might not have been able to adapt to the harsher conditions. One by one, species would have met their end. As is characteristic within the complex ecologic web, their demise would affect other organisms, resulting in waves of extinctions among dependent species.

Did a Combination of Factors Cause the Mass Extinctions?

Still other hypotheses account for the Late Cretaceous extinctions, but the examples provided above illustrate the complexity of the problem. The debate between those favoring sudden extraterrestrial causes and those supporting purely terrestrial hypotheses will continue for decades. Whatever the outcome, it is a fact that hard times near the end of the Mesozoic doomed the dinosaurs, pterosaurs, ammonites, and over three-fourths of the known species of marine plankton.

It is important to note that the extinctions did not happen simultaneously for all groups. Many groups died out sporadically over an interval of 0.5 to 5 million years. This is an argument against a sudden extraterrestrial catastrophe.

Perhaps the real cause of extinctions will be found in a combination of factors. The paleontologic evidence indicates that extinctions were widespread in the Late Cretaceous. The debilitated and diminished survivors of environmental hard times may have been dealt a *coup de grace* by an extraterrestrial event. Fortunately, small mammals, birds, lizards, crocodiles, turtles, many fish groups, many deciduous plants, and certain mollusks survived the hard times and expanded during the following Cenozoic Era.

SUMMARY

- Climates of the Mesozoic were generally mild and equable, except for occasional intervals of aridity and an episode of cooler conditions near the end of the era.
- In the widespread Mesozoic seas, coccoliths and diatoms flourished, as did invertebrate groups such as ammonoids, belemnites, oysters, and other bivalves, echinoderms, corals, and foraminifera.
- On land, gymnosperms (cycads and conifers) were common in Triassic and Jurassic forests. In the Cretaceous Period, the angiosperms (flowering plants) expanded and with them coevolved a multitude of modern-looking insects.
- Dinosaurs were the ruling vertebrates of the Mesozoic. Both carnivorous and herbivorous varieties occupied a variety of habitats. Based on difference in pelvic structure, we recognize two groups of dinosaurs, the Saurischia and Ornithischia. Certain dinosaur species, as well as the pterosaurs, may have been endothermic (warm-blooded). This is indicated by bone structure, posture, predator-prey relationships, and the presence of feathers on some theropods.
- Reptiles also were successful in the seas, where ichthyosaurs, plesiosaurs, mosasaurs, and large turtles competed successfully with sharks and bony fishes.

- The Mesozoic is noteworthy as the era during which mammals and birds appeared. The birds, amply feathered for insulation and flight, may have evolved from small carnivorous dinosaurs (theropods). The oldest unquestionable remains of a true bird are *Archaeopteryx* from the Jurassic.
- Primitive mammals debuted during the Triassic and had become common by Cretaceous time. One group of primitive prototherian mammals gave rise to marsupial and placental mammals during the Cretaceous.
- Like the Paleozoic Era, the Mesozoic Era closed with an episode of extinctions. Dinosaurs, pterosaurs, plesiosaurs, mosasaurs, ammonoid cephalopods, many groups of bivalves, and large numbers of planktonic foraminifera became extinct.
- An impressive body of evidence indicates that a large meteorite or asteroid struck Earth about 65.5 million years ago, and its effects may have caused or strongly contributed to the mass extinctions.
- Others argue that the extinctions can be attributed to extensive volcanic activity, coupled with loss of epicontinental seas, and resulting climatic changes.

KEY TERMS

adaptive radiation, p. 431

ammonite, p. 411

ammonoid, p. 411

Ammonoidea, p. 410

angiosperm, p. 442

ankylosaur, p. 428

archosaur, p. 415

belemnite, p. 412

bolide, p. 444

ceratite, p. 410

ceratopsian, p. 428

chloroplast, p. 440

coccolith, p. 440

coevolution, p. 443

convergence (evolutionary), p. 415

crustacean, p. 413

diatom, p. 441

dinosaur, p. 417

docodont, p. 438

ectothermic, p. 429

endothermic, p. 430

foraminifera, p. 413

frustule, p. 441

gastropod, p. 413

goniatite, p. 411

gymnosperm, p. 442

ichthyosaur, p. 434

ilium, p. 417

irregular echinoid, p. 410

ischium, p. 417

lepidosaur, p. 415

monotreme, p. 438

multituberculate, p. 438

Nautiloidea, p. 410

nodosaur, p. 428

nothosaur, pp. 433–434

Ornithischia, pp. 417, 423

ornithopod, p. 423

pachycephalosaur, p. 427

phytoplankton, p. 440

phytosaur, pp. 415, 416

placodont, p. 434

plankton, p. 440

plesiosaur, p. 434

prosauropod, p. 422

prototherian, p. 439

pubis, p. 417

radiolarian, p. 413

regular echinoid, p. 410

rhynchocephalian, p. 415

rudistid, p. 409

Saurischia, p. 417

sauropod, pp. 418, 422

sauropodomorpha, p. 418

septum (of cephalopod), p. 410

shocked quartz, p. 444

stishovite, p. 444

suture (of cephalopod), p. 410

symmetrodont, p. 438

tektites, p. 444

temnospondyl, p. 415

therian, p. 439

theropod, p. 418

triconodont, p. 438

QUESTIONS FOR REVIEW AND DISCUSSION

1. During what geologic period of the Mesozoic is chalk particularly abundant? What group of organisms provides skeletal remains for chalk deposition? Why are chalk formations rare in Paleozoic rocks?

2. What are diatoms? How do they differ in composition and morphology from coccolithophorids? How might the distribution of diatoms and coccolithophorids be related to the acidity or alkalinity of ocean water?

3. How did the terrestrial plant flora of the Jurassic differ from that of the Cretaceous? What environmental conditions might have driven the change in floras?

4. What are ammonoid cephalopods? What attributes of ammonoids result in their having special value as guide or index fossils?

5. What are foraminifera? Why are they of particular value to petroleum geologists involved in the correlation of subsurface strata?

6. Discuss the differences between the marine invertebrate faunas of the Mesozoic and those of the Paleozoic. What Paleozoic groups are not seen in the Mesozoic?

7. Discuss several lines of evidence that indicate that certain Mesozoic reptile groups were endothermic. Why would endothermy be less important for the giant saurischians than for small dinosaurs?

8. What attributes of Cretaceous mammals may have contributed to their survival during the biologic crisis at the end of the Mesozoic?

9. Discuss the differences between world geography at the end of the Permian as compared to the end of the Cretaceous.

10. What two classes of vertebrates appear for the first time during the Mesozoic Era?

11. Cite two examples of evolutionary convergence among animals living during the Mesozoic Era and today.

12. What is the particular evolutionary importance of the following: (a) basal archosaurs (b) *Archaeopteryx* (c) angiosperms?

13. Prepare a list of the reptilian groups that survived the mass extinction at the end of the Cretaceous Period.

14. Oceans cover about 71% of Earth's surface, yet evidence of impact craters on the ocean floor is rarely seen. Why?

15. What evidence at the boundary between the Cretaceous and Paleogene systems at many localities favors the bolide impact hypothesis for the extinction of the dinosaurs and many other animal groups? What arguments can be advanced against this popular hypothesis?

16. When were the Deccan Traps extruded? Discuss the possible role of this and other synchronous volcanism as a cause for mass extinctions.

WEB SITES

The Earth Through Time Student Companion Web Site (www.wiley.com/college/levin) has online resources to help you expand your understanding of the topics in this chapter. Visit the Web Site to access the following:

1. Illustrated course notes covering key concepts in each chapter;

2. Online quizzes that provide immediate feedback;

3. Links to chapter-specific topics on the web;

4. Science news updates relating to recent developments in Historical Geology;

5. Web inquiry activities for further exploration;

6. A glossary of terms;

7. A Student Union with links to topics such as study skills, writing and grammar, and citing electronic information.

Mount Hood in Oregon. *Mount Hood began to form during the Pleistocene. It is called a composite-cone volcano because it is composed of interbedded lava flows, ash beds, and cinder beds. Mount Hood's last major eruption was in the 1790s, but someday it will erupt again.*

Cenozoic Events

Many an aeon moulded earth before
her highest, man, was born,
Many an aeon too may pass when
earth is manless and forlorn.
Earth so huge and yet so bounded—
pools of salt and plots of land
Shallow skin of green and azure—
chains of mountains, grains of sand!
—Alfred, Lord Tennyson, "Locksley Hall:
Sixty Years After," 1866

OUTLINE

▶ THE TECTONICS–CLIMATE CONNECTION
▶ STABILITY AND EROSION ALONG THE
  NORTH AMERICAN EASTERN MARGIN
▶ GULF COAST: TRANSGRESSING AND
  REGRESSING SEA
▶ TECTONICS AND EROSION IN THE ROCKIES
▶ CREATING THE BASIN AND RANGE PROVINCE
▶ COLORADO PLATEAU UPLIFT
▶ COLUMBIA PLATEAU AND CASCADES
  VOLCANISM
▶ SIERRA NEVADA AND CALIFORNIA
▶ THE NEW WEST COAST TECTONICS
▶ MEANWHILE, DRAMA OVERSEAS . . .
▶ BIG FREEZE: THE PLEISTOCENE ICE AGE
▶ WHY DID EARTH'S SURFACE COOL?
▶ CENOZOIC CLIMATES
▶ SUMMARY
▶ KEY TERMS
▶ QUESTIONS FOR REVIEW AND DISCUSSION
▶ WEB SITES
▶ BOX 15–1 GEOLOGY OF NATIONAL PARKS
  AND MONUMENTS
▶ BOX 15–2 ENRICHMENT
▶ BOX 15–3 ENRICHMENT

Key Chapter Concepts

- The Cenozoic Era spans from 65.5 million years ago to present. We divide it into an older Paleogene Period and a younger Neogene Period. Major features of our modern Earth formed during the Cenozoic.

- The ridges and valleys of the present Appalachian Mountains are the result of repeated cycles of uplift and erosion.

- Eight marine transgressions and regressions caused by fluctuating sea levels are recorded in the Cenozoic sedimentary sequence of the Atlantic and Gulf Coastal Plains.

- Most geologic structures of the Rocky Mountains formed during orogenic events from Late Cretaceous to early Paleogene. Detritus from erosion of the ranges first filled intermontane basins. Then broad regional uplift inaugurated a new erosional cycle that spread sediment eastward and gave the Rocky Mountains their present form and relief.

- The Basin and Range Province was formed by normal faulting from tensional forces during the middle Cenozoic. Upfaulted blocks along a north-south trend formed elongate mountain ranges that supplied erosional detritus to downfaulted basins between the upfaulted blocks.

- The Colorado Plateau was uplifted, but not folded, during the Cenozoic. Its rejuvenated streams eroded the Grand Canyon and other deep canyons.

- The Columbia Plateau was created by Cenozoic basaltic lava flows. It was bordered on the west by Cascade Range volcanoes.

- The tectonic style changed during the Cenozoic. The earlier compressional style of collision–subduction–thrust-faulting was replaced by lateral movements of the American and Pacific Plates grinding past one another.

- The Tethys Sea closed and the Alpine–Himalayan mountain systems formed. Both were caused by the northward movement of Africa and the drifting of India toward Asia.

- African rift valleys and associated volcanoes and lakes were produced in the Late Cenozoic from tensional forces along the east side of the African continent.

- Continental glaciers covered a third of the Northern Hemisphere during the Pleistocene Epoch. The repeated advance and retreat of continental glaciers during the Pleistocene was related to Milankovitch cycles, periodic changes in Earth's rotational and orbital movement.

▶ **THE TECTONICS–CLIMATE CONNECTION**

The Cenozoic ("recent life") is our own era of geologic history. It is the era during which continents and landscapes acquired today's form, sea level reached its present position, and today's plants and animals evolved. We divide the Cenozoic Era into two periods, the **Paleogene** ("old born") and the **Neogene** ("new born"). Each is subdivided into various epochs (Table 15-1).

The Paris Basin is the type area for most of the Cenozoic epochs. In this basin there is a major unconformity, and in 2003 the International Commission on Stratigraphy chose it as the boundary between the Paleogene and Neogene. (Before 2003, the era was separated into Tertiary and Quaternary Periods, as shown in Table 15-1. You will often see these older terms used.)

The Cenozoic was busy with continued plate motion and seafloor spreading. In fact, *half of the present seafloor formed during the past 65.5 million years.* Much of this new ocean floor, extruded along the midoceanic ridges, was emplaced in the fast-expanding Atlantic and Indian Oceans. As this widening progressed, the Americas moved westward. The area that is now California came into contact with the northward-moving Pacific plate, thereby producing the San Andreas fault system. South America moved against the Andean trench, actually bending and displacing it.

TABLE 15-1 Geochronologic Terminology for Divisions of the Cenozoic Era.

Era	Periods	Epochs	Millions of Years Ago	Terminology Before 2003
Cenozoic	Neogene	Holocene		Quaternary
			0.01	
		Pleistocene		
			1.8	
		Pliocene		
			5.3	
		Miocene		
			23.0	
	Paleogene	Oligocene		Tertiary
			33.9	
		Eocene		
			58.8	
		Peleocene		
			65.5	

Throughout this discussion, please refer to the highlighted areas in Figure 15-1. Down the western backbone of the Americas, vigorous orogenic and volcanic activity formed the Isthmus of Panama, which today links North and South America. This little Panamanian land bridge had remarkable implications:

- It blocked the westward movement of the North Atlantic Current, forcing it to swing and flow northeastward as the Gulf Stream. This famous stream of warm water flows up the U.S. East Coast, across the Atlantic, to England and then to northern Europe (Fig. 15-2). Today's remarkably mild climate of the British Isles (which are at the same latitude as Newfoundland) and ice-free harbors in Norway (at the same latitude as glacial Greenland) are testament to the powerful warming influence of the Gulf Stream.
- The isthmus provided a pathway for plant, animal, and human migration between the Americas.
- The isthmus created a barrier for human travel between the Atlantic and Pacific that was finally breached by the 51-mile-long Panama Canal a hundred years ago.

There were two important continental break-ups during the Cenozoic. First, the North Atlantic rift extended toward the North Pole, separating Greenland from Scandinavia and severing the land connection between Europe and North America. And second, Australia separated from Antarctica, beginning its journey to today's location. Prior to this separation, Antarctica was warmed by ocean currents flowing toward it from warmer regions to the north.

However, by Oligocene time (about 30 million years ago), a frigid northward-flowing current developed in the widening rift between Antarctica and Australia. This current deflected warmer waters that had given Antarctica a milder climate. Around the now-isolated Antarctica, cold circumpolar currents were set in motion, and soon the continent began to assume its famous frigidity. Ocean water made dense by the extreme cold sank to the bottom surrounding Antarctica and began drifting northward along the ocean floor, exterminating benthic invertebrates that were adapted to warmer conditions. This transfer of frigid waters northward likely contributed to cooling conditions during the Late Eocene and Oligocene and to eventual development of the Pleistocene Ice Age.

The most dramatic Cenozoic tectonic event was the collision of Africa and India with Eurasia. This titanic smashup transformed much of the Tethys Sea into lofty mountain ranges, among which are the present Alps and Himalayas. Also during the Cenozoic, a branch of the Indian Ocean opened between Arabia and Africa, creating the Gulf of Aden and the Red Sea (Fig. 15-3).

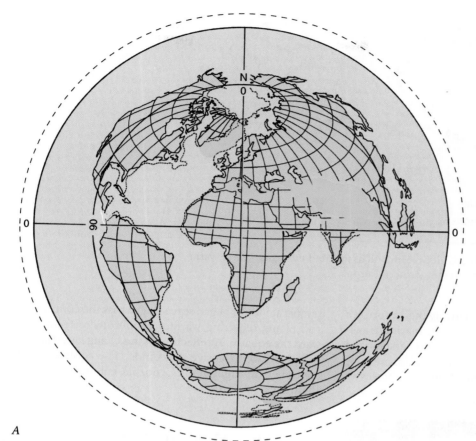

A

FIGURE 15-1 **Eocene global paleogeography.** (A) Major landmasses during the Eocene, about 50–45 million years ago. Note that Antarctica and Australia were still connected, and the Americas were not. (B) The world today. Compare the highlighted areas on both globes to note major tectonic changes. ▮ *Although Antarctica lay astride the South Pole, it did not experience extreme cold. How do we know this, and what would have caused its mild Eocene climate? (Answers to questions appearing within figure legends can be found in the Student Study Guide.)*

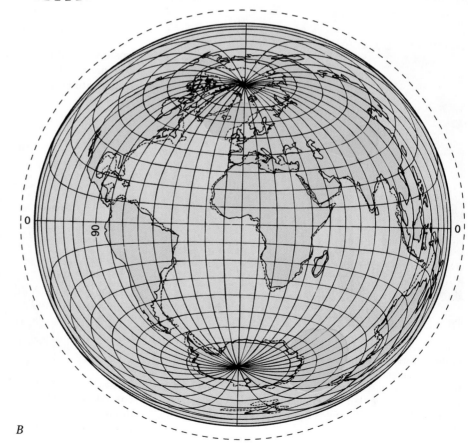

B

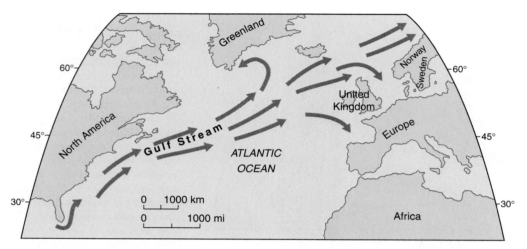

FIGURE 15-2 **Present trend of the Gulf Stream and associated currents.** These warm surface currents bring moderate climates to northern Europe.

Continental interiors stood relatively high above sea level during the Cenozoic. As a result, marine transgressions were limited. Warm climates characterized the early Cenozoic, as indicated by distribution of tropical and subtropical plants. By mid-Cenozoic, however, temperate floras began spreading across the continents, and plants requiring warmer conditions retreated toward the equator. A reflection of this change was the development of extensive grasslands. The cooling trend continued throughout the Cenozoic, culminating in the Pleistocene glaciation.

FIGURE 15-3 **The Red Sea.** The seaway opened about 30 million years ago when the Arabian Peninsula rifted away from Africa, and it continues to widen today. This view from the *Gemini* spacecraft is toward the south. The land wedge at the lower left is the Sinai Peninsula, bordered on the right by the Gulf of Suez and Egypt, and on the left by the Gulf of Aqaba and Saudi Arabia.

STABILITY AND EROSION ALONG THE NORTH AMERICAN EASTERN MARGIN

Erosion continued to rule the Appalachians, with little orogenic activity occurring along North America's eastern margin. Periodically, as the uplands were beveled by

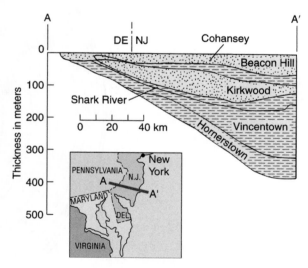

FIGURE 15-4 **Cenozoic strata across the New Jersey coastal plain.** The Cenozoic section of marine rocks is thinner near the Appalachian source areas and becomes thicker and less clastic offshore.

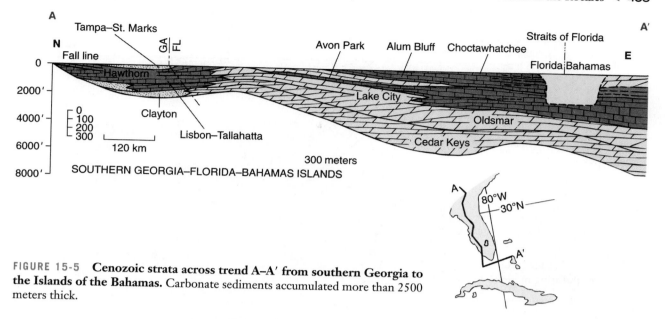

FIGURE 15-5 **Cenozoic strata across trend A–A′ from southern Georgia to the Islands of the Bahamas.** Carbonate sediments accumulated more than 2500 meters thick.

erosion, broad, gentle isostatic uplifts occurred. This revitalized streams, which sculpted a new generation of well-defined ridges and valleys. The most recent uplifts in the Appalachian belt were accompanied by gentle tilting of the Atlantic coastal plain and adjacent parts of the continental shelf.

The streams transported terrigenous clastic sediments from the highlands and deposited them on the plains. On occasion, these were reworked by waves and currents during marine transgressions. Because these seas were shallower along the western margins, the Cenozoic section of marine rocks is thinner near the Appalachian source areas and becomes thicker and less clastic offshore (Fig. 15-4).

Around Florida, which lacked both erodible highlands and major rivers, terrigenous clastics were sparse. Carbonate sediments accumulated to exceed 2500 meters thickness along subsiding, elongate coralline platforms that resembled today's Bahama Banks (Fig. 15-5). In the early Neogene, uplift along the northern end of this tract raised the land area of Florida above the waves, where it remains—so far.

GULF COAST: TRANSGRESSING AND REGRESSING SEA

The best record of Cenozoic strata in North America is found in the Gulf of Mexico coastal plain, where eight major transgressions and regressions are recorded (Fig. 15-6). The Paleocene transgression brought marine waters as far north as southern Illinois. Frequently, during marine regressions, nearshore deltaic sands were deposited above offshore shales.

The resulting interfingering of permeable sands and impermeable clays provided ideal conditions for the eventual entrapment of oil and gas. Much of the oil was trapped in structures around salt domes (see Fig. 13-9). Wells drilled in the Gulf Coast have penetrated more than 200 salt domes, and many more exist beneath the continental shelf. Gulf Coast Cenozoic rocks are famous for the petroleum they once yielded in great quantities. Although recovery continues in the Gulf, activity is much reduced today because of depletion.

"A wedge of sediments that thickens seaward" is a good description for the Cenozoic formations of the Gulf Coast. Geophysical measurements suggest that the thickness of Tertiary sediments beneath the northern border of the Gulf may exceed 10,000 meters. The area must have been subsiding rapidly in order to provide space for this great thickness of sediment.

TECTONICS AND EROSION IN THE ROCKIES

While marine sedimentation prevailed along the east coast, terrestrial deposition prevailed in the Rocky Mountain region (Fig. 15-7). Late Cretaceous and Paleogene deformation largely created the major structural features of the Western Cordillera. However, today's topography of this region is primarily erosional, following uplifts that began during the Miocene.

As always, erosion acting on tilted/folded hard and soft layers was the final factor in shaping the landscape. Following orogenesis in the Late Cretaceous to early

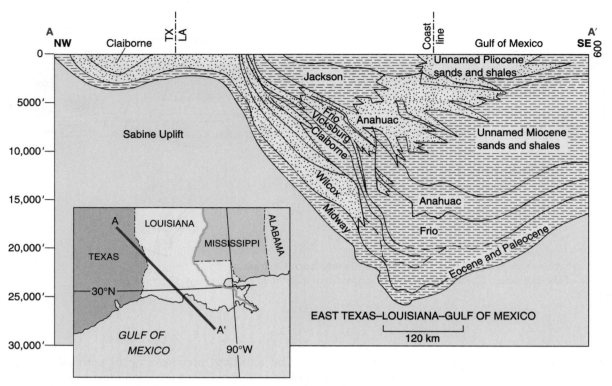

FIGURE 15-6 **Cenozoic strata across the Gulf coastal plain and Gulf of Mexico.** "A wedge of sediments that thickens seaward" is a good description for the Cenozoic formations of the Gulf Coast. ☑ *Marine invertebrate fossils indicate that most formations shown here accumulated at shallow-to-moderate water depths. How could 27,000 feet of sediment accumulate in water depths that rarely exceeded 1000 feet?*

Paleogene, erosional debris was trapped largely in basins between mountain ranges. These intermontane terrestrial sediments provide the record of the earlier Paleogene epochs.

Later uplifts and erosion resulted in the spectacular relief of the Rockies and caused the detritus of older basins and newer uplands to spread over plains to the east. The result was deposition of a vast apron of nonmarine Oligocene-through-Pliocene sands, shales, and lignites that underlie the western high plains. Beds of volcanic ash interspersed among the sediments provide useful isotopic dates.

Sediment and Mineral Wealth

In intermountain basins, continental sediments formed Lower Paleogene sedimentary rocks, including gray siltstones and sandstones, carbonaceous shales, lignites, and coal. These rocks are well represented in the Fort Union Formation, a unit recognized in several intermountain basins in the Rocky Mountain region.

The Fort Union is approximately 1800 meters (over a mile) thick. In its lower levels, it contains immense tonnages of coal with low sulfur content. This makes it very desirable as a "steam" coal (burned to generate electric power), because lower sulfur means less environmental pollution. Existence of these coal beds indicates widespread swampy conditions during Paleocene time.

Many of these basins had no outlet during the early Cenozoic, so water filled them to create large lakes. One formed in the Green River basin of southwestern Wyoming (Fig. 15-8). Sediments deposited in this basin comprise the Eocene Green River Formation (Fig. 15-9). Deposited here were more than 600 meters of freshwater limestones and fine, laminated shales.

The laminations are **varves**, each of which consists of a thin, dark winter layer and a lighter-colored summer layer. Counting varves reveals that more than 6.5 million years were required to deposit the Green River sediments. Fossils are abundant in Green River sedimentary rocks, including insects, plant fragments, and well-preserved fishes (Fig. 15-10).

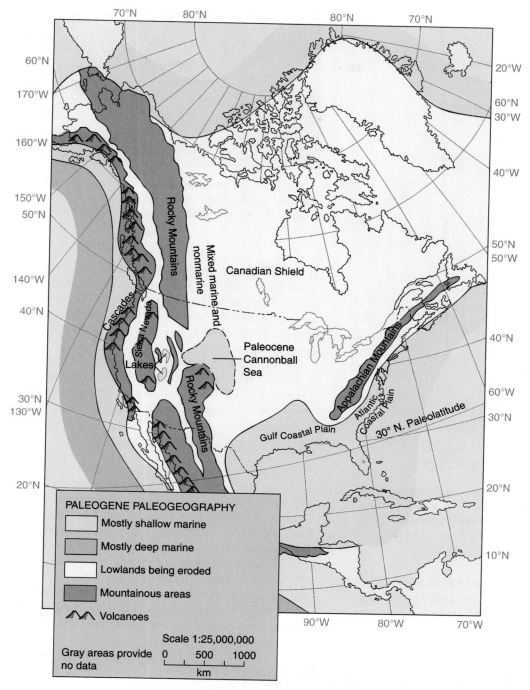

FIGURE 15-7 **Generalized paleogeographic map of North America during the Paleogene (65–23.5 million years ago).** Terrestrial deposition prevailed in the Rocky Mountain region, except for a solitary marine incursion called the Cannonball Sea. It probably was a vestige of the Late Cretaceous epicontinental sea. Paleocene strata reveal the Cannonball Sea's presence in dark shales of western North Dakota, with more than 150 marine invertebrate species.

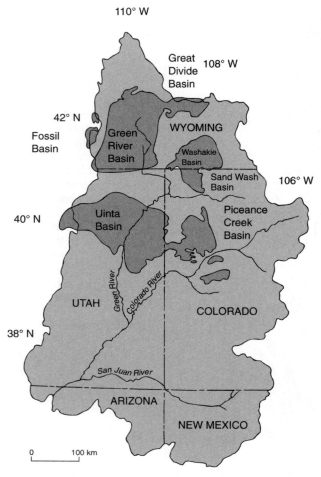

FIGURE 15-8 **Cenozoic basins containing important oil shale deposits in Colorado, Utah, and Wyoming.** The base map outlines the drainage basin of the Upper Colorado River (note the modern stream pattern).

In addition, the shales are rich in waxy hydrocarbons. For this reason, they are called **oil shales** and can be processed to yield petroleum (see Enrichment Box, "Oil Shale"). Several Wyoming fields pump oil

FIGURE 15-9 **The Green River Formation in a road cut near Soldier Summit, Utah.** The exposure here consists of dark shales and thin, lighter-colored limestones.

from the Claron Formation, an Eocene stream deposit that underlies and interfingers laterally with Green River beds. The source rock for the hydrocarbons was probably the Green River Shale. There is also plenty of colorful landscape geology in the region (Fig. 15-11).

To the south, in northeastern Utah, the Uinta basin is notable as the structurally deepest in the Colorado Plateau (Fig. 15-8). It received a particularly thick succession of Paleogene sediments (Fig. 15-12).

Remarkable Fossils

During the late Eocene and Oligocene, explosive volcanic activity blanketed with volcanic ash the region that today includes the San Juan Mountains and Yellowstone National Park (Fig. 15-13). For paleontologists, the more interesting rocks of this epoch are floodplain deposits of the White River Formation.

This famous formation contains entire skeletons of Oligocene mammals in extraordinary number, variety, and excellence of preservation. The animals apparently were victims of recent floods. The clays, silts, and ash beds of the White River Formation are the sediments

FIGURE 15-10 **Eocene freshwater fish (*Diplomystis*) from the Green River Formation, Wyoming.** Fossils are abundant in Green River sedimentary rocks, including insects, plant fragments, and well-preserved fishes like this specimen which is 8 cm long—approximately the size of a minnow or sardine.

ENRICHMENT

Oil Shale

The Eocene oil shale (Fig. A) that occurs in parts of Wyoming, Utah, and Colorado is rich in an oil-yielding organic compound known as kerogen. When heated to 480°C, the kerogen in oil shale vaporizes. This vapor can be condensed to form a thick oil. When enriched with hydrogen, it can be refined into gasoline and other products in much the same way as ordinary crude oil. The shale is heated in a retort, which resembles a giant pressure cooker that fuels itself with the very gases generated during heating.

Many oil shales yield between one-half and three-quarters of a barrel of oil (42 U.S. gallons) per ton of rock. If you mined only the known shale layers that are thicker than 10 meters, they would yield an impressive 540 billion barrels of oil. If this oil were produced during the next decade, it would appreciably reduce the U.S. dependency on foreign crude. Production of oil from oil shale, however, requires vast amounts of water in a region of North America where water is in short supply.

Two additional problems have prevented full-scale mining and retorting of oil shales. The first is the waste-disposal problem. In the process of crushing and retorting, the shale expands to occupy about 30% more volume than was present in the original rock. Geologists call this the "popcorn effect." Where could this great volume of light, dusty material be placed?

The second problem is air quality, for processing of huge tonnages of oil shale is likely to release large amounts of dust into the atmosphere. Unfortunately, the shales occur in arid regions, and there is little water available to suppress the dust and revegetate the land. Until these problems are solved, producing oil from oil shale will be slow. Perhaps this works to our advantage, for we may need this oil for processed goods a century or two from now, after the internal combustion engine has experienced its own mass extinction.

FIGURE A **An exposure of Eocene oil shale in Colorado.** The dark layers in these ancient lake deposits are rich in hydrocarbons.

FIGURE 15-11 **Eocene rocks at Cedar Breaks National Monument, Utah.** Here the Eocene Cedar Breaks Formation has been eroded into steep ravines, pinnacles, and razor-sharp divides. Early explorers used the term "breaks" to describe the change in topography where an elevated level area "breaks down" by eroding to a lower elevation. The formation is eroded into a magnificent badlands topography.

from which the Badlands of South Dakota have been sculpted (see Fig. A in "Badlands National Park, South Dakota").

Other interesting Oligocene beds are exposed in central Colorado, in the Florissant Fossil Beds National Monument. Volcanoes in this area produced voluminous ash that settled to the bottom of a neighboring lake, burying countless insects (Fig. 15-14), leaves, some fish, and even a few birds. Plant remains—including tree trunks in their original position, leaves, spores, and pollen—indicate subhumid conditions for the region and elevations between 300 and 900 meters.

The Miocene was a time of continued stream and lake sedimentation in intermountain basins as well as on the plains east of the mountains. By Miocene time, climates had cooled. Expanding grasslands supported Miocene camels, horses, rhinoceroses, deer, and other grazing mammals. In the central and southern Rockies, Miocene formations include beds of volcanic ash and lava flows, attesting to vigorous volcanic activity. The well-known gold deposits at Cripple Creek, Colorado, are mined from veins associated with a Miocene volcano.

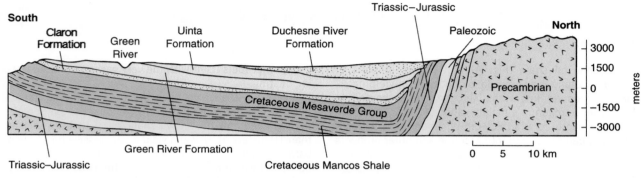

FIGURE 15-12 **A north–south view through the Uinta basin.** The basin received a thick succession of Paleogene sediments, including the Claron, Green River, Uinta, and Duchesne River formations. Paleogene units are above the Cretaceous Mesaverde Group. These formations can still be observed in exposures in the interior of the basin, where they have been affected only slightly by post-Eocene erosion.

Majestic Scenery

Regional uplift of the Rockies also began in the Miocene. This of course sharply increased erosion, and great sediment volumes filled intermountain basins. Sediment spread eastward, helping to construct the Great Plains.

Today's Rocky Mountain topography largely results from uplift and erosion that began in the Miocene. Sediment from the mountains covered the White River deposits and equivalent beds over extensive areas of South Dakota and Nebraska, continuing in the Pliocene.

FIGURE 15-13 **Yellowstone Falls and canyon, Yellowstone National Park, Wyoming.** Rocks in the canyon walls are Paleogene lava flows and volcanic ash, often altered to bright colors by hydrothermal activity.

Remains of plants and animals in some of these sediments indicate that the Pliocene was cooler and drier than the Miocene, a harbinger of an impending ice age.

Crustal movement continued throughout the remaining epochs of the Cenozoic, raising some of the highest peaks of the Rockies to spectacular elevations. Normal faulting and volcanism accompanying uplifts also produced some spectacular scenery (Fig. 15-15).

▶ CREATING THE BASIN AND RANGE PROVINCE

The Basin and Range Province is a large area of, as its name says, alternating basins and mountain ranges (Fig. 15-16). The province occupies a broad zone from Nevada and western Utah southward into central Mexico.

Most broadly, the basin-and-range pattern results from stretching of Earth's crust:

1. The region was up-arched during the Mesozoic Era.

2. Beginning in the Miocene, the arch subsided between great normal faults.

3. The uplifted blocks formed linear mountain ranges. These became sources of sediment, which filled adjacent downdropped basins.

4. Faulting opened fissures for the escape of molten rock from magmatic bodies below. The resulting volcanoes along the western side of the province produced extensive lava flows, while in the eastern region, the landscape was buried in ash and pumice.

5. Vigorous erosion of the upthrown fault blocks followed the volcanism. Coarse terrigenous clastics eroded from the mountains filled the downfaulted basins, clogged rivers, and caused lakes to develop behind debris dams.

6. Gypsum and salt layers formed when these lakes evaporated.

Badlands National Park, South Dakota

When you visit Badlands National Park, you enter an alien land of steep ravines and ethereal spires. You are amidst the world's finest example of badlands topography (Fig. A). Badlands are nearly devoid of vegetation, where erosion dissects the land into a labyrinth of chasms, steep ridges, and pinnacles. Bedrock of impermeable clays and shales, the absence of a protective plant cover, and infrequent but heavy rains are the conditions under which badlands form.

The rocks here are horizontal layers of clay, shale, and volcanic ash. Much of the ash has been weathered to a clayey sedimentary rock called bentonite, which erodes easily. Beds of river sandstones are more difficult to erode, and they often form a caprock atop buttes and provide an overhanging ledge, forming erosional features called "mushroom rocks."

Running water has been the most important agent in forming the badlands. Rain falling suddenly during cloudbursts cannot infiltrate the impermeable clays and shales. Instead, the rain runs off and becomes channeled into numerous small rivulets that have eroded a dense system of ever-enlarging, coalescing gullies and ravines. The impact of raindrops as they pelt the soft surfaces dislodges particles of rock, speeding the erosional processes.

Bentonite in the beds also facilitates erosion. The clay in bentonite

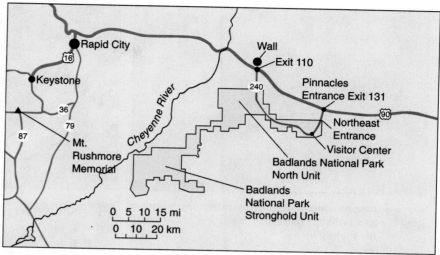

FIGURE B   Map of the park.

swells enormously when it becomes wet. On drying, bentonite crumbles and disintegrates, forming masses of easily eroded sediment. The White River takes its name from the whitish color imparted to the water by submicroscopic particles of bentonitic clays.

Formation of the badlands topography began with an episode of regional uplift late in the Pliocene Epoch. As a result of the uplift, major streams rejuvenated and entrenched themselves into the underlying poorly resistant beds. The White River became one of these entrenched streams. The process provided

steeper gradients to tributary streams that flowed southward into White River. The southern perimeter of the uplifted area that paralleled the White River Valley to the north became intricately gullied and dissected, forming an early badlands topography that subsequently eroded away. As tributary streams continued to extend their channels northward into the upland area, the badlands we view today began their development.

Badlands National Park has an interesting geologic history. Its oldest formation is the Pierre Shale, deposited in the shallow, extensive sea that covered much of western North America during the Late Cretaceous. The Cretaceous ended with regional uplift, and as the sea receded, erosion produced the unconformity on which the principal beds of the badlands were deposited. These are sediments of the Oligocene White River Group—primarily deposits of slow-moving streams that flowed across a landscape characterized by wide floodplains, marshlands, lakes, and ponds.

At this time, abundant plants nourished a rich fauna of turtles, lizards, alligators, huge titanotheres, aquatic rhinoceroses, three-toed horses, early camels, entelodonts, oreodonts, and tapirs. Many of these herbivores were prey for predatory members of the dog and cat families. Figure 16-46 provides a panoramic view of the environment and its inhabitants. The rich fossil discoveries in Badlands National Park prompted some geologists to call the Oligocene "The Golden Age of Mammals."

FIGURE A   **Badlands National Park, South Dakota.** This rugged terrain has been water-carved from flat-lying Eocene and Oligocene shales, mudstones, and ash beds. ◧ *What bedrock and climatic conditions favor the formation of badlands?*

FIGURE 15-14 **Fossil spider and insects from Oligocene beds near Florissant, Colorado.** Volcanoes in this area produced voluminous ash that settled to the bottom of a lake, burying countless insects and leaves. They are preserved in shale-like tuff, which is consolidated volcanic ash.

FIGURE 15-15 **Wyoming's lofty Teton Range.** The Teton Range was elevated along great normal faults, with displacement reaching nearly 6000 meters (3.7 miles). The magnificent eastern face is a fault scarp rising 2500 meters to an elevation exceeding 4000 meters. Following uplift, the range has been cut by water and ice erosion.

A

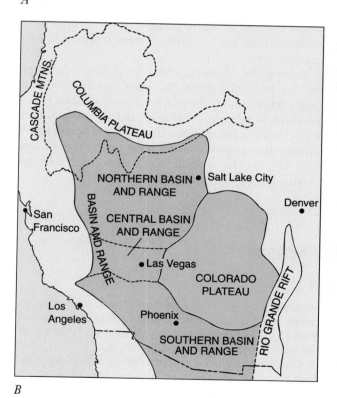

B

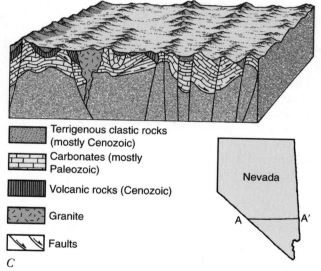

Terrigenous clastic rocks (mostly Cenozoic)

Carbonates (mostly Paleozoic)

Volcanic rocks (Cenozoic)

Granite

Faults

C

FIGURE 15-16 **Basin and Range Province.** (A) In the foreground is the top of Pioneer Mountain. In the distance is Idaho's Lost River Range. The valley is a downfaulted block (basin). (B)The Basin and Range Province relative to the Colorado and Columbia Plateaus. (C) The fault system responsible for Basin and Range in southern Nevada.

The cause of the crustal stretching—the tensional faulting—that produced the Basin and Range Province is much debated. Here are current hypotheses:

- When the westward-moving North America overrode part of the Pacific oceanic plate and a spreading center (the East Pacific rise) that was being subducted along the coast of California, the subducted spreading center caused uplift and stretching of the crust. But this has been questioned because there is too little progressive eastward deformation, which should occur during the passage of a spreading center beneath a continent.

- Normal faulting in the Basin and Range is simply how the crust adjusted to the change along the California coast when oblique shearing of the edge of the continent during the Miocene replaced the earlier subduction zone.

- This is extension and uplift of the crust from the remnants of an oceanic plate that was carried beneath the Basin and Range region by an earlier subduction episode. When subduction ceased, the oceanic slab may have formed a partially molten buoyant mass that pressed upward against the overlying crust, causing tensional faulting and escape of lava along fault and fracture zones.

- The crustal extension and tensional faulting is related to convectional movements beneath the continental plate similar to those that cause the breaking apart of continents.

Each hypothesis is supported by evidence, and the truth could be a combination of these ideas. As with so many things in science, more research is needed before we can conclusively explain the formation of the remarkable Basin and Range.

▶ COLORADO PLATEAU UPLIFT

The Colorado Plateau is a vast and magnificent region of uplift (Figs. 15-16B and 15-17). Here Paleozoic and Mesozoic rocks are relatively flat-lying, so we know the Plateau remained undeformed during the Mesozoic orogenies. The region formed a buttress around which folding and faulting produced highlands.

The plateau was repeatedly raised during early-to-middle Pliocene time (about 10 to 5 million years ago). Steep faults developed on the plateau during its rise, providing avenues for extrusion of lavas and volcanic ash (Fig. 15-18). The San Francisco Peaks near Flagstaff, Arizona, are a group of recent Colorado Plateau volcanoes and cinder cones (Fig. 15-19).

The best-known feature resulting from the linked processes of uplift and erosion on the Colorado Plateau is the Grand Canyon of the Colorado River (Fig. 15-20). This awesome monument to the forces of erosion exceeds a depth of 2600 meters (1.6 miles) in places. The river has eroded through a half-billion years of Phanerozoic strata into crystalline Precambrian basement rocks.

▶ COLUMBIA PLATEAU AND CASCADES VOLCANISM

The Columbia Plateau, named for the Columbia River, lies in the U.S. northwest (Fig. 15-16). Unlike the Colorado Plateau, which is constructed of layered sedimentary rocks, the Columbia Plateau was built by volcanic activity.

During the Miocene, about 15 million years ago, basaltic lavas like those that pour from midoceanic ridges erupted along deep fissures approximately parallel to the Washington–Idaho border. The liquid rock spread out and buried more than 500,000 square km of existing topography beneath layer upon layer of lava (Fig. 15-21). This is similar to the area covered by the basaltic Deccan Traps in India.

E N R I C H M E N T

Hellish Conditions in the Basin and Range Province

If the great Italian poet Dante Alighieri (1265–1321), author of *The Inferno*, had strolled the Basin and Range Province during the Oligocene and Early Miocene, his vision of the lower world might have included turbulent clouds of white-hot ash, jets of searing gases, and fiery explosions of molten rock. Such was the scene during the Mid-Cenozoic. Along the trails of mountain ranges in the province, evidence in the form of hardened ash falls and solidified lava beds is nearly everywhere underfoot. Volcanic rocks also lie between the mountain ranges, hidden by sediment layers.

Volcanic rocks of the Basin and Range Province are derived from high-silica magmas, which are highly viscous (thick). Such magmas produce explosive volcanism in which hot gases, steam, ash, and pyroclastics of all sizes erupt with enormous force. Fiery turbulent clouds of hot gas and ash called *nuée ardentes* accompany such eruptions and devastate the lands as they rush swiftly down slopes at speeds of up to 100 miles per hour. The volume of ash expelled was truly incredible. A single ash fall covered more than 10,000 square km (roughly the area of Florida's Everglades) to a depth exceeding 180 meters.

Commonly, high-silica magmas cool slowly at depth and form batholiths of granite and granodiorite. However, the magma of the Basin and Range Province breached the crust and produced the intense episode of volcanism. What might have been the cause?

During the Mid-Cenozoic, the North American plate was moving rapidly westward at about 15 cm per year. It overrode the Pacific plate so quickly that the plate was unable to descend into the mantle. Rather, the Pacific plate slid horizontally under North America. As the hot seafloor plate moved along beneath what is now the Basin and Range Province, it probably provided much of the heat responsible for volcanism.

FIGURE 15-17 **A Colorado Plateau Canyon.** The Dolores River near Grand Junction, Colorado.

West of the Columbia Plateau lies a belt that also was the site of extensive volcanic activity. Here, however, the outpourings of more viscous lavas generated the mountains of the Cascade Range. Volcanism in this region began about 4 million years ago and continues today, as was made dramatically obvious during the eruptions of Mount St. Helens in 1980 and 2004 (Fig. 15-22).

The recent activity of Mount St. Helens and the older eruptions that gave us the volcanic peaks of the Cascades are manifestations of an ongoing collision between the American plate and the small Juan de Fuca plate of the eastern Pacific. As the Juan de Fuca plate plunges beneath Oregon and Washington, molten rock rises to supply lava for the volcanoes (Fig. 15-23).

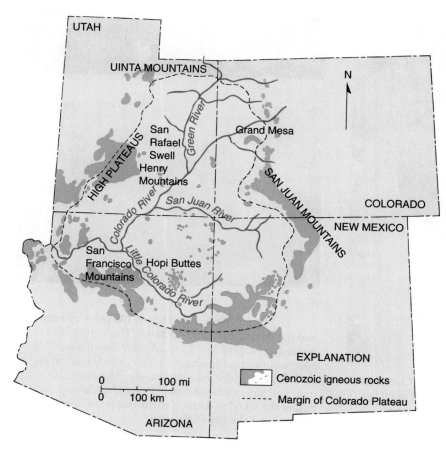

FIGURE 15-18 **Centers of Cenozoic igneous activity in and around the Colorado Plateau.**

FIGURE 15-19 **Bird's-eye view of a large cinder cone in the San Francisco volcanic field, northern Arizona.** The solidified flow from the cone is 7 km long and more than 30 meters thick. ◪ *Did the lava flow from the volcano before or after extrusion of the pyroclastics that built the cinder cone?*

FIGURE 15-21 **Basalt lava flows of the Columbia Plateau.** In some places, these low-viscosity lavas remained liquid and flowed 170 km (roughly 100 miles) from their source. Their combined thickness exceeds 2800 meters (1.7 miles). The Snake River is in the foreground. ◪ *What is the origin of the closely spaced vertical cracks seen on the cliff faces?*

A

FIGURE 15-20 **The Grand Canyon of the Colorado River.** This great gorge is largely the result of uplift accompanied by stream erosion during the Cenozoic Era.

B

FIGURE 15-22 **Mount St. Helens (A) prior to eruption and (B) during the 1980 eruption.** During the initial eruption, clouds of steam and ash were blown toward the northeast from the prominent plume, which reached about 20,000 meters (12.4 miles) in altitude. The eruption produced an ash volume roughly equivalent to that ejected during the A.D. 79 Mount Vesuvius eruption in Italy, which buried the city of Pompeii.

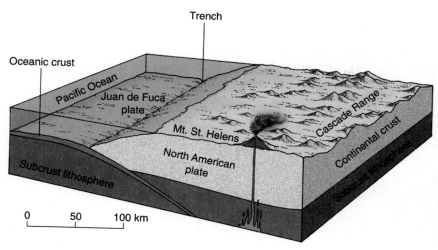

Trench

Oceanic crust

Pacific Ocean

Juan de Fuca plate

Mt. St. Helens

North American plate

Subcrust lithosphere

Cascade Range

Continental crust

Subcrust lithosphere

0   50   100 km

FIGURE 15-23 **The small Juan de Fuca plate plunges beneath Oregon and Washington.** As it does so, molten rock is generated and rises to supply the volcanoes of the Cascade Range.

Other mountains in the Cascades have important volcanic histories:

- California's Mount Lassen extruded lava and ash in 1914 and then exploded violently in 1915, producing a *nuée ardente* (French: "fiery cloud") that roared down the mountainside, incinerating everything in its path.

- Washington State's Mount Rainier had its last major eruption about 2000 years ago, but minor disturbances occurred frequently in the late 1800s.

- Oregon's Crater Lake (Fig. 15-24) formed from the volcanic cone of Mount Mazama after eruptions about 6000 years ago (Fig. 15-25).

FIGURE 15-24 **A lake in a volcano.** 6000 years ago, the volcano Mount Mazama erupted catastrophically, and its top collapsed to form a giant caldera (Spanish: *cauldron*) which today is filled with Crater Lake. The small cinder cone projecting above the lake surface is Wizard Island.

SIERRA NEVADA AND CALIFORNIA

The mountain ranges of the Sierra Nevada lie south of the Cascades (Fig. 15-7). The rocks of these mountains were folded and intruded during the Nevadan orogeny. For most of the Paleogene, the peaks were steadily eroded until their granitic cores lay exposed at the surface.

Then, during Pliocene time and continuing into the Pleistocene, the entire Sierra Nevada block was raised along normal faults (Fig. 15-26). Over time, its high eastern front was lifted an astonishing 4000 meters (more than two miles) above the California trough to the West. Streams and powerful valley glaciers eroded the block to form the magnificent landscapes of the present-day Sierra.

In the Paleogene, the region west of the Sierra Nevada was most directly affected by subduction tectonics. At about the beginning of Miocene time, this region underwent a change from subduction tectonics to strike-slip tectonics (large-scale lateral faulting). Fault movements created islands and intervening sedimentary basins in which fine-grained marine clastics, diatomites, and bedded cherts accumulated. Also during the Miocene, folding and uplift caused marine regressions, and by Pliocene time seas were restricted to a narrow tract along the western edge of California. Ultimately, uplifts erased this seaway as well.

THE NEW WEST COAST TECTONICS

During most of the Cenozoic, North America's western edge was an eastward-dipping subduction zone. This subduction created batholiths, compressional structures, volcanism, and metamorphism that accompanied Mesozoic and Cenozoic orogenies.

The building of ancient Mount Mazama. Magma chamber is filled and supplying magma to the volcano.

A

Often explosive eruptions begin to exhaust the reservoir of magma, leaving the upper part of the magma chamber empty.

B

Summit collapses into the vacated chamber below.

C

Minor eruptions build small volcanoes on the caldera floor, and the caldera gradually fills with water to form a lake.

D

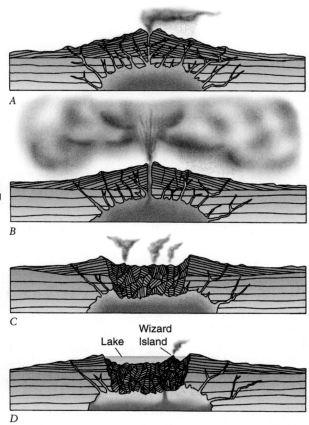

Lake   Wizard Island

FIGURE 15-25 **How Crater Lake's caldera formed.**

For most of Cenozoic time, the subducting Farallon plate (dense oceanic crust) was disappearing at the subduction zone faster than it was growing from its spreading center. As a result, most of the Farallon plate and part of the East Pacific rise that generated it were gobbled up at the subduction zone along the western edge of California (Fig. 15-27). Today, the only remnants of the Farallon plate are the small Juan de Fuca plate near the Oregon and Washington coast and the Cocos plate off the coast of Mexico.

With the loss of the Farallon plate near California, the North American plate was brought into direct contact with the Pacific plate, and a whole new set of plate motions commenced. Before making contact with the West Coast, the Pacific plate had been moving toward the northwest. As a result, when contact was made, the Pacific plate did not plunge under the continental margin but instead ground along laterally, like a moving car that sideswipes a parked car.

FIGURE 15-26 **Uplifted Sierra Nevada batholith, Yosemite National Park.** For most of the Paleogene, these peaks were steadily eroded until their granitic cores lay exposed at the surface.

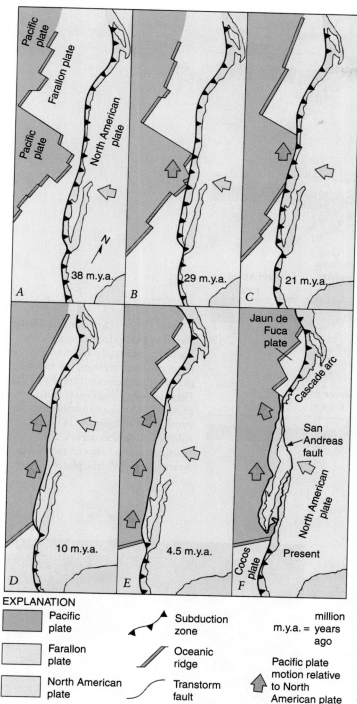

FIGURE 15-27 **The interaction of West Coast plates during the Cenozoic.** Note how the Farallon plate was largely subducted by Late Cenozoic time, leaving only remnants to the north (Juan de Fuca plate) and to the south (Cocos plate). The San Andreas and associated faults were caused by right-lateral movements that began about 29 million years ago.

This formed the San Andreas fault, with its strike-slip (lateral) motion (⇆).

Leave it to California to lead the nation in a change of "tectonic style"! No longer did the California sector of the West Coast have an Andean type of subduction zone (Fig. 15-28). Strike-slip movement now characterized the southwestern margin of the United States.

An important result of the shearing and wrenching of the West Coast was the tearing away of Baja California from the mainland of Mexico about 5 million years ago. Because it was located west of the San Andreas fault system, Baja California was sheared from the American plate and now accompanies the northward movement of the Pacific plate.

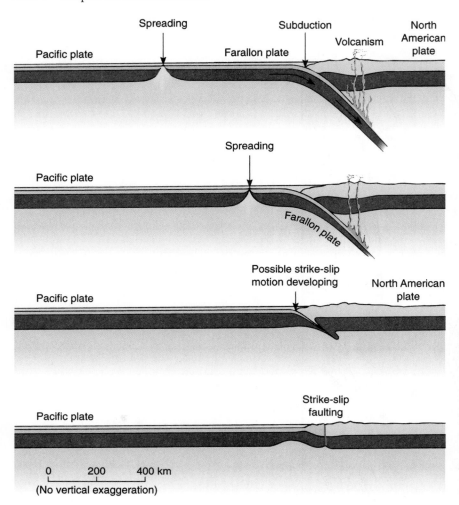

FIGURE 15-28 **Subduction of the Farallon plate.** In this model, the Pacific plate is considered to be fixed and overrun from the east by the North American plate. Even the Farallon spreading center (East Pacific Rise) moves to encounter the continental margin and is subducted. As this occurs, the relative motion between the North American and Pacific plates shifts to strike-slip.

MEANWHILE, DRAMA OVERSEAS . . .

One of the greatest events of the Cenozoic Era was the conversion of the **Tethys Sea** (between Eurasia from Gondwana) to a spectacular array of mountains (Fig. 15-29) and plateaus. A quiet ocean tract was transformed into a structurally complex mountain region that includes the Alps, Appennines, Carpathians, Caucasus, and Himalayas (Fig. 15-30).

At the beginning of the Cenozoic, the Tethys Sea lay south of Europe, its waters spilling onto the northern margin of Africa as an epicontinental sea. Tectonic activity began during the Eocene with large-scale folding and thrust-faulting as the northward-moving African block encountered the western underside of Europe and crumpled the strata that now compose the Pyrenees and Atlas Mountains.

Then, the Alpine region began to be squeezed. At first, marine conditions persisted between the emerging folds. Dark siliceous shales, poorly sorted sandstones, and cherts accumulated between elongate submarine banks. In the Alps, these marine sediments are referred to as *flysch* ("flish").

FIGURE 15-29 **Shivling Peak in the Himalayas.** What a history: These former sedimentary layers from the Tethys Sea bed were folded and uplifted, then eroded by mountain glaciers to form this sharp glacial horn.

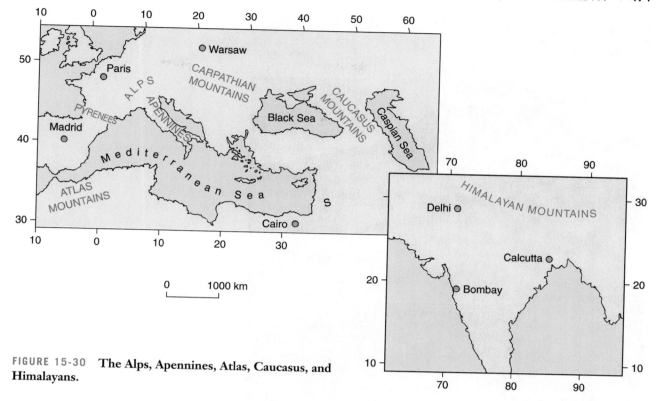

FIGURE 15-30 **The Alps, Apennines, Atlas, Caucasus, and Himalayans.**

By Oligocene time, compression from the south caused folds to rise as mountain arcs out of the seaway and to slide to the north along thrust faults. The great folds with faults along their undersides were pushed one atop another. North of these folds and thrust faults, a low-lying region received piedmont deposits eroded from the mountains. These terrestrial clastics are termed *molasse* (mo-LASS) by European geologists.

Compression continued, and during the Pliocene, further thrusts from the south carried the older folded belts northward over the molasse deposits and crumpled strata to form the Jura folds. Today these folded rocks form the northern front of the Alps.

East of the Alps, India moved northward and collided with Asia. During the Miocene, the collision was marked by extensive volcanism, folding, thrusting, and emplacement of granitic plutons. Regional uplifts brought plateaus and ranges to lofty elevations as the Indian plate moved northward beneath the southern edge of the Asian plate.

Northern Europe

North of the Tethys during the early Cenozoic, basaltic lavas were extruded in Scotland, Ireland, and in the Arctic Circle lands of Greenland, Baffin Island, and Norway's Svalbard islands. In Ireland, these lavas are vertically jointed and form the famous Giant's Causeway (Fig. 15-31). The separation of Greenland from

Europe provided the rift zone and associated fractures along which lavas rose to the surface.

Northeastern Europe underwent repeated early Cenozoic transgressions/regressions of the sea. Rocks deposited during these cycles have been thoroughly studied in the Paris basin, which is both a sedimentary basin and a structural basin. This is quickly apparent on a geologic map of the area (Fig. 15-32), which reveals younger Cenozoic deposits surrounded by older Mesozoic rocks in a classic basin structure. After the Oligocene, general uplift prevented further marine incursions.

Rifting Africa

The larger part of Africa that lay south of the Tethys was mostly emergent during the Cenozoic. Its most conspicuous changes were the rift valleys down the eastern side of the continent, formed during the Late Cenozoic by tensional faulting. Dozens of volcanoes formed along these faults, including Mount Kenya and Mount Kilimanjaro. In addition, elongate lakes formed within the downfaulted blocks (Fig. 15-33). Today, Lake Tanganyika and Lake Malawi are splendid examples of these fault-controlled inland water bodies.

Overall, Africa enjoyed a much quieter Cenozoic history than did southern Europe and Asia. In Libya and Egypt, the formations are flat or only moderately folded. The only severely folded mountain ranges are

FIGURE 15-31 **Striking hexagons of the Giant's Causeway, Northern Ireland.** This is one of Earth's best-known locations for columnar jointing in basaltic lava.

in western North Africa, which experienced pulses of orogeny throughout the Cenozoic.

Semitropical Antarctica

Antarctica had a semitropical climate during most of the early Cenozoic. During the Paleocene and Eocene, Australia had not yet separated from Antarctica, and the now-frigid continent was warmed by ocean currents that originated in temperate zones. By Miocene time, however, cold currents flowed through the oceanic tract that opened as the two landmasses drifted apart. Warm waters were forced northward, and snow began to accumulate on Antarctica.

As the continent became ice-bound, deep-marine clastics and volcanic rocks reflecting tectonic instability accumulated along its Pacific side. The crustal unrest and volcanism was the result of the eastern move-

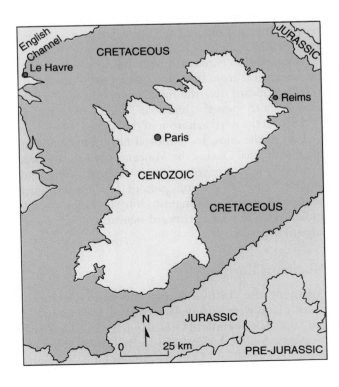

FIGURE 15-32 **Geologic sketch map of the Paris basin.** 🛈 *How does the outcrop pattern confirm that the structure is a basin rather than a dome?*

A

FIGURE 15-33 **African Rift Valleys show plate tectonics in action.** These valleys extend from the Red Sea southward for two-thirds the length of the continent. Some are occupied by lakes, notably Lake Tanganyika and Lake Malawi. (A) An African Rift Valley. (B) Locations of major rifts and associated lakes.

B

ment of an oceanic plate beneath the western margin of Antarctica.

BIG FREEZE: THE PLEISTOCENE ICE AGE

The two most recent epochs, the Pleistocene and the Holocene, represent only the past 2 million years of geologic time. But for humans, these are exceedingly significant. During the Pleistocene, primates of our own species evolved and rose to a dominant position on the globe. Also during the Pleistocene, about one-third of Earth's land surface became buried beneath more than 40 million cubic km of snow and ice (Fig. 15-34). This was the **Pleistocene Ice Age**.

This extensive cover of ice and snow had profound effects—not only for sculpting glaciated terrains (Fig. 15-35), but also for processes that occurred at great distances from the ice itself. Entire climatic zones in the Northern Hemisphere shifted southward, and arctic conditions prevailed across northern Europe and the United States. Mountains and highlands in the Cordilleran and Eurasian ranges were sculpted by spectacular mountain glaciers (Fig. 15-36). While the snow and ice accumulated and spread in higher latitudes, rainfall increased in lower latitudes, generally benefiting plant and animal life.

Even as late as the beginning of the present Holocene Epoch, presently arid regions in northern and eastern Africa were well-watered, fertile, and populated by nomadic tribes. Middle and late Pleistocene humans hunted along the fringes of continental glaciers, where game was abundant and the cold prevented meat from rapid spoiling. Animal furs provided warm clothing, and following the discovery of fire, caves were warmed against the arctic winds.

The Pleistocene also was a time of recurring crustal unrest. Active volcanoes erupted in New Mexico, Arizona, Idaho, Mexico, Iceland, north of the Arctic Circle, and the Pacific borders of North and South America. Crustal uplift proceeded in the Tetons, Sierra Nevada, central and northern Rocky Mountains, Alps, Himalayas, and all ranges in between.

FIGURE 15-34 Pleistocene continental glaciers in the Northern Hemisphere. During the Pleistocene, about one-third of Earth's land surface became buried beneath more than 40 million cubic km of snow and ice. Arrows show the direction of ice flow. Coastlines differ from today's because ice tied up seawater, lowering sea level about 75 meters below its present-day level.

FIGURE 15-35 Glaciated Canadian Shield north of Montreal, Canada. Imagine this scene covered with millions of tons of creeping ice, gouging and grinding away hills and depressions. ◀ *Draw an arrow on the photograph indicating which way the glacier moved.*

FIGURE 15-36 Mountain glaciers in Greenland. Mountain glaciers often are called "ice rivers," and you can see why. Their slow, powerful scouring action sharpens mountain peaks and converts V-shaped stream valleys into U-shaped glaciated valleys.

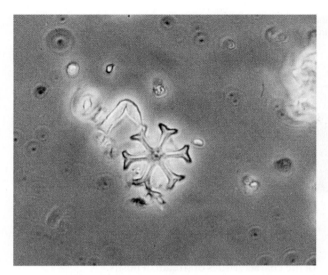

FIGURE 15-37 **Discoaster (*Discoaster challengeri*).** Magnified 1200 times through an optical microscope.

Pleistocene and Holocene Chronology

The Pleistocene Epoch did not begin with a sudden, simultaneous worldwide onslaught of frigid conditions. As with other epochs of the Cenozoic, the Pleistocene was defined by Charles Lyell in 1839 according to *the proportion of extinct to living species of mollusk shells* in layers of sediment. At the time, Lyell referred to rocks subsequently named Pleistocene as "Newer Pliocene."

Lyell's scheme for naming the epochs seems straightforward, but it is difficult to find suitably fossil-iferous sediments in various parts of the globe that can be confidently correlated to Lyell's type section from eastern Sicily. The beginning of the Pleistocene is dated at 1.8 million years ago. However, this date does not coincide everywhere with the beginning of glaciation (as indicated by glacial sediments). The most extensive glaciation appears to have begun about 1 million years ago. Also, the low temperatures and abundant precipitation needed for extensive continental glaciation occurred at different times in different places.

We use fossils to correlate Pleistocene marine beds. In some cases, a stratum that can be correlated to the standard section can be recognized by the earliest appearance of certain cooler-water mollusks and foraminifers. When studying cores of deep-sea sediments, the basal Pleistocene oozes can be recognized by the extinction point of tiny fossils called **discoasters** (Fig. 15-37). These calcareous, commonly star-shaped fossils are produced by golden-brown algae related to coccoliths. In continental deposits, the fossil remains of the modern horse *(Equus)*, the first true elephants, and particular species of other vertebrates are used to recognize the deposits of the lowermost Pleistocene.

Geologists divide the Pleistocene into four major glacial stages, with intervening interglacial stages (Table 15-2). However, they recognize up to 30 minor glacial advances over the past 3 million years. Widespread cold climates thus existed well before the beginning of the Pleistocene. Antarctica has had glaciers for at least the past 15 million years.

TABLE 15-2 Classic Nomenclature for Glacial and Interglacial Stages of the Pleistocene Epoch.

North America	Alpine Region	Years before Present
WISCONSINAN	Würm	—10,000 —75,000
Sangamon	Riss-Würm	—125,000
ILLINOISAN	Riss	265,000
Yarmouth	Mindel-Riss	—300,000
KANSAN	Mindel	—435,000
Aftonian	Günz-Mindel	500,000
NEBRASKAN	Günz	1800,000
Pre-Nebraskan	Pre-Günz	

In North America, the glacial stages are Nebraskan, Kansan, Illinoisan, and Wisconsinan. These terms correspond approximately to the Günz, Mindel, Riss, and Würm in Europe. The North American interglacial stages are Aftonian, Yarmouth, and Sanagamon.

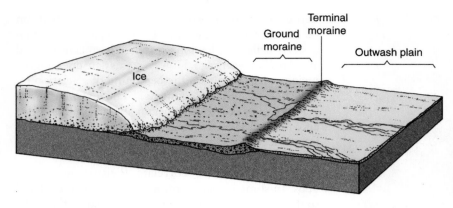

FIGURE 15-38 **Ground moraine, terminal moraine, and outwash plain.** ◢ *How might sediment of the outwash plain differ from sediment of the ground moraine?*

We think that the end of the Pleistocene/beginning of the Holocene was the time when ice sheets melted approximately to their present extent, with the associated rise in sea level. This would mean that the Pleistocene ended about 8000 years ago. However, there is some argument about the date of this boundary. Some geologists prefer to set the Pleistocene–Holocene boundary at the midpoint in warming of the oceans, in which case the Ice Age would have ended between 12,000 and 11,000 years ago.

Historical records and carbon-14 dating of old terminal moraines indicate that cold spells have recurred periodically into the Holocene. Climates are in delicate balance with atmospheric, geographic, and astronomic variables, and a change in any factor is likely to affect climate. For example, there is a good correlation between periods of minimum sunspot activity and episodes of colder conditions on Earth.

One well-documented period of cooler and drier conditions in the Northern Hemisphere occurred between A.D. 1540 and 1890, when temperatures were often 2° to 4°F cooler than today. These four centuries encompass the so-called **little ice age**. During the little ice age, frigid conditions periodically extended across most of Europe and in the United States down to the Carolinas. Arctic sea ice intruded far to the south of its present margin. The cold caused loss of harvests, with resulting famine, food riots, and warfare in Europe.

Stratigraphy of Terrestrial Pleistocene Deposits

Glacial deposits are difficult to correlate. They often consist of chaotic mixtures of coarse detritus completely lacking in fossil remains. Moving glaciers may carry away material, leaving only a bare surface of scoured and polished bedrock.

However, if the ice stalls for a time, a terminal moraine may be deposited in front of the glacier (Fig. 15-38). As the ice melts, a widespread ground moraine is left behind. Moraines may consist of **till**, an unsorted mixture of clay-sized to boulder-sized particles, dropped directly by the ice (Fig. 15-39). **Stratified drift**, on the other hand, has been washed by meltwater, is better sorted, and has many of the characteristics of stream deposits.

Whether the deposit is unsorted till or stratified drift, the stratigrapher is faced with similar-looking masses of debris that are woefully deficient in distinguishing characteristics. Several mutually substantiating criteria must be used in making correlations:

- The degree to which one sheet of glacial debris has been dissected by streams may indicate that it is older or younger than another sheet.
- The amount of chemical weathering may provide some estimate of the relative age of interglacial soils.
- Fossil pollen grains reflect fluctuations of climate and can be used to mark times of glacial advance and retreat.
- Varved clays deposited in lakes near the glaciers sometimes can be correlated to similar sediments of other lakes and may provide an estimate of the time required for deposition of the entire thickness of lake deposits.

FIGURE 15-39 **Glacial till in a valley glacier moraine, Kenai Peninsula, Alaska.**

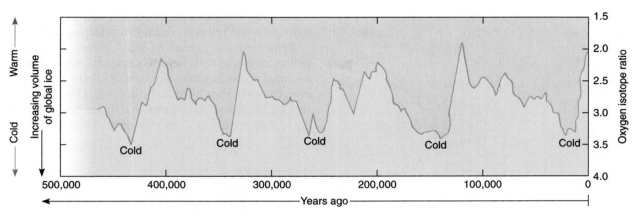

FIGURE 15-40 **Variations in global ice volume during the past 500,000 years.** Indirectly, these volume variations indicate paleotemperatures. Data are from radiometric dating and isotope measurements of Indian Ocean cores.

However, the most accurate means of dating and correlating Pleistocene sediments is to extract pieces of wood, bone, or peat and determine their age by radiocarbon dating techniques. Unfortunately, even this method has its limitations, the most significant of which is the relatively short half-life (5730 years) of carbon-14. This limits the method to materials less than about 100,000 years old and thus restricts the use of the carbon-14 technique to deposits of the most recent glacial stage.

Pleistocene Deep-Sea Sediments

Pleistocene seafloor sediments are easier to date and correlate than terrestrial deposits. The addition of sediment onto the ocean basins is relatively continuous, the deposits contain abundant fossil remains, and they can be dated by relating them to paleomagnetic data and to radiometric isotopes having short half-lives.

Continuous sections of deep-sea sediments are obtained by coring tools (basically, lengths of pipe driven into seafloor sediment). These provide cores more than 15 meters long. Oxygen isotope ratios from the shells of calcareous foraminifers in the cored sediment can be used to determine the volume of ocean water that was stored in glacial ice.

Oxygen-16 has a lower mass than oxygen-18. As a result, more oxygen-16 is evaporated and precipitated as snow. Hence, during an ice age when ice is accumulating globally, there will be greater percentages of oxygen-18 than oxygen-16 in seawater and in the shells of marine invertebrates. Plotted against depth in a deep-sea core, these isotope ratios indicate variations in ice volume and temperature with time (Fig. 15-40). Cooler conditions can then be correlated with lower sea level and marine regressions.

Some species of foraminifera are especially sensitive to temperature. For example, the tropical species

Globorotalia menardii (Fig. 15-41) lives in warmer water. The species is alternately present or absent within Pleistocene cores from the equatorial Atlantic. The absence of the species in part of a core is taken to indicate an episode of cooler climates and glaciation.

Another foraminifer, *Globorotalia truncatulinoides*, also indicates alternate cooling and warming of the oceans (Fig. 15-42). This species coils in a spiral with all the

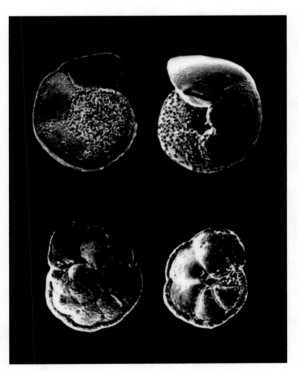

FIGURE 15-41 **Two well-known planktonic foraminifers used in correlating deep-sea sediments.** Top: both sides of *Globorotalia truncatulinoides*. Bottom: both sides of *Globorotalia menardii*. All images are magnified 50 times.

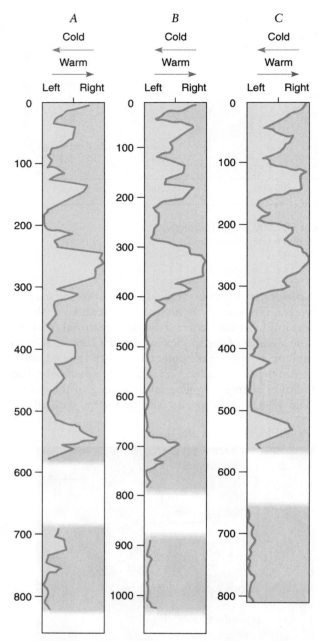

FIGURE 15-42 **Percentages of right-coiled and left-coiled *Globorotalia truncatulinoides*.** Data are from three cores (A,B,C) taken from Pleistocene sediments of the North Atlantic. In each column, the extreme left is 100% left-coiling cold-water forms, and the extreme right is 100% right-coiling warm-water forms. Numbers on the columns are deep-sea core depths in centimeters. White areas indicate that cores were not recovered.

whorls visible on one side, but only the last whorl visible on the other side (trochospirally coiled). Individuals that coil to the right dominate in warmer water, whereas left-coiled individuals prefer colder water. During glacial advances, when ocean temperatures dropped, the right-coiled populations in middle and low latitudes were replaced by left-coiled populations. The record of such coiling direction change is clear in the deep-sea cores.

Remanent magnetism in some cores of deep-sea sediments can be correlated to a magnetic reversal documented on land in volcanic rocks of known age. However, the analysis is likely to be in error if the sediment and its remanent magnetism has been disturbed by burrowing organisms (a process termed bioturbation).

Many Impacts of Pleistocene Glaciation

SHIFTING SEA LEVEL   During maximum glacial coverage, more than 40,000,000 cubic km of ice and snow buried the continents. Consider how much water would be required to create that much snow and ice. Removal of such an amount of water from the oceans strongly affected the Pleistocene environment.

One calculation is that sea level may have dropped at least 75 meters (225 feet) during maximum ice coverage. Extensive areas of the present continental shelves became exposed as dry land, many kilometers seaward. This exposed sea bottom became covered with forest and grasslands. The British Isles were joined to Europe, and a land bridge stretched from Siberia to Alaska.

With the drop in sea level, stream gradients increased and many streams cut deep canyons. During interglacial stages, marine waters returned to the low coastal areas, drowning the land flora and forcing terrestrial animals inland. With the rise in sea level, ocean water extended into the lower ends of stream valleys, forming estuaries like the Chesapeake Bay and Delaware Bay.

DEPRESSED CRUST REBOUNDS   The great weight of continental ice sheets depressed the crust in some regions by 200 to 300 meters below the preglacial position. With the melting of the last ice sheet, these formerly burdened areas of the crust gradually began to rebound, slowly returning to their former positions. The rebound is dramatically apparent in parts of the Baltic, the Arctic, and the Great Lakes region of North America, where former coastal features are now elevated high above sea level (Fig. 15-43).

REDIRECTING MIGHTY RIVERS   As the continental glaciers advanced, they obliterated old drainage channels and caused streams to erode new channels. These dislocations are especially evident in the north-central United States:

- Prior to the Ice Age, the northern segment of the Missouri River actually drained northward into the Hudson Bay.
- The northern part of the Ohio River flowed northeastward into the Gulf of St. Lawrence.

Those parts of the Missouri and Ohio rivers that once flowed toward the north were turned aside by the ice

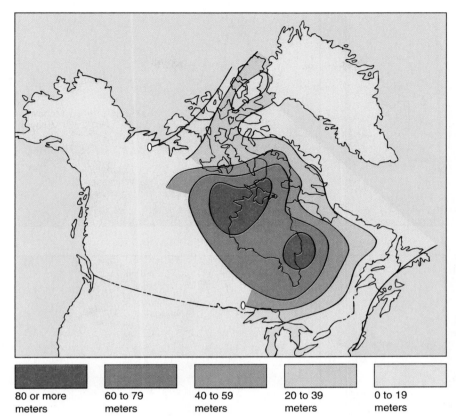

FIGURE 15-43 **Postglacial uplift (rebound).** The great weight of continental ice sheets depressed the crust in some regions by 200 to 300 meters. With the melting of the last ice sheet, these formerly burdened areas of the crust gradually began to rebound, slowly returning to their former positions. Uplift shown was determined by measuring the elevation of marine sediments 6000 years old.

80 or more meters

60 to 79 meters

40 to 59 meters

20 to 39 meters

0 to 19 meters

sheet and forced to flow along the fringe of the glacier until they found a southward outlet. Today's drainage of the Missouri and Ohio essentially follows the margin of the most southerly advance of the ice.

Niagara Falls, between Lake Erie and Lake Ontario, was formed when the retreating ice of the last glacial stage uncovered an escarpment formed by southwardly tilted resistant strata. Water from the Niagara River tumbled over the edge of the escarpment, which has the Silurian Lockport as its uppermost layer. Weak shales that lie beneath the Lockport are continuously undermined, causing southward retreat of the falls (see Fig. 10-14).

FORMING LAKES, GREAT AND SMALL Prior to the Pleistocene, there were no Great Lakes in North America. The present floors of these water bodies were lowlands. Glaciers moved into these lowlands, scouring them deeper. As the glaciers retreated, their meltwaters collected in the vacated depressions (Fig. 15-44).

South of the lake basins, huge ice blocks containing glacial debris melted and deposited their sediment to produce a hummocky topography composed of irregular hills and numerous small lakes called **kettles**. Somewhat later, sands deposited along the southern shores of Lake Michigan were carried eastward by gusty winds and deposited as dunes around the southeastern bend of the lake. Although many of these Pleis-

tocene dunes have been destroyed by developers, some remain and can be seen at Indiana Dunes National Lakeshore (Fig. 15-45).

Another large system of ice-dammed lakes covered a vast area of North Dakota, Minnesota, Manitoba, and Saskatchewan. The largest of these lakes has been named Lake Agassiz in honor of the great French naturalist who settled in North America and who had initially insisted on the existence of the Ice Age. Today, rich wheatlands extend across what was once the floor of the lake.

Other lakes developed during the Pleistocene, occupying basins that were not near ice sheets. These were formed as a consequence of the greatly increased precipitation and runoff that characterized regions south of the glaciers. Such water bodies are called **pluvial lakes** (Latin *pluvia* = "rain"). Pluvial lakes were particularly numerous in the northern part of the Basin and Range Province of North America, where faulting produced more than 140 lakes in downfaulted basins.

So-called pluvial (rainy) intervals, when lakes were most extensive, were generally synchronous with glacial stages. During interglacial stages, many lakes shrank to small saline remnants or even dried out completely. Lake Bonneville in Utah was one such lake. It once covered more than 50,000 square km and reached 300 meters deep in places. Parts of Lake Bonneville persist today as Utah's Great Salt Lake.

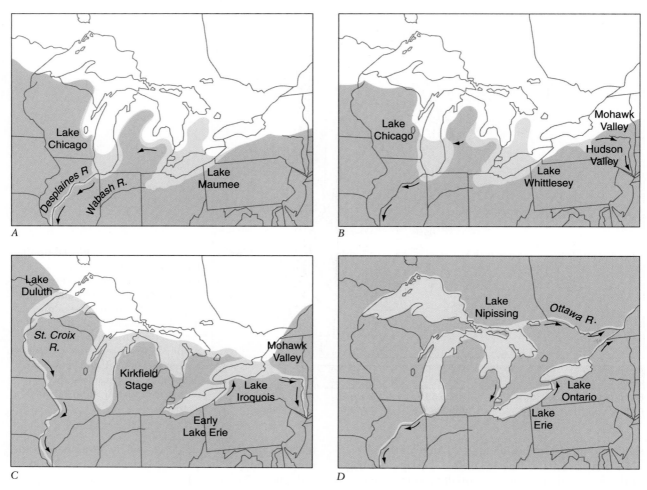

FIGURE 15-44 **How the Great Lakes formed.** (A) Continental glaciers moved into low-lying areas. (B) The glaciers scoured these depressions deeper. (C) As the glaciers retreated northward, meltwater filled the depressions. (D) The Great Lakes today.

WASHINGTON'S ALIEN LAND: THE CHANNELED SCAB-LANDS A spectacular Pleistocene lake event occurred about 18,000 years ago in the northwestern United States. A lobe of the southwardly advancing ice sheet blocked the Clark Fork River in northwestern Montana. The impounded water formed a long, narrow lake that cut diagonally across western Montana.

Called Lake Missoula, this vast impoundment contained an estimated 2000 cubic km of water, spread over 7800 square km (the size of Death Valley). With recession of the glacier, the ice dam broke repeatedly, sending unimaginable torrents of water across eastern Washington State. The extreme erosion left huge volumes of gravel, boulders, and cobbles and an alien landscape appropriately termed the **channeled scablands** (Figs. 15-46 and 15-47).

WINDBLOWN SEDIMENT The glacial conditions of the Pleistocene also affected soils. In many northern areas, fertile topsoil was stripped off bedrock and transported southward to form some of today's most productive farmlands.

Because of the flow of dense, cold air from the glaciers, winds were strong and persistent. Fine-grained glacial sediments that had been spread across outwash plains and floodplains were picked up, transported by the wind, and deposited as thick layers of windblown silt called **loess** ("luss"). Deposits of loess blanket large areas of the Missouri River valley (Fig. 15-48), central Europe, and northern China.

WHY DID EARTH'S SURFACE COOL?

First, note that we say Earth's *surface* cooled. It is a common misstatement to say "Earth cooled during the Ice Age," but it did not—the planet's massive body of rock was unaffected. The Ice Age was a surface phenomenon, chilling the atmosphere, oceans, lakes, rivers, soil, and living things. But the rocky Earth's internal temperatures continued as before.

FIGURE 15-45 **Indiana Dunes National Lakeshore.**
During the Pleistocene, sand carried to the southeastern
shores of Lake Michigan formed this beach and dunes.

FIGURE 15-47 **Conglomeratic scabland deposits,**
Washington State.

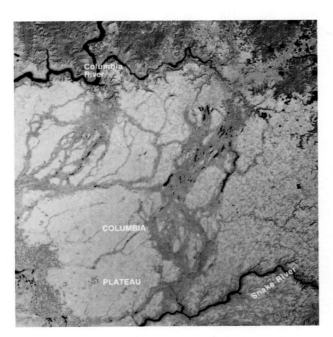

FIGURE 15-46 **Satellite image of Washington State's**
Channeled Scablands. Darker gray areas are channels
formed when floodwaters stripped away light-colored
surface sediment to expose the dark gray basaltic flows
beneath.

So why did the surface cool? World climates grew
progressively cooler from the Middle Cenozoic to the
Pleistocene, based on oxygen isotope data. But the culmination of this trend was not a single sudden plunge
into frigidity. It was an oscillation of glacial and interglacial stages. So, any hypothesis to explain the Ice Age
must consider:

1. The long-term decline in worldwide temperatures.

2. Repeated cycles of glacial and interglacial conditions.

Milankovitch Cycles

One widely accepted hypothesis for the temperature
fluctuations was developed by Yugoslavian mathematician Milutin Milankovitch (1879–1958). After 30 years
of careful study, he convincingly proposed that slow
cycles in Earth's movements and their influence on the
amount of solar radiation that Earth receives could account for Pleistocene glacial and interglacial stages.

Milankovitch based his calculations on three variables—Earth's *axial precession, orbital eccentricity,* and
axial tilt (Fig. 15-49):

- **Precession** is the way Earth's axis of rotation
  slowly "wobbles" once every 26,000 years. The
  effect is equivalent to tilting a rapidly spinning top
  (Fig. 15-49A). Like the variation in axial tilt, this
  varies the amount of solar radiation received toward Earth's poles.

- Earth's orbital path around the Sun varies about 2% over about 100,000 years. This variation (eccentricity) alternately brings Earth a bit closer to or farther away from the Sun, thus varying the solar radiation we receive.

- Earth tilts on its axis about 23.5°, which causes our seasons of the year. But Milankovitch recognized that this tilt actually varies slowly between about 21.5° and 24.5°, over a period of 41,000 years. This results in a corresponding variation in the seasonal length of days and in the amount of solar radiation received at higher latitudes (toward the poles).

According to Milankovitch's calculations, the effects of these three Earth motions combine. The combination periodically results in less solar radiation received by Earth's atmosphere, which causes cooling and recurrent glaciations.

The Milankovitch cycles correspond rather well to the timing of glaciation episodes over the past 100,000 years. The idea that Milankovitch cycles cause the alternation of glacial and interglacial stages is supported by oxygen isotope analyses of foraminifera shells (calcium carbonate) in deep-sea cores. These analyses confirm an interval of about 100,000 years between the times of coldest temperatures over the past 600,000 years.

However, if the Milankovitch cycles have been operating through most of the half-billion-year Phanerozoic eon, *why haven't there been Pleistocene-like glaciations continuously throughout geologic time?* Apparently, other factors are involved.

Earth's Albedo

Albedo is the specific fraction of solar energy reflected back into space from Earth, the moon, or other body. Earth is quite reflective, and regions overcast by cloud or covered by snow and ice are especially so. At present, Earth's average albedo is about 33%. If Earth's albedo increased, it could cause a lowering of atmospheric temperature and trigger an ice age.

Suppose that astronomic factors like those incorporated in the Milankovitch Effect brought cooler conditions. As these snow-covered areas increased, so too would the Earth's albedo, accelerating the rate of glacial growth. Just a 1% lessening of retained solar energy could bring a drop in average surface temperature up to 8°C. This would be sufficient to trigger a glacial build-up if ample precipitation were available over continental areas.

Other Factors

Other geologists speculate that absorption of solar energy was hindered by cloud cover, volcanic ash, and dust in the atmosphere or fluctuations in carbon dioxide. A decrease in carbon dioxide content would cause a corresponding decrease in the warmth-gathering greenhouse effect, for example. On the other hand, it is also possible that a buildup of carbon dioxide might trigger glaciation, for as warming occurred, there might be more rapid evaporation and an increase in highly reflective cloud cover.

Other hypotheses for Ice Age origin stress the need for preservation of ample snowfall on suitably located continental areas. Ample snowfall may have been produced by the northward deflection of the Gulf Stream after its westward movement was blocked by the formation of the Isthmus of Panama about 3.5 million years ago. A "conveyor belt" of precipitation resulted

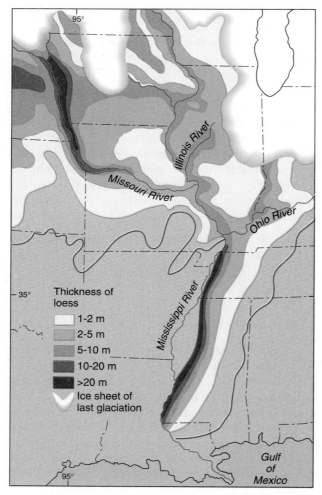

FIGURE 15-48 **Loess blankets a large area of the central United States.** Loess was deposited by winds beyond the limits of the Pleistocene ice sheets. Note how the thickness of the deposits thins downwind from the Mississippi River. This indicates that prevailing winds blew from the west toward the east. The red line indicates the downwind pinchout of the loess deposits.

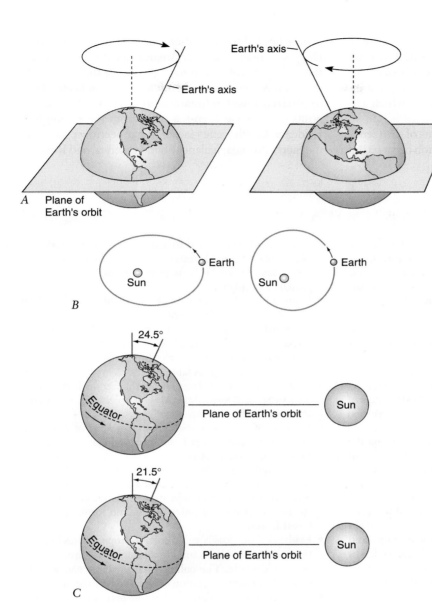

FIGURE 15-49 **Three astronomic cycles affecting global climate.**
(A) *Precession*: two positions in precession (wobble) of Earth's axis. At certain times in the precession, polar regions receive less of the sun's radiation, causing colder conditions. (B) *Orbital eccentricity*: when Earth's orbital path is more eccentric (less circular), the planet is farther from the sun in summer, so existing ice caps experience little melting. (C) *Axial tilt*: Earth's axial tilt varies between about 21.5° and 24.5°. It is presently 23.5°. When the tilt is smaller, polar regions receive less solar energy, enhancing glacier-producing cold.

when that warm, moist air became shunted northward. It fell on land areas as snow, accumulating to form continental glaciers.

Scientists are not ready to formulate a complete, unified theory for the cause of the Pleistocene's multiple glaciations. But it appears that they may be very near. Few deny that the Milankovitch cycles, albedo, and pole positions have played roles in creating Pleistocene climatic conditions. The problem is to determine the relative importance of each factor and how they relate to one another.

If Pleistocene-like climatic cycles are fundamentally the result of changes in Earth's orbital and axial geometry, it is possible to speculate about the future climate. Calculations indicate that the long-term climatic trend over the next 20,000 years is toward extensive Northern Hemisphere glaciation. But what effect will human

activities, such as the burning of coal, oil, and gas, and the associated increase of greenhouse gases have on this trend?

CENOZOIC CLIMATES

The worldwide cooling that culminated in the Pleistocene ice age actually began during the Early Cenozoic. However, the cooling was not entirely uniform. There were important warming trends during the late Paleocene and Eocene, as indicated by fossils of palm trees and crocodiles in Minnesota, Germany, and near London. Trees that are today characteristic of moist temperature zones thrived in Alaska, Norway, and Greenland. Coral reefs grew in regions 10° to 20° north of their present optimum habitat.

Cooling resumed during the Oligocene, and coral reefs slowly retreated toward the equator. On land, temperate and tropical forests also were displaced toward lower latitudes. Another pulse of warmer conditions occurred during the Miocene, after which climates grew steadily cooler again.

Although the most extensive glaciations of the Cenozoic were those of the Pleistocene, glaciation occurred on a more limited scale at other times during the Cenozoic. Deep-sea cores taken near Antarctica indicate that an ice sheet already had begun to form during the Eocene. In addition, Miocene glacial conditions are indicated both by glacial deposits in Antarctica and ice-rafted glacial deposits in Miocene deep-sea cores from the Bering Sea. Evidence for Pliocene glaciation has been recognized in the Sierra Nevada, Iceland, South America, and Russia.

SUMMARY

- Our world today was shaped during the Cenozoic era. Continents had drifted to their present locations, the Alpine-Himalayan mountain system formed, Australia separated from Antarctica, and uplifted regions of the American west were sculpted by erosion into their present form.
- The Appalachian Mountains were subjected to repeated cycles of isostatic uplift and erosion, producing their present topographic features.
- Along the Atlantic and Gulf coastal plains, repeated cycles of marine transgression reworked clastic sediments brought to the coast by streams. More than 9000 meters of sediment were deposited in the subsiding Gulf of Mexico basin.
- Continents stood at relatively high elevations during the Cenozoic, so that epicontinental and shelf seas were of limited extent. Non-marine deposition was dominant in western North America.
- In the Cordilleran region of the United States, earlier compressional forces ceased and were replaced by vertical crustal adjustments. Erosional debris worn from highland areas first filled the basins between ranges and then spread eastward to form the broad clastic wedge of sediments on the Great Plains. Lakes formed in intermountain lowlands sometimes served as collecting sites for oil shales.
- Crustal movements beginning in the early Neogene caused the elevation of the Sierra Nevada along a great fault producing the Basin and Range Province.

- Volcanic activity related to a subduction zone marks the Neogene of the northwestern U.S. Pacific coast. Cascades volcanoes are fed by lavas generated by the subducting Juan de Fuca plate.
- South of the Cascades, the Farallon plate and its East Pacific rise spreading center were overridden by the westward-moving North American plate. As a result, the formerly convergent plate boundary was transformed into a shear boundary, characterized by strike-slip motion, as seen in the San Andreas fault.
- Collision of tectonic plates during the Cenozoic closed the Tethys Sea and formed the Alps, Carpathians, Apennines, Pyrenees, and Himalayas.
- Advance of glaciers across more than 30% of Northern Hemisphere continents is the predominant event of the Pleistocene epoch. Among its effects were changes in stream behavior, alternating inundation/withdrawal of seas across coastal plains (related to oscillations in sea level), deposition of till and stratified drift, dislocation of streams, formation of pluvial lakes, and formation of the Great Lakes.
- Evidence from fossils and oxygen isotope studies indicate that Earth was on a fluctuating cooling trend through most of the Cenozoic. The culmination of that trend was the onset of extensive glaciation during the Pleistocene.
- The eccentricity of Earth's orbit, the precession or wobble of Earth's axis, and tilt of Earth's axis influence the amount of solar energy reaching the Earth. When extremes of these factors coincide, they may cause an ice age.

KEY TERMS

albedo, p. 482

channeled scablands, p. 480

discoasters, p. 475

kettle, p. 479

little ice age, p. 476

loess, p. 480

Neogene, p. 452

oil shale, p. 458

Paleogene, p. 452

Pleistocene Ice Age, pp. 473–483

pluvial lake, p. 479

precession, p. 481

stratified drift, p. 476

Tethys Sea, p. 470

till, p. 476

QUESTIONS FOR REVIEW AND DISCUSSION

1. Describe the manner in which each of the following major physiographic features developed:
   a. Mountains of the Basin and Range
   b. Great Plains
   c. Columbia and Snake River Plateau
   d. Red Sea
   e. Teton Range
   f. Cascade Range

2. What Eurasian mountain ranges resulted from the compression and upheaval of large areas of the Tethys Sea?

3. What epochs of the Cenozoic Era are included in the Paleogene Period and the Neogene Period?

4. What is the economic importance of the Green River Formation? the Fort Union Formation?

5. What is the origin of Lake Nyasa and Lake Tanganyika in eastern Africa?

6. What is the origin of the Great Salt Lake in Utah? Why is it so salty?

7. In a 30-meter-long deep-sea core that penetrated most of the Pleistocene section, how might you differentiate sediment that had been deposited during a glacial stage from that of an interglacial stage?

8. What land bridge, important in the migration of Ice Age mammals, resulted from the lowering of sea level during glacial stages?

9. What is the explanation for the gradual rise in land elevations that has occurred within historic time around Hudson Bay, the Great Lakes, and the Baltic Sea?

10. What changes in movement of tectonic plates along the western border of the United States were responsible for the development of the San Andreas fault?

11. What is the evidence that the Gulf of Mexico subsided simultaneously with deposition during the Cenozoic?

12. Discuss the advantages and disadvantages associated with the exploitation of Green River oil shales for their content of hydrocarbons.

13. Discuss conditions on Earth that might result in increased albedo. What are the possibilities that such conditions existed at the beginning of the Pleistocene?

14. What conditions during the Pleistocene favored the formation of extensive loess deposits? Where did this sediment come from? Why does loess consist of only very small grains (mostly 1/16 or 1/32 mm in diameter)?

15. The percentage of the isotope oxygen-18 relative to the percentage of oxygen-16 in the calcium carbonate shells of foraminifers increases during the colder glacial stages of the Pleistocene. Why?

WEB SITES

The Earth Through Time Student Companion Web Site (www.wiley.com/college/levin) has online resources to help you expand your understanding of the topics in this chapter. Visit the Web Site to access the following:

1. Illustrated course notes covering key concepts in each chapter;

2. Online quizzes that provide immediate feedback;

3. Links to chapter-specific topics on the web;

4. Science news updates relating to recent developments in Historical Geology;

5. Web inquiry activities for further exploration;

6. A glossary of terms;

7. A Student Union with links to topics such as study skills, writing and grammar, and citing electronic information.

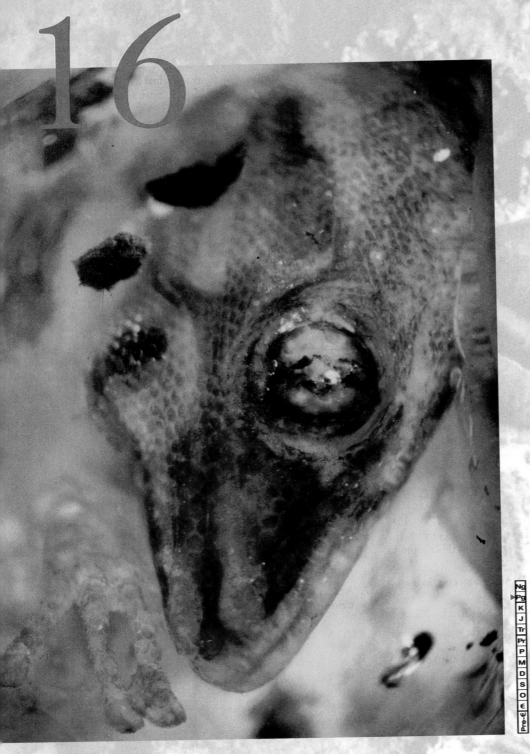

Ng
Pg
K
J
Tr
Pr
P
M
D
S
O
Є
Pre

Cenozoic lizard preserved in amber. *This specimen, 30–40 million years old, is from the Dominican Republic, La Toca Mine.*

Life of the Cenozoic

At the dawn of the Cenozoic Era the dinosaurs were extinct, the pterosaurs had disappeared, and the great marine reptiles were gone. Now the way was open for the mammals to begin their conquest of the world.
—Carl Dunbar, paleontologist, 1949

Key Chapter Concepts

- Modern plants and animals evolved during the Cenozoic Era.

- Expanding grasslands during the Cenozoic strongly influenced the evolution of plant-eating mammals as they adapted to this food source.

- Because continents were widely separated during the Cenozoic, distinctive faunas were able to evolve on each landmass, promoting greater global biological diversity.

- Corals constructed extensive reefs in the shallow, warm Cenozoic seas. In those seas, mollusks similar to those living today flourished, along with bryozoans and echinoderms. Sea urchins became particularly abundant.

- Abundant marine planktonic organisms of the Cenozoic included diatoms, coccolithophorids, radiolaria, and foraminifera.

- Bony fishes called teleosts achieved their highest level of evolution and diversity.

- Amphibians of the Cenozoic were small-bodied, smooth-skinned forms including frogs, salamanders, newts, and caecilians.

- Turtles, lizards, snakes, crocodilians, and the tuatara of New Zealand survived the mass extinction at the end of the Cretaceous and became common during the Cenozoic.

- Most modern families of birds had evolved by Neogene time.

- Early Paleogene mammals were small and unremarkable. However, they quickly expanded to dominate the Cenozoic world.

- Large plant-eating mammals populated Cenozoic prairies. We divide them into two groups: even-toed artiodactyls (cattle, sheep, goats, deer, camels) and odd-toed perissodactyls (horses, rhinos, tapirs).

- Many large plant-eating mammals became extinct at the end of the Pleistocene, possibly because of excessive harvesting by human hunters.

- We know more about life of the Cenozoic than earlier eras because Cenozoic fossils are stratigraphically topmost and have had fewer years in

O U T L I N E
- ▶ GRASSLANDS EXPAND, MAMMALS RESPOND
- ▶ PLANKTON
- ▶ MARINE INVERTEBRATES
- ▶ VERTEBRATES
- ▶ MAMMALS
- ▶ MONOTREMES
- ▶ MARSUPIALS
- ▶ PLACENTAL MAMMALS
- ▶ DEMISE OF THE PLEISTOCENE GIANTS
- ▶ SUMMARY
- ▶ KEY TERMS
- ▶ QUESTIONS FOR REVIEW AND DISCUSSION
- ▶ WEB SITES
- ▶ BOX 16-1 ENRICHMENT

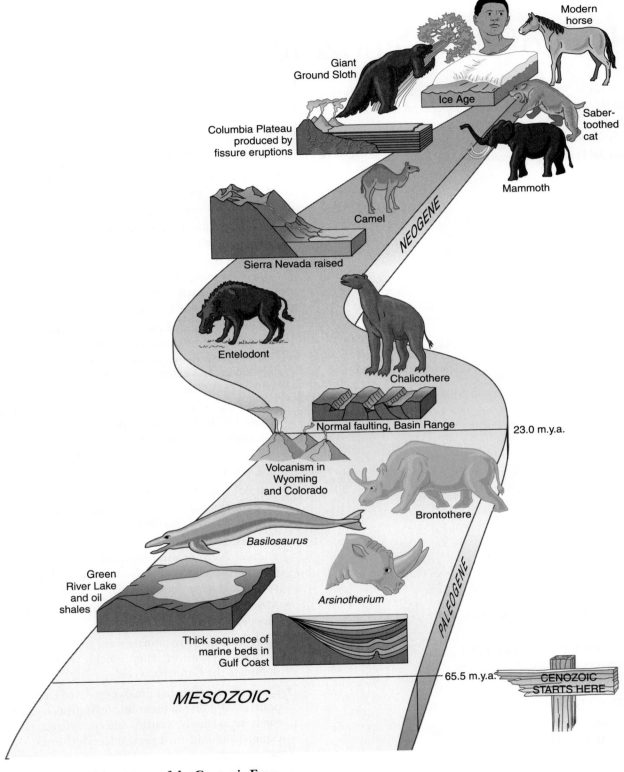

FIGURE 16-1 **Major events of the Cenozoic Era.**

which to be obliterated. They are better preserved and more accessible, and they more closely resemble life today. With these advantages, we are better able to see the effects of environmental and geographic changes on the evolution of Cenozoic animals and plants.

Separation of continents characterizes the Cenozoic. This isolation stimulated biologic diversity and resulted in distinctive faunal radiations on the separated landmasses, as well as in isolated marine basins (Fig. 16-1). In the seas flourished bivalves, gastropods, crustaceans, echinoids, scleractinid corals, bony fishes, and marine mammals.

On land, major mammal groups that have present-day relatives expanded. These included rodents, bats, carnivores, elephants, bison, camels, horses, and rabbits.

The Southern Hemisphere's continents—South America, Australia, and Antarctica—were separated from North America and Eurasia during most of the Cenozoic. As a result, a distinctive assemblage of mammals evolved, showing convergent evolution with Northern Hemisphere species.

When the Panamanian land bridge developed, many species of South American marsupials were driven to extinction by migrants from the north. To a lesser extent, some South American mammals migrated to North America and caused extinctions there.

Among the many evolutionary developments of the Cenozoic, none intrigue us more than the changes in primates. By the Neogene, primates had produced species believed to be ancestors of humans. It was during the Pleistocene that our own species appeared, *Homo sapiens*.

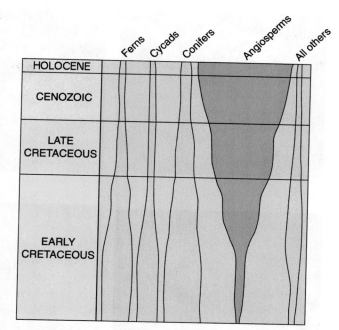

FIGURE 16-2 **Relative proportions of plant genera, Cretaceous to Holocene.**

GRASSLANDS EXPAND, MAMMALS RESPOND

Flowering plants (angiosperms) were the most recent plant group to evolve, but they now are the most widespread of all vascular land plants. Angiosperm floras did not begin to dominate until the mid-Cretaceous (Fig. 16-2). There were few spectacular floral innovations during the Cenozoic. Rather, this was a time of steady development toward today's complex plant populations. The Miocene is particularly noteworthy as the epoch during which grasses expanded and prairies spread widely over the lands. In response to the spread of grasses, grazing mammals began their remarkable evolution.

Many evolutionary modifications among herbivorous mammals are related to the development of extensive grasslands. One example is the change in their teeth. Grasses are abrasive because they grow close to the ground, and thus become coated with fine particles

of soil. They also contain siliceous secretions. To compensate for tooth wear, the major groups of herbivores evolved high-crowned cheek teeth that continue to grow at the roots during part of the animals' lives.

Enamel, the most resistant tooth material, became folded, so that as the tooth wore, a complex system of enamel ridges formed on the grinding surface. In general, incisors gradually aligned into a curved arc for nipping and chopping grasses. And to provide space for these high-crowned teeth, the overall length of the face in front of the eyes increased.

In the grassy plains, it was difficult to hide from predators. As a result, many herbivores evolved modifications that permitted speedy flight from their enemies. Limb and foot bones were lengthened, strengthened, and redesigned by forces of selection to prevent strain-producing rotation and permit rapid fore-and-aft motion. To achieve greater speed, the ankle was elevated, and, like sprinters, animals began to run on their toes. In many mammals, side toes were gradually lost. Hoofs developed as a unique adaptation for protecting the toe bones while the animals ran across hard prairie sod.

Some herbivores evolved a four-chambered stomach in response to selection pressures, which favored improved digestive mechanisms for breaking down tough grasses. The response of mammals to the spread of grasses provides a good example of how the environment influences evolution.

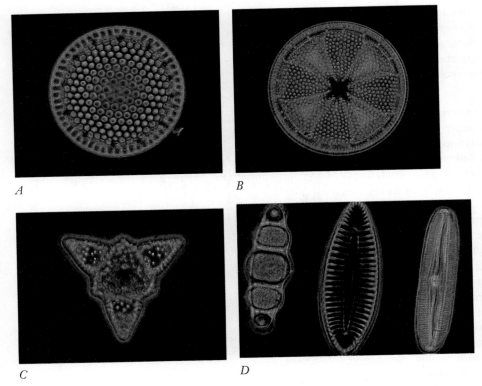

FIGURE 16-3 **Cenozoic diatoms showing various frustule shapes—discoidal, triangular, and spindle.**
(A) *Arachnoidiscus.*
(B) *Actinoptychus.*
(C) *Triceratum.* (D) *Biddulphia, Surirella,* and *Pinnularia.*
Images are magnified 50 times.
☙ *What is the composition of frustules of diatoms?*

PLANKTON

As noted in Chapter 14, entire families of planktonic organisms underwent extinction at the end of the Cretaceous Period. Only a few species in each major group survived and continued into the Paleogene. However, the survivors were able to take advantage of decreased competitive pressures and rapidly diversified.

In general, peaks in species diversity occurred in the Eocene and Miocene. A decrease in diversity has been recorded in the intervening Oligocene. The most abundant populations of Cenozoic marine phytoplankton were diatoms (Fig. 16-3), dinoflagellates (Fig. 16-4), and coccolithophorids (Fig. 16-5).

Among the zooplankton, Cenozoic seas contained dense populations of foraminifera and radiolaria (Fig. 16-6). The foraminifera included large numbers of benthonic as well as planktonic forms (Fig. 16-7).

Many benthonic foraminifera live in particular depths of water, so they are useful to geologists as indicators of ancient depths. They are called paleobathymetric indicators (Fig. 16-8).

Among the larger genera, foraminifera the size and shape of coins thrived in the clear warm waters of the

FIGURE 16-4 **Dinoflagellate viewed with scanning electron microscope.** From Paleocene sedimentary rocks of Alabama. Image is magnified 450 times.

FIGURE 16-5 **Coccolith of the coccolithophoroid *Coccolithus.*** Viewed with transmission electron microscope. Maximum diameter is 16 microns.

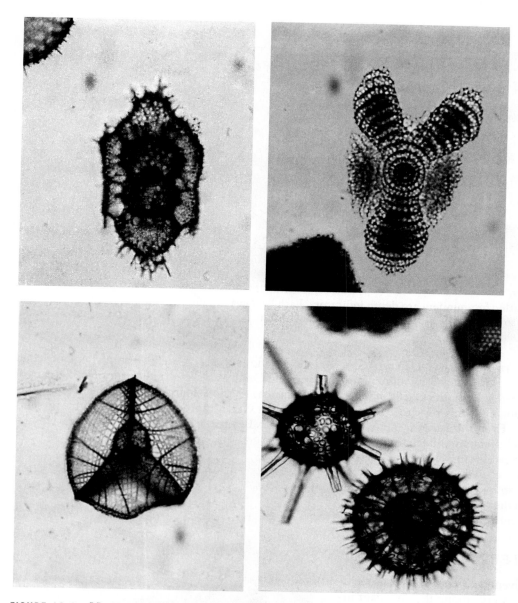

FIGURE 16-6   **Neogene radiolarians from the tropical Pacific Ocean.** Image is magnified 100 times.

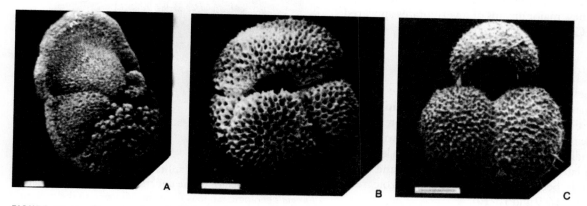

FIGURE 16-7   **Present-day planktonic foraminifera.** (A) *Globorotalia tumida*. (B) *Globigerinoides conglobatus*. (C) *Globigerinoides rubra*.

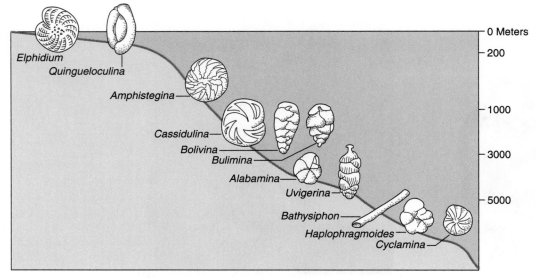

FIGURE 16-8 **Many benthic foraminifera are depth indicators.** Each lives in a specific depth range in the ocean, so they can be used to reveal ancient ocean depths.

U.S. Gulf Coast, western Atlantic, and Tethys Sea. The tests of these organisms, known as nummulitic foraminifera, accumulated to form thick beds of limestone. Ancient Egyptians used this rock to construct the Pyramids of Gizeh. Even the famous Sphinx was carved from a large block of nummulitic limestone.

Because of their incredible numbers and variety, Cenozoic foraminifera are used extensively by geologists for correlating strata penetrated during the drilling for oil. (60% of the world's total oil production comes from strata of Cenozoic age.)

MARINE INVERTEBRATES

Modern types of corals, plus diverse bryozoans, echinoids, and crustaceans, were among the abundant invertebrates of Cenozoic seas. The fauna acquired a decidedly modern appearance. Formerly successful groups were gone, such as ammonites and rudistid bivalves. No new major groups of invertebrates appeared during the Cenozoic.

Corals and South Pacific Atolls

Modern reef corals belong to an order of the phylum Cnidaria named the Scleractinia. **Scleractinids** began their evolution during the Mesozoic, but late in that era, their extent was limited by competition with reef-building rudistid bivalves. The rudistids were killed off during the late Cretaceous mass extinction, and scleractinid corals became the dominant Cenozoic reef-builders (Figs. 16-9 and 16-10).

Scleractinid corals were most extensively developed in parts of the Tethyan belt, West Indies, Caribbean, and

Indo-Pacific regions. Careful comparison of coral species in Cenozoic rocks on either side of the Isthmus of Panama has given geologists clues to when the Atlantic and Pacific Oceans were connected across this present-day barrier. Before the land barrier existed, coral popula-

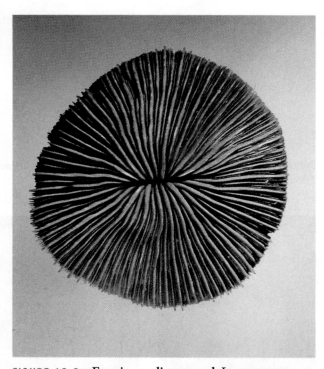

FIGURE 16-9 *Fungia,* **a solitary coral.** Its name comes from its resemblance to the underside of a mushroom cap. *Fungia's* geologic range is from the Miocene to the present, and it is common today in seas of the Indo-Pacific region. This specimen is about 9.0 cm (3.5 inches) in diameter.

FIGURE 16-10 **Living coral *Montastrea cavernosa*.** Its polyps are extended for feeding.

tions on the Atlantic side were similar to those on the Pacific side. But formation of the Isthmus of Panama prevented genetic interchange, so species became different.

Continuous formation of reef limestones occurred throughout the Cenozoic in the Pacific Ocean, progressively forming fringing reefs, barrier reefs, and atolls. **Atolls** are ringlike coral reefs that grow around a central lagoon. In 1842, Charles Darwin wrote an explanation of how atolls form around island volcanoes, summarized in Figure 16-11.

Darwin wrote that volcanic islands gradually subside because of their great weight. Corals growing around the fringe of the slowly sinking island grow upward at the same pace to maintain their living depth near the ocean's surface. Eventually, the volcano becomes submerged, but the encircling, living reef continues to grow upward.

Drilling into atolls has validated Darwin's hypothesis. A boring drilled at Eniwetok Island, an atoll in the Marshall Islands of the western Pacific, encountered the basaltic summit of the volcano at about 1200 meters depth. Eocene reefs lie directly above these igneous rocks.

It is likely that subsidence associated with some atolls is related to seafloor spreading rather than to the weight of the volcanic mass, as Darwin thought. Spreading begins along the crest of a midoceanic ridge, and as the plates move away from the ridge, they subside very slowly due to thermal contraction and isostatic adjustment. Thus, a volcano formed at the crest of

the midoceanic ridge is carried down the slight incline of the ocean plate until it is completely submerged. Coral growth that is able to keep pace with the subsidence will form **barrier reefs** and atolls.

Alternatively, scleractinid corals might build **fringing reefs** around the perimeters of volcanic islands that were erosionally truncated during a glacial period of low sea level. When sea level subsequently rises (as it did when the great Pleistocene ice sheets began to melt), the corals build the reef upward to stay at their optimum shallow living depth. Thus, some atolls may result from island subsidence, whereas others may be triggered by a rise in sea level. Still others probably result from a combination of sea-level rise and island subsidence.

Mollusks

Bivalves and gastropods are the dominant groups of Cenozoic mollusks (Fig. 16-12). Their range of adaptation is amazing. Cenozoic mollusk shells look very much like those found along coastlines today. Clams, oysters, mussels, and scallops were particularly abundant bivalves.

The climax of the evolution of chambered cephalopods passed with the demise of the ammonites. However, nautiloids similar to the modern chambered nautilus lived in Cenozoic seas, as they do today. Shell-less cephalopods, such as squid, octopi, and cuttlefish also were well represented in the marine environment, although their fossil record is sparse.

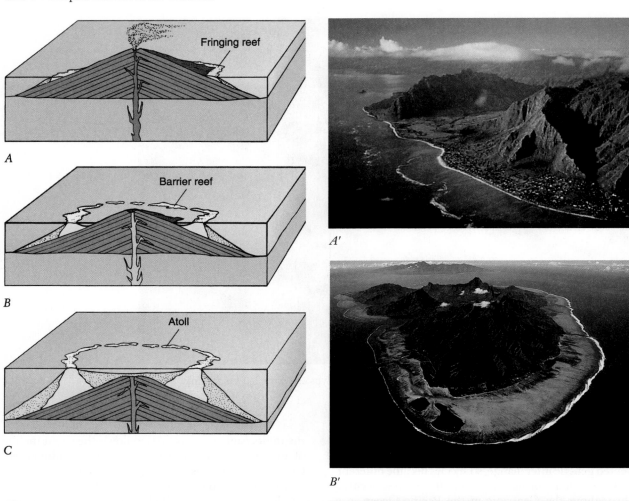

A

B

C

A'

B'

C'

FIGURE 16-11 **Stages in the growth of a fringing reef, barrier reef, and atoll.** (A) Initial stage of reef development around the margin of a volcanic island. (A') Fringing reef on the island of Oahu in the Hawaiian Islands. (B) Island begins to subside and corals build upward in order to stay in their optimum shallow-water life zone. The result is a barrier reef backed by lagoons. (B') Barrier reef encircling Moorea Island in the Society Islands. (C) With continued subsidence, center of island is completely flooded, leaving an irregular circle of reefs, creating an atoll. (C') Small atoll in the Society Islands. Plant-covered sandy areas have formed in the interior of this atoll.

Echinoderms, Bryozoans, Brachiopods, and Arthropoda

Many other invertebrates have continued successfully through the Cenozoic. Echinoderms were particularly prolific, mainly free-moving types like echinoids. Among the echinoids, sea biscuits (Fig. 16-13) and sand dollars appeared in the early Paleogene and spread rapidly. Their attractive flat shells (tests) are prized by shell collectors.

Among the arthropods, modern crustaceans became firmly established both in the salty oceans and in fresh water (Fig. 16-14).

FIGURE 16-12 **Common Cenozoic bivalves and gastropods.** Bivalves: (A) *Cardium*. (B) *Pecten*. (C) *Ostrea*. (D) *Mya*. Gastropods: (E) *Turritella*. (F) *Crepidula*—a small marine snail having a horizontal shelf that extends across the posterior of its shell.

FIGURE 16-13 **Cenozoic echinoid *Clypeaster*, commonly called a sea biscuit.** The view at the left shows the oral surface with its large central opening for the mouth and peripheral anal opening.

Brachiopods declined in abundance and diversity during the Cenozoic. Fewer than 60 genera survive today. They consist mostly of terebratulids (Fig. 16-15A), rhynchonellids, and such inarticulate brachiopods as *Lingula* (Fig. 16-15B). Bryozoa continued successfully into the Cenozoic. Many constructed attractive frond-like or laminar colonies attached to skeletal structures made by corals.

On land, insects thrived during the Cenozoic, just as they do today. More than 5000 fossil species have been described from Cenozoic strata. Wyoming's Eocene Green River shales contain well-preserved insect fos-

A

B

FIGURE 16-15 **Articulate and inarticulate brachiopods.** (A) Cluster of present-day articulate terebratulid brachiopods, *Terebratulina septentrionalis*. (B) *Lingula*, a persistent primitive inarticulate brachiopod with a thin shell of proteinaceous material and a long, fleshy, muscular pedicle.

sils, and splendidly preserved insects are recovered from Cenozoic amber deposits (see Fig. 6–7).

One of the world's best locations for fossil insects is the Florissant Fossil Beds National Monument, west of Colorado Springs, Colorado. Over 100,000 insect specimens and hundreds of spiders have been collected at this site, representing 1100 species. The fossils occur in finely textured volcanic ash. Particles of ash settled onto the floor of Oligocene Lake Florissant, burying insects and plant debris. Even delicate features like the veinlets of insect wings are preserved.

▼ **VERTEBRATES**

Fishes

Fishes with skeletons composed almost entirely of bone (instead of cartilage) are called **teleosts**. These advanced bony fishes were the dominant marine verte-

FIGURE 16-14 **Freshwater shrimp *Bechleja rostrata*, from Eocene Green River Formation.**

brates of the Cenozoic. Their enormous range of adaptive radiation produced such varied forms as perch, bass, snappers, seahorses, sailfishes, barracudas, swordfishes, flounders, flying fish, and numerous others. Wyoming's Green River Formation is well-known for beautifully preserved Eocene teleosts (see Fig. 15-10). Eocene strata at Monte Bolca in Italy also yield well-preserved teleost fossils (Fig. 16-16).

In addition to the teleost fishes, cartilaginous sharks were at least as common in the Cenozoic as they are today. Some of these sharks exceeded 12 meters (40 feet) in length and had teeth as big as a man's hand.

Amphibians

Throughout the Cenozoic, amphibians resembled modern forms. They were small-bodied, smooth-skinned creatures not at all like their often large Paleozoic ancestors. Relatively abundant were frogs, toads, newts, salamanders, and caecilians (strange-looking amphibians that resemble earthworms). The first frogs appeared during the Triassic, and by Jurassic time they were already completely modern in appearance. They have continued almost unchanged for more than 200 million years.

FIGURE 16-16 **Eocene flat-bodied fish.** *Gasteronomus* is from the Mount Bolla fossil locality near Verona, Italy. Specimen is 42 cm long.

Reptiles

By the beginning of the Cenozoic, the dinosaurs and most marine and flying reptiles had disappeared. The reptiles that survived and continue to the present include turtles, lizards, and snakes (Fig. 16-17). Another survivor is the tuatara, which inhabits islands off the coast of New Zealand (Fig. 16-18). The tuatara, formally known as *Sphenadon*, resembles a large lizard. It is the sole survivor of a group of ancient reptiles known as rhynchocephalians that evolved and diversified during the Triassic.

We can trace the lineage of Cenozoic turtles back to the Late Permian. The shell (carapace), an adaptation for protection that has been effective for more than 250 million years, is the distinctive trait of turtles. The ribs in turtles form broad, flat supports for the carapace. On the underside, a growth called the plastron provides additional protection. Turtles are toothless, with a beak that is used effectively in slicing through plants or flesh.

Both lizards and snakes belong to a group of reptiles known as the squamates. **Squamates** are by far the most varied and numerous of living reptiles. Lizards are the ancestors of snakes. In fact, snakes are essentially modified lizards in which the limbs are lost, the skull bones are modified into a highly flexible and mobile structure for swallowing prey, and the vertebrae

and ribs are greatly multiplied. Some snakes retain vestiges of rear limb and pelvic bones.

The evolution of limbless reptiles may have been driven by the habitat in which these animals lived. For example, a small tetrapod living in dense vegetation might have been impeded by having legs entangled by plants. Also, for vertebrates seeking refuge or food in small crevices, there is little space in which to use limbs.

Fossil snakes have been found in rocks as old as Early Cretaceous. They began to diversify in the Early Miocene, when poisonous snakes evolved. Specialized teeth or fangs are used by poisonous snakes to inject venom into their prey. One type of poison (neurotoxin) affects parts of the nervous system that control breathing and heart beat, whereas a second type (hemotoxin) destroys red blood cells and small blood vessels.

Crocodilians (crocodiles, alligators, gavials, and caimans) are archosaurians that began their evolution during the Triassic and were contemporaries of the dinosaurs throughout the Mesozoic. Modern crocodilians include the broad-snouted alligators, the narrow-snouted crocodiles, and the very narrow-snouted gavials. The nostrils of crocodilians are at the ends of their long snouts, so they are able to breathe by simply keeping the end of the snout above water.

A

B

C

D

FIGURE 16-17 **Surviving reptiles.** (A) Monitor lizard. (B) Three-toed box turtle.
(C) *Alligator mississippiensis.* (D) Hognose snakes emerging from their shells.

FIGURE 16-18 **Tuatara (*Sphenodon*), survivor of the late
Cretaceous extinction.** It still lives today in New Zealand.

Birds

The Cenozoic fossil record of birds is generally poor
because they are rarely preserved. However, the fragmentary record indicates that birds have been essentially modern in basic skeletal structure since the beginning of the Cenozoic. Distinctive skeletal features of
birds include fusion of bones of the "hand" to help support the wing, development of a vertical plate or keel on
the sternum for attachment of the large muscles leading
from the breast to the wings, and fusion of the pelvic
girdle and vertebrae to provide rigidity during flight.

Other characteristics include feathers, light and
porous bones, jaws in the form of a toothless horny
beak, a four-chambered heart, and constant body temperature. The extraordinary adaptive radiation of birds
has produced songbirds (such as robins), upland birds
(such as pheasants), forest birds (such as owls), oceanic

FIGURE 16-19 *Diatryma*, a large, ground-dwelling, predatory bird, evolved during the early Cenozoic. *Diatryma* stood 2 meters (6 feet) tall and had a huge, powerful beak.

birds (such as albatrosses), wading birds (such as plovers), flightless aquatic birds (such as penguins), and flightless land birds (such as ostriches).

The fossil record for large flightless terrestrial birds is better than for small flying varieties, because they generally have more robust skeletons. *Diatryma* (Fig. 16-19) from the Eocene of North America, was 2 meters tall and weighed nearly 300 pounds. It had massive legs, vicious claws, and a formidable beak.

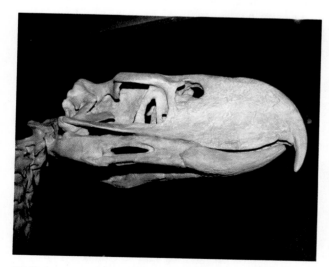

FIGURE 16-20 Skull of *Andalgalornis*, a giant flightless predatory bird from Pliocene rocks of Argentina.

Another large, flightless bird was *Andalgalornis* (Fig. 16-20), whose head was the size of a horse's. Some flightless birds were vegetarians. Among these, huge moas lived until relatively recent time in New Zealand. Some moas were over 3 meters tall and laid eggs with a 2-gallon capacity. Prior to settlement of New Zealand by Europeans, the moas were exterminated by native Maori hunters. The African ostrich, the South American rhea, and the emus and cassowaries of Australia are surviving flightless land birds.

Perhaps the most famous of Cenozoic flightless birds was the dodo, which lived on the island of Mauritius east of Madagascar until about A.D. 1700. At about that time, the dodo was either exterminated by sailors searching for provisions or was unable to survive the influence of animals brought to Mauritius by Europeans.

MAMMALS

During the Cenozoic, mammals came to dominate Earth much as reptiles had reigned during the Mesozoic. The evolution of mammalian traits had begun among the therapsids of the Permo-Triassic, 250 million years ago. For example, the Karoo beds of Africa contain bones of near-mammals that almost had made the transition from reptile to mammal.

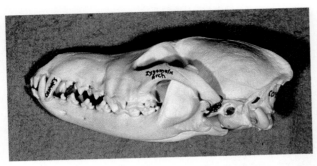

FIGURE 16-21 **Modern coyote skull.** One bone on each side, the dentary, forms the lower jaw. Like most mammals, its teeth are varied in shape: incisors at front of jaw, followed by daggerlike canines for stabbing, carnassials that shear past one another like scissors for cutting meat into smaller pieces, and more robust back teeth for crushing tougher objects and bone. Note the ample braincase.

Characteristics of Mammals

Mammals are warm-blooded vertebrates that suckle their young and have a body covering of hair. But neither body hair nor mammary glands are preserved in the fossil record, so they are of no help to the paleontologist, who relies on bones to recognize fossil mammals. Here are distinctive features of mammalian bones:

Lower jaw and teeth—Unlike the lower jaw in reptiles and birds, in mammals it consists of a single bone on each side, called the *dentary* (Fig. 16-21). The lower jaw is particularly useful in identification. The teeth are nearly always of different kinds (incisors, canines, molars), serving different functions in eating.

Ear bones—In other tetrapods, sound vibrations are transmitted by a single bone. But in mammals, a chain of three little bones conveys sound across the middle ear. (These ossicles are the malleus, incus, and stapes, which you might recall as the "hammer, anvil, and stirrup" taught in health classes.) Two of them—the malleus and incus—evolved from bones of the reptilian jaw. Unfortunately, these bones are so small that they rarely are preserved.

Neck bones—Typically, mammals have seven neck (cervical) vertebrae, regardless of neck length.

Large braincase—There is a relatively large braincase, compared to other tetrapods.

Secondary palate—To help maintain uniform body temperature, mammals have a well-developed secondary palate that separates the mouth cavity from the nasal passages. This makes simultaneous breathing and feeding possible. Without the secondary palate, an infant mammal could not suckle.

Earliest Mammals

The mammalian fossil record begins with rare, hard-to-find jaw fragments, tiny teeth, and shards of skull bone. However, these fossils are sufficient to reveal that the earliest members of our own taxonomic class were very small, and that they evolved from mammal-like reptiles such as those previously described from the late Paleozoic.

The cusp pattern of their teeth indicates they were insect-eaters. Evidence from brain casts also suggests that the parts of their brains dealing with smell and hearing were particularly well developed, as in animals that seek food at night.

Perhaps the most significant and ironic fact about these mammals is that they came on the scene at about the same time as the dinosaurs, which they were destined to succeed. By Cretaceous time, mammals as large as a medium-size dog existed. Some of these actually preyed on juvenile dinosaurs.

Cenozoic mammals benefited from efficient nervous and reproductive systems, reliable body temperature control, larger brains, and relatively high levels of intelligence. Armed with these attributes, mammals quickly expanded into habitats vacated by the dinosaurs and found many new pathways of adaptation (Table 16-1).

Mammal Groups and Their Features

The following sections present the remarkable mammalian groups that evolved to walk, swim, or fly in Earth's diverse environments. First, here is a short roadmap to the mammals we'll visit, with examples:

Monotremes (most primitive; lay eggs)—platypus

Marsupials (carry living young in external pouch)—kangaroo, opossum

Placental mammals (embryonic development within the mother's body)—you and I

- *Insectivores* (insect-eaters)—shrew, hedgehog of Europe
- *Edentates* ("toothless")—armadillo, tree sloth, South American anteater
- *Rodents* (gnawers—single pair of chisel-like incisors above and single pair below)—squirrel, mouse, rat, beaver
- *Rabbits* (gnawers—two upper pairs of incisors and single pair below)—"cottontail" rabbit, hare
- *Bats* (true powered flight)—Insect-eating bat, fruit bat
- *Carnivores* (meat-eaters)—cat, dog, bear, hyena, seal

TABLE 16-1   **Evolutionary Modifications of Mammals for a Variety of Habitats***

Habitat	Limbs	Teeth	Other Features
Land-Primitive walking (ambulatory)	Limbs generalized; fleet flat on ground	Incisors for grasping; canines for piercing; premolars for cutting; molars for grinding	Tongue, stomach, and intestine generalized; cecum small
Land-Running herbivores (cursorial) Horse, cow, deer	Limbs elongate; toes elongate and reduced in number; hoof or horse is one toenail; in cow and deer two toenails (cloven hoof)	Incisors for nipping grass; cheek teeth heavy and ridged to withstand wear of grass	Stomach (cow) or intestine (horse) large and complex; cecum long
Land-Running carivores (cursorial) Wolf, cheetah	Limbs elongate; often walk on toes; claws often long and sharp	Canines well developed; premolars sharp for cutting flesh	Stomach large; intestines short
Land-Heavy body Elephant	Bones in limbs massive; flat jointed; toes in circle around pad	Cheek teeth massive and ridged for grinding plant materials	Nose elongated into trunk to reach food on ground or in trees
Tree-Climbing (arborial) Tree squirrel, monkey	Limbs elongate; wide range of motion; toes long and flexible	Cheek teeth usually smooth-surfaced and flattened for crushing	Tail sometimes prehensile; diet usually fruit or vegetation
Land. Insect-eaters Anteaters, spiny echnida	Claws enlarged for digging after insects	Reduced or absent	Tongue long, sticky, and extensile
Burrowing (fossorial) Moles, marsupial moles Various rodents	Limbs short and stout; forefeet often enlarged, shovel-like; claws long and sharp	Teeth various, depending on diet (vegetation or insects and worms)	Pinna of ears usually short or absent; fur short; tail short
Aquatic Whales, seals, sea cows	Limbs shortened; feet enlarged and paddle shaped or absent	Often reduced to simple, conelike structures, sometimes absent	Tail often modified into swimming organ
Flying or gliding Flying phalangers, flying squirrels	Limbs extendible to sides to hold gliding membrane outstretched	Various but usually show herbivorous adaptations	Tail often flattened for use in balancing
Bats	Forelimbs and especially fingers of forelimbs elongate to support wing	Various for diets of insects, fruits and nectar, or other specific foods	Tongue, stomach, and intestine various; correlated with diet

*From Cockrum, E.L., and McCauley, W.J. 1965. *Zoology*, Philadelphia: W. B. Saunders Co.

- *Ungulates* (hoofed)—horse, cow, sheep, pig, deer, camel
- *Ruminants* (cud-chewing ungulates)—cow, sheep, goat, giraffe
- *Proboscideans* (trunks)—mastodon, mammoth, modern elephant
- *Cetaceans*— (blubber for insulation, tail and forelimbs modified for swimming)—whale, dolphin

▶ MONOTREMES

The most primitive living mammals are **monotremes**. These relics of an older time still lay eggs in the reptilian manner. However, unlike reptiles, monotremes nourish their young with milk from glands on the abdomen. The female monotreme has no nipples; milk is secreted from special glands onto hairs on the abdomen where the young can lick it up. The oldest monotreme fossils are lower Cretaceous. A few living monotremes exist today, such as the platypus of Australia and Tasmania and two species of spiny anteaters of New Guinea and Australia (Fig. 16-22).

▶ MARSUPIALS

Marsupials are mammals, but most nurture their young in an external pouch, or **marsupium** (Fig. 16-23). Most familiar to us today are the opossums of the Americas (Fig. 16-24) and some well-known Australian residents: kangaroos, wallabies, wombats, phalangers, bandicoots, and koalas (Fig. 16-25).

A

B

FIGURE 16-22 **Two modern monotremes.** (A) Duckbill platypus of Australia and Tasmania. (B) Spiny anteater (echnida) of Australia. ◀ *If monotremes lay eggs, why are they considered mammals?*

Marsupials had their greatest success in Australia and South America. Both continents were largely isolated during most of the Cenozoic, and marsupials were able to evolve with little competition from placental mammals. Although a few Paleogene marusupial fossils have been found, their fossils do not become common until the Pleistocene Epoch.

An older and more complete record exists in South America, where opossum-like marsupials evolved into many different carnivorous and herbivorous types. Many of these animals had an uncanny resemblance to plant-eaters and carnivores in North America. An example is the similarity between the South American saber-toothed marsupial *Thylacosmilus* (Fig. 16-26) and the North American saber-toothed placental *Smilodon*. There were also marsupial moles, as well as marsupials that were doglike, bearlike, and hippolike. These animals serve as excellent examples of **convergent evolution**, in which animals of widely different ancestry evolve similar structures and appearance in adapting to similar ways of life.

About 3 million years ago, the Central American land bridge formed, connecting North and South America. This permitted a major interchange of marsupials from the south and placentals from the

A

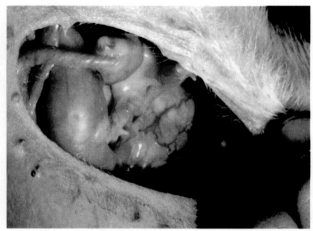

B

FIGURE 16-23 **Short-nosed bandicoot, a common Australian marsupial.** (A) Advanced offspring that have left the pouch return for nursing. (B) A fetal bandicoot attached to a teat in the mother's pouch.

FIGURE 16-24 **American opossum** *Didelphis*. This interesting marsupial still retains many of the dental and skeletal characteristics present in its Cretaceous ancestors.

north (Fig. 16-27). Entering South America were placental camels, elephants, deer, bears, peccaries, horses, tapirs, skunks, rabbits, cats, and dogs. In the opposite direction, entering North America were monkeys, opossums, rats, shrews, sloths, porcupines, and anteaters.

For a time, both marsupials and placentals flourished in their new locations, but eventually the marsupials began to decline. All of the hoofed marsupials became extinct, as did the placental ground sloths and heavily armored glyptodonts. Today, the marsupials are a dwindling group.

Marsupial "mouse"

Phalanger

Tasmanian "wolf"

Wombat

Bandicoot

Koala

Marsupial "mole"

Kangaroo

FIGURE 16-25 **Diverse Australian marsupial adaptations for various habitats.** The Tasmanian wolf (upper left) is thought to be extinct.

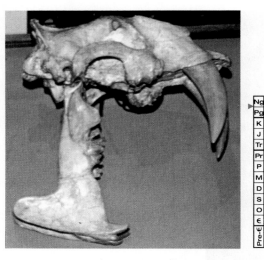

FIGURE 16-26 **Pliocene South American carnivorous marsupial,** *Thylacosmilus.* It is comparable to placental saber-toothed cats, with bladelike upper canine teeth about 18 cm long. These teeth were protected when the jaws were closed by a deep flange of bone in the lower jaw.

▶ **PLACENTAL MAMMALS**

Insectivores

Fossils of **placental mammals** in Cretaceous rocks are small, unspecialized insectivores. DNA evidence indicates they actually evolved earlier in the Jurassic, about 55 million years ago. The tiny tree shrew (Fig. 16-28) living today is anatomically very close to the ancestor of Cenozoic primates, rodents, bats, carnivores, and herbivores.

Edentates

Edentate mammals are "toothless," having no teeth at all, or very few. Living edentates include armadillos, tree sloths, and South American anteaters. Extinct edentates include ground sloths (Fig. 16-29) and **glyptodonts** (Fig. 16-30). Glyptodonts were roughly the size and shape of a Volkswagen Beetle. A spike-covered knob on *Glyptodon's* tail was used to bludgeon predators.

FIGURE 16-27 **North and South America connected by a land bridge, Late Pliocene.** This land bridge enabled many animals to migrate northward, including sloths, anteaters, caviomorph rodents, armadillos, porcupines, opossums, ground sloths, and glyptodonts. Conversely, headed south were jaguars, squirrels, rabbits, saber-tooths, elephants, deer, wolves, and horses. The event has been named "the great American interchange."

FIGURE 16-28 **Common tree shrew.** It most resembles the ancient insectivores that gave rise to the primates.

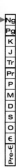

FIGURE 16-30 **Restoration of middle Pleistocene scene in Argentina.** The heavily armored animals are glyptodonts. At left is a giant ground sloth.

Cenozoic ground sloths included truly colossal Pleistocene beasts, such as *Megatherium*, which was 6 meters (20 feet) long. They were tall enough to browse on tree branches while resting on their hind legs. Dung and mummified body tissues of ground sloths have been found in association with human artifacts from caves in arid regions of the United States.

FIGURE 16-29 **Extinct great ground sloth** *Megatherium.* It lived during the Pleistocene and was nearly as large as a present-day elephant.

Rodents

Rodents, distinguished by their gnawing-nibbling behavior, have invaded nearly every habitat on Earth, and today they probably outnumber all other mammals. They have been exceptionally successful Cenozoic animals because their remarkable range of adaptations permits survival in diverse habitats, from arctic to tropics. A rapid breeding rate and small size also help. Their diversification has produced burrowers (chipmunks, marmots), partially aquatic animals (muskrat, beaver), desert-dwellers (kangaroo rats, hamsters, "gerbils"), and tree-dwellers (squirrels).

Rodents have distinctive adaptations of teeth and jaws to reflect their gnawing-nibbling specialization. They lack canines and have only two prominent and opposing pairs of incisors, which continuously grow. The outer surface of each incisor is made of enamel and is thus harder than the inner surface. When the rodent is chewing, the incisors are sharpened as the harder enamel layer persists to form a chisel edge. There is a toothless gap between the chisel incisors and the cheek teeth. The cheek teeth (molars) are long, ridged, and efficient tools for grinding coarse food such as grain.

The major groups of rodents are:

- **Protorogomorphs** originated in the early Paleogene. They are ancestors of the later true squirrels and Old World porcupines. They are represented in the fossil record by *Paramys*, which resembled a large squirrel (Fig. 16-31).

- Squirrels are the characteristic **sciuromorph** rodents, a group that also includes chipmunks and marmots.

FIGURE 16-31 **The Paleocene rodent *Paramys*.**

- Beavers and their relatives are **castorimorphs**. A Pleistocene member of the beaver family called *Paleocastor* left casts of peculiar corkscrew burrows, seen today in Nebraskan fossil beds. Some Ice Age beavers were as large as bears.

- Rodents called **myomorphs** evolved during the late Paleogene. Hamsters, those amiable pets, are myomorphs, and so are rats and mice.

Rats and mice have formed a one-sided partnership with humans. They invade our dwellings, eat our food, spread diseases, and voyage with us to the far corners of the world. Rodent persistence and tenacity are legendary.

Rabbits

Rabbits, hares, and the short-legged pika (rock rabbit or cony) were common throughout most of the Cenozoic. Many people are surprised to learn that rabbits are not rodents, since both groups have enlarged incisors. But rodents have a single pair of chisel-like incisors above, in opposition to a single pair below. Rabbits have two upper pairs and a single pair below.

Bats

Bats are the only mammals to have achieved true powered flight (others simply glide). Their greatly elongated finger bones support the membrane that forms the wing. Bats are adapted to varied diets:

- Insect-eating bats take flight at dusk and sweep the skies mostly at night, scooping insect prey from the air.

- Vampire bats (South America) feed on the blood of other mammals. Their front teeth are adapted to pierce the skin of their prey. They then lap the blood that oozes from the wound. (The association of bats with vampire legends, like those involving Count Dracula, began when Spanish explorers returned from South America and spread stories of blood-eating bats.)

- Some bats live on nectar from flowers or eat fruit.

Although many bats live in temperate zones, they are most abundant in the tropics. The well-preserved skeleton of a bat from Wyoming indicates that, by early Eocene time, bats were very similar to their modern relatives.

Flesh-Eaters

The earliest known flesh-eating placental mammals are Late Cretaceous in age. They are represented by a small weasel-like animal, *Cimolestes*. Although small, *Cimolestes* clearly was a predator. Cheek teeth called the **carnassials** were developed into blades able to shear

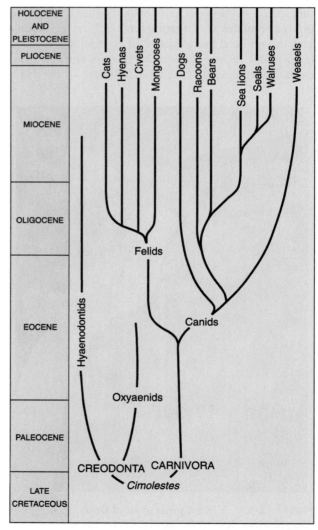

FIGURE 16-32 **General relationships of flesh-eating mammals.**

Ng
Pg
K
J
Tr
Pr
P
M
D
S
O
Є
Pre-Є

FIGURE 16-33 **Eocene creodont, *Patriofelis*.**

past one another and slice meat into small pieces. Enlarged canines were the principal killing weapon.

By the end of the Cretaceous, flesh-eating mammals had split into two independent orders, Creodonta and Carnivora (Fig. 16-32):

- **Creodonta** were relatively small-brained animals with short limbs, clawed toes, and long nails. *Patriofelis* is an example (Fig. 16-33). *Hyaenodon* is a creodont whose name was suggested by the animal's superficial resemblance to a hyena.

- **Carnivora** evolved larger brains than Creodonta. Also, their carnassial teeth were positioned farther to the front. By Miocene time, these more progressive predators had replaced the Creodonta.

The expansion of Carnivora accelerated during the remainder of the Cenozoic, producing a host of now-familiar cats, civets, mongooses, and true hyenas. Famous among the Pleistocene Carnivora was the large saber-toothed cat *Smilodon* (Fig. 16-34). This robust flesh-eater appears to have preyed on large herbivores, whereas the swift and agile biting cats probably sought their food among herds of grazing animals.

Dogs, raccoons, bears, and weasels were contemporaries of the cats. Indeed, wild dogs were doing very well long before humans came along to make companions of them. The modern dog *Canis* roamed widely during the Ice Age in the form of the hyena-like dire wolf.

As is true of carnivores today, Cenozoic carnivores were an essential element in the evolutionary process.

A   *B*

FIGURE 16-34 *Smilodon*, **the** *saber-toothed* **cat.** (A) Restoration. (B) Skull.

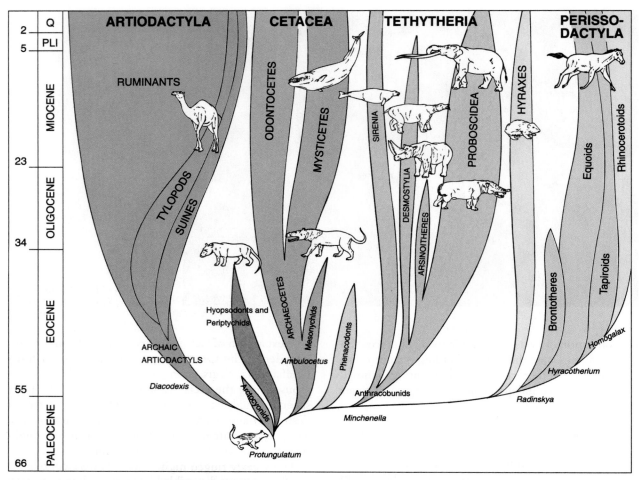

FIGURE 16-35 **Phylogenetic tree of major ungulate (hoofed) groups.**

To survive, their speed and cunning had to be equal to or better than that of their herbivore prey. The herbivores, in turn, responded to the carnivore threat by evolving adaptations for greater speed and defense. Then, as now, predators were not villains but necessary constituents of the total biologic scheme. They were able to cull out the weak, deformed, or sickly animals and thereby help to counteract the effects of degenerative mutation and overpopulation.

Not all the flesh-eaters of the Cenozoic are land dwellers. Seals, sea lions, and walruses obtain their food in, or at the edge of, the sea. As is evident from their sharp, pointed teeth, early Cenozoic seals and sea lions consumed flesh, in the form of fish. Walruses, however, feed upon mollusks, the shells of which they crush with their powerful jaws and broad, flat cheek teeth.

Ungulates—Hoofed Animals

Most animals with hoofs (or descendants of animals with hoofs) are called **ungulates**. These include horses, tapirs, rhinoceroses, cows, sheep, goats, hip-popotami, pigs, giraffes, antelopes, camels, and deer (Fig. 16-35).

Ungulates include some surprisingly different groups. Whales are ungulates because they descended from hoof-bearing mammals. **Proboscideans**, such as elephants and mastodons, are ungulates. The group also includes huge mammals called **arsinotheres** that sported an enormous pair of side-by-side horns over their skull and snout (Fig. 16-36).

Primitive ungulates were widespread by late Paleocene time. Their general characteristics are well represented by the sheep-sized herbivore *Phenacodus* (Fig. 16-37). *Phenacodus* had cheek teeth with broad grinding surfaces, rather like the teeth of primitive horses. It was flat-footed (plantigrade, like us), lacking the modifications needed to run on its toes for greater speed. This plantigrade stance also was characteristic of the enormous six-horned *Uintatherium* (Fig. 16-38A).

The two largest categories of today's hoofed animals are perissodactyls and artiodactyls, distinguished by the number of toes they have.

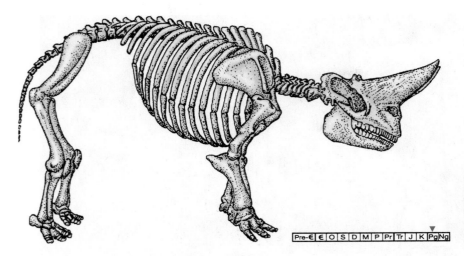

FIGURE 16-36 *Arsinotherium*, **huge browsing mammal of the Oligocene.** Its most striking feature was a pair of massive bony horns on its snout.

PERISSODACTYLS ("UNEVEN TOES"—ONE OR THREE)

Perissodactyls include modern horses, rhinoceroses, and tapirs, as well as extinct forms known as chalicotheres, brontotheres, and several less familiar groups (Fig. 16-39). Perissodactyl means "odd-numbered toes." Many have one or three toes on each foot. Tapirs and some extinct perissodactyls had four toes only on their forefeet.

In perissodactyls, the foot axis—along which body weight is primarily supported—lies through the third (middle) toe. Reduction of lateral toes is characteristic of perissodactyls, and in the modern horse, only the central toe remains (Fig. 16-40). Perissodactyls are digitigrade, meaning they run or walk on their toes to attain a longer stride and greater speed.

HORSE FAMILY TREE The horse family tree (see Fig. 6-24) has its roots in a small Paleocene herbivore from China named *Radinskya*. It is uncertain whether this primitive creature is a true perissodactyl or the pre-horse ancestor to Eocene browsing horses like *Protoro-hippus* and *Orohippus* (Fig. 16-38F). The latter were small horses only 40 cm (16 inches) tall, with four toes on the forefeet and three on the hind feet. The molar teeth were bluntly cusped for browsing and had not evolved the high crowns and ridged patterns that characterize modern grazers.

The rich fossil record of horses provides ample evidence of evolutionary change. You can easily follow the change from horses of small body size, short skulls, small cranial capacity, and low-crowned teeth to larger animals with fewer toes, longer skulls, larger brains, and complexly ridged high-crowned teeth for rendering grasses more digestible.

Mesohippus (Fig. 16-41) exhibits many of the general characteristics of Oligocene horses that lived in North America. Greyhound-sized, this horse had lost the fourth toe on its forefeet. Its teeth show an early pattern of ridges on their surfaces where they meet, but still were low-crowned. *Mesohippus* preferred a less-abrasive diet of succulent leaves rather than grasses. In Mesohippus and in subsequent horses, the

FIGURE 16-37 *Phenacodus*, **an Early Paleogene herbivorous mammal.** Like most Early Paleocene herbivores, *Phenacodus* walked on all five toes in a style called *plantigrade*. Speed was not essential to this forest-dwelling browser.

FIGURE 16-38 **Eocene mammals.** (A) *Uintatherium*, the large six-horned and tusked animal. (B) *Hyrachus*, a small, fleet rhinoceros. (C) *Trogosus*, a gnawing-toothed mammal. (D) *Mesonyx*, a hyenalike flesh eater. (E) *Stylinodon*, a gnawing-toothed mammal. (F) Three early members of the horse lineage (*Orohippus*). (G) *Machaeroides*, a saber-toothed mammal. (H) *Patriofelis*, an early carnivore. (I) *Palaeosyops*, an early titanothere. Restorations are based on skeletal remains from the Middle Eocene Bridger Formation of Wyoming.

front portion of the skull and lower jaws became deeper to make space for the long roots of the cheek teeth. In addition, the battery of molars was shifted toward the front of the muzzle, out of the way of the eye orbits and jaw articulation.

Most Miocene horses retained three toes, but the larger central one bore most of the animal's weight. *Merychippus* (Fig. 16-42) was a horse about a meter (3 feet) tall at the shoulder and was well adapted for life on the spreading grasslands. *Merychippus* lived on into the Pliocene alongside another group of three-toed horses called hipparions. Having evolved in North America, hipparions migrated into Eurasia and Africa. *Pliohippus* is one of several genera of horses present during the Late Miocene and Pliocene. The lateral toes in *Pliohippus* were reduced to useless vestiges.

The modern horse *Equus* appeared during the Pliocene and expanded across the prairies of North America, Eurasia, and Africa during the Pleistocene. Only a few thousand years ago, horses suffered extinction in North America. Some think the cause was a contagious disease; others speculate it was overkill by prehistoric human hunters. But the horse began to re-

populate North America when horses escaped from Spanish explorers in the 1500s.

Among the other surviving perissodactyls are the tapirs and rhinoceroses. Tapirs retain the primitive condition of four toes on the forefeet and three on the rear, as well as low-crowned teeth. They are forest-dwelling, leaf-eating animals whose fossil record begins in the Oligocene.

The first rhinoceros was a small animal rather like a modern tapir. However, by Oligocene time the lineage included some of the largest land mammals known. *Paraceratherium*, a colossal hornless rhinoceros, stood 5 meters (16 feet) tall at the shoulders (Fig. 16-43). Another branch contained water-dwelling rhinoceroses that lived and looked much like today's hippopotami.

The horns of rhinoceratids are unusual. Unlike most animals' horns, they are formed of tightly matted hair and thus are not preserved. A roughened area on the skull reveals where the horn was attached.

Two groups of perissodactyls known only from fossils are the **brontotheres** (Fig. 16-44G) and **chalicotheres** (Fig. 16-45A). Because of their large size and

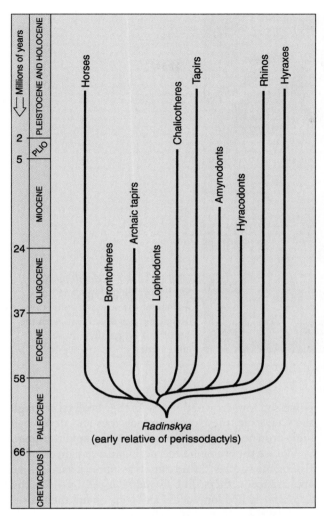

FIGURE 16-39 **Simplified evolutionary tree of Perissodactyls.**

huge, blunt, forked horns over their snouts, museum reconstructions of brontotheres are popular at natural history museums. Brontotheres likely used their horns for sexual display, possibly for jousting with other males for access to receptive females, or simply for defense.

Chalicotheres differed from all other Cenozoic perissodactyls in having three claws rather than hoofs. These odd creatures resembled a horse in head and torso, but their forelegs were considerably longer than the hind legs, so their back sloped rearward. They lived from Eocene into the Pleistocene.

ARTIODACTYLS ("EVEN TOES"—TWO OR FOUR) **Artiodactyls** have been more successful than the perissodactyls in survival, variety, and abundance. Those still living include the pig, peccary, deer, hippopotamus, goat, sheep, camel, llama, giraffe, and many kinds of cattle and antelope. Artiodactyls are important to us for meat, milk, and wool.

Fossils of the earliest artiodactyls have been recovered from Lower Eocene strata in Pakistan. This indicates that, like the perissodactyls, this great group of ungulates originated in the Old World before rapidly spreading around the globe.

The **oreodonts** and the **entelodonts** are extinct artiodactyls. Oreodonts once roamed the grassy plains of North America in great numbers (see Fig. 16-44H). Entelodonts were larger, some the size of an American bison. They were repulsive-looking, hoglike beasts. Their trademarks were bony protruberances that grew along the sides of the skull and jaws (Fig. 16-44D, 16-45F, and 16-46).

The radiation of artiodactyls that began in the Eocene produced such surviving groups as swine, camels, and ruminants. Of these three, the swine family

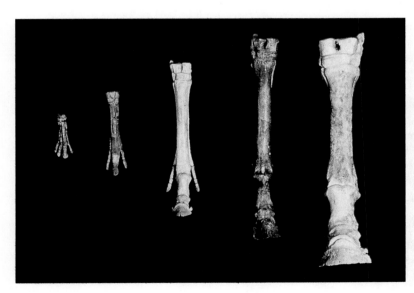

FIGURE 16-40 **Evolution of the lower foreleg in horses.** It starts at far left with an Eocene horse and continues to the modern horse at far right.

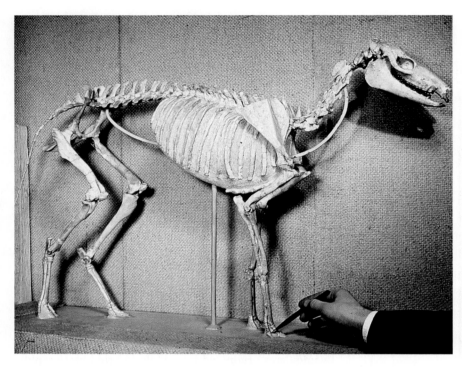

FIGURE 16-41 *Mesohippus*
Skeleton. This Oligocene horse
was about the size of a collie dog. It
had three toes on its feet, with the
center toe larger than those on
either side.

has remained the most primitive. They have kept all
four toes, even though most of the weight is supported
by the two middle toes. The group includes not only
pigs but peccaries, which are lighter framed and primarily South American. Hippopotami are the only
modern amphibious artiodactyls. Although their name
means "river horse," they are related to the pig family.

You might be surprised to learn that camels originated in North America, not Africa. Early members of
this group were present by Middle Eocene. They were
relatively small in stature and, like the early horses,

possessed low-crowned teeth. As they evolved through
the Cenozoic, they lost their side toes and developed
high-crowned teeth and longer legs for rapid running.

We see these changes occurring in a rich diversity of
Paleogene species. Some camels assumed a gazelle-like
appearance. Others, like *Oxydactylus* (Fig. 16-47) and
Aepycamelus had long necks and legs, giving them a superficial resemblance to modern giraffes. A particularly
unusual member of the camel lineage is *Synthetoceras*
(Fig. 16-48B). Synthetoceras had a pair of horns above
the eyes and a large Y-shaped horn on the snout.

FIGURE 16-42 *Merychippus,*
**the size of a small pony, was a
Miocene member of the horse
family.** It still had three toes, but
the side toes were reduced and of
little use.

FIGURE 16-43
***Paraceratherium*, a colossal, hornless rhinoceros.** It was 5 meters (16 feet) tall at the shoulder and weighed about 15 tons. This huge perissodactyl probably browsed on tree leaves.

FIGURE 16-44   Late Eocene mammals. This is based primarily on fossils from the Chadron Formation of the White River Group of South Dakota and Nebraska. (The White River Formation above the Chadron is Oligocene in age.) The flora is based on the somewhat younger plant fossils from the Florissant beds of Colorado. (A) *Trigonias*, an early rhinoceros. (B) *Mesohippus*, a three-toed horse. (C) *Aepinocodon*, a remote relative of the hippopotamus. (D) *Archaeotherium*, an enteledont. (E) *Protoceras*, a horned ruminant. (F) *Hyracodon*, a small rhinoceros. (G) *Brontops*, a brontothere. (H) Oreodonts named *Merycoidodon*. (I) *Hyaenodon*, a carnivore. (J) *Poëbrotherium*, an ancestral camel.

FIGURE 16-45 **Early Miocene mammals.** (A) The chalicothere *Moropus*. (B) The small artiodactyl *Merychyus*. (C) *Daphaenodon*, a large wolflike dog. (D) *Parahippus*, a three-toed horse. (E) *Syndyoceras*, an antelopelike animal. (F) *Dinohyus*, a giant, piglike entelodont. (G) *Oxydactylus*, a long-legged camel. (H) *Stenomylus*, a small camel. ◧ *Compare Figures 16-38, 16-44, 16-45, and 16-56. What trend in Cenozoic climate is indicated?*

RUMINANTS **Ruminants** are the most varied and abundant of modern-day artiodactyls. This group takes its name from the *rumen*, the first of four compartments in their multichambered stomach. Most are cud-chewers.

In ruminants, coarse vegetation is swallowed without much chewing and passed into the rumen and second compartment (reticulum). When the animal stops grazing, food is regurgitated from the reticulum to the mouth to be thoroughly chewed. The animal "chews its cud." When swallowed a second time, it passes into the final two stomach compartments for gastric digestion. Cud chewing aided digestion of coarse grasses. It may have had a survival advantage, in allowing the animal to crop large amounts of food in open areas and then move to safer locations to "ruminate."

FIGURE 16-46 *Dinohyus*, **giant piglike entelodont from the Miocene rocks in America.** The animal was about 3 meters (9 feet) long.

FIGURE 16-47 **Skeleton of Miocene camel *Oxydactylus*.** This small, graceful camel was one of a varied group that populated North American grasslands.

The earliest ruminants were small, delicate, four-toed animals called tragulids. An example is *Hyper-tragulus* (Fig. 16-49), whose remains are found in Lower Oligocene strata. Tragulids are represented today by the skinny-legged and timid mouse deer of Africa and Asia.

Ruminants include sheep, cattle, giraffes, and deer. Deer are primarily browsers that have made the forests their principal habitat ever since they first appeared in the Oligocene. Early deer were mostly small and horn-less, but later evolution increased their size and development of antlers. The culmination of this trend is represented by the Pleistocene *Megaloceros* (mistakenly called the "Irish Elk"), whose antlers measured more than 3 meters from tip to tip (Fig. 16-50).

The giraffe family probably branched off from the deer lineage sometime during the Miocene and became specialized in browsing on leaves of trees that grew rather sparsely in areas where most other herbivores were grazers. The exceptionally long neck by which we recognize modern giraffes was a late modification and is not typical of extinct giraffes, most of which had necks no longer than a deer's.

FIGURE 16-48 **Some early Pliocene mammals.** (A) *Amebelodon*, the shovel-tusked mastodon. (B) *Synthetoceras*. (C) *Cranioceras*. (D) *Merycodus*, an extinct pronghorn antelope. (E) *Epigaulis*, a burrowing, horned rodent. (F) *Neohipparion*, a Pliocene horse. (G) The giant camel *Megatylopus* and smaller *Procamelus*. (H) *Prosthennops*, an extinct peccary. (I) The short-faced canid *Osteoborus*.

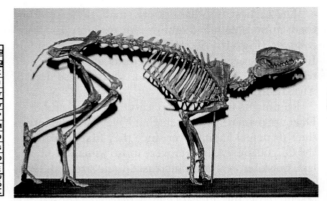

FIGURE 16-49 **Primitive ruminant *Hypertragulus* from Lower Oligocene beds of North America.** This tiny animal stood only about 30 cm high at the shoulder.

Bovids, including cattle, sheep, goats, muskoxen, and antelope, are presently the most numerous of ruminants. Miocene strata provide the earliest fossil record of bovids. Bison are bovids that hold a particular interest for Americans because of the vast numbers that once roamed our western prairies. Seven species of bison lived in North America during the Pleistocene. Some attained great size and had horns that measured 2 meters (6 feet) tip-to-tip.

Elephants and Kin

Animals with a trunk are called *proboscideans*, like elephants and the extinct **mastodons** and mammoths. The trunk functions in breathing and bringing food to the elephant's mouth. Other animals as tall as elephants reach food on the ground easily because of their long necks. Proboscideans, however, have short, muscular necks needed to support their massive heads. So they evolved their own unique anatomical solution to food-gathering.

Paleontologists are able to follow the development of the trunk in early proboscideans by noting the position of the external nasal opening at the front of the skull. Those openings recede toward the rear of the skull in sequential stages of trunk development. The trunk itself was formed by elongation of the upper lip and nose. The nasal passages that extend through the trunk are encased by over 500 muscles, which provide ample power and dexterity.

A second proboscidean trademark is tusks, which evolved by elongation of the second pair of incisors. Proboscideans use their tusks in combat and in browsing for food, using them to hold down branches pulled into reach by the trunk.

The earliest proboscidean remains were found in Morocco, in rocks that are 55 million years old. Named

FIGURE 16-50 ***Megaloceros*, the giant "Irish elk."** It actually was a deer whose remains are frequently found in Pleistocene-age peat bogs in Ireland. Tip-to-tip, its antlers were about 3.6 meters (12 feet).

Phosphatherium, the animal had distinctly proboscidean dental and cranial features. Another early proboscidean was *Moeritherium*, from Eocene fossil beds along Lake Moeris south of Cairo. *Moeritherium* was about the size of a pigmy hippopotamus. Upper and lower incisors were already formed into mini-tusks (Fig. 16-51).

From relatively primitive animals such as the moeritheres, proboscideans with trunks evolved and quickly divided into several branches. One branch led toward a group of Miocene and Pliocene animals called **dinotheres** (Fig. 16-52). The tusks of dinotheres were distinctive in that they were present only in the lower jaws and curved downward and backward, an orientation useful for uprooting plants and digging for roots and tubers.

Another branch of the proboscidean family tree produced mastodons and elephants. *Palaeomastodon*, in the Oligocene of North Africa, probably was representative of the group from which true mastodons evolved. The skull was long, the trunk was short, and the second incisors in both the upper and the lower jaws had developed into tusks.

ENRICHMENT

How the Elephant Got Its Trunk

Among the most interesting mammals to evolve during the Cenozoic are the proboscideans—mastodons, mammoths, and elephants—so-named for their most distinctive feature, their prominent trunk. The elephant's trunk is a muscular, much-elongated organ containing nostrils that extend through its entire length to the tip. How did this remarkable appendage develop?

In Rudyard Kipling's story *The Elephant's Child*, an overly inquisitive baby elephant tries to find out what the crocodile had for dinner. When the baby elephant lowers his head along the bank of the great, gray-green, greasy Limpopo River, the jaws of the crocodile close on his nose. There ensues a fierce tug-of-war in which the baby elephant's poor nose is stretched into the trunk we know today.

Nice children's story, but the elephant achieved its wonderfully long, grasping trunk by other means: evolutionary processes that operated over many generations, resulting in enormous elongation of the nose and upper lip. Scientists call such extraordinary growth of a body part *hypertrophy*. The elephant got its trunk by hypertrophy of the nose and upper lip. But why?

When we view the bones of proboscideans—beginning with the relatively small and primitive groups of the Eocene through the giant Pleistocene mastodons and mammoths—we can see several trends. One obvious trend was the size increase, of both the body and huge proboscidean head. Supporting the weight of the ponderous head required a short, powerful neck. As proboscideans grew taller, reaching food at ground level with a short neck would have been difficult. Natural selection paralleling that of the large head and short neck resulted in the development of the trunk.

The elephant's trunk is a structural adaptation for feeding that could be used not only for raising food from the ground, but also for stripping leaves from branches and stuffing them into the mouth. Along with the evolution of the trunk, the lower jaw became shorter, allowing the trunk to hang downward. Tusks in the lower jaw that were present in early proboscideans became smaller and gradually disappeared.

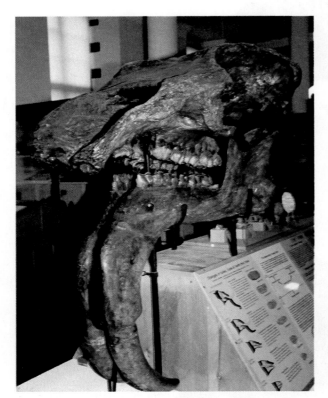

FIGURE 16-51 **Skull of *Dinotherium*, a Miocene proboscidean.** Skull length is about 1.2 meters.

By Miocene time, a larger proboscidean named *Gomphotherium* arrived in North America via the Bering isthmus into Alaska. Subsequent proboscidean evolution produced a variety of long-jawed mastodons, some with lower jaws almost 2 meters (6 feet) long.

The most bizarre proboscidean was the "shovel-tusked" *Amebelodon* (Fig. 16-48A) of the Pliocene. Tusks on its lower jaws were flattened, forming two sides of a broad, scooplike, ivory shovel. Mastodons with shorter jaws and massive tusks (Fig. 16-53) were common throughout North America during the Pleistocene and survived until comparatively recent times.

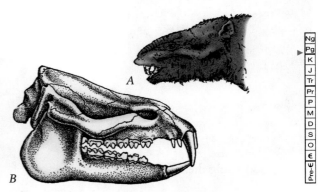

FIGURE 16-52 **Restoration of *Moeritherium*.** (A) Head. (B) Skull.

FIGURE 16-53 **Four-tusked mastodon, Late Miocene/Early Pliocene.** The mounted skeleton is in the collection of the Denver Museum of Natural History.

MAMMOTHS (ICE AGE ELEPHANTS) **"Mammoth"** loosely applies to Ice Age elephants of North America, Europe, and Africa. In their molar teeth, the cusps of earlier forms have merged to form transverse ridges of enamel (Fig. 16-54). The teeth also were higher-crowned and extended deep into the jaws.

Mammoths were a magnificent group of animals that included the famous woolly mammoths (Fig. 16-55) drawn by our own ancestors on the walls of their caves. The great imperial mammoth reached heights of 4.5 meters (15 feet) and ranged widely across California, Mexico, and Texas. The Columbian mammoth had immense spiral tusks that in older individuals overlapped at the tips, becoming useless for digging purposes.

Late Pleistocene proboscideans were hunted by humans. Then, about 8000 years ago, all but the African and Indian elephants became extinct. Their continued survival will depend on our good judgment and that of the governments we support.

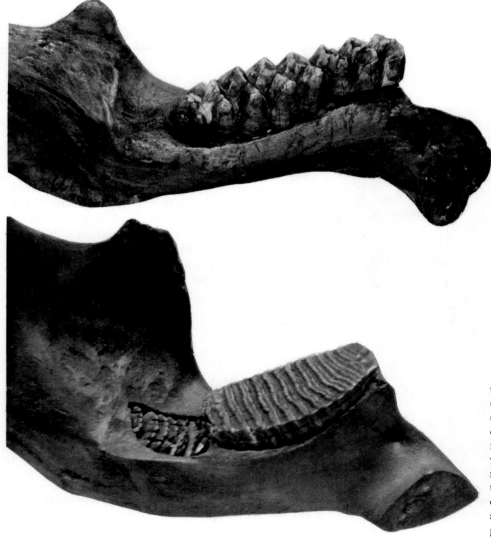

FIGURE 16-54 **Cheek teeth in jaws of (A) mastodon and (B) mammoth.** Paired cusps of mastodon teeth were used for crushing vegetation, as we typically see in browsing mammals. The mammoth tooth, with its parallel ridges of infolded enamel, was well-adapted for grazing on tough grasses and cereals.

FIGURE 16-55 **Woolly mammoth.** In the Late Pleistocene, these magnificent Ice Age elephants lived along the borders of the continental glaciers. Their remains have been found frozen in the tundra of northern Siberia.

CETACEANS (WHALES AND PORPOISES) Among all mammals, none have so completely adapted to life in the sea as the **cetaceans** (whales and porpoises). When we see their sleek bodies moving through the water, it is difficult to grasp that these mammals are descendants of the hoof-bearing artiodactyls! But this hypothesis has been confirmed by recent discoveries in Pakistan of a 55 million-year-old aquatic cetacean that is the direct ancestor of the first cetacean. The fossil has ankle bones characteristic only of artiodactyls.

An earlier discovery named *Pakicetus* also had the artiodactyl ankle structure. *Pakicetus* was a four-legged animal that probably ventured into streams, estuaries, and lakes to feed on fish, about 50 million years ago during the Eocene. Except for whale-like characteristics of its teeth and the ear region in the skull, *Pakicetus* did not look at all like a whale, as you can see (Fig. 16-56)!

But the link between artiodactyls and whales also is supported by DNA sequencing studies. As depicted on the cladogram (Fig. 16-57), molecular biologists think that hippopotami are closer to whales than any other group of artiodactyls.

In rocks only 10 million years younger, yet another ancestral whale named *Ambulocetus* has been found. *Ambulocetus* had webbed feet on its hind limbs and front flippers. The shape of its back vertebrae suggest it may have moved about on land with the up-and-down motion seen in today's seals. This motion may have been a sort of preadaptation for the up-and-down tail propulsion seen in modern whales.

In *Basilosaurus* (Fig. 16-58) from the Late Eocene of Egypt, we get a glimpse of the trend toward giganticism. This whale attained a length of 60 feet. Tiny hind limbs complete with toes were still present in *Basilosaurus*, but these rear limbs were small and too weak to provide strong propulsion.

Modern whales arose from the group that included *Basilosaurus* and divided into two lineages: toothed whales and whalebone (baleen) whales.

- Toothed whales include modern porpoises, killer whales, and sperm whales like author Herman Melville's famous Moby Dick.
- Whalebone whales include the titanic blue whale, right whale, humpback, gray, and Greenland whale. These plankton-feeding giants first appeared in the Miocene. Instead of teeth, they possess ridges of hardened skin that extend downward in rows from the roof of the mouth. The ridges are fringed with hair, which entangles the tiny invertebrates on which these cetaceans feed—thus, the paradox of the largest of all animals feeding on some of the smallest. The great blue whales far exceed in mass even the largest dinosaurs, for some have attained weights of over 135 metric tons (the weight of some locomotives).

FIGURE 16-56 **Ancestral whale *Pakicetus*, Eocene of Pakistan.** Because the remains of *Pakicetus* were found in stream-deposited sediments, it is doubtful that this primitive whale ever ventured into the open ocean. Except for whale-like characteristics of its teeth and the ear region in the skull, *Pakicetus* did not resemble a whale at all, as you can see.

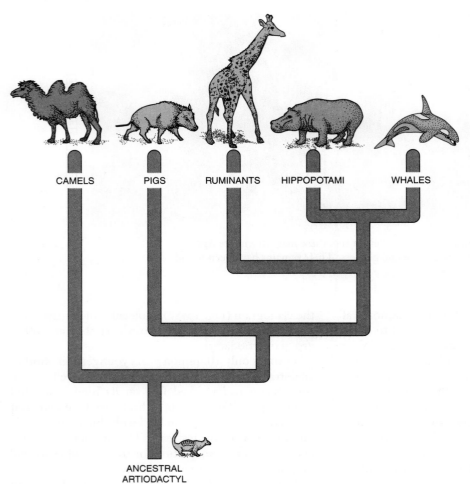

FIGURE 16-57 **Cladogram of major artiodactyl groups.** This diagram shows the close evolutionary relationship between artiodactyls and whales.

FIGURE 16-58 **Eocene whale *Basilosaurus*.** The cetacean tendency toward increased size already is evident in this early whale, more than 20 meters long (65 feet).

FIGURE 16-59 **Animals of central Alaska about 12,000 years ago, during the Late Pleistocene.** Fossil remains of this period are abundant. They indicate a fauna in which grazing animals predominated, with the rest composed of browsers and predators.

DEMISE OF THE PLEISTOCENE GIANTS

At the time of maximum continental glaciation approximately 17,000 years ago, the Northern Hemisphere supported abundant and varied large mammals—comparable to that which existed in Africa south of the Sahara a century ago (Fig. 16-59). There were giant beavers, mammoths, mastodons, elks, most species of perissodactyls, a variety of artiodactyls, and huge ground sloths.

Most of these great beasts maintained their numbers quite well during episodes of glaciation, but they experienced rapid decline and extinction beginning about 8000 years ago. What caused their extinction? There are two basic hypotheses. One proposes that human hunters killed them off. The second proposes that climatic change resulted in vegetation changes and associated disruption of food chains.

Human predation—Early humans had developed the ability to hunt in highly organized social groups, and they skillfully brought down large numbers of big animals on open prairies and tundras. Perhaps then as now, human predators may have killed in excess of their needs. Unlike wolves and lions, early human hunters probably did not seek the weak or sick but brought down large numbers at one time, including the best animals in the herds.

In support of this hypothesis, we know that most Pleistocene extinctions involved large terrestrial animals. Large herd animals were easily accessible and reproduced slowly. Large numbers could be driven off cliffs or into ravines, providing opportunities for slaughter of hundreds at the same time.

In contrast, marine genera, protected by the sea from human predation, continued to thrive. Small mammals also survived. Although frequently hunted, they were difficult to exterminate because of their more rapid breeding rate, their greater numbers, and the probability that they were killed only one at a time.

Climate change—Did climatic change cause the extinction of large Pleistocene mammals? The evidence is ambiguous. Large mammals were primarily adapted for living in extreme cold, yet they had successfully survived the warmer interglacial stages. Perhaps with the rapid retreat of the Wisconsin ice sheet they were unable to adapt to new sources of plant food, or perhaps their reproductive capacity was curtailed by warmer conditions.

SUMMARY

- The Cenozoic was a time of gradual change to present-day conditions. The warm, humid climates common during much of the Mesozoic persisted into the early Paleogene, but were replaced by cooler and dryer Neogene conditions. As a result, pervasive changes in land vegetation occurred as dense jungles gave way to deciduous forests, park-like woodlands, scrublands, and grasslands.

- Large discoidal foraminifera such as *Nummulites* were characteristic of areas in the Tethys Sea and the tropical western Atlantic. Smaller planktonic foraminifera proliferated and underwent a marked diversification during the Cenozoic.
- Marine invertebrates of the Cenozoic generally were similar to those living today. Reef-building corals, mollusks, echinoderms, and crustaceans were abundant.

- Teleost fishes with skeletons composed almost entirely of bone became varied and abundant.
- Dinosaurs did not survive into the Cenozoic. But other reptiles—turtles, crocodiles, lizards, snakes, the tuatara, and crocodiles—survived to modern times.
- Although the bird fossil record is poor, it is likely that most modern groups were present by Eocene time. Nearly all continents seem to have had one or more species of large, flightless, flesh-eating birds during the Cenozoic.
- The Cenozoic is the "Age of Mammals." During this era, mammals expanded rapidly into a multitude of vacated habitats. From ancestral shrew-like creatures, either directly or indirectly, the higher groups of placental mammals evolved.
- The adaptive radiation of mammals was as remarkable as that of the Mesozoic reptiles. In addition to adapting to living on and in the ground, mammals conquered the air (bats) and others returned to the sea (whales, seals, walruses), where vertebrate evolution had begun long ago.
- Cenozoic herbivores with hoofs—the ungulates—include two great categories: perissodactyls (odd-toed) and artiodactyls (even-toed). Perissodactyls include horses, rhinoceroses, and tapirs, as well as the extinct titanotheres and chalicotheres. Among artiodactyls, the large, piglike entelodonts and smaller oreodonts failed to survive beyond the Oligocene. However, living artiodactyls include swine, camels, hippopotami, and a diverse host of ruminants.
- Ungulates underwent specialization of teeth and limbs for grazing and running across relatively open plains. Gritty grass tends to wear teeth rapidly. To compensate, ungulates evolved teeth with complicated ridges of hard enamel and exceptionally high crowns.
- The earliest proboscidean fossils were discovered in Eocene beds in Egypt. Their radiation produced an impressive array of variously tusked and specialized mastodons and mammoths.
- The Pleistocene Epoch was a time of splendid diversity among terrestrial mammals. The fauna probably was more varied and impressive than that of Africa a century ago. The Pleistocene also was an age of giants: colossal ground sloths, beavers over 2 meters tall, bison, huge mammoths, and woolly rhinoceroses.
- Widespread extinction was the dominant biologic theme at the end of the Pleistocene. Either overkill by human hunters or climate change may have contributed to the demise of many giant Pleistocene mammals.

KEY TERMS

arsinothere, p. 509

artiodactyl, p. 512

atoll, p. 493

barrier reef, p. 493

bovid, p. 516

brontothere, p. 511

carnassial teeth, p. 506

Carnivora, p. 507

castorimorph, p. 506

cetacean, p. 519

chalicothere, p. 511

convergent evolution, p. 504

Creodonta, p. 506

dinothere, p. 517

edentate, p. 505

entelodont, p. 512

fringing reef, p. 493

glyptodont, p. 505

mammoth, p. 517

marsupial, p. 503

marsupium, p. 503

mastodon, p. 516

monotreme, p. 502

myomorph, p. 506

oreodont, p. 512

placental mammal, p. 505

perissodactyl, p. 509

proboscidean, p. 509

protorogomorph, p. 506

ruminant, p. 515

sciuromorph, p. 506

scleractinian coral, p. 492

scleractinid, p. 492

squamate, p. 498

teleost, p. 496

ungulate, p. 509

QUESTIONS FOR REVIEW AND DISCUSSION

1. What water depth do reef corals prefer? How does water depth control coral growth? Explain the development of atolls.

2. Which groups of marine phytoplankton proliferated during the Cenozoic Era?

3. How did the cephalopod faunas of the Cenozoic differ from those of the preceding era?

4. Which group of fishes is particularly characteristic of the Cenozoic?

5. What evidence suggests that whales are descendants of artiodactyls?

6. How might a paleontologist determine whether a fossil lower jaw containing teeth belonged to a mammal rather than a reptile?

7. What are ruminants? What advantages may be inherent in the ruminant type of digestion?

8. How is it possible to trace the development of the trunk or proboscis in the skeletal remains of proboscideans when the trunk itself is not preserved?

9. If evolutionary success can be measured in terms of diversity of a particular group of animals, how would you rate perissodactyls compared to artiodactyls?

10. Prepare a list of changes that have occurred in horses during their Cenozoic evolutionary history.

11. Describe the adaptations seen in Cenozoic herbivores that are related to the spread of prairies.

12. Cenozoic whales demonstrate a remarkable example of evolutionary convergence with the marine reptiles known as ichthyosaurs of the Mesozoic. What characteristics of whales, however, are distinctly *not* reptilian?

WEB SITES

The Earth Through Time Student Companion Web Site (www.wiley.com/college/levin) has online resources to help you expand your understanding of the topics in this chapter. Visit the Web Site to access the following:

1. Illustrated course notes covering key concepts in each chapter;

2. Online quizzes that provide immediate feedback;

3. Links to chapter-specific topics on the web;

4. Science news updates relating to recent developments in Historical Geology;

5. Web inquiry activities for further exploration;

6. A glossary of terms;

7. A Student Union with links to topics such as study skills, writing and grammar, and citing electronic information.

17

A band of Neandertals returns from a successful hunt.

Human Origins

It walked on its hind feet, like something out of the vanished Age of Reptiles. The mark of the trees was in its body and hands. It was venturing late into a world dominated by fleet runners and swift killers. By all the biological laws this gangling, ill-armed beast should have perished, but you who read these lines are its descendant.

—Loren Eiseley, *The Night Country,* 1971

O U T L I N E

▶ PRIMATES

▶ MODERN PRIMATES

▶ THE PROSIMIAN VANGUARD

▶ THE EARLY ANTHROPOIDS

▶ AUSTRALOPITHECINE STAGE AND THE EMERGENCE OF HOMINIDS

▶ THE *HOMO ERECTUS* STAGE

▶ FINAL STAGES OF HUMAN EVOLUTION

▶ HUMANS ARRIVE IN THE AMERICAS

▶ POPULATION GROWTH

▶ SUMMARY

▶ KEY TERMS

▶ QUESTIONS FOR REVIEW AND DISCUSSION

▶ WEB SITES

▶ BOX 17-1 ENRICHMENT

▶ BOX 17-2 ENRICHMENT

Key Chapter Concepts

- Primates evolved from small, arboreal, shrewlike insectivores.

- Adaptations among early primates that were important in the evolution of humans include hands capable of grasping and manipulating objects, an upright posture, stereoscopic vision, and enlargement of the brain.

- The two divisions of the Order Primates are the Suborders Prosimii and Anthropoidea. The Anthropoidea include humans, as well as monkeys and apes.

- Primate evolution began with the prosimians in the late Cretaceous. Prosimians were common in the Paleocene and Eocene forests.

- Australopithecines evolved primarily during the Pliocene and were the ancestors of early members of the genus *Homo.*

- *Homo erectus* lived primarily during the middle Pleistocene. *Homo sapiens* of the groups called Neandertals and Cro-Magnon people lived during the late Pleistocene.

- *Homo sapiens* reached the New World by migrating across the Bering Land Bridge during glacial stages of low sea level.

- The human population is growing at such a rapid rate that there is a danger it will exceed Earth's capacity to sustain it.

Evolution during the Cenozoic produced a large-brained mammal capable of shaping and controlling its own environment. That mammal was us: *Homo sapiens* (Latin: *intelligent human*). Unlike previous vertebrates, this remarkable creature profoundly changed the surface of the planet, modified the environment in ways both beneficial and destructive, and had such pervasive effects on populations of other creatures as to alter the entire biosphere. For these reasons, it is fitting that the final chapter in our history of Earth examines the evolution of humankind.

Homo sapiens shares many characteristics with other members of the Order Primates to which we belong. We resemble such primates as the great apes in basic body structure and biochemistry. On the other hand, we have become quite distinct from other primates in many ways:

- We have a larger, more complex brain.
- We stand and walk erect, and we have structural modifications of our vertebral column, legs, and pelvic bones to make such erect posture possible.
- Our face is flatter.
- Our teeth are less robust.
- We are capable of extraordinary manual dexterity.

These attributes have contributed to our ability to conceive, manufacture, and use sophisticated tools. Moreover, *Homo sapiens* exceeds all other primates in intelligence, which has led to language, culture, and aesthetic sensibility.

▶ PRIMATES

Primate Characteristics

What traits qualify a mammal as a member of the Order Primates? This question cannot be answered easily, for primates are structurally *generalized* compared to most other groups of mammals, which are *specialized* for a specific diet, running, swimming, or burrowing.

Primates retain the primitive number of five digits, have unspecialized teeth for eating either plant food or meat, and never developed such special features as hoofs, horns, trunks, or antlers. In the course of their evolution from shrewlike insectivores, primates have evolved large brains, modifications of the hand (Fig. 17-1), foot, and thorax, and binocular vision. Initially, these adaptations related to early life in trees and how primates obtained food. Ultimately, these adaptations were critical to the evolution of humans.

Snowball, the pig in George Orwell's *Animal Farm*, remarked that "the distinguishing mark of man is the hand, the instrument with which he does all his mischief." Although it expresses a biased point of view, the phrase is correct in suggesting the importance of the primate's grasping hand with its opposable thumb (Fig. 17-2). This characteristic not only permitted primates a firm grip on tree branches, but also allowed them to grasp, release, and manipulate food and other objects. Rotation of the ulna and radius upon one another in the forearm permits the hands to be turned at various angles and even reversed in position.

The development of the hand was accompanied by improvement in visual attributes. The eyes of primates became positioned toward the front of the face so that there was considerable overlap of both fields of vision. The result was an ability to see in "3-D," or to see depth—in other words, such eyes permit judging distance.

It would seem that grasping hands and feet and good stereoscopic vision are obvious adaptations for an animal

FIGURE 17-1 **Right hand of the Eocene prosimian** *Europolemus.* The hand shows key primate characteristics: broad fingertips and an opposable thumb (the thumb can be touched to each of the other fingertips). Thus the primate lineage extends back in time about 50 million years.

that leaps from branch to branch and seeks its food from precarious boughs. Yet many **arboreal** (tree-dwelling) animals such as squirrels, civets, and opossums lack the short face, close-set eyes, and opposable digits. And they get along very well. For this reason, anthropologists have recently suggested that the visual attributes of primates originated as adaptations that allowed early insect-eating primates to gauge accurately the distance to their prey without movement of the head. By being able to grasp narrow supports securely with its feet, the animal was able to use both of its mobile hands to catch the prey as it flew or scampered away.

Claws are used by animals such as squirrels in moving about on relatively wide branches. They would not be as advantageous for climbing about on thin boughs and vines, and might not provide a sufficiently secure hold for the quick catch of a moving insect. Thus, it seems that binocular vision and the grasping hand may not have been originally related to tree-swinging alone, but to precision and safety in the capture of visually located prey in the insect-rich canopy of tropical forests.

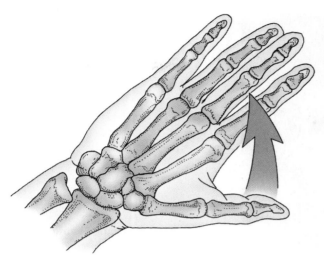

FIGURE 17-2 **Right hand of a human (palm up).** The human hand is not used in locomotion. It can be used to manipulate small objects between the fingers and thumb.

Other evolutionary modifications of primates were related primarily to changes in the eyes and limbs. On the outer margin of each eye orbit, a vertical ridge of bone developed (postorbital bar) that protected the eyes from bulging jaw muscles and accidental impact. As the

eyes became positioned more closely together, the snout was reduced so that the face became flatter. In response to a **brachiating** habit (swinging from branches), forelimbs and hindlimbs diverged in form and function, and a predisposition toward upright posture developed.

MODERN PRIMATES

Table 17-1 shows the two major primate groups:

- **Prosimii** (includes tree shrews, lorises, and tarsiers)
- **Anthropoidea** (the word means human-like, and includes monkeys, apes, and humans)

Prosimii

The Prosimii are more primitive than the Anthropoidea. For example, shrews (Fig. 16–28) possess clawed feet and a long muzzle with eyes on the sides of the head. Lemurs, confined largely to Madagascar, also have long snouts and lateral eyes, although more to the front than those in the tree shrew (Fig. 17-3). Some of the digits are clawed, whereas others have flattened nails. The East Indian tarsier has a relatively flat face (Fig. 17-4). As befits a **nocturnal** animal (feeds at night), the eyes are exceptionally large and positioned toward the front to provide stereoscopic vision.

TABLE 17-1

Order	Suborder	Superfamily	Common Names of Representative Forms	
Primates	Prosimii	*TUPAIOIDEA*	Tree shrew	
		LEMUROIDEA	Lemur	
		LORISOIDEA	Bush baby, Slender loris	
		TARSIOIDEA	Tarsier	
	Anthropoidea	*CEBOIDEA* (New World Monkeys)	Howler monkey, Spider monkey, Capuchin monkey, Common marmoset, Pinche monkey	
		CERCOPITHECOIDEA (Old World Monkeys)	Macaque, Baboon, Wanderloo, Rhesus monkey, Common langur, Mandrills, Proboscis monkey	
			Family ⇓	
		HOMINOIDEA	HYLOBATIDAE	Gibbon, Siamang
			PONGIDAE	Orangutan, Chimpanzee, Gorilla
			HOMINIDAE	Humans

FIGURE 17-3 **Ring-tailed lemur of Madagascar.** Note the prominent muzzle.

Anthropoidea

Monkeys are the most primitive members of the Anthropoidea. There are New World and Old World monkeys (Fig. 17-5). The **Ceboidea**, or New World monkeys, are an early branch not involved in the eventual evolution of humans. Included among the ceboids are the spider monkey and marmoset, as well as the familiar little capuchin, or "organ-grinder monkey." New World monkeys have flattish faces, widely separated nostrils, and **prehensile** tails that can be wrapped around branches. Most are small in comparison to their Old World cousins. Their oldest remains are found in Oligocene beds of South America.

The more advanced Old World monkeys, or **Cercopithecoidea**, are widely distributed in tropical regions of Africa and Asia. They include the familiar macaque, or rhesus monkey of laboratories and zoos; the Barbary ape of Gibraltar; langurs; baboons; and mandrills. In this group of monkeys, the nostrils are close together and directed downward (as in humans), and the tail is not prehensile.

Anthropoid apes (Fig. 17-6) are tail-less primates. Modern species evolved from the same ancestral stock that produced humans. DNA evidence indicates the divergence probably took place between about 7 and 5 million years ago. In this regard, comparative analysis of the DNA of humans, chimpanzees, gorillas, and other apes, as well as similarities in the proteins hemoglobin and myoglobin, indicates that the chimpanzee is our closest relative. There is a 98.4% correlation in DNA sequences between chimpanzees and humans.

The **Hylobatidae**, such as the gibbons, are the more primitive branch of the tail-less apes. Orangutans, chimpanzees, and gorillas are grouped within the **Pongidae**. Gorillas are essentially ground dwellers and spend only a small fraction of their time in the trees.

▶ THE PROSIMIAN VANGUARD

The early fossil record for primates includes a creature named *Purgatorius*, known only from a few teeth discovered in the Late Cretaceous Hell Creek Formation at Purgatory Hill in Montana. These finds indicate that the earliest primates were contemporaries of the last of the dinosaurs. (The fossil record for primates improves somewhat in the early epochs of the Cenozoic.)

Plesiadapis (Fig. 17-7), found in Paleocene beds in the United States and France, is the only primate genus (other than *Homo*) that has inhabited both the Old World and New World. This is one of many clues that the continents were not yet completely separated by the widening Atlantic during the Paleocene. *Plesiadapis* was a distinctive and specialized primate and represents a sterile offshoot of the primate family tree. Its incisors were rodentlike and separated from the cheek teeth by a toothless gap (diastema). Fingers and toes terminated in claws rather than nails. The rodentlike characteris-

FIGURE 17-4 **Philippine Tarsier (*Tarsius syrichta*).** This tree-dwelling prosimian has large, forward-directed eyes.

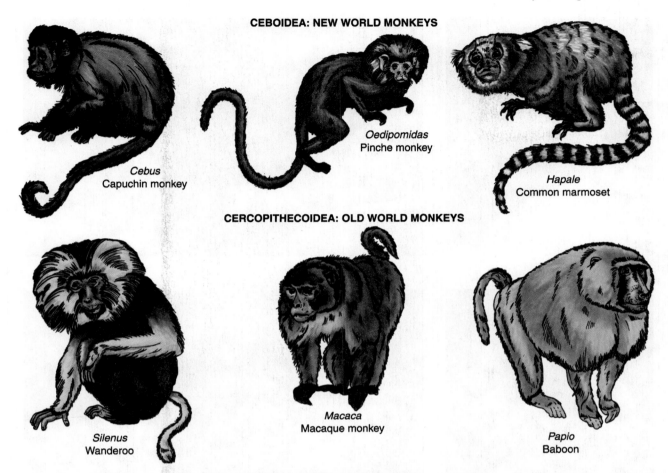

CEBOIDEA: NEW WORLD MONKEYS

Cebus
Capuchin monkey

Oedipomidas
Pinche monkey

Hapale
Common marmoset

CERCOPITHECOIDEA: OLD WORLD MONKEYS

Silenus
Wanderoo

Macaca
Macaque monkey

Papio
Baboon

FIGURE 17-5 **Representative New World and Old World monkeys.** Most New World monkeys have prehensile tails (adapted for grasping) that function almost as effectively as another limb.

tics and habits of these prosimians may have contributed to their extinction by late Eocene time.

General trends in prosimian evolution during the Eocene Epoch (55–37 million years ago) involved these changes:

- Reduction in muzzle length
- Increase in brain size
- Shifting of eye orbits to a more forward position
- Development of a grasping big toe

These trends are evident in the fossil remains of *Cantius* from the Wind River Basin of Wyoming. In the Wyoming stratigraphic sequence, you can trace the evolution of successive species of *Cantius*. Near the top of the seqence, a new genus, *Notharctus* (Fig. 17-8), occurs as the apparent direct descendant of *Cantius*.

Notharctus has been known from other localities for over a century. In general form, it appears to be just the kind of lemurlike animal from which monkeys and apes evolved. Prosimian populations during the Late Paleocene and Eocene also included tarsiers. *Tetonius* was

an Eocene form with tarsioid traits of closely spaced large eyes and a shortened muzzle.

Both tarsiers and lemurs were abundant and widely dispersed on Northern Hemisphere continents during the Eocene. However, with the advent of cooler Oligocene climates, they virtually deserted North America. In the Eastern Hemisphere they were forced southward into the warmer latitudes of Asia, Africa, and the East Indies. Surviving prosimians are much reduced in variety and number. They probably did not do well in competition with growing numbers and varieties of monkeys.

▶ THE EARLY ANTHROPOIDS

Anthropoids (the higher primates: monkeys and apes) were the next evolutionary step. Discoveries in the Fayum region south of the Nile River delta in Egypt give us a wealth of information about early anthropoids. More than a hundred specimens of Fayum fossils occur in several Oligocene horizons. Many of the

A

B

C

D

FIGURE 17-6 **Anthropoid apes.** (A) Gibbons are considered by many scientists to be the most perfectly arboreal of the anthropoid apes. (B) Orangutans are solitary apes that seldom leave the protection of the trees. (C) A gorilla knuckle-walking. (D) Chimpanzees live in groups and have complex social behavior.

Pre-Є | Є | O | S | D | M | P | P | Pr | Tr | J | K | Pg | Ng

FIGURE 17-7 **Paleocene prosimian *Plesiadapis*.** ◗ *What features of its skull are rodentlike?*

FIGURE 17-8   **Eocene prosimian** *Notharctus*.

skull fragments and teeth retain subtle vestiges of prosimian ancestry, but none of the remains are prosimians. They are fossils of primates that reached the monkey stage of organization.

One primate discovered at Fayum is *Aegyptopithecus* (Fig. 17-9), a relatively robust arboreal anthropid with monkeylike limbs and tail, a brain larger than that of *Notharctus*, and eye orbits rotated to the front of the skull. Bone fragments of *Aegyptopithecus zeuxis* have been dated at 33–34 million years old, indicating that the prosimian-anthropoid transition had taken place by Oligocene time. Egypt was not as arid during the Oligocene as it is now. It was a well-watered region covered with lush tropical forests.

Evolution is a continuing process in which each animal is a transitional link between older and younger species. Thus, it is difficult to assign the fragment of a fossil jaw or tooth to either the monkey or the ape category. One of many clues used to make this assignment is the teeth—specifically, the cusp pattern on molar teeth. In Old World monkeys, certain molars have four cusps. Among apes and humans, these same molars have five cusps, with an intervening Y-shaped trough. *Aegyptopithecus* molars display the Y-5 pattern (Fig. 17-10), which is one reason why anthropologists consider the genus possibly to be the very early ancestor of Miocene apes. Evidence indicates that the transition from prosimians to

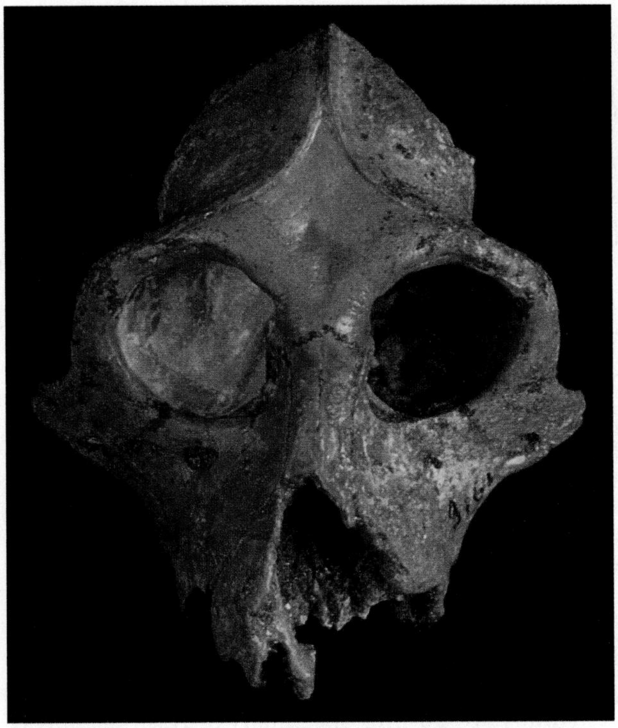

FIGURE 17-9   **Early ape, *Aegyptopithecus zeuxis*, from Oligocene fossil beds near Cairo.** *Aegyptopithecus* is considered an early ape that preceded the Miocene apes exemplified by *Dryopithecus*.

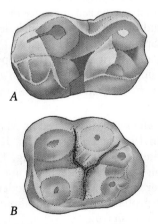

A

B

FIGURE 17-10   **General pattern of cusps on the molars of Old World monkeys.** (A) Baboon lower molar showing a cusp at each corner. (B) Chimpanzee lower molar showing the "lazy Y-5" pattern characterized by a Y-shaped depression that separates five cusps.

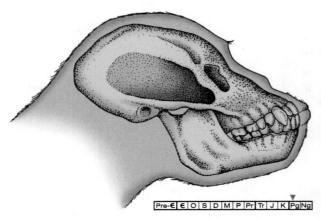

Pre-€ | € | O | S | D | M | P | Pr | Tr | J | K | Pg|Ng

FIGURE 17-11 **Skull of *Proconsul* (a dryomorph) from Lake Victoria, Kenya.**

anthropoids, as well as the differentiation of apes from monkeys, occurred during the Oligocene.

The Miocene was an important epoch in primate evolution, and plate tectonics played a role. The unfragmented block containing Africa and Arabia drifted northward, ultimately colliding with Eurasia. This event blocked east-west circulation of tropical currents across the former Tethys Sea, so East Africa became cooler and dryer. The dense forests waned, replaced by extensive grass-covered plains. These environmental changes created selective pressures that stimulated adaptive radiation of Old World primates. During the Late Miocene, primates appeared that gave rise to the pongids (chimpanzee, gorilla, and orangutan family) on the one hand, and the hominids on the other.

New players during the Miocene included:

• **Dryomorphs** are named from *Dryopithecus fontani*, a species first discovered in France in 1856. Variable in size and appearance, they include forms from France, Spain, Greece, Hungary, Turkey, India, Pakistan, and Africa. Many anthropologists consider the important species *Proconsul africanus* to be either a very early dryomorph or the immediate dryomorph ancestor.

• *Proconsul africanus*, discovered by Mary and Louis Leakey in 1948 at Lake Victoria, Kenya, has an apelike skull, jaws, and teeth (Fig. 17-11). However, it also has a monkeylike long trunk, arms, and finger bones. Much like modern monkeys, *Proconsul* probably scampered about on all four legs and lived on fruit.

By about 18 million years ago, after Africa converged on Eurasia, monkeys and apes migrated into Eurasia. The Miocene apes *Ramapithecus, Sivapithecus,* and

Gigantipithecus established themselves. Loosely termed **ramamorphs**, these primates are considered the ancestral stock for both later apes and hominids.

Miocene primates lack an orderly sequential change along a single trend. The primate family tree is a complex of parallel and diverging branches. It is more of a family bush than a family tree. Tracing the ascent of humans through the many splits and dead ends is difficult. Paleontologists must reinterpret the story each time new fossil material is uncovered.

▶ THE AUSTRALOPITHECINE STAGE AND THE EMERGENCE OF HOMINIDS

The story of identifying human (**hominid**) emergence begins in 1924. Anthropologist Raymond Dart (1893–1988) discovered fossil remains of an immature primate in a South African limestone quarry. He named the fossil *Australopithecus africanus* (Fig. 17-12). In succeeding years, others found many additional *Australopithecus* skeletal fragments. East African fossil sites have yielded hundreds of hominid bones. These fossils provide an unsurpassed record of human evolution over the past 4 million years. Some of the eastern African sites, such as Olduvai Gorge (Fig. 17-13), are famous as a result of lifelong research programs of the Leakeys.

Volcanic eruptions were frequent along the eastern margin of Africa during the Cenozoic, and many of the fossil sites have interspersed layers of volcanic ash and lava flows. This has greatly helped paleoanthropologists, for ash can be dated by the potassium-argon method. Dates from two succeeding ash beds provide a close age estimate for intervening fossil-bearing sediment layers.

The oldest known hominid fossils are in the central African nation of Chad, just west of Sudan (Fig. 17-13). Named *Sahelanthropus tchadensis*, this upright, four-foot-tall primate lived 7 to 6 million years ago. It was an unexpected discovery, for previous early hominids had been found only in eastern and southern Africa. Also, the Chad fossils are older than east African hominids. Prior to the discovery of *Sahelanthropus*, the oldest known hominid was the Ethiopian *Ardipithecus ramidus*, dated at 5.2 to 5.8 million years (Fig. 17-14).

Each new field season discovers specimens younger than *Ardipithecus ramidus*. These discoveries have provided new information about the complex phylogenetic "bush" of early hominids. Among recent finds are 20-plus specimens of *Australopithecus anamensis* from Kenya's Lake Turkana (formerly Lake Rudolf). *Australopithecus anamensis* lived from about 4.2 to 3.9 million years ago, and thus existed

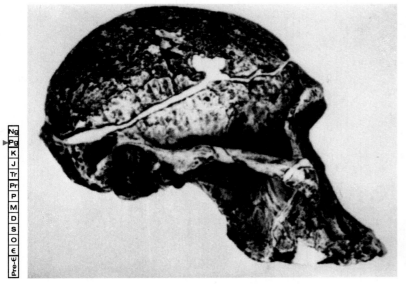

FIGURE 17-12 *Australopithecus africanus* **from the Transvaal of South Africa.** Anthropologist Raymond Dart discovered its remains in a South African limestone quarry.

more recently than *Ardipithecus ramidus*. The species appears to show intermediate evolution between *Australopithecus ramidus* and *Australopithecus afarensis*, the famous "Lucy."

"Lucy" (*Australopithecus afarensis*)

"Lucy" was discovered by Donald Johanson and nicknamed after the Beatles song, "Lucy in the Sky with Diamonds." Later evidence shows that Lucy may have been male, but the nickname has stuck (Fig. 17-15).

The pelvic and leg bones tell us that *Australopithecus afarensis* walked erect. Further evidence of bipedalism came with the discovery of footprints at Laetoli (Fig. 17-16). We can trace these footprints over 9 meters (30 feet), and they were made by two contemporaries of Lucy who walked side-by-side over a layer of soft,

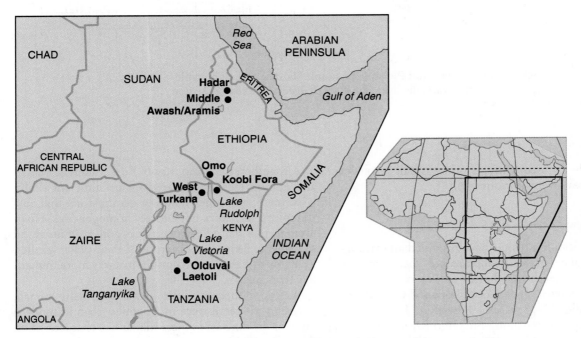

FIGURE 17-13 **Rich hominid fossil sites in East Africa.** Beds of volcanic ash between fossil beds facilitate radioisotope dating in this region, and dry conditions limit the amount of vegetation covering bone beds.

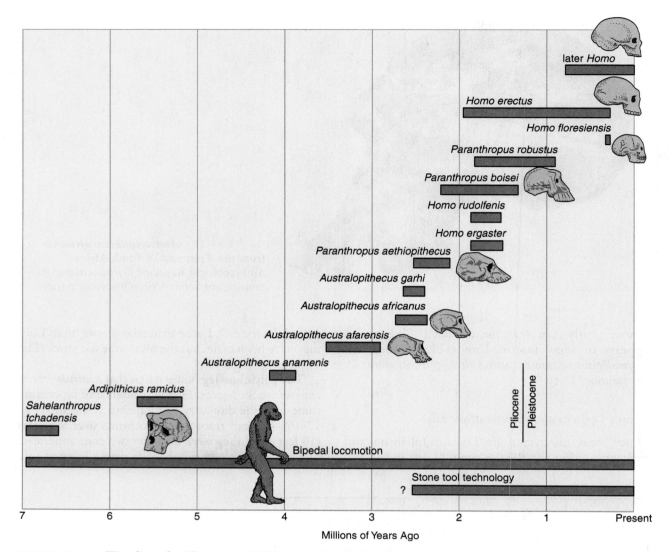

FIGURE 17-14 Timelines for Pliocene and Pleistocene hominids. The earliest hominid presently known is *Sahelanthropus tchadensis*, discovered in 2002. It lived in the central African nation of Chad, for which it is named.

volcanic ash. (Subsequently, another ashfall formed a protective seal over the footprints.)

Rich African Fossil Sites

Eastern African fossil sites have provided an extraordinary number of **australopithecine** specimens, as well as a few teeth and bones of our own species, *Homo*. Such sites include Koobi Fora and Omo (location map, Fig. 17-13; Koobi Fora, Fig. 17-17). The Omo digs are located along the Omo River in remote southwestern Ethiopia. Here, a nearly continuous section of sediments and volcanic ash beds span 2.2 million years.

Altogether, the Omo, Koobi Fora, Olduvai, Hadar, and Laetoli localities have given us sufficient skulls, jaws, teeth, and other bones to demonstrate that australopithecines were not entirely homogeneous. However, certain shared characteristics appear in nearly all specimens. From the structure of their pelvic girdles, we know that they stood upright in a way more humanlike than apelike.

Australopithecine dentition was essentially human, although the teeth were more robust than in modern humans. In contrast to these humanlike characteristics, the australopithecine cranial capacity of about 600 cubic cm was far less than the 1400–1600 cubic cm in modern humans.

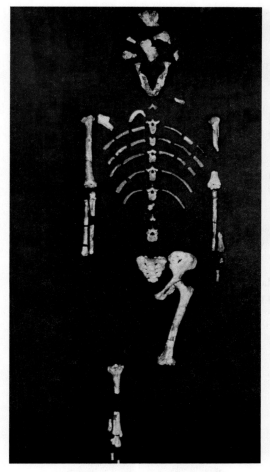

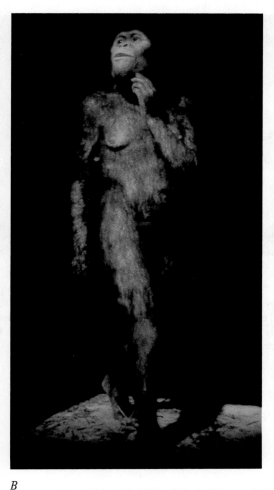

A   *B*

FIGURE 17-15 **"Lucy," formally named *Australopithecus afarensis*.** Lucy was a young, erect-walking hominid who lived in East Africa about 3.5 million years ago. Naming this fossil Lucy may have been premature, for anthropologists at the University of Zurich recently proposed that Lucy might have been male. (A) The specimen shown is the oldest-known, most complete skeleton of an erect-walking human ancestor. (B) A reconstruction of Lucy on display at the St. Louis Zoo.

In the past, paleoanthropologists distinguished two kinds of australopithecines:

- "Gracile" ("gracefully slender") australopithecines had lighter, smaller teeth and somewhat smaller body size. Lucy is an example.
- "Robust" australopithecines were heavier, larger-toothed types. Cheek teeth had large grinding surfaces, suggesting a diet that included coarse plant food. Species include *Australopithecus boisei* and *Australopithecus robustus* (Fig. 17-18). Both appear to be evolutionary side branches that went nowhere.

Genus *Homo*

Interestingly, the transition from gracile australopithecines to hominids of the genus *Homo* is not marked by striking anatomical differences. Compared to australopithecines, early *Homo* species had a higher cranial vault and a somewhat larger cranial capacity. The opening at the base of the cranium, through which the spinal cord joins the brain (the foramen magnum), moved slightly forward, indicating more erect posture. Premolars were narrower, with a shorter row of cheek teeth. Notably, crude stone tools are found in association with *Homo* fossils.

A   *B*

FIGURE 17-16   **"Footprints in the ashes of time," evidence for bipedalism.** (A) Footprints probably made by *Australopithecus afarensis* about 3.2 million years ago as a pair of these hominids walked across wet volcanic ash. The footprints confirm skeletal evidence that the species had a fully erect posture. (B) A female and male stroll the Late Pliocene landscape of eastern Africa, leaving telltale footprints in the ash.

FIGURE 17-17 **The Koobi Fora site in eastern Africa provides an exceptional record of some steps in hominid evolution.** Both australopithecine remains and the skull of an early member of the genus *Homo* have been found here. Paleontologists also found pebble choppers and flake tools associated with the bones of a hippopotamus in deltaic deposits. Finding the tools and bones together suggests that this was a butchering site. We infer that about 1.8 million years ago, some australopithecines came across a recently deceased hippopotamus and feasted, using the sharp edges of flaked chert to get the meat.

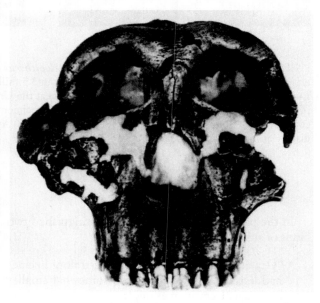

FIGURE 17-18 *Ausralopithecus robustus.* Because these hominids had larger molars and premolars than *Australopithecus africanus*, some paleoanthropologists think they were primarily plant-eaters and taxonomically distinct from *A. africanus*. Recent interpretations favor inclusion of these creatures under the generic name *Australopithecus*. This specimen derived from Bed I on the south side of the main ravine at Olduvai Gorge, Tanzania.

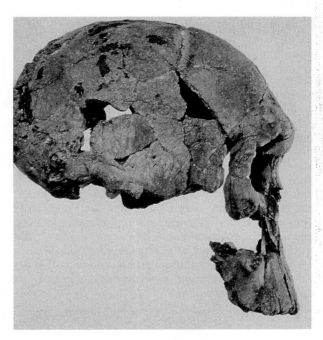

FIGURE 17-19 **By about 2 million years ago, *Homo rudolfensis* roamed the African plains.**

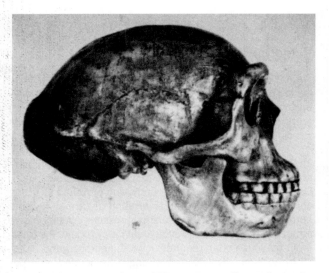

FIGURE 17-20 **Replica of *Homo erectus*.** It was formerly known as *Pithecanthropus erectus* from Java and *Sinanthropus pekinensis* from China.

The oldest *Homo* remains are nearly 2.5 million years old. Jaw fragments and teeth occur near stone tools of the same age. By about 2 million years ago, *Homo rudolfensis* (Fig. 17-19) and *Homo ergaster* (also called *Homo habilis*) were roaming the African plains.

What conditions triggered the evolution of *Homo* from *Australopithecus*? We now have ample evidence that about 2.7 million years ago, Africa began to grow cooler and drier. As a result, rainforests gave way to broad expanses of grasslands. In this more open and difficult environment, hominids would have experienced selective pressures that could have led to improved bipedalism, greater intelligence, and the development of a stone tool technology.

THE *HOMO ERECTUS* STAGE

The next stage in hominid evolution is represented by *Homo erectus* (Fig. 17-20). This is the first hominid known to have moved out of Africa into Eurasia. Noteworthy discoveries include:

- A lower jaw with teeth from the Caucasus Mountains (1.8 million years old)
- A largely complete skeleton from the western Turkana fossil site (1.5 million years old)
- A skull cap from Bed II of Olduvai Gorge (750,000 years old) (Fig. 17-21)

The Turkana skeleton is that of a boy, aged 11–13, who perished in a marsh and was quickly covered with mud

so his body was not shredded by scavengers. The skeleton reveals a long-limbed, tall, slender individual. Adults would be about the same size or larger than many natives of eastern Africa today. Below the neck, the Turkana boy had a remarkably modern skeleton.

Although the skeleton of *Homo erectus* was generally similar to that of modern humans, the skull was not. Cranial capacity ranged from about 775 cubic cm to nearly 1300. In comparison, brain capacity in modern *Homo sapiens* is 1400–1600 cubic cm. Thus, the maximum brain size of *Homo erectus* approaches near the bottom range of modern people. Most would agree that *Homo erectus* represents a stage in hominid evolution

FIGURE 17-21 **Upper skull portion of *Homo erectus* specimen from Olduvai Gorge.** This specimen was found associated with stone implements.

ENRICHMENT

Being Upright: Good News, Bad News

We are unique among primates in several ways. One is by having a body structured for standing fully erect and walking on two legs. Our nonhuman ancestors walked on all fours. But the evolutionary transition from those ancestors to hominid bipedalism had its challenges.

For example, the human vertebral column is somewhat imperfect as a vertical support. That imperfection is one reason why so many people are plagued by lower back pain. Under the weight of the upper body, and provoked by lifting and unusual movements, intervertebral discs may become herniated and protrude, causing pressure against nerve structures. Severe pain and disability may follow, the treatment for which often requires delicate neurosurgery.

In addition, evolutionary changes associated with our upright stance have increased the distance between the lowest ribs and the top of the pelvis. This gives us our dis-tinctive waist, but also weakens the abdominal wall. Thus, we are particularly prone to hernias (ruptures).

But for survival, the advantages of erect posture far outweigh the disadvantages. An upright stance and bipedalism freed our hands for the precise manipulation that led to tool manufacture and the advent of technology. For humans, hands are not merely motor organs. They also are sensory organs that can investigate by touch, "seeing" into places the eye may not reach. Our hands explore and react to what has been discovered. They can gesture, give instruction, indicate directions, and even express emotion.

If we merely walked on our hands, what chance would we have had to evolve the complex human brain that provides for abstract thought, symbolic communication, and the development of culture?

---

during which relatively rapid increase in brain size had begun. No doubt the expansion of the brain involved the reshaping of both the cranium and the birth canal to accommodate fetuses with larger heads.

Aside from its larger cranial capacity, the *Homo erectus* skull was massive and rather flat. Above the eyes were heavy, bony, supraorbital ridges (Fig. 17-20). The forehead sloped, and the jaw jutted forward at the tooth line (a condition termed **prognathous**). A jutting chin was lacking, and the nose was broad and flat. These were rather primitive traits. However, the teeth and dental arcade in *Homo erectus* were essentially modern, except for their being somewhat more robust.

There is ample evidence that *Homo erectus* made good use of his larger brain. From the bones of other animals found where they lived, it is clear these hominids were excellent hunters. They also were skilled at making simple implements of flint and chert (Fig. 17-22), such as axes and scrapers. Some also appear to have engaged in cannibalism, either for food or as a part of a ritual.

Paleoanthropologists do not know if *Homo erectus* spoke a distinctive language, wore clothes, or built dwellings. There is vague evidence that they hunted together in bands. What appear to be slaughter sites have been found in Europe. Unfortunately, these sites did not yield any bones of *Homo erectus*, to confirm their presence.

A few fossil localities in Europe and China contain traces of carbon. However, this is not indisputable evidence that *Homo erectus* had learned to use fire, for the carbon could have come from natural fires, commonly triggered by lightning.

Homo erectus fossils of differing ages and locations show some variability. Some paleoanthropologists consider this reason enough to divide *Homo erectus* into two or more species, whereas others consider the variability no greater than what we see today among our own species.

FINAL STAGES OF HUMAN EVOLUTION

The Neandertals

From the *Homo erectus* stage of the Middle Pleistocene, it is a short step to Late Pleistocene hominids called **Neandertals** (Fig. 17-23). Some paleoanthropologists regard Neandertals as a variety or subspecies of *Homo sapiens*, which they have designated *Homo sapiens neandertalensis*.

Radiocarbon dating of a jawbone from Croatia indicates that Neandertals were in central Europe as recently as 28,000 years ago (Figure 17-24). This means that Neandertals and early modern humans coexisted for thousands of years, with the inevitable result: interbreeding. Evidence for interbreeding came in 2002 with the discovery of a jawbone and skull fragments from a cave in the Carpathian Mountains. The bones possess a mix of Neandertal and modern human traits.

The first Neandertal specimen was found in Germany, in the Neander River Valley, giving the species its name. Subsequently, many additional fossils have been found, indicating that Neandertal people ranged across the entire Old World. With their heavy brow ridges and prognathous (jaws projecting forward from the face), chinless jaws, Neandertals have become the stereotypical "cave man." Indeed, their face seems a brutish carryover from the Middle Pleistocene.

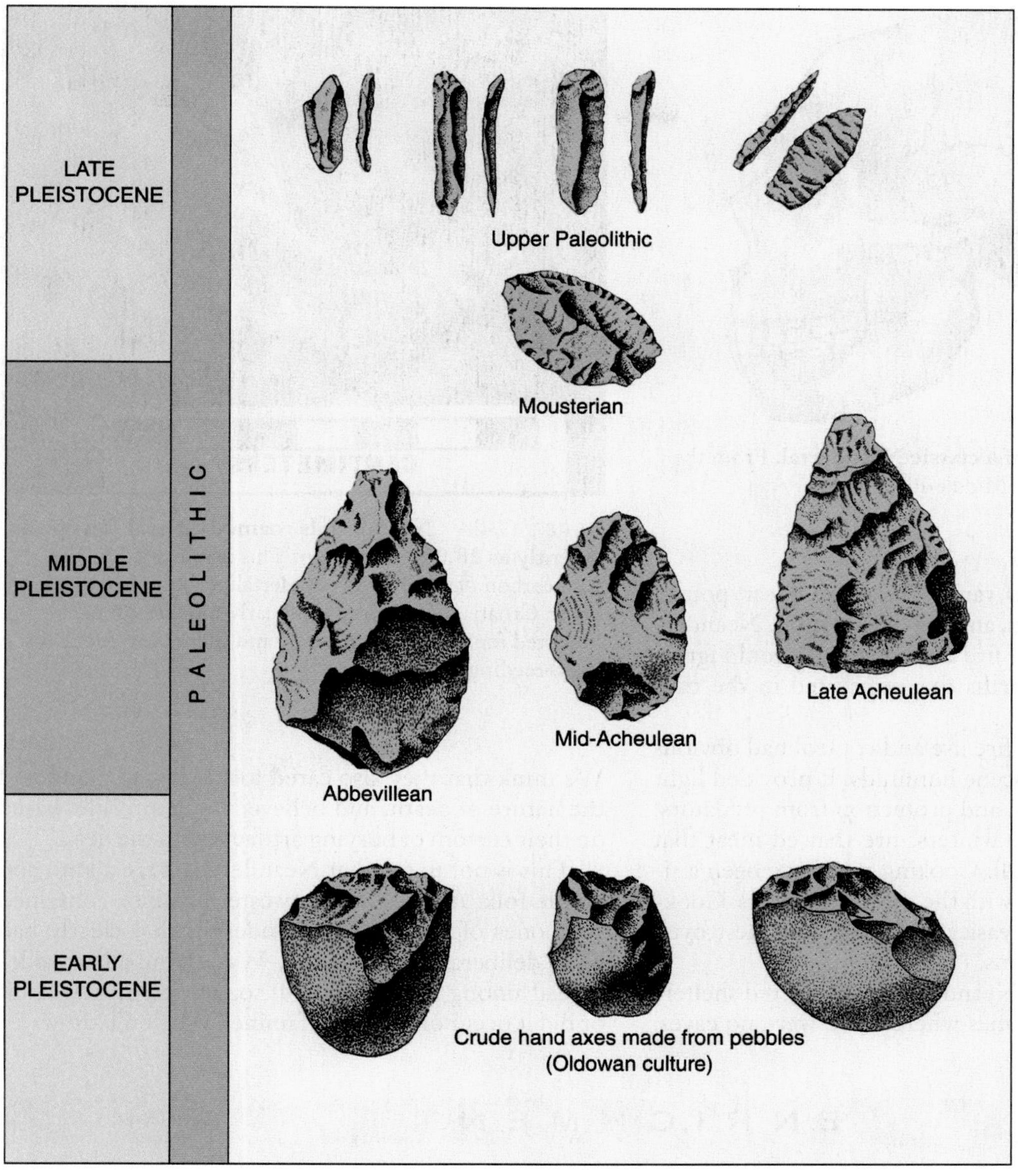

FIGURE 17-22 **Progressive improvement in tool-making from stone during the Pleistocene.** The crude stone tools of the Early Pleistocene (bottom) were produced by australopithecines. *Homo erectus* produced the better-shaped tools of the Middle Pleistocene. The Upper Paleolithic tools included carefully chipped blades and points. The next stage, not shown here, is the Neolithic, with refined and polished tools of many kinds.

Below the neck, Neandertal skeletons were somewhat more robust than our own. They were rather bulky around the middle because of a more flared rib cage and pelvis (Fig. 17-25). Their brain size equaled or even exceeded that of present humans. Thus, the once popular depiction of Neandertals as bent-kneed, flatfooted, bullnecked brutes with curved backs is incorrect. In fact, much of that early interpretation was from restoration of an elderly Neandertal skeleton that had severe osteoarthritis.

Most "classic Neandertals" were sturdy people of small stature. They apparently adapted well to chilly living near the edge of the glacial ice sheet, in a very cold climate. Relatively short limbs and a bulky torso may have helped them conserve body heat. Their large nasal cavities also may have been an adaptation for warming frigid air as it was inhaled.

Many Neandertals lived in caves (Fig. 17-26). They hunted mammoth, woolly rhinoceros, reindeer, bison, and fierce ancestors of modern cattle known as aurochs.

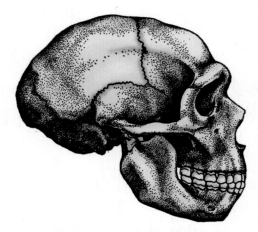

FIGURE 17-23 **Skull of a classic Neandertal.** From the *Homo erectus* stage of the Middle Pleistocene.

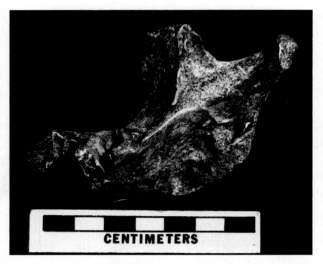

CENTIMETERS

FIGURE 17-24 **Neandertals roamed central Europe as recently as 28,000 years ago.** This is indicated by the radiocarbon date for this Neandertal jawbone from a cave site in Croatia. Neandertals and early modern humans coexisted for thousands of years, and there is evidence for interbreeding.

They manufactured a variety of stone spear points, scrapers, borers, knives, and saw-edged tools. Neandertals made ample use of fire and apparently could ignite one at will in the hearths they excavated in the cave floors.

The technology of fire use and control had obvious advantages for Pleistocene hominids. It provided light in caves, gave warmth, and protection from predators. During frigid Ice Age winters, fire thawed meat that froze quickly after a kill. Cooking may have been a simultaneous discovery with the thawing process. Cooking foods promoted easier digestion and destroyed harmful microorganisms.

In addition to fire, Neandertals constructed shelters of skins, sticks, and bones where there were no caves.

We think that they also cared for their sick, pondered the nature of death, and believed in an afterlife, based on their custom of burying artifacts with the dead.

This is not to say that Neandertals were a kind and gentle folk. A cave in southwestern France contained the bones of at least six Neandertals that clearly had been deliberately butchered. Was cannibalism widespread among Neandertals? If so, was it a dietary habit or did it occur only during famine? We don't know.

ENRICHMENT

Neandertal Ritual

The Neandertals lived from about 125,000 to about 40,000 years ago. Their culture included chipped flint tools, crude carvings, the use of fire, and burial of their dead in carefully prepared graves. Some burial sites contain indications of religious beliefs. For example, consider the La Chapelle fossil site in southwestern France. It contains a Neandertal male, placed in ritual position within a shallow grave, with a bison leg on his chest. Flint tools also are in the grave, possibly in the belief that they could be used in an afterlife.

At the Ferrassie Neandertal site in France, an apparent family cemetery was discovered within a rock shelter. Two adults, presumed to be a father and mother, are buried head to head. Nearby, three small children and a newborn infant are buried. Once again, flint flakes and bone splinters were placed in the grave of the adult male. A heavy, flat stone was placed over his head and shoulders. Perhaps the

stone was to protect the body, or possibly to restrain the deceased from returning to life.

That Neandertals had a special attitude about death also is evident at a fossil site in Uzbekistan, near Russia. Anthropologists working there uncovered an array of goat horns surrounding the buried body of a 9-year-old child.

Yet another interesting Neandertal fossil site is in Shanidar, Iraq. Researchers uncovered the remains of nine Neandertals, four of whom were deliberately buried. One burial contains assorted pollen grains, suggesting that flowers may have been placed in the grave.

Although not related to funerary ritual, the Shanidar site also includes the remains of a 30-year-old male who had been born with a crippled arm. The arm was amputated below the elbow, and this hardy Neandertal survived the surgery. The excessive wear on his teeth suggests that he used them to compensate for his missing limb.

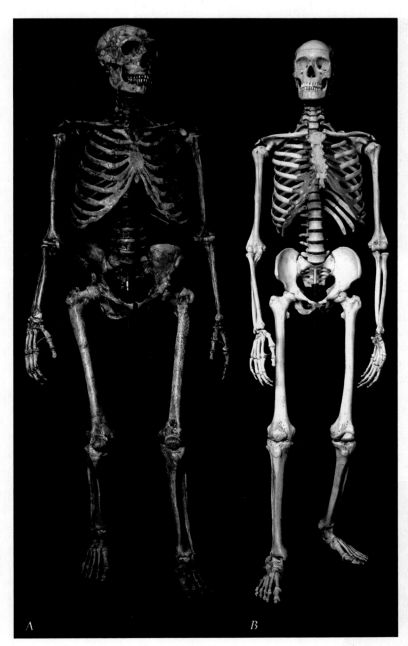

A   B

FIGURE 17-25 **(A) Neandertal skeleton assembled from actual skeletal bones compared with skeleton of a modern human.** (B) Note the flaring ribcase and pelvis of the Neandertal skeleton. The Neandertal reconstruction was produced by Blaine Maley at Washington University and Gary Sawyer at the American Museum of Natural History.

Little People of the South Pacific

In 2004, a team of Australian and Indonesian scientists announced a remarkable discovery: a new human species. Digging in a cave on the Indonesian island of Flores, they hoped to find evidence of *Homo erectus*. Instead, they unearthed stone tools and bones of a new species, *Homo floresiensis*.

The site produced bones of seven small individuals, including the complete skeleton of an adult female. Adults were only three feet tall, prompting the popular press to call them "hobbits of the South Pacific,"

after the short people in J.R.R. Tolkien's *Lord of the Rings*.

Although the ancestry of *H. floresiensis* is uncertain, anthropologists working at the site believe that *Homo erectus* reached Flores by about 800,000 years ago, and evolved into the smaller species as a result of living on an island with few large predators and limited food.

Homo floresiensis lived as recently as 13,000 years ago, and was a contemporary of Neandertals and modern humans. They were a clever people who knew how to hunt, use fire, and make carefully crafted points, blades, and awls from stone.

FIGURE 17-26 **Reconstruction of a Neandertal family group.** The technology of fire use and control had obvious advantages for Pleistocene hominids.

Cro-Magnon People

About 34,000 years ago, during the fourth glacial stage, humans closely resembling modern Europeans moved from Africa into regions inhabited by the Neandertals. They quickly were assimilated or replaced the Neandertals through tribal warfare and competition for hunting grounds.

These early members of our own species are called **Cro-Magnon people** (Figs. 17-27 and 17-28). They were mostly taller than Neandertals, had a more vertical

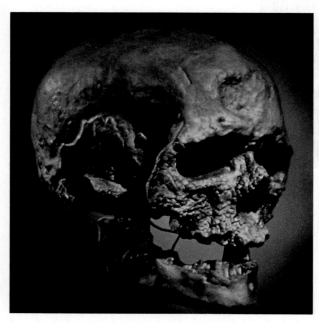

FIGURE 17-27 **Skull of Cro-Magnon human (*Homo sapiens sapiens*) about 30,000 years old.** As in present-day humans, the face of Cro-Magnon was vertical rather than prognathous (jaws projecting forward from the face). The heavy supraorbital ridges of the European Neandertals are not present. Cro-Magnon had a prominent chin.

brow, and had a decided chin projection (Fig. 17-29C). In short, Cro-Magnon's bones were modern, and anthropologists have recognized definite Cro-Magnon skull types among today's western and northern Europeans.

Cro-Magnon continued and further developed the cultural traditions of the Neandertals. Finely crafted spear points, awls, needles, scrapers, and other tools are found in Cro-Magnon caves. Handsome paintings and drawings were made on cave walls and ceilings. Engravings and sculptures include mammoths, horses, and women (Fig. 17-30). These were produced from fragments of bone or ivory. The statues of women probably were used in fertility rites.

Evidence is that Cro-Magnon people enjoyed wearing body ornaments and frequently fashioned necklaces from pieces of ivory, shells, and teeth. Burial of the dead became an elaborate affair. Hunters were buried with their weapons and children with their ornaments. This apparent concern for an afterlife and the sense of self-awareness that resulted in art and complex ritual suggest that the beginning of the age of the philosopher had arrived.

Beginnings of Recorded History

Through most of the species' early history, *Homo sapiens* was a wandering hunter and gatherer of wild edible plants. However, about 15,000 to 10,000 years ago, near the beginning of the Holocene Epoch, tribes began to domesticate animals and cultivate plants. They learned to grind their tools to unprecedented perfection and to make utensils of fired clay. With more reliable sources of food, permanent settlements developed. Individuals were spared the continuous demand of searching for food, and so were able to build and improve their cultures. Languages improved, and symbols developed into forms of writing. The era of recorded history had begun.

FIGURE 17-28 **Cro-Magnon hunters.**

HUMANS ARRIVE IN THE AMERICAS

The record for hominids in the New World is discontinuous and often ambiguous. Too often, anthropologists must trace human presence through discarded fragments of tools or weapons, rather than actual skeletal remains. Actual dating can be difficult and flawed by contamination. For these reasons, it still is not definitely known when the first human set foot in America.

A classic hypothesis suggests that bands of humans entered North America by way of a land bridge spanning the present trend of the Bering Straits between Asia and Alaska. During glacial stages, sea level would have dropped as much as 100 meters, exposing a rich tract of grazing land hundreds of miles wide. Migrating herds of reindeer, elk, and bison, as well as mastodons and mammoths, lived along the land bridge, enticing human hunters from Siberia. Humans entered the New World with much the same motivation as other predators. They followed the source of their sustenance.

Once they had crossed the land bridge, the migrating clans were able to follow an ice-free corridor down into Canada, the United States, and even South America.

Earlier, the same route had been used by mammoth, reindeer, the dire wolf, bear, and other mammals.

Asia lay at the west end of the land bridge, and the first arrivals to America were certainly Asian. Linguistic studies and mitochondrial DNA clearly link native Americans to Asian populations. But precisely when these migrations occurred is difficult to determine and hotly debated:

- There is unassailable evidence that human populations lived in North America about 12,000 years ago. That evidence consists mainly of stone tools and weapons made by people of the **Clovis culture**, named for an archeological site in New Mexico. Clovis artifacts, however, are known throughout the New World from Alaska down into South America.

- At a site in western Pennsylvania known as the Meadowcroft Shelter, stone weapon points and tools have been uncovered that indicate human occupation since about 19,600 years ago. However, there is concern that the radiocarbon used in dating materials at the site might have been contaminated by nearby coal deposits.

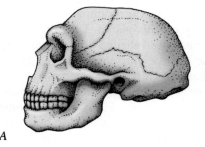
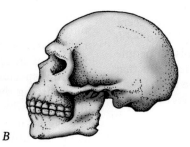
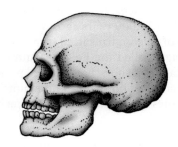

A   B   C

FIGURE 17-29 **Skull comparison.** (A) Neandertal. (B) Skull from a rock shelter on the slope of Mount Carmel (Israel) that appears to show both Neandertal and Cro-Magnon features. (C) Cro-Magnon. The Mount Carmel skull is intermediate in both form and age between Neandertal and Cro-Magnon.

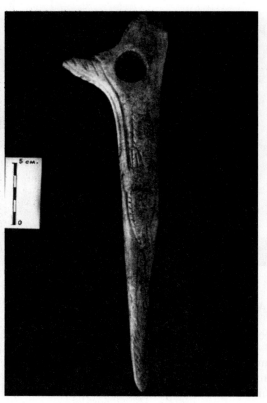

A　　　　　　　　　　　　　　　　　　B

FIGURE 17-30　Prehistoric art by Late Pleistocene *Homo sapiens*. (A) "The Venus of Willendorf"—probably a fertility figure—found in Austria. (B) A tool called a thong-stropper, used either to work thongs made from hide or to straighten arrow shafts. The tool is made from an antler and is intricately carved.

- There are several South American contenders for pre-Clovis humans in America, some yielding dates as old as 13,000 years.

Numerous more recent sites yield indirect evidence of **paleoindians**. In the western United States, projectile points are found in definite association with extinct Late Pleistocene elephants and bison. Sites near Clovis and Sandia, New Mexico, have yielded spear points and tools of two separate cultures that existed from about 13,000 to 11,000 years ago. A somewhat more recent group of paleoindians lived from about 11,000 to 9000 years ago, called the Folsom Culture. Folsom people manufactured short, finely flaked projectile points that were mounted on shafts. These flints have been found in kill sites in association with extinct species of bison.

POPULATION GROWTH

One of the many lessons of historical geology is that the advent of *Homo sapiens* represents only one re-

cent and momentary event along the sinuous, branching, 500-million-year evolution of vertebrates. We humans are linked by similarities of structure and body chemistry to lungfish struggling out of stagnating Devonian pools, to small, shrew-like precursors of the Primate Order, and to axe-carrying hunters that wandered along the margins of great ice sheets.

The descendants of these and other Pleistocene hunters have emerged as the dominant species of higher life presently on this planet. Like other animals, *Homo sapiens* has been shaped by the combined powers of genetic change and environment. However, our species has quickly become a pervasive force in modifying the very physical and biologic environment from which it was shaped. Humanity has come to rely heavily on science to improve its lot, but it has had great difficulty in finding ways to manage the resulting technology and to control the burgeoning problems arising from too few resources for too many humans.

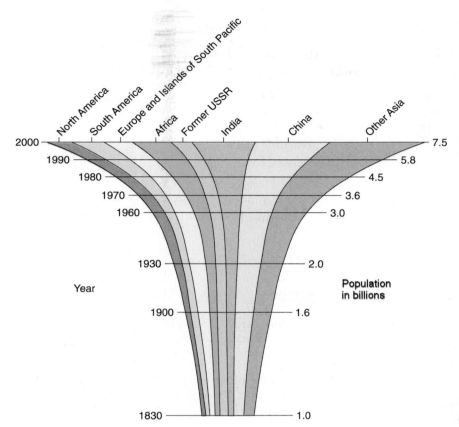

FIGURE 17-31 **World population growth to the year 2000.**

Figure 17-31 depicts world population growth just since 1830. How fast has Earth's human population expanded?

- About 10,000 years ago (a geologic eye blink), there were more than 6 million humans on Earth. For perspective, that is the present population of Arizona, or Massachusetts, or Tennessee.

- By A.D. 1, the number jumped 50-fold to 300 million (near the present U.S. population).

- In 1970, the number expanded over 10-fold to 3.6 billion.

- By 1995, Earth supported 5.7 billion people.

- Our numbers are expected to double to about 10–12 billion by 2050. All will require food, shelter, clean air, clean water, and energy for survival.

Can Earth sustain such an exploding human population? Resources are limited, and environments are easily damaged.

- Already, we have squandered about a fifth of the topsoil needed to grow food.

- Over the past five decades we have lost a third of the globe's forested areas.

- About an eighth of all agricultural land has been altered to barren desert, partly as a result of overgrazing associated with the demand for meat.

- Some soil productivity has been lost due to accumulation of mineral salts caused by extensive irrigation.

- Water shortages now are critical in many regions.

- Climate-altering greenhouse gases have increased by more than a third, with a 5% reduction of the ozone layer that protects all life from lethal ultraviolet radiation.

- Our mineral resources are also finite. As they are consumed and become rare or require more expensive extraction and refining, costs rise and standards of living decline in even the richest of nations.

With concerns like these, the survival of our species, *Homo sapiens*, will depend on our ability to stabilize world population, protect agricultural lands, reforest Earth, conserve and recycle mineral resources, and halt pollution of our environment. This task is vast and complex.

If we fail, other animals may replace our species and add their own distinctive chapters to the history of life on Earth. Hopefully, they will be able to record it, as we can.

KEY TERMS

Anthropoidea, p. 527

arboreal, p. 526

australopithecines, p. 534

brachiating, p. 527

Ceboidea, p. 528

Cercopithecoidea, p. 528

Clovis culture, p. 543

Cro-Magnon people, p. 542

dryomorphs, p. 532

hominids (Hominidae), p. 532

Hylobatidae, p. 528

Neandertals, p. 540

nocturnal, p. 527

paleoindians, p. 544

Pongidae, p. 528

prehensile (tail in primates), p. 528

primates, p. 526

prognathous, p. 538

Prosimii, p. 527

ramamorphs, p. 532

SUMMARY

- Arboreal prosimians were the dominant primates during the Paleocene and Eocene. By the Oligocene, monkeys and apes had appeared.
- The general evolutionary changes in apes during the Cenozoic included the following:
  - Loss of the long snout, development of a flatter face, reshaping of the head to a more rounded form.
  - Relocation of the eyes to a more forward position for improved binocular vision.
  - Development of hands capable of grasping and manipulating objects.
  - Skeletal modifications accompanying a trend toward erect posture.
- Apes known as dryomorphs and ramamorphs were common during the Miocene when climate change resulted in less forest area and more prairie. The ramamorphs were the probable ancestors of both apes and humans.
- Australopithecines were fully erect, primarily Pliocene hominids.
- During the middle Pleistocene, *Homo erectus* began to populate much of Africa, Asia, and Europe. These erect-walking hominids had larger brains than the Australopithecines and improved tool-making abilities.
- The late Pleistocene was marked by the appearance of Neandertal peoples.
- Anatomically modern *Homo sapiens* spread widely during the late Pleistocene. An early group known as Cro-Magnon replaced the Neandertals.
- Humans may have come to the Americas by migrating across the Bering land bridge.

QUESTIONS FOR REVIEW AND DISCUSSION

1. In general, how do prosimians differ from anthropoids? What kinds of primates are *hominoids* (members of the Superfamily Hominoidea)? What kinds of primates are *hominids* (members of the Family Hominidae)?

2. Early arboreal primates such as *Notharctus* had close-set eyes and grasping hands. Other than facilitating movement among tree branches, what function might these attributes have provided?

3. During what geologic period do we find the earliest known remains of primates? When does the first member of the genus *Homo* appear?

4. What physical properties of chert and flint render these rocks useful in the manufacture of spear points, axes, and scrapers by early humans?

5. Why is it easier to find fossils and date them in the eastern African fossil sites than in many other parts of the world?

6. What distinctly humanlike traits were possessed by *Australopithecus*? What apelike characteristics were retained?

7. What features of Cro-Magnon people differentiate them from Neandertal people?

8. Discuss the general changes in the skulls of fossil primates as one progresses from an animal such as *Plesiadapis* to modern humans.

9. Discuss the route by which *Homo sapiens* may have entered the New World. Would the migrations have been easier during a glacial or interglacial epoch? Why?

10. Prepare text for a debate in which you first argue for and then against the premise that life will be more difficult for humans during the year 2050 than it is today.

WEB SITES

The Earth Through Time Student Companion Web Site (www.wiley.com/college/levin) has online resources to help you expand your understanding of the topics in this chapter. Visit the Web Site to access the following:

1. Illustrated course notes covering key concepts in each chapter;

2. Online quizzes that provide immediate feedback;

3. Links to chapter-specific topics on the web;

4. Science news updates relating to recent developments in Historical Geology;

5. Web inquiry activities for further exploration;

6. A glossary of terms;

7. A Student Union with links to topics such as study skills, writing and grammar, and citing electronic information.

Classification of Living Things

The classification presented below recognizes six kingdoms of organisms. Two of these, the Archaeobacteria and Eubacteria, are components of the superkingdom Prokaryota. The remaining Protista, Fungi, Plantae, and Animalia are grouped under the superkingdom Eukaryota. The classification has been simplified in that some groups lacking a fossil record or not mentioned in the text are omitted.

SUPERKINGDOM PROKARYOTA

Kingdom Archaeobacteria   Prokaryotes with primitive genetic codes, including methane-producing bacteria (methanophiles), bacteria tolerant of extremely salty water (extreme halophiles), and bacteria tolerant of high temperatures (extreme thermophiles). In the three-domain classification, archaeobacteria are equivalent to the archaea.

Kingdom Eubacteria   Advanced prokaryotes that include digestive and disease-causing bacteria as well as photoautotrophs like the cyanobacteria, formerly called blue-green algae. Cyanobacteria are represented in the fossil record by **stromatolites**. (Archean to Holocene).

SUPERKINGDOM EUKARYOTA

Kingdom Protista   Solitary or colonial unicellular eukaryotes that do not form tissues.

Phylum Euglenophyta   Euglenoid organisms. Members of this phylum illustrate the impossibility of distinguishing "plants" from "animals" among the Protista.

Phylum Chrysophyta   Golden-brown algae. This phylum includes the **diatoms** and **coccolithophorids**, both of which are abundant and useful microfossils. (Triassic to Recent)

Phylum Chlorophyta   Green algae. Includes the stoneworts and charophytes, as well as *Halimeda*, which grows in modern carbonate-depositing environments. (Cambrian to Holocene)

Phylum Pyrrophyta   Dinoflagellates and cryptomonads.

Phylum Sarcodina   Protozoans with pseudopodia, including the living *Amoeba* as well as important microfossil groups such as **Foraminifera** and **Radiolaria**. (Cambrian to Holocene)

Kingdom Fungi   Multicellular eukaryotes that exist mostly as decomposers, such as molds and mushrooms. Some are parasites that infest plants. Definite fungal remains occur in rocks as old as Devonian, although some questionable remains have been reported from the Proterozoic. (Proterozoic? to Holocene)

Kingdom Plantae

Phylum Rhodophyta   Red algae, usually marine, multicellular.

Phylum Phaeophyta   Brown algae, often with large bodies, as in seaweed and kelp.

Phylum Bryophyta   Liverworts, mosses, and hornworts. (Devonian to Holocene)

Phylum Psilophyta   Primitive, spore-bearing, leafless, rootless, vascular land plants. (Mid-Paleozoic to Holocene)

Phylum Lycopodophyta (= Lycopsida)   Spore-bearing vascular plants with true roots and leaves, including the scale trees (lycopsids like *Lepidodendron* and *Sigillaria*) of the Carboniferous. (Silurian to Holocene)

Phylum Equisetophyta (= Sphenopsida)   Horsetails, scouring rushes, including the Carboniferous **sphenopsids** such as *Calamites* and *Annularia*. (Devonian to Recent)

Phylum Polypodiophyta   The true ferns, or **pteropsids**. (Devonian to Recent)

Phylum Pinophyta (= Gymnospermophyta)   The "**gymnosperms**," including conifers, cycads, and many evergreen plants that do not develop true flowers. (Middle Paleozoic to Holocene)

　　Class Pteridospermophyta   Seed ferns, known from such Carboniferous fossil plants as *Neuropteris* and *Glossopteris*. (Devonian to Holocene)

　　Class Bennettitopsida   The extinct **cycadeoids**. (Triassic to Early Cretaceous)

　　Class Cycadopsida   **Cycads**. (Triassic to Holocene)

　　Class Ginkgopsida   **Ginkgoes**. (Mesozoic to Holocene)

　　Class Pinopsida   **Conifers**, as well as the extinct **Cordaites**. (Late Paleozoic to Holocene)

Phylum Magnoliophyta   The flowering plants, or "**angiosperms**." Seeds enclosed in an ovary; fertilization involving flowers. (Cretaceous to Holocene)

Kingdom Animalia

Phylum Porifera   **Sponges**, multicellular aquatic invertebrates that lack tissue level of organization, gather their food from the flow of water through numerous pores and canals, and possess special flagellated collared cells called choanocytes. (Proterozoic to Holocene)

  Class Demospongiae   Skeleton of siliceous spicules, an organic substance called spongin, or a combination of these materials. (Proterozoic to Holocene)

  Class Hyalospongea   The "glass sponges," so named because their skeleton is composed of siliceous (opaline) spicules. (Cambrian to Holocene)

  Class Sclerospongea   Skeleton of siliceous spicules and spongin underlain by an encasement of aragonite. The class includes the **stromatoporoids**, Paleozoic reef-building poriferans that build structures composed of calcareous lamellae supported by connecting pillars and partitions. (Cambrian to Holocene)

  Class Calcispongea (= Calcarea)   Sponges that secrete calcareous spicules. (Devonian to Holocene)

  Phylum Archaeocyatha   Cup- or vase-shaped fossils of extinct sponge-like organisms. Important as Cambrian reef-builders and widely used in correlation of Cambrian strata. (Cambrian)

  Phylum Cnidaria (= Coelenterata)   Jellyfishes, anemones, and coals; radially symmetric, aquatic invertebrates with specialized cells called cnidocysts that produce capsules used in capturing prey. True tissues are developed. (Latest Proterozoic to Holocene)

  Class Hydrozoa   The *Hydra* and hydra-like cnidarians.

  Class Scyphozoa   True jellyfishes. The Scyphozoa include **conularids**, jellyfish that build pyramidal shells. (Proterozoic to Holocene)

  Class Anthozoa   Corals and sea anemones. Included are the extinct **tabulate** and **rugose** corals and the extant **scleractinid** corals (also called hexacorals).

Phylum Bryozoa   Tiny, colonial, filter-feeding, mostly marine invertebrates with lophophore and digestive tract with mouth and anus. (Ordovician to Holocene)

Phylum Brachiopoda   Lophophore-bearing marine invertebrates with shell composed of two unequal valves. The larger valve is called the pedicle or ventral valve, and the smaller the brachial or dorsal valve. Often attached to substrate by a flexible stalk called a pedicle. (Cambrian to Holocene)

  Class Inarticulata   Shell lacking teeth and sockets and without hinge line. Most composed of chitinophosphatic material. (Cambrian to Holocene)

  Class Articulata   Shell with definite hinge, usually bearing teeth and sockets. (Cambrian to Holocene)

  Order Orthida   Unequally biconvex shells with straight hinge line. (Cambrian to Permian)

  Order Strophomenida   Wide, straight hinge line, one valve concave. (Ordovician to Jurassic)

  Order Pentamerida   Large, biconvex shells, overhanging beak, short hinge line. (Cambrian to Devonian)

  Order Rhynchonellida   Strongly ribbed (plicate), strongly biconvex shells with prominent beak. (Ordovician to Holocene)

  Order Spiriferida   Interior of brachial valve with spiral lophophore supports. (Ordovician to Jurassic)

  Order Productida   Often large concavo-convex shells with elaborate development of spines over shell surface. (Devonian to Permian)

  Order Terebratulida   Interior of brachial valve with loop-shaped lophophore supports. (Devonian to Holocene)

Phylum Phoronida   Lophophore-bearing, slender, worm-like marine animals that secrete and live within a chitious, leathery tube. (Devonian to Holocene)

Phylum Annelida   Segmented worms. The phylum includes polychaetes (marine sandworms and tubeworms), earthworms, and leeches. (Proterozoic to Holocene)

Phylum Arthropoda   Animals with paired, jointed appendages, an armor-like exoskeleton, segmented body, and respiratory systems consisting of gills for many aquatic forms and tracheae for terrestrial forms. (Cambrian to Holocene)

  Subphylum Trilobita   Invertebrates with body divided into three parts (cephalon, thorax, and pygidium) and divided longitudinally by a central lobe and two lateral or pleural lobes. (Cambrian to Permian)

  Subphylum Chelicerata   Arthropods having claw-like feeding appendages (chelicerae) and no antennae. Includes horseshoe crabs and arachnids (spiders, scorpions, ticks, and mites). (Cambrian to Holocene)

  Subphylum Crustacea   Arthropods, which like trilobites, have biramous appendages. Lobsters, crabs, barnacles, and ostracodes are all included in the Crustacea. (Cambrian to Holocene)

  Subphylum Uniramia   Arthropods with unbranched (uniramous) appendages and a single pair of antennae. (Cambrian to Holocene)

  Class Onycophora   Living and ancient velvet worms, including *Aysheia* from the Burgess Shale fossil locality.

  Class Myriapoda   Uniramids (centipedes and millipedes) having head and elongate trunk bearing either one pair of legs on each segment (most centipedes), or two pair (most millipedes). (Carboniferous to Holocene)

  Class Hexapoda   Insects, all characterized by a body consisting of head, thorax, and abdomen, six legs, and tracheal tubes for respiration. (Devonian to Holocene)

Phylum Mollusca   Like the annelids and arthropods, mollusks are protostome animals (animals in which the blastopore develops into the mouth and the anus forms secondarily). The shell characteristic of most mollusks is secreted by the mantle, which also covers the visceral mass. (Cambrian to Holocene)

Class Monoplacophora   Primitive mollusks with cap- or spoon-shaped shells, multiple pairs of gills. Monoplacophorans are thought to have evolved from annelid worms and may be ancestral to all other classes of mollusks. The living form **Neopilina** and the Silurian **Pilina** are examples. (Cambrian to Holocene)

Class Polyplacophora   Chitons, marine mollusks with shells composed of eight articulating calcarous plates. (Cambrian to Holocene)

Class Scaphopoda   Tusk shells; mollusks with curved tubular shells open at both ends. (Ordovician to Holocene)

Class Gastropoda   Mollusks having a twisted body and a shell that, when present, is coiled. Includes snails, slugs, and abalones. (Cambrian to Holocene)

Class Bivalvia (= Pelecypoda)   Shells composed of two valves (right and left) that are hinged dorsally. Includes clams, mussels, oysters, and scallops. (Cambrian to Holocene)

Class Cephalopoda   Mollusks with foot divided into tentacles that surround the central mouth of the large head. Large, well-developed, image-forming eyes. Includes modern *Nautilus*, squid, cuttlefishes, and octopods. (Cambrian to Holocene)

   Subclass Nautiloidea   Cephalopods with external shell divided into chambers by septa and having simple suture lines. (Cambrian to Holocene)

   Subclass Ammonoidea   Cephalopods with external shell divided into chambers by septa with fluted margins, resulting in complexly folded suture lines. (Devonian to Cretaceous)

   Subclass Coleoidea   Cephalopods that lack a shell or have an internal shell. Includes belemnites (Mississippian to Tertiary) as well as living and fossil squid, cuttlefish, and octopods (Mississippian to Holocene).

     Order Belemnoidea   Skeleton consisting of small chambered structure surrounded by a thick rostrum. Shell developed internally like the shell of a cuttlefish. The **belemnites**. (Mississippian to Early Tertiary)

     Order Sepioidea   Cuttlefishes. (Jurassic to Holocene)

     Order Teuthoidea   Squids. (Jurassic to Holocene)

     Order Octopoda   Octopods. (Cretaceous to Holocene)

Phylum Hyolitha   Marine, bilaterally symmetrical, solitary metazoans, typically with conical shells closed by a lid. (Cambrian to Devonian)

Phylum Echinodermata   Spiny-skinned marine invertebrates that are radially symmetrical as adults (bilateral in larval stage) and have calcareous, spine-bearing skeletal plates and unique water vascular systems. Echinoderms are **deuterostomes**, in that the anus develops from the blastopore. (See Phylum Mollusca above for comparison to *protostome*.) (Cambrian to Holocene)

Class Asteroidea   Starfishes. Animals with a central disk from which radiate five or more arms. (Ordovician to Holocene)

Class Ophiuroidea   Brittle stars and serpent stars having a central disk from which abruptly radiate five or more slender arms that lack ambulacral grooves. (Ordovician to Holocene)

Class Echinoidea   Sea urchins and sand dollars. Have solid shell without arms; often covered with spines. (Ordovician to Holocene)

Class Holothuroidea   Sea cucumbers. Echinoderms with elongate, flexible bodies and a mouth surrounded by a circle of modified tube feet serving as tentacles. (Cambrian to Holocene)

Class Crinoidea   Sea lillies and feather stars. Many sessile and attached to sea floor by stem and "root." (Cambrian to Holocene)

Class Blastoidea   Blastoids commonly have bud-shaped compact thecae constructed of 13 or 14 plates. (Silurian to Permian)

Classes Rhombifera and *Diploporita*   Component groups of the former Class Cystoidea. These are stalked echinoderms with thecae characterized by distinctive pore patterns.

Phylum Hemichordata   As suggested by their name, which means "half chordates," hemichordates share certain features with chordates that may include a pharynx perforated by gill slits or pores and a hollow dorsal nerve cord. The phylum includes the extinct **Class Graptolithina**, as well as living acorn worms that have larval forms similar to those of echinoderms.

Phylum Chordata   Bilaterally symmetric animals with notochord, dorsal hollow neural tube, and gill clefts in the pharynx. Include sea squirts, lancelets, and possibly conodont-bearing animals.

Subphylum Urochordata   Sea squirts or tunicates. Larval forms have notochord in tail region.

Subphylum Cephalochordata   *Branchiostoma* ("amphioxus"); small marine animals with fish-like bodies and musculature, notochord, and dorsal neural tube.

Subphylum Vertebrata   Animals with a backbone composed of vertebrae, well-defined head, ventrally located heart, and well-developed sense organs. Includes fishes, amphibians, reptiles, synapsids, birds, and mammals.

Class Agnatha   Jawless fishes, including living lampreys and hagfishes, as well as extinct armored jawless **ostracoderms**. (Cambrian to Holocene)

Class Acanthodii   Earliest group of fishes with jaws. Often called "spiny sharks" because of their many spine-supported fins. (Middle to Late Paleozoic)

Class Placodermi   Extinct jawed fishes with heavy bony armor and a specialized joint at rear of head. (Middle Paleozoic)

Class Chondrichthyes   Fishes with cartilaginous skeletons, including sharks, rays, chimaeras, and skates. (Middle Paleozoic to Holocene)

Class Osteichthyes   Fishes in which bone becomes the most abundant skeletal material. (Devonian to Holocene)

Subclass Actinopterygii   Bony fishes in which the fins are supported by bony rays, hence "ray-finned" fishes. Actinopterygians are by far the dominant fishes in the modern world. (Devonian to Holocene)

Subclass Sarcopterygii   Lobe-finned fishes. Stout, muscular fins are supported by bony skeletal elements. Many with lungs and the ability to breath air as well as features of the skull and teeth that indicate their relationship to early tetrapods. Include modern **lungfishes** and **coelacanths**. (Devonian to Holocene)

Order Crossopterygii   Lobe-finned fishes most closely related to tetrapods.

Order Dipnoi   Lungfishes, including the Devonian fish *Dipterus* and living lungfish of South America, Africa, and Australia. (Devonian to Holocene)

Class Amphibia   Amphibians; the Earth's earliest tetrapods (vertebrates with four limbs.) Include living frogs, toads, salamanders, and newts as well as prominent Carboniferous labyrinthodontic amphibians known as **temnospondyls**. *Eryops*, pictured in the text, has the broad flat skull typical of temnospondyls.

Amniota   Reptiles, birds, and mammals are all amniotes in that they develop an egg with membranes around the embryo (allantois, chorion, and amnion) that serve to retain water and provide for gas exchange, support, and protection. Thus amniotes do not require water bodies for reproduction. Amniotes evolved during the Carboniferous.

Tetrapoda   Vertebrates with two pairs of walking limbs. The term tetrapods is derived from two Greek words meaning "four-footed." (Carboniferous to Holocene)

Class Reptilia   Amniotes that include turtles, crocodiles, and many ancient vertebrates with distinctive elements of the skull. Excluded from the Reptilia are the so-called mammal-like reptiles (synapsids), mammals, and birds.

Subclass Anapsida   The Anapsida are reptiles that lack special openings in the side of the skull to accommodate jaw muscles. This group includes many extinct reptiles of the Carboniferous to Triassic, as well as *Mesosaurus*, the geographic distribution of which provides evidence of drifting continents.

Subclass Testudinata   Turtles, which also have anapsid skull structure, but in addition have shells and unique postcranial skeletal features. (Triassic to Holocene)

Subclass Diapsida   Reptiles with two temporal openings in the skull. (Carboniferous to Holocene)

Infraclass Lepidosauromorpha   Lizards, snakes, the tuatara (*Sphenodon*), and their extinct relatives. Relatives of early lepidosauromorphs may include such Mesozoic marine reptiles as the **ichthyosaurs** and **plesiosaurs**.

Infraclass Archosauria   The archosaurs include the dinosaurs, the pterosaurs, and the crocodilians. They differ from lepidosauromorphs in having an opening or fenestra in front of the eye orbit, a fenestra in the lower jaw, laterally compressed and serrated teeth (in flesh eaters), no teeth on the palate, and limbs which supported the body in semi-upright or upright posture. Included within the Archosauria are archaic ancestral archosaurs, the more advanced **Ornithischia** (ankylosaurs, stegosaurs, ceratopsians, ornithopods) and **Saurischia** (sauropods, theropods), and according to cladistic analyses, birds (**Aves**).

Class Synapsida   The synapsids include Late Paleozoic amniotes such as the finbacks and other Late Permian and Early Triassic vertebrates erroneously termed "mammal-like reptiles." (They are not reptiles). Mammals, the warm-blooded synapsids with body hair and mammary glands in females for nourishing offspring, range from Triassic to Holocene.

Subclass Pelycosauria   Archaic synapsids of the Carboniferous and Permian, the most familiar of which are the carnivorous and herbivorous finbacks of the Permian Period.

Subclass Therapsida   Progressive synapsids approaching the mammalian stage of evolution. Mammals, for example, have one dentary bone on either side of the lower jaw, and therapsids show enlargement of the dentary at the expense of other jaw bones. They also show development of a secondary palate and differentiation of teeth. (Permian to Triassic)

Subclass Mammalia   Synapsids in which each half of the lower jaw consists of only a single bone, the dentary; also possess body hair and mammary glands. (Triassic to Holocene).

Superorder Monotremata (Eutheria)   Monotremes are egg-laying mammals. Representatives are the platypus and echidna (spiny anteater).

Superorder Marsupialia (Metatheria)   Marsupials are pouched mammals, including kangaroos, koalas, and opposums. (Cretaceous to Holocene)

Superorder Eutheria (= placental mammals)   Mammals that retain their young for a long period within the body of the female, where they are nourished by means of a complex structure called the placenta. (Cretaceous to Holocene)

Order Insectivora   Primitive insect-eating placental mammals, including moles and shrews. (Cretaceous to Holocene)

Order Chiroptera   Bats. (Early Tertiary to Holocene)

Order Primates   Prosimians (e.g., lemurs, tarsiers), New World and Old World monkeys, apes, and humans. (Early Tertiary to Holocene)

Order Edentata   Living armadillos, anteaters, and tree sloths, as well as extinct **glyptodonts** and **ground sloths**. (Early Tertiary to Holocene)

Order Rodentia   Squirrels, mice, rats, beavers, and porcupines. (Early Tertiary to Holocene)

Order Lagomorpha   Rabbits, hares, and pikas. (Early Tertiary to Holocene)

Order Creodonta   Extinct, archaic, carnivorous placental mammals. (Paleocene to Pliocene)

Order Carnivora   Progressive carnivorous placentals, including living and ancient dogs, cats, bears, hyenas, seals, sea lions, and walruses. (Early Tertiary to Holocene)

Order Perissodactyla   Odd-toed ungulates (hoofed mammals), including horses, rhinoceroses, and tapirs, as well as the extinct **titanotheres** and **chalicotheres**. (Early Tertiary to Holocene)

Order Artiodactyla   Even-toed ungulates, including antelopes, cattle, deer, giraffes, camels, llamas, hippopotamuses, and pigs, as well as extinct **oreodonts** and **entelodonts**. (Early Tertiary to Holocene)

Order Proboscidea   Elephants and extinct **mastodons** and **mammoths** (Early Tertiary to Holocene).

Order Cetacea   Whales and porpoises, both of which are descendents of ungulates. (Paleocene to Holocene)

Physiographic Provinces of the United States

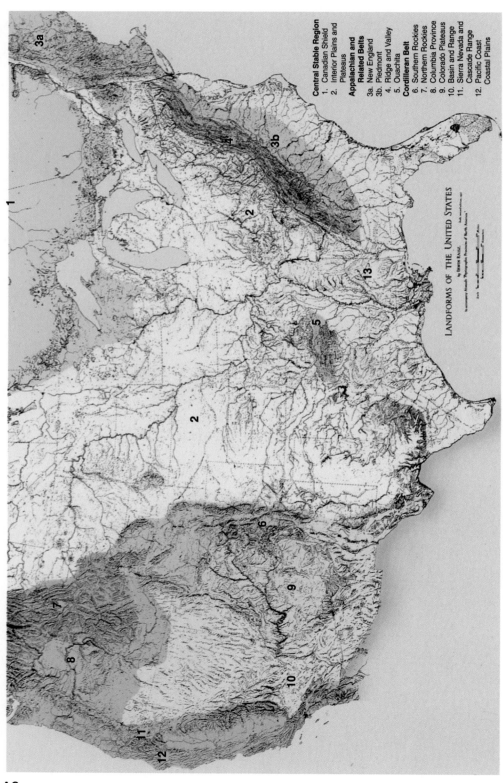

Central Stable Region
1. Canadian Shield
2. Interior Plains and Plateaus

Appalachian and Related Belts
3a. New England
3b. Piedmont
4. Ridge and Valley
5. Ouachita

Cordilleran Belt
6. Southern Rockies
7. Northern Rockies
8. Columbia Province
9. Colorado Plateaus
10. Basin and Range
11. Sierra Nevada and Cascade Range
12. Pacific Coast
13. Coastal Plains

LANDFORMS OF THE UNITED STATES
by ERWIN RAISZ

World Political Map

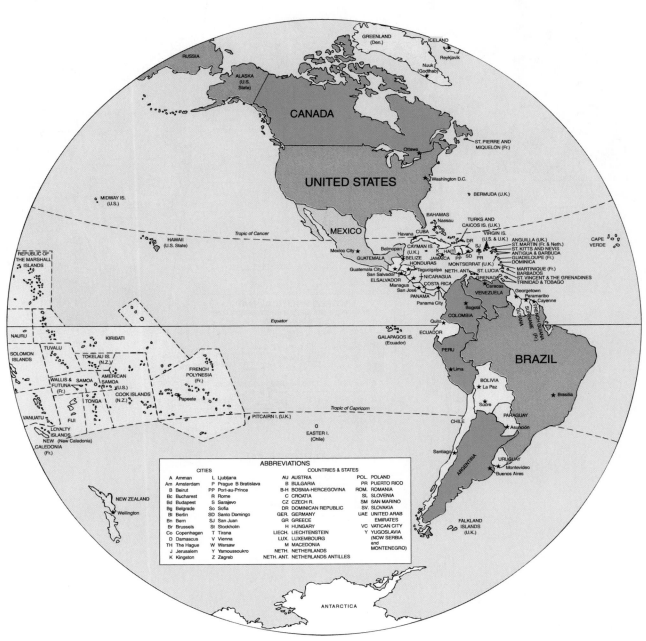

WESTERN HEMISPHERE

ABBREVIATIONS

CITIES

A	Amman	L	Ljubljana
Am	Amsterdam	P	Prague B Bratislava
B	Beirut	PP	Port-au-Prince
Bc	Bucharest	R	Rome
Bd	Budapest	S	Sarajevo
Bg	Belgrade	So	Sofia
Bl	Berlin	SD	Santo Domingo
Bn	Bern	SJ	San Juan
Br	Brussels	St	Stockholm
Co	Copenhagen	T	Tirana
D	Damascus	V	Vienna
TH	The Hague	W	Warsaw
J	Jerusalem	Y	Yamoussoukro
K	Kingston	Z	Zagreb

COUNTRIES & STATES

AU	AUSTRIA	POL.	POLAND
B	BULGARIA	PR	PUERTO RICO
B-H	BOSNIA-HERCEGOVINA	ROM.	ROMANIA
C	CROATIA	SL	SLOVENIA
CZ	CZECH R.	SM	SAN MARINO
DR	DOMINICAN REPUBLIC	SV.	SLOVAKIA
GER.	GERMANY	UAE	UNITED ARAB
GR	GREECE		EMIRATES
H	HUNGARY	VC	VATICAN CITY
LIECH.	LIECHTENSTEIN	Y	YUGOSLAVIA
LUX.	LUXEMBOURG		(NOW SERBIA
M	MACEDONIA		and
NETH.	NETHERLANDS		MONTENEGRO)
NETH. ANT.	NETHERLANDS ANTILLES		

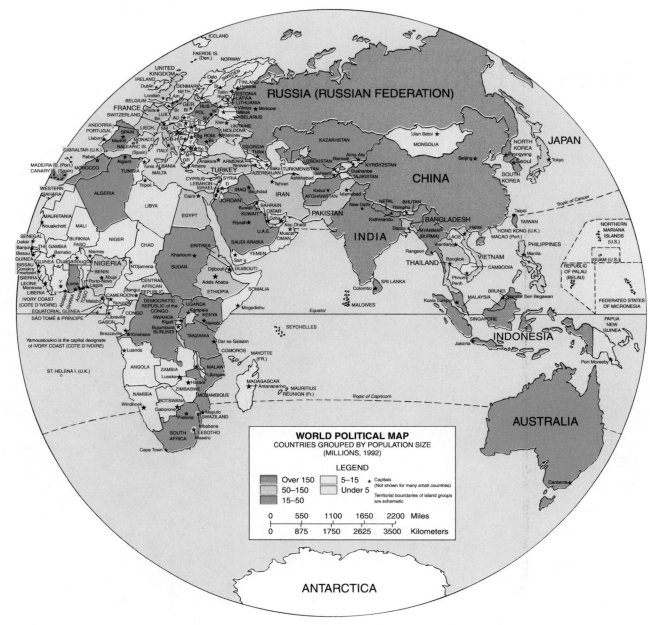

WORLD POLITICAL MAP
COUNTRIES GROUPED BY POPULATION SIZE
(MILLIONS, 1992)

LEGEND

Over 150
50–150
15–50
5–15
Under 5

★ Capitals
(Not shown for many small countries)

Territorial boundaries of island groups
are schematic

| 0 | 550 | 1100 | 1650 | 2200 | Miles |
| 0 | 875 | 1750 | 2625 | 3500 | Kilometers |

EASTERN HEMISPHERE

Periodic Table and Symbols for Chemical Elements

Scientists generally group or organize things that are similar to better undestand them and determine how they relate to one another. An important step in the organization of elements was made in 1869 by Dimitri Mendeleev. Mendeleev showed that when elements are arranged in order of their increasing atomic weights, their physical and chemical properties tend to be repeated in cycles. The arrangement of elements was depicted on a chart called the periodic table of elements, a modern version of which is shown in Table D–1. In the periodic table, each box contains the symbol of the element and its atomic number. Except for two long sequences set apart at the bottom, the elements appear in the increasing order of their atomic numbers. The vertical columns are called *groups* and contain elements with similar properties.

For example, in column VIIA, one finds fluorine, chlorine, bromine, and iodine. All of these elements are colored and highly reactive and share other similarities. Fluorine, however, is chemically the most active, chlorine somewhat less active, bromine still less active, and iodine the least active of the four. Except for hydrogen, all of the elements in group IA are soft, shiny metals and very reactive chemically. The horizontal rows on the table are called *periods* and contain sequences of elements having electron configurations that vary in characteristic patterns. It is thus apparent that the periodic table shows relationships among elements rather well and certainly better than an arbitrary listing.

Table D–2 provides the chemical names for the symbols of the elements used in the periodic table.

TABLE D-1 Periodic Table of the Elements

Key:

26	— Atomic number (Z)
Fe	— Element symbol
55.845	— Atomic mass of naturally occurring isotopic mixture; for radioactive elements, numbers in parentheses are mass numbers of most stable isotopes

Metals
Semimetals
Nonmetals

Transition elements

I	II												III	IV	V	VI	VII	VIII
1 **H** 1.0079																		2 **He** 4.0026
3 **Li** 6.941	4 **Be** 9.0122												5 **B** 10.811	6 **C** 12.011	7 **N** 14.007	8 **O** 15.999	9 **F** 18.998	10 **Ne** 20.180
11 **Na** 22.990	12 **Mg** 24.305												13 **Al** 26.982	14 **Si** 28.086	15 **P** 30.974	16 **S** 32.066	17 **Cl** 35.453	18 **Ar** 39.948
19 **K** 39.098	20 **Ca** 40.078	21 **Sc** 44.956	22 **Ti** 47.87	23 **V** 50.942	24 **Cr** 51.996	25 **Mn** 54.938	26 **Fe** 55.845	27 **Co** 58.933	28 **Ni** 58.693	29 **Cu** 63.546	30 **Zn** 65.39		31 **Ga** 69.723	32 **Ge** 72.61	33 **As** 74.922	34 **Se** 78.96	35 **Br** 79.904	36 **Kr** 83.80
37 **Rb** 85.468	38 **Sr** 87.62	39 **Y** 88.906	40 **Zr** 91.224	41 **Nb** 92.906	42 **Mo** 95.94	43 **Tc** (98)	44 **Ru** 101.07	45 **Rh** 102.91	46 **Pd** 106.42	47 **Ag** 107.87	48 **Cd** 112.41		49 **In** 114.82	50 **Sn** 118.71	51 **Sb** 121.76	52 **Te** 127.60	53 **I** 126.90	54 **Xe** 131.29
55 **Cs** 132.91	56 **Ba** 137.33	71 **Lu** 174.97	72 **Hf** 178.49	73 **Ta** 180.95	74 **W** 183.84	75 **Re** 186.21	76 **Os** 190.23	77 **Ir** 192.22	78 **Pt** 195.08	79 **Au** 196.97	80 **Hg** 200.59		81 **Tl** 204.38	82 **Pb** 207.2	83 **Bi** 208.98	84 **Po** (209)	85 **At** (210)	86 **Rn** (222)
87 **Fr** (223)	88 **Ra** (226)	103 **Lr** (262)	104 **Rf** (261)	105 **Db** (262)	106 **Sg** (263)	107 **Bh** (262)	108 **Hs** (265)	109 **Mt** (268)	110 **Uun** (269)	111 **Uuu** (272)								

Lanthanide series

57 **La** 138.91	58 **Ce** 140.12	59 **Pr** 140.91	60 **Nd** 144.24	61 **Pm** (145)	62 **Sm** 150.36	63 **Eu** 151.96	64 **Gd** 157.25	65 **Tb** 158.93	66 **Dy** 162.50	67 **Ho** 164.93	68 **Er** 167.26	69 **Tm** 168.93	70 **Yb** 173.04

Actinide series

89 **Ac** (227)	90 **Th** 232.04	91 **Pa** 231.04	92 **U** 238.03	93 **Np** (237)	94 **Pu** (244)	95 **Am** (241)	96 **Cm** (247)	97 **Bk** (247)	98 **Cf** (251)	99 **Es** (252)	100 **Fm** (257)	101 **Md** (258)	102 **No** (259)

TABLE D-2 **Elements and Their Chemical Symbols**

Actinium	Ac	Erbium	Er	Mercury	Hg	Scandium	Sc
Aluminum	Al	Europium	Eu	Molybdenum	Mo	Seaborgium	Sg
Americium	Am	Fermium	Fm	Neodymium	Nd	Selenium	Se
Antimony	Sb	Fluorine	F	Neon	Ne	Silicon	Si
Argon	Ar	Francium	Fe	Neptunium	Np	Silver	Ag
Arsenic	As	Gadolinium	Gd	Nickel	Ni	Sodium	Na
Astatine	At	Gallium	Ga	Niobium	Nb	Strontium	Sr
Barium	Ba	Germanium	Ge	Nitrogen	N	Sulfur	S
Berkelium	Bk	Gold	Au	Nobelium	No	Tantalum	Ta
Beryllium	Be	Hafnium	Hf	Osmium	Os	Technetium	Tc
Bismuth	Bi	Hassium	Hs	Oxygen	O	Tellurium	Te
Bohrium	Bh	Helium	He	Palladium	Pd	Terbium	Tb
Boron	B	Holmium	Ho	Phosphorus	P	Thallium	Tl
Bromine	Br	Hydrogen	H	Platinum	Pt	Thorium	Th
Cadmium	Cd	Indium	In	Plutonium	Pu	Thulium	Tm
Calcium	Ca	Iodine	I	Polonium	Po	Tin	Sn
Californium	Cf	Iridium	Ir	Potassium	K	Titanium	Ti
Carbon	C	Iron	Fe	Praseodymium	Pr	Tungsten	W
Cerium	Ce	Krypton	Kr	Promethium	Pm	Ununnilium	Uun
Cesium	Cs	Lanthanum	La	Protactinium	Pa	Unununium	Uuu
Chlorine	Cl	Lawrencium	Lr	Radium	Ra	Uranium	U
Chromium	Cr	Lead	Pb	Radon	Rn	Vanadium	V
Cobalt	Co	Lithium	Li	Rhenium	Re	Xenon	Xe
Copper	Cu	Lutetium	Lu	Rhodium	Rh	Ytterbium	Yb
Curium	Cm	Magnesium	Mg	Rubidium	Rb	Yttrium	Y
Dubnium	Db	Manganese	Mn	Ruthenium	Ru	Zinc	Zn
Dysprosium	Dy	Meitnerium	Mt	Rutherfordium	Rf	Ziconium	Zr
Einsteinium	Es	Mendelevium	Md	Samarium	Sm		

Convenient Conversion Factors

To Convert from	To	Multiply by*
Centimeters	Feet	0.0328 ft/cm
	Inches	0.394 in/cm
	Meters	0.01 m/cm
	Microns (micrometers)	10,000 μm/cm
	Miles (statute)	6.214×10^{-6} mi/cm
	Millimeters	10 mm/cm
Feet	Centimeters	30.48 cm/ft
	Inches	12 in/ft
	Meters	0.3048 m/ft
	Microns (micrometers)	304,800 μm/ft
	Miles (statute)	0.000189 mi/ft
	Yards	0.3333 yd/ft
Kilometers	Miles	0.6214 mi/km
Gallons (U.S., liquid)	Cubic centimeters	3785 cm^3/gal
	Cubic feet	0.133 ft^3/gal
	Liters	3.785 L/gal
	Quarts (U.S., liquid)	4 qt/gal
Grams	Kilograms	0.001 kg/g
	Micrograms	1×10^6 μg/g
	Ounces	0.03527 oz/g
	Pounds	0.002205 lb/g
Inches	Centimeters	2.54 cm/in
	Feet	0.0833 ft/in
	Meters	0.0254 m/in
	Yards	0.0278 yd/in
Kilograms	Ounces	35.27 oz/kg
	Pounds	2.205 lb/kg
Meters	Centimeters	100 cm/m
	Feet	3.2808 ft/m
	Inches	39.37 in/m
	Kilometers	0.001 km/m
	Miles (statute)	0.0006214 mi/m
	Millimeters	1000 mm/m
	Yards	1.0936 yd/m
Miles (statute)	Centimeters	160,934 cm/mi
	Feet	5280 ft/mi
	Inches	63,360 in/mi
	Kilometers	1.609 km/mi
	Meters	1609 m/mi
	Yards	1760 yd/mi
Ounces	Grams	28.3 g/oz
	Pounds	0.0625 lb/oz
Pounds	Grams	453.6 g/lb
	Kilograms	0.454 kg/lb
	Ounces	16 oz/lb

*Values are generally given to three or four significant figures.
Source: Turk, J., and Turk, A. 1977. *Physical Science*. Philadelphia: W. B. Saunders Company.

Exponential or Scientific Notation

Exponential or scientific notation is used by scientists all over the world. This system is based on exponents of 10, which are shorthand notations for repeated multiplications or divisions.

A positive exponent is a symbol for a number that is to be multiplied by itself a given number of times. Thus, the number 10^2 (read "ten squared" or "ten to the second power") is exponential notation for $10 \cdot 10 = 100$. Similarly, $3^4 = 3 \cdot 3 \cdot 3 \cdot = 81$. The reciprocals of these numbers are expressed by negative exponents. Thus $10^{-2} = 1/10^2 = 1/(10 \cdot 10) = 1/100 = 0.01$.

To write 10^4 in longhand form you simply start with the number 1 and move the decimal four places to the right: 10000. Similarly, to write 10^{-4} you start with the number 1 and move the decimal four places to the left: 0.0001.

$$\underbrace{1\ 000\ 000}_{6 \text{ places}} = 10^6.$$

Similarly, the decimal place of the number 0.000001 is six places to the left of 1:

$$\underbrace{0.000001}_{6 \text{ places}} = 10^{-6}.$$

Rock Symbols

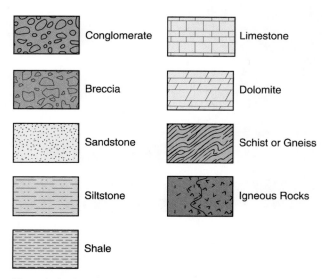

Conglomerate

Breccia

Sandstone

Siltstone

Shale

Limestone

Dolomite

Schist or Gneiss

Igneous Rocks

The standard geologic symbols used by geologists and in diagrams in this book are indicated above.

Bedrock Geology of North America

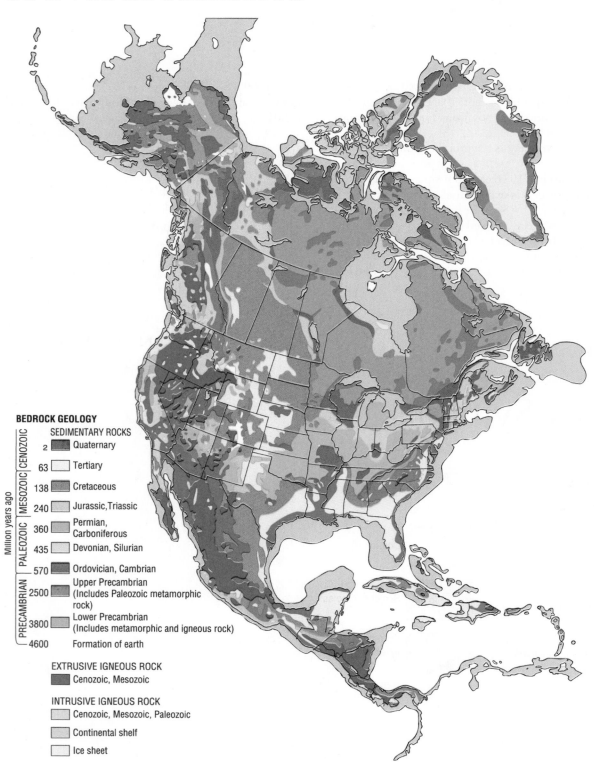

BEDROCK GEOLOGY

Million years ago

CENOZOIC		SEDIMENTARY ROCKS
	2	Quaternary
	63	Tertiary
MESOZOIC	138	Cretaceous
	240	Jurassic, Triassic
PALEOZOIC	360	Permian, Carboniferous
	435	Devonian, Silurian
	570	Ordovician, Cambrian
PRECAMBRIAN	2500	Upper Precambrian (Includes Paleozoic metamorphic rock)
	3800	Lower Precambrian (Includes metamorphic and igneous rock)
	4600	Formation of earth

EXTRUSIVE IGNEOUS ROCK
Cenozoic, Mesozoic

INTRUSIVE IGNEOUS ROCK
Cenozoic, Mesozoic, Paleozoic

Continental shelf

Ice sheet

Common Rock-Forming Silicate Minerals

Silicate Mineral	Composition	Physical Properties
Quartz	Silicon dioxide (silica, SiO_2)	Hardness of 7 (on scale of 1 to 10)*; will not cleave (fractures unevenly); specific gravity: 2.65
Potassium feldspar group	Alluminosilicates of potassium	Hardness of 6.0–6.5; cleaves well in two directions; pink or white; specific gravity: 2.5–2.6
Plagioclase feldspar group	Aluminosilicates of sodium and calcium	Hardness of 6.0–6.5; cleaves well in two directions; white or gray; may show striations on cleavage planes; specific gravity: 2.6–2.7
Muscovite mica	Aluminosilicates of potassium with water	Hardness of 2–3; cleaves perfectly in one direction, yielding flexible, thin plates; colorless; transparent in thin sheets; specific gravity: 2.8–3.0
Biotite mica	Aluminosilicates of magnesium, iron, potassium, with water	Hardness of 2.5–3.0; cleaves perfectly in one direction, yielding flexible, thin plates; black to dark brown; specific gravity: 2.7–3.2
Pyroxene group	Silicates of aluminum, calcium, magnesium, and iron	Hardness of 5–6; cleaves in two directions at 87° and 93°; black to dark green; specific gravity: 3.1–3.5
Amphibole group	Silicates of aluminum, calcium, magnesium, and iron	Hardness of 5–6; cleaves in two directions at 56° and 124°; black to dark green; specific gravity: 3.0–3.3
Olivine	Silicates of magnesium and iron	Hardness of 6.5–7.0; light green; transparent to translucent; specific gravity: 3.2–3.6
Garnet group	Aluminosilicates of iron, calcium, magnesium, and manganese	Hardness of 6.5–7.5; uneven fracture; red, brown, or yellow; specific gravity: 3.5–4.3

Ferromagnesian minerals (Biotite mica through Olivine)

*The scale of hardness used by geologists was formulated in 1822 by Frederich Mohs. Beginning with diamond as the hardest mineral, he arranged the following table:

10 Diamond	8 Topaz	6 Feldspar	4 Fluorite	2 Gypsum
9 Corundum	7 Quartz	5 Apatite	3 Calcite	1 Talc

For terms not included in this glossary, students may wish to consult the *Glossary of Geology* (1997, edited by J. A. Jackson, 4th ed., Falls Church, Virginia, American Geological Institute).

Absaroka sequence A sequence of Permian–Pennsylvanian sediments bounded both above and below by a regional unconformity and recording an episode of marine transgression over an eroded surface, full flood level of inundation, and regression from the craton.

Absolute geologic age The actual age, expressed in years, of a geologic material or event.

Acadian orogeny An episode of mountain building in the northern Appalachians during the Devonian Period.

Acanthodians The earliest known vertebrates (fishes) with a movable, well-developed lower jaw, or mandible; hence, the first jawed fishes.

Accretionary terrane A block of continental crust having fault boundaries that is geologically distinct from surrounding terranes.

Adaptation A modification of an organism that better fits it for existence in its present environment or enables it to live in a somewhat different environment.

Adaptive radiation The diversity that develops among species as each adapts to a different set of environmental conditions.

Adenosine diphosphate (ADP) A product formed in the hydrolysis of adenosine triphosphate that is accompanied by release of energy and organic phosphate.

Adenosine triphosphate (ATP) A compound that occurs in all cells and that serves as a source of energy for physiologic reactions such as muscle contraction.

Aerobic organism An organism that uses oxygen in carrying out respiratory processes.

Age The time represented by the time-stratigraphic unit called a stage. (Informally, may indicate any time span in geologic history, as "Age of Cycads.")

Agnatha The jawless vertebrates, including extinct ostracoderms and living lampreys and hagfishes.

Algae Any of a large group of simple plants (thallophyta) that contain chlorophyll and are capable of photosynthesis.

Allegheny orogeny The late Paleozoic episodes of mountain building along the present trend of the Appalachian Mountains. (Also termed the Appalachian orogeny.)

Alluvium Unconsolidated, poorly sorted detrital sediments ranging from clay to gravel sizes and characteristically fluvial in origin.

Alpha particle A particle equivalent to the nucleus of a helium atom, emitted from an atomic nucleus during radioactive decay.

Alpine orogeny In general, the sequence of crustal disturbances beginning in the middle Mesozoic and continuing into the Miocene that resulted in the geologic structures of the Alps.

Amino acids Nitrogenous hydrocarbons that serve as the building blocks of proteins and are thus essential to all living things.

Ammonites Ammonoid cephalopods having more complex sutural patterns than either ceratites or goniatites.

Ammonoids An extinct group of cephalopods with coiled, chambered conch(s) and having septa with crenulated margins.

Amniotic egg An egg produced by reptiles, birds, and monotremes in which the developing embryo is maintained and protected by an elaborate arrangement of shell membranes, yolk, sac, amnion, and allantois.

Amphibians "Cold-blooded" vertebrates that utilize gills for respiration in the early life stages but that have air-breathing lungs as adults.

Amphibole A ferromagnesium silicate mineral that occurs commonly in igneous and metamorphic rocks.

Anaerobic organism An organism that does not require oxygen for respiration but rather makes use of processes such as fermentation to obtain its energy.

Andesite A volcanic rock that in chemical composition is intermediate between basalt and granite.

Angiosperms An advanced group of plants having floral reproductive structures and seeds in a closed ovary. The "flowering plants."

Anthropoidea The suborder of primates that includes monkeys, apes, and humans.

Anticline A geologic structure in which strata are bent into an upfold or arch.

Antler orogeny A Late Devonian and Mississippian episode of mountain building involving folding and thrusting along a belt across Nevada to southwestern Alberta.

Archaea The domain that includes thermophylic (heat-dependent), halophylic (salt-dependent), methane-producing, and sulfur-producing bacteria. (The remaining two domains are the Bacteria and Eurkarya.)

Archaeocyatha A group of Cambrian marine organisms having double, perforated, calcareous walls and conical-to-cylindric skeletal form. **Archaeocyathids** lived during the Cambrian.

Archean Pertaining to the division of Precambrian time beginning 3.8 billion years ago and ending 2.5 billion years ago.

Archosaurs Advanced reptiles of a group called diapsids, which includes thecodonts, "dinosaurs," pterosaurs, and crocodiles.

Arcoids A group of bivalves (pelecypods) exemplified by species of Arca.

Artiodactyl Hoofed mammals that typically have two or four toes on each foot.

Asteroid One of numerous relatively small planetary bodies (less than 800 kilometers in diameter) revolving around the Sun in orbits lying between those of Mars and Jupiter.

Asthenosphere The zone between 50 and 250 kilometers below the surface of the Earth, where shock waves of earthquakes travel at much reduced speeds, perhaps because of less rigidity. The asthenosphere may be a zone where convective flow of material occurs.

Atoll A ring-like island or a series of islands formed by corals and calcareous algae around a central lagoon.

Atom The smallest divisible unit retaining the characteristics of a specific element.

Atomic fission A nuclear process that occurs when a heavy nucleus splits into two or more lighter nuclei, simultaneously liberating a considerable amount of energy.

Atomic fusion A nuclear process that occurs when two light nuclei unite to form a heavier one. In the process, a large amount of energy is released.

Atomic mass A quantity essentially equivalent to the number of neutrons plus the number of protons in an atomic nucleus.

Atomic number The number of protons in the nuclei of atoms of a particular element. (An element is thus a substance in which all of the atoms have the same atomic number.)

Aulocogen The failed arm of a triple junction rift.

Australopithecines A general term applied loosely to Pliocene and early Pleistocene primates whose skeletal characteristics place them between typically ape-like individuals and those more obviously human.

Autotroph An organism that uses an external source of energy to produce organic nutrients from simple inorganic chemicals.

Bacteria Unicellular, prokaryotic microorganisms belonging to the Kingdom Monera. One of the three great domains of organisms that include also the Archaea and the Eukarya.

Basin A depressed area that serves as a catchment area for sediments (basin of deposition). A structural basin is an area in which strata slope inward toward a central location. Structural basins tend to experience periodic downsinking and thus receive a thicker and more complete sequence of sediments than do adjacent areas.

Belemnites Members of the molluscan Class Cephalopoda, having straight internal shells.

Benioff seismic zone An inclined zone along which frequent earthquake activity occurs and that marks the location of the plunging, forward edge of a lithospheric plate during subduction.

Benthic Pertaining to a bottom-dwelling organism.

Bentonite A layer of clay, presumably formed by the alteration of volcanic ash.

Beta particle A charged particle, essentially equivalent to an electron, emitted from an atomic nucleus during radioactive disintegration.

Bioturbation The disturbance of sediment by burrowing, boring, and sediment-ingesting organisms.

Bivalvia A class of the Phylum Mollusca (also known as the Class Pelecypoda).

Blastoids Sessile (attached) Paleozoic echinoderms having a stem and an attached cup or calyx composed of relatively few plates.

Bolide A meteorite, asteroid, or comet that explodes on striking the Earth.

Brachiating Swinging from branch to branch and tree to tree by using the limbs, as among monkeys.

Brachiopod A bivalved (doubled-shelled) marine invertebrate. Brachiopods were particularly common and widespread during the Paleozoic and persist in fewer numbers today.

Breccia A clastic sedimentary rock composed largely of angular fragments of granule size or larger.

Bryozoa A phylum of attached and incrusting colonial marine invertebrates.

Burgess Shale fauna A beautifully preserved fossil fauna of soft-bodied Cambrian animals discovered in 1910 by Charles Walcott in Kicking Horse Pass, Alberta, Canada.

Caledonian orogeny A major early Paleozoic episode of mountain building affecting Europe that created an orogenic belt, the Caledonides, extending from Ireland and Scotland northwestward through Scandinavia.

Carbon-14 A radioactive isotope of carbon with an atomic mass of 14. Carbon-14 is frequently used in determining the age of materials less than about 50,000 years old.

Carbonate A general term for a chemical compound formed when carbon dioxide dissolved in water combines with oxides of calcium, magnesium, potassium, sodium, and iron. The most common carbonate minerals are calcite (which forms the carbonate rock limestone) and dolomite.

Carbonization The concentration of carbon during fossilization.

Cast (natural) A replica of an organic subject, such as a fossil shell, formed when sediment fills a mold of that object.

Catskill delta A build-up of Middle and Upper Devonian clastic sediments as a broad, complex clastic wedge derived from the erosion of highland areas formed largely during the Acadian orogeny.

Ceboidea The New World monkeys, characterized by prehensile tails, and including the capuchin, marmoset, and howler monkeys.

Centrifugal force The apparent outward force experienced by an object moving in a circular path. Centrifugal force is a manifestation of inertia, the tendency of moving things to travel in straight lines.

Ceratites One of the three larger groups of ammonoid cephalopods having sutural complexity intermediate between goniatites and ammonites.

Ceratopsians The quadrupedal ornithischian dinosaurs characterized by the development of prominent horns on the head.

Cercopithecoidea The Old World monkeys (of Asia, southern Europe, and Africa), including macaques, guenons, langurs, baboons, and mandrills.

Cetaceans The group of marine mammals that includes whales and porpoises.

Chalicotheres Extinct perissodactyls having robust claws rather than hoofs.

Chalk A soft, white, fine-grained variety of limestone composed largely of the calcium carbonate skeletal remains of marine microplankton.

Chert A dense, hard sedimentary rock or mineral composed of submicrocrystalline quartz. Unless colored by impurities, chert is white, as opposed to flint, which is dark or black.

Chlorophyll The catalyst that makes possible the reaction of water and carbon dioxide in green plants to produce carbohydrates. Photosynthesis is the reaction.

Choanichthyes   That group of fishes that includes both the dipnoans (lungfishes with weak pelvic and pectoral fins) and crossopterygians (lungfishes with stout lobe-fins).

Chondrichthyes   The broad category of fishes with cartilaginous skeletons that is exemplified by sharks, skates, and rays.

Chondrites   Stony meteorites that contain rounded silicate grains or chondrules. Chondrules are believed to have formed by crystallization of liquid silicate droplets.

Chromosome   A thread-like, microscopic body composed of chromatin. Chromosomes appear in the nucleus of the cell at the time of cell division. They contain the genes. The number of chromosomes is normally constant for a particular species.

Chromosphere   One of the concentric shells of the Sun, lying above the photosphere and telescopically visible as a thin, brilliant red rim around the edge of the sun for a second or so at the beginning and end of a solar eclipse.

Cladistic phylogeny   An approach to biologic classification in which organisms are grouped according to similarities that are derived from a common ancestor.

Clastic texture   Texture that characterizes a rock made up of fragmental grains such as sand, silt, or parts of fossils. Conglomerates, sandstones, and siltstones are clastic rocks; the individual clastic grains are termed clasts.

Clastic wedge   An extensive accumulation of largely clastic sediments deposited adjacent to uplifted areas. Sediments in the wedge become finer and the section becomes thinner in a direction away from the upland source areas. The Queenston and Catskill "deltas" are examples of clastic wedges.

Coccolithophorids   Marine, planktonic, biflagellate, golden-brown algae that typically secrete coverings of discoidal calcareous platelets called coccoliths.

Colorado mountains   Highlands uplifted in Pennsylvanian time in Colorado. Sometimes inappropriately termed the "ancestral Rockies."

Conodonts   Small, tooth-like, stratigraphically useful fossils composed of calcium phosphate and found in Cambrian to Triassic rocks. Conodonts are elements of an apparatus used by primitive chordates in capturing and chewing food.

Contact metamorphism   The compositional and textural changes in a rock that result from heat and pressure emanating from an adjacent igneous intrusion.

Convergence (in evolution)   The process by which similarity of form or structure arises among different organisms as a result of their becoming adapted to similar habitats.

Cordaites   A primitive order of tree-like plants with long, blade-like leaves and clusters of naked seeds. Cordaites were in some ways intermediate in evolutionary stage between seed ferns and conifers.

Core   The central part of the Earth that lies beneath the mantle.

Coriolis effect   The deflection of winds and water currents to the right in the Northern Hemisphere and to the left in the Southern Hemisphere as a consequence of the Earth's rotation.

Cosmic rays   Extremely high-energy particles, mostly protons, that move through the galaxy and frequently strike the Earth's atmosphere.

Craton   The long-stable region of a continent, commonly with Precambrian rocks either at the surface or only thinly covered with younger sedimentary rocks.

Creodonta   Primitive, early, flesh-eating placental mammals.

Crinoids   Stalked echinoderms with a calyx composed of regularly arranged plates from which radiate arms for gathering food.

Cross-bedding (cross-stratification)   An arrangement of laminae or thin beds transverse to the planes of stratification. The inclined laminae are usually at inclinations of less than 30A7 and may be either straight or concave.

Crossopterygii   That group of choanichthyan fishes ancestral to earliest amphibians and characterized by stout pectoral and pelvic fins as well as lungs.

Crust (Earth)   The outer part of the lithosphere; it averages about 32 kilometers in thickness.

Crustacea   A subphylum of the phylum Arthropoda that includes such well-known living animals as lobsters and crayfishes.

Curie temperature   The temperature at which a cooling mineral acquires permanent magnetic properties that record the surrounding magnetic field orientation and strength at the time of cooling. Magnetic properties of a mineral above the Curie temperature will change as the surrounding field changes. Below the Curie temperature, magnetic characteristics of the mineral will not alter.

Cyanobacteria   Prokaryotic, photosynthetic microorganisms that possess chlorophyll and produce oxygen. Formerly termed blue-green algae.

Cycadales   A group of seed plants that were especially common during the Mesozoic and were characterized by palm-like leaves and coarsely textured trunks marked by numerous leaf scars.

Cyclothem   A vertical succession of sedimentary units reflecting environmental events that occurred in a constant order. Cyclothems are particularly characteristic of the Pennsylvania System.

Cystoids   Attached echinoderms with generally irregular arrangement and number of plates in the calyx and perforated by pores or slits.

Deccan traps   A thick sequence (3200 meters) of Upper Cretaceous basaltic lava flows that cover about 500,000 kilometers^2 of peninsular India.

Décollement   The feature of stratified rocks in which upper formations may become "unstuck" from lower formations, deform, and slide thousands of meters over underlying beds.

Diatoms   Microscopic golden-brown algae (chrysophytes) that secrete a delicate siliceous frustule (shell).

Differentiation   The process by which a planet becomes internally zoned, as when heavy materials sink toward its center and light materials accumulate near the surface.

Dinoflagellates   Unicellular marine algae usually having two flagella and a cellulose wall.

Diploid cells   Cells having two sets of chromosomes that form pairs, as in somatic cells.

Dipnoi   An order of lungfishes with weak pectoral and pelvic fins; not considered ancestral to land vertebrates.

Disconformity A variety of unconformity in which bedding planes above and below the plane of erosion or non-deposition are parallel.

DNA (deoxyribonucleic acid) The nucleic acid found chiefly in the nucleus of a cell that functions in the transfer of genetic characteristics and in protein synthesis.

Docodonts A group of small, primitive Late Jurassic mammals possibly ancestral to the living monotremes.

Dolomite (or dolostone) A carbonate sedimentary rock of which more than 50 percent is the mineral dolomite $CaMg(CO_3)_2$.

Domain A major taxonomic division ranking higher than a kingdom. The three domains are the Archaea, Bacteria, and Eukarya.

Dome An upfold in rocks having the general configuration of an inverted bowl. Strata in a dome dip outward and downward in all directions from a central area. An example is the Ozark dome.

Dryopithecine In general, a group of lightly built primates that lived during the Miocene and Pliocene in mostly open savannah country and that includes Dryopithecus, a form considered to be in the line leading to apes.

Dynamothermal (regional) metamorphism Metamorphism that has occurred over a wide region, caused by deep burial and high temperatures associated with pressures resulting from overburden and orogeny.

Echinoderms The large group (phylum Echinodermata) of marine invertebrates characterized by prominent pentamerous symmetry and a skeleton frequently constructed of calcite elements and including spines. Cystoids, blastoids, crinoids, and echinoids are examples of echinoderms.

Edentata An order of placental mammals that includes extinct ground sloths and gyptodonts, as well as living armadillos, tree sloths, and South American anteaters.

Ediacaran fauna The late Proterozoic fauna of multicellular animals first discovered in Australia but subsequently found in rocks about 600 million years old in many continents.

Endemic population The native fauna of any particular region.

Entelodonts A group of extinct artiodactyls bearing a superficial resemblance to giant wild boars.

Eon A major division of the geologic time scale. The Phanerozoic Eon comprises all of the geologic periods from the Cambrian to the Holocene. The term is also sometimes used to denote a span of 1 billion years.

Epifaunal organisms Organisms living on, as distinct from in, a particular body of sediment or another organism.

Epoch A chronologic subdivision of a geologic period. Rocks deposited or emplaced during an epoch constitute the series for that epoch.

Era A major division of geologic time, divisible into geologic periods.

Eukaryote A cell containing a true nucleus, enclosed within a nuclear membrane, and having well-defined chromosomes and cell organelles.

Eurypterids Aquatic arthropods of the Paleozoic, superficially resembling scorpions and probably carnivorous.

Eustatic Pertaining to worldwide simultaneous changes in sea level, such as might result from change in the volume of continental glaciers.

Evaporites Sediments precipitated from a water solution as a result of the evaporation of that water. Evaporite minerals include anhydrite gypsum ($CaSO_4$) and halite (NaCl).

Evolution The continuous genetic adaptation of organisms or species to the environment.

Facies A particular aspect of sedimentary rocks that is a direct consequence of sedimentation in a particular depositional environment.

Fault A fracture in the Earth's crust along which rocks on one side have been displaced relative to rocks on the other side.

Felsic Term used to describe light colored igneous rocks having the general composition of granite.

Fermentation The partial breakdown of organic compounds by an organism in the absence of oxygen. The final product of fermentation is alcohol or lactic acid.

Filter-feeders Animals that obtain their food, which usually consists of small particles or organisms, by filtering it from the water.

Fissility That property of rocks that causes them to split into thin slabs parallel to bedding. Fissility is particularly characteristic of shale.

Flood basalts Regionally extensive layers of basalt that originated as low-viscosity lava pouring from fissure eruptions. The lavas of the Columbia Plateau and the Deccan Plateau are flood basalts.

Fluvial Pertaining to sediments or other geologic features formed by streams.

Flysch Thick sequences of rapidly deposited, poorly sorted marine clastics.

Focus (earthquake) The location at which rock rupture occurs that generates seismic waves.

Foliation A textural feature especially characteristic of metamorphic rocks in which laminae develop by growth or realignment of minerals in parallel orientation.

Foraminifera An order of mostly marine, unicellular protozoans that secrete tests (shells) that are usually composed of calcium carbonate.

Formation A mappable, lithologically distinct body of rock having recognizable contacts with adjacent rock units.

Fossil The remains or indications of an organism that lived in the geologic past.

Fractional crystallization The separation of components of a cooling magma by sequential formation of particular mineral crystals at progressively lower temperatures.

Fusulinids Primarily spindle-shaped foraminifers with calcareous, coiled tests divided into a complex of numerous chambers. Fusulinids were particularly abundant during the Pennsylvanian and Permian periods.

Galaxy An aggregate of stars and planets, separated from other such aggregates by distances greater than those between member stars.

Gamete Either of two cells (male or female) that must unite in sexual reproduction to initiate the development of a new individual.

Gamma rays   Very high-frequency electromagnetic waves.

Garnet   A family of aluminosilicates of iron and calcium that are particularly characteristic of metamorphic rocks.

Gene   The unit of heredity transmitted in the chromosome.

Gene pool   All the genes present in a species population.

Genus   The major subdivision of a taxonomic family or subfamily of plants or animals, usually consisting of more than one species.

Geochronology   The study of time as applied to Earth and planetary history.

Geologic range   The geologic time span between the first and last appearance of an organism.

Geology   The science of the Earth, including the materials that the planet is made of, the physical and chemical changes that occur on and within the Earth, and the history of the Earth and its inhabitants.

Glacier   A large mass of ice, formed by the recrystallization of snow, that flows slowly under the influence of gravity.

Glauconite   A green clay mineral frequently found in marine sandstones and believed to have formed at the site of deposition.

***Glossopteris* flora**   An assemblage of fossil plants found in rocks of late Paleozoic and early Triassic age in South Africa, India, Australia, and South America. The flora takes its name from the seed fern *Glossopteris*.

Gondwanaland   The great Permo-Carboniferous Southern Hemisphere continent, comprising the assembled present areas of South Africa, India, Australia, Africa-Arabia, and Antarctica.

Goniatites   One of the three large groups of ammonoid cephalopods with sutures forming a pattern of simple lobes and saddles and thus not as complex as either the ceratites or the ammonites.

Gowganda Conglomerate   An apparent tillite of the Canadian Shield. The Gowganda rests on a surface of older rock that appears to have been polished by glacial action.

Graptolite facies   A Paleozoic sedimentary facies composed of dark shales and fine-grained clastics that contain the abundant remains of graptolites and that are associated with volcanic rocks.

Graptolites   Extinct colonial marine animals considered to be hemichordates. Graptolites range from the late Cambrian to the Mississippian.

Gravity anomalies   The differences between the observed value of gravity at any point on the Earth and the calculated theoretic value.

Greenhouse effect   A process in which incoming solar radiation that is absorbed and reradiated cannot escape back into space because the Earth's atmosphere is not transparent to the reradiated energy, which is in the form of infrared radiation (heat).

Greenstones   Low-grade metamorphic rocks containing abundant chlorite, epidote, and biotite and developed by metamorphism of basaltic extrusive igneous rocks. Great linear outcrops of greenstones are termed greenstone belts and are thought to mark the locations of former volcanic island arcs.

Guide fossil   A fossil with a wide geographic distribution but narrow stratigraphic range and thus useful in correlating strata and for age determination.

Gutenberg discontinuity   The boundary separating the mantle of the Earth from the core below. The Gutenberg discontinuity lies about 2900 kilometers below the surface.

Gymnosperms   An informal designation for flowerless seed plants in which seeds are not enclosed (hence, "naked seeds"). Examples are conifers and cycads.

Hadrosaurs   The ornithischian duck-billed dinosaurs of the Cretaceous.

Half-life   The time in which one-half of an original amount of a radioactive species decays to daughter products.

Haploid cell   A cell having a single set of chromosomes, as in gametes (see diploid cells).

Hercynian orogeny   The major late Paleozoic orogenic episode in Europe that formed the ancient Hercynian mountains. Today, only the eroded stumps of these mountains are exposed in areas where the cover of Mesozoic and Cenozoic strata has been removed by erosion.

Heterotroph   An organism that depends on an external source of organic substances for its nutrition and energy.

Holocene   A term sometimes used to designate the period of time since the last major episode of glaciation. The term is equivalent to Recent.

Homologous organs   Organs having structural and developmental similarities due to genetic relationship.

Hydrosphere   The water and water vapor present at the surface of the Earth, including oceans, seas, lakes, and rivers.

Hylobatidae   A group of persistently arboreal, small apes that are exemplified by the gibbons and siamangs.

Hyomandibular bone (in fishes)   The modified upper bone of the hyoid arch, which functions as a connecting element between the jaws and the braincase in certain fishes.

Ichnology   The study of trace fossils (tracks, trails, burrows, borings, castings, etc.).

Ichthyosaurs   Highly specialized marine reptiles of the Mesozoic, recognized by their fishlike form.

Ictidosauria   A group of extinct mammal-like reptiles or therapsids whose skeletal characteristics are considered to be very close to those of mammals.

Igneous rock   A rock formed by the cooling and solidification of magma or lava.

Infaunal organisms   Organisms that live and feed within bottom sediments.

Ion   An atom that, because of electron transfers, has excess positive or negative charges.

Island arcs   Chains of islands, arranged in arcuate trends on the surface of the Earth. The arcs are sites of volcanic and earthquake activity and are usually bordered by deep oceanic trenches on the convex side.

Isopachous map   A map depicting the thickness of a sedimentary unit.

Isotopes   Atoms of an element that have the same number of protons in the nucleus, the same atomic number, and the same chemical properties but different atomic masses because they have different numbers of neutrons in their nuclei.

Karoo system A sequence of Permian to Lower Jurassic rocks, primarily continental formations, which outcrop in Africa and are approximately equivalent to the Gondwana system of peninsular India.

Kaskaskia sequence A sequence of Devonian-Mississippian sediments, bounded above and below by regional unconformities and recording an episode of transgression followed by full flooding of a large part of the craton and by subsequent regression.

Komatiite An ultramafic volcanic rock prevalent in Archean terranes.

Lacustrine Pertaining to lakes, as in lacustrine sediments (lake sediments).

Lagomorph The order of small placental mammals that includes rabbits, hares, and pikas.

Laramide orogeny In general, those pulses of mountain building that were frequent in Late Cretaceous time and were in large part responsible for producing many of the structures of the Rocky Mountains.

Lateral fault A fault in which the movement is largely horizontal and in the direction of the trend of the fault plane. Sometimes called a strike-slip fault.

Laurasia A hypothetic supercontinent composed of what is now Europe, Asia, Greenland, and North America.

Limestone A sedimentary rock consisting mainly of calcium carbonate.

Lithofacies map A map that shows the areal variation in lithologic attributes of a stratigraphic unit.

Lithosphere The outer shell of the Earth, lying above the asthenosphere and comprising the crust and upper mantle.

Litopterns South American ungulates whose evolutionary history somewhat paralleled that of horses and camels.

Logan's line A zone of thrust-faulting produced during the Taconic orogeny that extends from the west coast of Newfoundland along the trend of the St. Lawrence River to near Quebec and southward along Vermont's western border. (Named after the pioneer Canadian geologist Sir William Logan.)

Lophophore An organ located adjacent to the mouth of brachiopods and bryozoans that bears ciliated tentacles and has as its primary function the capture of food particles.

Low-velocity zone The interior zone of the Earth, characterized by lower seismic wave velocities than the region immediately above it.

Lunar maria Low-lying, dark lunar plains filled with volcanic rocks rich in iron and magnesium.

Lycopsids Leafy plants with simple, closely spaced leaves bearing sporangia on their upper surfaces. They are represented by living club mosses and vast numbers of extinct late Paleozoic "scale trees."

Magnetic declination The horizontal angle between "true" (geographic) north and magnetic north, as indicated by the compass needle. Declination is the result of Earth's magnetic axis being inclined with respect to the Earth's rotational axis.

Magnetic inclination The angle between the magnetic lines of force for the Earth and the Earth's surface; sometimes called "dip." Magnetic inclination can be demonstrated by observing a freely suspended magnetic needle. The needle will lie parallel to the Earth's surface at the equator but is increasingly inclined toward the vertical as the needle is moved toward the magnetic poles.

Mammoth The name commonly applied to extinct elephants of the Pleistocene Epoch.

Marsupials Mammals of the Order Marsupialia. Female marsupials bear mammary glands and carry their immature young in a stomach pouch.

Mass spectrometer An instrument that separates ions of different mass but equal charge and measures their relative quantities.

Mastodonts The group of extinct proboscideans (elephantoids), early forms of which were characterized by long jaws, tusks in both jaws, and low-crowned teeth.

Medusa The free-swimming, umbrella-shaped jellyfish form of the phylum Cnidaria.

Meiosis That kind of nuclear division, usually involving two successive cell divisions, that results in daughter cells having one-half the number of chromosomes that were in the original cell.

Mélange A body of intricately folded, faulted, and severely metamorphosed rocks, examples of which can be seen in the Franciscan rocks of California.

Mesosphere The zone of the Earth's mantle where pressures are sufficient to impart greater strength and rigidity to the rock.

Metamorphic rock A rock formed from a previously existing rock by subjecting the parent rock to high temperature and pressure but without melting.

Metamorphism The transformation of previously existing rocks into new types by the action of heat, pressure, and chemical solutions. Metamorphism usually takes place at depth in the roots of mountain chains or adjacent to large intrusive igneous bodies.

Metazoa All multicellular animals whose cells become differentiated to form tissues (all animals except Protozoa).

Meteorites Metallic or stony bodies from interplanetary space that have passed through the Earth's atmosphere and hit the Earth's surface.

Meteors Generally small particles of solid material from interplanetary space that approach close enough to the Earth to be drawn into the Earth's atmosphere, where they are heated to incandescence. Sometimes called "shooting stars." Most disintegrate, but a few land on the surface of the Earth as meteorites.

Micrite Pertaining to a texture in carbonate rocks that, when viewed microscopically, appears as murky, fine-grained calcium carbonate. Micrite is believed to develop from fine carbonate mud or ooze.

Milankovitch effect The hypothetic long-term effect on world climate based on positional changes between the Earth and the Sun. The changes provide a possible explanation for repeated glacial to interglacial climatic swings.

Miliolids A group of foraminifers with smooth, imperforate test walls and chambers arranged in various planes around a vertical axis. Miliolids are common in shallow marine areas.

Mineral A naturally occurring element or compound formed by inorganic processes that has a definite chemical composition or range of compositions, as well as distinctive properties and form that reflect its characteristic atomic structure.

Mitosis The method of cell division in which each of the two daughter nuclei receives exactly the same complement of chromosomes as had existed in the parent nucleus.

Mobile belt An elongate region of the Earth's crust, characterized by exceptional earthquake and volcanic activity, tectonic instability, and periodic mountain building.

Mohorovičić discontinuity A plane that marks the boundary separating the crust of the Earth from the underlying mantle. The "moho," as it is sometimes called, is at a depth of about 70 kilometers below the surface of continents and 6 to 14 kilometers below the floor of the oceans.

Molasse Accumulations of primarily nonmarine, relatively light-colored, irregularly bedded conglomerates, shales, coal seams, and cross-bedded sandstones that are deposited subsequent to major orogenic events.

Mold An impression, or imprint, of an organism or part of an organism in the enclosing sediment.

Mollusk Any member of the invertebrate Phylum Mollusca, including bivalves (pelecypods), cephalopods, gastropods, scaphopods, and chitons.

Monoplacophorans Primitive marine molluscans with simple cap-shaped shells.

Monotremes The egg-laying mammals.

Morganucodonts Early mammals found in Triassic beds of Europe and Asia and characterized by their small size and their retention of certain reptilian osteologic traits.

Mosasaurs Large marine lizards of the Late Cretaceous.

Multituberculates An early group (Jurassic) of Mesozoic mammals with tooth cusps in longitudinal rows and other dental characteristics that suggest they may have been the earliest herbivorous mammals.

Mutation A stable and inheritable change in a gene.

Mytiloids Bivalvia having rather triangular shells that are most commonly identical but in some forms unequal. The edible mussel *Mytilus* is a representative form.

Nappe A large mass of rocks that have been moved a considerable distance over underlying formations by overthrusting, recumbent folding, or both.

Nektonic Pertaining to swimming organisms. (Some texts shorten the term to nektic.)

Neogene A subdivision of the Cenozoic that encompasses the Miocene and Pliocene.

Neutron An electrically neutral (uncharged) particle of matter existing along with protons in the atomic nucleus of all elements except the mass 1 isotope of hydrogen.

Nevadan orogeny In general, those pulses of mountain building, intrusion, and metamorphism that were most frequent during the Jurassic and Early Cretaceous along the western part of the Cordilleran orogenic belt.

Newark series A series of Upper Triassic, nonmarine red beds (shales, sandstones, and conglomerates), lava flows, and intrusions located within downfaulted basins from Nova Scotia to South Carolina.

New World monkeys Monkeys whose habitat is today confined to South America. They are thoroughly arboreal in habit and have prehensile tails, by which they hang and swing from tree limbs.

Niche The full range of physical and biologic conditions under which an organism can live and reproduce.

Nonconformity An unconformity developed between sedimentary rocks and older plutons or massive metamorphic rocks that had been exposed to erosion before the overlying sedimentary rocks were deposited.

Normal fault A fault in which the hanging wall appears to have moved downward relative to the footwall; normally occurring in areas of crustal tension.

Nothosaurs Relatively small early Mesozoic sauropterygians that were replaced during the Jurassic by the plesiosaurs.

Notochord A rod-shaped cord of cartilage cells forming the primary axial structure of the chordate body. In vertebrates, the notochord is present in the embryo and is later supplanted by the vertebral column.

Notungulates A group of ungulates that diversified in South America and persisted until Plio-Pleistocene time.

Novaculites A term applied originally to rocks suitable for whetstones and, in America, to white chert found in Arkansas. Now applied to very tough, uniformly grained cherts composed of microcrystalline quartz.

Nuclear fission tracks Submicroscopic "tunnels" in minerals produced when high-energy particles from the nucleus of uranium are forcibly ejected during spontaneous fission.

Nucleic acid Any of a group of organic acids that control hereditary processes within cells and make possible the manufacture of proteins from the amino acids ingested by the cells as food.

Nuclides The different weight configurations of an element caused by atoms of that element having differing numbers of neutrons. Nuclides or isotopes of an element differ in number of neutrons but not in chemical properties.

Nummulites Large, coin-shaped foraminifers, especially common in Tertiary limestones.

Offlap A sequence of sediments resulting from a marine regression and characterized by an upward progression from offshore marine sediments (often limestones) to shales and finally sandstones (above which will follow an unconformity).

Oil shale A dark-colored shale rich in organic material that can be heated to liberate gaseous hydrocarbons.

Old World monkeys Monkeys of Asia, Africa, and southern Europe that include macaques, guenons, langurs, baboons, and mandrills.

Onlap A sequence of sediments resulting from a marine transgression. Normally, the sequence begins with a conglomerate or sandstone deposited over an erosional unconformity and followed upward in the vertical section by progressively more offshore sediments.

Oölites Limestones composed largely of small, round or ovate calcium carbonate bodies called oöids.

Ophiolite suite An association of radiolarian cherts, pelagic muds, basaltic pillow and flow lavas, gabbros, and ultramafic rocks such as periodotite regarded as surviving masses of former oceanic crust largely destroyed in former subduction zones.

Oreodonts North American artiodactyls of the middle and late Tertiary.

Ornithischia An order of dinosaurs characterized by bird-like pelvic structures and including such herbivores as the ornithopods, stegosaurs, ankylosaurs, and ceratopsians.

Orogenic belt Great linear tracts of deformed rocks, primarily developed near continental margins by compressional forces accompanying mountain building.

Orogeny The process by which great systems of mountains are formed. (Orogenesis means mountain building.)

Ostracoderms Extinct jawless fishes of the early Paleozoic.

Ostracodes Small, bivalved, bean-shaped crustaceans.

Ostreoids The "oyster family" of bivalves (pelecypods).

Outcrop An area where specific rock units are exposed at the Earth's surface or occur at the surface but are covered by superficial deposits.

Palatoquadrate Upper jaw element of primitive fishes and chondrichthyes.

Paleoecology The study of the relationship of ancient organisms to their environment.

Paleogene A subdivision of the Cenozoic that encompasses the Paleocene, Eocene, and Oligocene epochs.

Paleogeography The geography as it existed at some time in the geologic past.

Paleolatitude The latitude that once existed across a particular region at a particular time in the geologic past.

Paleomagnetism The Earth's magnetic field and magnetic properties in the geologic past. Studies of paleomagnetism are helpful in determining positions of continents and magnetic poles.

Paleontology The study of all ancient forms of life, their interactions, and their evolution.

Pangea In Alfred Wegener's theory of continental drift, the supercontinent that included all present major continental masses.

Panthalassa The great universal ocean that surrounded the supercontinent Pangea prior to its break-up.

Paraconformity A rather obscure unconformity in which no erosional surface is discernible and in which beds above and below the break are parallel.

Partial melting The process by which a rock subjected to high temperature and pressure is partially melted, and the liquid fraction moved to another location. Partial melting results from the variations in melting points of different minerals in the original rock mass.

Pectenoids Bivalvia exemplified by the scallops. They have generally subcircular shells and straight hinge lines.

Pelycosaurs Early mammal-like reptiles exemplified by the sail-back animals of the Permian Period.

Period A subdivision of an era.

Perissodactyl Progressive, hoofed mammals characteristically having an odd number of toes on the hind feet and usually on the front feet as well.

Permineralization A manner of fossilization in which voids in an organic structure (such as bone) are filled with mineral matter.

Petrification The process of conversion of organic structures, such as bone, shell, or wood, into a stony substance, such as calcium carbonate or silica.

Phanerozoic The eon of geologic time during which the Earth has been populated by abundant and diverse life. The Phanerozoic Eon followed the Cryptozoic (or Precambrian) Eon and is divided into the Paleozoic, Mesozoic, and Cenozoic eras.

Phenotypic trait A trait that is the observable expression of an organism's genes.

Phosphorite A sediment composed largely of calcium phosphate.

Photosynthesis The process of synthesizing carbohydrates from carbon dioxide and water, utilizing the radiant energy of light captured by the chlorophyll in plant cells.

Phyletic gradualism Gradual evolutionary change of one species population into another.

Phytoplankton Microscopic marine planktonic plants, most of which are various forms of algae.

Phytosaurs Extinct aquatic, crocodile-like thecodonts of the Triassic.

Pillow lava Type of lava that is extruded under water and in which many pillow-shaped lobes break through the chilled surface of the flow and solidify. (Resembles a pile of pillows.)

Pinnipeds Marine carnivores such as seals, sea lions, and walruses.

Placer deposit An accumulation of sediment rich in a valuable mineral or metal that has been concentrated because of its greater density.

Placoderms Extinct primitive jawed fishes of the Paleozoic Era.

Placodonts Extinct walrus-like marine reptiles that fed principally on shellfish.

Plankton Minute, free-floating aquatic organisms.

Plate tectonics The theory that explains the tectonic behavior of the crust of the Earth in terms of several moving plates that are formed by volcanic activity at oceanic ridges and destroyed along great ocean trenches.

Platform That part of a craton covered thinly by layered sedimentary rocks and characterized by relatively stable tectonic conditions.

Plesiosaurs The group of extinct Mesozoic marine reptiles (sauropterygians) characterized by large, paddle-shaped limbs and broad bodies, with either very long or relatively short necks.

Pluvial lake A lake formed in an earlier climate when rainfall was greater than at present.

Polyp The hydra-like form of some cnidaria in which the mouth and tentacles are at the top of the body.

Population (species population) A group of individuals of the same species occupying a given area at the same time.

Porphyry A textural term used to describe an igneous rock in which some of the crystals, called phenocrysts, are distinctly larger than others.

Precambrian Pertaining to all of geologic time and its corresponding rocks before the beginning of the Paleozoic Era.

Primary earthquake waves Seismic waves that are propagated through solid rock as a train of compressions and dilations. Direction of vibration is parallel to direction of propagation.

Principle of cross-cutting relations The principle that states that geologic features such as faults, veins, and dikes must be younger than the rocks or features across which they cut.

Principle of biologic succession The principle that states that the observed sequence of life forms has changed continuously through time, so that the total aspect of life (as recognized by fossil evidence) for a particular segment of time is distinct and different from that of life of earlier and later times.

Principle of temporal transgression The principle that stipulates that sediments of advancing (transgressing) or retreating (regressing) seas are not necessarily of the same geologic age throughout their lateral extent.

Proboscidea The elephants and their progenitors.

Productid An articulate brachiopod usually with a flat or concave brachial valve and a convex pedicle valve. Productids were present from Devonian to Permian.

Prokaryotes Organisms that lack membrane-bounded nuclei and other membrane-bounded organelles.

Prosimii The less advanced primates, such as lemurs, tarsiers, and tree shrews.

Proteinoids Extra-large organic molecules containing most of the 20 amino acids of proteins and produced in laboratory conditions simulating those found in nature.

Proteins Giant molecules containing carbon, hydrogen, oxygen, nitrogen, and usually sulfur and phosphorus; composed of chains of amino acids and present in all living cells.

Proton An elemental particle found in the nuclei of all atoms that has a positive electric charge and a mass similar to that of a neutron.

Pterosaur A flying reptile of the Jurassic and Cretaceous.

Punctuate equilibrium The model of evolution proposing that long periods of little or no evolutionary change are punctuated by short episodes of rapid change.

Pyroclastics Fragments of volcanic debris that have usually been found fragmented during eruptions.

Pyroxene group A group of dark-colored iron and magnesium-rich silicate minerals.

Queenston delta A clastic wedge of red beds shed westward from highlands elevated in the course of the Taconic orogeny.

Radioactive decay The spontaneous emission of a particle from the atomic nucleus, thereby transforming the atom from one element to another.

Radiolaria Protozoa that secrete a delicate, often beautifully filigreed skeleton of opaline silica.

Recumbent fold A fold in which the axial plane is essentially horizontal; a fold that has been turned over by compressional forces so that it lies on its side.

Red beds Prevailing red, usually clastic sedimentary deposits.

Regoliths Any solid materials, such as rock fragments or soil, lying on top of bedrock.

Regression A general term signifying that a shoreline has moved toward the center of a marine basin. Regression may be caused by tectonic emergence of the land, eustatic lowering of sea level, or prograding of sediments, as in deltaic build-outs.

Relative geologic age The placing of an event in a time sequence without regard to the absolute age in years.

Replacement A fossilization process in which the original skeletal substance is replaced after burial by inorganically precipitated mineral matter.

Reverse fault A fault formed by compression in which the hanging wall appears to move up relative to the foot wall.

Rhynchonellids A group of brachiopods having pronounced beaks, accordion-like plications, and triangular outlines.

Rift valley A valley formed by faulting, usually involving a central fault block that moves downward in relation to adjacent blocks.

Rudists Peculiarly specialized Mesozoic bivalvia often having one valve in the shape of a horn coral, covered by the other valve in the form of a lid.

Rugosa The large group of solitary and colonial Paleozoic horn corals.

Ruminant A herbivorous, cud-chewing ungulate.

Salt dome A structural dome in sedimentary strata resulting from the upward flow of a large body of salt.

Sarcopterygii Lobe-finned bony fishes, including air-breathing crossopterygian fishes.

Sauk sequence A sequence of upper Precambrian to Ordovician sediments bounded both above and below by a regional unconformity and recording an episode of marine transgression, followed by full flooding of a large part of the craton and ending with a regression from the craton.

Saurischia An order of dinosaurs with triradiate pelvic structures, including both the gigantic herbivorous sauropods and the carnivorous theropods.

Scleractinid coral Coral belonging to the order Scleractinia, which includes most modern and post-Paleozoic corals.

Sea-floor spreading The process by which new sea-floor crust is produced along midoceanic ridges (divergence zones) and slowly conveyed away from the ridges.

Secondary earthquake wave A seismic wave in which the direction of vibration of wave energy is at right angles to the direction the wave travels.

Sedimentary rock A rock that has formed as a result of the consolidation (lithification) of accumulations of sediment.

Sessile Pertaining to the bottom-dwelling habit of aquatic animals that live continuously in one place.

Series The time-rock term representing the rocks deposited or emplaced during a geologic epoch. A series is a subdivision of a system.

Shelly facies In general, sedimentary deposits consisting primarily of carbonate rocks containing the abundant fossil remains of marine invertebrates.

Siliclastic sedimentary rock. A sedimentary rock such as a conglomerate, sandstone, or shale that is composed largely of particles of silicate minerals such as quartz, feldspar, and mica.

Silicoflagellates Unicellular, tiny, flagellate marine algae that secrete an internal skeleton composed of opaline silica.

Silicon tetrahedron An atomic structure in silicates consisting of a centrally located silicon atom linked to four oxygen atoms placed symmetrically around the silicon at the corners of a tetrahedron.

Sonoma orogeny Middle Permian orogenic movements, the structural effects of which are most evident in western Nevada.

Sorting A measure of the uniformity of the sizes of particles in a sediment or sedimentary rock.

Spar carbonate As viewed microscopically, the clear, crystalline carbonate that has been deposited in a carbonate rock as a cement between clasts or has developed by recrystallization of clasts.

Species A unit of taxonomic classification of organisms. In another sense, a species is a population of individuals that are similar in structural and functional characteristics and that in nature breed only with one another.

Sphenodon (tuatara) Large, lizard-like reptiles that have persisted from Triassic to the present and now inhabit islands off the coast of New Zealand.

Sphenopsids A group of sponge-bearing plants that were particularly common during the late Paleozoic and were characterized by articulated stems with leaves borne in whorls at nodes.

Spiracle In cartilaginous fishes, a modified gill opening through which water enters the pharynx.

Spontaneous fission Spontaneous fragmentation of an atom into two or more lighter atoms and nuclear particles.

Spore A usually asexual reproductive body, such as occurs in bacteria, ferns, and mosses.

Stage The time-rock unit equivalent to an age. A stage is a subdivision of a series.

Stapes The innermost of the small bones in the middle ear cavity of mammals; also recognized in amphibians and reptiles.

Stasis A long interval of little evolutionary change.

Strain Deformation of a rock mass in response to stress.

Stratification The layering in sedimentary rocks that results from changes in texture, color, or rock type from one bed to the next.

Stratified drift Deposits of glacial clastics that have been sorted and stratified by the action of meltwater.

Stratigraphy The study of rock strata, with emphasis on their succession, age, correlation, form, distribution, lithology, fossil content, and all other characteristics useful in interpreting their environment of origin and geologic history.

Stratophenic phylogeny The traditional model of evolution in which organisms are arranged in treelike fashion, with the most recently evolved groups on the upper branches and the older, ancestral groups on the lower branches and trunk.

Stromatolites Distinctly laminated accumulations of calcium carbonate having rounded, branching, or frondose shape and believed to form as a result of the metabolic activity of marine algae. They are usually found in the high intertidal to low supratidal zones.

Stromatoporoids An extinct group of reef-building organisms now believed to have affinities with the Porifera and noted for the large, often laminated masses constructed by the colonies.

Subaerial Formed or existing at or near a sediment surface significantly above sea level.

Subduction zone An inclined planar zone, defined by high frequency of earthquakes, that is thought to locate the descending leading edge of a moving oceanic plate.

Sublittoral zone The marine bottom environment that extends from low tide seaward to the edge of the continental shelf.

Surface earthquake waves Seismic or earthquake waves that move only about the surface of the Earth.

Symmetrodonts A group of primitive Mesozoic mammals characterized by a symmetric triangular arrangement of cusps on cheek teeth.

Syncline A geologic structure in which strata are bent into a downfold.

System The time-rock unit representing rock deposited or emplaced during a geologic period.

Taconic orogeny A major episode of orogeny that affected the Appalachian region in Ordovician time. The northern and Newfoundland Appalachians were the most severely deformed during this orogeny.

Taxon (pl. taxa) Any unit in the taxonomic classification, such as a phylum, class, order, or family.

Taxonomy The science of naming, describing, and classifying organisms.

Tectonics The structural behavior of a region of the Earth's crust.

Teleosts The most advanced of the bony fishes, characterized by thin, rounded scales, completely bony internal skeleton, and symmetric tail. Teleosts range from Cretaceous to Recent.

Terebratulids A group of Silurian to Recent, mostly smooth-shelled brachiopods having a loop-shaped attachment for the lophomore. Terebratulids were most abundant during the Jurassic and Cenozoic.

Terrane A three-dimensional block of crust having a distinctive assemblage of rocks (as opposed to terrain, which implies topography, such as rolling hills or rugged mountains).

Tethys seaway A great east-west trending seaway lying between Laurasia and Gondwanaland during Paleozoic and Mesozoic time and from which arose the Alpine-Himalayan Mountain ranges.

Thecodonts An order of primarily Triassic reptiles considered to be the ancestral archosaurians.

Therapsids An order of advanced, mammal-like reptiles.

Thermal plume A "hot spot" in the upper mantle believed to exist where a huge column of upwelling magma lies in a fixed position under the lithosphere. Thermal plumes are thought to cause volcanism in the overlying lithosphere.

Thermoremanent magnetism Permanent magnetization acquired by igneous rocks as they cool past the Curie point while in the Earth's magnetic field.

Theropods The carnivorous saurischian dinosaurs.

Thrust fault A low-angle reverse fault, with inclination of fault plane generally less than 45A7.

Till Unconsolidated, unsorted, unstratified glacial debris.

Tillite Unsorted glacial drift (till) that has been converted into solid rock.

Time-stratigraphic unit (chronostratigraphic unit) The rocks formed during a particular unit of geologic time. Also called time-rock unit.

Tippecanoe sequence A sequence of Ordovician to Lower Devonian sediments bounded above and below by regional unconformities and recording an episode of marine transgression, followed by full flooding of a large region of the craton and subsequent regression.

Titanotheres Large, extinct perissodactyls (odd-toed ungulates) that attained the peak of their evolutionary development during the Early Eocene. Brontotherium is a widely known titanothere.

Trace fossils Tracks, trails, burrows, and other markings made in now-lithified sediments by ancient animals.

Transcontinental arch An elongate, uplifted region extending from Arizona northeastward toward Lake Superior. During the Cambrian, the arch was emergent, as indicated by the observation that Cambrian marine sediments are located on either side of the arch but are missing above it. (Along the crest of the arch, post-Cambrian rocks rest on Precambrian basement.)

Transform fault A strike-slip fault bounded at each end by an area of crustal spreading that tends to be more or less perpendicular to the trace of the fault.

Triconodonts A group of primitive Mesozoic mammals recognized primarily by the arrangement of three principal cheek tooth cusps in a longitudinal row.

Trilobites Paleozoic marine arthropods of the class Crustacea, characterized by longitudinal and transverse division of the carapace into three parts, or lobes. Trilobites were especially abundant during the early Paleozoic.

Tuff Volcanic ash that has become consolidated into rock.

Turbidites Sediment deposited from a turbidity current and characterized by graded bedding and moderate to poor sorting.

Turbidity current A mass of moving water that is denser than surrounding water because of its content of suspended sediment and that flows along slopes of the sea floor as a result of that higher density.

Tylopod The artiodactyl group to which camels and llamas belong.

Unconformity A surface separating an overlying younger rock formation from an underlying formation and representing an episode of erosion or nondeposition. Because unconformities represent a lack of continuity in deposition, they are gaps in the geologic record.

Ungulate Pertaining to four-legged mammals whose toes bear hoofs.

Uniformitarianism A principle that suggests that the past history of the Earth can be interpreted and deciphered in terms of what is known about present natural laws.

Vagile Pertaining to the bottom-dwelling habit of aquatic animals capable of locomotion.

Varve A thin sedimentary layer or pair of layers that represent the depositional record of a single year.

Vascular plants Plants, including all higher land plants, that have a system of vessels and ducts for distributing moisture and nutrients.

Veneroids A group of bivalves exemplified by the common clam, *Mercenaria*.

Vestigial organ An organ that is useless, small, or degenerate but representing a structure that was more fully developed or functional in an ancestral organism.

Vindelician arch A highland area believed to have formed a barrier separating the Germanic and Alpine depositional areas during the Triassic and Jurassic.

Volatile element A chemical element that generally occurs as a gas at moderate to high temperatures.

Williston basin A large structural basin extending from South Dakota and Montana northward into Canada; well known for the petroliferous Devonian formations deposited therein.

Zechstein sea An arm of the Atlantic that extended across part of northern Europe during the late Permian and in which were deposited several hundred meters of evaporites, including the well-known potassium salts of Germany.

Zircon A zirconium silicate mineral that is useful in uranium-lead isotopic dating.

Zone A bed or group of beds distinguished by a particular fossil content and frequently named after the fossil or fossils it contains. More formally known as a biozone.

Zygote The cell formed by the union of two gametes. Thus, a zygote is a fertilized egg.

In this index, an italic *f* following the page number indicates a *figure*, and an italic *t* after the page number indicates a *table*. Entries from the appendices are preceded by a capital A.

Aa lava, 200
Absaroka sequence, 274*f*, 275, 301
Absolute age, 7
Abyssal environment, 148
Abyssal plain, 174*f*
Acadia National Park, 322–323
Acadian orogeny, 267, 297, 308–309
Acanthodians, 356
Accretion, 210, 220*f*
Accretionary prism, 188
Accretionary tectonics, 378–380
Acquired trait, 128, 129*f*
Acritarchs, 256–258
Actinopterygians, 357
Actual geologic dating, 28, 36–43
Actualism, 17
Adaptations (defined), 133
Adaptive radiation, 132–134, 132*f*, 431
Aegyptopithecus zeuxis, 531, 531*f*
Aepinocodon, 513*f*
Aepycamelus, 512
Aerobic organisms, 235
Afar triangle (Ethiopia), 191, 192*f*
Africa:
    in Cenozoic Era, 471–472
    in Mesozoic Era, 400–401
    rift valley in, 473*f*
Agassiz, Louis, 22
Agate, 51
Age:
    absolute, 7
    of earth, 43
    relative, 7
Aglaophyton, 365, 365*f*
Agnathids, 354–356
Alabaster gypsum, 55*f*
Albedo, 482
Albite, 51
Aldanella, 330
Alien terrane, 322–323
Alighieri, Dante, 464
Alleghenian orogeny, 297, 308, 312–313
Allegheny-Cumberland Plateau, 84
Allegheny orogeny, 267
Alligator mississippienis, 498*f*
Allosaurus, 420*f*
Alluvial fans, 82, 83*f*
Alluvium, 16
Almandine garnet, 49*f*
Alpha particles, 35
Altai Mountains, 15
Aluminosilicates, 51. *See also* Feldspar(s)
Alvarez, Walter, 443
Amazon River (Brazil), 84
Amber, 13, 13*f*, 122–123

Ambulocetus, 139, 519
Amebelodon, 515*f*, 517
Amitsoq gneiss (Greenland), 229*f*
Ammonites, 12*f*, 411
Ammonoids, 410–412
Ammonoid cephalopods, 344, 346*f*
Ammonoidea, 410
Amniotes, 362–363
Amniotic egg, 329, 354*f*, 362
Amniotic membrane, 353
Amniotic vertebrates, 353
Amphibians, 415, 497
Amphiboles, 53
Anabarites, 330, 330*f*
Anaerobic organisms, 235
Andesite, 58*f*, 60
Andesitic magna, 64
Andesitic melt, 64*f*
Andes mountains, 160*f*
Andrews, Roy Chapman, 431
Angiosperms, 154, 364, 442–443
Angular unconformity, 17, 106, 106*f*, 107*f*
Anhydrite, 55
Animals:
    arboreal, 526
    history of, 154–156
    taxonomic classification of, 126
Animalia kingdom, 126, 127*f*
Animikiea, 256, 257*f*
Animikie group, 249
Ankylosaurs, 428, 429*f*
Anning, Mary, opp. p. 1
Annularia, 367*f*
Anomalocaris, 332*f*
Anorthite, 51
Anorthosites, 226
Antarctica, 472–473
Anthropoidea, 527–532
Antiarchs, 356, 357*f*
Anticline(s), 170, 172*f*–174*f*, 391*f*
Antler orogeny, 317
Anus, 351
Apatosaurus, 422, 426
Apes, 528, 530*f*. *See also* Anthropoidea
Apollo space mission, 194
Appalachian Mountains, 108*f*, 454–455
Appalachian orogenic belt, 312–313
Apparent polar wandering paths, 180–181
Aragonite, 54–55
Arboreal animals, 526
Arches:
    Cincinnati, 272*f*
    Kolob, 382
    Transcontinental, 271*f*, 276
Archacotherium, 513*f*
Archaea domain, 127, 127*f*
Archaebacteria kingdom, 126, 127*f*
Archaeocyathids (cup animals), 335, 335*f*
Archaeopteryx, 437
Archean Eon, 28, 208, 208*f*, 220–239
    atmosphere in, 222–224

Earth's crust in, 220–221
    fossils, 236, 239
    hydrologic cycle in, 224, 225*f*
    and origin of life, 230–239
    and origin of Precambrian rocks, 224–228
    sedimentation in, 228–229
Archelon, 435, 437
Archimedes, 141, 141*f*
Archosaurs, 415, 416*f*, 431–433
Arduino, Giovanni, 15
Argon-40, 39–40
Arkarua, 349
Arkoses (sandstone), 95, 95*f*–97*f*
Arsinotheres, 508
Arsinotherium, 509*f*
Arthrodires, 356
Arthropods, 345–348, 494, 496*f*
Articulate brachiopods, 340, 341*f*
Artiodactyls, 511–514, 514*f*, 515*f*, 520*f*
Asaphia kowalewskii, 133, 133*f*
Assemblage zone, 143
Assilina-Heterostegina concurrent range zone, 143, 143*f*
Asteroids, 443–444
Asteroid belt, 217–218
Asteroidea (starfish), 348
Asthenosphere, 7, 8*f*, 165*f*, 166–167, 181
Astraeospongea, 336*f*
Astrapis, 354, 355*f*
Athabasca glacier (British Columbia), 282*f*
Atmosphere:
    in Archean Eon, 222–224, 222*f*
    development of, 263–264, 263*f*
Atolls, 493, 494*f*
Atom(s), 34–35
Atomic mass, 34, 34*f*, 35
Atomic number, 34, 34*f*
Auditory ossicles, 144, 144*f*
Augite, 53
Australopithecine(s), 532–537
Australopithecus afarensis (Lucy), 533–535, 535*f*, 536*f*
Australopithecus africanus, 532, 533*f*
Australopithecus boisei, 535
Australopithecus robustus, 535, 536*f*
Autotrophs, 234–235
Avalonia, 323
Avalon terrane, 308
Awramic, Stanley, 254
Axial precession of Earth, 482, 483*f*
Axial tilt of Earth, 482, 483*f*
Azurite, 48*f*

Bacillus infernus, 234
Back-arc basin, 396*f*
Back-reef facies, 309
Bacteria domain, 127
Badlands National Park, 461
Bahama Banks, 98, 98*f*
Bakker, Robert, 430

Baltica, 269
Banded iron formations (BIFs), 223, 223*f*, 249, 249*f*, 250
Bandicoots, 501, 502*f*, 503*f*
Barred basin, 278, 281*f*
Barrier islands, 85, 85*f*
Barrier reefs, 493
Baryonyx, 424*f*
Basalt, 58, 61, 61*f*
   Deccan Trap, 401, 401*f*
   in Hawaiian volcanoes, 200
   in oceanic crust, 167
   pillow, 167*f*
Basaltic lava, 61, 179, 466*f*
Basaltic magma, 64
Basilosaurus, 519, 520*f*
Basin(s), 171
   back-arc, 396*f*
   barred, 278, 281*f*
   in Cenozoic Era, 458*f*
   in Early Paleozoic Era, 269, 272*f*, 273–274
   foreland, 391–396
   Green River, 456, 458*f*
   Gulf of Mexico, 380*f*
   Karoo, 400
   opening of ocean, 246
   Oquirrh, 318*f*
   Paradox, 303, 307*f*
   Paris, 471, 472*f*
   Permian, 304–306
   Uinta, 458, 460*f*
Basin and Range province, 460, 463–464, 463*f*
Bats, 500, 506
Batholith(s):
   Coast Range, 384, 389*f*
   Idaho, 384, 389*f*
   of Mesozoic Era, 389*f*
   Sierra Nevada, 384, 389*f*
   of Sierra Nevada, 468
Bathyal environment, 148
Bechleja rostrata, 496*f*
Becquerel, Henri, 33, 35
Belemnites, 152, 153*f*, 412–413
Benthic foraminifera, 490, 492*f*
Benthic marine ecosystem, 147–149
Bentonite, 36, 284, 393, 461
Bering Strait land bridge, 150, 151*f*, 543
Beta particles, 35
Biblical calculation of earth's age, 32–33
BIFs, *See* Banded iron formations
Big Bang hypothesis, 213
Binomial nomenclature, 125
Bioclasts, 68
Biodiversity:
   in Cenozoic Era, 489
   in Paleozoic Era, 333–335
Biofacies, 100
Biology, and evidence for evolution, 137–139
Biosphere, development of, 263*f*
Biostratigraphic correlation, 104
Biostratigraphic zones, 142–143
Biotite mica, 61, 65
Bioturbation, 147–148, 334
Biozone(s), 142–143
Bipedal dinosaurs, 417–421, 417*f*, 418*f*, 423–427

Bipedal hominids, 533–534, 536*f*
Birds, 437, 498–499
Bivalves:
   in Cenozoic Era, 493, 495*f*
   in Mesozoic Era, 409
Bivalvia, 341, 344*f*
Blastoidea (blastoids), 348, 350–351, 351*f*
Blastopore, 351
Blastula, 351
Blueschists, 189
Body waves, 162–163
Bolide(s), 443–446
Bones, 360
   of crossopterygian, 359*f*
   in cynodonts, 363
   dinosaur pelvic, 417, 418*f*
   of forelimb, 138*f*
   hyomandibular, 144, 144*f*, 360–361
   incus (ear), 144, 144*f*
   isotope analysis of, 430
   malleus (ear), 144, 144*f*
   of mammals, 500
   stapes (ear), 144, 144*f*
Boney fish, 357–359, 415
Bovids, 516
Bowen, Norman L., 61
Bowen's Reaction Series, 61–62, 62*f*
Box turtle, 498*f*
Brachiopods, 133*f*, 339–341, 341*f*–343*f*, 496, 496*f*
Brachiosaurus, 422, 422*f*
Bragdon formation, 319*f*
Braincase of mammals, 500
Branchiostoma, 355*f*, 359
Breccia, 66, 67*f*
Bright angel shale, 113, 114*f*
Brongniart, Alexander, 18–19
Brontops, 513*f*
Brontosaurus, See Apatosaurus
Brontotheres, 510–511, 513*f*
Bryophytes, 364
Bryozoans, 338, 339, 340*f*, 496
Burgess shale, 330*f*, 331–333

Cacops, 362, 362*f*
Calamites, 367*f*
Calcareous oozes, 148–149
Calcite, 49*f*, 54–55, 55*f*, 349
Caldera, 468*f*
Caledonian orogenic belt, 288, 290
Caledonian orogeny, 269, 297
California (Cenozoic Era), 467
Calliops, 349
Cambrian explosion, 330–333
Cambrian Period:
   invertebrates in, 154–155
   North America in, 271*f*
Cambrian/Precambrian boundary, 273
Cambrian System, 30
Camels, intercontinental migration of, 151*f*
Camptosaurus, 425
Canada, 56, 73
Canadian Shield, 225, 238, 474*f*
Cannonball sea, 457*f*
Cantius, 529
Canyonlands National Park (Utah), 14*f*
Canyons of the Escalante, 394

Carbon-12, 41
Carbon-14, 38, 41–43, 41*f*
Carbonaceous chondrites, 211, 212*f*
Carbonate(s), 54
   clasts, 67–68
   sedimentary rock, 65, 67–68
   spar, 67, 68
Carbonate compensation depth (CCD), 149
Carbon dioxide, 41–42
Carboniferous Period:
   and equator/continent relationship, 321*f*
   major land/sea regions in, 150*f*
Carboniferous plants, 365–366
Carboniferous System, 32
Carbonization, 121, 121*f*
Carcharodontosaurus, 420
Carnassials (cheek teeth), 506, 508, 509*f*, 528
Carnegie, Andrew, 426
Carnivora, 507
Carnivores, 500. *See also* Flesh-eaters
Carnivorous reptiles, 363*f*
Carolina trough, 381*f*
Cascade Range volcanoes, 465, 467, 467*f*
Cassiar orogeny, 318
Casts, 122
Castorimorphs, 506
Catastrophism, 18–19
Catskill clastic wedge, 309–310, 312*f*, 313*f*
Caudipteryx, 430
CCD (carbonate compensation depth), 149
Ceboidea, 528
Cells, 130–131, 131*f*
Cell division, 130–131, 131*f*, 351, 353
Cell membrane, 231
Cement (bonding material in sedimentary rocks), 87
Cenozoic Era, 28, 32, 150, 452–484, 488–521
   Africa in, 471–472
   Antarctica in, 472–473
   biodiversity in, 489
   climate in, 483–484
   and eastern North America, 454–455
   evidence of continental drift in, 178
   grassland expansion in, 489
   horses in, 136
   major events of, 488
   mammals in, 499–521
   marine invertebrates in, 492–496
   northern Europe, 471
   plankton in, 490–492
   Pleistocene Ice Age of, 473–483
   and tectonic/climate connection, 452–454
   Tethys Sea in, 470–471
   vertebrates in, 496–499
   and western North America, 455–470
Central American land bridge, 502–503, 504*f*
Cephalochordata, 355*f*
Cephalodiscus graphtoloides, 352
Cephalopods, 155*f*, 342–345, 346*f*
   coiling in, 133
   evolutionary change in, 137, 137*f*
   in Mesozoic Era, 410–413
   suture patterns of, 345*f*, 346*f*, 410–412, 412*f*
Ceratites, 345, 411
Ceratopsians, 428–429, 429*f*

Cercopithecoidea, 528
Ceresiosaurus, 435*f*
Cetaceans, 501, 519, 520*f*
Chalicotheres, 510–511, 514*f*
Chalk, 32, 68, 378, 385*f*
Chalk Marl formation, 399
Channeled scablands, 480, 481*f*
The Channel Tunnel, 399
Charniodiscus, 259
Chattanooga shale, 298–301
Checkerboard Mesa, 383, 391*f*
Cheek teeth, *See* Carnassials
Cheirolepis, 357, 358*f*
Chemosynthesis (hyperthermophiles),
    232–234
Chengjiang (China), 333
Chert, 51, 52*f*, 68, 69*f*
Chicxulub structure, 445–447, 445*f*
Chief Mountain (Montana), 253*f*
China, 269, 333
Chinle formations, 113*f*, 115, 382, 383,
    387*f*, 394
Chlorite, 70
Chlorophytes (green algae), 154, 364
Chloroplasts, 440
Chondrichthyes, 356–357
Chondrites, 211, 212*f*
Chondrules, 211
Chordates, 333, 351–352
Chromosomes, 130–131, 131*f*
Chronostratigraphic correlation, 104–105
Chronostratigraphic units, 28. *See also*
    Time-rock units
Chuar group (Grand Canyon), 255
Cibolites waageni, 137*f*
Cidaris, 411*f*
Cimolestes, 506–507
Cincinnati arch, 272*f*
Cinder cone, 466*f*
Cladogram, 134–135, 136*f*, 520
Cladoselache, 357, 357*f*
Claron formation, 458
Class (taxonomy), 126
Classification system for life forms, 125–128
Clasts, 87
Clastics, molasse, 471
Clastic sedimentary rock, 66–67
Clastic wedge, 286*f*
    Catskill, 309–310, 312*f*, 313*f*
    Queenston, 278, 285, 287*f*
Clay, 54
Clay minerals, 53–54
Cleavage:
    in calcite, 55*f*
    embryologic, 353
    in halite, 55*f*
    in minerals, 49
Climate(s):
    astronomic cycles affecting, 483*f*
    in Cenozoic Era, 483–484
    in Early Paleozoic Era, 290–291
    fossil relationship to past, 152–156
    in Late Paleozoic Era, 320–321
    in Mesozoic Era, 407–408
    in Proterozoic Eon, 263–264, 263*f*
    and tectonics in Cenozoic Era, 452–454
Climatius, 356*f*

Cloning, dinosaur, 424
Cloud, Preston, 262
Cloudina, 262*f*, 330
Clovis culture, 543–544
Clypeaster (sea biscuit), 496*f*
Cnidaria, 81*f*, 337
Cnidocyte cells, 337
Coal, 69, 312, 321
Coal-bearing cyclothems, 305*f*
Coast Range batholith, 384, 389*f*
Coccolith, 441*f*
Coccolithophorid, 148*f*, 440, 490*f*
Coccolithus, 490*f*
Coccolithus pelagicus, 148*f*
Cochlea, 144
Cocos plate, 468, 469*f*
Coelacanths, 359, 360*f*
Coelophysis, 419, 419*f*
Coelum, 261
Coelurosauravus, 431
Coevolution, 443
Coiling, cephalopods, 133
Collins, Desmond, 332*f*
Color:
    and depositional environment of
        sedimentary rocks, 86
    in minerals, 48
Colorado National Monument, 387*f*
Colorado Plateau, 464, 465*f*
Colorado River, 23, 26*f*, 56
Columbia River Plateau, 58, 63, 63*f*, 464,
    466*f*
Composition (igneous rock), 59–60, 60*f*
Conception group (Newfoundland), 258*f*
Concurrent range zone, 143, 143*f*
Conglomerates, 66, 67*f*
Conifers, 366
Conodont fossils, 142, 142*f*, 359–360
Contact metamorphism, 69–70
Continent(s):
    Cenozoic Era separation of, 489
    geographic fit of, 175, 176*f*
Continental collisions, plate tectonic model
    for, 314*f*
Continental-continental crust convergence,
    187, 188*f*
Continental crust, 165*f*, 167, 199, 201–202,
    202*f*, 221*f*, 225–226
Continental drift, 174–178, 269
    early hypotheses, 174–175
    evidence for, 175–178
Continental environments, and sediment
    deposition, 82–83
Continental margin, passive, 183
Continental-oceanic crust convergence:
    in plate margins, 188–189, 188*f*, 190*f*
    and rocks, 188–189
Continental passive margin, 183
Continental shelves, 80–81, 81*f*
Continental slope, 81–82
Continent/equator relationship, 321*f*
Continuous series (term), 62
Convection cells, 189
Convergence, evolutionary, 415, 417
Convergent evolution, 502
Convergent plate boundaries, 7, 8*f*, 185,
    187–189, 187*f*

Cooksonia, 365, 365*f*
Cooling rate, as geologic dating method, 33
Cope, Edwin D., 23
Copper, 249–250
Coral, 337–338
    in Cenozoic Era, 492–493
    in Mesozoic Era, 409–410
    Montastrea cavernosa, 493*f*
    rugose, 339*f*
Coralliochama orcutti, 409*f*
Coral reef(s), 338
    and atolls, 493, 494*f*
    in Wallowa Mountains, 202
Cordaites, 366, 368*f*
Cordillera, North American, 390*f*
Cordilleran belt, 278–281, 316–318
Core (of earth), 164–166
Correlation methods (stratigraphy),
    104–105, 141–142
Corythosaurus, 425*f*
Cosmic rays, 41
Cosmopolitan species, 141
Cranioceras, 515*f*
Crater(s), 445–446
Crater Lake, 467, 467*f*, 468*f*
Craton(s), 79, 225
    in Early Paleozoic Era, 269, 273, 273*f*
    in Late Paleozoic Era, 298–302
    Pennsylvanian Period, 306*f*
Cratonic sequence(s):
    Absaroka, 274*f*, 275, 301
    in Early Paleozoic Era, 274–278
    Kaskaskia, 274*f*, 275, 298–301
    Sauk, 274*f*, 275–277
Creodonta, 507
Cretaceous Period, 123, 377–378
    climate cooling in, 408
    epicontinental sea of, 389
    limestone(s), 389, 393*f*
    paleogeographic map, 384*f*, 408*f*
    sandstone, 372*f*
    tectonic features of, 393*f*
Cretaceous System, 32
Crichton, Michael, 424
Crinoids, 300*f*, 351, 352*f*, 410*f*
Crinoidea (crinoids), 348, 352*f*
Crocodile skull (Jurassic), 434*f*
Crocodilians, 497
Cro-Magnons, 542, 544*f*
Cross-bedding, 91, 92*f*, 94
    in Checkerboard Mesa, 383, 391*f*
    of Devonian sandstone, 313*f*
    in Navajo sandstone, 386, 391*f*
    of sandstone, 303*f*
Cross-cutting relationships, principle of,
    19–20, 19*f*
Crossing over (in chromosomes), 131
Crossopterygians, 358–359
Cross-sections, 107
Crust (of Earth), 7, 8*f*, 167–168, 221*f*
    chemical elements in, 50*t*
    continental, 165*f*, 167, 199, 201–202,
        202*f*, 221*f*
    felsic, 225–226
    mafic, 225
    oceanic, 58, 165*f*, 167, 168, 202
    postglacial rebound of, 478, 479*f*

Crustaceans (Mesozoic Era), 413
Crystal form (of minerals), 49–50
Crystalline (term), 51
Crystallization (igneous rock), 61–62
Cup animals, *See* Archaeocyathids
Curie point, 179
Cuvier, Georges Léopold, 18–19
Cyanobacteria, 153, 249, 255*f*, 364, 364*f*
Cyclomedusa, 259
Cyclothems, 301–302, 305*f*
Cynodonts, 363
Cynognathus, 178, 363, 363*f*
Cystoidea (cystoids), 348, 351*f*
Cystoids, 350, 351*f*

Dakota group, 393
Daphaenodon, 514*f*
Dart, Raymond, 532
Darwin, Charles, 7–8, 18, 20–21, 21*f*, 128, 493
Daspletosaurus, 421*f*
Dating, geologic, *See* Geologic dating
Daughter element, 35
David (Michaelangelo), 73*f*
Day, length of, 247
Dead Horse Point (Utah), 26*f*
Deccan lava plateau (India), 58, 63
Deccan Traps, 401, 401*f*
DeChelly sandstone, 318, 319*f*
Deep marine environment, 82
Deep time, 7
Deformation, sevier, 390*f*
Deformed beds (Franciscan formation), 388*f*
Deinonychus, 419, 421*f*
Deiphon, 133, 133*f*
Deltas, sediment accumulation in, 83–84
Density, mineral, 49
Deoxyribonucleic acid (DNA), 8, 125, 130, 130*f*, 139, 231
Detrital grain (in sedimentary rock), 36
Deuterostomes, 351
Devonian Period, 296*f*, 297
    extinctions in late, 367–369
    forest, 366*f*
    paleogeographic map, 299*f*
Devonian System, 32, 113*f*, 115
Diatoms, 146–147, 441
    of Cenozoic Era, 490*f*
    modern marine, 148*f*
Dicellomus, 341
Dickinsonia costata, 260, 260*f*
Didelphis, 503*f*
Differentiated planet, 219, 220*f*
Dimetrodon, 362, 363*f*, 429–430
Dinoflagellate(s), 257, 440, 441*f*, 490*f*
Dinohyus, 514*f*
Dinosaur(s), 32, 156, 156*f*, 417–431
    cloning of, 424
    footprint, 375, 378*f*
    plant-eating, 422–429
    reproduction of, 430–431
    trackways, 125*f*
Dinosaur National Monument, 426–427, 426*f*
Dinosaur Quarry, 426*f*, 427
Dinotheres, 516, 517*f*

Dinotherium skull, 517*f*
Diorite, 60
Dip (strata), 170, 171*f*, 172*f*
Diploid cells, 130, 131*f*
Diplomystis, 458*f*
Dipnoans, 358
Dipnoi (term), 358
Diptera, 122, 123
Dipterus, 358*f*
Discharge (term), 82
Discoasters, 475
Discoaster challengeri, 475*f*
Discoidal fossil(s), 259*f*, 260*f*, 261
Disconformity, 106, 106*f*
Discontinuities, 164, 165*f*
Discontinuous series (term), 62
Divergent plate boundaries:
    earthquakes and, 181
    and plate tectonics, 7, 181, 183, 184*f*, 188*f*
DNA, *See* Deoxyribonucleic acid
Docodonts, 438
Dolomite, 47, 54, 67, 68
Domains (taxonomy), 126, 127–128, 127*f*
Domes, 171, 172*f*, 269, 272*f*, 273–274
Douglas, Earl, 426
Dropstones, 251
Drowned coastline, 322
Dryomorphs, 532
Dryopithecus fontani, 532
Duck-billed dinosaurs, *See* Hadrosaurids
Duck-billed platypus, 502*f*
Duluth gabbro, 250
Dunkard group, 313
Dunkleosteus, 356*f*

Ear, middle, 361
Earbones:
    evolution of, 144, 144*f*
    of mammals, 500
Early Paleozoic, 269–291
    basins/domes in, 269, 272*f*, 273–274
    and caledonian orogenic belt, 288, 290
    and Cambrian/Precambrian boundary, 273
    climate in, 290–291
    cordillera region in, 278–281
    cratonic sequences in, 274–278
    cratons in, 269, 273, 273*f*
    Earth's paleogeography in, 270*f*
    eastern North America in, 281–288
    events of, 273–274
    North America in, 271*f*
    ocean expansion/contraction in, 288, 290
    sediment deposition in, 275–290
    western North America in, 279–281
Earth:
    age of, 43
    albedo of, 482
    axial precession of, 482, 483*f*
    axial tilt of, 482, 483*f*
    core of, 164–166
    crust of, 167–168, 220–221
    differentiation of, 219–220, 220*f*
    formation of, 219–221
    geologic structures on, 168–171
    interior divisions of, 181*f*

    internal structure of, 162–171
    mantle of, 164, 166–167
    moon of, 215, 215*f*
    orbital eccentricity of, 482, 483*f*
    paleogeography configurations of, 270*f*
    precession of, 481–482
    and seismic waves, 162–163
    slowing rotation of, 247
    surface cooling of, 480–483
    temperature range of, 214
Earthquake(s):
    distribution of, 182*f*, 196*f*
    and divergent boundaries, 181
    in India, 163*f*
East African rift, 181
Eastern North America:
    in Cenozoic Era, 454–455
    in Early Paleozoic Era, 281–288
    in Late Paleozoic Era, 306, 308–316
    in Mesozoic Era, 375–378
Echinoderm(s), 348–353*f*, 411*f*
    in Cenozoic Era, 494, 496*f*
    and chordate relationship, 351–352
    Clypeaster, 496*f*
    crinoids, 300*f*, 351, 352*f*
    graptolites, 352
    living representatives, 349*f*
Echinoids (sea urchins), 350, 351, 353*f*, 410, 411*f*
Echinoidea, 348
Ecology (defined), 143
Ecosystem(s):
    defined, 143
    marine, 145–149
Ectothermic animals, 429–430
Edaphosaurus, 362
Edentates, 500, 504–505
Ediacaran fauna, 258–262
Edmontonia, 429*f*
Edrioasteroid (*Edrioaster bigsbyi*), 350*f*
Edrioasteroidea (crinoids), 348
Egg(s):
    amniotic, 329, 354*f*
    dinosaur, 430–431
    fertilized, 131, 131*f*
"Egg Mountain," 430
Elasmosaurus, 434
Electrons, 34
Elements, 34, 35
Elephants, 517
Elephant Rocks State Park (Missouri), 57*f*
Embryologic cleavage, 353
Embryos, vertebrate, 139*f*
Endemic species, 141
Endothermic, 430
Entelodonts, 511, 513*f*, 514*f*
Entrada formation, 395
Environments:
    abyssal, 148
    bathyal, 148
    classification of marine, 145*f*
    of deposition, 80, 80*f*
    fossil relationship to past, 143–149
    hadal, 148
    marine, 80–82, 145*f*
    and paleoecology, 143, 145
    and paleogeography, 149–152

Eoastrion, 256, 257*f*
Eocene Epoch:
    global paleogeography in, 453*f*
    mammals of late, 513*f*
    oil shale, 459, 459*f*
    rocks of, 459*f*
Eoentophysalis, 255*f*
Eomaia, 439
Eons, 28
Eoradiolites davidsoni, 381*f*
Eoraptor, 419
Eosphaera, 256, 257*f*
Eozostrodon, 437
Epeiric (defined), 103
Epicontinental inland seas, 245
Epicontinental seas:
    of Cretaceous Period, 389
    and seafloor spreading, 388
Epifaunal animals, 334
Epifaunal organisms, 147
Epigaulis, 515*f*
Epochs, 28
Equator:
    geographic, 179*f*
    geomagnetic, 179*f*
Equator/continent relationship, 321*f*
Equus, 136*f*, 510
Eras, 28, 30*f*
Erosion, 7
    of Appalachians in Cenozoic Era,
      454–455
    of Rocky Mountains in Cenozoic Era,
      455–460
Erosional unconformities, 106*f*
Erratics, glacial, 22
Erwin, Douglas, 369
Estuaries, 86
Eubacteria kingdom, 126, 127*f*
Eukarya domain, 127, 127*f*
Eukarya superkingdom, 126
Eukaryotes, 235–236, 236*f*, 237*f*, 256–258
Euoplocephalus, 429*f*
Eureka Valley (California), 84*f*
Europe:
    in Cenozoic Era, 471
    in Late Paleozoic Era, 319
    orogenic belts of, 273*f*
    paleogeographic map of, 398*f*
Europolemus, 526*f*
Eurypterids, 291*f*, 347–348, 348*f*
Eurypterus, 291*f*
Eustachian tube, 361
Eusthenopteron, 358, 359*f*
Evaporites, 69, 177, 303–304
Evolution, 7–8, 20–21, 124–139, 355*f*
    and adaptive radiation, 132–134, 132*f*
    of auditory ossicles, 144, 144*f*
    and classification of life forms, 124–128
    evidence for, 136–139
    of fish, 355*f*, 361*f*
    of fossils, 33
    and heredity, 130
    of horse foreleg, 511*f*
    of jaw, 355–356
    models for, 134
    and modification in mammals, 501*f*
    of North American cordillera, 390*f*

    and phylogeny, 134–136
    and reproduction, 130–132
    and speciation, 132
    theories of, 128–129
Evolutionary convergence, 415, 417
Exogyra arietina, 409*f*
Exotic terranes, 199, 201–202, 202*f*,
    378–379
Exploration geologists, 3
External molds (fossils), 121–122
Extinctions, mass, *See* Mass extinctions
Extraterrestrial life, 156–157
Extremophiles, 126
Extrusive igneous rock, 57, 58
Eyes, trilobite, 349

Facies, 100–102
    Guadalupe Mountain, 310*f*
    reef, 309
Failed arm (thermal plumes), 191, 192*f*
Family (taxonomy), 126
Farallon plate, 468, 469*f*, 470*f*
Fault(s), 169–170, 169*f*
    formation of San Andreas, 469
    San Andreas, 169, 170*f*
Fault plane, 169
Favosites coral, 337, 338*f*
Feathered theropoda, 430
Feel (of minerals), 50
Feldspar(s), 50–52, 65
Felis, 125
Felis domesticus, 125
Felis leo, 125
Felis onca, 125
Felis thomas, 125
Felsic continental crust, 225–226
Fermentation, as nutrition source, 234
Ferric iron oxide, 86
Ferromagnesian mineral, 53
Ferrous iron oxide, 86
Fertilization, 130, 131*f*
Fertilized egg, 131, 131*f*
Filter feeders, 334
Finger Lakes (New York), 22
Fish, 354–359
    acanthodians, 356
    agnathids, 354–356
    boney, 357–359, 415
    in Cenozoic Era, 496–497
    chondrichthyes, 356–357
    evolution of, 355*f*, 361*f*
    osteichthyes, 357–359
    placoderms, 356
Fission tracks, 42–43
Flagella, 336
Flesh-eaters, 362, 506–508, 506*f*. *See also*
    Carnivores
Flexibility (in minerals), 50
Flexicalymene, 348*f*
Flint, 51
Florissant Fossil Beds National Monument,
    459
Flysch sediments, 470
Folds (rock strata), 170–171, 171*f*, 172*f*
Foliated metamorphic rocks, 71, 72
Foliation, 71*f*
Fool's gold, *See* Pyrite

Footwall, 169, 169*f*, 170, 170*f*
Forams, 413
Foraminifera, 4*f*, 334–335, 482*f*, 490, 491*f*,
    492*f*
    fossils, 134, 143*f*
    geologic ranges of, 143*f*
    living planktonic, 146*f*
    in Mesozoic Era, 413
    planktonic, 152, 152*f*, 414*f*
Foreland basin, 391–396
Foreleg, 511*f*
Fore-reef facies, 309
Forest (in Devonian Period), 366*f*
Formations, 99, 99*f*, 100*f*, 110*f*, 111, 114*f*
    Bragdon, 319*f*
    Chalk Marl, 399
    Chinle, 382, 383, 387*f*, 394
    Claron, 458
    Entrada, 395
    Fort Union, 456
    Franciscan, 388*f*
    Gowganda, 248
    Green River, 456, 458*f*, 497
    Kayenta, 382, 383, 395
    Moenave, 382
    Moenkopi, 113*f*, 115, 382, 383, 394
    Morrison, 387, 389, 392*f*
    Navajo, 383
    Niobrara, 393, 396, 397*f*
    Phosphoria, 324
    Purslane, 314*f*
    Rockwell, 314*f*
    Sundance, 387
    Temple Gap, 383
    White River, 458, 459
Formation sequence (of rock units), 105*f*
Fort Union formation, 456
Fossil(s), 12–13, 120–124, 139–145,
    149–156
    Archean Eon, 236, 239
    casts, 122
    of Chengjiang China, 333
    in chronostratigraphic correlation, 104
    conodont, 142, 142*f*
    and continental drift evidence, 177–178
    earliest evidence of, 153
    earliest land plant, 364, 365
    fusulinid, 334*f*
    geologic dating based on evolution of, 33
    geologic ranges of, 140–141, 140*f*
    graptolites, 280, 284*f*
    guide, 141*f*, 142
    horses, 135*f*–137
    index, 142
    molecular, 236, 239, 256
    of Neoproterozoic era, 258–262
    of Oligocene Epoch, 458, 459, 462*f*, 496
    and paleoecology, 143, 145
    paleogeography and, 149–152
    past climates and, 152–156
    preservation of, 120–123
    prokaryote, 257*f*
    record, 124
    reworked, 142
    sassafras leaf, 442*f*
    shell-bearing, 262, 262*f*, 327–330, 329*f*
    sites for hominid, 533*f*

Fossil(s), (*Continued*)
  small-shelly, 329–330
  spider of Oligocene Epoch, 462*f*
  and stratigraphy, 139–143
  stromatolite, 254, 254*f*
  trace, 123–124
  wasp, 120*f*
Fossil succession, principle of, 18
Fountain Formation (Colorado), 306*f*
Fractional crystallization (term), 64
Franciscan formation, 388*f*
Fringing reefs, 493, 494*f*
Frondlike fossils, 259
Frustule, 441
Fuller, Buckminster, 369
Fullerenes, 369
Fungia (coral), 492*f*
Fungi kingdom, 126, 127*f*
Fusion, 210
Fusulinid fossils, 334*f*
Future, predicting the, 2

Gabbro, 58*f*, 60–61
  Duluth, 250
  in oceanic crust, 167
Galileo (spacecraft), 218
Gametes, 130, 131*f*
Gamma radiation, 35
Garnet, 49*f*, 70
Gases, in Earth's primitive atmosphere,
    222–223, 222*f*
Gasteronomus, 497*f*
Gastropods, 341, 342, 344*f*
  in Cenozoic Era, 493, 495*f*
  in Mesozoic Era, 413
Genes, 130–131
Gene pool, 132
Genetic mutation, 21
Genetic variation, 130–132
Genus, 125, 126
Geochronologic units, 28. *See also* Time units
Geochronology, 27. *See also* Geologic dating
Geographic equator, 179*f*
Geographic poles, 179, 179*f*
Geography, global, 175–176*f*
Geologic columns, 107
  Grand Canyon National Park, 114*f*
  Niagara Falls, 279*f*
  Zion National Park, 386*f*
Geologic dating, 27–28
  actual, 28
  early attempts at, 32–33
  by radioactivity, 35–43
  relative, 27–28
Geologic maps, 107, 109–110, 112–115
Geologic ranges:
  of foraminifera, 143*f*
  of fossils, 140–141, 140*f*
  of invertebrate animals, 155*f*
  of plants, 154*f*
  of species, 140–141
Geologic sequence of events, 20*f*
Geologic structures of earth, 168–171
  basins, 171
  domes, 171
  faults, 169–170
  folds, 170–171

Geologic time scale, 27–33
  chart, 29*f*
  development of, 30–32
  divisions in, 28–30
  Proterozoic/Cambrian, 329*f*
Geology, 2–3
  development of, as science, 12–23
  historical, 2, 13
  physical, 2
  in United States, 22–23
Geomagnetic equator, 179*f*
Geomagnetic poles, 179, 179*f*
Geosaurus, 434*f*
Giant's Causeway (Northern Ireland), 471,
    472*f*
Gigantosaurus, 420
Ginkgo biloba, 366, 442*f*
Ginkgoes, 366
Giraffes, 129*f*
Glacial conditions:
  and Great Lakes, 250
  in Late Paleozoic, 320
  major episodes of, 252–253, 252*f*
  in Neoproterozoic Era, 251–252
Glacial deposits (of sediment), 83, 84*f*,
    476–477
Glacial erratics, 22, 238
Glacial moraines, 22
Glacial striations, 22
Glacial till, 476*f*
Glaciers:
  mountain, 473, 474*f*
  Pleistocene, 322, 474*f*
Global geography, 175–176*f*
Globorotalia menardii, 477, 477*f*
Globorotalia truncatulinoides, 152, 152*f*,
    477–478, 477*f*, 478*f*
Glomar Challenger, 5, 6, 6*f*, 196*f*
Glossopleura, 107, 108*f*
Glossopteris, 177–178, 177*f*, 365–366
Glyptodonts, 504
Gneiss, 71, 71*f*
  Amitsoq, 229*f*
  Monson, 72*f*
  Precambrian, 238*f*
Gomphotherium, 517
Gondwanaland, 175, 176, 176*f*, 269
  correlations for, 179*f*
  ice caps of Ordovician Period, 367
  in Late Paleozoic Era, 320
  in Mesozoic Era, 399–401
  reconstruction of, 177*f*
Goniatites, 345, 411
Gowganda formation, 248
Graded bedding, 91, 93, 93*f*, 94
Grain:
  orientation in sedimentary rocks, 90
  orientation study, 91*f*
  shape in sedimentary rocks, 89
  size of, in sedimentary rocks, 87–88
  size of igneous rocks, 57–58
Grand Canyon, 23, 56, 464, 466*f*
  Cambrian strata in, 276*f*
  Precambrian rocks, 254–255
Grand Canyon National Park (Arizona),
    100*f*, 113–115, 114*f*
Grand Canyon Supergroup, 113

Grand Staircase-Escalante National
    Monument, 394–395
Granite, 21*f*, 59, 59*f*, 60, 65
  weathering of, 66*f*, 67*f*
  Zoroaster, 113, 254*f*
Granodiorite, 60, 389*f*
Granulite rocks, 226, 238
Graptolites, 280, 284*f*, 352, 354*f*
Grassland expansion (Miocene Epoch), 489
Gravity:
  and plate tectonic theory, 194, 196
  variation over sea trench, 197*f*
Gravity anomaly, 194
Graywackes (sandstone), 95, 95*f*–97*f*
Great Britain, Paleozoic strata in, 31*f*
Great Lakes, 250, 480*f*
Greenhouse effect, 321
Green River, 456, 458*f*, 497
Green River shale, 458
Greenstone association, 72, 228, 229*f*, 230*f*,
    238
Grendelius, 436*f*
Grenville orogen, 251, 251*f*
Groups (in rock formations), 100
  Animikie, 249
  Chuar, 255
  Conception, 258*f*
  Dakota, 393
  Dunkard, 313
  Monongahela, 313
  Pocono, 310, 312
  Supai, 113*f*, 115
  Tonto, 100
  Torrowangee, 153
Guide fossils, 141*f*, 142
Gulf of Aden rift, 181
Gulf of Mexico basin, 380*f*
Gulf of Suez, 83
Gulf Stream, present currents of, 454*f*
Gunflint Chert, 249, 257*f*
Gutenberg discontinuity, 164, 165*f*
Guyots:
  origin of, 184*f*
  and plate tectonics, 183–185
Gymnosperms, 154, 154*f*, 364, 442–443
Gypsum, 55, 55*f*

Hadal environment, 148
Hadrosaur, 425*f*
Hadrosaurids, 425, 430–431
Haikouichthys, 354
Half-life, 7, 38
Halimeda, 98, 98*f*
Halite, 55, 55*f*
Hall, James, 22, 78
Halley, Edmund, 33
Hallucigenia, 332*f*
Hand, 526
  of *Europolemus*, 526*f*
  of human, 527*f*
Hanging wall, 169, 169*f*, 170, 170*f*
Haploid cells, 130–131, 131*f*
Hardness (minerals), 49
Hasmark Formation (Montana), 108*f*
Hawaiian Islands, 56, 57*f*, 58, 63*f*, 198*f*
  formation of, 197, 199
  and hotspots, 197–199

and plate tectonics, 197–200
volcanoes in, 62–63
Hawaii Volcanoes National Park, 200, 200f
Hayden, Ferdinand V., 22–23, 22f
Heart, three-chambered, 360
Heavy minerals, 298
Helicoplacus, 350f
Heliophyllum halli, 247f
Heliotropic stromatolites, 254
Hemiaster, 411f
Hemicyclaspis, 355, 356f
Hercynian orogeny, 297, 319
Heredity, evolution and, 130
Hermatite, 86
Hermit shale, 113f, 115
Herodotus, 12
Herrerasaurus, 419, 419f
Hesperonus, 398f, 437
Hesperornis, 398f
Hesperosuchus, 415, 416f
Hess, Harry H., 183, 191
Heterotrophs, 235
Himalayan Mountains, 13f, 190f
Historical geology, 2, 13
Hogbacks, 393, 397f
Holocene Epoch, 123, 476
Holocene Series, 32
Hominid(s), 532–545
australopithecine stage of, 532–537
Australopithecus afarensis, 533–535, 535f, 536f
and Bering Strait land bridge, *543*
bipedal, 533–534, 536f
Cro-Magnons, 542
fossil sites of, 533f
Homo erectus stage of, 537–538
Homo sapiens, 1, 125, 525–526, 537, 538–545
Neandertals, 538–540, 540f
oldest known, 532
population growth of, 544–545
stone tools of, 539f
timelines for, 534f
Homo, 535, 537
Homo erectus, 537–538, 537f
Homo ergaster, 537
Homo floresiensis, 541
Homologous (term), 137–138
Homo rudolfensis, 537, 537f
Homo sapiens, 1, 125, 525–526, 537, 538–545
Hooker, Joseph, 21
Horizontal strata, 15
Hornblende, 53, 53f, 61, 62
Horner, John R., 430
Hornfels, 72
Horses:
in Cenozoic Era, 136
fossil, 135f–137
phylogenetic tree of, 135f, 509
Hotspots, 198f
and Hawaiian Islands, 197–199
Mammoth Geyser, 199f
and plate tectonics, 196–202
and Yellowstone National Park, 199
Hsu, Kenneth J., 5
Hudsonian orogen, 249
Humboldt, Alexander von, 32

Hutton, James, 10, 16–18, 16f
Huxley, Thomas, 430
Hyaenodon, 507, 513f
Hydrologic cycle, 224, 225f
Hydrothermal vents, 223, 224f
Hylobatidae, 528
Hylonomus, 362
Hyomandibular bone, 144, 144f, 360–361
Hyperthermophiles, 232–234
Hypertragulus, 515, 516f
Hypotheses, 4
of atoll formation, 493
Big Bang, 213
for origin of Pleistocene ice age, 480–483
oscillating-Universe, 213
steady-state, 213
Hyrachus, 510f
Hyracodon, 513f

Iapetus Ocean, 269
Icarosaurus, 431
Ice, variations in global volume of, 477, 477f
Ice age(s), 22
evidence for first, 248
hypotheses for origin of Pleistocene, 480–483
little, 476
Ordovician, 290
Pleistocene, 42, 473–483
Ice caps, 367
Iceland, 192f
Ichthyosaurs, 434, 436f
Ichthyostega skeleton, 361f
Ichthyostegids, 361
Idaho batholith, 384, 389f
Igneous rock(s), 15, 20f, 56–65. *See also specific types, e.g.:* Granite
composition of, 59–60, 60f
crystallization of, 61–62
extrusive, 57
formation of, 55
geologic dating based on, 36–38, 37f
grain size of, 57–58
intrusive, 56
and paleomagnetic determinations, 180
texture of, 56–59, 58f, 60f
and volcanic eruption, 62–64
Iguanodon, 424f, 425
Ilium, dinosaur, 417, 418f
Inarticulate brachiopods, 340, 341f
Inclusions, 20, 21f
Incus (ear), 144, 144f
Index fossils, 142
Index minerals, metamorphic, 70
India:
earthquake in, 163f
in Mesozoic Era, 401
northward migration of, 189f
Indiana Dunes National Lakeshore, 479, 481f
Infaunal animals, 334
Infaunal organisms, 147, 147f
Insect:
fossils of Oligocene Epoch, 462f, 496
preservation in amber, 122, 123
Insectivores, 500, 504, 505f

Internal molds (fossils), 121, 122
Intraplate volcanoes, 200
Intrusive igneous rock, 56, 57
Invertebrate(s), 155f
in Cambrian Period, 154–155
in Cenozoic Era, 492–496
geologic ranges of, 155f
in Mesozoic Era, 413–414, 439–441
of Mesozoic Era, 409–414
in Paleozoic Era, 328, 335–352
Iridium, 443–445, 445f
Iron, 249. *See also* Banded iron formations
Iron meteorites, 212, 212f
Iron ore, 290f
Iron oxides, 86
Irregular echinoids, 410
Ischium, dinosaur, 417, 418f
Isochron (straight line), 40–41, 40f
Isopach (thickness) maps, 111, 111f, 280f, 288f
Isostasy, 167–168, 168f
Isostatic adjustment, 168f
Isotope(s):
analysis of bones, 430
defined, 35
and partial melting of Earth, 220
for radioisotope dating, 39f
rate of decay, 35
Isotope ratio of oxygen, 152
Isotopic dating, 35–43
and igneous rocks, 36–38
methods of, 38–43

Jasper, 51
Jasper National Park (British Columbia), 282–283
Jaw:
of crossopterygian, 359f
evolution of, 355–356
of mammals, 500
Johanson, Donald, 533
JOIDES Resolution, 196f
Joly, John, 33
Juan de Fuca plate, 465, 467f, 468, 469f
Jupiter, 218
Jurassic Period, 375–376, 383–387
paleogeographic map, 379f, 392f, 408f
sandstone, 372f
Jurassic System, 32

Kaibab limestone, 113f, 115, 382
Kaiparowits plateau, 394
Kakabekia, 256, 257f
Kangaroos, 501, 503f
Kant, Immanuel, 209
Kaolinite, 54f
Karaurus, 415
Karoo basin, 400
Kaskaskia sequence, 274f, 275, 298–301
Kayenta formation, 382, 383, 395
Kazakhstania, 269
Kelvin, William Thomson, 33
Kenoran orogeny, 238
Kettles (small lakes), 479
Keweenawan rocks, 249–250
Kilauea volcano, 200, 201f
Kimberella, 260–261, 261f

Kimmswick limestone (Missouri), 99
Kingdoms (taxonomy), 126, 127f
Koalas, 501, 503f
Kolob Arch, 382
Komatiites, 221
Koobi Fora dig, 534, 536f
Krakatoa Island (Indonesia), 63

Labradorite, 53f
Labrador Trough, 248f, 249
Labyrinthodonts, 361–362
Lacustrine deposits (of sediment), 83
Lagoons, 85, 85f
Lakes:
    Pleistocene ice age formation of, 479–480
    pluvial, 479
Lake deposits (of sediment), *See* Lacustrine
        deposits
Lamarck, Jean Baptiste de, 128
Lamarck's theory of evolution, 128, 129f
Lambeosaurus, 425f
Land bridges, 149–150
    Bering Strait, 150, 151f, 543
    Central American, 502–503, 504f
Landmasses (Neoproterozoic Era), 269f
Land plants, 441–443
Lapworthella, 330, 330f
Laramide orogeny, 385–386
Lascaux cave (France), 42
Late Paleozoic, 296–324
    climates of, 320–321
    cratons in, 298–302
    and eastern North America, 306, 308–316
    Europe in, 319
    Gondwana in, 320
    mineral products of, 321, 324
    paleogeography of, 296f
    periods in, 296–297
    sedimentation/orogeny in Cordilleran
        belt, 316–318
    and southwestern North America,
        303–306, 307f
Lateral continuity, principle of, *See* Original
        lateral continuity, principle of
Lateral faults, 169, 169f
Lateral moraines, 84f
Latimeria, 360f
Laurasia, 175, 176f, 297
Laurentia, 245, 251, 251f, 269
Lava, 16, 56, 56f, 59
    aa, 200
    basaltic, 61, 466f
    Keweenawan, 249–250
    pahoehoe, 200
    pillow, 238
    tubes, 200
Laws, natural, 16–17
Lead-206, 39f
Leaf, fossil, 442f
Leakey, Louis, 532
Leakey, Mary, 532
Lemur, 528f
Leperditia fabulites, 348f
Lepidodendron, 365, 366f
Lepidosaurs, 415
Life forms:
    in Cambrian Period, 154–155

earliest traces of, 153–154
history of animal, 154–156
history of plant, 154
mass extinctions of, 156
in Mesozoic Era, 156
origin of, 230–239
in Paleozoic Era, 155–156
possibility of extraterrestrial, 156–157
in Proterozoic Eon, 255–262
Limacina, 148f
Limestone(s), 20f, 21f, 65, 67, 68
    Cambrian, 282f
    Cretaceous Period, 389, 393f
    Kaibab, 382
    kaibab, 113f, 115
    kimmswick, 99
    late Jurassic Solnhofen, 118f
    muav, 113, 114f
    oölitic, 99f, 301f
    redwall, 113f, 115, 115f
    and sediment deposition, 98
    temple butte, 113f, 115
    textures of, 68f
Lindow Man, 123, 124f
Lingula, 496f
Linnaean classification system, 125
Linnaeus, Carolus, 125
Lithic (sandstone), 95f–97f
Lithification (sedimentary rocks), 65
Lithofacies, 100, 101f
Lithofacies map(s), 111–112, 112f
    lower Tippecanoe, 277f
    Upper Cambrian, 275f
Lithographic limestone, 68
Lithosphere, 7, 8f, 165f, 167, 181, 183f,
        263f
Lithostratigraphic correlation, 104, 105f
Lithostratigraphic units, *See* Rock units
Lithotrophs, 234
Little ice age, 476
Littoral zone, 147
Lizard(s):
    of Cenozoic Era, 486f
    monitor, 498f
Loess silt, 480, 482f
Lo'ihi seamount, 197, 199, 200
Long Island contour map, 81f
Louann salt bed, 375
Lower Carboniferous System, 32
Lower Cretaceous Period, 123
Lower Silurian rocks, 112f
"Lucy," *See Australopithecus afarensis*
Lunar highlands, 215
Lungfish, *See Eusthenopteron*
Luster, 48
Lycopsids, 365
Lyell, Charles, 19, 19f, 33, 69, 475
Lystrosaurus, 178

Machaeroides, 510f
McMenamin, Mark, 262
Mafic crust, 225
Mafic silicate mineral, 53
Magma(s), 56, 57, 63, 167
    andesitic, 64
    basaltic, 64
    of Basin and Range province, 464

Magma ocean, 220–221
Magnetic field, 172, 178–180, 180f
    dipole model of, 179f
    normal, in seafloor, 192, 192f, 193f
    polarity in, 192–193, 194f
    reversals in, 193f
    reversed, in seafloor, 192, 192f, 193f
Magnetic intensity, 191–192
Magnetic minerals, 48
Magnetic poles, 179
    mobile locations of north, 181f
    and paleomagnetism, 180–181
Magnetism, 50
Magnetite, 50f, 180f
Magnetometer, 172
Malachite, 48f
Malign Lake (British Columbia), 2f
Malleus (ear), 144, 144f
Malthus, Thomas, 21
Mammals:
    of Cenozoic Era, 499–521
    characteristics of, 500
    earliest, 500
    of early Miocene Epoch, 514f
    of early Pliocene Epoch, 515f
    evolutionary modifications of, 501f
    of late Eocene Epoch, 513f
    marsupials, 501–504f
    of Mesozoic Era, 156, 437–439
    monotremes, 501
    placental, 439, 504–521f
Mammoths, 123f, 516, 518, 519f
Mammoth Geyser (Yellowstone National
        Park), 199f
Mantle (of earth), 7, 8f, 164, 165f, 166–167
Maps (geologic), 107, 112–115
Maps (isopach), 111, 111f, 280f, 288f
Maps (paleogeographic), 110, 110f, 149
    of Cretaceous Period, 384f, 408f
    of Devonian Period, 299f
    of Europe, 398f
    of Jurassic Period, 379f, 392f, 408f
    of Mississipian Period, 302f
    of North America in Paleogene Epoch,
        457f
    of Ordovician North America, 287f
    of Triassic Period, 376f, 408f
Marble, 71, 72f
Marginifera ornata, 133, 133f
Maria (term), 215
Marine ecosystems, 145–149
Marine environments:
    classification of, 145f
    and sediment deposition, 80–82
Marine invertebrates:
    in Cenozoic Era, 492–496
    in Mesozoic Era, 439–441
    in Paleozoic Era, 335–352
Marine life (Mesozoic Era), 439–441
Marl (term), 399
Marl rocks (Cretaceous Period), 389,
        393f
Marrella, 332f
Mars, 216–217, 216f, 217f
Marsh, Othniel C., 23
Marsupials, 500, 501–504f
Marsupuim, 501, 502f

Mass extinctions, 156
  cause of, 284, 368–369
  hypotheses for, 443–448
  of large Pleistocene animals, 521
  major, 368f
  in Mesozoic Era, 443–448
  in Paleozoic Era, 329, 366–369
Mass spectrometer, 35–36
Mastodons, 516, 517, 518f
Matrix (in sedimentary rocks), 87
Matthews, Drummond, 192
Maturity (sandstone forming sediments), 95
Mauna Loa volcano, 200
Mawsonites fossil, 154f
Mechanisms (in plate tectonics), 189–191
Mediterranean seafloor, 4–6
Medusa (cnidaria), 337, 338f
Megaloceros, 515, 516f
Megatylopus, 515f
Megazostrodon, 437
Meiosis, 130, 131f
Mélange, 188, 384
Melonechinus, 353
Members (in rock formations), 100
Mendel, J. Gregor, 21, 128, 130
Mercury, 214, 214f
Merychippus, 136f, 510, 512f
Merychyus, 514f
Merycodus, 515f
Merycoidodon, 513f
Mesohippus, 509–510, 512f, 513f
Mesonyx, 510f
Mesoproterozoic era, 243, 249–251
Mesosaurus, 178, 178f
Mesozoic Era, 28, 113f, 115, 374–401,
    406–448
  climate in, 407–408
  dinosaurs of, 417–431
  eastern North America in, 375–378
  and Gondwana, 399–401
  invertebrates of, 409–414
  land plants in, 441–443
  major events of, 406
  mammals in, 437–439
  marine life in, 439–441
  mass extinction in, 443–448
  molars of mammals, 439f
  organisms in, 156
  and Pangea breakup, 374–375
  reptiles, 431–437
  rock, and continental drift evidence, 178
  and Tethys Sea, 396–398
  vertebrates of, 415–416
  western North America in, 378–396
Metallic core of Earth, 165–166
Metallic luster, 48
Metamorphic index minerals, 70
Metamorphic rock(s), 15, 36, 48, 69–73.
    See also specific types, e.g.: Slate
  foliated, 71, 72
  formation of, 55
  nonfoliated, 71, 72
Metamorphism, 69–70
Metazoan embryo, 259f
Metazoans, 258–262
Meteors, 210
Meteor crater, 446

Meteorites, 3f, 43, *166* f, 210–212
  and earth's core, 165
  and Solar System, 210–212
Mica, 53, 53f, 61, 65
Mica schist, 72f
Michaelangelo, 73f
Michigan basin, 272f
Micrite (textural term), 67
Microcontinents (defined), 201
Microplates (oceanic crust), 202
Mid-Atlantic ridge, 174f, 181
Middle ear, 361
Midoceanic ridges, 175f, 183, 183f,
    232–234, 233f
Midoceanic trenches, 175f, 183, 183f
Migration:
  of camels, 151f
  of Indian land mass, 189f
  intercontinental, 149–151f
  and paleogeography, 149–151
Milankovitch, Milutin, 481–482
Milankovitch cycles, 481–482
Milky Way galaxy, 209f
Miller, Stanley, 231–232
Minerals, 48–55
  clay, 53–54
  defined, 48
  heavy, 298
  magnetic, 48
  metamorphic index, 70
  nonsilicate, 54–55
  properties of, 48–50
  silicate, 50–54
Mineral deposits:
  in Cenozoic Era, 456, 458
  of Late Paleozoic Era, 321, 324
Miocene Epoch, 459–460
  camel skeleton, 515f
  mammals of early, 514f
Mischoptera, 349
Mississippian Period, 113f, 115, 297
  deposition rates, 314
  infaunal organisms of, 147f
  paleogeographic map, 302f
  strata, 310, 312
Mississippian System, 32
Mississippi River Delta, 84, 85f
Mitosis, 130, 131f
Mjølnir crater, 445
Moenave formation, 382
Moenkopi formation, 113f, 115, 382, 383,
    394
Moeritherium, 516, 517f
Mohorovicˇicˊ, Andrija, 164
Mohorovicˇicˊ discontinuity, 164, 165f, 167,
    167f
Mohs, Frederick, 49
Mohs scale, 49f, *136*
Molars, 136f, 439f
Molasse clastics, 471
Molds:
  external, 121–122
  fossils as, 121–123
  internal, 121, 122
  in preservation of fossils, 121–123
Molecular fossils, 236, 239, 256
Mollusca, 343f

Mollusks, 341–345
  in Cenozoic Era, 493, 495f
  of Cretaceous Period, 381f
Monitor lizard, 498f
Monkeys, 528, 529f. *See also* Anthropoidea
Monoclines, 170, 172f
Monongahela group, 313
Monoplacophorans, 341, 344f
Monotremes, 438, 439, 500, 501, 502f
Monson Gneiss (New Hampshire), 72f
Montastrea cavernosa, 493f
Moon, 215, 215f
Moon rocks, 43, 216f
Moraines, 22, 84f, 476, 476f
Morganucodon, 437, 438f
Moropus, 514f
Morphology (teeth), 438–439, 439f
Morrison formation, 387, 389, 392f
Mosasaurs, 435, 437f
Mountain building, *See* Orogeny(-ies)
Mountain glaciers, 473, 474f
Mountain Lakes Formation (Canada), 92f
Mount Hood (Oregon), 450f
Mount Lassen, 467
Mount Mazama, 467, 467f, 468f
Mount Rainier, 467
Mount St. Helens, 3f, 63, 465, 466f
Mouth, 351
Muav limestone, 113, 114f
Mud cracks, 91, 91f, 94
Multituberculates, 438, 439
Munising Formation (Michigan), 93f
Murchison, Roderick Impey, 32
Muscovite mica, 53f
Mutation, 131–132
Myllokunmingia, 333f, 354
Myomorphs, 506

Natural laws, 16–17
Natural radioactivity, as geologic dating
    method, 33
Natural selection, 8, 21, 128
Nautiloid cephalopods, 344, 346f
Nautiloidea, 410
Nautilus, 342–344, 345f
Navajo formation, 383
Navajo sandstone, 382–383, 386, 391f
Neandertals, 538–540, 540f, 541f
Nebula hypothesis, 208–210, 212f
Nekton (swimmers), 147
Neogene Period, 452, 452f
Neogene Series, 32
Neohipparion, 515f
Neoproterozoic era, 243, 251–253
  fossil record, 258–262
  landmasses during, 269f
Neptune, 219, 219f
Neptunists, 16
Neritic zone, 145
Neutrons, 34, 34f
Nevadan orogeny, 383–384
Niagara Falls geologic section, 279f
Niger Delta (Africa), 84, 85f
Nile River delta (Egypt), 83
Niobrara formation, 393, 396, 397f
Nitrogen-14, 42
Nocturnal animals, 527

Nodosaurs, 428
Nomenclature, binomial, 125
Non-amniotic vertebrates, 353
Nonconformity, 106, 106f, 107f
Nonfoliated metamorphic rocks, 71, 72
Nonsilicate minerals, 47, 54–55
Norell, Mark A., 431
Normal faults, 169, 169f
North America. *See also* Eastern North
America; Western North America
in Cambrian Period, 271f
cratonic sequences of, 274f
orogenic belts of, 273f
Silurian paleogeography of, 289f
southwestern, in Late Paleozoic Era,
303–306, 307f
North American cordillera, 390f
North American plate, 465, 467f, 468, 469f,
470f
Northosaurus, 435f
Notharctus, 529, 531f
Nothosaurs, 433–434, 435f
Notochord, 333
Novaculites, 314
Nucleic acids (defined), 231
Nucleotide base pairs (in DNA), 139
Nuée ardentes, 464, 467

Obsidian, 58–59, 59f
Ocean(s):
in Archean Eon, 224
contraction in Early Paleozoic, 290
expansion in Early Paleozoic, 288, 290
floor mapping, 172, 174f, 175f
Iapetus, 269
magma, 220–221
transgressions of, 274–279
Ocean basin(s):
opening of, 246
in Paleozoic Era, 273–274
Oceanic crust, 58, 165f, 167, 168, 202, 221f
basaltic, 174f
mafic, 225
Oceanic-oceanic crust convergence,
187–188, 187f, 188f
Oceanic sediments, 194
Oceanic zone, 145
Ocean salinity, 33
Offlap sequence (in facies), 100, 102f
Offspring, genetic variation in, 130–132
Oil shales, 458, 459, 459f
Olduvai Gorge, 532, 533f
Oligocene, 123
Oligocene Epoch fossils, 458, 459
Olivine, 53, 54f, 62, 166f
Olivine crystals, in Hawaiian volcano basalt,
200
Onlap sequence (in facies), 100, 101f
On the Origin of Species (Charles Darwin), 21
Oöids, 68, 98, 99f
Oölites (textural term), 67
Oölitic limestone, 99f, 301f
Oozes:
calcareous, 148–149
siliceous, 148
Opabinia, 332f
Ophiolite suite (defined), 188, 191f

Ophiuroidea (brittle stars), 348
Opossums, 501, 502f
Opportunity (Mars rover), 3f, 216
Oquirrh basin, 318f
Orbital eccentricity (of Earth), 482, 483f
Order (taxonomy), 126
Ordovician Period:
extinctions in late, 366–367
ice age, 290
marine limestones, 37f
North America paleogeographic map, 287f
sedimentary rock (isopach map), 288f
sedimentation, 285f
shale beds, 286f
trilobite, 133f
Ordovician System, 32
Oreodonts, 511, 513f
Organs, vestigial, 138, 139f
Organelles, 230
Organic evolution, 7–8. *See also* Evolution
Organic phosphorous compounds, 231
Organisms:
aerobic, 235
anaerobic, 235
Origin(s):
of Earth, 219–221
of life, 230–239
of Universe, 213
Original horizontality, principle of, 14, 14f
Original lateral continuity, principle of,
14–15, 15f, 103
Oriskany sandstone, 298, 300f
Ornithischia dinosaurs, 417–418, 417f, 418f
Ornithomimus, 420f
Ornithopods, 423–428
Orogenic belt(s), 79, 273, 273f
Appalachian, 312–313
Caledonian, 288, 290
of Europe, 273f
of North America, 273f
Ouachita, 314, 316
Wopmay, 245, 245f
Orogeny(-ies), 245
Acadian, 267, 297, 308–309
Alleghenian, 267, 297, 308, 312–313
Antler, 317
Caledonian, 269, 297
Cassiar, 318
Grenville, 251, 251f
Hercynian, 297, 319
Hudsonian, 249
Kenoran, 238
Laramide, 385–386
in Late Paleozoic Era, 316–318
Nevadan, 383–384
Palisades, 377f
Sevier, 385
Sonoma, 318, 381, 383
Taconic, 267, 269, 283–284
Trans-Hudson, 247–248, 248f
Wopmay, 245–247, 245f
Orohippus, 509, 510f
Orthoclase feldspars, 51–52, 52f
Orthograptus quadrimucronatus, 354f
Oscillating-Universe hypothesis, 213
Osteichthyans, 357
Osteichthyes, 357–359

Osteoborus, 515f
Ostracods, 347, 348f
Ostracoderms, 354
Ouachita-Marathon trough, 287–288
Ouachita Mountains (Oklahoma), 14f
Ouachita orogenic belt, 314, 316
Outgassing, 222
Outwash plain, 476f
Oviraptor, 431
Oxides, 47
Oxydactylus, 512, 514f–515f
Oxygen (Proterozoic Eon), 263–264, 263f
Oxygen-16/oxygen-18:
isotope ratio and ancient seawater
temperature, 152
in Pleistocene deep-sea sediments, 477

Pachycephalosaurs, 427
Pachycephalosaurus, 427, 427f
Pacific plate, 468, 469f, 470f
Pahoehoe lava, 200
Pakicetus, 519, 519f
Palaeomastodon, 516
Palaeosyops, 510f
Palate, secondary, 500
Paleocene Epoch, sediments in, 456–458
Paleoclimatology, 176–177
Paleoecology, 143, 145
Paleogene Period, 452, 452f, 457f
Paleogeographic map(s), 110, 110f
of Cretaceous Period, 384f, 408f
of Devonian Period, 299f
of Europe, 398f
of Jurassic Period, 379f, 392f, 408f
of Mississipian Period, 302f
of North America in Paleogene Period,
457f
of Ordovician North America, 287f
of Triassic Period, 376f, 408f
Paleogeographic relations of continents, 195f
Paleogeography:
configurations of Earth's, 270f
in Eocene Epoch, 453f
and fossil distribution, 149–152
of Late Paleozoic Era, 296f
mapping, 149
migration and, 149–150
of Silurian North America, 289f
and species diversity, 150–152
Paleoindians, 544
Paleomagnetism:
magnetic poles and, 180–181
and plate tectonics, 178–181
and rocks, 179–180
Paleontology, and evidence for evolution,
136–137
The Paleontology of New York (James Hall), 22
Paleoproterozoic Era, 243, 245–249
glaciation in, 248
length of day in, 247
rocks in, 249
Wilson cycle in, 246–248
Paleothyris, 362
Paleozoic Era, 28, 113, 267–291, 296–324,
327–369. *See also* Early Paleozoic;
Late Paleozoic
biodiversity in, 333–335

Cambrian explosion of, 330–333
life forms in, 155–156
major events of, 268f
marine invertebrates in, 335–352
mass extinctions of, 329, 366–369
plants in, 363–366
protistans, 334–335
sauk sequence in, 275–277
and shell-bearing fossils, 327–330
tippecanoe sequence in, 278
vertebrates in, 329, 354–363
and western North America, 279–281
Paleozoic strata (Great Britain), 31f
Palisades (Hudson River), 378f
Palisades orogeny, 377f
Pallas, Peter Simon, 15
Pander, Christian, 359
Pangea, 175, 297
breakup of, 374–375
break-up of, 176f
and Permian extinctions, 369
Panthalassa, 175, 176f, 251
Paraceltites elegans, 137f
Paraceltites rectangularis, 137f
Paraceratherium, 510, 513f
Paradox basin, 303, 307f
Parahippus, 514f
Parasaurolophus cyrtocristatus, 427f
Parent element, 35
Paris basin, 471, 472f
Partial melting:
defined, 64
of Earth, 220
of peridotite, 64f
Parvancorina, 260f
Passive continental margin, 183
Patriofelis, 507, 507f, 510f
Pecopteris, 367f
Pelagic marine ecosystem, 145–147
Pelecypoda, 341, 344f
Pelycosaurs, 362
Pelycosaurian sail, 362, 363f
Penicillus, 98, 98f
Pennsylvanian Period, 297, 306f
Pennsylvanian System, 32, 312, 317f
Pentacrinus, 410, 410f
Peridot, 200
Peridotite rock, 53, 64, 166, 166f
Periods, 28
Perissodactyls, 509–511
brontotheres, 510–511, 513f
chalicotheres, 510–511, 514f
phylogenetic tree of, 511f
Permian Period, 113f, 115, 297, 297f
basin, 304–306
extinctions in late, 369
reconstruction of Gondwana in, 177f
rock, and continental drift evidence, 178
stages of, 307f
Permian System, 32
Permineralization, 120, 121f
Petrifaction, 120, 121f
Petrified Forest National Park, 383, 387f
Petroleum, 321, 324
Petroleum geologists, 3
Phacops rana, 349
Phalangers, 501, 503f

Phanerozoic Eon, 28
Phenacodus, 508, 509f
Phenocrysts, 58f, 59, 200
Phenotypic traits, 126–127
Phosphatherium, 516
Phosphoria formation, 324
Phosphorite, 320f
Photochemical dissociation, 222–223
Photosynthesis:
interdependence with respiration, 146f
as oxygen-generating mechanism, 223
Phycodes, 273f
Phyletic gradualism, 134, 134f
Phyllite, 71
Phylogenetic tree, 134, 135f
of horses, 135f, 509
of perissodactyls, 511f
of ungulates, 508f
Phylogeny, 134–136
Phylum (taxonomy), 126
Physical geology, 2
Phytoplankton, 146–147, 440, 440f
Phytosaurs, 415, 416f
Pierre shale, 461
Pilaia, 333, 333f
Pilina, 344f
Pillow basalt, 167f
Pillow lava, 238
Placental mammals, 439, 500, 504–521f
Placoderms, 356
Placodonts, 433–434, 435f
Placodus, 435f
Placophorans, 341, 344f
Plagioclase, 61, 62
Plagioclase feldspar, 51–52, 53f, 60
Planets, 43
properties of, 210
size comparison of, 211
in Solar System, 214–219
Plankton, 146–147, 440, 490–492
Plants:
genera in Cenozoic Era, 489, 489f
geologic ranges of, 154f
history of, 154
land, 441–443
in Mesozoic Era, 441–443
in Paleozoic Era, 363–366
plankton, 146–147
sea, 439–440
taxonomic classification of, 126
vascular, 154, 154f, 329, 364–365
Plantae kingdom, 126, 127f
Plant-eating dinosaurs, 422–429
Plantigrade, 508
Plates, tectonic, *See* Tectonic plates
Plate margins, 187–189
continental-continental crust
convergence in, 187, 188f
continental-oceanic crust convergence in,
188–189, 188f
oceanic-oceanic crust convergence in,
187–188, 187f, 188f
Plate movement, rate of, 195f, 197
Plate tectonics, 7, 171–202
and continental drift, 174–178
convergent boundaries and, 185
current theory of, 181–186

defined, 171
development of, 171–172
divergent boundaries and, 181, 183
driving force in, 189–191
early theories, 226
force of, 285f
guyots and, 183–185
and Hawaii, 197–200
and hotspots, 196–202
margins and, 187–189
as model for continental collisions, 314f
in Northern Appalachians, 310f
and paleomagnetism, 178–181
in Paleoproterozoic Era, 245–248
theory development for, 171–172
theory verification for, 191–196
transform boundaries and, 185
Platforms, 79, 225
Platypus, duck-billed, 502f
Playfair, John, 17–18
Pleistocene Epoch:
beginning of, 475
Bering land bridge in, 150
extinctions of large terrestrial animals in,
521
stages of, 475f
Pleistocene glaciers, 322
Pleistocene Ice Age, 42
of Cenozoic Era, 473–483
deep-sea sediments of, 477, 478
glacial deposits of, 476–477
major impacts of, 478–480
Pleistocene Series, 32
Plesiadapis, 528, 530f
Plesiosaurs, 434, 436f
Pliocene Epoch:
early mammals of, 515f
land bridge of late, 504f
Pliohippus, 510
Pluto, 219
Plutons, 57, 57f
Plutonic igneous rock, 56–57
Plutonists, 16
Pluvial lakes, 479
Pocono group, 310, 312
Poëbrotherium, 513f
Polarity:
in magnetic fields, 192–193
reversals of, 194f
Polar wandering paths, apparent, 180–181
Poles:
geographic, 179, 179f
geomagnetic, 179, 179f
Polymorphs (term), 55
Polyp (cnidaria), 337, 338f
Polyplacophorans, 341
Pongidae, 528
Population:
defined, 132
growth of hominid, 544–545
Porifera (sponges), 335–337, 336f
Porphyritic texture, 59
Potassium-40, 38, 39–40
Potassium-argon method of isotopic dating,
39–40
Potassium feldspar, 52, 52f, 58f
Powell, John, 23

Precambrian/Cambrian boundary, 273
Precambrian Period, 28
  and living organisms, 153–154
  time divisions of, 226f
Precambrian provinces, 225, 228f, 245
Precambrian rocks, 224–228
  exposed, 225, 227f
  greenstone, 228, 229f, 230f
Precambrian Superior province, 238
Precession (of Earth), 481–482, 483f
Prehensile tails, 528
Preservation of fossils, 120–123
  in amber, 123
  common processes in, 120–121
  and fossil record, 124
  molds and casts in, 121–123
  soft tissue in, 122–123
Primary waves, 162, 162f, 163, 163f, 165 f
Primate(s), 525–545
  anthropoidea, 527–532
  australopithecine stage of, 532–537
  characteristics of, 526–527
  groups of, 527–528, 527f
  hominid, 532–545
  Homo erectus stage of, 537–538
  Homo sapiens, 538–542, 544–545
  prosimii, 527, 528–529, 530f, 531f
Principles of Geology (Charles Lyell), 19
Prionodinium alveolatum, 441f
Proboscideans, 501, 508, 516–518
Procamelus, 515f
Proconsul africanus, 532, 532f
Productid brachiopod, 340, 343f
Prognathous, 538
Prokaryota superkingdom, 126
Prokaryotes, 235, 236f, 237 f, 255–256
Promissum, 360
Prosauropods, 422
Prosimii, 527, 528–529, 530f, 531f
Prosthennops, 515f
Protarchaeopteryx, 430
Protein, 231
Proterozoic Eon, 28, 208, 208f, 243–264
  life in, 255–262
  major events of, 244f
  Mesoproterozoic era of, 243, 249–251
  Neoproterozoic era of, 243, 251–253
  oxygen/climate changes in, 263–264, 264f
  Paleoproterozoic era in, 243, 245–249
  rocks in, 253–255
Protista kingdom, 126, 127f
Protistans:
  in Mesozoic Era, 413
  in Paleozoic Era, 334–335
Protoceras, 513f
Protoceratops, 431
Protons, 34, 34f
Protoplanets, 210
Protorogomorphs, 505
Protorohippus, 509, 510f
Protostomes, 351
Protostomes, 351
Prototherians, 439
Protozoa, 257–258
Provinces:
  Basin and Range, 460, 463–464, 463f
  physiographic, of eastern United States,
    315f

Pseudohizostomites, 260f
Psittacosaurus, 439
Pteranodon, 432, 433f
Pterichthyodes, 357f
Pterobranchs, 352
Pterodactylus, 118f
Pterodausto, 432, 434f
Pteropod, 148f
Pterosaurs, 431–433, 434f
Pubis, dinosaur, 417, 418f
Pumice, 59
Punctuated equilibrium, 134, 134f
Purgatorius, 528
Purslane formation, 314f
P-waves, See Primary waves
Pyrite (fool's gold), 49f, 121
Pyroxene crystals, 61, 62, 166f

Quartz, 50–51, 52f, 65
  grains of, 90
  sandstone, 95, 95f–97f
  shocked, 444, 445f
Quartzite, 71, 73f, 176f
Queenston clastic wedge, 278, 285, 287f
Quetzalcoatlus northropi, 432, 433f

Rabbits, 500, 506
Radial cell division, 351, 353
Radiation:
  gamma, 35
  UV, 231
Radinskya, 136, 136f
Radioactive dating, 35–43
  accuracy of, 35–36
  and carbon-14, 41–42
  fission track method, 42–43
  and half-life, 38
  potassium-argon method, 39–40
  rubidium-strontium method, 40–41
  and type of rock, 36–38
  uranium-lead methods, 38–39
Radioactive decay:
  discovery of, 33
  for uranium-238, 38f
Radioactive decay series, 36f
Radioactive isotopes, See Isotope(s)
Radioactivity, 7, 35
Radiolaria, 148f, 335, 490, 491f
Radiolarians, 413, 414f
Ramamorphs, 532
Range zone, 142–143
Rate (plate movement), 195f, 197
Receptaculids, 364
Recrystallization (minerals), 69
Red beds, 86, 263
Red Sea, 454f
Red Sea rift, 181
Red shift, 213
Red tide, 257
Redwall formation (rock unit), 100
Redwall limestone, 113f, 115, 115f
Reefs, 309
  barrier, 493, 494f
  in Cretaceous Period, 378
  fringing, 493
Reef core, 309
Regional metamorphism, 70

Regular echinoids, 410
Relative age, 7
Relative geologic dating, 27–28
Repenomamus giganticus, 438
Repenomamus robustus, 438–439, 439f
Replacement (in fossil preservaton), 121,
  121f
Reproduction:
  of dinosaurs, 430–431
  and evolution, 130–132
Reptiles, 362
  of Cenozoic Era, 497–498f
  mammal-like, 363f
  in Mesozoic Era, 431–437
Respiration, and photosynthesis, 146f
Reticulum, 514
Reverse faults, 169, 169f, 170
Reworked fossils, 142
Rhipidistians, 359
Rhynchocephalians, 415
Rhynia, 365, 365f
Rhyolite, 59, 59f, 60
Ribonucleic acid (RNA), 126–127, 231
Ridge(s):
  mid-Atlantic, 174f, 181
  midoceanic, 175f, 183, 183f, 232–234,
    233f
Ridge-push model, 189–190, 191f
Rifts:
  Gulf of Aden, 181
  major, 473f
Rift valleys, 181, 473f
Rift zone, 250
Ring of Fire, 200
Ripple marks, 91, 93, 93f, 94f
Rivers, Pleistocene ice age redirection of,
  478–479
RNA, See Ribonucleic acid
Rock(s), 55–73. See also specific headings, e.g.:
  Igneous rock(s)
  of Catskill clastic wedge, 313f
  color of, and sediment deposition, 86
  and continental-oceanic convergence,
    188–189
  determining age of, See Geologic dating
  of Eocene Epoch, 459f
  in exotic terranes, 202f
  geologic cycle of, 55–56, 56f
  Keweenawan, 249–250
  marl, 389, 393f
  moon, 43
  origin of Precambrian, 224–228
  and paleomagnetism, 179–180
  Paleoproterozoic Era, 249
  in Paleoproterozoic Era, 249
  Pennsylvanian System, 312
  peridotite, 53
  Proterozoic Eon, 253–255
  texture of, and sediment deposition, 86–90
  Upper Devonian sedimentary, 311f
Rocknest Formation (Canada), 246f
Rock sequences, and continental drift
  evidence, 178
Rock stratum, 12
Rock units, 98–100
  correlation of, 108f
  correlation with fossils, 137, 141–142

in Grand Staircase-Escalante National Monument, 395
in Hasmark Formation (Montana), 108*f*
in Paleoproterozoic Era, 246*f*
in Zion National Park, 382
Rockwell formation, 314*f*
Rocky Mountains, 2*f*
Canadian, 281*f*
in Cenozoic Era, 455–460
exotic terranes in, 202
Rodents, 500, 505–506, 506*f*
Rodinia, 251, 251*f*, 267, 269
Rounding, 89
Rubidium-87, 38, 40–41
Rubidium-strontium method of isotopic dating, 40–41
Rudistids, 378, 409, 409*f*
Rugose coral, 337, 339*f*
Ruminants, 501, 514–516, 516*f*
Rutiodon, 416*f*
Ryan, William B. F., 5

Saber-toothed cat, 502, 507, 507*f*
Sahelanthropus tchadensis, 532
"Sails" (pelycosaurs), 362, 363*f*
St. Petersandstone, 278, 278*f*
Salinity, ocean, 33
Salt domes, 375–376, 380*f*
Samoa, 58
San Andreas fault, 169, 170*f*, 186*f*, 469
Sandstone, 20*f*, 21*f*, 65, 66, 76*f*
catagories of, 95, 96*f*, 97
at Colorado Springs, 294*f*
Cretaceous, 372*f*
cross-bedding of, 303*f*
DeChelly, 318, 319*f*
Devonian cross-bedding of, 313*f*
Jurassic, 372*f*
Navajo, 382–383, 386, 391*f*
Oriskany, 298, 300*f*
sorting in, 88*f*
St. Peter, 278, 278*f*
Tapeats, 113, 114*f*, 276, 276 f
and tectonic setting, in sediment deposition, 95–97
Wingate, 383, 387*f*
Sarcopterygians, 357–358
Sassafras leaf fossil, 442*f*
Satellites, 194
Satin spar gypsum, 55*f*
Saturn, 218–219, 219*f*
Sauk sequence, 274*f*, 275–277
Saurischa dinosaurs, 417–421, 417*f*, 418*f*
Sauropods, 422–423, 423*f*
Sauropodomorpha, 418, 422–423
Schist, 71
mica, 72*f*
Vishnu, 113, 114*f*, 254–255, 254*f*, 255*f*
Scientific laws, 4
Scientific method, 3*f*
Sciuromorph, 505
Scleractinids, 492
Scour marks, 94–95
Sea biscuit, 496*f*
Seafloor:
determining age of, 193
magnetic fields in, 192, 192*f*, 193*f*

Seafloor spreading, 181, 183, 184*f* . *See also* Divergent plate boundaries
in Cenozoic Era, 452
and epicontinental seas, 388
rates of, 193–194
Seafloor volcanoes, flat-topped, 183–185
Sea level:
changes in, 274–275
changes in, and geologic history, 102–103, 103*f*
in Cretaceous Period, 377
Pleistocene ice age shifting of, 478
Seamounts (submarine volcanoes), 174*f*, 183, 197, 199, 200
Sea plants, 439–440
Sea regions, in Carboniferous Period, 150*f*
Sea turtles, 435, 437
Seawater temperature, ancient, 152
Secondary palate (of mammals), 500
Secondary waves, 162, 162*f*, 163, 163*f*, 165 f
Sedgwick, Adam, 30, 31*f*
Sediment(s), 12–13
age of midoceanic, 196*f*
deep-sea of Pleistocene Ice Age, 477, 478
flysch, 470
glacial deposits of, 83, 84*f*, 476–477, 480
of Paleocene Epoch, 456–458
and plate tectonic theory, 194
Sedimentary quartz, 51
Sedimentary rock(s), 15, 26*f*, 36, 47, 65–69, 78. *See also specific types, e.g.:* Sandstone
carbonate, 67–68
of Catskill clastic wedge, 313*f*
clastic, 66–67
depositional environment and color, 86
formation of, 55
structure of, 90–95
texture of, 86–90
tidal rhythmites, 247
of Upper Devonian, 311*f*
Sedimentation, 101*f*, 102, 102*f*
in Archean Eon, 228–229
in Devonian Period, 298
in Late Paleozoic Era, 316–318
Ordovician, 285*f*
Sediment deposition, 78–97
in Early Paleozoic Era, 275–290
environments for, 80–86
and limestone, 98
rate of, as geologic dating method, 33
rates in Mississipian Period, 314
rock color, 86
rock structure, types of, 90–95
rock texture in, 86–90
sandstone and tectonic setting in, 95–97
tectonic setting, 78–79
Sediment feeders, 334
Sediment particle shape, 90*f*
Sedwick, Adam, 32
Seilacher, Adolf, 261
Seismic activity, and plate tectonic theory, 194
Seismic waves, 162–163, 165*f*, 166–167
Seismograph, 162, 162*f*
Seismograph(s), 172
Seismusaurus, 423*f*
Septa (in coral), 337, 338*f*

Septa patterns, 410–412
Septum, 410
Sequence:
of geologic events, 20*f*
stratigraphic, in greenstone belt, 230*f*
Series (term), 30
Sevier orogeny, 385
Sevier-type deformation, 390*f*
Seychelles Bank, 199, 199*f*, 201
Shadow zone, 164, 165*f*
Shale, 20*f*, 21*f*, 54, 65, 66
bright angel, 113, 114*f*
Chattanooga, 298–301
Green River, 458
hermit, 113*f*, 115
metamorphism of, 70*f*
oil, 458, 459, 459*f*
Ordovician, 286*f*
Pierre, 461
Soom, 360
Shale bed, 36–38
Shape, of sediment particles, 90*f*
Sharks, 357, 357*f*
Sharovipteryx, 431, 432*f*
Shear zones, 252*f*
Sheep Mountain anticline, 391*f*
Shell-bearing fossils, 329*f*
Cloudina, 262, 262*f*
in Paleozoic Era, 327–330
Shield, 79, 225
Shield volcanoes, 200
Shinarump conglomerate, 383
Shivling peak (Himalayas), 470*f*
Shocked quartz, 444, 445*f*
Shrew, tree, 505*f*, 527, 528*f*
Shrimp, 496*f*
Siccar Point (Scotland), 10*f*, 17
Sierra Nevada, 60
batholith, 384, 389*f*, 468
in Cenozoic Era, 467
Signature fossil, 12
Silica, 59–60, 63
Silicate minerals, 50–54
Siliceous oozes (sediments), 148
Silicoflagellates, 440, 441*f*
Sillimanite, 70
Silt, loess, 480
Siltstone (Mississippian age), 147*f*
Siltstones, 66, 76*f*
Silurian:
North America paleogeography, 289*f*
sedimentary deposits, 286–288
trilobite, 133*f*
Silurian System, 32
Silver Hill Formation (Montana), 108*f*
Sinosauropteryx, 430
Skeleton:
of *Hypertragulus*, 516*f*
of *Ichthyostega*, 361*f*
of *Megaloceros*, 516*f*
of *Mesohippus*, 512*f*
of Miocene Epoch camel, 515*f*
of Neandertal, 541*f*
Skull:
of *Aegyptopithecus zeuxis*, 531*f*
comparison of hominid, 543*f*
of Cro-Magnon, 542*f*

Skull: (*Continued*)
  of crossopterygian, 359*f*
  of *Dinotherium*, 517*f*
  of *Homo erectus*, 537*f*
  of *Homo rudolfensis*, 537*f*
  of Jurassic crocodile, 434*f*
  of *Moeritherium*, 517*f*
  of Neandertal, 540*f*
  of ornithischian dinosaur, 418*f*
  of *Proconsul africanus*, 532*f*
  of saber-toothed cat, 507*f*
  of saurischian dinosaur, 418*f*, 420*f*
Slab-pull model, 189–190, 191*f*
Slate, 71
Sloth, ground (*Megatherium*), 505*f*
Small-shelly fossils, 329–330
Smilodon (saber-toothed cat), 502, 507, 507*f*
Smith, William, 12, 18, 18*f*, 20, 98–99, 139
Snakes, 497, 498*f*
Snake River Plateau, 58, 63
Snider, A., 174
Sodium-calcium aluminosilicates, 51
Solar nebula, 209. *See also* Nebula
  hypothesis
The Solar System, 208–221
  earth's formation in, 219–221
  meteorites and, 210–212
  nebula hypothesis and, 208–210
  planets in, 214–219
Solar wind, 210
Sonett, Charles, 247
Sonoma orogeny, 318, 381, 383
Sonoma terrane, 381
Soom shale, 360
Sordes pilosus, 433
Sorting of sedimentary rocks, 88–89
Sound, and tympanic membrane, 144, 144*f*
Southwestern North America (Late
  Paleozoic Era), 303–306, 307*f*
Speciation, 132
Species, 125–126
Species diversity, 150–152, 151*f*
Sphenodon, 443, 497, 498*f*
Sphenosids, 365, 367*f*
Sphericity, 89
Spider fossil (Oligocene Epoch), 462*f*
Spinal column, 360
Spiracle, 361
Spiral cell division, 351
Spiriferid brachiopods, 340, 343*f*
Sponges, *See* Porifera
Spore-bearing plants, 364
Spore bodies (plants), 364
Spreading center (plate tectonics), 181
Spriggina floundersi, 260, 261*f*
Squamates, 497
Stage (term), 30
Stapes (ear), 144, 144*f*, 361
Starfish, 350*f*
Steady-state hypothesis, 213
Stegosaurus, 427–428, 428*f*, 430
Stelleroidea, 411*f*
Steno, Nicolaus, 13–15
Stenomylus, 514*f*
Stensen, Niels, 13–15
Stishovite, 444

Stone tools (hominid), 539*f*
Stony-iron meteorites, 212
Strachey, John, 15
Straight line, *See* Isochron
Strata, 15, 65, 98–103
  Cambrian, 276*f*
  of Cenozoic Era, 454*f*, 456*f*
  facies, 100–102
  Late Paleozoic, 316*f*
  rock units, 98–100
  and sea level changes, 102–103
Strata Identified by Organized Fossils (William
  Smith), 18
Stratified drift, 476
Stratigraphic cross-section, 107
Stratigraphic sequence, 230*f*
Stratigraphy, 14, 15, 103–107
  correlation methods, 104–105
  and fossils, 139–143
  of Pleistocene deposits, 476–478
  unconformities, 105–107
Streak, 48
Streak plate, 48
Stream deposits (of sediment), 82
Striations, glacial, 22
Strike (fold direction), 170, 171*f*
Strike-slip faults, 169, 169*f*
Stromatolites, 153, 153*f*, 236, 239*f*,
  255–256, 256*f*, 363–364
  Neoproterozoic, 253*f*
  in Proterozoic rocks, 254, 254*f*
Stromatoporoids, 336–337, 337*f*
Strontium-87, 40–41
Structural cross-section, 107, 108*f*
Studies of Glaciers (Louis Agassiz), 22
Stylinodon, 510*f*
Subduction, 470*f*
Subduction zone(s), 185, 186*f*, 379–380
  in Mesozoic Era, 383–384
Sublittoral zone, 147, 147*f*
Submarine fans (sediment deposits), 82, 82*f*
Sudbury, Ontario, 40
Suess, Eduard, 174–175
Sundance formation, 387
Sundance sea, 387, 392*f*
Supai group, 113*f*, 115
Superior province (Precambrian), 238
Superkingdoms (taxonomy), 126
Superposition, principle of, 13–14, 13*f*, 93
Supersaurus, 423*f*
Supratidal zone, 147
Surface waves, 162, 162*f*
Suture patterns (cephalopods), 345*f*, 346*f*,
  410–412, 412*f*
Suture zone, 187
S-waves, *See* Secondary waves
Symmetrical ripple marks, 93
Symmetrodonts, 438
Synapsids, 362
Synclines, 170, 172*f*, 173*f*
Syndyoceras, 514*f*
Synthetoceras, 512, 515*f*
Systems, 28, 30

Tabulae (coral), 337
Tabulate coral, 337.338*f*

Taconic orogeny, 267, 269, 283–284
Taconite, 250
Tahiti, 58
Tails, prehensile, 528
Tapeats sandstone, 113, 114*f*, 276, 276* f*
Tarbosaurus, 421*f*
Tarsier, 527, 528*f*
Tarsius syrichta, 528*f*
Tasmanian "wolf," 503*f*
Taste (of minerals), 50
Taum Sauk Mountain (Missouri), 107*f*
Taxonomy, 126–128
Tectonics, 7
  accretionary, 378–380
  and climate in Cenozoic Era, 452–454
  of Rocky Mountains in Cenozoic Era,
    455–460
  term, 78
  thin-skinned, 313
Tectonic collage, 380
Tectonic parts (of North America), 79*f*
Tectonic plates, 184*f*
  Cocos, 468, 469*f*
  Farallon, 468, 469*f*
  Juan de Fuca, 465, 467*f*, 468, 469*f*
  major, 182*f*, 184*f*
  North American, 465, 467*f*, 468, 469*f*, 470*f*
  Pacific, 468, 469*f*, 470*f*
  subduction of Farallon, 470*f*
Tectonic plate boundaries, 7
  convergent, 7, 8*f*, 185, 187–189, 187*f*
  divergent, 7, 181, 183, 184*f*, 188I*f*
  events/features at, 188*f*
  transform, 7, 185, 188*f*
Teeth:
  of *Aegyptopithecus zeuxis*, 531*f*
  of australopithecines, 534
  carnassials, 506, 508, 509*f*, 528
  of crossopterygian, 359*f*
  in cynodonts, 363
  of labyrinthodonts, 361
  of mammals, 500
  of mammoth, 518*f*
  of mastodon, 518*f*
  morphology of, 438–439, 439*f*
Tektites, 444
Teleosts, 496–497
Temnospondyls, 415
Temperature, of ancient seawater, 152
Temple butte limestone, 113*f*, 115
Temple Gap formation, 383
"Temples" (Navajo sandstone), 382–383
Temporal transgression, 277
Terebratulina septentrionalis, 496*f*
Terminal moraines, 84*f*
Terranes:
  alien, 322–323
  Avalon, 308
  exotic, 199, 201–202, 202*f*, 378–379
  Sonoma, 381
Terrestrial invertebrates, of Mesozoic Era,
  413–414
Tethys Sea, 187, 202
  in Cenozoic Era, 470–471
  in Mesozoic Era, 396–398
Tetonius, 529

Teton Range, 462f
Tetrads (plants), 364
Tetrapods, 329, 360–362
Texture:
    of igneous rock, 56–59, 58f, 60f
    porphyritic, 59
    of sedimentary rock(s), 86–90
Theca (coral), 337, 338f
Theories, 4
Theory of the Earth (James Hutton), 17
Therapsids, 329, 362–363
Therians, 439
Thermal plumes, 191, 191f
Theropoda, 418–421, 430
Thin-skinned tectonics, 313
Three-chambered heart, 360
Thrust faults, 169f, 170
Thylacosmilus, 502, 504f
Tidal flats, 85
Tidal rhythmites, 247
Till, glacial, 476f
Tillites, 176, 248
Till particles, 476
Time, deep, 7
Time-rock units, 28, 30f, 99, 112f
Time units, 28, 30f
Tippecanoe sequence, 274f, 275, 277f, 278
Tissue preservation, soft, 122–123
Tonto group (rock unit), 100
Toroweap formation, 113f, 115
Torrowangee group (Australia), 153
Trace fossils, 123–124, 125f, 262, 273f
Tracheophytes, 364
Trailing edge (plates), 181
Traits:
    acquired, 128, 129f
    phenotypic, 126–127
Transcontinental Arch, 271f, 276
Transform faults, 185, 185f
Transform plate boundaries, 7
    and plate tectonics, 7, 185, 188If
    and San Andreas fault, 186f
Transgressions (advances of oceans),
    274–279, 455
Trans-Hudson orogen, 247–248, 248f
Transitional depositional environments,
    83–86
Transitional (shoreline) environments,
    83–86
Trees, 177. *See also* Phylogenetic tree
Tree shrew, 505f, 527, 528f
Trenches, midoceanic, 175f, 183, 183f
Triadobatrachus, 415, 415f
Triassic Period, 375, 380–383
    paleogeographic map, 376f, 408f
    stages of, 377f
Triassic System, 32
Tribrachidium, 259, 260f
Triceratops, 429f, 430
Triconodonts, 438, 439
Trigonias, 513f
Trilete spores, 364
Trilobite(s), 346–347, 347f, 348f
    eyes of, 349
    Ordovician age, 133f
    Silurian, 133f

Triple junction (thermal plumes), 191, 192f
Trogosus, 510f
Trunk (proboscideans), 516, 517
Tuatara, 497, 498f
Tuff (term), 59
Turbidites, 82
Turbidity currents, 82, 91, 93
Turquoise, 48f
Turtles, sea, 435, 437
Tympanic membrane, 144, 144f
Tyrannosaurus, 419–421f
Tyrannosaurus rex, 404f

Uinta basin, 458, 460f
Uintacrinus, 410, 410f
Uinta formation (Utah), 76
Uintatherium, 508, 510f
Ultramafic (term), 228
Ultramafic rock, 53
Ultrasaurus, 423f
Ultraviolet radiation, and origin of life, 231
Umcompahgre Mountains (Colorado), 303
Unconformities, 17–18, 20f
    angular, 17
    in Grand Canyon Supergroup, 254f, 255
    in stratigraphic continuity, 105–107
Ungulates, 501, 508, 508f
Uniformitarianism, 16–19
United States, geology in, 22–23
Universe, origin of, 213
Upper Carboniferous System, 32
Upper Devonian, 311f
Upper ordovician formations
    (Pennsylvania), 111f
Ural Mountains, 15
Uranium-235, 7, 38
Uranium-238, 35, 36f, 38
Uranium-lead method of isotopic dating,
    38–39
Uranus, 219, 219f
Ussher, James, 32

Varangian glaciation, 251
Varves, 248, 456
Vascular plants, 154, 154f, 329, 364–365
Vascular system, 348
Velociraptor, 419
Vendoza controversy, 261–262
Vent, hydrothermal, 223, 224f
Venus, 214–215, 215f
Vertebral column, 353
Vertebrates, 333, 353
    in Cenozoic Era, 496–499
    and continental drift evidence, 178
    of Mesozoic Era, 415–416
    in Paleozoic Era, 329, 354–363
Vertebrate embryos, 139f
Vesicles, 59
Vestigial gill slit, 361
Vestigial organs, 138, 139f
Viking mission to Mars, 157
Vine, F.J., 192
Vishnu schist, 113, 114f, 254–255, 254f, 255f
Volcanic eruption:
    and igneous rock, 62–64
    and mass extinctions, 447

Volcanic islands, 187f, 196–197
Volcanoes:
    of Cascade Range, 465, 467, 467f
    of Colorado plateau, 464
    of Columbia plateau, 464
    flat-topped seafloor, 183–185
    intraplate, 200
    Kilauea, 200
    Mauna Loa, 200
    shield, 200
    submarine, 174f, 183
Voyageurs National Park (Ontario), 238

Wadati-Benioff seismic zone, 194
Walcott, Charles D., 331–332
Walking erect, 538
Wallabies, 501, 503f
Wallace, Alfred R., 8, 128
Wallace, Russel, 21
Wallowa Mountains (Oregon), 202
Walther, Johannes, 101
Walther's principle, 101, 102f
Wasp, fossil, 120f
Water vascular system, 348
Wave(s):
    body, 162–163
    primary, 162, 162f, 163, 163f, 165 f
    secondary, 162, 162f, 163, 163f, 165 f
    seismic, 162–163, 165f, 166–167
    surface, 162, 162f
Wegener, Alfred, 175, 176f, 199, 199f
Wells, John, 247
Werner, Abraham Gottlob, 15–16
Western North America:
    in Cenozoic Era, 455–470
    in Cretaceous, 393f
    in Early Paleozoic Era, 279–281
    in Mesozoic Era, 378–396
Whales, vestigial organs from, 138, 139f
White River formation, 458, 459
Whittington, Harry B., 332–333
Wilson, J. Tuzo, 189, 246
Wilson cycle(s), 189, 245f, 246–248, 250
Wind deposits (of sediment), 83
Wingate sandstone, 383, 387f
Wombats, 501, 503f
Wopmay orogen, 245–247, 245f

Xenacanthus, 357, 357f

Yellowstone National Park, 199, 460f

Zinsmeister, William, 446
Zion National Park, 382–383
    cross-bedding in, 92f
    geologic column in, 386f
Zircon, 36, 43, 208f
Zones:
    biostratigraphic, 142–143
    subduction, 185, 186f, 379–380, 383–384
Zooid, 338
Zooplankton, 147
Zoroaster granite, 113, 254f

Highlights in the History of Life

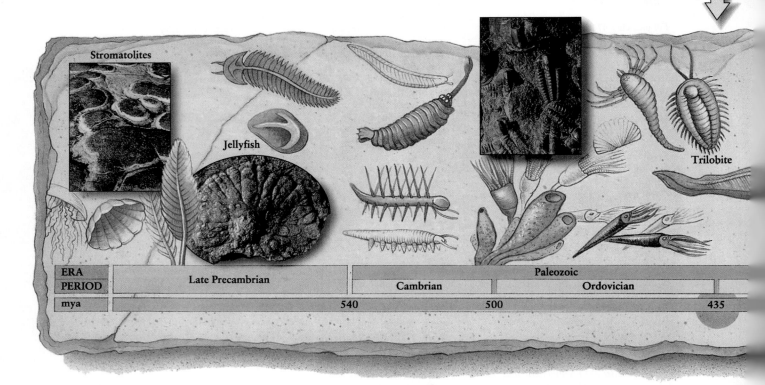

Late Ordovician Mass Extinction
22% of families of invertebrates perish. Possible causes: global cooling, growth of continental glaciers, and loss of epeiric seas.

Stromatolites

Jellyfish

Trilobite

ERA			Paleozoic		
PERIOD	Late Precambrian		Cambrian	Ordovician	
mya		540	500		435

Cambrian Invertebrates:
trilobites, inarticulate brachiopods, monoplacophorans, hyolithids, archaeocyathids, early echinoderms, Burgess Shale invertebrates.

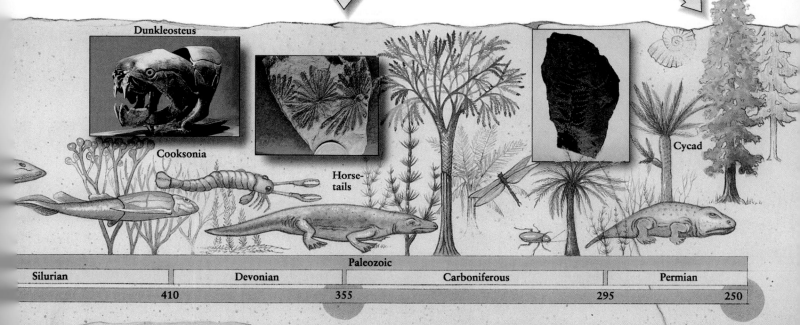

Late Devonian Mass Extinction
21% of families of marine animals perish. Possible causes: global cooling, lethal fluctuations in sea level, bolide impact.

Late Permian Mass Extinction
57% of families of marine animals perish. Possible causes: severe climatic change resulting from assembly of Pangea, global increase in volcanic activity.

Dunkleosteus

Cooksonia

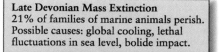

Horse-tails

Cycad

Paleozoic

Silurian	Devonian	Carboniferous	Permian
410	355	295	250

Paleozoic Invertebrates:
articulate brachiopods, bryozoa, rugose and tabulate corals, stromatoporoids, cephalopods, crinoids, graptolites, ostracodes, and new species of trilobites.

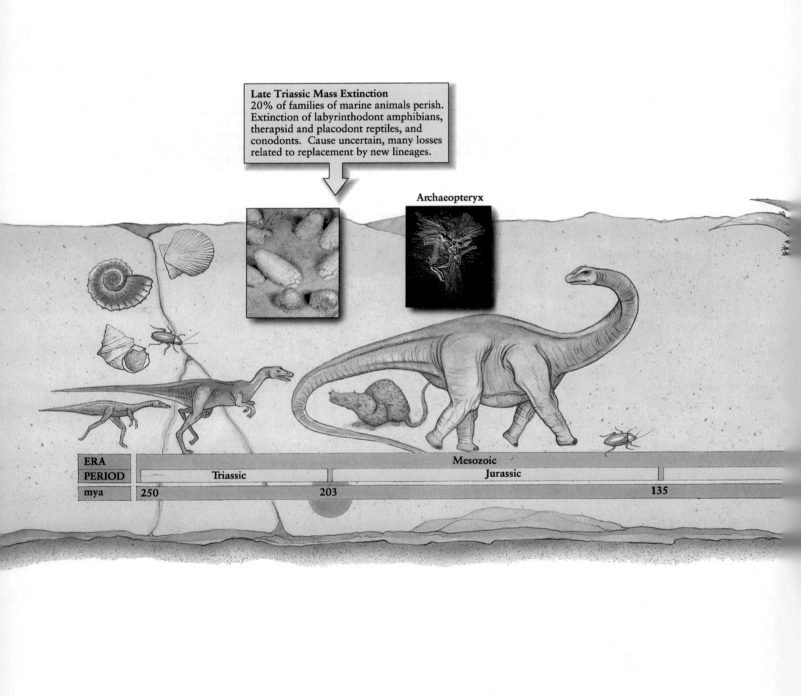

Late Triassic Mass Extinction
20% of families of marine animals perish. Extinction of labyrinthodont amphibians, therapsid and placodont reptiles, and conodonts. Cause uncertain, many losses related to replacement by new lineages.

Archaeopteryx

ERA	Mesozoic		
PERIOD	Triassic	Jurassic	
mya	250	203	135

Highlights in the History of Life

Late Cretaceous Mass Extinction
Dinosaurs become extinct, as well as 15% of families of marine invertebrates. Possible causes: bolide impact (and/or volcanic actiyity global diminution of epeiric seas).

The Coming Mass Extinction
By the year 2030, rates of extinction are expected to reach several hundred species a day if preventive programs fail. Cause: human activity, including destruction of forests and other habitats.

Sassafras leaf

Mosasaur

Mesozoic	Cenozoic	
Cretaceous	Tertiary	Quaternary →
65		1.75

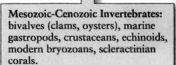

Mesozoic-Cenozoic Invertebrates: bivalves (clams, oysters), marine gastropods, crustaceans, echinoids, modern bryozoans, scleractinian corals.